Advanced Engine Technology

Heinz Heisler MSc, BSc, FIMI, MIRTE, MCIT

Head of Transport Studies
The College of North West London
Willesden Centre
London, UK

Edward Arnold
A member of the Hodder Headline Group
LONDON   MELBOURNE   AUCKLAND

First published in Great Britain 1995 by
Edward Arnold, a division of Hodder Headline PLC,
338 Euston Road, London, NW1 3BH

British Library Cataloguing in Publication Data

ISBN 0 340 568224

1 2 3 4 5   95 96 97 98 99

Typeset in 10/11pt Times by Wearset, Boldon, Tyne and Wear.
Printed and bound in Great Britain by The Bath Press, Avon.

To my long-suffering wife, who has
provided support and
understanding throughout the
preparation of this book.

Contents

Preface

Advanced Vehicle Technology has been written with the aim of presenting an up-to-date, broad-based and in-depth treatise devoted to the design, construction and maintenance of the engine and its auxiliary equipment in a clear and precise manner.

The intention is to introduce and explain fundamental automotive engineering concepts so that the reader can appreciate design considerations and grasp an understanding of the difficulties the manufacturer has in producing components which satisfy the designers requirements and the production engineer and maintenance assessibility. This will enable the mechanic or technician to develop an enquiring mind to search for and reason out symptoms leading to the diagnosis and rectification of component faults.

Line drawings are used to illustrate basic principles and construction details. Photographs of dismantled or assembled components are deliberately excluded because these can be obtained from any manufacturer's repair manual and do not adequately explain the information which is being sought.

This follows on from *Vehicle and Engine Technology* and covers the engine technology for levels 2 and 3 of CGLI 383 Repair and Servicing of Road Vehicles, underpinning knowledge for level 3 NVQ Motor Vehicle Engines, and the engine technology for BTEC National Certificate/Diploma Motor Vehicle Engineering. In addition, the content and depth of this text encompasses the engine technology modules for both the BTEC Higher National Certificate/Diploma in Motor Vehicle Engineering and the Institute of the Road Transport Engineers.

Heinz Heisler
1994

Acknowledgements

The author would like to express his thanks to the many friends and colleagues who provided suggestions and encouragement in the preparation of this book; to Diane Leadbetter-Conway and Sarah Jenkin of Edward Arnold (Publishers) Ltd who scrutinized and edited the text; and to his daughter Glenda who typed the manuscript.

1
Valves and Camshafts

1.1 Engine valve timing diagrams

(Figs 1.1, 1.2 and 1.3)

In its simplest form the inlet valve is open and the exhaust valve is closed at top dead centre (TDC) on the piston's outward induction stroke (Fig. 1.1). As the piston initially accelerates and then decelerates down its stroke, it creates a depression which induces the fresh charge to enter and fill the expanding cylinder space. Towards the end of its induction stroke the piston slows down and momentarily comes to rest at bottom dead centre (BDC).

At the end of the induction stroke the piston reverses its direction of motion and, at the same time, the inlet valve is made to close, the piston

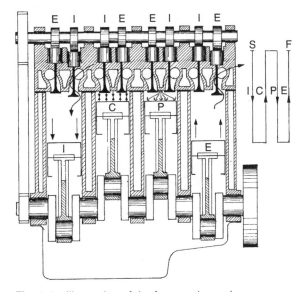

Fig. 1.1 Illustration of the four-stroke cycle

then moves inward towards the cylinder head on its compression stroke so that the cylinder volume is reduced with a corresponding rise in the cylinder charge pressure.

As the piston approaches the end of its compression stroke, it slows down, comes to a halt and then changes its direction of motion. The cylinder charge is then ignited by either a spark if it is a petrol engine or by the heat generated during the compression stroke in the case of a diesel engine, which accordingly causes the mixture of air and fuel to burn. The resulting pressure rise forces the piston outwards on its power stroke.

When the piston reaches the end of its power stroke and comes to a standstill, the exhaust valve opens, the piston then reverses its direction of movement and sweeps up the cylinder bore, thus clearing out all the burnt gas products on its exhaust stroke ready for the next induction stroke.

The clearing out of the exhaust gases from the cylinder is improved by opening the exhaust valve before the piston has completed its power stroke and then delaying the closure of this valve until the piston has swept the cylinder on its exhaust stroke. The piston then reverses its direction of motion and commences its induction stroke (Fig. 1.2(a and b)).

Similarly, to improve the filling of the cylinder with a fresh charge, the inlet valve is designed to open just before the piston reaches TDC on its exhaust stroke. The inlet valve then remains open for the full induction stroke and the early part of the piston's compression stroke (Fig. 1.2(a and b)).

The angular crank movement which occurs when the inlet or exhaust valve opens before TDC or BDC, is referred to as the valve lead

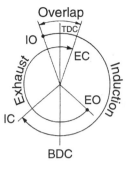
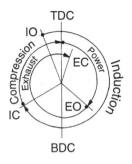

(a) Circular valve
timing diagram

(b) Spiral valve
timing diagram

Fig. 1.2 Valve timing diagrams

angle, and the crank-angle movement, after an inlet or exhaust valve closes after TDC or BDC, is called the valve lag angle. The total angular movement when both inlet and exhaust valves are open simultaneously in the TDC region is then known as the overlap period.

When the inlet and exhaust valve opening periods are combined as two semi-circles, then the resulting configuration is known as a circular valve timing diagram (Fig. 1.2(a)).

If an overall picture of the four operating periods—induction, compression, power and exhaust—is required, then they can be shown altogether on a spiral valve timing diagram (Fig. 1.2(b)).

A better appreciation of the progressive opening and then closing of both the induction and exhaust valves and their overlap is well illustrated by a linear valve opening area timing diagram (Fig. 1.3). The diagram shows that for the first 20° opening and the last 20° closing of the valves, there is very little valve opening area, but between these two extremes the valve opening area changes rapidly.

1.1.1 Exhaust valve opening and closing periods
(Figs 1.2(a and b), 1.4 and 1.5)

To maximize the expulsion of the exhaust gases from the cylinder the cam opens the exhaust valve as the piston slows down on its approach to BDC towards the end of its power stroke (before BDC). Consequently, when the exhaust valve opens, the bulk of the gases that may still be at a pressure of 3–4 bar will be expelled by their own kinetic energy through the exhaust system to the atmosphere (Fig. 1.4). This rejection of the exhaust gases is sometimes referred to as exhaust blowdown. It only remains for the piston to sweep the residual gases out on the inward moving exhaust stroke with the minimum of negative work. The early expulsion of the majority of gases by their own energy speeds up the exhaustion of the gases from the cylinder and minimizes the work done on the exhaust stroke by the piston. However, the early opening of the exhaust valve on the piston's power stroke is not such a great loss since the exhaust valve only opens when the

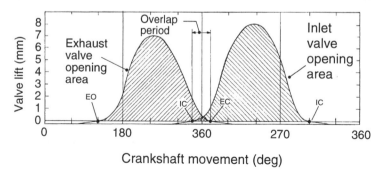

Fig. 1.3 Valve opening area timing diagram

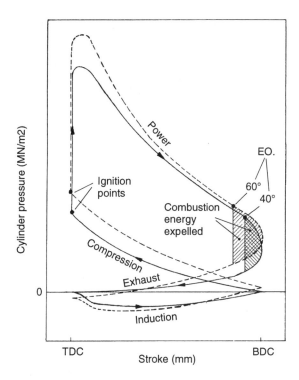

Fig. 1.4 The effects of exhaust valve early opening greatly exaggerated

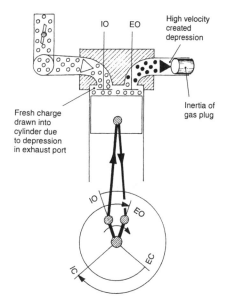

Fig. 1.5 Effects of valve overlap at high-speed large throttle opening

1.1.2 Inlet valve opening period
(Figs 1.2, 1.3 and 1.6)

To induce as much fresh charge as possible into the cylinder, the inlet valve starts to open towards the end of the piston's exhaust stroke when the outgoing column of burnt gas in the exhaust port still has sufficient velocity to form a depression in its wake (behind it in the exhaust port and combustion chamber). As a result, the fresh charge in the induction port will be drawn in the direction of the escaping exhaust gases, so that, in effect, it fills and occupies the combustion chamber space as it sweeps out the remaining products of combustion.

If the inlet valve opens very early towards the end of the exhaust stroke, some of the exhaust gases will actually be pushed back through the inlet valve into the induction manifold instead of passing out through the exhaust port. This may occur particularly under part throttle conditions, when the average depression in the induction manifold may be far greater than in the cylinder, as the piston is approaching TDC towards the end of the exhaust stroke. The inlet valve then remains open for the complete outward movement of the piston on its induction stroke and for some of the return compression stroke.

The crank-angle lag after BDC before the inlet valve closes is there to utilize the inertia of the

piston is reducing its speed and the crankshaft is revolving in the ineffective crank-angle region; that is, the piston approaches BDC with the effective perpendicular crank-pin offset at a minimum.

The exhaust valve is kept open for the whole of the piston's inward exhaust stroke movement as well as at the beginning of the piston's induction stroke. The delay in closing the exhaust valve until the crankshaft has moved past TDC at the beginning of the outward piston induction stroke makes use of the outrushing gases leaving behind a partial vacuum in the exhaust port and the combustion chamber (Fig. 1.5). This encourages the fresh charge to enter the combustion chamber space when the inlet valve has only partially opened even though the piston has slowed down almost to a standstill and therefore cannot create the vacuum pump effect above the piston. If the exhaust valve closing lag is very large then, at low speed, the fresh charge will chase the last of the exhaust gases across the combustion chamber and may actually follow these spent gases out of the exhaust port into the exhaust manifold.

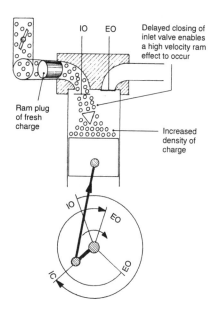

IO EO

Delayed closing of inlet valve enables a high velocity ram effect to occur

Ram plug of fresh charge

Increased density of charge

IO

EO

EO

IC

Fig. 1.6 Effect of inlet valve late closing at high-speed large throttle opening

fresh charge moving through the induction manifold tract and valve port. It therefore provides more time towards the end of the induction phase when the momentum of the fresh charge has built up a slight positive pressure so that the column of charge rams itself into the cylinder, thereby raising the density of the cylinder charge (Fig. 1.6).

Delaying the closure of the inlet valve until after BDC, when the piston has commenced its compression stroke, improves the cylinder volumetric efficiency in the medium to high engine-speed range. However, at lower engine speeds there is insufficient density and momentum of air, or air and petrol mixture, to pile into the cylinder to boost the charge mass. In fact, the inward movement of the piston on its compression stroke will oppose and actually push back some of the newly arrived charge through to the induction manifold with the result that the effective cylinder charge is much reduced. For example, if the engine is rotating at its minimum speed with a wide-open throttle and the inlet valve closes 60° after BDC, then roughly one-fifth of the cylinder's swept volume will be returned to the induction port and manifold before the effective part of the compression stroke commences. Hence, not only will there be a 20% volumetric efficiency reduction, but the shortened compression stroke would lower a nominal compression ratio of 10:1 down to something like 7:1.

1.1.3 Effect of inlet valve closing lag
(Figs 1.7 and 1.8)

The volumetric efficiency variation characteristic over an engine's speed range represents a measure of the torque the engine is capable of producing; therefore, induction and exhaust tuning of the valve timing for the highest possible overall volumetric efficiency is fundamental to maximize and spread out the engine's response and performance.

Delaying the closing of the inlet valve means that the piston pushes back a portion of the fresh charge at low engine speed as the piston moves up its compression stroke before the inlet valve closes. However, as the engine speed rises the momentum of the air, or air and petrol mixture, increases as it rushes through the induction tract and port on its way into the cylinder. Subsequently, the charge molecules move closer together so that when they enter the cylinder they ram into the expanded and less dense charge filling the cylinder (Fig. 1.6). This condition continues throughout the induction stroke and on the early part of the compression stroke causing the continuous fast-moving pulses to pressurize and pump more and more charge into the cylinder even when the piston on its return stroke reduces the cylinder volume.

The benefits of inertia ramming of a moving column of induction charge, as opposed to flow-back caused by the reduction in cylinder space on the return stroke before the inlet valve closes, becomes counter-productive at low engine speeds. This is because the low induction velocity is insufficient to generate a high enough pressure pulse for cylinder cramming and, therefore, instead of extra charge being contained in the cylinder, some of it will actually be expelled back to the induction port and manifold.

Consequently, if there is a large inlet valve lag, say 80°, (Fig. 1.7) there will be a fall-off in volumetric efficiency as the engine speed decreases due to the reduction in charge velocity ramming, permitting an increasing amount of inlet valve flow-back to occur. Conversely, if the inlet valve closing lag is small and is well within the ineffective crank-angle movement, say 40°, (Fig. 1.7) the volumetric efficiency will be at a maximum at low engine speeds, but this falls progressively as the speed rises because there is an insufficient inlet valve late closing ram effect to compensate for the shortening cylinder filling time. Neither the large 80° lag or the small 40°

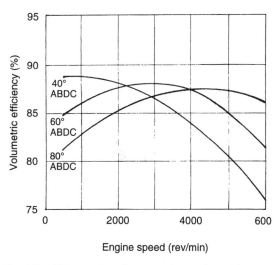

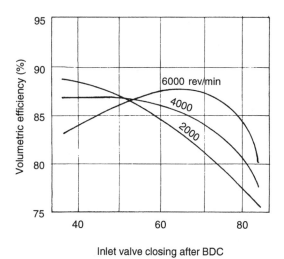

Fig. 1.7 Effects of engine speed on volumetric efficiency for different inlet valve closings after BDC

Fig. 1.8 Effects of inlet valve closing after BDC on volumetric efficiency for different engine speeds

inlet valve closing lag satisfies the overall speed range of either a petrol or diesel engine suitable for car or commercial vehicle applications. Thus, by compromising and selecting a moderate inlet valve lag, say, 60°, (Fig. 1.7), the resultant induction positive ram charge and the negative cylinder flow-back provides the most effective volumetric efficiency improvement over the mid-speed range. There will be some degree of inlet valve flow-back response reducing the cylinder filling at low speed, but this will be contrasted by the inlet valve delayed closure inertia response preventing too much decline in cylinder breathing capability at high engine speeds.

The relationship between volumetric efficiency and inlet valve closing lag, at different constant engine speeds, is very well illustrated in Fig. 1.8. The graphs show that at a low engine speed of 2000 rev/min, there is a considerable reduction in volumetric efficiency as the inlet valve closing lag is prolonged, indicating that at low engine speeds the inertia impulse pressure charging is ineffective. However, at a high engine speed of 6000 rev/min the inertia ram charge improves the volumetric efficiency when the inlet valve closing lag is extended (Fig. 1.8). Eventually, the flow-back factor opposing the inertia ram pressure charging predominates, thus causing the volumetric efficiency to decrease fairly rapidly when the inlet valve delaying closure exceeds 65°. In the mid-speed range of 4000 rev/min the volumetric efficiency remains consistent as the inlet valve closing lag increases from 40° to about 60°, after that, there is a noticeable decline in

volumetric efficiency caused by the flow-back response of the piston, pumping a proportion of the fresh charge back into the induction port before the bulk of the charge is trapped in the cylinder by the closure of the inlet valve. Nevertheless, the decline in breathing capacity in the mid-speed range of 4000 rev/min is superior to that when the engine is running at 2000 rev/min, but inferior to the engine operating at 6000 rev/min (Fig. 1.8). These volumetric efficiencies demonstrate the necessity of engine speed and, correspondingly, induction charge velocity for boosting inertia ram charging in conjunction with increasing inlet valve closing lag.

1.1.4 Valve overlap limitations
(Figs 1.9, 1.10 and 1.11)

The benefits of delaying the exhaust closing while opening the inlet valve earlier, so that both inlet and exhaust opening periods overlap each other, are better cylinder clearing and filling in the mid to upper speed range of the engine.

The improvement in cylinder volumetric efficiency owing to the extended valve overlap is caused by the high exit velocity of the exhaust gases establishing a depression in the exhaust port and manifold branches, this greatly assists in drawing in fresh air or mixture, for diesel and petrol engines respectively, from the induction manifold even before the piston has completed its exhaust stroke in the ineffective piston-stroke region where the piston is not able to perform as a vacuum pump.

Unfortunately, the advantage of opening the inlet valve early and closing the exhaust valve late has various detrimental side effects which are not compatible with the minimization of exhaust pollution.

Exhaust valve closing lag induces a fresh charge to enter the combustion chamber and cylinder during the ineffective part of the piston stroke— at the end of the exhaust stroke and the beginning of the induction period—when the engine is running in the higher speed band. However, as the engine speed is reduced, some of the fresh charge will not only enter the cylinder but will actually be carried out with the fast moving burnt gases into the exhaust system. The loss of fresh charge to the exhaust will become more pronounced as the exhaust valve lag is extended and the engine speed is reduced (Fig. 1.9). Consequently, this will show up in an increasing amount of unburnt and partially burnt exhaust gas in the form of larger quantities of hydrocarbon and carbon monoxide being present in the exhaust composition.

The inlet valve opening lead provides an opportunity for fresh charge to commence entering the cylinder early, providing there is a difference of pressure across the partially opened inlet valve sufficient to force the fresh charge into the combustion chamber and cylinder space. This is not only possible with a wide open throttle but it is

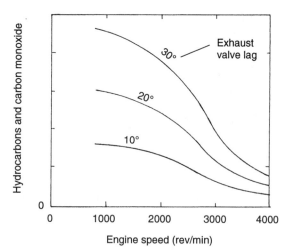

Fig. 1.9 Effects of exhaust valve delayed closing on expelled exhaust gas composition

also an effective way of initiating the beginning of induction. However, as the throttle opening is progressively reduced, the manifold depression rises until there may be a reversal of conditions within the cylinder and manifold, and, in fact, the mean depression in the induction manifold may, at the end of the exhaust stroke, be greater than in the cylinder. Under these part throttle conditions, some of the exhaust gas escaping from the cylinder will not only go out of the exhaust port, but will also be drawn back through the induction port into the induction manifold where it originated before being burnt (Fig. 1.10).

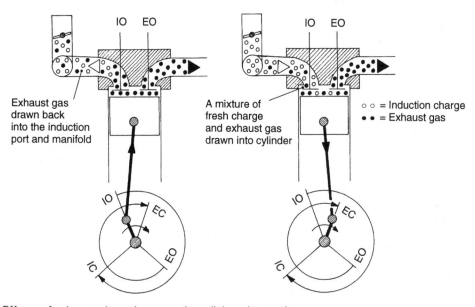

Fig. 1.10 Effects of valve overlap at low-speed small throttle opening

6

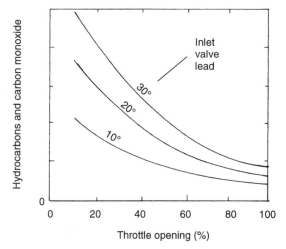

Fig. 1.11 Effects of early inlet valve opening on expelled exhaust gas composition

Inlet valve opening before TDC	10°–30°
Inlet valve closing after BDC	40°–75°
Exhaust valve opening before BDC	40°–75°
Exhaust valve closing after TDC	10°–30°

Consequently, the fresh charge in the induction manifold will be diluted by the exhaust gas, making it difficult for the new charge, when it enters the cylinder, to burn effectively in the time available. Therefore, a proportion of exhaust gas subsequently expelled through the exhaust port will not have completed the combustion process, this being reflected by a higher hydrocarbon and carbon-monoxide content. Note the increasing amount of exhaust emission as the inlet valve lead becomes larger (Fig. 1.11).

Survey of valve timing diagrams
(Fig. 1.12(a, b, c and d))
Popular crank angle displacements for both inlet and exhaust valve leads and lags may vary considerably for engines designed for similar operating conditions.

Broadly speaking, engine applications can be broken up into four groups.

  i Saloon cars powered by either petrol or diesel high-speed engines.
 ii Commercial goods and passenger public-service vehicles powered by slow to medium-speed diesel engines.
iii Industrial and marine low-speed diesel engines.
 iv Competition and racing petrol engines.

A survey of a large number of petrol and diesel engine valve timing diagrams suitable for both on and off road applications (Fig. 1.12(a)) fell within the following ranges.

A typical valve timing diagram (Fig. 1.12(b)) for a saloon car has an inlet opening valve lead and closing lag of 18° and 64° respectively, whereas the exhaust valve lead and lag is 58° and 22° respectively. This, therefore, provides a 40° valve overlap which promotes good final clearance of exhaust gases from the cylinder and an early effective commencement of the induction period.

The general concept of utilizing valve overlap to improve the engine breathing is effective when there is an unhindered supply of fresh charge in the induction manifold, but under certain speed and part throttle conditions small amounts of exhaust gas dilute the fresh charge, thereby slowing down the combustion process. In addition, fresh charge and incomplete combustion products due to the contamination of the fresh charge may be expelled through the exhaust port causing high levels of hydrocarbon and carbon monoxide to come out of the exhaust system. Consequently, there has been a move towards reducing the valve overlap to counteract the mixing of exhaust gas with the fresh charge, and the loss of fresh charge out from the exhaust port. A compromise of a more modest inlet valve opening, 10°, and exhaust valve closing of something like 15° providing an overlap of 25° tends to satisfy the pollution problem without greatly affecting performance (Fig. 1.12(b)).

Diesel engines which do not have throttled induction and only inhale fresh air do not suffer pollution caused by valve overlap and, therefore, can benefit by late closing of the exhaust valve. Supercharged diesel (turbo or positively driven) can benefit from a large inlet valve opening lead and, similarly, large exhaust valve closing lag since these engines only have to deal with air as a blow-through media. Therefore, the valve overlap can be extended, but only up to a point where, under certain driving conditions, the exhaust gas is expelled into the induction manifold, or a very late closing of the exhaust valve interferes with the ram filling of the cylinder through the induction manifold. Large diesel engines which have been purposely designed for supercharging would therefore have their valve timing (Fig. 1.12(c)) comforming approximately to the following.

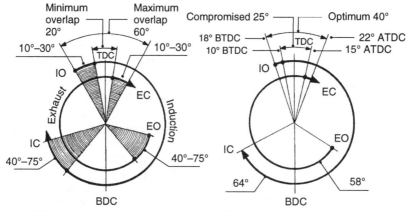

(a) Range of Valve Leads and Lags
Suitable for Petrol and Diesel Engines

(b) Typical saloon car petrol and diesel
naturally aspirated or supercharged engines

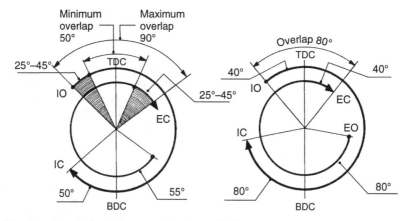

(c) Diesel engines designed for supercharging

(d) Petrol racing engine naturally aspirated

Fig. 1.12 Survey of valve timing diagram

Inlet valve opening before TDC	25°–45°
Inlet valve closing after BDC	50°
Exhaust valve opening before BDC	55°
Exhaust valve closing after TDC	25°–45°

Competition naturally-aspirated engines, which operate at high speeds over a relatively narrow speed range in conjunction with a close ratio gear box, can usually take advantage of very large inlet and exhaust valve leads and lags (Fig. 1.12(d)) without having to consider fuel economy and exhaust emission.

Valve timing periods for these operating conditions would roughly be as follows.

Inlet valve opening before TDC	40°
Inlet valve closing after BDC	80°
Exhaust valve opening before BDC	80°
Exhaust valve closing after TDC	40°

1.1.5 Variable inlet valve timing
(Fig. 1.13(a, b, c and d))

The pros and cons of variable inlet valve timing
(Fig. 1.13(a))
Engine valve timing is designed to produce a valve overlap period between the inlet valve commencing to open at the start of the induction period before TDC and the closure of the exhaust valve at the end of the exhaust period just after TDC.

Valve overlap is designed to make use of the outgoing exhaust gas momentum leaving behind a depression that induces a fresh charge to enter the combustion chamber when the piston is near TDC in the ineffective part of the piston's stroke

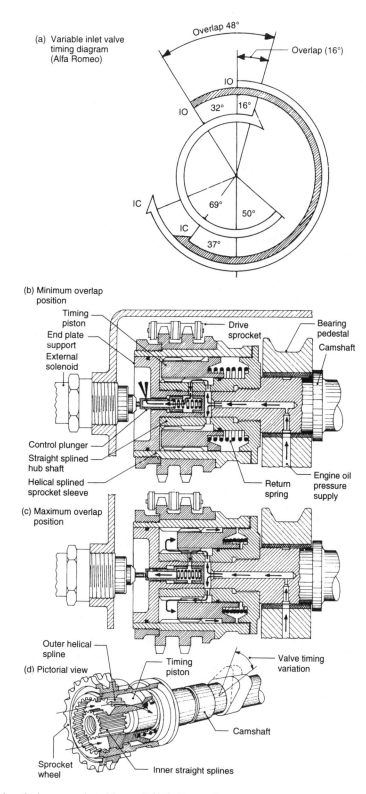

(a) Variable inlet valve timing diagram (Alfa Romeo)

Overlap 48°

Overlap (16°)

IO

IO 32° 16°

IC 69° 50°

IC 37°

(b) Minimum overlap position

Timing piston
End plate support
External solenoid
Drive sprocket
Bearing pedestal
Camshaft
Control plunger
Straight splined hub shaft
Helical splined sprocket sleeve
Return spring
Engine oil pressure supply

(c) Maximum overlap position

Outer helical spline
Timing piston
Valve timing variation
(d) Pictorial view
Camshaft
Sprocket wheel
Inner straight splines

Fig. 1.13 Variable valve timing sprocket drive unit (Alfa Romeo)

9

and, therefore, it cannot contribute to the induction process.

The benefits gained by increasing the valve overlap are only realized in the upper speed and load range, and not in starting and low-speed running conditions. With a large valve overlap, some fresh unburnt charge will be expelled into the exhaust manifold which, accordingly, increases the hydrocarbon exhaust emission. Conversely, residual exhaust gas may actually be drawn into the induction manifold under part throttle and idling conditions and therefore it contaminates and dilutes the fresh mixture entering the cylinder.

Partially to meet the conflicting engine requirements, twin camshaft engines can have a variable timing drive sprocket mounted on the front end of the inlet camshaft which is able to advance or retard the inlet valve angular opening and closing points relative to the fixed exhaust camshaft to crankshaft timing. The characteristics of this variable inlet valve opening point mechanism are that the total induction period remains constant but the start and finish points are variable (Fig. 1.13(a)). One drawback of this type of variable inlet timing mechanism is that when the inlet valve opening lead and valve overlap are increased, it is at the expense of the inlet valve closing lag, which is correspondingly reduced by the same amount. Thus, at high speed and wide-open throttle conditions, when a delayed inlet valve closure is desirable to take advantage of the inertia of the fresh charge ramming itself into the cylinder towards the end of the induction period just after BDC, the reverse happens—the inlet valve point of closure is made to occur earlier, thereby cutting short the final cylinder filling when the piston inward movement is still small relative to the crank angle displacement.

The overall gain in expelling the last of the exhaust gases from the unswept combustion space and encouraging an early inducement of fresh charge to enter the cylinder, outweighs that lost by reducing the inlet valve closure lag after BDC and, at the same time, the larger inlet valve closing lag reduces the effective compression ratio at idling speed, which is beneficial for good low-speed tick-over stability.

Operation of variable timing sprocket drive unit
(Fig. 1.13(b, c and d))
The main components of the sprocket drive unit are a timing piston with internal straight splines and external helical splines, an internal helical splined sleeve sprocket, an external straight-splined hub shaft mounted on the camshaft, a control plunger and an external control solenoid.

Minimum valve overlap
(Fig. 1.13(b))
When the engine is running and the solenoid stem is withdrawn, engine lubrication under pressure enters the camshaft from the bearing pedestal, flows axially towards the hub end of the camshaft, it is then diverted radially to the surface of the hub shaft and the bore of the timing piston. Oil then passes between the clearance space formed between the meshing piston and hub shaft straight splines, it is then directed radially inwards, passes through the centre of the control plunger and finally escapes to the sump via the relief hole in the side of the plunger.

Maximum valve overlap
(Fig. 1.13(c and d))
At some predetermined speed and load condition the solenoid is actuated by the electronic petrol injection and ignition management system, this extends the solenoid stem, pushing the control-plunger into the end plate until it covers the relief hole in its side. Immediately, the oil flow from the plunger is blocked, which therefore pressurizes the stepped side of the timing piston. As a result, the timing piston is forced to move towards the flanged end of the hubshaft, and as it does so, the external helical piston splines progressively screw the matching internal helical sleeve splines, thus causing the sleeve and sprocket to rotate in a forward direction relative to the hubshaft which is attached to the camshaft. Accordingly, the inlet valve lead will be increased and the inlet valve lag will decrease by the amount of relative angular twist imposed on the sleeve and sprocket as the piston slides along the external sleeve splines from the left-hand to the right-hand end.

1.2 Valve operating conditions

1.2.1 Heat absorption and dissipation of an exhaust valve
(Fig. 1.14)

The quantity of heat absorbed by the valve when heat is released from the products of combustion amounts roughly to 70% by conduction through the valve head when it is closed, and 30% by conduction through the fillet underside of the valve when it is open and the exhaust gas is escaping through the exhaust port (Fig. 1.14).

When the valve is closed, about 70% of the total heat input from the heat released by combustion is transferred by conduction to the cylinder head coolant by way of the valve seat, and only 24% of the total heat input to the valve is conveyed through the valve stem to the cylinder head coolant via the valve guide.

The majority of heat taken in by the valve is transferred to the coolant by conduction through the relatively large circumferential valve-seat heat path when the valve is closed. However, when the valve opens, the exhaust gas sweeps across the throat of the valve on the way out and the only way the heat can then be rejected from the valve head is along the slender valve to the valve guide and cylinder head. Thus, the hottest region of the valve is generally not the centre of the valve head, but below the valve head where the valve fillet and parallel stem merge.

1.2.2 Valve operating temperature strength and durability
(Figs 1.15 and 1.16)

With rising temperature, the tensile strength of the exhaust-valve alloy steel and nickel base alloys decreases initially slowly and then at an increasing rate as the temperature exceeds 600°C (see Fig. 1.15).

However, although the short-term tensile strength may appear to be adequate, its creep strength or hot fatigue strength may be insufficient in the valve's working temperature range of between 700 and 900°C, Fig. 1.15 shows the hot fatigue strength considerably lower over the same temperature range. The inadequacy of most valve materials is primarily due to the combination of three kinds of loads imposed between the crown

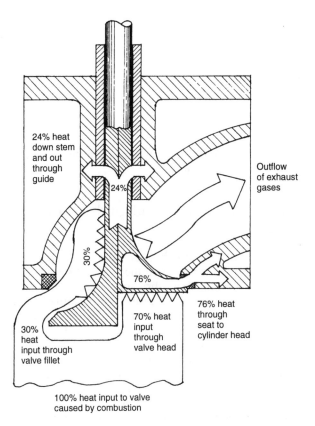

24% heat down stem and out through guide

24%

Outflow of exhaust gases

30%

76%

30% heat input through valve fillet

70% heat input through valve head

76% heat through seat to cylinder head

100% heat input to valve caused by combustion

Fig. 1.14 Heat absorption and dissipation of an exhaust valve

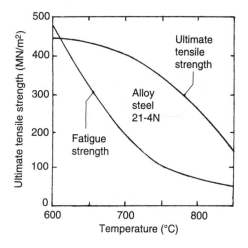

Fig. 1.15 Effect of temperature on strength of an exhaust valve alloy steel

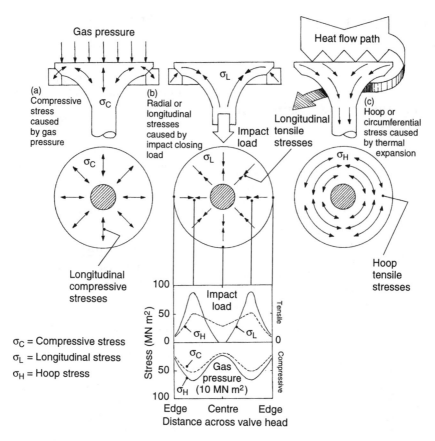

Fig. 1.16 Poppet valve working stresses

and neck of the valve (Fig. 1.16(a, b and c)), the stresses created are as follows:

1 the gas pressure compressive stress imparted to the valve crown during every power stroke (Fig. 1.16(a));
2 the radial (longitudinal) impact tensile stress produced between the valve crown and neck every time the valve closes (Fig. 1.16(b));
3 the hoop (circumferential) tensile stress around the rim of the valve head caused by the cyclic temperature expanding and contracting the valve head (Fig. 1.16(c)).

If the combined dynamic and thermal cyclic load of the valve head is excessive, then this will cause the valve head to distort permanently and eventually to collapse.

Unless the head section thickness is substantial, the alternating deflection and cupping of the head disc each time the valve closes may reduce the heat transfer due to the poor seating of the valve and, accordingly, cause the operating temperature of the valve to rise.

1.2.3 Valve stem and guide clearance
(Fig. 1.17)

For optimum sliding between the valve stem and guide the operating clearance should be just sufficient to establish a lubrication film between the rubbing pair under operating temperature conditions. A small clearance is desirable since the smaller the stem-to-guide clearance, the greater the amount of heat transference from the stem to the cylinder-head via the valve guide, this therefore enables the valve head to operate at lower temperatures (Fig. 1.17). If the stem-to-guide clearance is large, not only will this act as a heat conduction barrier but it will produce scuffing of the stem and a side thrust rocking motion caused by the rocker arm contact on the valve tip.

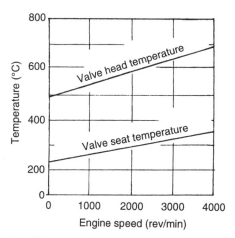

Fig. 1.17 Effect of valve stem-to-guide clearance on exhaust valve temperature

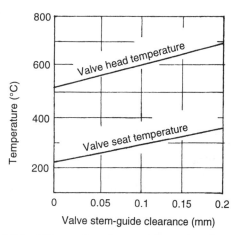

Fig. 1.18 Effects of engine speed on exhaust wave temperature

Consequently, if there is a large clearance, the guide and stem will wear, causing bell mouthing of the guide bore at its ends and localized wear on opposite sides of the valve stem where the tilting stem makes contact with the guide bore.

1.2.4 Effects of engine speed and ignition advance on exhaust valve temperature
(Figs 1.18 and 1.19)

Rising engine speed increases the amount of heat transferred from the valve to the cylinder-head coolant in a given time, this then causes the heat flow path from the valve head through the valve seat and valve stem to become saturated. It therefore becomes more difficult to dissipate the heat absorbed by the valve head, and consequently the valve head and seat temperatures will rise, see Fig. 1.18.

The influence of ignition timing on exhaust valve temperature is most marked; and, generally, advancing the point of ignition reduces the mean exhaust valve temperature (Fig. 1.19). The reason for this temperature reduction is because more of the energy released by combustion is put into doing useful work. Also, less heat is therefore absorbed by the valve neck and conducted away via the valve stem when the exhaust gases are expelled.

Over-advancing the engine, however, will produce the reverse results; that is, there will be a marked rise in exhaust valve temperature.

1.2.5 Sodium filled hollow exhaust valves
(Fig. 1.20)

To improve the cooling of the head and throat (underside) of exhaust valves operating at high output or heavy duty conditions, a hollow valve stem and, in exceptional circumstances, hollow valve heads may be utilized in preference to conventional solid-type valves. During manufacture, the hollow valve is partially (40–50%) filled with sodium crystals which melt at approximately 98°C and which have a boiling point of 883°C, the hollow valve stem is then sealed with a low scuffing steel plug, which also becomes the valve stem tip and therefore must resist the rocker arm rubbing contact motion.

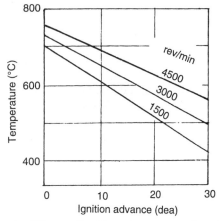

Fig. 1.19 Effects of ignition advance on exhaust valve clearance

Metallic sodium has the thermal ability that, when in its liquid state, it can transmit roughly nine times more heat than the austenitic steels, which are relatively poor conductors and are commonly used for exhaust valves. In its liquid state sodium readily spreads and wets the surfaces of most metal alloys it contacts.

While the engine operates, the opening exhaust valve flings the liquid sodium towards the valve head, where it absorbs a proportion of the heat which it transfers from the escaping exhaust gas to the surface of the valve head and its underside. When the valve closes, the hot sodium liquid is flung towards the valve tip and the heat stored in the sodium liquid is then readily dissipated radially outwards from the wetted liquid to the valve stem hollow wall interface to the much cooler cylinder-head coolant via the valve guide and cylinder-head casting.

Through these means, a much larger quantity of heat is transferred from the valve head to the valve guide than would be possible with a solid valve stem. Consequently, the valve crown's, and fillet (underside) region's, operating temperature gradients are much reduced. As can be seen in Fig. 1.20 the maximum temperature in the centre

of the valve crown and at the valve throat (fillet) is reduced at full load by roughly 100°C for a hollow stem valve, and this can be reduced further by almost the same amount if the valve head itself is hollow and sodium filled.

1.3 Mechanism of valve head corrosion

(Figs 1.21 and 1.22)

The composition of the combustion products consists of the following hot gases—carbon dioxide, carbon monoxide, hydrocarbons, water vapour and sulphur dioxide; in addition, there are some solid deposits which may contain mixtures of oxides, halides and sulphates and, in some cases, phosphates and vanadates. Other deposits which may attach themselves to the valve head, derived from the lead anti-knock compounds, are lead sulphate, lead monoxide, lead chloride, lead bromide and barium sulphate. To some extent the amount of these deposits will depend upon the effectiveness of the type of lead anti-knock compound scavenger used.

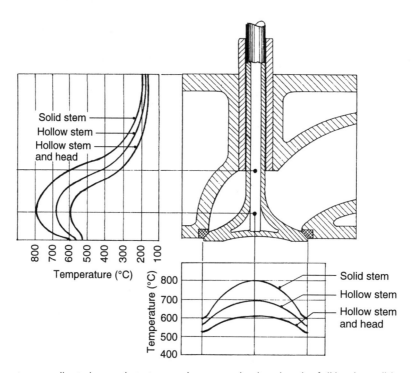

Fig. 1.20 Temperature gradient along valve stem and across valve head under full load conditions

14

The corrosion resistance at elevated temperatures of any valve material is dependent upon the formation of a firmly adherent continuous protecting oxide film. Good corrosion resisting alloys rely on the oxide film rapidly healing itself if it is ruptured in service when exposed to the burning gases. If the ruptured film does not immediately repair itself, severe localized corrosion results.

The mechanism of oxidation is that a chromium-rich oxide consisting of Cr_2O_3 for austenitic steel alloys, $NiO-Cr_2O_3$ for Nimonic alloys and $NiO-Cr_2O_3-Fe_2O_3-Mn_2O_3$ for steel alloys is readily formed.

The corrosion rate will depend to a large extent upon the conditions existing at the surface of the valve head; that is, the operating temperature of the valve, the cyclic change in gas composition and the nature of the surface deposits.

Local concentration of carbon may be established on the valve head, which may result in different oxidation rates of iron and carbon in the alloy. The presence of iron, nickel and cobalt oxides under certain temperatures and pressure conditions can catalytically cause the carbon monoxide to change to carbon dioxide and carbon. Local diffusion of the free carbon may take place, and carbon may also be produced on the metal surface by the breakdown of hydrocarbon products in the hot gas. Therefore, at the high operating temperature of the valve, carburization and then oxidation of the alloy surface occurs. Chromium carbides are then probably formed which cause the depletion of the surrounding surface of chromium (Fig. 1.21). Consequently, without the even and continuous distribution of chromium at the surface of the valve head, the network of chromium oxide will be inadequate to protect the valve surface from high temperature corrosion.

Fortunately, the recovery of chromium by diffusion is generally rapid at the depleted zones so that carburization and oxidation almost overlap each other as one and then the other reaction occurs. Corrosion of this kind occurs only where small spots of carbon have formed on the valve seat face and, accordingly, internal oxidation, intergranular penetration and pitting will result (Fig. 1.22).

The formation of the internal carbides and oxides produces a large increase in volume of the protection film, the film then becomes highly strained so that the film, in effect, bursts. Consequently, cracks appear at the surface of the film which cannot repair themselves whilst this kind of

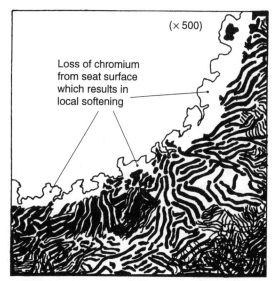

(× 500)

Loss of chromium from seat surface which results in local softening

Fig. 1.21 Chromium denudation of failed valve seat surface

corrosion persists. As a result, severe corrosion takes place at these exposed openings, the oxides then convert into a brittle scale which tends to dislodge itself from the valve head surface.

Consequently, if the oxide film is disrupted on any part of the seat face, a gap will be left at the interface of the valve and its seat when the valve closes. Accordingly, there will be a local reduction in the heat transfer from the valve head to the cylinder head and, at the same time, the hot

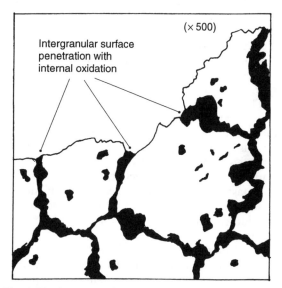

(× 500)

Intergranular surface penetration with internal oxidation

Fig. 1.22 Intergranular attachment of failed valve seat surface

gases escaping through the gap will also cause a rise in the valve head temperature. The outcome will be a rapid rise in the local rate of corrosion followed by considerable gas blow-by, valve seat erosion and seat burning.

The cause of valve guttering and burning can be considered in four stages.

1 At high working temperatures a black magnetic scale will form all over the valve head.
2 This black magnetic scale acts as a catalyst causing the hot gases to react with the surface of the valve so that there is a high rise in local surface temperature of the valve head.
3 As a result the surface scale becomes very brittle and crumbly so that chromium denudation of the surface takes place (Fig. 1.21) followed by intergranular surface penetration (Fig. 1.22).
4 Eventually the intergranular surface penetration leads to guttering that is a concave circumferential burning away of the valve seat which, if severe, will lead to valve failure.

1.3.1   The influence of combustion products on valve corrosion

The reduction of sulphate deposits to sulphides produces a low melting point sulphide–oxide mixture and to sulphide penetration of the metal.

Sodium hydroxide is produced by the reduction of sodium sulphate in the presence of water vapour, this causes local breakdown of the protective scale due to fluxing.

Chloride, in the presence of sulphate deposits, causes a considerable increase in the corrosion rate.

The corrosion rate of iron in sulphur dioxide is considerable compared with iron in a carbon dioxide–carbon monoxide atmosphere. This is because the iron surface forms a sulphide which occupies a much larger volume, which tends to crumble so that the protective scale is disrupted.

Catastrophic corrosion occurs when nickel-based alloys are exposed to sulphates and sulphur dioxide caused by the creation of a low melting point eutectic.

Lead monoxide or vanadium pentoxide forming part of the surface deposits considerably raises the rate of oxidation due to fluxing of the protective scale.

Lead monoxide exposed to high chromium

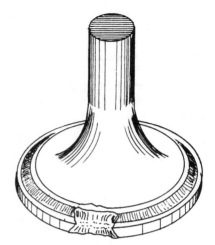

Fig. 1.23   Local rim burning caused by valve seat distortion

alloys disrupts the protective Cr_2O_3 layer through the formation of $Pb_3Cr_2O_9$.

The corrosion rate can usually be slowed down by increasing the chromium or aluminium content, which then preferentially reacts with the environment to form a coherent protective sulphide scale.

1.4   Identification of the causes of valve failure

1.4.1   Valve rim burning
(Fig. 1.23)

High operating temperatures due to incorrect ignition timing and mixture strength raise the valve head temperature until the valve head becomes the focal point for preignition. A portion of the valve head then becomes the point of ignition which again leads to further heating of the valve head until eventually the valve seat distorts. Consequently, the poor seating of the valve during the combustion phase provides an escape route for some of the burning gases and, as a result, causes excessive local overheat and burning in the valve rim and seat region.

1.4.2   Valve seat guttering
(Fig. 1.24)

If large quantities of combustion products accumulate around the valve seat they will event-

16

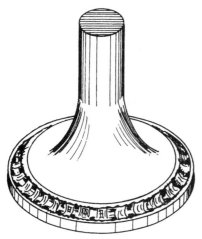

Fig. 1.24 Guttering caused by deposits flaking away from the valve face and seat

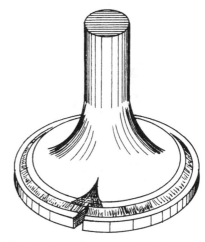

Fig. 1.25 Split rim failure caused by thermal fatigue

ually become dislodged due to the repeated valve closing impact, some of the break-away deposits will then provide exit passages for the burning gases—which are still subjected to high pressure and temperatures. The partial radial tunnelling around the valve seat when it is closed will therefore cause the valve seat to overheat as the gas escapes, since the heat flow path around the head is interrupted. This may result in severe carburization and oxidation of the valve seat surface, and ultimately it may burn a semi-concave groove into and around the valve seat. The name 'guttering' is derived from the top face contour of house guttering.

A second cause of valve face guttering may be caused by insufficient valve clearance which does not permit the valve to close fully during the power combustion stroke periods; consequently, the leaking valve overheats and encourages carburization and oxidation to take place, which finally results in a burnt circumferential concave valve seat contour.

1.4.3 Valve rim splitting
(Fig. 1.25)

Owing to the large amount of heat absorbed by the neck of the exhaust valve as the hot gases flow around and across its stem when it is open, and the relatively good heat transfer path of the valve seat when the valve is closed, the temperature distribution over the valve head will be uneven. The hottest part of the valve head will be roughly just less than half way inwards from the valve rim.

Therefore, as the temperature rises, the valve head will expand circumferentially, this expansion being greatest in the annular zone approximately mid-way between the valve rim and stem where the valve head temperature is at its highest.

Consequently, the tendency for the valve material just in from the valve seat to expand circumferentially more than the rim causes internal hoop stresses to be created within the disc-shaped head. If the temperature differential across the head is relatively large, then the repeated heating and cooling of the valve head may cause large cyclic thermal stresses to be established in the disc part of the valve which can, in severe operating situations cause the valve head to fracture in the form of a radial vee split, with its origin at the valve's rim.

1.4.4 Valve stem bending fracture
(Fig. 1.26)

A distorted valve seat insert possibly caused by overheating due to a fault in the cooling system's excessive detonation or preignition or due to a solid combustion deposit build-up over a narrow section of the valve seat, will prevent the valve head fully seating. The valve will therefore be subjected to repeated tilting of the head each time the valve closes. If this condition is permitted to continue, the gas leakage between the valve head and its seat insert will raise the valve head and neck temperature. At the same time, the cyclic bending moment produced will progressively weaken the valve material in the stress raising

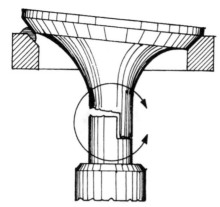

Fig. 1.26 Bending stresses caused by uneven valve seating

region where the reducing cross-section of the neck meets the parallel stem section, until the valve fractures at this point.

1.4.5 Valve head cupping
(Fig. 1.27)

Continuous high-speed and heavy load driving may abnormally raise the valve head temperature and, at the same time, the valve head absorbs the high frequency impact hammer-like blows every time the valve closes. The impact load is created by the combination of the valve mass inertia and the valve spring return force.

The effects of the repeated loading of the valve at high elevated temperatures produces radial and longitudinal tensile stresses in the valve head and neck respectively, which, if severe, will tend to stretch the valve stem and dish the head disc at its centre. Eventually, if the rigidity of the valve

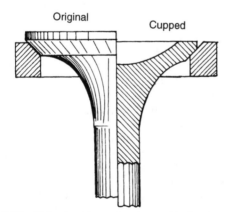

Original Cupped

Fig. 1.27 Valve head cupping caused by inadequate creep strength

head section and the creep strength of the valve alloy are inadequate, the centre of the valve collapses and takes on a permanent cupped set and, as a result, will cause the valve's tapered circumferential seat to move further into the valve seat ring insert. Excessive cupping of the valve may be followed by distorted seating of the valve causing the valve seat to overheat and burn.

1.5 Hydraulic operated tappets

1.5.1 The reason for incorporating hydraulic tappet follower

The hydraulic tappet follower incorporates three main operating characteristics which are as follows.

1 The hydraulic tappet follower is able to take up the extra valve train slack when the engine is cold and during the warm-up period or if the static tappet clearance is abnormally large due to wear or incorrect adjustment, thereby eliminating tappet valve train noise.
2 The hydraulic tappet eliminates static tappet clearance once the engine has started, this ensures that the camshaft's designed valve-timing periods are maintained during normal engine operation.
3 The initial (static) tappet clearance is not critical and therefore the engine's top-end cylinder-head assembly tolerance does not normally need the additional setting of the tappet clearance. Consequently, some manufacturers do not consider it necessary to set the static (stationary engine) tappet clearance after the engine has been disturbed and, therefore, throughout the engine's life no periodic tappet clearance setting is required.

1.5.2 Push-rod operated hydraulic tappet follower
(Fig. 1.28(a and b))

Description of follower tappet unit
The hydraulic tappet follower assembly consists of an outer follower tappet very similar to the conventional solid cylindrical follower but bored out to house an inner plunger. Both the follower

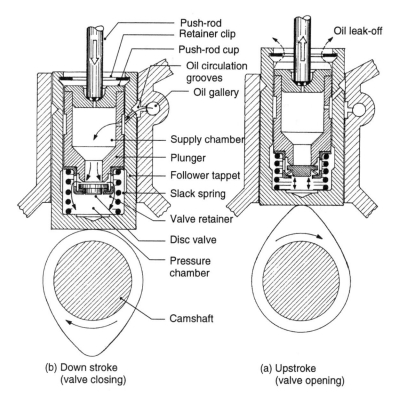

| Push-rod |
| Retainer clip |
| Push-rod cup |
| Oil circulation grooves |
| Oil gallery |
| Oil leak-off |
| Supply chamber |
| Plunger |
| Follower tappet |
| Slack spring |
| Valve retainer |
| Disc valve |
| Pressure chamber |
| Camshaft |

(b) Down stroke
(valve closing)

(a) Upstroke
(valve opening)

Fig. 1.28   Push-rod hydraulic tappet follower

body and plunger have annular grooves machined on their outside and each groove has a radial drilling so that there is always some degree of passage alignment between the oil gallery and the centre region of the plunger—known as the supply chamber. A push-rod cup supports the lower end of the push-rod and at the same time seals the upper end of the plunger's supply chamber.

The large central passage in the base of the plunger connects the supply chamber with the space between the plunger and the blank end of the follower cylinder, this being known as the pressure chamber. A valve retainer supports a disc check valve on the underside of the plunger which is held in position by a slack spring which reacts between the plunger and cylinder so that the plunger is constantly tending to extend outwards to take up the tappet clearance slack.

Engine oil is continuously delivered under pressure from the oil gallery's vertical slot, through a vee groove and hole in the side of the follower body to the supply chamber, via a second groove and hole in the plunger. The oil then passes through the check valve passage to fill up the pressure chamber. The purpose of the oil gallery vertical slot and the circumferential grooves on both the follower body and plunger is to ensure that, at all times, the oil passages between the oil gallery supply and the plunger supply chamber align, irrespective of the follower's lift or fall position. A small amount of oil will also be metered through the central hole in the base of the push-rod cup, and with some designs through the push-rod central passage to lubricate the rocker, its pivot, and adjustment assembly.

Upward stroke valve opening
(Fig. 1.28(a))
As the rotating cam approaches the beginning of valve opening, the follower's rubbing contact changes from basic-circle to opening flank contact as it moves through the opening ramp angular interval. During this time the follower will commence to lift. The plunger will therefore be forced further into the follower cylinder causing a reduction in the pressure chamber volume and a corresponding rise in the oil pressure. Immediately, the

pressure increase compels the disc-valve to snap closed, thereby trapping the bulk of the oil in the pressure chamber. Since the oil sealed in the pressure chamber is practically incompressible, the rising follower produces an equal amount of plunger and push-rod lift (in fact the plunger does compress approximately 0.1 mm during the valve lift period and so allows the valve to seat fully). The clearance between the sliding plunger and the follower cylinder permits the deliberate escape of small quantities of oil from the pressure chamber, this leakage therefore automatically compensates for the expansion of the valve train as the engine temperature increases with rising speed and load.

Thus, the plunger's inward or outward position in the follower cylinder continuously adjusts itself so that it takes up the valve train slack to compensate for the expansion or contraction of the valve train and cylinder head, which takes place as the engine operating conditions change. In effect, the valve train tappet slack is eliminated and, at the same time, both inlet and exhaust valves are able to seat fully during their closing periods.

Down stroke (valve closing)
(Fig. 1.28(b))
When the valve begins to close and the cam profile-to-follower contact changes from closing flank to base-circle the compressive load will be removed from the valve train, this permits the slack-spring to extend, thus causing the follower's body to bear against the cam and for the plunger to take up the valve train slack. The slight increase in the pressure chamber volume reduces the oil pressure below that existing in the supply chamber, this causes the disc-valve to snap open and, accordingly, oil from the supply chamber feeds into the pressure chamber to compensate for any oil leakage losses between the plunger and follower cylinder. In other words, the tappet follower adjusts and corrects its length every time the valve closes so that zero clearance is maintained under all operating speeds and working conditions.

1.5.3 Direct acting overhead camshaft hydraulic tappet follower
(Fig. 1.29(a, b, and c))

Description of follower tappet unit
The hydraulic tappet follower comprises an inverted bucket follower body which incorporates a plunger guided by a sleeve formed by a steel pressing, which also retains the engine's oil supply in the upper half of the follower. A slack-spring, ball-check-valve and sliding thrust-pin are housed inside the plunger, the thrust-pin is permitted to slide inside the plunger and, similarly, the plunger itself slides in the mouth of the steel sleeve pressing. An oil gallery is drilled alongside the follower in the cylinder head parallel to its length. Each follower bore has an oil passage joining it to the parallel oil gallery and, to enable oil to flow into the upper half of the follower, a shallow groove is machined around the follower about half-way down.

Downstroke (valve opening)
(Fig. 1.29(a))
Rotation of the camshaft causes the cam opening flank to commence pushing the follower body downwards, immediately the pressure chamber volume tends to decrease and the oil pressure increases above that of the oil supply pressure. Instantly, the ball-valve snaps onto its seat, thereby sealing the pressure chamber. Further downwards movement of the follower body will also displace the plunger and valve-stem an equal amount since the trapped oil in the pressure chamber reacts as a solid incompressible column. Thus, there is no lost motion between the thrust-pin and plunger and, as a result, the follower body, thrust-pin and valve-stem assembly are moved downwards as a whole, thereby opening the valve with the same lift as the cam profile predicts.

Upward stroke (valve closing)
(Fig. 1.29(b))
Once the cam closing flank contact with the flat face of the follower changes to cam base-circle contact, the compressive load is removed from the pressure chamber allowing the slack-spring to expand so that the thrust-pin pushes the follower body against the cam base-circle and the plunger simultaneously presses down onto the valve stem. Consequently, the pressure chamber volume marginally increases and the pressure of the trapped oil decreases with the result that the slightly higher oil pressure in the supply chamber dislodges the ball valve from its seat, thus permitting the oil in the supply chamber to enter the pressure chamber to compensate for any loss of oil which is deliberately leaked between the thrust-pin and plunger wall's clearance. Hence, zero cam-to-

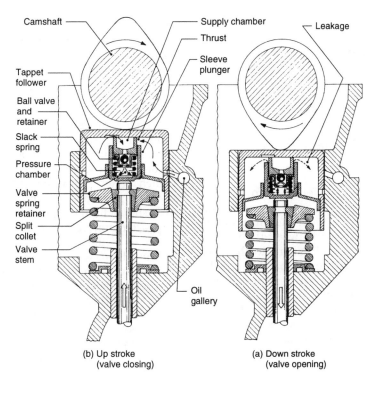

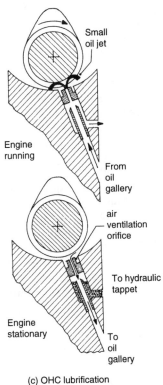

(b) Up stroke
(valve closing)

(a) Down stroke
(valve opening)

(c) OHC lubrification

Fig. 1.29 Direct acting overhead camshaft hydraulic tappet follower

21

follower take-up clearance is maintained when the valve is in its closed phase, fully seated and gas tight. During the period of valve closure any expansion or contraction of the valve-stem, follower assembly length and cylinder-head height is automatically taken into account by the amount the thrust-pin moves in or out of the plunger.

Overhead cam (OHC) hydraulic cam follower lubrication
(Fig. 1.29(c))
For the hydraulic tappet follower to operate immediately when the engine is started after being stationary for some time, the oil supply in the horizontal feed passage between the vertical supply passage from the oil gallery to the hydraulic tappet follower and cam bearing must be prevented from draining down the vertical passage when the engine is stopped. This is achieved by enlarging the top end of the vertical oil supply drilling (passage) and positioning a protruding tube in the base of the enlarged hole, a ventilation orifice then partially seals the vertical passage.

Engine stopped
When the engine is stopped, oil in the vertical passage will drain back to the oil gallery, but the oil in the horizontal supply passage between the vertical passage and the hydraulic follower and camshaft bearing will be prevented from returning due to the protruding tube top end being above the oil level. Even so, a syphoning action might take place due to the emptying vertical passage leaving behind a vacuum in the enlarged space if the top end of the vertical drilling was completely enclosed. Therefore, a small ventilation orifice/jet is used to block partially the top exit from the vertical drilling. Thus, when the oil drains down, the vertical drilling air will enter and vent the emptying vertical drilling space via the restriction orifice, thereby preventing any tendency for the oil to be syphoned from the horizontal supply and storage passage.

Engine started
When the engine is started the supply of oil from the horizontal passage to the hydraulic follower and camshaft bearing is instantly available and, within a short span of time, oil will be pumped up from the oil gallery to continue the supply. At the same time a small oil jet will be projected onto the camshaft.

1.5.4 Hydraulic valve clearance compensating pivot post
(Fig. 1.30(a and b))

Description of compensating pivot post element
This type of zero tappet clearance mechanism utilizes the rocker follower end pivot as the hydraulic automatic compensating element. The hydraulic pivot end-post consists of a body unit, which screws into the cylinder head, and an integral pivot post and plunger located vertically as a sliding fit in the pivot-post body. The plunger has a circumferential groove machined near the top end of the plunger, with two radial holes to permit the passage of oil from the oil gallery into the plunger supply chamber, this takes place via a vertically machined slot intersecting the pivot-post screw thread cut in the cylinder head and an inclined hole drilled in the upper wall of the pivot-post body. A ball-check valve and retainer attached to the plug sealing the supply chamber in the base of the plunger isolates the pressure chamber formed between the plunger and pivot-post cylinder from the supply chamber, while the thrust-pin to rocker follower slack clearance is continuously being taken up by the extending tendency of the slack-spring. The clearance fit between the plunger and pivot-post cylinder walls is designed to provide controlled leakage of oil from the pressure chamber when the valve is in its opening period and the rocker follower and pivot post is under load.

Downstroke (valve opening)
(Fig. 1.30(a))
When the cam lobe pushes down the rocker follower to open the valve, a reaction load is applied to the pivot post which tends to reduce the pressure chamber space created by the gap between the base of the pivot-post cylinder and plunger. Instantly, the oil pressure rises in the pressure chamber causing the ball-valve to close rapidly. Consequently, the trapped oil now acts as an incompressible substance thereby preventing the plunger moving further downwards. As a result, cam lift motion is relayed to the valve-stem via the rocker follower without any loss of slack clearance take-up. The minute initial inward movement of the plunger before valve lift commences, ensures that the valve fully seats during its closed period.

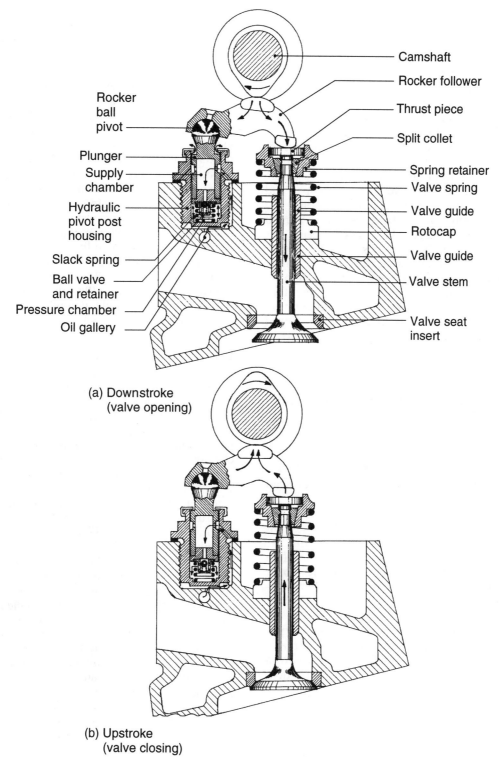

Fig. 1.30 Direct acting overhead camshaft end pivoted rocker follower with hydraulic valve clearance compensating element

Upstroke (valve closing)
(Fig. 1.30(b))

After the cam lobe has completed its valve lift period the cam contact changes from flank closing to one of base-circle, the reaction load is therefore removed from the pivot post, thereby allowing the slack-spring to move the plunger upwards so that full contact between rocker follower to cam and pivot-post is maintained. The very small increase in the pressure chamber space therefore removes the load pressurization of the trapped oil to a level equal to or just below supply pressure.

As a result, the ball-valve is forced to open and oil from the supply chamber thus flows into the pressure chamber to compensate for the small amount of oil leakage which hence occurs between the plunger and pivot-post cylinder walls.

Subsequently, there is a recuperative filling phase during each valve closure period following the valve opening period load phase, it therefore provides the automatic valve train take-up to compensate for the expansion and contraction of the valve actuating layout between cold start to fully warmed-up operating conditions.

1.5.5 Rocker arm operated overhead camshaft hydraulic tappet
(Fig. 1.31(a and b))

Description of hydraulic rocker arm tappet

The normal rocker arm tappet screw is replaced by the hydraulic tappet unit.

It comprises a tappet cylinder body bored out to house a tappet end cap, fixed plunger, non-return ball-valve and a slack expansion-spring.

Oil is supplied from the rocker-shaft central oil passage to an enlarged upper part of the tappet cylinder housed in the rocker-arm at the valve-stem end.

A valve-stem cap with an upper spherical depression and lower flat contact face is incorporated between the tappet and valve-stem to maintain a uniform distributed load transference from the tappet to the valve as the rocker-arm articulates.

Downstroke valve opening
(Fig. 1.31(a))

When the engine is running the revolving cams periodically lever-up the contact shoe situated at the end of the rocker-arm, thus compelling the rocker to pivot accordingly. This motion is relayed in a clockwise direction to the tappet and valve. Thus, when the cam profile commences to lift the rocker, the tappet will instantly be forced against the valve-stem and the opposing valve-spring load. Consequently, the pressure chamber space tends to decrease causing a corresponding pressure rise, immediately the ball-valve snaps onto its seat, the trapped oil then acts as a rigid member so that the full downward motion of the rocker is transferred to the valve-stem, thereby opening the valve. For operational purposes the high oil pressure build-up in the pressure chamber causes a small deliberate seepage of oil between the plunger and tappet cylinder walls and, as a result, the tappet compresses up to a maximum of 0.1 mm during the valve lift period. This leakage feature is necessary as it enables a very small degree of slackness to be created between both the cam-base circle and the valve-stem tip, which is essential for hydraulic compensating action to take place.

Up stroke valve closing
(Fig. 1.31(b))

As the camshaft rotates, the cam lobe moves beyond the contact shoe of the rocker arm thus permitting the valve to close, this enables the slack-spring to expand by pushing apart the tappet cylinder and the plunger until the tappet clearance has been eliminated. Immediately, the pressure chamber volume increases, but conversely this decreases the oil pressure in the chamber. As a result, the supply chamber oil pressure snaps open the ball-valve permitting the supply oil to transfer rapidly into the pressure chamber to compensate for the oil lost.

The actual amount of oil entering the pressure chamber will depend on the expansion and contraction in the valve actuating train which is directly related to the varying tappet clearance as the engine's driving conditions change.

The cycle of pressure chamber oil-loss and top-up respectively, over the opening and closing periods of the valve automatically forms a variable load bearing hydraulic element which eliminates the normal tappet clearance and yet permits the valve to be fully seated when it is operating during its closed period.

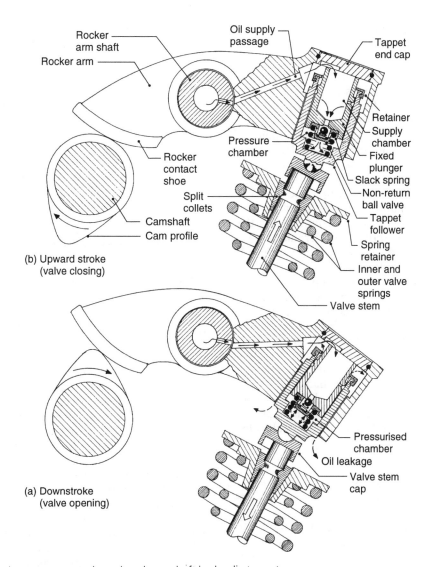

Fig. 1.31 Rocker arm operated overhead camshaft hydraulic tappet

1.6 Cams and camshafts

The purpose of the cams and their shaft is to actuate and control the opening and the closing intervals of the inlet and the exhaust poppet valves relative to each piston position along its respective stroke.

A cam is a rotating member which imparts reciprocating motion to a second member (known as a follower) in a plane at right angles to the axis of the cam by forming with it a slip rolling pair.

The type of cam normally used to actuate the engine valves is referred to as a radial-plate cam, it resembles a circular disc (plate) with a radially protruding profile, called a lobe, which occupies about one-third of its circumferential periphery. The cam profile is suitably contoured to provide a smooth rise and fall for the predicted motion of the follower which, accordingly, is then relayed to the individual poppet valves.

Each pair of inlet and exhaust cams may be cast or forged in position integrally on a single camshaft or twin camshafts can be adopted—one for the phasing of the inlet cams and, similarly, the other for the exhaust cams.

25

1.6.1 Rotational speed ratio between camshaft and crankshaft

The completion of one four-stroke cycle is equivalent to two revolutions of the crankshaft. During this cycle the inlet valve opens to allow the cylinder to be filled with charge, after which it closes and, likewise, the exhaust valve opens to expel the burnt gases and then closes. Therefore, each cam must open and close its respective inlet or exhaust valve once every two revolutions of the crankshaft. Thus, the camshaft must be made to complete one revolution or one opening and closing cycle for every two revolutions of the crankshaft. Consequently, the crankshaft-to-camshaft drive is arranged so that the camshaft rotates at half crankshaft speed.

Now, each cam-lobe must keep the inlet or exhaust valve open for its whole crank angle opening period, this being roughly equivalent to 240° crank-angle movement which can be seen from any valve timing diagram. Since the camshaft only completes one revolution for two crankshaft revolutions, the camshaft opening angular movement will only be half that of the crankshaft opening valve period. Hence, for an inlet or exhaust valve crank-angle opening period of 240°, the cam-lobe lift equivalent angular displacement would be 120°.

1.6.2 Cam profile phases
(Fig. 1.32)

There are two sides of the cam profile: these are the follower lift or valve opening side and the follower fall or valve closing side (Fig. 1.32). Each side of the profile may be considered as taking place in three phases. On the valve opening side there is an initial transition opening ramp phase which joins the base-circle to the cam-lobe flank with a very small lift rate. Secondly, there is the flank opening phase which accelerates the follower lift to a maximum velocity, this position on the cam profile is the point of inflection at which the flank concave curve meets the nose convex curve, and thirdly, there is the nose opening phase which decelerates a maximum follower lift velocity to zero velocity as the follower approaches its full lift position. The deceleration force on the third nose phase is provided by the valve-spring load.

The second-half of the cam profile is concerned with the return of the follower from full to no lift.

This cam angular movement commences with the closing nose phase causing the follower to accelerate from zero velocity to its maximum velocity, under the influence of the valve-spring, the cam profile then decelerates the follower's fall during the closing flank phase to almost a standstill. The follower's fall then changes into a gentle ramp closing phase to the point where it blends in with the base-circle dwell period during which time the follower remains in its lowest position, the valve will then be fully closed until the next opening period occurs.

1.7 Cam design

1.7.1 Three-arc cam
(Figs 1.33 and 1.34)

The early manufactured cams were simply formed from a base circle and a smaller nose circle which has an offset or eccentricity equal to E_N (Fig. 1.33). The two circles are then joined together by flank arcs of radius or eccentricity E_F. The maximum flank acceleration is proportional to E_F and the maximum nose deceleration is proportional to E_N. Normally, the ratio of flank-to-nose eccentricity $= E_F/E_N$ varies between 2 and 3 as this provides the best compromise for cam flank follower lift acceleration and deceleration and cam nose spring controlled acceleration and deceleration (Fig. 1.34).

To provide a gradual and smooth initial follower take-up and final return to base circle, additional arc ramps are introduced between the base circle and both opening and closing flank sides. These low-lift ramps therefore ensure that the initial rise and fall of the follower occurs with the minimum of acceleration and deceleration, respectively.

1.7.2 Constant velocity cam
(Fig. 1.35)

Profile geometry and characteristics
A better insight to the construction of the cam profile may be obtained by stretching out the lift-angular region of the cam into a straight line to produce a base and having the cam rise as the vertical axis (Fig. 1.35).

The simplest of cam profiles is a straight-line

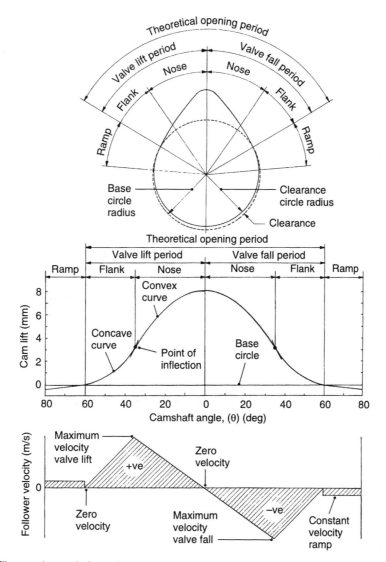

Fig. 1.32 Cam profile opening and close phases

outstroke rise end on to an intermediate full-lift dwell period followed by a straight-line return stroke. Because the cam profile slopes are uniform, both the lift and fall of the follower take place at constant velocity.

The shortcoming of this type of profile is the jerky change from the dwell on the base circle to the constant velocity opening flank and, likewise, the sudden change from the rising flank to the full-lift dwell interval. Similarly, at the beginning of the return stroke the follower motion abruptly changes from full-lift dwell to the constant velocity closing flank return slope, the closing flank slope then again abruptly merges into the fully lowered base-circle dwell interval. Accordingly, there is an instantaneous infinitely high acceleration (∞) and deceleration (∞) at the start and end of the opening flank, and similarly, there is an equally high instantaneous follower acceleration and retardation at the beginning and at the end of the closing flank. Since the rise and fall cam slopes are uniform, there is no acceleration and deceleration in between the starting and finishing interval of both the opening and closing flanks.

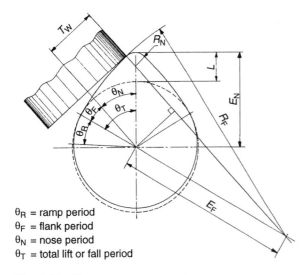

θ_R = ramp period
θ_F = flank period
θ_N = nose period
θ_T = total lift or fall period

Fig. 1.33 Three-arc cam geometry

Summary
(Fig. 1.35)

With a constant velocity cam profile there is no follower acceleration or deceleration during its rise and fall on the opening and closing cam flanks respectively. However, the follower is subjected to very high impact inertia forces at the commencement of its rise and at the end of its return, whereas at the end of its lift and at the beginning of its fall the follower is subjected to very large separating inertia forces.

Consequently, cams with this simple profile operating at even low speeds would subject the follower and cam to very high impact stress with its associated noise, surface wear and damage. At the same time, this form of profile would require a very large spring load to control the follower jump-off when the valve has just reached full lift and is about to commence its return stroke.

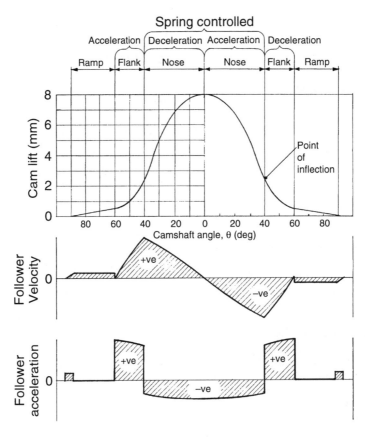

Fig. 1.34 Three-arc cam characteristics

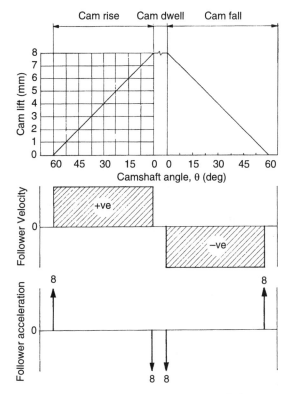

Fig. 1.35 Constant velocity cam characteristics

1.7.3 Constant acceleration (parabolic) cam
(Fig. 1.36)

Profile geometry and characteristics
Sometimes this cam is referred to as parabolic since the profile is derived from two parabolic arcs.

The cam construction is obtained by dividing the angular displacement of the cam and the lift of the follower into the same number of equal divisions and then drawing vertical and horizontal lines through these points (Fig. 1.36). Diagonal lines are then projected from the start and end lift points of the cam profile towards each of the intersecting horizontal lines onto the vertical mid-lift line. Points are then plotted along each diagonal line where they intersect the corresponding vertical division lines, these points are then joined by a smooth curve to form the constant acceleration cam profile.

As can be seen, the cam rise profile consists of two parabolic curves meeting at the point of inflection. At the commencement of lift on the concave curve part of the profile, the rate of the follower rise is small but then quickly builds up to a maximum (greatest gradient) at the point of inflection. The second half of the profile forms a convex curve starting with its steepest slope (highest lift rate) which then reduces to a gentle slope as the follower approaches full lift.

Consequently, the follower lift velocity rises directly in proportion to the angular displacement of the cam up to a maximum at the point of inflection and then decreases at a similar rate to zero as the follower reaches its full outstroke position.

Since the follower outward velocities increase and decrease uniformly on the concave and convex halves of the follower lift profile respectively, the corresponding positive acceleration and negative deceleration of the follower during lift will be constant.

Summary
(Fig. 1.36)
The constant acceleration curve profile provides the least maximum acceleration and retardation compared with all other predicted curve profiles considered, i.e. constant velocity, simple harmonic motion, cycloidal and multi-sinewave cams.

With this profile there is an abrupt rise in the follower's inertia force when the base circle meets both the opening and closing flanks and, at full lift, if there is a maximum opening dwell period. Likewise, at the transition point of inflection where the steepness of the cam profiles is changing from increasing to decreasing, the follower's inertia force instantaneously switches from a positive impact to a negative separating force. This condition therefore requires a considerably large spring load to maintain the follower-to-cam contact.

1.7.4 Simple harmonic motion (SHM) cam
(Fig. 1.37)

Profile geometry and characteristics
This profile is obtained as follows. Draw a semi-circle of diameter equal to the follower stroke on the vertical axis and divide it into any convenient number of equal parts. Divide the angular displacement during the outward stroke and inward stroke of the follower into the same number of equal parts, and then draw lines vertically along

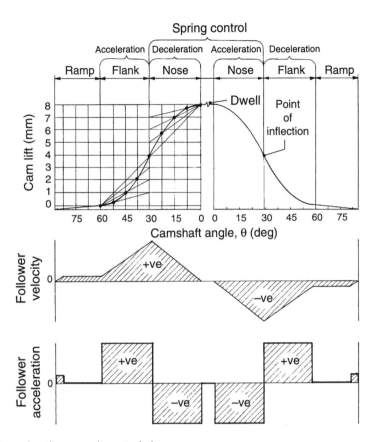

Fig. 1.36 Constant acceleration cam characteristics

the angular displacement axis and horizontally from the angular divisions on the semicircle. Points are then plotted on the intersecting lines, and these are joined by a smooth curve to produce the SHM cam profile (Fig. 1.37).

With the simple harmonic motion profile, the velocity of the follower on the first half of lift, rises to a maximum at the point of inflection, and then decreases to zero over the second half of the lift in a sine-wave fashion. Conversely, since the velocity curve slope is greatest at the start and finish of the rise profile, and is a minimum at mid-stroke (point of inflection), the positive acceleration and negative deceleration is at a maximum at the start and finish of the follower lift. Conversely, there is no acceleration or deceleration at the point of inflection, this being a point on the cam profile where it is transformed from a concave curve to a convex one.

The acceleration and deceleration over the profile's outward and inward strokes with respect to the angular movement of the cam assumes a cosine form. Since the velocity curve slope is zero at mid-stroke, there is no rate of change of velocity in this position, therefore the acceleration and deceleration reduces to zero at the profile's inflection point.

Summary
(Fig. 1.37)
The simple harmonic motion velocity curve rises steeply at the start of follower lift but then steadily bends over towards the point of inflection. Thus, where the flank and nose curves meet, the slope will be horizontal. The velocity then gently decreases, finally changing into a steep decline to zero as the follower approaches full lift.

The rapid increase and decrease in follower velocity at the beginning and end of the lift or fall dictates that the initial positive inertia force and the final negative inertia force will be sudden and high. Conversely, the very gradual rise and then fall in follower velocity on either side of the point

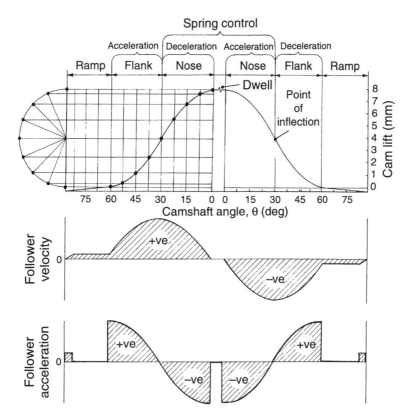

Fig. 1.37 Simple harmonic motion cam characteristics

of inflection where the flank and nose curves meet provides a progressive change from a positive impact inertia force to a negative separating inertia force. As a result there will be relatively high impact jerks when the follower commences to rise and when it has completed its fall. In addition, there will be a similar jump-off and return jerk at full lift if there is a dwell period at maximum lift.

However, due to the gradual change of follower inertia force from a positive compressive force to a negative jump-off force at the point of inflection, only a relatively light spring load will be needed to control the follower motion during the opening and closing nose intervals.

1.7.5 Cycloidal cam
(Fig. 1.38)

Profile geometry and characteristics
The cycloid is the curve traced by a point on the circumference of a circle that rolls without slipping on the vertical follower lift axis (Fig. 1.38).

The cycloid is constructed as follows. Make the circumference of the rolling circle equal to the follower lift, and draw one circle at the start of the cam lift. Divide the follower lift and the circle into the same number of equal parts, in this case eight, and then erect perpendiculars through each division on the horizontal line and draw vertical lines through each division on the lift circle. Now, draw the remaining eight circles using each centre in turn. The cycloidal locus is then plotted at points where each circle cuts the division circle's vertical lines, a smooth curve is then drawn through each plotted point.

Finally, the cycloidal cam profile is produced by projecting horizontal lines from each cycloidal plotting across the angular displacement vertical lines marking the intersecting points and drawing a smooth curve through these points.

During the first part of the follower lift, its velocity increases slowly, it then moves rapidly and then slows down as the velocity reaches its peak value at the point of inflection on the cam profile. At the beginning of the second half of the follower lift, its velocity is reduced slowly, and

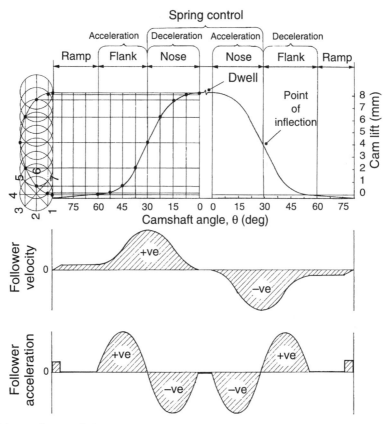

Fig. 1.38 Cycloidal cam characteristics

then at a faster rate, and finally it decreases very slowly to zero.

The follower acceleration and deceleration during its lift is the rate of change of the follower outstroke velocities. The acceleration of the follower at first rises fairly steeply, then rises more slowly as it reaches its peak value, which occurs at the point of inflection on the velocity diagram's rising side. The follower outward acceleration then reduces at an increasing rate to zero as the maximum follower velocity is reached, corresponding to the cam profile inflection point. The follower then decelerates initially quickly and then more slowly as the maximum deceleration is approached, which is also the point of inflection on the velocity diagram's declining side. Finally, the follower lift deceleration reduces at an increasing rate until full follower lift is reached—at which point the deceleration will be zero.

Summary
(Fig. 1.38)
The characteristics of the cycloidal profile is its very low initial and final follower rise rate—this cam profile is difficult to generate accurately by grinding without using equipment which can work to very fine tolerances.

During the flank lift, the follower's inertia produces a positive impact force which rises to a peak over half the flank interval and then reduces to zero over the second half flank interval. The change-over from flank to nose contour converts the follower's inertia positive impact force to a negative separating force as the follower's velocity approaches a maximum. This negative separating force will increase to a maximum at the mid-nose interval, it then decreases to zero as full lift is approached.

Thus, it can be seen that as the follower approaches the start of the opening flank a positive impact force is created which rapidly, but not instantly, increases; and towards the end of the

32

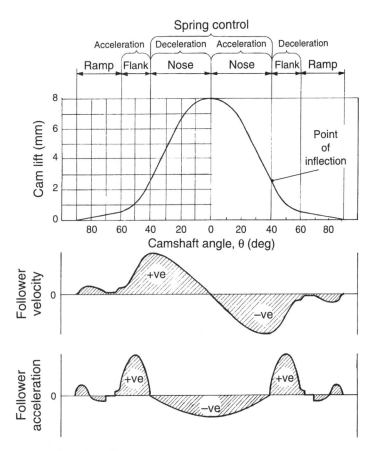

Fig. 1.39 Multi-sine-wave cam characteristics

closing flank interval the follower's inertia positive impact force rapidly, but not instantly, decreases to zero.

Similarly, as the follower moves towards the end of the opening nose interval, and when it begins the closing nose interval, its inertia—which is of a negative separating nature—rapidly (but not instantly) reduces to zero and then, at the commencement of its return stroke, rises at the same rate as it decreased.

In addition, the change in follower inertia force from positive impact to a negative separating one produces a smooth and gradual transition which passes through zero at the point of inflection where the flank and nose curves meet.

The cycloidal cam profile minimizes the initial and final impact loads imposed on the opening and closing flanks and enables good spring control to be achieved over the nose opening and closing intervals.

1.7.6 Multi-sine-wave cam
(Fig. 1.39)

Profile geometry and characteristics

As with all the other profiles examined, the multi-sine-wave profile is derived from a concave curve which commences with the follower lift and increases its velocity to a maximum. At this point, the profile changes to a convex curve, this then reduces the follower velocity to zero as full lift is approached (Fig. 1.39).

The first flanked interval raises the follower according to the function $L = -a_1 \sin b_1 \theta + c\theta$, whereas the second interval completes the lift by obeying the law $L = a_2 \sin b_2 \theta$ where

L = follower lift
θ = angular movement of camshaft
a_1, a_2 = amplitude of the half sine wave of the first and second interval respectively
b_1, b_2 = the number of periods per revolution of

33

camshaft for the first and second intervals respectively

c = the constant slope added to the negative half-period sine wave comprising the first interval

During the first interval up to the point of the inflection, the follower's acceleration increases to a peak and then decreases to zero as maximum velocity is reached. In the second lift interval, from the inflection point to maximum profile lift, the follower's outward velocity is decelerated at an increasing rate to a standstill.

Summary
(Fig. 1.39)
A study of the multi-sine-wave cam profile would show a relatively flat flank-to-nose contour compared with the three-arc, the constant acceleration and simple harmonic motion cams which tend to have a more rounded lift contour.

Excluding the quietening ramps and having a two-to-one nose-to-flank angular interval ratio, the follower positive velocity rises steeply and then curves over as it peaks at the point of inflection where the short flank curve joins the much longer nose curve. The follower's velocity then gently decreases over the much larger nose interval to zero as full lift is reached, it then takes on a negative value as the follower commences its return stroke until the maximum inward velocity is reached at the closing nose-to-flank point of inflection. Finally, the follower's inward velocity begins to decrease fairly rapidly from its peak negative value, to zero, as the follower moves down the closing flank to the beginning of the quietening ramp.

It is worthwhile at this stage reminding ourselves that acceleration and deceleration are derived as the rate of change of velocity with respect to an increment of time or angular cam displacement, and that the inertia force produced by the follower is equivalent to its mass times the outward or inward-moving follower's acceleration or retardation, at any point on the cam profile.

As can be seen, the follower's acceleration growth and then the deceleration decay, over the opening and closing flank intervals, resemble a steeply rising arch, and they also represent the magnitude of the follower's positive impact inertia force as it is raised and lowered by the cam flanks. Conversely, when the follower motion is predicted by the opening and closing nose profile the follower inertia force takes on a negative

separating (jump-off) nature. Thus, the inertia forces created due to the deceleration of the follower during the open and closing nose interval will gently increase to a maximum at full lift and then decrease to zero at a similar rate over the closing nose interval.

In effect, this cam is subjected to moderate opening and closing flank jerk loads and, during the transition to nose lift and fall, the jump-off inertia forces are relatively low, enabling good spring control to be achieved at high engine speeds.

1.7.7 Cam quietening ramps
(Figs 1.32, 1.40 and 1.41)

Most cams these days have a bridging (transitional) lift contour between the base-circle upon which the opening and closing flanks start and end, respectively, and the clearance-circle: these joining contours are known as quietening ramps (Fig. 1.32).

The purpose of a quietening ramp is to ensure that the follower-to-clearance circle radial clearance is taken up only on the ramp portion of the cam and that the initial lift will always commence at the same, or almost the same, relatively low velocity provided that the tappet clearance is maintained between the upper and lower tappet clearance tolerance.

Therefore, the follower slack clearance take-up and the start of lift commences with the minimum of follower jerk.

It is usual for the tappet clearance to be eliminated and a very slow initial lift to begin at mid-ramp height so that a slight tappet clearance variation—due to expansion or contraction of the valve actuating train, incorrectly adjusted clearance or valve train wear—shifts the nominal follower to the ramp contact point to an earlier or later contact position but still within the ramp interval.

Constant velocity ramps
(Fig. 1.40)
The simplest of ramps have uniform rising and falling slopes and therefore provide a constant velocity follower lift within their angular displacement interval, but they suffer from high instantaneous accelerations of the follower at the instant it passes from the clearance-circle to the ramp contour (Fig. 1.40).

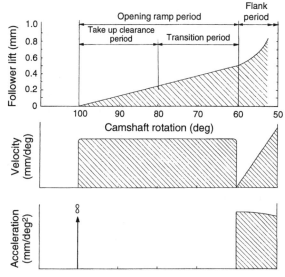

Fig. 1.40 Constant velocity (eccentricity) ramp

The slope of these ramps usually ranges between 0.0125 and 0.035 mm per camshaft degree and the total lift caused by the ramp may be anything between 0.5 mm and 0.65 mm. These maximum ramp rises correspond roughly to angular camshaft displacements of 20° to 50°.

When hydraulic cam followers are incorporated which automatically maintain a very small but constant clearance, much lower ramp heights are used, typical values being 0.08 mm for the opening ramp and 0.25 mm for the closing ramp.

Variable velocity ramps
(Figs 1.41 and 1.42)
The tappet clearance is taken up at a variable velocity as the follower is lifted by the quietening ramp. Initially the velocity is very low but quickly rises. The follower lift velocity is then steadily reduced until all the clearance has been eliminated, and the valve itself starts to lift with a velocity, which is roughly only one-third of that of a conventional constant velocity ramp (Fig. 1.41).

The transition from the tappet clearance take-up and initial low-velocity lift to the actual flank profile lift is achieved in two stages. Firstly, there is a constant velocity interval during which the ramp rise actually begins, and secondly there is a constant acceleration ramp interval which merges

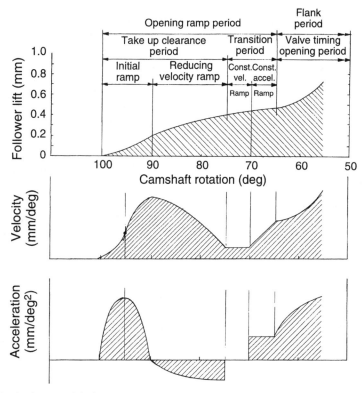

Fig. 1.41 Variable velocity (eccentricity) ramp

35

the low shift constant velocity ramp portion with the steeply rising flank contour.

At low engine speed the tappet clearance is taken up at the beginning of the constant velocity ramp interval so that a very low-lift rate commences covering the constant velocity interval, which then leads onto a constant acceleration lift interval. The follower lift, having passed through the transition from tappet clearance take-up at relatively low velocity and acceleration, is then subjected to the much higher lift-rate of the flank interval.

As the engine speed rises the whole valve train becomes increasingly compressed and strained due to the reciprocating inertia forces of the valve train so that the commencement of the follower lift is progressively pushed back from the beginning of the constant velocity ramp interval. Thus, in effect, the effective lift interval of the constant velocity portion of the ramp becomes shorter or, putting it another way, the follower does not begin to lift until the compressive deflection of the valve train has been taken up further back but still within the low constant velocity ramp interval (Fig. 1.42). Accordingly therefore, the start and

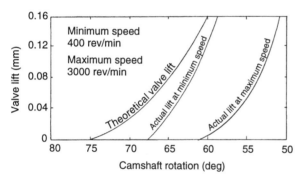

Fig. 1.42 Variation of valve left at low and high engine speeds due to the inertia loading of the actuating mechanism and cam

end of the follower lift and fall transition on the opening and closing ramps will always take place at an approximately similar low velocity throughout the engine's speed range.

1.7.8 Valve timing
(Figs 1.41 and 1.42)

Valve opening and closing are assumed to take place at the nominal points where the opening and closing quietening ramps join their corres-

ponding opening and closing flank contours. In fact, the valve timing is designed so that the tappet clearance has been eliminated at roughly mid-ramp height—this could mean the cam would rotate anything from 10° to 20° further or later before the follower actually contacts the much faster rising flank profile (Figs 1.41 and 1.42). However, this is not a real problem because the rate of valve opening or closing during the ramp intervals is very low indeed. Therefore, for all intents and purposes the valve opening and closing can be said to commence at the start or end of the opening and closing flank. This is because a follower rise or fall between the angular movement between mid-ramp and flank may be no more than between 0.15 and 0.25 mm, which will hardly influence the fresh charge or gas flow effectiveness caused by an early start or the later completion of the induction or exhaust periods.

Consequently, when the engine valves are timed to coincide with the relevant crank angle position, the inaccuracy of determining the actual opening point of the valve when the follower has taken up the tappet clearance on the ramp, has made engine manufacturers provide a much larger tappet clearance just for timing purposes. Thus, the timing tappet clearance is chosen so that the clearance is just taken up at the beginning of the flank lift at the point where the slightest forward angular movement of the cam will cause the valve to lift. The point at which the valve actually commences to open can be sensed by rotating the push-rod between finger and thumb until it becomes tight.

After the valve timing has been checked and set, the tappet clearance for the valve which has been timed (usually the last inlet valve) must be readjusted to its normal operating tappet clearance.

1.7.9 Valve train inertia forces versus cam profile and spring control
(Figs 1.34, 1.36, 1.37, 1.38 and 1.39)

Any to and fro motion of the valve train (valve, rocker push-rod, follower etc) produces inertia forces which may be positive or negative. Positive forces are regarded as those which tend to press the cam follower onto the cam contour, whilst negative forces tend to separate the two. Negative forces are the critical forces, as they must be opposed by the valve return-springs. If these negative inertia forces are large, stiff return-springs will become necessary to counteract the undesirable jump-off response to the negative forces. Unfortunately, the same spring load is added to the inertia load during periods of positive acceleration and deceleration which will increase cam and follower stress and, accordingly, will cause wear and damage to the heavily loaded cam flanks and follower contact faces.

The motion of the follower is only controlled positively by its inertia bearing onto the cam contour opening and closing flanks during the first part of its outstroke and by the second part of its return stroke, whereas the opening and closing nose-to-follower contact on the second half of the outstroke and the first part of the return stroke is maintained solely by the valve return-spring stiffness.

The energy needed to accelerate the follower's initial lift on the cam flank must be equal to the energy dissipated in retarding the follower on the final part of the nose lift. The enclosed areas above and below the zero line on the acceleration and deceleration diagram are equal as they represent the energy absorbed in lifting and lowering the follower.

For the valves to operate effectively the cam must satisfy two basic requirements:

1 a rapid opening and closing of the valves to maximize the breathing and the exhausting of the cylinders;
2 a low deceleration as the follower approaches full lift on its outstroke and a low acceleration when it commences its return stroke so that the valve return-spring force is minimized.

These two desired requirements are partially achieved by shortening the flank interval and extending the nose interval within the valve opening or closing periods (Fig. 1.39). Accordingly, the point of inflection is moved to make the angular nose interval approximately twice that of the flank angular interval. By these means the follower has a higher initial flank acceleration and, correspondingly, a higher final flank deceleration to make the valves open and close rapidly, and at the same time, the reduced retardation and acceleration on both sides of the cam nose enables the follower to maintain its contact with the cam contour with a relatively small valve-spring force.

1.8 Valve actuating mechanism considerations

1.8.1 Criteria for good design of valve actuating mechanisms

1 Minimum valve train inertia.
2 Minimum rubbing velocity between cam and follower as this can cause lubrication problems.
3 Minimum contact pressure between cam and follower as this can produce surface scuffing.
4 Maximum rigidity of camshaft bearing mounting.
5 Maximum stiffness of the cam, follower, push-rod and rocker arm etc.
6 Minimum length of push-rod by adopting a high mounted camshaft.
7 Minimum length of rocker-arm or rocker-follower to minimize the rocker moment of inertia.
8 Minimum valve lift necessary to exhaust and fill the cylinder as this will help to keep down cam acceleration and deceleration.
9 Minimum cam profile lift acceleration and deceleration to enable similar movement to be faithfully transferred from the cam to the valve.
10 Minimum spring load necessary to prevent valve jump at high engine speed.

1.8.2 A comparison of direct and indirect cam operated poppet valves
(Figs 1.29(a and b), 1.30, 1.43(a and b) and 2.41(a and b))

A high spring rate is no substitute for compensating for poor cam and actuating mechanism design and only serves to over-stress the components and cause rapid wear. Cam opening and closing designed lift motion may be accurately relayed to the valve at relatively low camshaft speeds, but due to the elasticity in the actuating valve mechanism, at high speeds the dynamic forces will distort the transmitted motion from the cam to the valve so that the actual valve lift will not be a true copy of the original cam profile.

With a rocker arm and push-rod valve lift mechanism (Fig. 2.41(a and b)) the cam lift is multiplied by the rocker arm leverage ratio whereas with the direct acting OHC using an

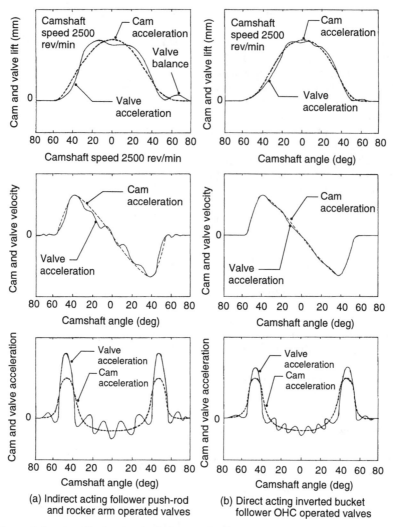

Fig. 1.43 Comparison of direct and indirect actuated poppet valve

inverted bucket follower (Fig. 1.29(a and b)) the valve lift can only equal the cam lift. Therefore, a larger cam lobe must be used with direct acting overhead cams compared with a push-rod actuated valve if similar maximum valve lifts are to be obtained. Consequently, the rubbing velocity between the cam and follower will be higher for OHC relative to the push-rod operated valve mechanism. As a result of having to use a bigger cam lobe a larger diameter inverted bucket follower is necessary so that its contact face extends to the full rubbing length of both the opening and closing sides of the cam profile, thus the OHC follower will be relatively heavier than a push-rod follower used to obtain a similar valve lift.

The inertia of the rocker arm could be considerable, but fortunately it is the moment of inertia about the rocker shaft from which it pivots that forms part of the valve train inertia and this is normally very small. A very popular OHC valve actuating mechanism uses a horizontal rocker arm or beam follower, this pivots on a ball stud mounted on the cylinder head at one end whereas the other free end rests on top of the valve stem (Fig. 1.30). The cam which is mounted over the arm is positioned roughly two-thirds from the pivot end and therefore provides something like a 1.4–1.8 leverage ratio, it therefore enables a smaller cam lobe to be employed for a given valve lift compared with the direct acting inverted bucket follower OHC arrangement. Therefore, this rocker follower OHC layout not only reduces the inertia of the valve actuating mechanism to that of the moment of inertia of the rocker follower about its pivot, but also reduces the size and rubbing velocity of the cam lobe. The moment of inertia for the end pivoting OHC follower arm is larger than for the centrally pivoting push-rod rocker arm layout.

Thus, comparing the magnitude of the inertia levels involved with push-rod and OHC-operated valve mechanisms, the inertia forces of the two OHC arrangements described may be very slightly lower than the push-rod layout but the difference is far too small to justify the extra cost of adopting an OHC arrangement on these grounds alone.

The real benefits of the OHC-operated valve mechanisms exist in the increased rigidity provided by the direct acting inverted bucket follower and the end pivoting rocker follower when transferring the cam lift motion to the valve stem, as compared with the more flexible push-rod and rocker arm valve actuating layout. A very stiff, rigid, lift and compact valve operating mechanism permits each valve to be opened and closed faster and yet precisely without creating undue vibration which would cause the follower and valve to bounce and rebound from the cam lobe profile and from the valve seat in the cylinder head.

Thus, a very stiff and rigid OHC valve actuating configuration enables the follower and valve to follow the designed opening and closing cam lift and fall profile more accurately under the varying acceleration and deceleration conditions, whereas with the more elastic push-rod arrangement the small deflection which occurs in the push-rod and rocker arm is sufficient to set up some degree of superimposed oscillating movement into the designed cam lift-and-fall motion. Generally, a more precise control of the opening and closing of the valve periods improves the exhausting and filling of the cylinder and, therefore, smaller valve overlaps can be chosen to improve exhaust emission without sacrificing performance.

An ideally designed valve opening and closing mechanism should be able to rely on identical acceleration and deceleration patterns to the valve, as produced by the cam profile. This is never quite possible due to the inertia of the moving components and the small amount of elasticity and lack of rigidity of the mechanism which must always exist.

The reduction in elasticity and the improvement in rigidity of the inverted bucket follower

OHC as opposed to the push-rod operated layout can be compared by observing the acceleration and deceleration oscilloscope pattern traces at various camshaft speeds of both cam and valve-lift over their opening and closing periods (Fig. 1.43(a and b)). It can be seen that neither the OHC or push-rod operated valve actuating mechanisms are able to provide a perfectly relayed acceleration and deceleration motion from the cam profile to the valve. In fact, the push-rod arrangement produces a considerable increase in maximum opening and closing acceleration and that, instead of the acceleration curve rising and decaying smoothly, there are many secondary oscillations superimposed onto the main acceleration and deceleration curve—due basically to the elasticity and flexibility within the mechanism. As can be seen from Fig. 1.43(a and b) with the push-rod and OHC arrangements, the push-rod secondary oscillations of the valve are of much larger amplitude and are of a lower frequency than for the more rigid OHC mechanism. The more flexible push-rod mechanism relative to the inverted bucket follower OHC layout is reflected in the much greater deviation in the valve lift and fall over the opening period than that of the designed cam lift and fall (Fig. 1.43(a and b)). These graphs also illustrate the extended valve opening at the end of the designed opening period due to valve bounce, this being much more pronounced with the push-rod layout as opposed to the direct-acting OHC layout.

If the rocker arm follower-type OHC mechanism valve opening and closing response is compared with both the push-rod and direct acting OHC, then the ability of the rocker arm follower to provide the minimum valve acceleration and deceleration and to follow closely the designed cam profile lift movement would lie somewhere between the two extremes; that is, between the push-rod and direct acting OHC arrangements.

1.8.3 Valve spring control
(Figs 1.44 and 1.45)

The stiffness of the valve return springs plays an important part in making the valve lift and fall closely follow the designed lift and fall of the cam. If the rate is very high (strong) large stresses will be imposed on the actuating components and the severe pressure between the rubbing surfaces of the cam and follower may cause scuffing due to a breakdown in the separating film or lubricant.

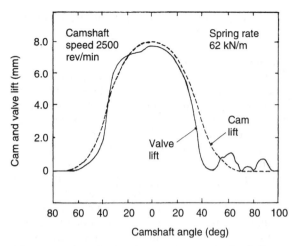

Fig. 1.44 Standard spring rate cam and valve left movement

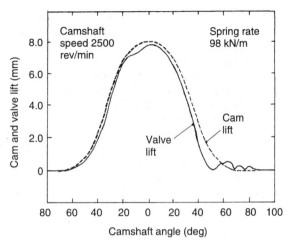

Fig. 1.45 High spring rate cam and valve left movement

Conversely, if the spring rate is very low (weak) there can be two side-effects at high speed. Firstly, the spring will not be able to keep the follower in contact with the cam profile as the follower motion changes from acceleration to deceleration in the mid lift-to-nose region of the cam. Secondly, when the valve closes the relatively small spring load will be insufficient to prevent the impact of the valve seating, making it bounce and rebound (Fig. 1.44) at least once or twice before it fully closes. Therefore, the engine designer has to select a spring stiffness (spring rate) (Fig. 1.45) which is able to control the lift-and-fall motion of the valve without causing excessive cam and follower surface wear.

2
Camshaft Chain Belt and Gear Train Drives

2.1 Camshaft timed drives

The timed rotational drive from the crankshaft to the camshaft may be transmitted by one of four methods:

1 roller chain;
2 inverted tooth chain;
3 toothed belt;
4 gear train.

2.1.1 Roller chains

Construction
(Fig. 2.1(a and b))
The roller chain consists of a series of inner and outer side plate links which are coupled together by pin and bush joints. The opposed inner links are held together by seamed bushes which are forced into pairs of holes punched in each link-plate, whereas the outer opposing plates are clamped together by bearing pins which slip through the inner plate bearing bushes and are then forced through the outer plate holes before being swagged at their ends. Articulation of each joint can then take place by each bearing pin rolling inside its respective bearing bush. To reduce chain-to-sprocket tooth friction, loose fitting seamed rollers slip over each bush when the chain is initially assembled so that when the chain is driving, the rollers revolve about their bearing bushes as they roll over the teeth profiles. The bearing pins and bushes and the rollers are case hardened to reduce wear between the rubbing surfaces.

Roller chain drive considerations
(Fig. 2.1(a, b and c))
Roller chains can be used to drive complicated and awkward-positioned camshaft and auxiliary shafts which may be a short distance or a considerable distance apart. Roller chains are particularly suitable for camshaft drive layouts subjected to relatively large crankshaft-to-camshaft centre distance variations, due to cylinder-head and block expansion and contraction caused by changing operating conditions. Roller chains are mounted on sprocket wheels which are attached to the crankshaft and either to a single or twin camshaft, or sometimes to a third shaft. This third shaft may drive the distributor, injection pump, oil pump, coolant pump, hydraulic power steering pump, exhauster, air compressor, generator etc.

An automatic chain tensioner is usually required to take up slackness in the chain caused by the chain stretching and sprocket wear and, for more positive drive control between long chain spans, restraint guide mechanisms are recommended. In some designs a jockey idler sprocket may be incorporated to increase the angle of chain wrap of the crankshaft sprocket or camshaft sprocket wheels in the case of twin camshafts. Generally, the chain should provide a minimum of 95°, preferably 120°, angle of lap for the smaller crankshaft sprocket and a minimum of 85°, preferably 95°, for the larger camshaft sprocket wheels and if there is a jockey idler

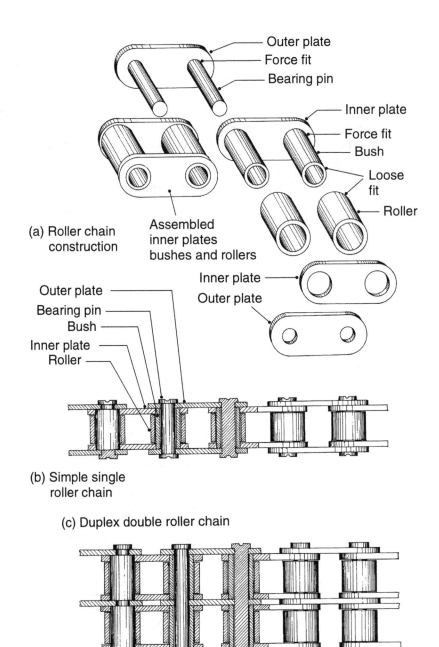

Fig. 2.1 Timing drive roller chain

sprocket, there should be at least three teeth in engagement with the chain, and the spacing between the jockey sprocket and adjacent drive sprockets should be at least four pitches of the chain apart or more.

The recommended minimum number of crankshaft sprocket teeth for low shock loading, wear and noise level is 23, but due to the usual dimensional limitations of chain drive arrangements, a minimum of 19 teeth for engines up to 1.6 litres and 20 teeth for engines having cylinder capacities of above 1.6 litres may be adopted.

The use of single, double (Fig. 2.1(b and c)) or triple roller chains will depend largely on engine type—petrol or diesel, the number of cylinders and their capacity, the maximum engine speed and the engine's applications. Normally, for petrol engines with capacities up to 2.0 litres, a simple (single) chain can cope, but for 2.0 litre to 5.0 litre engines a duplex (double chain) is necessary; whereas the increased driving torque for an injection pump on a diesel engine justifies a duplex (double) chain for all engines up to 5.0 litres and triplex (triple) chains for engine capacities above 5.0 litres.

The maximum number of sprocket teeth should be kept below 125 otherwise the smallest amount of pitch elongations between each link, due to wear, will cause the chain to ride and jump the sprocket teeth long before the chain is worn out. In practice, this situation is unlikely to happen since the maximum camshaft sprocket teeth generally never exceed 50.

For optimum chain performance the centre distance between shafts should be between 30 to 50 times the chain pitch: 40 times the chain pitch is about normal and 80 times the chain pitch is a maximum.

It is usual for a chain drive to have an even number of pitches in the chain; therefore, by using a crankshaft sprocket with an odd number of teeth, the times when the same sprocket teeth and rollers come into contact with each other are considerably extended, this then helps to produce a more uniform wear distribution over both chain and crankshaft sprockets.

Roller chain drive efficiencies are in the region of 98% to 99% under correctly tensioned and well-lubricated working conditions.

The life expectancy of a roller timing chain under favourable running conditions may span 160 000 km, 100 000 miles or even more.

Roller chain wear should not be allowed to vary the ignition distributor or injection pump timing by more than $\pm\frac{1}{2}°$ if low emission is to be maintained.

Roller chain chordal action
(Figs 2.1(a, b and c), 2.2(a and b) and 2.3)
The chain links form a polygon on the sprocket wheel with each adjacent side of the polygon intersecting at one of the chain joint centres (Fig. 2.2(a and b)). The pitch circle will therefore be

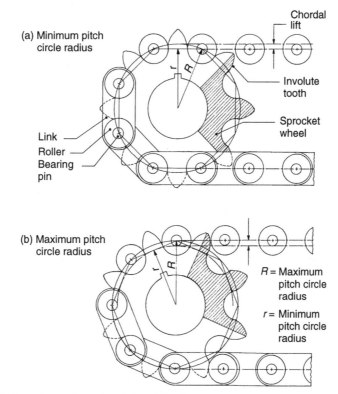

Fig. 2.2 Roller timing chain and sprocket wheel chordal action

43

equal to a circle drawn through each joint centre surrounding the sprocket and the chordal distance between adjacent joint centres is equal to the chain pitch. When the sprocket rotates one chain roller will engage a sprocket tooth on the drive side and equally disengage a sprocket tooth on the non-drive slack side; for an eight-toothed sprocket (Fig. 2.2(a and b)) this would occur every one-eighth of a revolution.

When a sprocket tooth is at right angles (at the top of the sprocket) to the lead in of the chain (Fig. 2.1(a)) the effective driving radius of the chain will be the distance from the sprocket centre to the chord, 'r', but as the sprocket revolves so that a roller becomes positioned at the highest point (top) of the sprocket, the effective driving radius of the chain increases to the distance between the sprocket and roller joint centres, 'R'. Now, the linear speed of the chain is directly related to the effective driving radius of sprocket, that is

$$V_1 = 2\pi r N \quad \text{m/min}$$
$$V_2 = 2\pi R N \quad \text{m/min}$$

where V_1 and V_2 = minimum and maximum chain linear velocity (m/min);
  r and R = minimum and maximum effective chain drive radii (m);
  N = constant sprocket angular speed (rev/min).

Therefore, if the sprocket rotates at a constant speed, the cyclic difference in effective chain driving radius $R - r$ will cause the chain speed to fluctuate correspondingly. The percentage speed fluctuation may therefore be given by

$$V = \frac{V_2 - V_1}{V_2} \times 100$$

where V = percentage speed variation. Thus, the chordal rise and fall of each chain pitch as it contacts a sprocket tooth is known as chordal action and results in repeated variations in linear chain speed. The fewer the number of sprocket teeth, the greater will be the cyclic speed fluctuation and vice versa—this can be seen in Fig. 2.3 which shows the relationship between the number of sprocket teeth and the cyclic percentage speed variation. It can be seen that the percentage speed variation decreases rapidly as the number of sprocket teeth increases. Above 25 teeth or more the speed fluctuation becomes almost insignificant.

A large chordal rise and fall will produce a

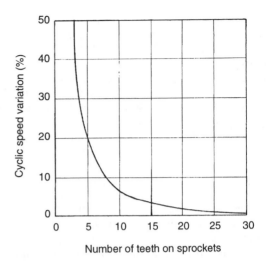

Fig. 2.3 Relationship between number of sprocket teeth and cyclic speed variation

correspondingly large acceleration and deceleration pulsation in the chain which can, in the extreme, produce a very jerky chain drive, noise and rapid wear.

Sprocket wheels
(Fig. 2.2(a and b))
The roller chain sprocket wheels have involute tooth profiles permitting the rollers to ride farther out on the teeth as the chain is stretched under load or the pitch increases as wear occurs, this profile thus distributes the load over a larger portion of the tooth surface.

For best crankshaft sprocket and chain compatibility a 0.6% carbon steel hardened to Brinell Hardness 180–220 is recommended whereas the larger camshaft sprocket wheel, which has a greater number of teeth in contact with the chain, can be made from close grained cast iron or even case hardened mild steel.

2.1.2 Inverted tooth chain (Morse HY-VO link)

Construction
(Figs 2.4 and 2.5)
The chain comprises a series of inverted (inward facing teeth) horseshoe-shaped steel links hinged together by pin and rocker joints (Fig. 2.4). The width of the chain is made up from adjacent inverted links alternately stacked with outer guide

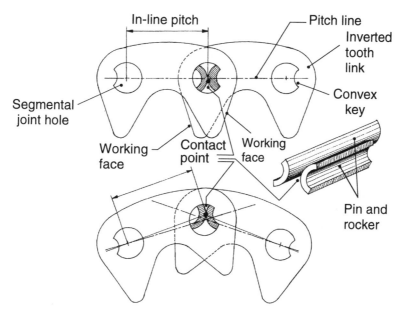

Fig. 2.4 Inverted tooth chain pin and rocker joint articulation

links on either side, all held together by the riveted ends of the pins forming the pin and rocker joint (Fig. 2.5).

The riveted link plates have pairs of segmental holes punched through them at either end, the swing joints are then formed by pairs of identical semi-circular pins passing through the alternately stacked adjacent links.

Link joint articulation is provided by the concave back of each joint pin fitting snuggly against a convex key formed in each segmental joint hole; thus, the pins are prevented from rotating relative to the inverted link to which they are keyed, but the semi-circular pin and rocker faces contact each other and are therefore able to roll together as the links wrap themselves around the sprocket wheels.

Concentric pin-rocker compensating joint
(Fig. 2.6(a and b))

The load bearing contact of the inverted link joints is provided by the pin and rocker curved working faces. When the chain links are in-line, that is stretched out straight, the contact point will be just below the pitch line (Fig. 2.6(a)) of the chain. However, as adjacent links hinge about their pivots, the rolling contact point of the two curved surfaces moves upwards, thereby marginally increasing the pitch between adjacent articulated links (Fig. 2.6(b)). The extended pitch is designed to equal that needed for the chain to wrap itself around the sprocket along the pitch circle. It is through this method of elongating the articulated link pitch in combination with the inverted link teeth engaging the involute sprocket

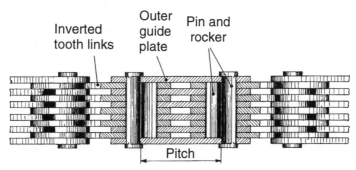

Fig. 2.5 Plan view of inverted tooth chain

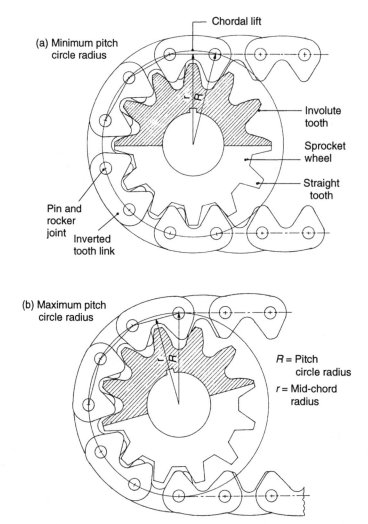

Fig. 2.6 Inverted tooth timing chain (Morse HY Vo) and sprocket wheel chordal action

teeth, that the chordal action is reduced. This automatic adjustment of the link pitch, so that the chain closely follows the circular pitch with very little variation in pitch circle radius is known as chordal compensation.

A reduction in chordal action with inverted link chains decreases chain pulsation, vibration, wear and noise.

Inverted-tooth chain considerations
(Fig. 2.6(a and b))

High strength alloy steel is used for the pins and links and heat-treatment is controlled so that the hardness tolerance is only 3 points on the Rockwell C scale; in addition, the pins and links are shot peened and pre-stressed for greater load capacity and fatigue resistance.

The pin and rocker combination provides a rolling action between the curved bearing contact faces, this minimizes the sliding friction and joint galling—that is, the rubbing away of one surface by the other when there is very little relative movement, as happens with the pin and bush joint of a conventional roller chain.

No chain tensioner or damper of any kind should be used with the inverted-teeth type chain, which is in contrast to the roller chain where some sort of tensioning to compensate for the stretch/wear of the chain is essential.

Wear in an inverted-link chain elongates the pitch between joints, this forces the chain further out from the sprocket teeth thereby creating a larger pitch circle (Fig. 2.6(a and b)). This results in the chain teeth engaging the involute flanks of the sprocket teeth further out. Consequently, as the chain wears, the chain teeth contact on the sprocket teeth, move to the unworn portion of involute sprocket teeth, and the increased pitch circle of the chain automatically takes up any slack tendency of the chain spanning the drive and driven sprocket wheels.

Sprocket wheels suitable for supporting inverted link chain drives are best made from a cast pearlitic malleable iron surface, induction hardened to 60 Rockwell 'C' scale. Sometimes, timing camshaft sprocket wheels are made from fibre-reinforced nylon.

2.1.3 Synchronous toothed-belt drive

A synchronous belt drive is an internally cogged (teeth) endless strap which engages identically shaped grooves formed on the drive and on the driven pulley wheel circumferences, so that a positive non-slip timed drive is provided.

Construction
(Fig. 2.7)
Toothed belts are basically composed of four components each having a specific function:

1 load carrying cords;
2 moulded drive teeth;
3 cord support backing;
4 wear resistant facing.

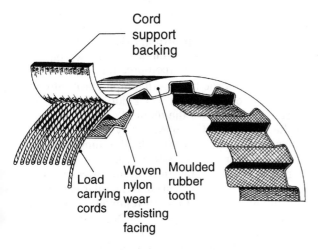

Fig. 2.7 Synchronous timing toothed belt drive

Load carrying cords
(Fig. 2.7)
The belt member transferring the torque load from the drive to the driven pulley is made of very strong flexible glass-fibre cords. These cords are dimensionally stable ensuring that the length of these belts never changes; that is, they do not stretch and the belt teeth easily and continuously mesh with the pulley grooves, which provide the spacing for the belt teeth. The flexibility, stability and compactness of the cords make them particularly suitable for forming the backbone of the camshaft belt drive.

Moulded drive teeth
(Fig. 2.7)
Belt grip is provided by internally spaced, tough, shear-resistant Neoprene rubber cogs or teeth mouldings that are bonded to the inside layer of the tensile load bearing cords to a close tolerance on pitch and length. This enables the belt to mesh perfectly with the pulley teeth so that engagement and release occurs, without drag, in a progressive and smooth manner.

Cord support backing
(Fig. 2.7)
The support backing is bonded to the load-carrying cords and is made from oil and age resistant synthetic rubber, which is both flexible and durable. This rubber backing, which is smooth on the outside, positions and supports the glass-fibre cords so that they all share the drive load and pull together smoothly and evenly. At the same time, the smooth outer surface of the backing reduces vibration when the belt is in contact with the jockey idler pulley.

Wear resistant facing
(Fig. 2.7)
The toothed side of the belt is protected by lining it with a tough woven nylon fabric that has been chemically treated and is impregnated with abrasion oil and ageing resistant synthetic rubber. As a result, the belt-to-pulley interface friction is reduced and the teeth are able to deform slightly under load or by torsional oscillation without prematurely cracking in the roots of the teeth.

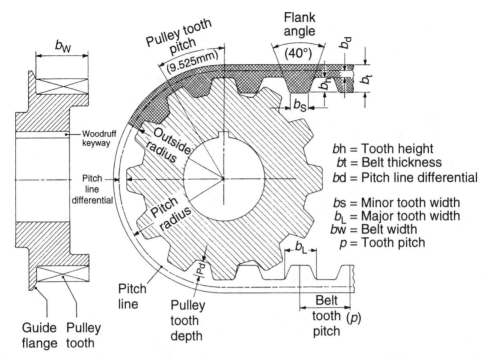

Fig. 2.8 Straight tooth profile timing belt and sprocket pulley

In the figure:

- Woodruff keyway
- Outside radius
- Pitch radius
- Pitch line differential
- Pitch line
- Guide flange
- Pulley tooth
- Pulley tooth depth
- Pulley tooth pitch (9.525mm)
- Flank angle (40°)
- Belt tooth (p) pitch

b_h = Tooth height
b_t = Belt thickness
b_d = Pitch line differential

b_s = Minor tooth width
b_L = Major tooth width
b_w = Belt width
p = Tooth pitch

Belt and pulley operating conditions
(Figs 2.8 and 2.9(a and b))

The belt and pulley teeth generally have a pitch of 9.5 mm and they may have a shallow straight-sided profile (Fig. 2.8). However, the deep involute tooth profile (Fig. 2.9(a and b)) is becoming more popular since this profile tends to reduce belt scuffing, noise, improve belt life and make it more difficult for tooth jumping under cold starting conditions. To sustain the average load transmitted by a camshaft and auxiliary driven equipment, a 19 mm or 25 mm belt width standard is adopted (Fig. 2.8). The tensile strength of the belt cords is very high and more than adequate to cope with the combined steady and fluctuating torque load, therefore the most common belt weaknesses are tooth wear and tooth jumping.

If the pulley grooves are roughly finished or are burred, the nylon lining quickly wears away and, once this happens, fairly rapid rubber tooth moulding wear occurs until the load is too high for the remaining section of the tooth, which then shears. Another cause of rapid belt wear is due to pulley misalignment, which results in unequal tension across the belt width so that cyclic distortion occurs between the load carrying cords and moulded rubber teeth. Extreme edge wear therefore takes place, the rubber surrounding the cords disintegrates and, with inadequate support for the cords, the belt will eventually fracture due to cord fatigue.

A minimum pulley size of 19 grooves, and a minimum of six teeth in mesh with the crankshaft drive pulley, is considered necessary. In general, for large torque consuming drives like diesel injection pump drives, the angle of wrap should be increased so that more teeth are in engagement with both the drive and driven pulleys than is absolutely necessary for a simple two-pulley crankshaft-to-camshaft drive. A further precaution to prevent the belt teeth jumping over any of the pulley teeth is a snubber guide which can be placed directly above the point of entry of the belt to the pulley, and approximately 0.75 mm from the smooth backing of the belt. Belt jump is more likely to occur during engine cranking when the engine is cold. For injection pump drive, torque occurs periodically over a small angular camshaft movement of about 10° with relatively high peak values of something like 40 Nm (Fig. 2.10). Under these fluctuating load operating conditions, involute deep-tooth profile belts and pulleys become essential to avoid any of the belt

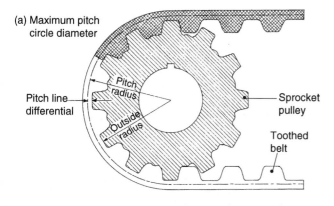

(a) Maximum pitch circle diameter

Pitch line differential

Pitch radius

Outside radius

Sprocket pulley

Toothed belt

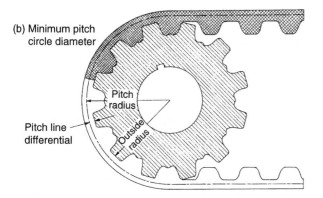

(b) Minimum pitch circle diameter

Pitch radius

Pitch line differential

Outside radius

Fig. 2.9 Involute tooth profile timing belt and sprocket pulley drive

teeth jumping the matched pulley teeth.

To prolong the life of the belt it should be kept free of grease or oil contamination as this tends to counteract the smooth grip of the belt as it engages and disengages the pulley circumferential toothed profile. At the same time any grease or oil may react with the lining fabric protecting the teeth and so cause it to deteriorate.

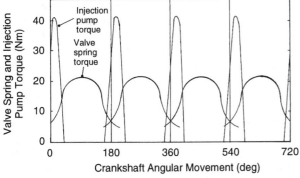

Fig. 2.10 Illustration of cyclic spread valve spring torque and instantaneous injection pump torque for a four-cylinder engine

Toothed-belt tension

Engine temperature may vary in the UK from something like 0°C before starting on a cold winter's day to a running temperature of 90°C, and in cold climates the initial temperature may drop as low as −30°C.

The engine block and cylinder head will therefore expand or contract something like 0.1 to 0.2% depending upon the engine's component materials and the amount the mean engine temperature changes.

The coefficient of linear expansion for the belt will be smaller than for the material used for the engine and, in addition, the timing belt will operate at a much lower temperature than that of the block and cylinder head. Therefore, the belt tension will increase as the engine temperature rises from the off-load to the full-load working conditions.

A belt tensioner device serves only to adjust initially and set the belt tension correctly since, practically, it does not stretch during service. In contrast, a roller chain tensioner device is provided to take up the chain slack caused by the stretching of the chain as wear takes place during the engine's operating life.

For easy removal and assembly of a toothed timing belt, a sideway deflection between pulleys should be of the order of 4 to 5 mm, whereas for adjustment of the belt tension only 1 to 2 mm movement is generally required.

The belt must not be overstretched when its tension is adjusted between the crankshaft pulleys and any other auxiliary drive pulley. For belts fitted to 1.5 to 3.0 litre diesel car engines a typical tension would be about 150 N. A more precise belt tension for both cold starting and normal running would be a minimum of 50 N and a maximum just above 250 N respectively, but an automatic variable tensioning mechanism has not been considered necessary up to now.

Overtensioning of these timing belts considerably shortens belt life and simultaneously increases the pulley bearing radial loads—subsequently increasing bearing wear.

Pulleys and flanges
(Fig. 2.8)
Pulleys can be made from carbon steel forgings and iron castings but the machined finish and final heat treatment to harden the pulley grooved surfaces is expensive. A better alternative is sintered steel powder preformed to improve the

surface finish and to work-harden the material around the teeth profile surface. A very low-cost solution for large volume production is injection moulding with thermoplastic polymide resins such as Nylon 66 glass filled from 20% to 30%, and anti-friction additives such as graphite or molybdenum bisulphate from 1% to 5%. These plastic compounds are moulded onto steel hubs to provide the pulley with adequate dimensional stability and support. The major problem with plastic pulleys has been to predict with accuracy the shrinkage of the mould pulley—which may vary under different moulding conditions and therefore produce pulleys outside the recommended dimensional tolerances.

Toothed timing belts have an inherent tendency to generate a slight side thrust when in motion, therefore some sort of guide flange (Fig. 2.8) is necessary to prevent the belt winding itself off the edge of the pulley. In general, for drives with two pulleys, there should be one flange on opposite sides on each pulley or on each side of the drive pulley only. With three pulleys, at least one should be flanged on two sides, while for drives with more than three pulleys at least two pulleys should be flanged on two sides or more than two on opposite sides.

Merits of toothed-belt drives

Wide speed range capability The lightness and flexibility of the belt enables it to accelerate and decelerate smoothly and to operate at very high speeds and over a wide speed range without any side effects.

Silent belt drive system The drive provides a low noise level due to the absence of metal-to-metal contact between the belt and pulley teeth as they move into and out of mesh.

Belt chatter and vibration is eliminated Speed is transmitted uniformly; this is because there is no chordal rise and fall of the pitch line, no belt creep or slippage and therefore no belt chatter or vibration.

Low heat build-up and high mechanical efficiency There is very little heat build up due to the very low belt-to-pulley surface friction and relatively low tensile tension. The thin belt cross-section makes the belt very flexible and there will be negligible hysteresis. As a result, the belt-to-pulley drive has an unusually high mechanical efficiency.

No belt drive maintenace required The absence of metal-to-metal contact and the relatively low belt tension eliminates the need for lubricating the belt and pulley system and, with the low radial pulley bearing load, periodic belt-and-pulley and pulley-bearing adjustment is eliminated.

Low-cost drive system A belt drive arrangement can be arranged to cope with complicated drive configurations with relatively low-cost grooved pulleys, a jockey idler tensioner and a toothed belt.

Limitations of toothed belt drive

The life expectancy of a toothed belt is about one-third of a chain drive used for a similar application. The recommended belt replacement period is normally 60 000 km (36 000 miles). However, there is a new generation of toothed belts which claim a life expectancy of 160 000 km (100 000 miles).

There is no warning indication when a belt is about to fracture and when it does, valves in the cylinder head may bump against the piston crown necessitating their replacement.

Cold start instantaneous belt loads can cause the belt teeth to jump over engaging pulley teeth, thereby altering the valve, ignition or injection pump timing.

2.1.4 Gear train drive

When high loads or high speeds are to be sustained for long operating periods, then the most reliable method of driving both camshafts and injection pump shafts is the involute gear train. Helical involute gears are generally used to ensure continuous tooth contact and to overcome tooth pitting and wear, which could result in increased backlash, and to provide minimal valve and injection-pump timing variation throughout the life of the engine.

Conflicting reasons for not using gears are that accurate distances between shaft centres under all operating temperatures are required, a large number of gears may be needed to make up a train set suitable for driving overhead cam layouts, and the backlash between each pair of gears connected in series becomes accumulative.

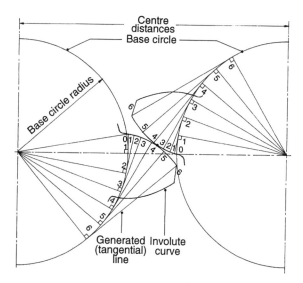

Fig. 2.11 Sliding action of meshing involute teeth

Construction of an involute tooth profile
(Fig. 2.11)

The involute of a circle can be considered as the curve traced by the end of a cord being unwound from the circle.

The involute tooth profile is constructed by drawing a base-circle of diameter D mm, say 140 mm (Fig. 2.11). The circumference is then divided into a convenient number of parts, say intervals of 10°. Each division on the circumference is equal to

$$\frac{\pi D}{36} = \frac{\pi 140}{36} = 12.2 \text{ mm}.$$

Tangents are next drawn at each marked point on the circumference of the base circle.

The length of the first tangent line
           is equal to $\pi 140/36 = 12.2$ mm
The length of the second tangent line
           is equal to $12.2 \times 2 = 24.4$ mm
The length of the third tangent line
           is equal to $12.2 \times 3 = 36.6$ mm
The length of the fourth tangent line
           is equal to $12.2 \times 4 = 48.8$ mm
The length of the fifth tangent line
           is equal to $12.2 \times 5 = 61.0$ mm
The length of the sixth tangent line
           is equal to $12.2 \times 6 = 73.2$ mm

Finally, the involute curve is generated by drawing a line joining up the ends of each generated tangential line.

Involute tooth sliding and rolling
(Fig. 2.11)

When teeth of an involute form are in mesh the action is one of rolling and slipping. For every 10° interval around the base circle, projected tangential lines of increasing length equal to the unwrapped base circle arc from the start point are produced, the joined-up ends then form the involute tooth profile (Fig. 2.11).

The unwrapped tangential lines, as they move away from the starting point on the base circle, increase in length but at a decreasing rate, thus: the second tangential line length has increased 100% relative to the first; the third tangential line length has increased 50% relative to the second; the fourth tangential line length has increased $33\frac{1}{3}$% relative to the third; the fifth tangential line length has increased 25% relative to the fourth; the sixth tangential line length has increased 20% relative to the fifth.

The incremental distance between tangential line ends making up the involute curve becomes larger and larger as the curve moves outward from the root to the tip of the tooth (Fig. 2.11).

Hence, for equal base-circle intervals, the incremental involute curve length 5–6 is greater than distance 4–5 and 4–5 is, similarly, greater than distance 3–4 and, accordingly, distance 3–4 is greater than 3–2, and so on.

If the gears are revolving at a uniform speed, then each angular interval, and therefore involute incremental distance, must be completed in the same time. Consequently, when two teeth are in mesh and, say, increment 3–4 on one tooth contacts say, increment 4–5 on the other, then the smaller increment distance has to be completed in the same time as the larger increment length; thus, this can only be achieved by relative slippage between the contacting profiles of each tooth.

Maximum sliding occurs where the largest increment length, say 5–6, on one tooth rubs against the smallest increment distance, say 1–2, on the adjacent tooth, whereas intertooth rolling only occurs when equal length involute profiles face each other near or at the pitch point. Thus, the further the engaging teeth move away from the pitch point, the greater will be the rubbing action.

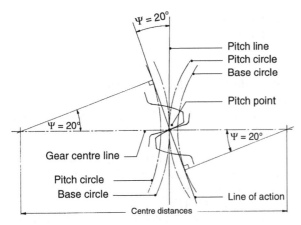

Fig. 2.12 Pressure angle and line of contact on spur gears

Line of action and pressure angle
(Fig. 2.12)

Turning effort is transferred from the driver to the driven gear wheel along a line of action which is perpendicular to the teeth at their point of contact and which passes through the pitch point to form common tangents to both base circles (Fig. 2.12).

The angle made between the line of action in a direction in which the pressure is exerted by one tooth on the other, and a line perpendicular to the centre line of these gears, which also passes through the pitch point, is known as the pressure angle, which is normally 20°.

When two teeth of any pair of gears having involute profiles engage, their point of contact lies somewhere along this line and, as these gear wheels rotate, adjacent meshing-teeth contact points move continuously along the line of action.

Gear teeth definitions
(Fig. 2.12)

Pitch circle A circle, the radius of which is equal to the distance from the gear axis to the pitch point.

Pitch point The point of contact and tangency of the pitch circle or the point where the centre line of mating gears intersects the pitch circles.

Pitch line A line drawn through the pitch point tangential to both pitch circles.

Line of action That portion of the common tangent to the base circles along which contact between mating involute gear teeth occurs.

Circular pitch Length of the arc of the pitch circle between the centres or other corresponding points of adjacent teeth.

Pressure angle (ψ) The pressure angle of a pair of mating involute gears is the angle between the line of action and a line perpendicular to the centre line of these gears.

Tooth engagement and disengagement
(Fig. 2.13)

In order that a pair of gears may give a constant angular velocity it is essential to design the teeth so that before one pair has disengaged another pair has taken up the load. Thus, to produce a smooth and continuous drive each pair of teeth must come into contact before the preceding pair has broken contact; hence, the greater the over-lap interval of tooth contact the smoother the

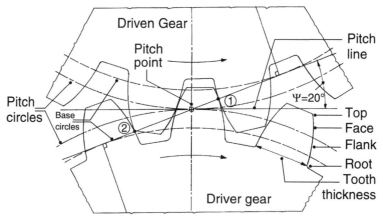

Fig. 2.13 Meshing teeth load distribution

action. To achieve a smooth transition of load from one gear to another the average number of pairs of teeth in contact must be greater than one.

Under these conditions, where one particular tooth makes contact near either its outer edge or base circle, a second tooth of the same gear is also in contact with a corresponding tooth. The whole turning thrust load is taken by a single tooth only when the tooth contact is near the pitch line. The engagement and disengagement process can be explained in the following way, as shown in Fig. 2.13.

The fully meshed central driver gear tooth right-hand face contacts the driven gear left-hand flank of the adjacent tooth point (1) along the line-of-action; whereas, a second pair of teeth, just coming into mesh, contact each other on the flank near the base-circle on the driver gear which aligns with the outer face tip on the driver gear tooth point (2), again along the line-of-action. Immediately before the second pair of teeth touch, it is the driver gear central tooth alone that absorbs the transmitted load point (1), this being the outermost point of its contour at which it will carry the full tooth load. Further rotation of the gears causes the driver gear profile contact to continue to move outwards and along the line-of-action until the tip of the tooth is reached. Thus, during the period when the second pair of teeth progressively come into contact, an increasing proportion of the load is taken by the second driver gear tooth as its point of contact moves from the base circle to the pitch-circle. Therefore, a driver gear tooth is subjected to the greatest bending stress when it passes through the pitch point just before the next pair of teeth mesh and contact each other.

The cycle of tooth engagement commences with the tip of the driven tooth making contact with the flank of the driver tooth and terminates when the tip of the driver loses contact with the flank of the driven gear.

Helical tooth gears

When gears are used to connect parallel shafts in the same plane and are inclined to the axis of the shafts forming part of the helix they are known as single helical gears. For these helical gears to mesh together they must have the same helix angle but be of the opposite hand. Helix angles may range from 25° to 45° but a more typical value would be 30°. The limitation to large helix

angles is that the end thrust rises in proportion to this angle.

The engagement of a pair of helical teeth takes place with contact commencing at one end and steadily extending over the whole tooth width. The helix angle and the face width are proportional so that the leading edge and the tooth come in contact at the pitch-line before the trailing edge of the preceding tooth has passed out of contact at the pitch-line. By these means, the approach and recession of mating teeth enables the transference of power to be continuous and smooth.

It is interesting to note that with straight involute teeth, if the circle pitch and tooth profiles were perfectly machined and no deformation of the teeth occurs under load, then the smooth transference of power from the driver to the driven gears would be of equal quality to that of the helical gearing, but unfortunately this is never achieved.

Tooth wear

Failure of gear teeth is frequently due to the non-uniform load distribution across the tooth width caused by the deflection of the shaft and the support housing under load so that stress concentration occurs somewhere near the tooth root causing the tooth to fracture. Excessive compressive loads show up in the form of pitting and scoring which normally occurs near the pitch line. Eventually the involute contour is destroyed in the region so that the engagement becomes jerky and noisy. Therefore, under extreme high-load operating conditions the impact transference load caused by the damaged surfaces rapidly accelerates tooth profile wear further.

2.2 Camshaft pulley, sprocket and gear wheel attachments

2.2.1 Camshaft timing belt toothed pulley wheel attachment
(Fig. 2.14(a))

The toothed pulley central bore is a forced fit over the overhang portion of the camshaft. Positive drive is provided by the segmental woodruff key located between the external semi-circular

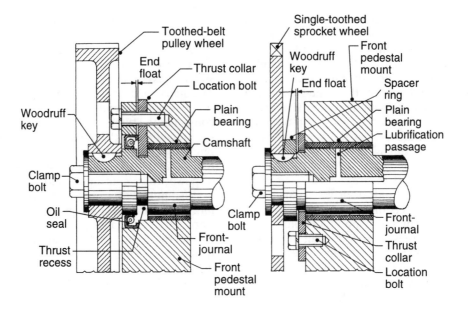

(a) Camshaft timing belt toothed-pulley
    wheel attachment

(b) Camshaft timing chain single-toothed
    sprocket wheel attachment

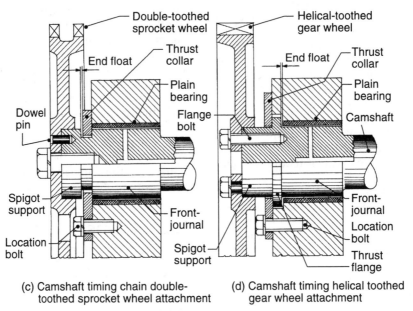

(c) Camshaft timing chain double-
    toothed sprocket wheel attachment

(d) Camshaft timing helical toothed
    gear wheel attachment

Fig. 2.14(a–d) Camshaft timing drive mounting with end float control

machined groove setbolt and the internal parallel groove formed in the pulley hub. A central axial setbolt draws the pulley hub via the large washer over the reduced diameter portion of the camshaft until it is positioned hard against its shoulder.

Camshaft end float is controlled by a thrust collar which is located in a slot machined in the front pedestal mount. When correctly positioned the semi-circular side of the collar fits in the groove next to the cylindrical reduced diameter of the recess machined on the camshaft.

A further feature of this pulley attachment is that an oil seal is located in the enlarged mouth of the camshaft support bearing to prevent oil escaping from the front of the camshaft and spilling onto the pulley and belt—which must be kept dry at all times.

2.2.2 Camshaft timing chain single-toothed sprocket wheel attachment
(Fig. 2.14(b))

A single-toothed sprocket wheel attachment for light-to-medium duty applications is shown in Fig. 2.14(b); here, the sprocket wheel is a single flat disc with a central hole mounted in the reduced diameter portion of the overhanging camshaft. A woodruff key is then located between a flat and semi-circular groove machined in both sprocket wheel and shaft respectively.

The positioning of the sprocket wheel on the end of the camshaft is determined by the thickness of the spacer ring which is sandwiched behind the sprocket. A central setbolt, when tightened, pulls the sprocket wheel and spacer ring hard against the first step on the shaft through the overlapping flat washer.

End thrust, which develops in operation, is absorbed by a thrust collar situated between the spacer ring and the second step on the camshaft.

The removal of the camshaft is achieved by aligning holes in the sprocket wheel with the thrust collar location bolts which can then be unscrewed, this then permits the thrust collar to be withdrawn.

2.2.3 Camshaft timing chain double-toothed sprocket attachment
(Fig. 2.14(c))

A more substantial method of mounting the sprocket wheel onto the camshaft provides the hub of the sprocket wheel with a recess which fits directly over the spigot support at the end of the shaft. Drive is then transferred by a dowel-pin which is pressed into the offset aligned holes in both the camshaft and sprocket wheel, and the attachment is then completed by a central setbolt which draws the sprocket wheel recess and camshaft end flat faces together.

Camshaft end thrust reaction is then provided by a thrust-collar which slips into the grooved

recess machined on the camshaft and is held in place by a pair of location bolts (only one shown).

2.2.4 Camshaft timing helical-toothed gear wheel attachment
(Fig. 2.14(d))

For heavy duty applications where straight or helical timing gears are used a central recess is machined into the gear hub on the side facing the camshaft. The gear wheel can then be attached to the end of the camshaft by pushing the recessed hub over the spigot support provided by the camshaft. The drive is then transferred from the gear to the camshaft through four or more set-bolts passing through holes drilled in the flanged part of the gear hub.

End-thrust of the camshaft in this example is provided by the flanged portion of the camshaft fitting in the enlarged bore formed in the front pedestal mounting, and outward movement of the camshaft is prevented by the thrust-collar encompassing half the flanged shoulder of the camshaft.

Removal or replacement of the thrust-collar is made possible by aligning holes in the side of the gear wheel with the thrust-collar location bolts.

2.2.5 Camshaft end thrust and float
(Fig. 2.15(a, b, c and d))

The transmission of motion from one shaft to another by means of a belt chain or gear drive will always produce some degree of end thrust on the camshaft even if the pulley sprocket or straight-toothed gear appears to be parallel to the shafts.

This is usually due to some misalignment between driver and driven wheel under operating conditions and by a cam and follower offset interaction when opening and closing the cylinder-head valves.

End-float and thrust control is normally achieved with an end-plate, or a collar, which is shaped to fit into or over a recessed groove or a shouldered portion of the camshaft, respectively.

With the recessed groove, the thrust-collar reacts against the end-thrust in either direction (Fig. 2.15(a and d)), whereas, with the flanged camshaft, end thrust is taken either by the thrust collar (Fig. 2.15(b)) or by the end-plate (Fig. 2.15(c)) which, in the latter case, provides an adjustment to the end float by altering the thickness of the end-plate flange gasket.

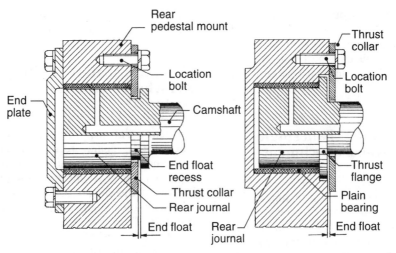

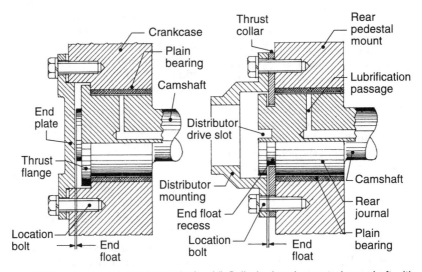

Fig. 2.15(a–d) Camshaft end float control

The amount of end float can generally be checked by inserting a feeler gauge between the thrust collar and grooved side (Fig. 2.15(a)) or flanged shoulder (Fig. 2.15(b)), alternatively a dial gauge can be placed against one of the camshaft journals or cam shoulders, the end float can then be read from the clock gauge as the camshaft is levered to and fro.

2.3 Camshaft chain drive arrangements

2.3.1 Cylinder-block mounted camshaft chain drive
(Fig. 2.16)

For low mounted camshafts, the simplest of drives is a short chain joining together the two-to-

56

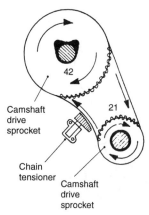

Fig. 2.16 Cylinder-block mounted camshaft chain drive (Renault)

one speed step-down ratio crankshaft and camshaft sprocket wheels, with a slipper-head chain tensioner situated mid-way between the sprockets on the chain slack side. A typical sprocket wheel size would be 21 and 42 teeth for the crankshaft and camshaft sprockets respectively.

2.3.2 Cylinder-block camshaft, injection pump and generator chain drive with two idler sprockets
(Fig. 2.17)

A well-proven chain sprocket and idler configuration, which has been adopted by Gardner diesel engines for many years, is shown in Fig. 2.17.

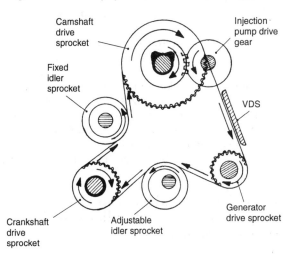

Fig. 2.17 Cylinder-block camshaft injection pump and generator chain drive with two idler sprockets (Gardner Diesel)

This layout not only drives the low-mounted camshaft and generator directly, but indirectly drives the injection pump via a pair of meshing gear wheels driven by way of the camshaft sprocket wheel. Strict control of two sides of the triangular chain layout is obtained by a pair of idler sprockets, one fixed and the other with an eccentric adjuster, whereas experience has shown that the third side only requires a vibration damper strip to minimize any chain whip and vibration. A chain drive of this type was basically chosen as it could readily cope with the relatively large expansion and contraction occurring between crankshaft and camshaft centre distances caused by the aluminium alloy crankcase.

2.3.3 Overhead camshaft chain drive
(Fig. 2.18)

Overhead camshaft drives require a much longer chain to span the relative large centre distance between the small crankshaft and large camshaft sprocket wheels; therefore, these chains must be supported not only by a chain tensioner but additional vibration damper strips which guide and absorb whip over the extensive chain span.

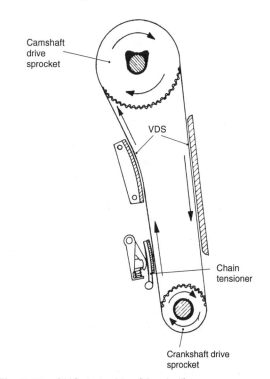

Fig. 2.18 OHC chain drive (Vauxhall)

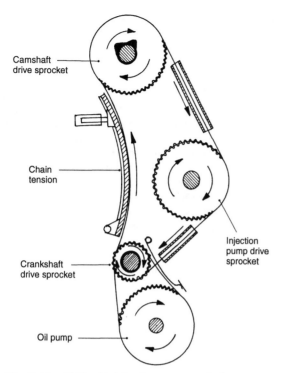

Camshaft drive sprocket

Chain tension

Crankshaft drive sprocket

Injection pump drive sprocket

Oil pump

Fig. 2.19 OHC with injection pump and oil pump chain drive (Mercedes)

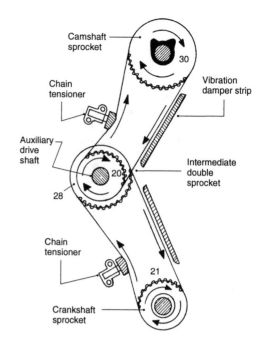

Camshaft sprocket

30

Vibration damper strip

Chain tensioner

Auxiliary drive shaft

20

28

Intermediate double sprocket

Chain tensioner

21

Crankshaft sprocket

Fig. 2.20 OHC with two-stage chain drive with auxiliary drive shaft (Rover)

2.3.4 Overhead camshaft with injection pump and oil pump chain drive
(Fig. 2.19)

A three-cornered chain configuration is used for this car diesel overhead camshaft engine, this layout uses a long curved and hinged rail and automatic chain tensioner for the slack chain side and a pair of double-sided vibration damper guides to give closer chain control on the drive side of the chain. A second short chain is used to drive the oil pump and, at the same time, to reduce its speed relative to the crankshaft, thereby conserving power to drive the pump in the upper speed range.

2.3.5 Overhead camshaft two-stage chain drive with auxiliary shaft
(Fig. 2.20)

A single chain looped between the crankshaft sprocket and the overhead camshaft sprocket

forms a relatively large chain span which can be the cause of excessive chain whip noise and wear.

A convenient approach to reduce the centre distance between chain sprockets is to employ an intermediate double sprocket wheel and two short timing chains, this not only halves the length of each chain span but also provides a drive for an auxiliary shaft utilized to drive the distributor, oil pump, fuel pump etc.

Each endless chain between sprocket wheels forms a single drive stage, the primary stage is therefore between the crankshaft sprocket and the large intermediate sprocket while the secondary stage completes the drive from the small intermediate sprocket to the camshaft sprocket.

The sprocket teeth sizes are so chosen that the two-to-one speed reduction is obtained in two stages, this enables a smaller camshaft sprocket to be used compared with a single chain drive when using the same number of teeth on the crankshaft.

2.3.6 Twin overhead camshaft two-stage chain drive with idler sprocket between camshafts
(Fig. 2.21)

To reduce the chain span between the crankshaft and camshaft sprockets it is usual on twin over-

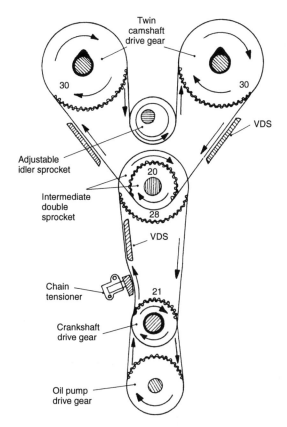

Fig. 2.21 Twin OHC two-stage chain drive with tensioner between camshaft drive sprockets (Jaguar)

Let crankshaft sprocket size $= 21$ teeth
large intermediate sprocket size $= 28$ teeth
small intermediate sprocket size $= 20$ teeth
camshaft sprocket size $= 30$ teeth

speed reduction ratio =

$$= \frac{\text{primary output}}{\text{primary input}} \times \frac{\text{secondary output}}{\text{secondary input}}$$

$$= \frac{\text{large intermediate sprocket}}{\text{crankshaft sprocket}}$$

$$\times \frac{\text{camshaft sprockets}}{\text{small intermediate sprocket}}$$

$$= \frac{28}{21} \times \frac{30}{20} = 2:1$$

With this arrangement the camshaft chain lap angle has been extended by fitting a deep-mounted central idler jockey tensioner, and chain vibration is absorbed by the two vibration damper strips located either side of the secondary chain between the intermediate and camshaft sprocket wheels. Primary chain slackness is supported by a slipper-head chain tensioner on the outside, just above the crankshaft, whereas a vibration damper strip is positioned on the inside of the chain just below the intermediate sprockets. A third chain is used to provide a speed step-down and drive for the lubrication pump.

2.3.7 Twin overhead camshaft and auxiliary equipment two-stage chain drive
(Fig. 2.22)

A second example of a two-stage chain drive is shown in Fig. 2.22: such an arrangement provides a speed reduction in each chain drive stage, thus bringing down the camshaft revolutions so that they revolve at half crankshaft speed. A triple-sprocket triangular-chain drive forms the primary stage, the crankshaft sprocket providing the input drive to the larger intermediate double sprocket wheel and the auxiliary shaft sprocket. Likewise, the second-stage triple-sprocket wheel-chain lay-out is triangular, taking its input drive from the smaller of the intermediate double sprocket wheels; this sprocket relays motion via the chain to the twin camshaft sprocket wheels. No idler jockey sprocket is used as in Fig. 2.21, there-fore chain slackness is controlled entirely by the

head camshaft engines to deliver the drive in two stages employing an intermediate double sprocket wheel to transfer the drive from the primary to the secondary stage. The overall two-to-one speed reduction between crankshaft and camshaft can therefore be carried out in two parts, hence smaller camshaft sprocket wheels can be used which make the cylinder-head twin camshaft drive more compact as compared with a single chain drive. The primary chain provides a parallel chain drive conveying motion from the crankshaft sprocket of, say, 21 teeth to the larger of the two intermediate sprockets of, say, 28 teeth. The secondary chain then transfers the chain drive from the smaller intermediate sprocket wheel of, say, 20 teeth upwards and outwards to the camshaft sprockets having, say, 30 teeth and then downwards between the adjacent sprockets to loop around the centrally mounted eccentric adjustable idler jockey sprocket.

The speed reduction can therefore be derived from a popular set of sprocket wheel sizes as follows.

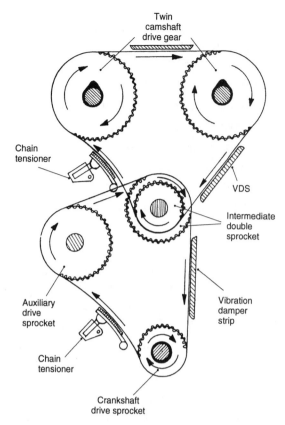

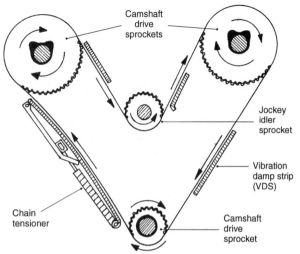

Fig. 2.23 Vee 8 and 12 single OHC single-stage chain drive (Jaguar and Mercedes)

Fig. 2.22 Twin OHC and auxiliary equipment two-stage chain drive (Jaguar)

hinged rail tensioner and vibration damper guide strips.

2.3.8 Vee 8 and 12 single overhead camshaft single-stage chain drive
(Fig. 2.23)

A very popular Vee 8- or 12-cylinder engine chain drive arrangement for a single overhead camshaft, uses only a single chain, four sprockets, a curved strip chain tensioner and three vibration damper guide strips (Fig. 2.23).

The chain layout is basically a vee configuration with the crankshaft and idler jockey sprockets forming the apex and the camshaft sprockets and chain span representing the arms. The deep mounted idler jockey sprocket provides a large angle of wrap for both camshaft sprocket wheels. Chain slack control is achieved by the long curved strip chain tensioner and chain vibration is restrained by the normal vibration damper guide strips. A separate oil pump chain drive can be dispensed with by incorporating a crescent-type oil pump driven directly off the crankshaft.

Provided twin camshafts are not essential, the single-chain drive weighs far less, and costs are reduced, compared with the twin overhead camshaft two-stage arrangement shown in Fig. 2.24, which utilizes four chains and 12 sprockets. Also, the noise level is lowered and the chain drives smoother with twice the number of cam profiles on each shaft.

2.3.9 Vee 8 and 12 twin overhead camshaft two-stage chain drive
(Fig. 2.24)

A twin camshaft chain-drive double cylinder-bank arrangement, such as is used by vee 8 and 12 engines, is shown in Fig. 2.24.

This layout relays the drive in two steps via intermediate double sprocket wheels. The primary chain first speed-reduction stage conveys the drive from the crankshaft via a pair of intermediate double-sprocket wheels mounted on each of the cylinder blocks, a large angle of wrap being provided by incorporating an idler jockey sprocket between the intermediate double sprockets. The second speed-reduction stage then continues the drive from each intermediate sprocket to its corresponding twin camshafts' sprockets via the usual triple-sprocket triangular chain drive layout. Chain slackness and vibration control for each independent triangular chain drive is then catered for in the normal way with both slipper-

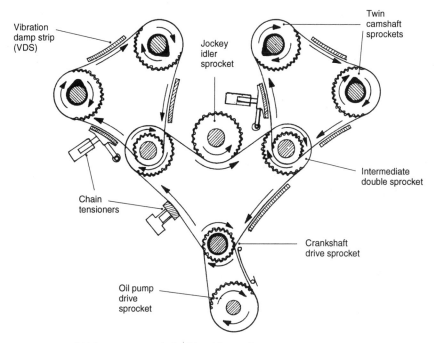

Fig. 2.24 Vee 8 and 12 twin OHC two-stage chain drive (Jaguar)

head and hinged-rail chain tensioners for the primary and secondary stages respectively.

Chain vibration relief on all the non-drive sides of the chain are absorbed by conventional vibration damper strips. A fourth simple short-chain drive rotates the oil pump, a large oil-pump sprocket being used to reduce the oil-pump speed so that the power consumed by the pump is minimized.

2.4   Chain tensioners and vibration damper strips

(Fig. 2.25(a and b))

With a chain drive, the torque transmitted from the driver to the driven sprocket wheel is conveyed by the taut side of the chain, which alone takes the load, whereas the slack side of the chain only provides a means of returning the chain links unwound on the driver sprocket.

Thus, when the engine is accelerating or steadily pulling, the slack side of the chain is free to flap or whip to and fro between the sprocket wheels, within relative safety (Fig. 2.25(a)). However, as the chain and sprockets wear and the chain stretches, the amount of slackness of the chain on the non-drive side may eventually cause the rol-

lers to jump over the camshaft sprocket teeth, thereby altering the valve timing.

Conversely, when the engine decelerates from a high speed, momentarily the camshaft sprocket becomes the driver and the crankshaft sprocket the driven (Fig. 2.25(a)). Thus, the chain moving from the crankshaft to the camshaft sprocket becomes the taut drive side and the opposite chain spans the slack side. However, generally, the deceleration period is only of a short duration and is therefore not a major problem.

Consequently, chain tensioners are incorporated on the normally slack side of the chain to take up any chain slack, thereby restricting chain whiplash (Fig. 2.25(b)). In severe operating conditions guide strips may be provided on the drive side of the chain span between the sprockets.

2.4.1   Heavy-duty twin-ratchet automatic chain tensioner
(Fig. 2.26)

When relatively heavy pulsating loads are to be chain driven then an idler-jockey sprocket-tensioner is more durable and suitable than a slipper or rail-friction type tensioner.

The tensioner comprises a flanged mounting

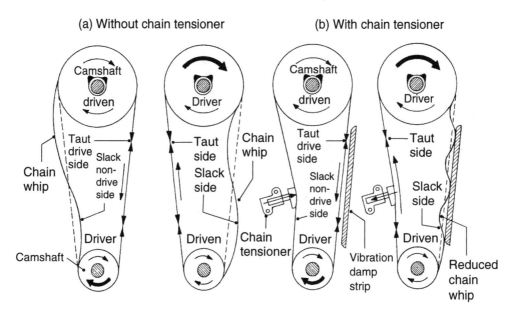

(a) Without chain tensioner   **(b) With chain tensioner**

Fig. 2.25 Timing chain response-under operating conditions

which forms part of a protruding rectangular sectioned block. A cylindrical slider-block with a rectangular elongated central slot fits over the flanged mounting (Fig. 2.26). The cylindrical outer-profile of the slider-block provides the bearing surface to support the sprocket-wheel and a compression spring assembled inside a bored hole on one side of the rectangular section of the flanged mounting. The spring therefore moves the slider-block and sprocket-wheel assembly hard against the chain, thereby taking up any chain slack which may exist. A rectangular twin pawl housing is screwed onto the end of the flanged mounting, whereas a pair of segmental ratchet plates are screwed on either side of the pawl housing against the slider-block. This, therefore, restricts both the slider-block and sprocket-wheel end-float respectively. When assembled, the ratchet-pawls are forced outwards by the central bias-spring until they both engage the ratchet-plate teeth.

In service, as the chain stretches, the compression-spring situated between the flanged mounting and the slider-block pushes the slider block and sprocket wheel assembly further towards the chain to compensate for the increased amount of slackness in the chain. As the slider-block moves progressively over, both toothed pawls ride over the segmental ratchet-plate teeth until they align and drop into place with the next set of meshing teeth. Consequently, the tensioner sprocket-

wheel assembly is permitted to move in one direction towards the chain but is prevented from retracting rearwards under chain backlash conditions.

2.4.2 Ratchet strip and block automatic chain tensioner
(Figs 2.27 and 2.28)

Tensioning of the chain with these tensioners utilizes a fixed and hinged arm, compression spring and guide pin, and a nylon toothed ratchet strip and toothed block (Figs 2.27 and 2.28).

As the chain stretches, the light compression spring thrust is relayed to the hinged arm causing it to swing the slipper head (Fig. 2.27) or tensioning rail (Fig. 2.28) towards the chain to take up the extra slack.

The automatic tensioning is achieved by the toothed block which is wedged between the ratchet strip arm and hinged arm by the thrust of the compression spring. The gap between the fixed and hinged arms is bridged by the block's curved profile and inclined arm, on one side, and the block's trailing ratchet and ratchet strip arm on the other. Thus, as the two arms move further apart with increased chain slackness, the block slides further between the inclined arm and over the ratchet strip to engage the next adjacent teeth. If chain backlash attempts to force the

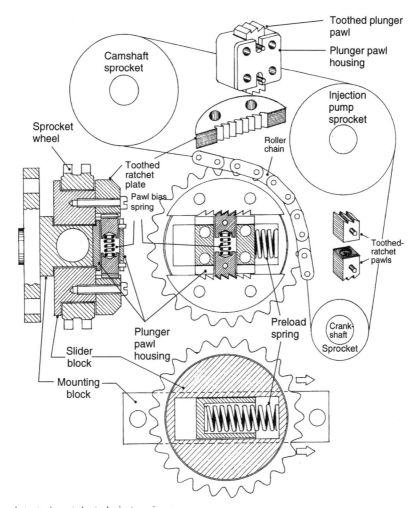

Fig. 2.26 Heavy-duty twin-ratchet chain tensioner

adjuster arms together, then the squeeze reaction of the arms causes the matching block and strip teeth to slot together and lock. Tensioners of this type do not rely upon a heavy spring load and therefore they eliminate excessive chain tensioning.

These tensioners can be adapted to operate via a rubber slipper head or indirectly through a hinged rubber-lined rail.

2.4.3 Cylindrical spiral ratchet automatic chain tensioners (Renold)
(Fig. 2.29(a, b and c))

The adjuster consists of an oil abrasion resistant Neoprene rubber slipper head bonded to a steel plate mounted on the end of a hollow plunger which slides in the cylinder housing. Located inside the plunger is a ratchet sleeve which has a slot spiralling from top to bottom with a series of stepped semi-circular notches formed along one edge of the spiral (a helix). A light spring inside the ratchet sleeve combined with engine oil pressure pushes the plunger out from the housing so that the slipper head bears against the chain, thereby restricting the chain slackness to a minimum and at the same time applying the correct chain tensioner (Fig. 2.29(a)). Whip-back of the chain is prevented by the limit-peg—protruding inside the plunger near its mouth—being pushed back with the plunger and slipper head until it aligns and sits in one of the ratchet-stepped semi-circular notches (Fig. 2.29(b)).

As the chain stretches in service, the slipper head and plunger extend further outward to take

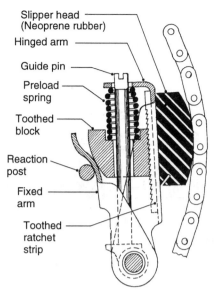

Fig. 2.27 Ratchet strip and block slipper-head chain tensioner

up the chain slack and, in so doing, the limiting-peg contacts the smooth side of the spiral slot, this causes the ratchet sleeve to rotate partially (Fig. 2.29(c)). The limiting-peg will therefore have moved further around the spiral slot so that when the chain whips back, the slipper head and plunger's rearward movement is restricted by the limiting-peg aligning with a new semi-circular

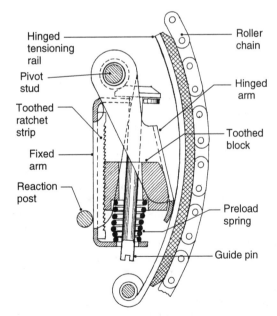

Fig. 2.28 Ratchet strip and block hinged rail chain tensioner

notch on a portion of the spiral which is further out from the blank end of the ratchet sleeve. By these means the plunger and slipper head automatically move outwards, whereas the ratchet sleeve revolves as the chain and sprocket teeth wear. Because of this, the backlash retraction of the plunger remains constant as the plunger progressively extends from its cylinder housing to compensate for the increasing chain slackness.

Before the plunger is inserted into its cylinder housing during assembly, the ratchet sleeve, using a 3 mm Allen-key fitted in the hexagonal hole in the base of the sleeve, is screwed inwards against the spring tension until it moves beyond the limiting peg—this action retains the sleeve and spring within the cylinder. After the tensioner assembly has been bolted to the engine block, the spring-loaded plunger is released by removing the back end-plug, inserting the Allen-key in the base of the ratchet sleeve and turning it in a clockwise direction until a distinct click is heard, the slipper head will then freely move outwards until it contacts the chain.

Lubrication oil from the engine is supplied to the automatic ratchet mechanism via a drilling in the backplate, and it passes out through a hole in the slipper head. The body of oil tends to damp down any violent to and fro movement of the plunger and also lubricates the slipper head and chain.

2.4.4 Hydraulic plunger-controlled automatic chain tensioner
(Fig. 2.30)

With this type of chain tensioner the outward thrust of the plunger, tensioning the chain, is maintained constant.

The tensioner assembly consists of a cylinder housing screwed into the cylinder head, an end cap nut which supports the inlet non-return ball-valve, a thrust-plunger, compression-ring, and a pressure-relief ball-valve assembly, all of which contribute in converting the hydraulic pressure into a tensioning load acting perpendicular to the chain (Fig. 2.30).

Initially, the compression spring exerts an axial thrust against the plunger alone thereby forcing it to load the friction rail lightly against the chain.

When the engine is running, oil from the engine's lubrication system passes into the cylinder housing, via passages from the cylinder head,

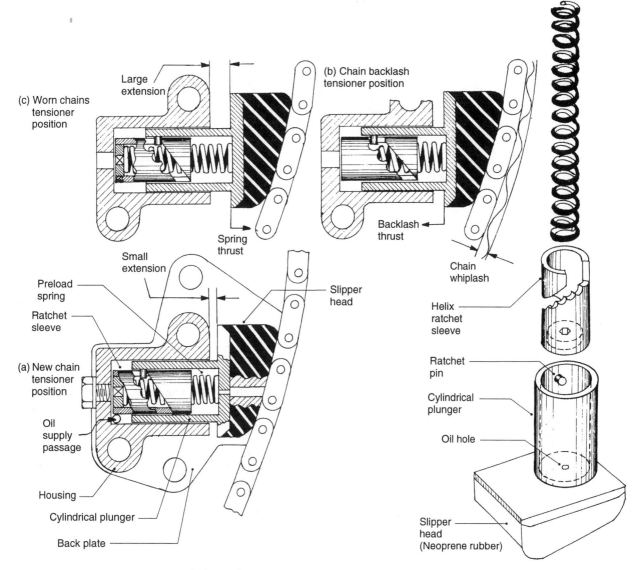

Fig. 2.29 Cylindrical helix ratchet chain tensioner

tensioner cylinder housing, cap nut and finally through the inlet non-return ball-valve.

The initial oil pressure build-up in the cylinder forces the plunger out, thus causing the tensioning friction rail to swing over against the chain. When the maximum pressure is reached the relief-valve opens and spills the excess oil onto the friction rail.

Lubrication between the plunger and cylinder housing is obtained by a permanent leakage of oil through the sliding clearance between them.

Automatic tensioning of the chain, as stretch takes place by the plunger progressively moving outwards, keeps the chain slackness to a minimum and any chain backlash is absorbed by the oil trapped between the cylinder and plunger.

A volume-reducer rod is placed inside the compression spring to reduce the initial volume formed between the cylinder and hollow plunger, it thereby provides faster venting during the filling operation.

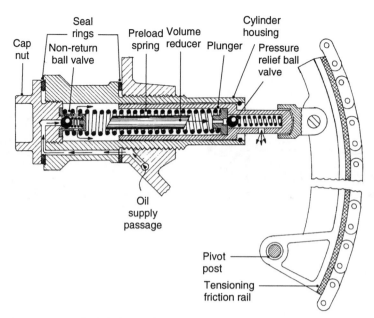

Fig. 2.30 Hydraulic plunger-controlled chain tensioner

2.4.5 Ratchet pawl and plunger automatic chain tensioner
(Fig. 2.31)

Controlled automatic chain tensioning for heavy duty conditions may be achieved by incorporating an idler jockey sprocket wheel which relays the thrust from the sliding cylinder and fixed-plunger tensioner to the chain (Fig. 2.31).

A compression-spring preloads the jockey-sprocket thrust towards the chain with the assistance of the engine hydraulic oil pressure which is conveyed to the cylinder via the central plunger passage. Excess oil pressure is spilled through the relief ball-valve which then drips and lubricates the jockey sprocket and chain assembly.

A spring-loaded ratchet-arm hinges on the plunger base and contacts a fixed pawl mounted on the side of the cylinder body.

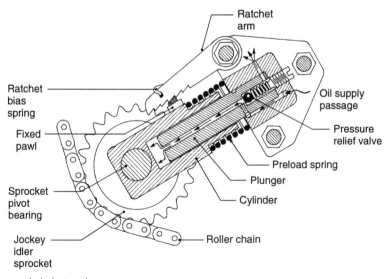

Fig. 2.31 Ratchet pawl chain tensioner

As the amount of chain slack increases, the cylinder moves out due to the combined effort of the compression-spring and oil pressure, and, in so doing, the pawl slides along the underside of the spring-loaded ratchet-arm, clicking into each sawtooth notch as the chain stretch and wear permits.

Therefore, the cylinder and jockey-sprocket can move outwards but are prevented from moving rearwards by the ratchet and pawl locking when the chain tries to whip back.

Note that the permanent outward movement of the jockey-sprocket assembly occurs in steps equal to the pitch of the ratchet teeth, so that the chain slack builds up from a minimum, when a new ratchet tooth engages, to a maximum just before the pawl engages the next tooth.

2.4.6 Chain vibration damper strips (Snubbers)
(Fig. 2.25(b))

Over an engine's speed range, torsional vibration (or cyclic speed variation) of the crankshaft and camshaft occurs at one or more critical speeds. Generally, these vibrations are partially absorbed or detuned by the crankshaft vibration damper but they may still persist with a much reduced magnitude and the critical speed or speeds may shift slightly to another part of the engine's speed range.

If the engine, when passing through one of these critical speeds, should linger for any length of time, then the wind and unwind oscillating motion imposed on the revolving crankshaft drive sprocket will be relayed to the chain and camshaft. This will cause the chain tension to alternate repeatedly between taut and slack on the drive side of the chain at a relatively high frequency, and if this becomes severe long unsupported chain spans can be excited into vibrating.

In the majority of cases these vibrations disappear when the engine speed increases or decreases above or below the critical speed. However, for some engines, these critical or resonant speeds occur in the normal-use driving speed range. It is usual, therefore, to minimize these vibrations by supporting and guiding the chain spans with synthetic rubber-lined guides, known as vibration damper strips (snubbers) (Fig. 2.25(b)). The effects of fitting these strips is to absorb some of the chain whip deflection energy and therefore damp down the vibration tendency

at the critical speeds when these vibrations build up.

2.5 Camshaft toothed-belt drive arrangements

2.5.1 Overhead camshaft belt drive
(Fig. 2.32(a))

The simplest of toothed belt drives consists of a driver and driven grooved or toothed pulley and jockey pulley tensioner layout. The two to one speed ratio step down for the camshaft is obtained with something like a 21 tooth crankshaft driver pulley relaying motion to a 42 tooth camshaft pulley via the toothed belt.

The correct working belt tension then being provided by a jockey tensioner situated on the smooth non-drive side of the belt approximately mid-way between the two pulleys, the jockey pulley being made to contact and roll against the smooth side of the belt.

2.5.2 Overhead camshaft with coolant pump on the non-drive side of the belt drive
(Fig. 2.32(a and b))

A triple-grooved pulley and jockey tensioner belt drive has a very similar configuration to a two-grooved pulley drive with the exception that a third grooved pulley, which is mounted on the coolant pump spindle, is positioned on the non-drive side of the belt, engaging the belt teeth between the crankshaft pulley and jockey tensioner. With this arrangement the belt passes from the outside of the coolant pulley to the inside of the smooth jockey pulley, thereby increasing the belt angle of contact with both the coolant pump and jockey tensioner pulleys. Likewise, the position of the jockey tensioner winds the belt further around the return side of the camshaft pulley.

2.5.3 Overhead camshaft with coolant pump on the drive side of the belt drive
(Fig. 2.32(c))

An alternative triple-grooved pulley and jockey tensioner belt drive configuration, which drives

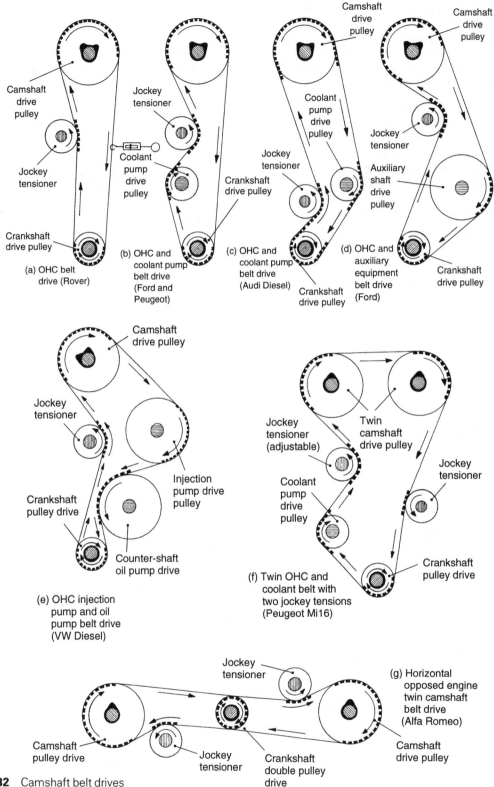

Fig. 2.32 Camshaft belt drives

both the camshaft and coolant pump, mounts the coolant pump on the drive side of the belt to the right of the crankshaft and camshaft centres, and brings the jockey pulley tensioner lower down and roughly in line with the crankshaft and camshaft centres. This layout has the benefit of increasing the angle of belt contact for the toothed crankshaft and coolant pump pulleys, and also for the plain jockey pulley.

2.5.4 Overhead camshaft and auxiliary equipment belt drive
(Fig. 2.32(d))

A further triple-grooved pulley and jockey tensioner arrangement uses a large third pulley, mounted low down to the right of the crankshaft and camshaft centres, to drive the auxiliary shaft at half crankshaft speed (distributor, petrol pump and oil pump take off). The jockey pulley tensioner, however, is positioned just below the camshaft pulley, again a little to the right of the crankshaft and camshaft centres. This configuration is chosen to maximize belt contact with each grooved pulley and jockey tensioner.

2.5.5 Overhead camshaft with injection pump and oil pump belt drive
(Fig. 2.32(e))

A quadruple-grooved pulley toothed-belt drive uses a pair of large pulleys mounted adjacent to each other on the drive side of the belt, to the right of the crankshaft and camshaft centres, to drive both the injection pump and the oil pump. The jockey tensioner pulleys as before, is positioned high up on the non-drive side of the belt. The belt is then looped on opposite sides of the injection pump and oil pump pulleys to extend their angle of belt wrap. Note that the oil pump pulley is driven on the smooth side, this being possible since it does not have to be timed.

2.5.6 Twin overhead camshafts and coolant pump belt drive with two jockey pulleys
(Fig. 2.32(f))

A further example of a twin overhead camshaft quadruple-grooved pulley toothed-belt drive us-ing a pair of jockey pulleys (one fixed and the other adjustable) is shown in Fig. 2.32(f).

With this arrangement the coolant pump pulley is offset to the left of the crankshaft with a jockey pulley on both the non-drive and drive sides of the belt. Motion is transferred from the crankshaft to the coolant pump and both camshafts, via the toothed belt, which is partially wound over each grooved pulley. The jockey pulley positioned between the left-hand camshaft and coolant pump, and between the crankshaft and right-hand camshaft, bends the belt inwards on either side, thereby extending the angle of pulley contact, and sets the belt tension.

2.5.7 Horizontally opposed twin camshaft engine belt drive
(Fig. 2.32(g))

The belt drive for a horizontally opposed twin camshaft engine consists of two separate belt drives which are driven from a crankshaft double pulley, each belt is tensioned by its own jockey pulley tensioner mounted on the non-drive side of each belt, the tension being created by the plain jockey pulley surface rolling against the smooth side of each belt.

2.6 Belt tensioners

2.6.1 Spring-loaded pivot plate jockey pulley tensioner
(Fig. 2.33(a, b and c))

The difficulty of accurately measuring the tension of a belt while levering the pivot plate and pulley against the smooth side of the belt is overcome by spring loading the tensioner so that a constant torque is applied to the pivot plate causing it to deflect the belt until the correct tension is achieved. The spring force which produces the pivot swing action may be provided by one of the following types of spring: (1) compression (Fig. 2.33(a)); (2) tension (Fig. 2.33(b)); or (3) torsion (Fig. 2.33(c)). Tensioning of the belt is obtained by slackening both the pivot-post and the elongated hole locknuts, this then permits the preload spring force to swing the pivot-plate and jockey-pulley hard against the smooth side of the belt, the relayed spring load therefore applies to the

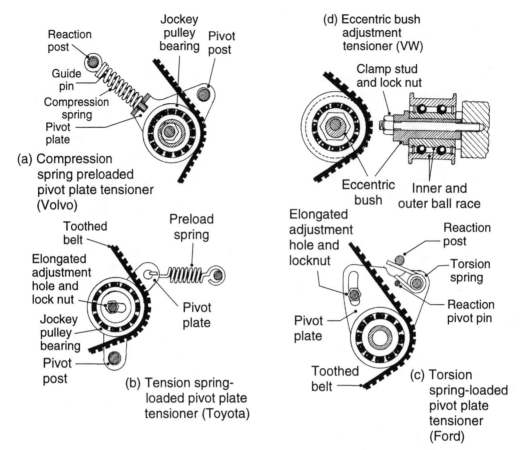

Fig. 2.33 Jockey idler pulley belt tensioners

belt, via the jockey-pulley, the correct tension. Both pivot-plate locknuts are then tightened, and the crankshaft pulley is then rotated two complete revolutions, enabling the belt to settle and accurately align itself between pulley-wheels. Finally, the pivot plate locknuts should again be fully released and then tightened to permit the pivot-plate to establish its optimum tensioning position.

2.6.2 Eccentric bush adjustment tensioner
(Fig. 2.33(d))

This method of tensioning the belt makes use of an eccentric bush supporting the jockey pulley bearing inner race, which is itself mounted on a clamp stud. Tensioning of the belt is achieved by rotating the eccentric bush with a spanner fitted over the hexagonal-shaped bush head until the belt can just be twisted 90° when held between the

finger and thumb halfway between the crankshaft and camshaft. The eccentric bush is then held in this adjusted position while the clamp stud lock nut is fully tightened.

2.6.3 Spring-loaded slide plate jockey pulley tensioner
(Fig. 2.34(a and b))

With this belt tensioner (Fig. 2.34(a)) the jockey-pulley is mounted on a sliding-plate via a bearing pulley post. Tensioning of the belt is obtained by slackening the bolts or nuts securing the elongated holes thereby enabling the preload compression-spring to expand until the pulley is pushed against the smooth side of the belt with a predetermined force, the locknuts are then re-tightened to maintain the slide plate in its adjusted position.

One design of belt jockey pulley tensioner

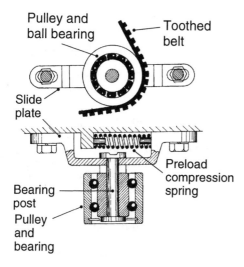

(a) Compression spring
and sliding plate
tensioner (Ford)

Pulley and
ball bearing

Toothed
belt

Slide
plate

Preload
compression
spring

Bearing
post

Pulley
and
bearing

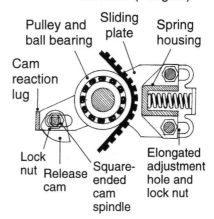

(b) Cam and slide plate
tensioner (Peugeot)

Pulley and
ball bearing

Sliding
plate

Spring
housing

Cam
reaction
lug

Lock
nut

Release
cam

Square-
ended
cam
spindle

Elongated
adjustment
hole and
lock nut

Fig. 2.34 Jockey idler pulley belt tensioner

incorporates a cam to release the tension in the belt before removing it so that it can easily be slipped off or pushed on the grooved pulleys without causing damage to the belt fabric (Fig. 2.34(b)). Thus, before removing the belt, the three slide-plate locknuts are slackened, and the square ended cam spindle is then rotated clockwise until the cam compresses the preload-spring and, at the same time, moves the slide-plate away from the belt. The cam-spindle locknut is then tightened. The belt can then be withdrawn easily from the jockey-pulley. Once the belt has been replaced, the cam spindle locknut is again slackened, and the square end cam-spindle is then rotated anti-clockwise until the cam is horizontal. This releases the preload-spring so that it is again free to move the slide-plate towards the belt until the jockey-pulley loads and tensions the belt. The crankshaft should then be rotated two full revolutions in the normal direction of rotation to settle the belt. Finally, the slide-plate locknuts should be slackened and then retightened so that the slide-plate and pulley assembly have the opportunity to reposition after the belt has been realigned.

2.7   Camshaft gear train drive arrangement

2.7.1   Camshaft direct gear train drive
(Fig. 2.35(a))

For vee 8 engines with push-rod operated valves, the central-mounted camshaft can simply be driven by a pair of meshing helical toothed gears. Typical gear sizes for the crankshaft driver and the camshft driven gear would be 26 and 52 teeth respectively, this will then drive the camshaft at the required half crankshaft speed.

2.7.2   Camshaft and injection pump with intermediate idler gear train drive
(Fig. 2.35(b))

With an in-line engine using short push-rod operated valves, a cylinder-block high-mounted camshaft necessitates an additional idler-gear so that it can be positioned high enough and to one side of the cylinders. By these means, motion will be relayed from the crankshaft to the camshaft via the intermediate idler-gears. The injection-pump

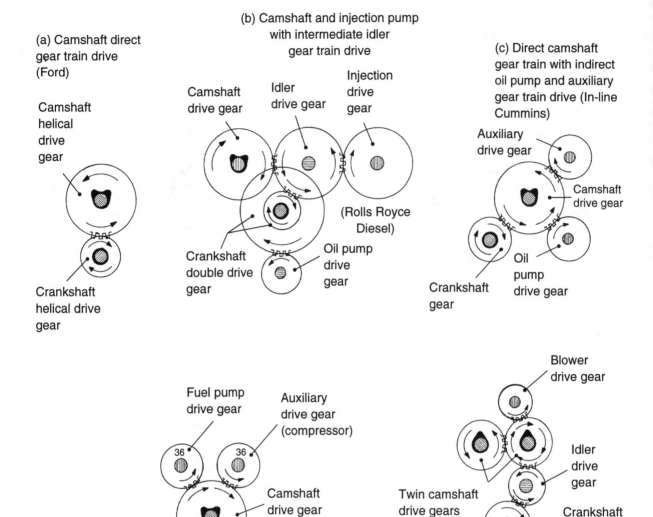

Fig. 2.35 Camshaft gear train drives

shaft gear is also driven off the idler-gear at half crankshaft speed, whereas the oil pump direct-drive small gear steps up its speed relative to the crankshaft to compensate for a low revving engine.

2.7.3 Camshaft direct gear train with indirect oil pump and auxiliary gear train drive
(Fig. 2.35(c))

If an in-line engine with a low mounted cylinder block camshaft is to be used with long push-rod operated valves, then a simple pair of meshing

72

gears can be utilized, with the camshaft positioned to the right-hand side of the cylinders. In turn, the auxiliary and oil pump shafts can be driven off the camshaft gear using smaller gears to bring these shafts up to crankshaft speed again, this being required for Cummins injection pumps and to compensate for the engine's low operating speed range.

2.7.4 Camshaft, injection pump and auxiliary drive via intermediate idler gear train drive
(Fig. 2.35(d))

This gear train configuration provides the drive to the camshaft, fuel pump shaft and auxiliary shaft. The centrally mounted camshaft actuates the push-rods for both cylinder banks and, because the distance between the crankshaft and the camshaft centres is relatively large, an extra idler-gear is used to span the gap and to transfer motion from the small crankshaft gear to the large camshaft gear. The camshaft gear is also utilized to relay motion to the fuel-pump shaft and auxiliary equipment shaft, mounted above and on either side of the camshaft.

2.7.5 Two-stroke engine with twin camshaft gear train drive
(Fig. 2.35(e))

A well-proven gear train, adopted on the Detroit Diesel two-stroke vee 4, 6 and 8 cylinder engines, provides a gear train drive to a pair of camshafts mounted high up between the cylinder banks.

The gear drive is relayed from the crankshaft to the right-hand camshaft via the idler-gear and the drive then continues to the left-hand camshaft by way of the direct gear engagement between the camshaft; similarly, the blower is driven at a faster speed directly from the right-hand camshaft gear.

2.7.6 Overhead camshaft gear train drive
(Fig. 2.36)

This overhead camshaft and auxiliary equipment drive is driven by two completely separate gear trains, the rear train at the flywheel end drives the

camshaft, compressor and injection pump, while the front gear train drives the water pump and steering hydraulic pump. In Fig. 2.36 there are two gear trains appearing to be driven off the same crankshaft gear, but, in fact, there are two separate crankshaft gears, one at the front and the other one at the rear of the crankshaft.

The rear gear train drive transfers motion from the crankshaft's 36 tooth gear to the camshaft via a pair of lower and upper idler gears each having 60 teeth, and the compressor and injection-pump drive 72-toothed gear is taken off the lower idler gear to the left-hand side of the crankshaft, whereas the oil pump 72-toothed gear is driven directly off the rear crankshaft gear.

For convenience the water pump and steering hydraulic pump 28-toothed gear is driven from the front crankshaft 36-toothed gear through the 52-toothed idler gear.

Thus, camshaft gear reduction
$$= \frac{60}{36} \times \frac{60}{60} \times \frac{72}{60} = 2:1$$

Injection pump gear reduction
$$= \frac{60}{36} \times \frac{72}{60} = 2:1$$

Water pump and hydraulic pump gear step-up
$$= \frac{52}{36} \times \frac{28}{52} = 0.78:1$$

2.8 Valve and camshaft timing

Inlet and exhaust valve timing are governed by the relative positions of the camshaft and the crankshaft and, as these are coupled via chain, belt or gears directly together, the timing cannot alter in service unless the chain stretches or jumps a tooth, the belt fractures or jumps a tooth or, with a gear train drive, a gear tooth shears.

The majority of engines have some sort of timing marks for alignment, such as centre punch marks or line marks; however, there are occasions when a deeper study of valve timing becomes necessary to carry out the valve timing procedure. This will now be discussed and explained.

2.8.1 Camshaft (valve) timing using a valve timing diagram

Camshaft timing generally only requires the inlet valve opening point to be observed but there are a

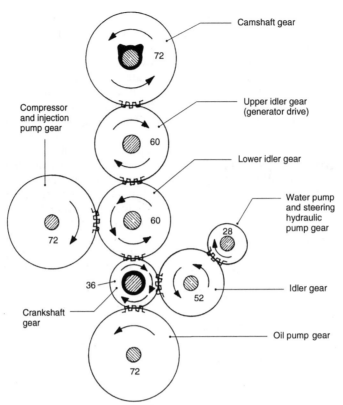

Fig. 2.36 OHC gear train drive (Leyland diesel engine)

few instances where the camshaft timing is carried out on the exhaust valve closing point instead.

Procedure for camshaft timing
(Fig. 2.37(a and b))

1 Adjust the tappet clearance of the inlet and exhaust valve of the rear cylinder—that is, number 4 or 6 cylinder—to the recommended timing tappet clearance. This is usually greater than the normal cold tappet clearance.

   Having a slightly larger tappet clearance for valve timing permits the commencement of valve lift or the completion of valve fall to occur on the relatively steep part of the opening and closing sides of the cam flanks and not on the low lift rate ramp portion of the cam profile. Thus, the increased tappet clearance enables the first point of valve opening or closing to be accurately determined.

2 Rotate the crankshaft so that 1 and 4 or 1 and 6 cylinder pistons are at TDC, then mark the flywheel rim with chalk or pencil opposite a fixed datum pointer or at some other convenient position on the bell housing.

3 If the inlet or exhaust valve opening or closing positions are not already marked on the flywheel rim then they must be derived from the valve timing diagram and converted from degrees to linear measurements on the flywheel rim. The conversion of valve lead or lag from degrees to flywheel rim measurements can be obtained in the following way:

$$\frac{x}{\pi D} = \frac{\theta}{360}$$

therefore $x = \dfrac{\theta \pi D}{360}$

where x = linear flywheel rim distance (mm)
  θ = angular crankshaft displacement (deg)
  D = diameter of flywheel (mm)

For example, with a 420 mm diameter flywheel and with the inlet valve opening 18° before TDC, determine the equivalent distance on the flywheel rim:

then $x = \dfrac{\theta \pi D}{360} = \dfrac{18 \times 3.142 \times 420}{360} = 66\,\text{mm}$

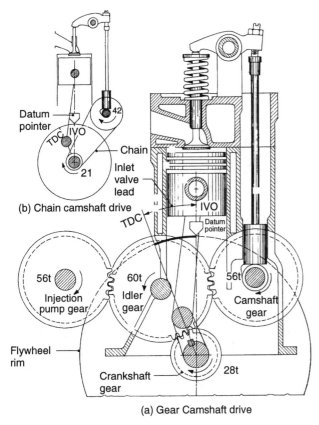

(b) Chain camshaft drive

(a) Gear Camshaft drive

Fig. 2.37 Valve timing fundamentals

4 Measure the inlet valve opening position on the rim of the flywheel on the lead side of the datum pointer or mark (Fig. 2.38(a)), or on the lag side of the datum pointer or mark (Fig.

2.38(c)) if the exhaust valve closing position is being measured.

5 Turn back the flywheel in the opposite direction of normal rotation (Fig. 2.37(a and b)) until the inlet valve opening (IVO) position on the flywheel rim aligns with the datum pointer or mark. If timing the exhaust closing position, move the flywheel forward in the direction of normal rotation until the exhaust closing (EVC) mark on the rim of the flywheel is opposite the datum pointer or mark on the bell housing.

6 Rotate the camshaft slowly in the normal direction of rotation and with first finger and thumb twist the rear cylinder inlet valve push-rod until the first sign of tightness, this will be the point of inlet valve opening, at this point stop rotating the camshaft. Recheck the inlet valve opening position by moving the camshaft backward and then come forward until the first point of opening is established again. Similarly, if the exhaust valve closing point is to be found, apply a steady rotary effort on the rear cylinder exhaust valve push-rod and rotate the camshaft in the direction of rotation until the push-rod is released.

7 The two major ingredients for camshaft timing relative to the crankshaft have now been completed, these are:

i positioning the crankshaft either at the inlet lead position before TDC or at the exhaust closing lag position after TDC;

ii positioning the camshaft so that the rear cylinder inlet valve is about to open or so that the rear cylinder exhaust valve is about to close.

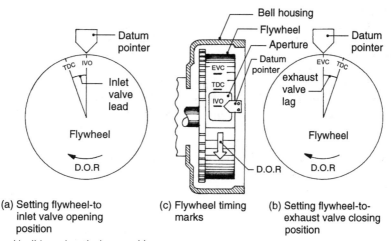

(a) Setting flywheel-to inlet valve opening position

(c) Flywheel timing marks

(b) Setting flywheel-to-exhaust valve closing position

Fig. 2.38 Flywheel and bell-housing timing markings

8 At this stage, the drive train between the crankshaft and camshaft, be it chain, belt or gear, is ready to be assembled.

In the case of the chain drive remove the camshaft sprocket-wheel, place the chain over both crankshaft and camshaft sprockets and then replace the camshaft sprocket-wheel onto the camshaft.

With the belt-drive, slacken off the jockey tensioner and slip the belt over both attached crankshaft and camshaft pulleys, then tension the belt.

When assembling a helical gear train an allowance must be made for the extra angular movement of the camshaft gear when it is pushed into mesh with its drive gear.

After the camshaft drive has been assembled re-check the valve timing.

For a 21/42 tooth drive combination sprocket, pulley or gear, if the timing is one tooth out then the camshaft timing will be $360/42 = 8.5°$ out or, on a 420 mm diameter flywheel rim, 31 mm out. This is a considerable amount and would show up immediately when the camshaft timing is being checked.

2.8.2 Valve timing check
(Fig. 2.37(a and b))

1 Adjust the rear cylinder inlet valve tappet to the correct timing clearance.
2 Measure and mark the inlet valve lead before TDC on the rim of the flywheel (if not already marked).
3 Rotate the crankshaft until pistons 1 and 4 or 1 and 6 are approaching TDC with number one cylinder on its compression stroke.
4 With finger and thumb, twist the rear cylinder inlet valve push-rod and slowly continue to rotate the crankshaft until the push-rod tightens, i.e. the valve commences to open.
5 At this point, stop rotating the crankshaft and observe the flywheel rim inlet valve opening mark, this should now be opposite the datum pointer on the bell housing.
6 The tolerance of the inlet valve opening mark and datum pointer alignment should be within $±2.5°$ camshaft movement or $±10$ mm on a 420 mm diameter flywheel.

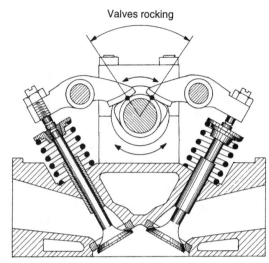

Fig. 2.39 OHC with twin rocker shafts and arms illustrating valve rock

2.8.3 Camshaft (valve) timing by the approximate method

Background to valve overlap and valve rock
(Fig. 2.39)

When examining a typical valve timing diagram, it will be seen that the piston is at TDC at the end of the exhaust stroke and, at the beginning of the induction stroke, the inlet valve has just begun to open and the exhaust valve is about to close. This period when both inlet and exhaust valves are partially open is known as valve overlap. Generally, the inlet valve lead and the exhaust valve lag, which make up valve overlap, are roughly equal although there is a tendency for the exhaust valve lag to be slightly larger than the inlet valve lead.

Therefore, when piston numbers 1 and 4 or 1 and 6 are at TDC and the rear piston (4 or 6) is at the end of its exhaust stroke, then for all practical purposes, both valves in the rear cylinder will be rocking. This means that, when turning the camshaft slowly backward and forward in the TDC region, the camshaft will rock open and close both inlet and exhaust valves approximately the same amount.

Procedure for camshaft timing
(Fig. 2.39)

1 Adjust the rear cylinder inlet and exhaust valve tappets to their normal operating clearance.

2 Rotate the crankshaft until pistons 1 and 4 or 1 and 6 are at TDC with the camshaft drive (chain, belt or gears) disconnected.
3 Rotate the camshaft until the rear inlet and exhaust valves are equally rocking; that is, the inlet valve is just opening and the exhaust valve is near closing.
4 Re-check the crankshaft and camshaft timing positions and then couple the chain belt or gears.

Valve timing check
(Fig. 2.39)
To check the engine valve timing when the camshaft drive is assembled, just rotate the crankshaft until the rear cylinder inlet valve is just opening and the exhaust valve is about to close (back two valves rocking) and note the position of the crankshaft. If the valve timing is correct, cylinders 1 and 4 or 1 and 6 should be at TDC and cylinder number 1 is about to begin its power stroke.

2.8.4 Vernier hole camshaft valve timing adjustment

The vernier hole gear camshaft drive is employed when accurate timing is necessary, where several gears are involved or a very long chain is used and when timing may alter in service due to chain wear and stretch.

Construction and description of vernier hole adjustment
(Fig. 2.40)
A vernier hole camshaft drive gear consists basically of a flanged hub mounted on the end of the camshaft and secured by a woodruff key, and the shoulder or spigot on the hub supports a separately attached gear wheel. Relative angular movement between the hub and gear wheel can take place when the two vernier setbolts are withdrawn from the aligned holes forming part of the vernier hole ring. The flanged hub (Fig. 2.40) has 20 holes of equal pitch spacing and the gear wheel has 18 holes of equal pitch and of the same pitch circle as the hub holes. This means that only two diametrically opposed holes will align when the hub and gear flanges are placed together.

A fine angular adjustment is achieved in the following manner.

Angular pitch between the flanged hub 18 holes

$$= \frac{360}{18} = 20°$$

Angular pitch between the gear wheel 20 holes

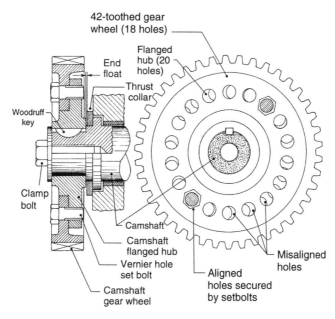

Fig. 2.40 Vernier hole camshaft valve timing adjustment

77

$$= \frac{360}{20} = 18°$$

Therefore, the smallest angle of camshaft can be varied

$$= 20 - 18 = 2°$$

This may be compared with a 42-toothed camshaft gear or sprocket wheel in which the angular pitch between teeth will be 360/42 = 8.5°.

Procedure for camshaft timing
(Fig. 2.40)

1 Set the rear cylinder inlet and exhaust timing tappet clearance making sure that the tappets are on the backs of the cams when doing so.
2 Rotate the flywheel until pistons 1 and 4 or 1 and 6 are at TDC; that is, the TDC mark on the flywheel rim aligns with the timing datum pointer looking through the bell housing timing aperture (Fig. 2.38(c)).
3 Remove the two setbolts securing the vernier camshaft gear wheel to its hub.
4 Rotate the camshaft until the rear cylinder inlet and exhaust valve just rock; that is, the inlet valve is just opening and the exhaust valve is just closing.
5 Insert and screw in the two setbolts in the two sets of diametrically opposed aligning holes.
6a Now turn the flywheel very slowly in the direction of rotation and apply a rotary effort to the last inlet valve push-rod until it tightens, the inlet valve will be at opening point, and observe if the flywheel inlet valve opening (IVO) lead point aligns with the datum pointer in the bell housing aperture (Fig. 2.38(c)).
6b Rotate the flywheel further in the normal direction of motion and try to spin the last exhaust valve push-rod. When it is released the exhaust valve will be closed. The flywheel exhaust closing (EVC) lag point should then align with the bell housing datum pointer. The check on exhaust valve closing is not always necessary and therefore need not be carried out unless there is any doubt about the inlet valve opening point timing.
7 If the inlet or exhaust valve opening points are incorrect and do not align with the datum pointer within ±10 mm (2.5°) on, say, a 420 mm diameter flywheel rim, remove both vernier hole setbolts and rotate the camshaft to advance or retard as required, then refit and screw tight the setbolts. Again check the inlet valve timing.

In most engines, if the timing is more than two teeth out the engine cannot be turned as the valves will bump the piston crown.

After timing the camshaft, reset the larger timing tappet clearances to their normal operating clearance.

3
Engine Balance and Vibration

3.1 The concept of balance and vibration

A perfectly balanced engine is one in which the relative motion of the component parts do not set up an accumulation of forces that tend to make the engine shake and rock. Hence, if the perfectly balanced engine were to be suspended freely in space no vibration or other movement would be observed and, therefore, in theory, such an engine could be attached directly to its support frame. Conversely, a partially balanced engine requires some sort of suspension mounting to isolate the engine from its support frame to prevent any of the unbalanced reaction movement being transmitted through to the vehicle's chassis and body.

Engine vibrations can be broadly divided into two unwanted modes:

1 vibration of the engine and its rigid components as a whole, in which there is no elastic yielding of the components—these vibrations being caused by the imbalance of the rotating and reciprocating components;
2 vibration of engine components due to the elastic deformation of the material of the component parts under the influence of the periodic combustion impulses causing torsional and lateral oscillation of the crankshaft and camshaft.

3.1.1 Balancing a single-cylinder crankshaft
(Figs 3.1 and 3.2)

Rotating masses in a single-cylinder engine may be completely balanced but reciprocating compo-

nent parts are normally only partially balanced. Considering the rotating masses only, if the crankpin and connecting-rod big-end of a single-cylinder crankshaft of mass (m) rotate at a constant velocity (ω) about some fixed centre (main journal) at a radius (crank throw) (R) the mass will experience an outward pulling centrifugal force of magnitude (F)

$$F = m\omega^2 R \tag{1}$$

where F = centrifugal force (N)
      m = crankpin and connecting-rod big-end mass (kg)
      ω = angular velocity (rad/s)
      R = crank throw (m)
      F_r = reaction force (N)

The reaction (F_r) of the generated centrifugal (F) force on the axis of the main journal is a force equal and opposite in direction (Fig. 3.1).

The disturbing force pulling radially outward from the centre remains constant but its direction is changing continuously as it moves in its circular path.

In order that the centre of rotation is in equilibrium, an equal and opposite force must be applied. This may be obtained by extending the big-end crankpin webs to the other side of the crankshaft, by a distance (R_b), so that half the equivalent out-of-balance masses ($\frac{1}{2}m_b$) are situated opposite and to each side of the crankpin, thereby causing the centrifugal pivoting reaction to become zero (Fig. 3.2).

Thus, the two centrifugal forces must be equivalent and therefore can be equated:

revolving mass centrifugal force = balance mass centrifugal force.

$$m\omega^2 R = m_b \omega^2 R_b \tag{2}$$

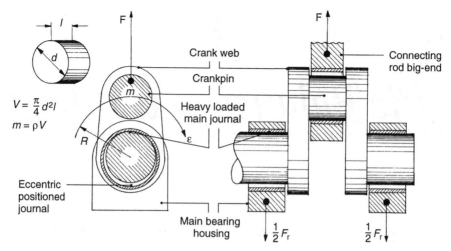

Fig. 3.1 Unbalanced single-cylinder crankshaft

Since the angular velocities are common on both sides of the equation the relationship becomes:

$$mR = m_b R_b \tag{3}$$

where mR is known as the mass moment un-balanced effort due to the off-centre rotating mass (m).

Thus, the mass moments of the two half-balance masses can be equated as shown.

crankpin and big-end mass moment
= balance mass moments

$$mR = \left(\frac{m_b}{2} + \frac{m_b}{2}\right)R_b \tag{4}$$

3.1.2 Rotating components of crankshaft and connecting rod
(Fig. 3.2)

The rotating components of a crankshaft are basically the crankpin and the big-end of the connecting-rod (Fig. 3.2). The revolving mass of

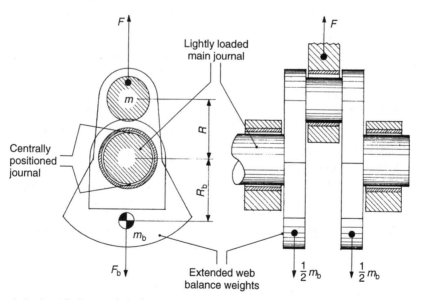

Fig. 3.2 Balanced single-cylinder crankshaft

the crankshaft itself is derived from the mass of the offset crankpin and its supporting side webs but, because these webs are generally made to extend to the opposite side of the main journal, the rotating web imbalance is cancelled so that their masses can be ignored. As a result, only the crankpin mass is considered when examining the balance of the crankshaft.

Crankpin journal mass = material density
$$\times \text{ crankpin volume}$$

$$m_J = \rho V = \rho \frac{\pi}{4} d^2 L \text{ (kg)} \tag{5}$$

where m_J = crankpin journal mass (kg)
ρ = material density (kg/m^2)
d = diameter of crankpin (m)
L = length of crankpin (m)
for steel $\rho = 7900$ kg/m^3

It is usual to divide the connecting-rod mass into two components, a rotating mass component and a reciprocating mass component. The rotating mass component of the connecting-rod is obtained from the big-end portion of the rod while the reciprocating mass component is derived from the rod's small-end. Experiments have shown that approximately two-thirds of the connecting-rod mass contribute to the rotating mass component of the rods, whereas only one-third of the rod's mass effectively reciprocates, thus:

rotating connecting-rod mass
$$= \text{two-thirds of the con-rod mass}$$

$$m_c = 2/3C \tag{6}$$

where m_c = rotating con-rod mass (kg)
C = con-rod mass (kg)

Consequently, the total rotating inertia components acting on the crankshaft are equal to the sum of the crankpin mass and the rotating big-end mass component, thus:

total rotating mass = crankpin mass + big-end mass

$$m = m_J + m_c$$

$$= \rho \frac{\pi}{4} d^2 L + 2/3C \tag{7}$$

3.1.3 Centrifugal moment of a couple
(Figs 3.3 and 3.4(a and b))

If two crankpins each of mass (m) are spaced a distance (a) apart (Fig. 3.3) on the opposite sides of the main journals at a radius (r) from the centre axis of the rotating crankshaft, then the centrifugal force ($F = m\omega^2 r$) exerted on each big-end and crankpin, will produce a moment of a couple, its magnitude being proportional to the square of the speed of rotation. The moment of a couple can be expressed as follows:

$$C = F\frac{a}{2} + F\frac{a}{2} = 2\left(F\frac{a}{2}\right) = Fa \text{ (Nm)}$$

The directional rock caused by the couple will continuously change as the crankshaft rotates in the vertical plane when crankpin No. 1 is at TDC and crankpin No. 2 is at BDC (Fig. 3.4(a)). The crankshaft will tend to produce a clockwise twist or rocking movement, but a further half a revolution of rotation moves crankpins Nos 1 and 2 to BDC and TDC positions, respectively, thereby altering the sense of the couple to an anticlock-

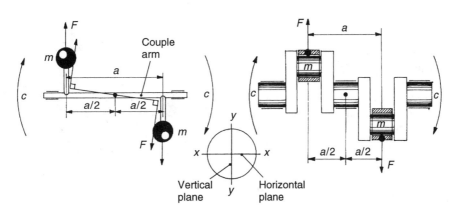

Fig. 3.3 Illustration of a moment of a couple

81

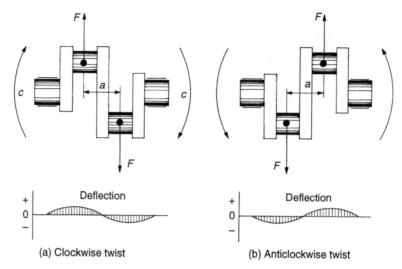

Fig. 3.4(a and b) Change in sense of couple and deflection as twin-cylinder crankshaft rotates

wise rock (Fig. 3.4(b)). Likewise, in the horizontal plane the sense of the moment of the couple continuously reverses as crankpins Nos 1 and 2 move towards and from their mid-stroke positions of 90° and 270°.

Due to the reaction of the main bearings the outward pull on the crankpins deflects the crankshaft in a sine-wave fashion as it rotates (Fig. 3.4(a and b)).

The magnitude of the crankshaft deflection will be greatly intensified in the vertical plane when the reciprocating component's inertia effects are compounded with the rotational inertia forces as the crankpins approach either the outer or inner dead centres. Thus, the rotational inertia effects are constant but continuously changing their sense throughout one revolution of the crankshaft whereas the reciprocating component inertia effects are only prominent as TDC or BDC is approached.

3.1.4 Static crankshaft balance
(Figs 3.2, 3.5 and 3.6)

Static balance of a single-cylinder crankshaft can be achieved by positioning half the equivalent balancing mass opposite the big-end crankpin on either side so that the product of these masses $2(m_b/2)$ and its distance from the centre of rotation (R_b) is equal to the mass of the big-end crankpin (m) times its distance from the centre of the shaft (R) (Fig. 3.2) that is

$$mR = m_b R_b$$

The static balance of the single-cylinder crankshaft can then be checked by mounting it on a pair of horizontal knife edges which are effectively frictionless and slowly rolling the shaft in steps of 45° until one revolution is completed. At the end of each 45° interval the shaft should be held still and then released. If the shaft is perfectly balanced then it will remain stationary in any position in which it is stopped; conversely, if the shaft is unbalanced it will tend to roll one way or the other until the heaviest part of the crankshaft is in its lowest position.

Multi-cylinder engines are statically balanced by positioning the opposing crank-throws so that they cancel out the individual centrifugal forces created by the crankpin and web structure. Thus,

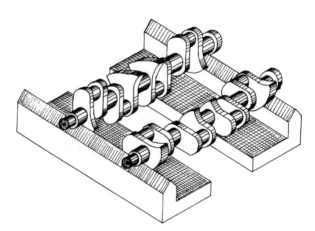

Fig. 3.5 Static balance check on knife edges

82

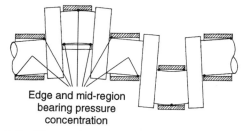

Edge and mid-region
bearing pressure
concentration

Fig. 3.7 Distorted crankshaft caused by rotating crankpin inertia force

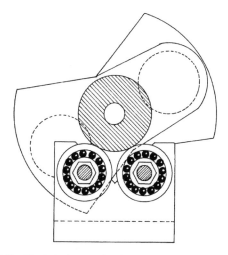

Fig. 3.6 Static balance check on ball race bearings

if a multi-cylinder statically balanced crankshaft is rolled on knife edges (Fig. 3.5) or, alternatively spun on ball race bearing supports (Fig. 3.6) then the crankshaft should remain stationary in any position.

3.1.5 Dynamic crankshaft balance
(Fig. 3.4(a and b))

Unfortunately, with multi-cylinder engines, the individual crankpins are offset to each other so that, as the crankshaft commences to rotate, the centrifugal inertia forces act in different planes and will therefore tend to produce rocking moments which make the crankshaft twitch in each direction as it completes each revolution. This condition is known as *dynamic imbalance*, which is caused by the opposing centrifugal inertia forces being spaced apart, an example of this is shown by a twin cylinder crankshaft (Fig. 3.4(a and b)). Thus, a crankshaft which is balanced statically will produce dynamic imbalance as it rotates unless provision for dynamic balance has been built into the shaft, whereas a dynamically balanced crankshaft will always be statically balanced.

3.2 Multi-cylinder crankshaft balancing of rotating masses
(Fig. 3.7)

Most multi-cylinder crankshaft configurations are

arranged so that their crankpins oppose and therefore neutralize any static out-of-balance of the individual crankpins and their webs. However, as soon as the crankshaft commences to rotate, internal couples will be created by the crankpin centrifugal inertia force, which tends to bend the crankshaft laterally. An exaggerated illustration of a distorted crankshaft (Fig. 3.7) concentrates extreme pressure at the edges and the mid-region of the bearing lengths. The magnitude of the crankshaft deflection increases with rising speed and, if the crankshaft stiffness is inadequate, the excessive bearing pressures will produce early fatigue failure.

To reduce the lateral deflection of the crankshaft when it rotates, balance masses in the form of extended webs can be positioned opposite and on either side of each crankpin. In practice, normally only one of the crankpin side webs is extended to the opposite side of the main journal, but for high-performance engines both side webs may be extended to smooth out the bending of the crankshaft, thereby relieving the bearing edge loading so that a more uniform pressure distribution exists over the working bearing surface.

3.2.1 Twin-cylinder crankshaft balance
(Figs 3.8 and 3.9)

Twin-cylinder crankshafts have offset diametrically opposed crankpins supported on a pair of outer cranked arms and by a common central arm which joins together both crankpins. With the in-line 180° out-of-phase twin (Fig. 3.8) the central arm is diagonally inclined to increase the offset between cylinders to accommodate the piston diameter width, the thickness of the cylinder walls and the coolent jacket. However, with the horizontally opposed twin-cylinder crankshaft (Fig. 3.9) the crankpin offset between cylinders is kept to a minimum by having the common central

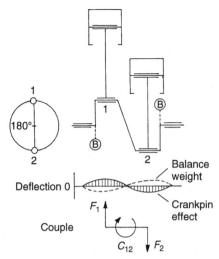

Fig. 3.8 180° out of phase twin-cylinder crankshaft balance

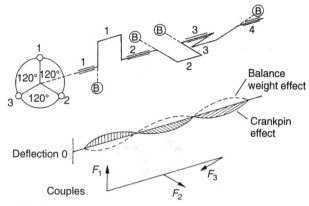

Fig. 3.10 Three-cylinder in-line crankshaft balance

3.8 and 3.9) thereby neutralizing the offset crankpin and web couple moment.

3.2.2 Three-cylinder in-line crankshaft balance
(Fig. 3.10)

The three crankpins are spaced at intervals of 120° and their crank arms are supported on four main journals. The even angular spacing of the crankpins statically balances the rotating masses but the longitudinal offset of each crankpin creates a dynamic imbalance which can only be cancelled by attaching balance weights opposite the individual crankpin webs (Fig. 3.10). Without balance weights, internal couples will cause the shaft to deflect laterally and, simultaneously, a resultant external couple will be generated which will rock the crankshaft within its main bearings. The normally adopted counterweight system extends both central crankpin webs to the opposite side of the shaft's main journals, whereas the first and last crankpins only have counterweights extended on their outer webs.

3.2.3 Four-cylinder in-line crankshaft balance
(Fig. 3.11(a and b))

With this single-plane crankshaft two internal couples of opposite sense are created by the opposing crankpins and their supporting webs. These couples are absorbed by the stiffness of the crankshaft material and cancel each other out, but the rigidity of the crankshaft structure is generally insufficient to prevent a small amount of

arm perpendicular to the crankshaft axis since there is no piston interference problem with opposing piston layouts. Consequently, a larger couple is produced with 180° out-of-phase twin-cylinder crankshafts relative to the horizontal opposed twin-engine.

In both crankshaft arrangements the out-of-balance opposing crankpins and webs counterbalance. This provides an inherent static balance for the shaft, but dynamic balance can only be achieved by extending the outer crank arm webs (B) to the opposite side of the main journals (Figs

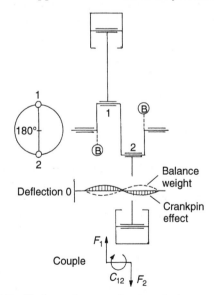

Fig. 3.9 Horizontal opposed twin-cylinder crankshaft balance

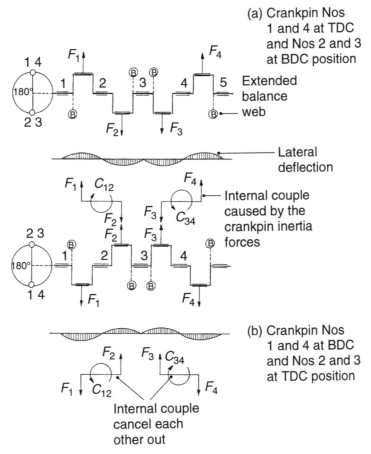

Fig. 3.11 Four-cylinder in-line crankshaft with extended balance webs illustrating the effects of internal couples

lateral deflection when the crankshaft rotates. Therefore, to improve the crankshaft alignment four extended webs are incorporated, two on the outside webs of the first and last crankpins and a second pair opposite the inner crankpins adjacent to each other (Fig. 3.11(a and b)). These extended balance webs produce dynamic couples which counteract each of the internal couples caused by the big-end crankpins so that the crankshaft deflection is minimized and the bearing load is spread more evenly over a large area. This crankshaft configuration is dynamically balanced with or without the extra balance masses, the extended web masses are there only to reduce the crankshaft lateral deflection and therefore minimize bearing pressure at high rotational speeds.

3.2.4 Five-cylinder in-line crankshaft balance
(Fig. 3.12)

A straight five-cylinder crankshaft has five separate big-end crankpins equally spaced around the crankshaft axis of rotation at intervals of 72° in a consecutive firing order of 12453 (Fig. 3.12). Each crankpin and its two side webs are separated and supported by one of six main journal bearings. With this arrangement, there are no pairs of opposing crankpins to balance out the inertia effects of each crankpin; therefore, each crankpin and big-end portion of its connecting rod is individually counterbalanced by balance weights which are provided by both crankpin side webs being extended to the opposite side of the main journals. This method of obtaining both static and dynamic crankshaft balance of the rotating com-

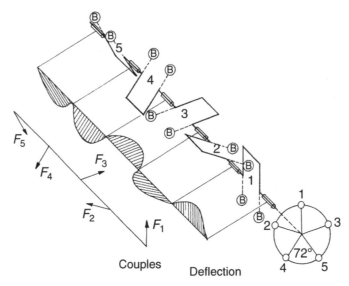

Fig. 3.12 Five-cylinder in-line crankshaft balance

Couples Deflection

ponents also eliminates any internal couples which might have been generated at its source, so that crankshaft lateral deflection and therefore bearing loading will be at a minimum. The penalty of increasing the inertia wind-up effect produced by having the full complement of counterweights, is compensated by having relatively large-diameter short-length journal surfaces which considerably improve the crankshaft's rigidity.

3.2.5 Six-cylinder in-line crankshaft balance
(Fig. 3.13)

Six-cylinder crankshafts have either four or seven main journal bearings. Four main bearing crankshafts were very popular until the 1960s for petrol engine capacities up to 3.5 to 4.0 litres because of their low manufacturing costs. However, owing to higher crankshaft speeds, compression ratios and better cylinder charge filling since then, the seven main bearing crankshaft has become necessary to provide greater lateral crankshaft support, thereby reducing running roughness and bearing loading caused by crankshaft bending.

Both four and seven main bearing crankshafts generate pairs of front and rear internal couples due to the inertia forces of the rotating big-end crankpins, these couples and their resulting lateral crankshaft deflection are shown with the crank-

shaft in three different angular positions in Fig. 3.13. The generated couples act in the opposite sense to each other and therefore tend to become neutralized, but this is generally at the expense of the crankshaft bending in several places along its length, which can cause excessive main journal deflection and very high bearing loads. A partial remedy is to attach counterweights in the form of extended crankpin webs to one, or in some cases both, sides of each big-end crankpin so that rotating crankpin inertia forces are balanced, or at least partially balanced, thus, limiting very high bearing loads caused by lateral tilting of the crankshaft.

3.2.6 60° vee six-cylinder crankshaft balance
(Fig. 3.14)

A 60° vee cylinder banked angle enables the width of an engine to be reduced by approximately 25% at the expense of something like a 20% increase in engine height, as compared with a 90° vee engine of the same capacity. Whereas the overall engine length is reduced by roughly 35% relative to a similar capacity straight six-cylinder engine, this reduction in crankshaft length considerably increases the rigidity of the crankshaft and, therefore, practically eliminates any torsional vibration.

The crankshaft has six separate big-end crank-

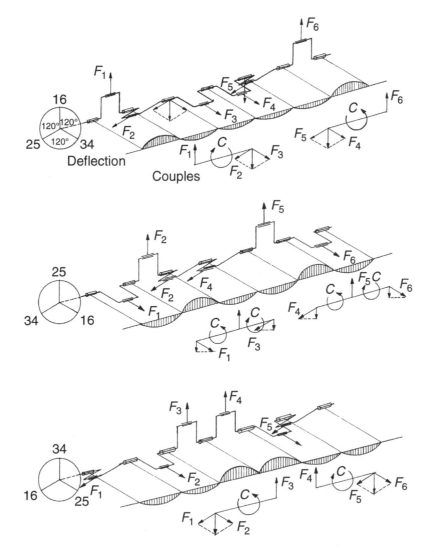

Fig. 3.13 Six-cylinder in-line crankshaft illustrating crankshaft lateral deflection caused by the internal couples

pins equally spaced at intervals of 60° arranged in pairs between the four main journal bearings (Fig. 3.14).

The centrifugal inertia forces of the crankpins and connecting-rod big-ends cancel out due to there being three diametrically opposing pairs of crankpins 1 and 6, 4 and 2 and 3 and 5, but these generate three longitudinally active couples which create a dynamic imbalance to the crankshaft. These three couples produce a resultant longitudinal couple in the plane of crankpins 1 and 6 and this is normally cancelled by providing balance weights directly opposite and on each side of crankpin Nos. 1 and 6. In some long stroke engines a third additional pair of smaller balance

weights are positioned in the same plane and direction as the other counterweights, but they are attached to the inner webs of crankpins 3 and 4 as this arrangement tends to relieve more of the load on the two inner main bearings.

3.2.7 90° vee eight two-plane or cruciform crankshaft balance
(Fig. 3.15 (a–d))

This is a two-plane five main bearing crankshaft with four big-end crankpins each supporting a pair of connecting-rods.

A pair of rotating couples is generated by the

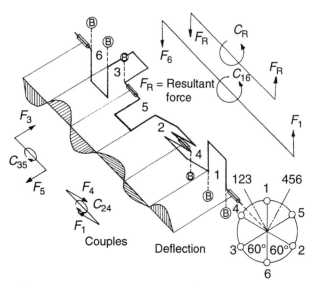

Fig. 3.14 60° vee six-cylinder crankshaft balance

two outer diametrically opposing big-end crank-pins 1 and 4 in the vertical plane (Fig. 3.15(a)) and the two opposing inner crankpins 2 and 3 in the horizontal plane, both couples thus being at right-angles to each other. The plane of the resultant couple occurs at approximately 18° to the plane of crankpin No. 1.

To neutralize the unbalanced couples (C_V and C_H) caused by the rotating crankpins, balance weights (B) are positioned directly opposite the crankpins on webs 2, 3, 6 and 7, the mass moment of each of these balance weights being equivalent to something like half to two-thirds of the purely rotating mass moment of each crankpin, its sup-port webs and the crankpin ends of the connect-ing-rods. A further two balance weights (B) are also positioned on the outer webs of crankpins 1 and 4. Their mass moment and angular disposi-tion relative to No. 1 crankpin (18°) then com-pletely counterbalance the remaining resultant couple produced by the combination of the rotat-ing crankpin and webs and the rotating compo-nent of the connecting-rods.

3.2.8 Crankshaft dimensional proportions
(Figs 3.16, 3.17, 3.18 and 3.19)

The crankshaft dimensions will be greatly in-fluenced by the magnitude of the piston inertia and gas loads imparted to the big-end crankpin

and the load-carrying capacity of the shell bear-ings.

The load-carrying capacity of the bearing will be determined basically by the big-end journal projected area and the maximum pressure the bearing material can sustain.

Early crankshafts adopted long slender dia-meter big-end and main journals, thereby keeping the rubbing velocities between the journal and bearing to a minimum (Fig. 3.16). This design produced a long crankshaft with very low torsion-al and lateral stiffness; however, modern trends have been to increase the journal diameter and to shorten the length of each journal without sac-rificing any of the projected bearing area. This considerably raises the torsional stiffness and the lateral rigidity of the crankshaft, but at the ex-pense of increasing the surface rubbing velocity between bearing and journal.

However, high rubbing velocities do not seem to be a major problem now that a much better understanding of hydrodynamic lubrication exists since this has produced great advancements in bearing materials and design and, therefore, im-provements in engine lubrication systems.

Crankshafts were originally all made from forged steel; however, due to the development of suitable nodular iron and the improvements in foundary techniques using shell mouldings, cast crankshafts which are much cheaper to produce have mainly replaced the forged shafts for small and medium-sized engines. Only for high per-formance petrol engine applications and for large heavy-duty diesel engines is the forged crankshaft still prominent.

With increased gas pressure and inertia loading of the crankshaft it has become necessary to provide substantial main bearing support on either side of each big-end crankpin instead of every second big-end crankpin, particularly when cast crankshafts are used which tend to need more lateral support.

Big-end and main journals are sometimes cast (Fig. 3.17) with hollow cores since the metal in the central portion of the journals contributes very little to the torsional and lateral stiffness of the crankshaft. Thus, for a given weight of crank-shaft, the metal which would normally occupy the journal cores can then be used to enlarge the journal and web dimensions, where it is more effective in raising the torsional stiffness and the lateral rigidity of the crankshaft.

Long stroke under-square engine crankshafts (Fig. 3.18) have very little journal overlap, and

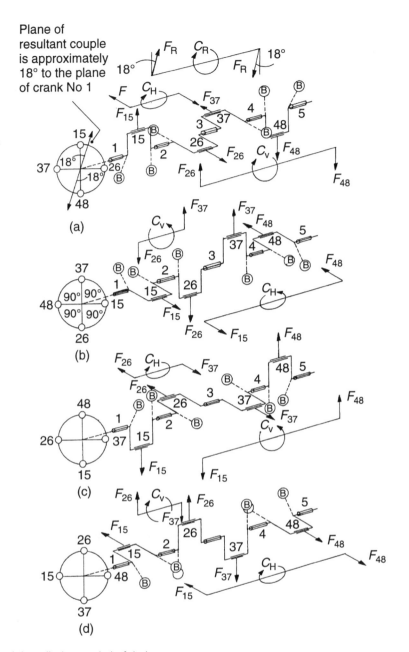

Fig. 3.15 90° vee eight-cylinder crankshaft balance

therefore the shaft relies entirely on the web section thickness for bending strength. On the other hand, over-square engine crankshafts (Fig. 3.19) have relatively short crank-throws, which provide adjacent journals with a certain amount of lateral overlap. Consequently, this overlap contributes somewhat to the lateral rigidity of the crankshaft.

Sometimes engines are designed to have two or more cylinder capacities. This can be achieved by increasing the cylinder diameter. However, this may be difficult, or even impossible, with some cylinder-block castings since the thickness of the cylinder walls would not permit additional boring. However, a relatively straightforward method of altering the cylinder's capacity without introduc-

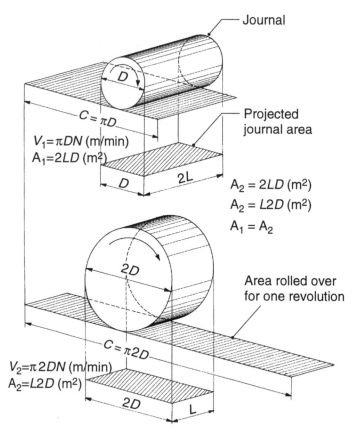

Fig. 3.16 Relationship between rubbing velocity and journal diameter

ing any major changes is to have crankshafts with two or more crank-throw sizes (Figs 3.18 and 3.19). An interesting feature which results when different strokes are used with a common connecting-rod length, is that the engine with the smaller swept volume (shorter crank-throws) will have a larger connecting-rod to crank-throw ratio, i.e. $n = 1/r$. Therefore, the reciprocating

secondary inertia forces will be greater for the larger swept volume engine, which has a longer crankthrow. Thus, with an in-line four-cylinder engine there will be an increase in secondary force vertical shake as the engine cylinder capacity is raised.

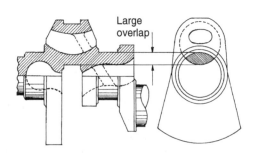

Fig. 3.17 Short stroke cored crankshaft with large journal overlap

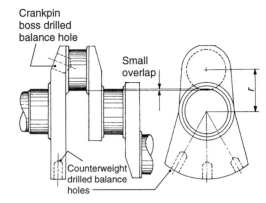

Fig. 3.18 Long stroke crankshaft with very little journal overlap

90

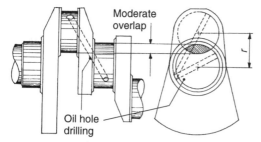

Fig. 3.19 Normal stroke solid crankshaft with moderate journal overlap

3.3 Primary and secondary piston movement

3.3.1 Primary piston movement
(Fig. 3.20(a))

Primary piston motion may be studied by constructing a crankpin circle with a radius (R) to represent to a predetermined scale, the length of the crank throw (Fig. 3.20(a)). Next, draw the line-of-stroke through the centre of the circle, divide one half of the crankpin circle into a number of equal parts at, say, 30° intervals.

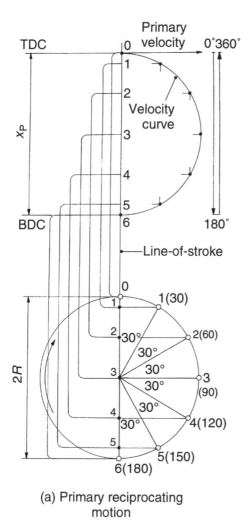

(a) Primary reciprocating motion

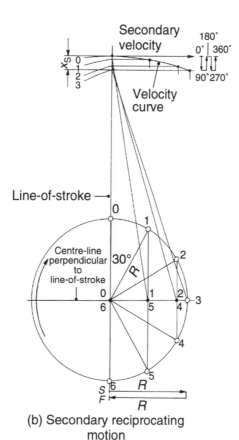

(b) Secondary reciprocating motion

Fig. 3.20 Illustration of primary and secondary motion

Project lines from each crankpin position to intersect the line-of-stroke. Set a compass to the scaled connecting-rod length and mark off the small-end gudgeon pin positions along the extended line-of-stroke for corresponding crankpin big-end angular dispositions. Draw perpendicular lines to the line-of-stroke for each incremental position of the small-end gudgeon-pin, also draw a second set of lines parallel to the line-of-stroke passing through the crankpin points until they intersect the corresponding perpendicular lines for each gudgeon pin position. Mark off the perpendicular and horizontal intersection point and then draw a smooth curve through these points.

The distance between gudgeon-pin points along the line-of-stroke then represents the piston's movement for a corresponding crank-angle displacement. Primary piston movement may be defined as the piston's travel caused by the rotating crankpin's projected position along the line-of-stroke. The curve drawn between plottings represents a graph of the piston's velocity with the base as the line-of-stroke while the perpendicular distance from the base (line-of-stroke) to the curve represents the piston velocity at any point between TDC and BDC.

Analysing the motion for the first and last 30° crankpin angular intervals, there will be an equal but relatively small corresponding piston movement at each end of the piston's stroke. The next 30° crankpin inward intervals, from TDC and BDC, i.e. 30–60° and 150–120°, will again be equal but their corresponding piston travel will be much greater. Finally, the third 30° crankpin inward intervals from TDC and BDC, i.e. 60–90° and 120–90°, meeting at the mid-stroke position will produce similarly equal but only slightly increased piston travel compared with the previous crankpin intervals. Hence, the primary piston movement x_p between TDC and mid-stroke and between BDC and mid-stroke will be identical; however, the effective primary movement will be greatest in the mid-stroke region, and the minimum as the piston approaches or departs from either end of its stroke. Thus, the piston motion near either dead centre is known as the ineffective crank angle movement.

3.3.2   Secondary piston movement
(Fig. 3.20(b))

The relative piston-to-crankpin secondary movement can be shown by constructing a crankpin circle to scale, of radius (R) (crankthrow) (Fig. 3.20(b)). Draw a line-of-stroke and a perpendicular to the line-of-stroke passing through the centre of rotation. Divide one half of the crankpin circle into a number of equal parts in, say, 30° intervals. Draw projected lines through each crankpin centre interacting the perpendicular centre-line and mark these points 0,6 1,5 2,4 and 3. Set the compass to the scaled connecting-rod length and mark off the piston gudgeon-pin position along the line-of-stroke for each intersected point on the perpendicular centre-line.

Secondary piston movement may be defined as the piston travel caused by the rotating crankpin's projected positions on the perpendicular centre-line.

The secondary movement velocity graph can then be produced by drawing lines perpendicular to the line-of-stroke through each gudgeon-pin position and then intersecting these lines with the parallel lines to the line-of-stroke which pass through the crankpin centre's perpendicular projection points. The horizontal and vertical intersections in the gudgeon-pin region are then joined together with a smooth curve which represents the piston's secondary motion velocity. The axes of the graph are the line-of-stroke as the base representing the piston displacement, and the perpendicular height to the line-of-stroke as the measure of the piston's velocity at any instant. The construction also shows that the piston changes its direction of motion every quarter of a revolution and, therefore, completes two forward and backward secondary strokes every revolution.

3.3.3   Relationship between primary and secondary piston movement
(Fig. 3.20 (a and b))

Due to the obliquity of the connecting-rod—that is, the connecting-rod inclination angle to the line-of-stroke—there is a projected crankpin outward and inward movement perpendicular to the line-of-stroke as the piston moves to and fro between its dead centres (Fig. 3.20(b)). This additional piston displacement produced due to the perpendicular crankpin outward and inward projection movement is known as the secondary movement. Thus, as the crankpin on its first half angular displacement moves from TDC to the mid crankpin position, the connecting-rod big-end will move perpendicularly outward and away from the line-of-stroke, therefore causing the piston to have an additional movement away from TDC in the same direction as the primary piston movement. However, when the crankpin has reached its mid-position and commences the second half of its angular displacement towards BDC, the big-end perpendicular projected movement will be towards the line of stroke. Hence, the piston movement caused by the perpendicular projected crankpin's inward movement reverses its direction and pushes it in the opposite direction to the main primary movement (Fig. 3.20(a)).

Therefore, as the crankpin turns through its first quadrant from TDC to mid-stroke position the connecting-rod big-end moves perpendicularly outwards and away from the line-of-stroke and, as a result, additional movement in the same direction as the basic primary movement occurs. After the crankpin has reached its mid-crank angle position, it then moves into the second quadrant, the big-end of the connecting-rod now reverses its perpendicular outward movement and commences to move inward towards the line-of-stroke. Correspondingly, the piston movement caused by the perpendicular projected movement of the crankpin will now move in the opposite direction to the primary movement which still continues to move towards BDC.

Hence, for the first quadrant (0–90°) crankpin movement, from TDC to mid-crankpin position, both primary and secondary movement are in the same direction and it is therefore additive. However, during the second quadrant (90–180°), from mid-crankpin position to BDC, the secondary movement has reversed its direction and cancels out a portion of the primary movement. The consequences of these events means that the resultant piston movement from TDC to mid-crankpin position is greater due to the addition of the secondary movement, i.e. $x_p + x_s$, whereas the resultant piston movement from mid-crankpin position to BDC is smaller due to the subtraction of the secondary movement from the primary movement, i.e. $x_p - x_s$.

When the piston reaches the end of the outward stroke and commences its return inward stroke from BDC to TDC, the relative displacements of the primary and secondary movements are similar to those on the outward-going stroke. Thus, when moving from BDC to mid-crankpin position the secondary movement reduces a portion of the primary movement and, with further movement from mid-crankpin position to TDC,

the secondary movement then, in effect, extends the primary movement. The overall result is that the piston travel from TDC for the crank-angle inner stroke quadrants, 0–90° and 270–360°, is greater than for the outer stroke crank-angle quadrants, 90–180° and 180–270°.

Subsequently, for the crankshaft to revolve at a uniform speed, the piston's acceleration and deceleration between TDC and mid-crankpin position must be higher (due to the increased distance moved) than the deceleration and acceleration which occurs between BDC and mid crankpin position, where the relative piston travel is less.

3.3.4 Combined primary and secondary reciprocating movement
(Figs 3.21 and 3.22)

The crank–piston reciprocating mechanism relative motion may be examined by constructing a circle with a radius representing, to a predetermined scale, the length of the crank-throw. One half of the circle is then divided into any number of equal parts, say 15° intervals (Fig. 3.21). Set the compass to the scaled connecting-rod length and mark off the small-end gudgeon-pin positions along the line-of-stroke for corresponding crankpin big-end angular dispositions. Draw straight lines between the big-end and small-end centres and for the first crank-angle quadrant (0–90°). From TDC extend back these lines until they intersect the big-end circle centre perpendicular line and mark off these points. Draw perpendiculars along the line-of-stroke for each incremental gudgeon-pin position and then draw lines parallel to the line-of-stroke passing through the projected crankpin intersecting points on the perpendicular crankpin circle line. The resulting velocity graph is then produced by drawing a smooth curve through the perpendicular and parallel intersecting points. With the piston stroke as the base the magnitude of the piston velocity at any position of the gudgeon-pin is presented by the perpendicular distance between the line-of-stroke and the curve.

Thus, the piston travel for the first crankpin quadrant (0–90°) is larger compared with that of the second crankpin quadrant (90–180°), and the piston movement for equal crankpin intervals in the TDC and BDC region is greater at TDC; that is, the piston movement at BDC is relatively ineffective compared with similar crankpin movements at TDC. However, the piston's movement

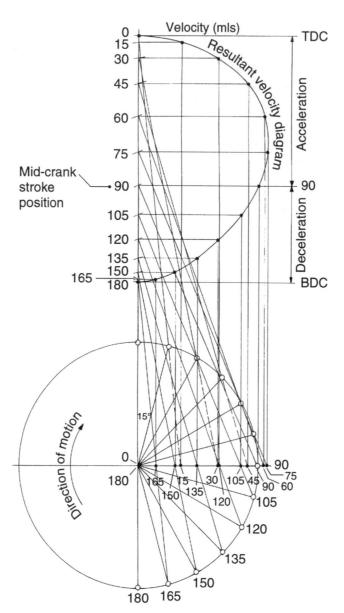

Fig. 3.21 Resultant primary and secondary reciprocating motion

at both TDC and BDC is very ineffective compared with the 60–75° crankpin interval which provides the most effective (greatest) piston movement over the entire half revolution of the crankpin. The velocity graph also shows that the piston velocities are zero at either dead centres and that the piston velocity rises rapidly for the first 45° crank angle movement from TDC, the steepness of the velocity rise then reduces until the maximum velocity is reached roughly 80° after

TDC. The piston velocity rise then commences to decrease slowly until the crankpin has been displaced approximately 90° from TDC, after this the piston's velocity steadily declines until, at BDC, it becomes zero.

The increased distance travelled by the piston for the first half of the crankpin angular movement compared with that moved in the second half (Fig. 3.22) of the piston's stroke implies that the piston's rate of speed increase or decrease

94

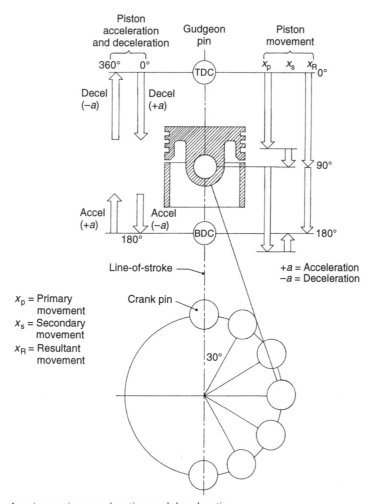

Fig. 3.22 Illustration of reciprocating acceleration and deceleration

(acceleration or deceleration) has to be greater between TDC and the mid-crankpin position than when the piston is moving in its second half of its stroke and the crankpin is somewhere between its mid-position and BDC. Consequently, there is a relative imbalance of acceleration and deceleration when the piston changes its direction of motion at either dead-centres.

3.3.5 Reciprocating piston connecting-rod and crank dynamics
(Fig. 3.23)

An analytical determination of how the piston's velocity, acceleration and inertia forces vary as the piston travels between the inner and outer dead-centres can be derived for any crank angle movement.

Let x be the displacement of the piston from the inner dead-centre position (Fig. 3.23) then l = length of the connecting-rod, r = crank throw, ϕ = connecting-rod angle to the line-of-stroke, θ = crank angle movement from TDC, ω = angular velocity of crank.

Thus, it can be proved that

$$\text{piston velocity } (v) = \omega r\left(\sin\theta + \frac{\sin 2\theta}{2n}\right) \text{ m/s}$$

$$\text{piston acceleration } (a) = \omega^2 r\left(\cos\theta + \frac{\cos 2\theta}{n}\right) \text{ m/s}^2$$

$$\text{piston inertia force } (F) = m\omega^2 r\left(\cos\theta + \frac{\cos 2\theta}{n}\right) \text{ N}$$

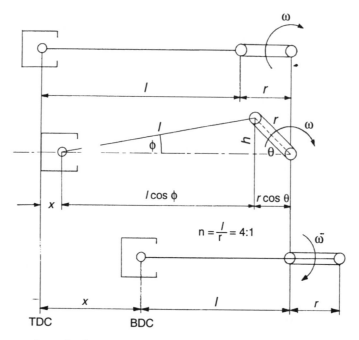

Fig. 3.23 Reciprocating crank mechanism geometry

Multiplying out the brackets

$$F = m\omega^2 r \cos\theta \pm m\omega^2 r \frac{\cos 2\theta}{n} \text{ N}$$

The first term $m\omega^2 r \cos\theta$ is called the primary inertia force and reaches a maximum value of $m\omega^2 r$ twice per revolution, when $\theta = \theta$ and 180°.

The second term $m\omega^2 r (\cos 2\theta/n)$ is called the secondary inertia force and reaches a maximum value of $m\omega^2/n$ four times per revolution when $\theta = \theta°, 90°, 180°$ and 270°, thus inertia piston force = primary force ± secondary force.

The primary force is the inertia force produced by the piston mass due to the rotating crankpin's projected movement along the line-of-stroke being relayed to the piston via the connecting-rod.

The secondary force is the inertia force produced by the piston mass due to the rotating crankpin's outward and inward projected movement perpendicular to the line-of-stroke relaying motion to the piston via the inclined connecting-rod.

In practice when the crankpin rotates, its projected movements for each crank-angle position, both along the line-of-stroke and perpendicular to the line-of-stroke occur simultaneously, this then causes a resultant acceleration or deceleration motion to the piston and hence a corresponding inertia force will be created by the reciprocating piston's mass.

3.3.6 Primary and secondary piston acceleration curves
(Fig. 3.24)

From the derived formula for the piston acceleration for any crank shaft angular displacement it can be seen that both primary and secondary piston acceleration and deceleration between the two extreme dead-centres follow a cosine waveform (Fig. 3.24). It can be seen that the primary piston acceleration and deceleration cycle takes place between the outward and inward stroke displacing the crankshaft through 360°, whereas the secondary piston motion completes two acceleration and deceleration cycles for one complete crankshaft revolution. In other words, the secondary piston movement operates at twice the frequency of the piston's primary movement. However, its peak magnitude is only between a quarter to one-third of the primary acceleration depending upon the connecting-rod length to crank-throw ratio. Maximum primary piston acceleration and deceleration occurs at TDC and

96

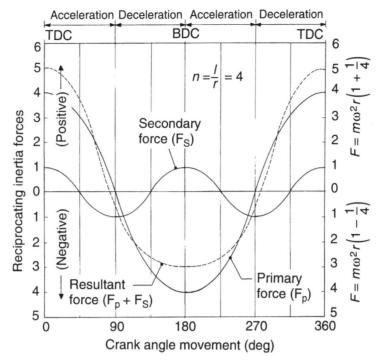

Fig. 3.24 Primary, secondary and resultant inertia forces required to move piston over one revolution of the crankshaft

BDC respectively while at mid-crankpin positions 90° and 270° the acceleration or deceleration is zero. However, the secondary piston periodic acceleration and deceleration occurs at 0° (TDC), 90°, 180° (BDC), 270° and 360° crankpin movement, whereas zero acceleration occurs at 45°, 135°, 225° and 315°.

Comparing both the primary and secondary piston motion over one complete revolution, it can be observed that both primary and secondary piston movements, moving away from TDC, are accelerating together, whereas when BDC is approached the primary piston motion is decelerating while the secondary piston movement is accelerating. Algebraically combining both the positive and negative accelerations of the primary and secondary piston movements over one complete revolution produces an increase in piston acceleration as it moves out from TDC—this being shown by the height and steepness of the resultant curve for the first crank-angle quadrant. In contrast, the resultant deceleration of the piston as BDC is approached is reduced, this being shown by the almost flat plateau at the base of the resultant acceleration curve. A further feature of the resultant, primary and secondary

piston acceleration curve is that the piston's zero acceleration has shifted to just before and just after 90° and 270° crank angle movement respectively due to the secondary piston movement's interaction with that of the primary piston motion.

3.4 Balancing of the reciprocating parts of a single cylinder engine

(Fig. 3.25)

If the reciprocating masses of the piston and connecting-rod are completely balanced in the vertical plane by extending backward the crank-arm webs on each side of the crankpin then, as the crankshaft rotates (Fig. 3.25), the opposing reciprocating and counterweight mass vertical-force component cosine-curve phasings are such that they become the mirror image of each other and therefore cancel out. Thus, the vertical reciprocating and counterweight mass force components on the outward stroke decline together from a maximum at TDC to zero during the first

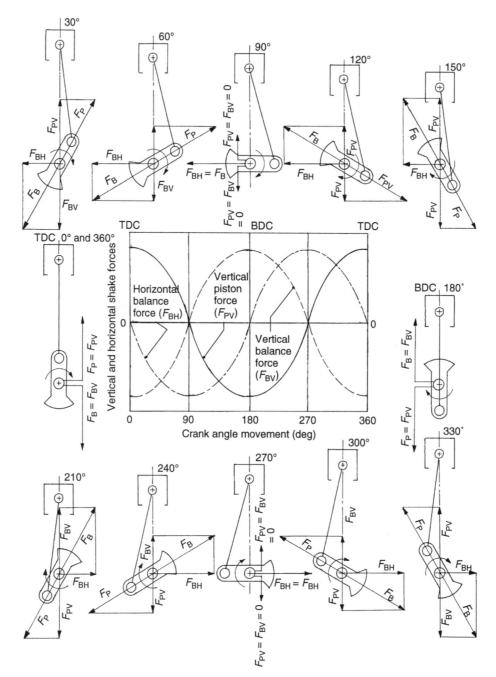

Fig. 3.25 Illustration of single-cylinder engine cyclic reciprocating force and counter weight balance in both vertical and horizontal plane

quadrant 0–90°; then, they again increase from zero at the mid-crankpin position to a maximum at BDC during the second-quadrant 90–180° movement. On the return inward stroke the opposing vertical force components repeat the half-cycle of events. Hence, in any position in the

vertical plane (along the line-of-stroke) the piston's reciprocating inertia force will be completely neutralized by the counterweight's vertical force component. However, in the horizontal plane there is no reciprocating force component to be balanced by the two half counterweights so

98

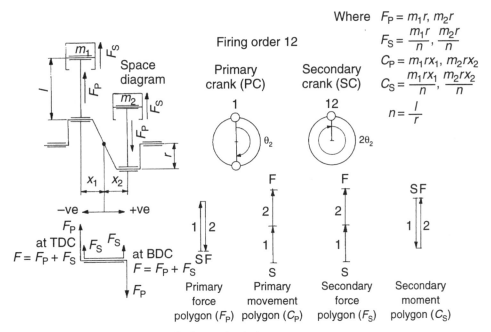

Where $F_P = m_1r, m_2r$

$$F_S = \frac{m_1r}{n}, \frac{m_2r}{n}$$

$$C_P = m_1rx_1, m_2rx_2$$

$$C_S = \frac{m_1rx_1}{n}, \frac{m_2rx_2}{n}$$

$$n = \frac{l}{r}$$

Firing order 12

Space diagram

Primary crank (PC)

Secondary crank (SC)

$F = F_P + F_S$ at TDC

$F = F_P + F_S$ at BDC

Primary force polygon (F_P)

Primary movement polygon (C_P)

Secondary force polygon (F_S)

Secondary moment polygon (C_S)

Fig. 3.26 In-line 180° out-of-phase twin cylinder engine balance

that the latter produce their own imbalance at right-angles to the line-of-stroke. Thus, if the counterweight masses completely balance the reciprocating masses in the vertical plane then there will be a maximum out-of-balance horizontal force at crankshaft positions where the reciprocating vertical force components become zero; that is, at 90° and 270° crank angle positions, which are equal in magnitude to the reciprocating inertia forces at TDC and BDC. Thus, all that has been achieved is that the maximum out-of-balance forces have been shifted a quarter of a revolution from TDC and BDC positions, to the mid-crankpin positions, 90° and 270°.

As a compromise, it is usual for the counterweights to provide only a 50% reduction in reciprocating imbalance. Consequently, the crankshaft will move through four instead of two equal peak out-of-balance positions every revolution—that is, at TDC and BDC in the vertical plane caused by the reciprocating masses and at 90° and 270° in the horizontal plane, caused by the counterweight masses—but the magnitude of each peak imbalance will amount to only a half of the maximum reciprocating inertia force.

3.4.1 Primary and secondary reciprocating inertia force balance
(Figs 3.26 and 3.27)

Before examining the balance of different crankshaft and cylinder configurations it is worthwhile reminding ourselves in which direction the primary and secondary forces point along the line-of-stroke relative to the extreme dead-centres and mid-stroke (Figs 3.26 and 3.27).

Due to the obliquity of the connecting-rod the inertia forces are greater at the TDC crank position than at the BDC position, which necessarily results in an imbalance of forces that cannot be balanced simply by attaching counterweights on the opposite side of the crankpins.

As a result, both primary and secondary forces act in the same direction at TDC (0° and 360°), that is $F_p + F_s$, whereas at BDC (180°) the secondary force opposes the primary force—that is $F_p - F_s$. The secondary forces at TDC and BDC therefore point in the same direction towards TDC and are referred to as being positive.

Conversely, when the crankshaft approaches mid-stroke (90° and 270°) positions the piston primary inertia force decreases to zero due to the piston velocity in this position being approximately constant. However, the crankpin's projected perpendicular movement to the line-of-stroke

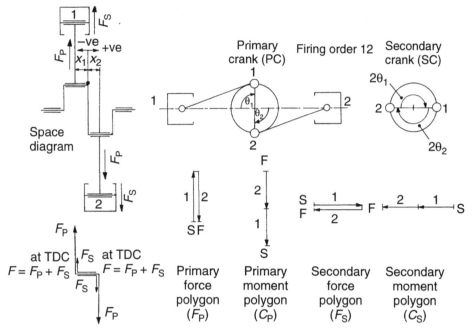

Fig. 3.27 Horizontal opposed twin-cylinder engine balance

changes its direction, at mid-stroke, and therefore there will be a maximum negative secondary piston force which is in the opposite sense to that produced at either dead centres.

3.4.2 Method of solving the state of balance for primary and secondary forces and couples for various crankshaft and cylinder configurations
(Figs 3.26, 3.27 and 3.32)

Analysing the dynamic balance of an engine in terms of primary and secondary forces and couples with single-plane crankshafts, such as are used in two and four-cylinder engines, is relatively simple. However, when the crank-throws are arranged in several planes, it becomes very difficult, and therefore a more precise method of studying the state of engine balance is needed.

The method chosen in this text uses what are known as a primary and secondary cranks, and from their angular configurations, primary and secondary vector force and couple polygons are drawn. From these, the balance or imbalance of the system can be derived.

Using the in-line four-cylinder engine as an example (Fig. 3.32) the procedure for obtaining the force and couple polygons is described as follows.

1 Draw a plan or pictorial view of the crankshaft with vertical lines representing the crankpin centres.

2 Mark off the reference plane, usually mid-way along the crankshaft length, using the notation that anything to the left of the plane is positive and to the right is negative.

3 Draw dimension lines parallel to the crankshaft length and insert distances x_1, x_2, x_3 etc from the reference plane.

4 Draw the primary crank, this is the end view of the crankshaft showing each crank throw (arm) with its angular displacement (θ) from its piston's respective line-of-stroke.

5 Draw a secondary (imaginary) crank, this is the end view of the crankshaft with each crank-throw having been rotated twice the primary crank displacement (2θ) from the piston's respective line-of-stroke.

6 Draw both primary and secondary force-vector polygons. Primary and secondary force vectors are treated as acting radially outwards from crank centres.

7 Draw both primary and secondary moment vector polygons. All positive primary and secondary moment vectors are treated as

acting radially outwards and the negative vectors radially inwards.

3.4.3   In-line 180° out-of-phase twin-cylinder engine balance
(Fig. 3.26)

With the in-line 180° out-of-phase twin-cylinder engine, when one piston is at inner-dead-centre the other will be at outer-dead-centre; therefore, since the primary forces of both pistons act in opposition to each other they are neutralized. However, the secondary force on piston No. 1, when at TDC, acts in the same direction as the primary force; that is, towards the cylinder head, whereas the secondary force on piston No. 2, when at BDC, acts in the opposite direction to the primary force; that is, away from the cylinder head.

Consequently, the offset between the two pistons causes the primary forces (which are parallel but point in opposite directions to each other) to produce a moment of a couple, this tends to rock the crankshaft alternately clockwise and anticlockwise. In contrast, both secondary forces act in the same direction and therefore they produce a vertical shake.

3.4.4   Horizontal opposed twin-cylinder engine balance
(Fig. 3.27)

With this crankshaft and cylinder configuration, both pistons arrive at or depart from TDC or BDC simultaneously, therefore the primary forces created are directed outwards when TDC is reached or point inwards when approaching BDC. Likewise, both secondary forces are directed outward at TDC and inwards at BDC.

Due to the offset of the primary and secondary forces there will be both a primary and secondary couple which rocks the engine in both a clockwise and anticlockwise direction. However, the couple–arm length is smaller with the opposed twin compared with the in-line twin since there is no piston overlap interference when the pistons are on either side of the crankshaft. There is no secondary shake as both of the secondary forces act in opposition to each other.

3.4.5   90° vee twin-cylinder engine balance
(Figs 3.28 and 3.29(a–d))

With this basic crankshaft arrangement the primary inertia force of each piston is completely out-of-balance, whereas the secondary inertia forces produce an approximate 70% secondary force imbalance (Fig. 3.28). However, by attaching counterweights to each web on either side of the crankpin on the opposite side of the cranks, the primary inertia force may be completely balanced but, unfortunately, the secondary inertia forces still remain unaffected and unbalanced.

Primary force balance by attaching balance weights to the crankarms will now be described (Fig. 3.29). With the crankpin in the TDC position (Fig. 3.29(a)) relative to the left-hand No. 1 piston and cylinder, the primary inertia force is completely balanced by the centrifugal force set up by the counterweights, and the right-hand piston will be in the mid-stroke position where there is no primary inertia force. However, there will be a secondary inertia force on piston No. 1 when it is at TDC and on piston No. 2 when it is at mid-stroke position, which produces a resultant secondary inertia force $F_{SR} = 1.414F_S$.

When the crankpin has turned through 90° (Fig. 3.28(b)) the right-hand piston No. 2 will now be at TDC and the primary inertia force will be completely neutralized by the counterweights' centrifugal force, which is now diametrically opposite piston No. 2. Correspondingly, piston No. 1 will have moved to its mid-stroke zero primary inertia force position. Consequently, piston No. 2 in its mid-stroke position also produces a maximum secondary inertia force along its line-of-stroke; both of these forces being at right-angles to each other and, therefore, they produce a resultant secondary force F_{SR} at 45° to each individual secondary inertia force.

As the crankpin revolves through 180° (Fig. 3.29(c)) relative to piston No. 1's line-of-stroke, piston No. 1 will be at BDC with the crankshaft counterweights aligned in the opposite direction. Also, piston No. 2 will again be in a mid-stroke zero primary inertia force position. In this crank-angle-position, piston No. 1 at BDC and piston No. 2 at mid-stroke position produce a maximum secondary inertia force along their respective lines-of-stroke, which has a resultant secondary force, F_{SR}.

A further crankpin angular movement from

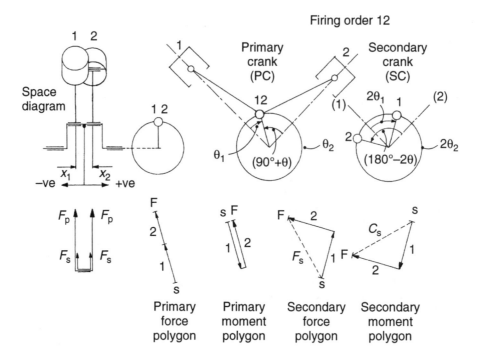

Fig. 3.28 90° vee twin-cylinder engine balance

piston No. 1's line-of-stroke of 270° (Fig. 3.29(d)) positions piston No. 2 at BDC so that its primary inertia force is exactly cancelled by the centrifugal force of the counterweights, while piston No. 2 will be in its mid-stroke zero primary inertia force position. Under these conditions the secondary inertia forces acting on each piston are at a maximum and will again produce a resultant secondary force of F_{SR}.

3.4.6   In-line three-cylinder engine balance

(Figs 3.30(a–f) and 3.31)

A three-dimensional view of a three-cylinder engine crankshaft with its pistons attached is shown in Fig. 3.30. With this simple crankshaft arrangement both primary and secondary inertia forces are balanced as shown by the two force polygons (Fig. 3.31). In contrast, there are both primary and secondary couples which cannot be simply balanced by counterweights since they would then introduce additional couples at right-angles to the cylinder axis. The out-of-balance couples are portrayed by the open primary (C_p) and secondary (C_s) moment polygons as shown (Fig. 3.31). The couples produced as the crankshaft rotates will be

of varying magnitudes, as explained in the following.

When piston No. 1 is at TDC, its upwardly directed inertia force in the vertical plane is equal to the combined downward inertia force components of pistons Nos 2 and 3, which are both 60° from BDC. The effect of these opposing offset forces along the crankshaft is to create a vertical clockwise pitching couple with the front-end tilting up and the rear-end pressing downwards (Fig. 3.30(a)).

Rotating the crankshaft 60° from piston No. 1's position brings piston No. 2 to BDC, causing its maximum inertia force to be directed downward, this is counteracted by pistons No. 1 and No. 3, which are both 60° from TDC. They, therefore, produce opposing upward inertia force components which equal the middle piston's maximum downward inertia force. Accordingly, with the crankshaft in this position, there will be no pitch couple (Fig. 3.30(b)).

Rotating the crankshaft 120° from piston No. 1's TDC position brings piston No. 3 to TDC and its maximum upward inertia force position, while pistons Nos 1 and 2 move to 60° either side of the BDC position. The lower pistons, therefore, produce downward inertia force components equal to that produced by piston No. 3's

102

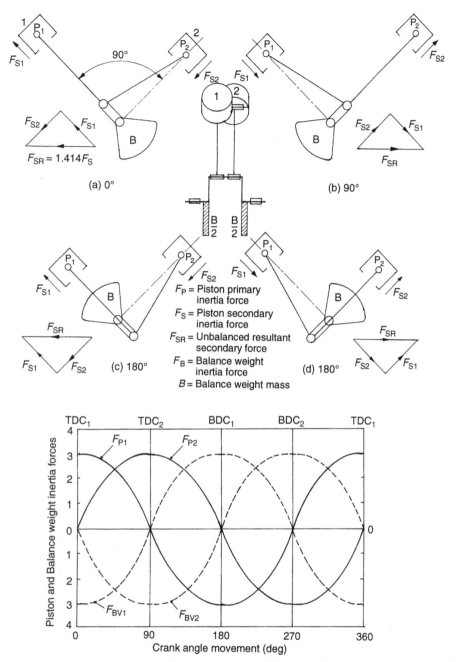

Fig. 3.29 90° vee twin-cylinder engine with crankweb counter weight primary force balance principle

upward force. However, due to all the pistons being in different planes the downward force of the first two pistons is opposed by the upward force of the rear piston, the result being that an anticlockwise pitch couple is produced which tends to tilt the engine downward at the front and upward at the rear (Fig. 3.30(c)).

Rotating the crankshaft 180° from piston No. 1's TDC position brings piston No. 1 to BDC and pistons Nos 2 and 3 to 60° either side of TDC (Fig. 3.30(d)). Accordingly, the maximum downward inertia force of piston No. 1 is counterbalanced by the upward inertia force components of pistons Nos 2 and 3, but their offset along the

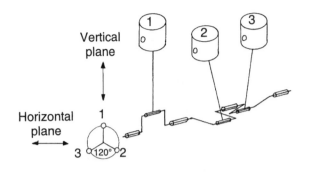

Vertical plane

Horizontal plane

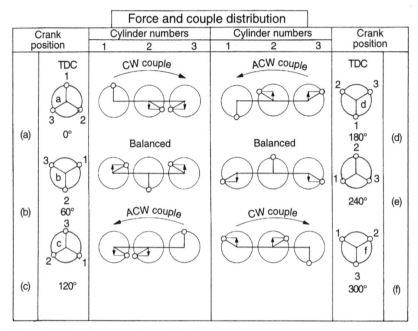

	Crank position	Cylinder numbers			Cylinder numbers			Crank position	
		1	2	3	1	2	3		
(a)	TDC 1 ... 3 2 0°	CW couple			ACW couple			TDC 2 3 ... d 1 180° 2	(d)
(b)	3 ... 1 b 2 60°	Balanced			Balanced			1 ... 3 240°	(e)
(c)	2 3 ... c 2 1 120°	ACW couple			CW couple			1 ... 2 f 3 300°	(f)

Fig. 3.30 In-line three-cylinder primary and couple balance

crankshaft length produces a similar anticlockwise pitch couple to the previous 120° crankshaft position (Fig. 3.30(d)).

Rotating the crankshaft 240° from piston No. 1's TDC position brings piston No. 2 to the TDC position and pistons Nos 1 and 3 move to 60° either side of BDC. In this position, the upward force of piston No. 2 is balanced by pistons Nos 1 and 3's downward force components. Therefore, since the downward force components are at an equal distance from the middle piston's upward force there will be no pitch couple in this position (Fig. 3.30(e)).

Rotating the crankshaft still further to 300° from piston No. 1's TDC position brings piston No. 3 to BDC and pistons Nos 1 and 2 to 60° before and after TDC position. The resulting offset upward force components and the max-

imum downward piston force along the crankshaft length now produces a clockwise pitch couple which tends to tilt the front end up and the rear end down (Fig. 3.30(f)).

Rotating the crankshaft a further 60° to 360° from piston No. 1's TDC position brings piston No. 1 to TDC again, whereas pistons Nos 2 and 3 move to 60° either side of BDC. Thus, the cycle of events is repeating (Fig. 3.30(a)).

3.4.7 In-line four-cylinder engine balance
(Fig. 3.32)

The in-line four-cylinder engine can be treated as a pair of in-line 180° out-of-phase twin crank-

104

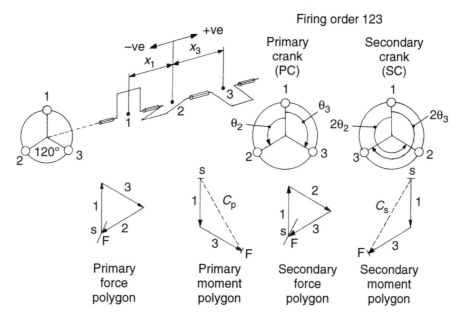

Firing order 123

Fig. 3.31 In-line three-cylinder engine balance

shafts joined together end on (Fig. 3.32). Thus, all the crank throws are in a single plane (in-line) so that when pistons Nos 1 and 4 are at TDC, pistons Nos 2 and 3 will be at the opposite ends of their stroke at BDC. Accordingly, the primary forces caused by the pistons' upward movement in cylinders Nos 1 and 4 will be balanced by the primary forces produced by pistons Nos 2 and 3 on their downward movement and, correspondingly, when the outer pistons are on their downward stroke approaching BDC and the inner pistons are on their upward stroke towards TDC then the opposing directional sense of the inner and outer primary forces also cancel each other out (Fig. 3.32).

It is worthwhile reminding ourselves of the relationship between the primary and secondary forces as the piston reciprocates between TDC and BDC. When the piston approaches TDC the secondary force produced acts upwards in the same direction as the primary force $(F_p + F_s)$, but when the piston approaches BDC the secondary force always acts upwards in opposition to the primary force $(F_p - F_s)$ which is now directed downwards.

With this knowledge it is easy to visualize that the secondary forces for both inner and outer pairs of pistons when approaching either TDC or BDC always point upwards, that is, at TDC the secondary forces act in the same direction as the primary forces, but at BDC the secondary forces react in opposition to the primary forces.

The overall result is that a pair of primary couples will be created at the front between pistons Nos 1 and 2 and at the rear between pistons Nos 3 and 4 and, since one has a clockwise rotation and the other an anticlockwise rotation and their magnitudes are equal, then both are neutralized by each other through the internal stresses created within the crankshaft material. Thus, the couples are absorbed internally by the crankshaft's rigidity, thereby eliminating all external rocking.

However, all the pistons' secondary forces are directed upwards, and therefore the crankshaft is subjected to a vertical secondary shake.

3.4.8 Horizontally opposed flat four-cylinder engine balance
(Fig. 3.33)

This crankshaft arrangement can be considered as a pair of horizontally opposed twin-crankshafts joined together end on (Fig. 3.33). The engine configuration consists of two horizontally opposed cylinder banks, the left-hand bank is formed by cylinders Nos 1 and 3 and the right-hand bank includes cylinders Nos 2 and 4. Thus, the first two crank throws (Nos 1 and 2) will be at

105

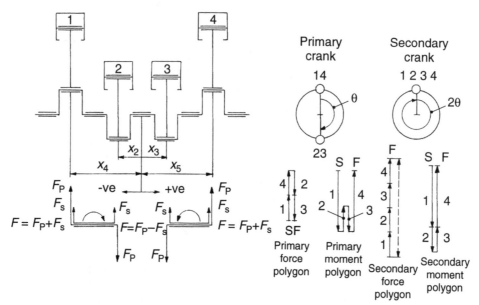

Fig. 3.32 In-line four-cylinder engine balance

TDC or BDC when the rear two crank throws (Nos 3 and 4) are at the opposite ends of their strokes, at BDC or TDC respectively.

Observing the primary forces produced when the four pistons are in motion, as can be seen, the primary forces of piston No. 1 at TDC and piston

No. 4 at BDC both point towards the left-hand bank, whereas piston No. 2 at TDC and piston No. 3 at BDC point in the opposite direction, towards the right-hand bank; therefore, the primary forces are balanced (Fig. 3.33).

Inspecting the secondary forces for pistons Nos

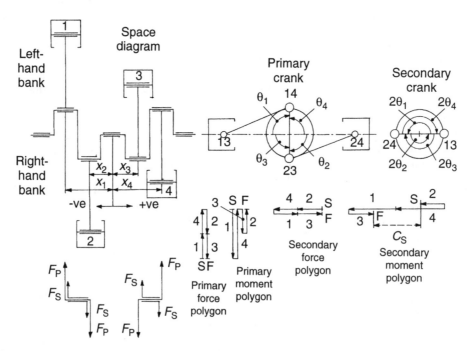

Fig. 3.33 Horizontally opposed four-cylinder engine balance (flat four)

1 and 2, since both pistons are at TDC they point outwards in the same direction as their respective primary forces. In contrast, the secondary forces imposed on pistons Nos 3 and 4 act in opposition to their respective primary force since both of these pistons are at BDC and therefore their sense of direction must be towards their respective TDC positions.

As a result, a clockwise couple for both primary and secondary forces is generated between pistons Nos 1 and 2, whereas only a primary anti-clockwise couple is generated between pistons Nos 3 and 4, this being partially counteracted by a clockwise secondary couple, in the opposite sense to the rear primary couple.

The overall state of balance is that both primary couples are neutralized internally, but there now exists an unbalanced external secondary couple because the front and rear couple both tend to rotate in the same direction.

3.4.9 60° vee four-cylinder engine balance

(Figs 3.34, 3.35 and 3.36(a, b and c))

This four-throw two-plane crankshaft has throw

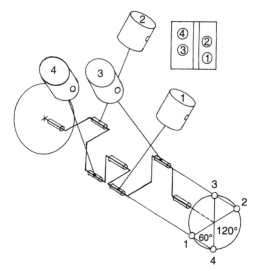

Fig. 3.34 60° vee four-cylinder engine crank and piston layout

intervals of 60°, 120°, 60° and 120° respectively. The left-hand cylinder-bank pistons are connected to one of the crankthrow planes while the right-hand cylinder-bank pistons are connected to the crankpins of the other plane (Fig. 3.34). Thus, when one piston of either cylinder bank is at TDC

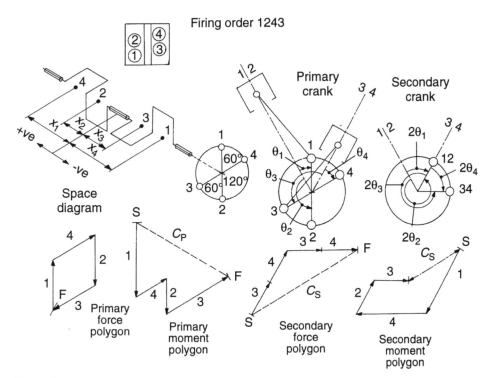

Fig. 3.35 60° vee four-cylinder engine balance

107

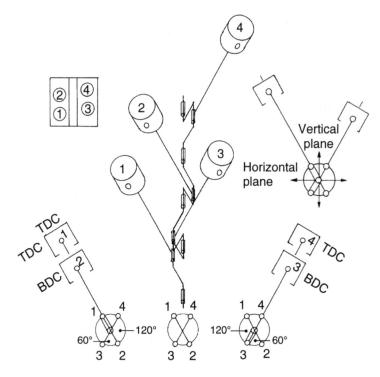

Fig. 3.36 60° vee four-cylinder primary force and couple balance

the second piston from the same bank will be at BDC. Power impulse intervals occur every 180°. Thus, it can be seen from the force and moment polygon diagrams that only the primary forces are balanced and that the primary couples, secondary forces and secondary couples are unbalanced (Fig. 3.35). Examining the state of primary piston inertia forces and couples (Fig. 3.36(a)) and considering the left-hand cylinder-bank; when

piston No. 1 is at TDC, piston No. 2 will be at BDC. Likewise, piston No. 4 will be at TDC and piston No. 3 will be at BDC in the right-hand cylinder-bank. A plan view of these piston forces shows that a horizontal clockwise couple is generated in both the left- and right-hand cylinder banks which produces a resultant compounded horizontal clockwise couple.

Rotating the crankshaft 120° from piston No. 1

TDC position in the left-hand cylinder-bank brings crankpin No. 3 to TDC and crankpin No. 4 to BDC on the left-hand cylinder-bank, whereas crankpins Nos 1 and 2 will be in a horizontal position (Fig. 3.36(b)). In this crankshaft position the pistons in the left-hand cylinder-bank generate a small clockwise couple, and this results in an overall small horizontal anticlockwise (yaw) couple.

Rotating the crankshaft 240° from crankpin No. 1's TDC position in the left-hand cylinder-bank, brings crankpin No. 2 to TDC and crankpin No. 1 to BDC on the right-hand cylinder-bank, whereas crankpins Nos 2 and 4 will be in a horizontal position (Fig. 3.36(c)). In this crankshaft position the pistons in the left-hand cylinder-bank generate a small horizontal clockwise couple while the pistons in the right-hand cylinder-bank will generate a large horizontal anticlockwise couple. A resultant horizontal anticlockwise couple is produced of moderate magnitude.

Thus, it can be seen that there will be a resultant horizontal couple which changes both its direction and magnitude every revolution of the crankshaft. The unbalanced primary couples cannot be simply balanced by attaching counter-weights to the crankshaft webs, but they can be neutralized by incorporating a counterbalance-shaft parallel to the crankshaft.

3.4.10 In-line five-cylinder engine balance
(Figs 3.37 and 3.38(a–e))

With this crankshaft configuration five separate crank throws are equally spaced 72° apart. Each successive crank throw angular interval (commencing at the front and moving to the rear) occurs in the order 12453 (Fig. 3.37).

The construction of the primary and secondary force and couple polygons, enabling the state of balance to be determined, is fully described for the in-line five-cylinder engine as follows.

Procedure for solving primary and secondary force and couple polygons
(Fig. 3.37)
Draw a plan or pictorial view of the crankshaft and mark off a reference plane mid-way along the crankshaft, this being No. 3 crankpin centre.

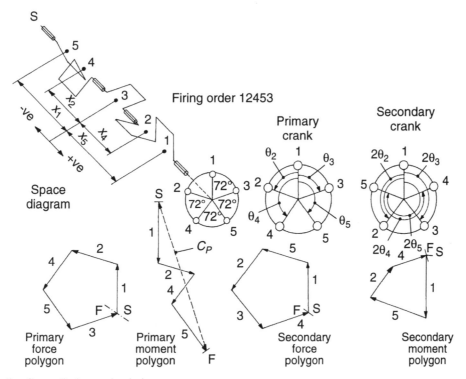

Fig. 3.37 In-line five-cylinder engine balance

Draw dimension lines parallel to the crankshaft length and insert distances x_1 and x_2 to the left of the reference plane and x_4 and x_5 to the right. When drawing the primary and secondary moment polygons use the notation that pistons to the left of the reference plane are negative and pistons to the right are positive. Draw the primary crank (PC), which is the end view of the crankshaft, showing each crank-throw arm with its angular displacement (θ) from the line-of-stroke. Finally, draw the secondary (imaginary) crank (SC), this being the end view of the crankshaft with each crankpin having been rotated twice the primary-crank (PC) displacement (2θ) from the line-of-stroke.

Primary and secondary force polygon
(Fig. 3.37)
Choose a suitable scale for the force polygon and, starting from the point(s), draw to scale a line parallel to the primary crank arm (1) representing the piston primary force (1) and mark an arrow pointing in the outward direction from the primary crank centre. From the end of the first line draw to scale a parallel line to the crank arm (2) representing the piston primary force (2) and insert an arrow pointing again in an outward direction from the centre of the primary crank. From the end of the second line draw to scale a parallel line to the crank arm (3) representing the piston primary force (3) and again insert an outward pointing arm from the primary crank centre. Repeat the foregoing procedure for primary forces (4) and (5) marking the finish of the last vector line with (F).

If the start (S) and finish (F) points meet, then the primary forces balance, whereas if the primary reciprocating forces are unbalanced there will be a gap proportional to the imbalance between the first and last vector-line's open ends.

The secondary force polygon is constructed similarly to the primary force polygon, but this time parallel lines are drawn from the secondary-crank instead of the primary-crank.

Primary and secondary moment polygon
(Fig. 3.37)
Select a suitable scale for the primary moment polygon and then draw to scale a line proportional in length to the distance from crankpin No. 1 to the reference line parallel to the primary crank arm (1). Mark an arrow pointing in the outward

direction from the primary crank centre for crankpins on the positive side of the reference plane and, conversely, point an arrow towards the centre of the primary crank if the crankpin is on the negative side of the reference plane.

From the end of the first moment line draw to scale a second parallel line to the primary crank arm (2) representing the primary moment (2) making it proportional to its distance from the reference plane, i.e. x_1, x_2, x_3. Repeat this procedure for primary moment vectors 3, 4, and 5.

If the start (S) and finish (F) points of the moment polygon do not meet, then the distance between these points represents the out-of-balance primary moment.

The secondary moment polygon is constructed similarly to the primary moment polygon except that it is developed from the secondary crank.

Summary of primary and secondary force and couple balance

Referring to Fig. 3.37 both primary and secondary piston inertia forces are balanced: this is indicated by the force polygon being closed. It can also be seen that the secondary moment polygon is closed so that there is no unbalanced secondary couple. Unfortunately, the primary moment polygon does not close and, in fact, the large gap (C_p) between the start and finish of the vector moment indicates a relatively large unbalanced primary couple.

One advantage of the five-cylinder engine is that there are separate angular phasings for each crankpin and therefore no two pistons reach or depart from either dead centres at any one time. Consequenttly, the resultant vertical primary force of a five-cylinder engine is less than for a six-cylinder engine where pairs of pistons approach either TDC or BDC together.

Primary force and couple distribution

The primary couple produced as the crankshaft rotates will be of varying magnitude as illustrated in Fig. 3.38(a–e).

When piston No. 1 is at TDC the upward maximum inertia force of piston No. 1 and the smaller upward inertia force components of pistons Nos 2 and 3 are counter-balanced by the

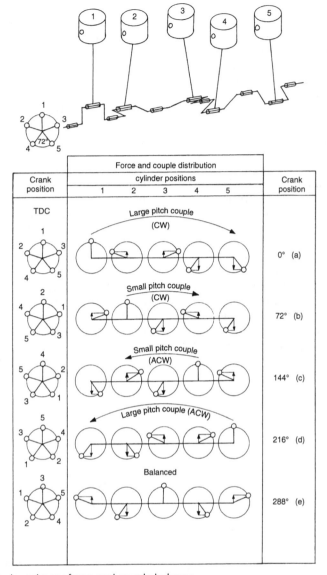

Fig. 3.38 In-line five-cylinder primary force and couple balance

downward inertia force components of pistons Nos 4 and 5, but the offset planes of the individual pistons cause a relatively large clockwise pitching couple to occur (Fig. 3.38(a)).

Rotating the crankshaft 72° from piston No. 1's TDC position brings piston No. 2 to TDC, causing the maximum inertia force to be directed upward; likewise, pistons Nos 1 and 4 have small upward inertia force components, whereas pistons Nos 3 and 5 have medium downward inertia force components (Fig. 3.38(b)). Consequently, the upward and downward inertia forces are equal, but, due to their offset to each other along the crankshaft's length, they create a resultant clockwise pitching couple which is of medium magnitude.

Rotating the crankshaft 144° from piston No. 1's TDC position brings piston No. 4 to TDC in its maximum upward inertia force position and, together with pistons Nos 2 and 5's small upward inertia force components, they balance the downward inertia force components for pistons Nos 1 and 3 (Fig. 3.38(c)). However, due to the offset of each piston along the crankshaft an anticlock-

111

wise pitching couple is now produced of medium magnitude.

Rotating the crankshaft 216° from piston No. 1's TDC position brings piston No. 5 to TDC in its maximum upward inertia force position and, together with pistons Nos 3 and 4, which also provide small upward inertia force components, they counteract the medium downward inertia force components of pistons Nos 1 and 2 (Fig. 3.38(d)). Thus, due to the staggering of the pistons over the crankshaft span, a large anti-clockwise pitching couple is produced.

Rotating the crankshaft further, to 288° from piston No. 1's TDC position, brings piston No. 3 to TDC and to its maximum inertia force position and, together with the small upward inertia force components of pistons Nos 1 and 5, they neutral-ize the evenly spaced downward inertia force components of pistons Nos 2 and 4, which are also evenly spaced on either side of the middle piston (Fig. 3.38(e)). This combination of vertical in-ertia forces (which are equally spaced and are of equal magnitude) cancel out any likely pitch couple. Hence, there is no pitch couple while the crankshaft remains in this position.

To remove the unwanted pitching and there-fore improve the smoothness of this type of engine, twin counter rotating balance shafts are sometimes installed which are able to cancel out the inherent unbalanced primary couples.

3.4.11 In-line six-cylinder engine balance
(Figs 3.39 and 3.40)

This in-line six-cylinder crankshaft has its six crank-throws arranged in pairs, the three pairs being equally spaced 120° apart with their crank-pins usually paired off in the order 1–6, 2–5 and 3–4 (see Fig. 3.39). The usual firing order for this configuration is 153624.

The angular phasing between the first and last three crankpins is 120° to one another; thus, the front three and rear three crank-throw configura-tions are the mirror image of each other and therefore constitute a pair of in-line three-cylinder crankshafts joined together end on (Fig. 3.39).

The three-cylinder crankshaft is balanced for

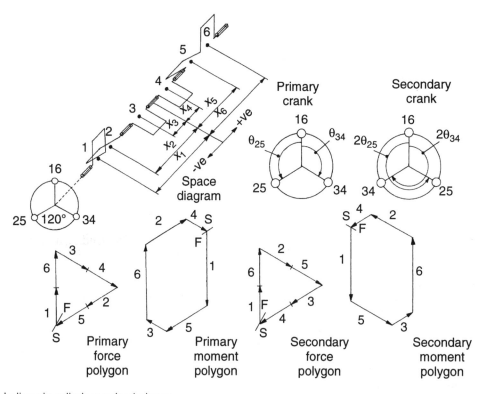

Fig. 3.39 In-line six-cylinder engine balance

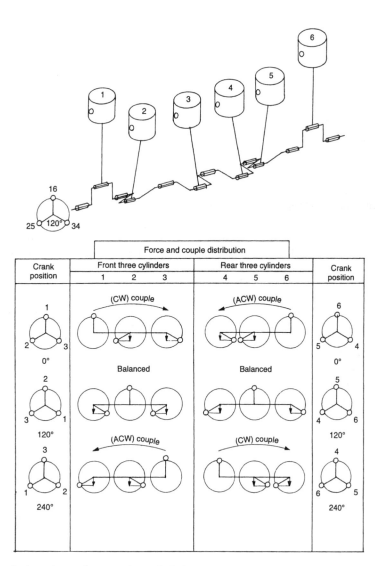

Fig. 3.40 In-line six-cylinder primary force and couple balance

both primary and secondary shake forces, but because these vertical forces are in different planes along the crankshaft, rocking couples will be introduced which are difficult to balance. However, when two of these three-cylinder crankshafts are joined together they may be treated as the mirror image of each other with the first and last (1–6), adjacent middle (3–4) and in between front and rear (2–5) crankpins paired (Fig. 3.39). Consequently, the unbalanced pitch couple for the front three cylinders will be counter-balanced by the rear three-cylinder pitch couple since both vertical couples will always act in opposition to each other and therefore they will

cancel each other out. This is demonstrated by the closed primary and secondary moment polygon (Fig. 3.40). The bending moment and stresses are absorbed by the rigidity of the crankshaft material itself and therefore there is no external rocking couple load imposed on the crankshaft's main bearings and its supporting crank-case.

A similar treatment may be given to the secondary couples created in both the front and rear halves of the crankshaft, these being again neutralized internally by the crankshaft due to the double in-line three-cylinder crankshaft arrangement which causes the couples created at both ends to balance out.

The in-line six-cylinder crankshaft is therefore perfectly self-balanced for primary and secondary shake forces and primary and secondary rocking couples and, with its 120° cylinder filling intervals, it makes a very popular engine for large cars and commercial vehicles where engine length is not at a premium.

3.4.12 Horizontally opposed flat six-cylinder engine balance
(Figs 3.41 and 3.42)

This cylinder configuration can be considered as a 180° vee six-cylinder engine with both cylinder banks sharing a common six-throw crankshaft (Fig. 3.41). Successive pistons in each cylinder bank are linked via the connecting-rods to every second crankpin, the angular intervals between the crank-throws for each cylinder bank are 120° and the crank-throws for both cylinder banks alternatively intermesh along the crankshaft's length and are out-of-phase with each other by 60°. Therefore, the left- and right-hand cylinder banks are staggered to accommodate and align the connecting-rods with their respective crank-pins, the left-hand bank being slightly forward relative to the right-hand bank.

In effect, the left- and right-hand cylinder banks consist of a pair of three-cylinder in-line engines with each set of crank-throws being a mirror image of the other.

As a result, the unbalanced couples produced by each cylinder bank, from the imaginary pair of three-throw crankshafts, exactly counter-balance each other for all crank angle positions (Fig. 3.41).

Thus, this cylinder and crankshaft configuration provides perfect self-balance for primary and secondary shake forces and, similarly, all primary and secondary rocking couples which can be seen by the closed force and moment polygon diagrams (Fig. 3.42).

3.4.13 60° vee six-cylinder engine balance
(Figs 3.43, 3.44 and 3.45)

These six-cylinder engines have two cylinder banks of three cylinders inclined at 60° to each other (Fig. 3.43). Three evenly spaced crank-

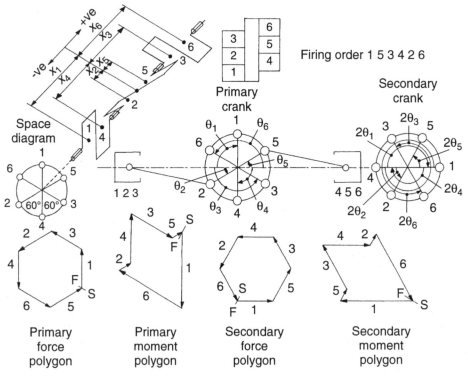

Firing order 1 5 3 4 2 6

Space diagram

Primary crank

Secondary crank

Primary force polygon

Primary moment polygon

Secondary force polygon

Secondary moment polygon

Fig. 3.41 Horizontal opposed six-cylinder engine balance (flat six)

114

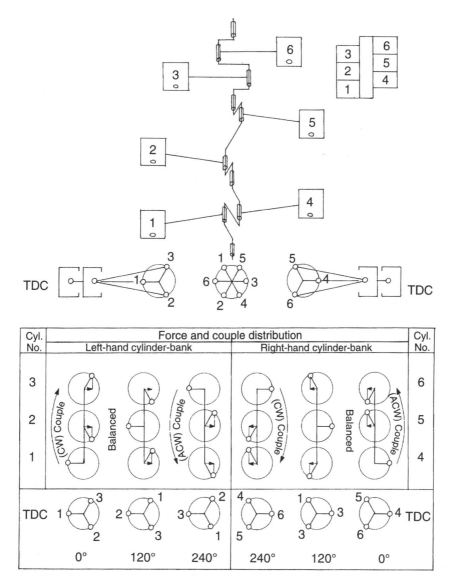

Fig. 3.42 Horizontal opposed six-cylinder primary force and couple balance

throws (with their pins aligned with each cylinder plane in the left-hand bank) and, similarly, three evenly spaced crankpins, align with their respective cylinders in the right-hand bank. The angular interval between crankpins for each bank is 120°; however, when both sets of crank-throws merge together to form a single crankshaft, the relative spacing of the left and right-hand bank crank-throws is such that the three prongs of each set of crank-throws intermesh with each other so that the angular phasing between both sets of crank-throws is 60°. The left-bank is numbered from the front to the rear 1, 2 and 3 and the right-hand

bank also follows this pattern with numbers 4, 5 and 6. Thus, the left-bank cylinders align with the first, third and fifth crankpin whereas the right-hand bank cylinders align with the second, fourth and sixth crankpin, counting from the front of the crankshaft. Consequently, the engine should be visualized as two three-cylinder in-line engines inclined 60° to each other (Fig. 3.44) with the crank throws for each cylinder bank alternately staggered along the crankshaft length.

Therefore, the primary and secondary forces for this double three-cylinder in-line engine are balanced, this is illustrated by both force polygons

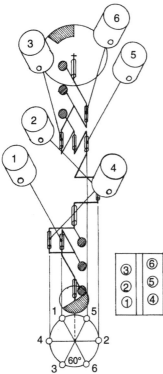

Fig. 3.43 60° vee six-cylinder engine crank and piston layout

closing (Fig. 3.45). Moreover, there is no direct secondary couple, again shown by there being no resultant on the secondary moment polygon. Conversely, however, there is a large primary couple created as shown by the open primary moment polygon (Fig. 3.45).

The inertia forces created by each piston react on their respective crank-throw pins, each being offset along the length of the crankshaft; therefore, each cylinder bank produces its own primary couple (Fig. 3.44). Therefore, as the crankshaft revolves, the directional sense and magnitude of both sets of couples are continuously changing and they are not necessarily in phase. Consequently, there will be angular periods when the direction and magnitude of these couples do not cancel each other out, this then causes periodic external couples to be generated.

With counterweights forming part of the fly wheel and rear three extended crank webs, and diagonally opposed counterweights imposed on the front pulley and first three extended crank webs, primary couples can be neutralized (Fig. 3.43).

3.4.14 90° vee eight-cylinder engine balance with single-plane crankshaft
(Figs 3.46, 3.47 and 3.48)

The crankshaft is similar in shape to an in-line four-cylinder engine which has two up and two down crank-throws all in the same plane, except that each crankpin, instead of supporting one connecting-rod big-end, accommodates a pair, one from each cylinder bank (Fig. 3.46). This means that at any one time, when the crankshaft is at TDC or BDC relative to one cylinder bank's line-of-stroke, pairs of pistons will be either at TDC or BDC simultaneously, while the pistons in the other cylinder bank will all be in their mid-stroke positions. With power impulses every 90° of the crankshaft movement a typical firing order would be 18364527.

Thus, the primary inertia forces of one pair of pistons at TDC or BDC will always be counter-balanced by the other pair of pistons in the same cylinder bank (Fig. 3.47). Conversely, the pistons in the other cylinder bank in their mid-stroke position do not generate any primary inertia force and therefore do not influence the state of primary force balance.

The internal couples, which are created by the offset distance between the first and second pairs of pistons when the outer and inner dead centres are reached, are counteracted by both couples opposing and neutralizing each other.

However, the inherent imbalanced secondary forces in each four-cylinder bank are cancelled out in the vertical plane due to the 90° inclination to each other of both banks, but these secondary forces combine in the horizontal plane to produce a secondary force disturbance equal to $\sqrt{2}$ (1.414) times the total secondary inertia force on one bank of cylinders (Fig. 3.48).

Thus, both primary forces and couples are balanced (Fig. 3.48). This is shown by the closed polygon force and moment diagrams. However, the secondary force is unbalanced as the force diagram does not close, whereas that of the secondary moment polygon closes, indicating that the secondary couples are neutralized.

3.4.15 90° vee eight-cylinder engine balance with two-plane crankshaft
(Figs 3.49, 3.50(a–d) and 3.51)

These two-plane 90° vee eight-cylinder crankshafts have four spaced crank-throws which are

116

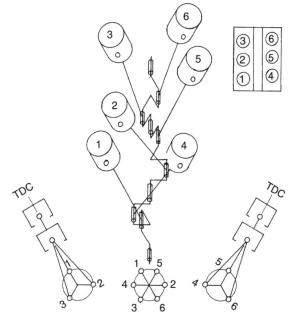

Crank position	Left-hand cylinder-bank			Right-hand cylinder-bank			Crank position
	1	2	3	4	5	6	
TDC 1	(CW) Couple			Balanced			TDC 5
0° 3	(ACW) Couple			(CW) Couple			0° 4
120° 1	Balanced			(ACW) Couple			120° 6
240°							240°

Force and couple distribution

Fig. 3.44   60° vee six-cylinder primary force and couple balance

set at angular intervals of 90° to each other (Fig. 3.49). A pair of connecting-rods, one from each cylinder bank, are linked to each crankpin, and therefore the engine can be considered as two four-cylinder engines sharing a common crankshaft. There is a 90° interval between power pulses which adapts readily to one of several firing orders, such as 15486372.

Considering the left-hand cylinder bank (Fig.

3.50(a)) when piston No. 1 is at TDC, piston No. 4 will be at BDC and pistons Nos 2 and 3 are in mid-stroke. However, on the right-hand cylinder bank, piston No. 6 will be at TDC while piston No. 7 is at BDC and pistons Nos 5 and 8 are in mid-stroke positions. Thus, when the crankshaft arrives in this position a large clockwise couple is created by the left-hand cylinder bank of pistons, whereas the right-hand cylinder

117

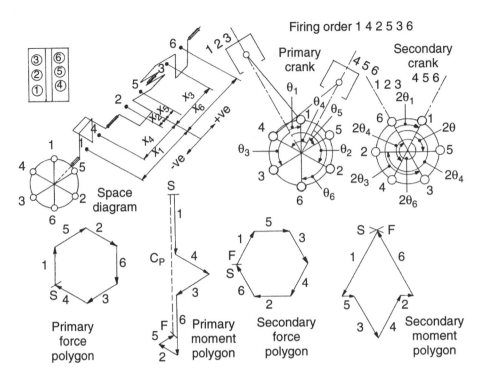

Firing order 1 4 2 5 3 6

Space diagram

Primary crank

Secondary crank

Primary force polygon

Primary moment polygon

Secondary force polygon

Secondary moment polygon

Fig. 3.45 60° vee six-cylinder engine balance

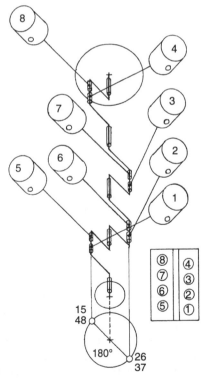

Fig. 3.46 90° vee eight-cylinder engine with a single-plane crankshaft crank and piston layout

bank also produces a clockwise couple, but of half its magnitude. Hence, the two individual sets of couples for each cylinder bank acting 90° to each other produce a compounded relatively large resultant couple which reacts through the main bearings onto the crank case.

When the crankshaft has rotated a further 90° (Fig. 3.50(b)) the left-hand cylinder bank pistons Nos 3 and 4 will now be at TDC and BDC respectively, while pistons Nos 1 and 4 are now at mid-stroke positions, whereas the right-hand cylinder bank pistons Nos 5 and 8 have moved to TDC and BDC respectively and pistons Nos 6 and 7 are in their mid-stroke positions. The effects are a large clockwise couple in the right-hand cylinder bank, which is opposed by an anticlockwise couple of half the others' magnitude on the left-hand cylinder bank. These couples which are at right-angles to each other therefore produce a relative moderate unbalanced resultant couple.

Rotating the crankshaft 180° (Fig. 3.50(c)), from piston No. 1's TDC position now produces a large and small anticlockwise couple in the left- and right-hand cylinder bank respectively, this causes a relatively large resultant couple to be created which will tend to pitch the engine in an anticlockwise direction.

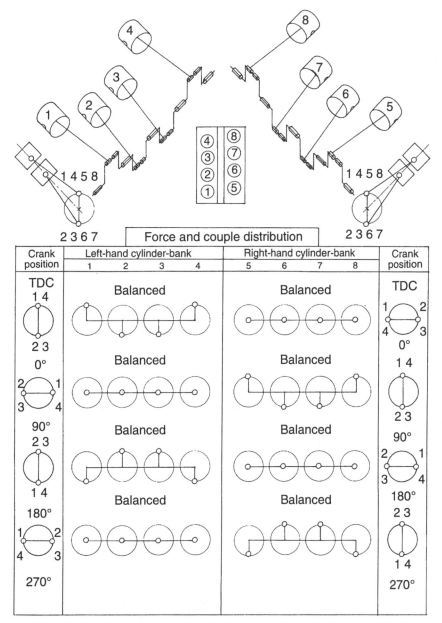

Fig. 3.47 90° vee eight-cylinder primary force and couple balance with single-plane crankshaft

Further rotation of the crankshaft to 270° (Fig. 3.50(d)), from piston No. 1's TDC position now generates a large anticlockwise couple in the right-hand cylinder bank which is partially counteracted by a small clockwise couple coming from the right-hand cylinder bank; consequently, there will again be an anticlockwise resultant couple, but of a smaller magnitude.

A further rotation of 90° will bring the crankshaft to its original position in which the resultant couple has again changed its sense and now pitches the engine in a clockwise direction (Fig. 3.50(a)). However, all the primary and secondary shake forces are balanced and there is no secondary rocking couple (shown by the closed force and moment polygon diagrams, Fig. 3.51). The

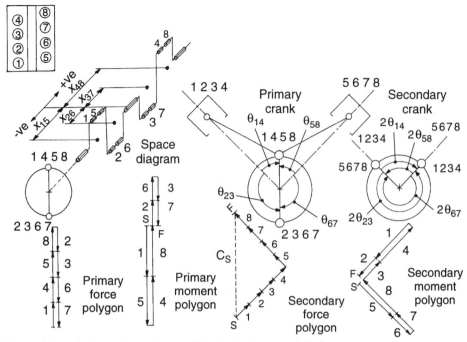

Fig. 3.48 90° vee eight-cylinder engine balance with single-plane crankshaft

magnitude of the imbalance of the primary couples is illustrated by the relatively large gap in the primary moment polygon (Fig. 3.51).

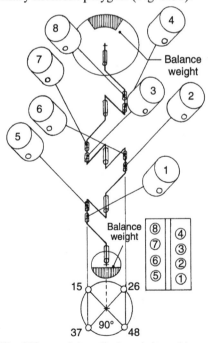

Fig. 3.49 90° vee eight-cylinder engine with a two-plane crankshaft crank and piston layout

A much easier approach to understanding and achieving reciprocating inertia balance with a two-plane four-throw crankshaft is to consider the engine as four transverse 90° vee twin-cylinder engine, with each pair of connecting rods sharing a common crankpin and each pair of crank-throws joined to the next pair via the main journals.

The crank-throws are arranged with the first and fourth crank-throws in the same plane but with their crankpins diametrically opposite each other, while the second and third crank-throws are both in a second plane and again opposite each other, but situated at right-angles to the first crank-throw plane.

Primary piston inertia force balance can be achieved by placing a counterweight diametrically opposite one of each pair of crank-throw arms, and equal to the piston and connecting-rod reciprocating mass (Fig. 3.29).

Thus, when any of the pistons are at TDC or BDC their primary inertia force will be exactly balanced by the counterweight mass, while the adjacent piston on the opposite cylinder-bank will be in its mid-stroke position where there is no primary piston inertia force. Since each primary piston force is balanced by a counterweight virtually in the same plane, there will now be no

120

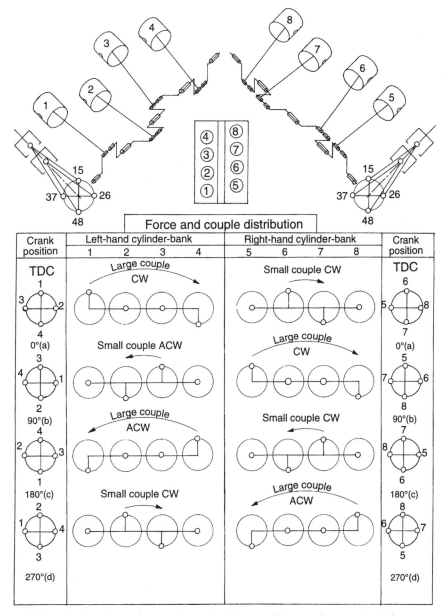

Fig. 3.50 90° vee eight-cylinder primary force and couple balance with two-plane crankshaft

primary couples. However, the secondary inertia force of each piston at mid-stroke is negative (see Fig. 3.24) and is at a maximum so that it points downward along its line-of-stroke. Conversely, when the piston is at either TDC or BDC the secondary force is positive and at a maximum and points upwards along its line-of-stroke. This is because when a piston is approaching TDC the secondary inertia force reinforces the primary force and, therefore, points in the same direc-

tion—that is, upwards—whereas when the piston moves towards BDC the secondary inertia force opposes the primary force and therefore also points upwards along its line-of-stroke, while the primary force is directed downwards.

Accordingly, the negative secondary inertia forces which are produced at mid-stroke from pairs of pistons in each cylinder bank are neutralized by the positive secondary inertia forces produced at TDC and BDC by the other pair of

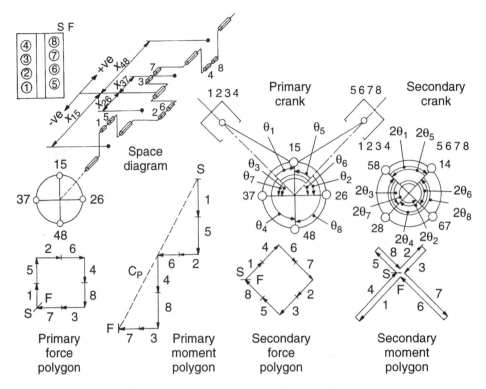

Fig. 3.51 90° vee eight-cylinder engine balance with two-plane crankshaft

pistons. Therefore, the secondary reciprocating inertia forces and couples are balanced.

The single plane crankshaft provides a filling and exhausting interval between adjacent transverse cylinders of 180°, whereas the two-plane crankshaft produces an uneven induction and exhaust interval between cylinder banks.

The two-plane cruciform crankshaft is more complex and heavier, due to the second plane and the counterweights, than the single-plane crankshaft configuration. Therefore, it can be seen that, with the use of counterweights, all primary and secondary forces and couples are eliminated (Fig. 3.49).

Single-plane vee-eight crankshafts have been used for racing engines where a slightly higher level of vibration is tolerated. However, the two-plane crankshafts are preferred for saloon car applications that demand an engine to be balanced not only for primary forces and couples, but also secondary forces and couples so that the engine does not experience shake and rock and therefore will run smoothly over its normal operating speed range.

3.4.16 90° vee ten-cylinder engine balance
(Figs 3.52 and 3.53)

This double five-cylinder banked engine with each line-of-stroke set at right-angles to each other has adopted an in-line five-cylinder engine crankshaft which has five crank-throws evenly spaced 72° apart (Fig. 3.52). Pairs of traversely located pistons are attached to each crankpin via their connecting-rods. If the engine is considered as two in-line five-cylinder engines sharing a common crankshaft, then the inherent imbalance of secondary inertia forces, primary and secondary couples can be observed by the open force and moment polygons (Fig. 3.53).

Fortunately, these unbalanced primary couples can be balanced by treating each pair of transverse cylinders as a 90° vee twin-cylinder engine and extending the crank-throw webs to the opposite side to form counterweights, equal in magnitude to one piston and connecting-rod reciprocating mass (Fig. 3.29).

Hence, when either piston is at TDC or BDC the counterweights balance the primary piston

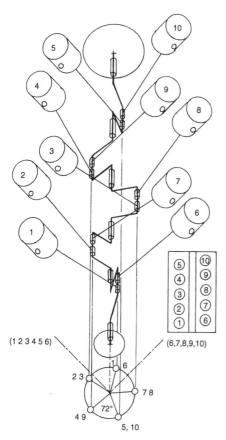

(1 2 3 4 5 6)   (6,7,8,9,10)

5, 10

Fig. 3.52 90° vee ten-cylinder engine crank and piston layout

3.4.17 60° vee 12-cylinder engine balance
(Figs 3.54, 3.55 and 3.56)

This 60° double-banked 12-cylinder engine adopts a conventional in-line six-cylinder crankshaft with paired crank-throws evenly spaced at 120° intervals (Fig. 3.54). Each crankpin supports a pair of connecting-rod big-ends which are connected to adjacent pistons in each transverse cylinder bank. This crankshaft and cylinder arrangement provides a power impulse every 60° of crank movement with the induction and exhaust occurring at equal intervals of 120° from alternative cylinder banks. The two cylinder banks can be treated as a pair of separate six-cylinder engines sharing a common crankshaft.

The crankshaft has three pairs of throws evenly spaced 120° apart so that the rear half of the crankshaft is a mirror image of the front half (Fig. 3.55). This symmetry gives perfect self-balance for both the primary and secondary forces and ensures that the internal couples produced by the front three cranks and pistons cancel out the couples produced by the rear three cranks and pistons.

Hence, the engine is perfectly balanced for primary and secondary inertia forces and couples (Fig. 3.56) without having to resort to adding counterweights, although counterweights are used to provide dynamic balance of the crank-throws and pins themselves. This crankshaft and cylinder configuration therefore provides the very best smoothness and acceleration response of all the engines examined.

inertia force, whereas the piston from the opposite cylinder bank is in its mid-stroke position during which time its primary force is zero and therefore does not contribute in any way to the state of balance. As the crankshaft rotates, the piston which was initially at mid-stroke, now approaches TDC or BDC and the other piston, which was originally at one of the dead-centres, now moves to the mid-stroke position. Thus, the piston which is now in line with the crank-throw is balanced and, since the cycle of events is periodic, this system will always cancel out unwanted primary inertia imbalance.

The question of secondary piston inertia force shake and secondary rocking couples still remains, but these effects are relatively small and can be tolerated with good engine mount design and for commercial vehicle applications.

3.5 Primary and secondary piston movement

3.5.1 Primary inertia force balancing of a single-cylinder engine using twin countershafts
(Fig. 3.57(a–e))

The primary inertia force, created by the piston and that part of the connecting-rod that is considered to reciprocate, produces a considerable amount of vertical shake which might not be tolerated. Normally, if the cylinder capacity is to be increased the solution is to have more cylinders and thereby make each piston and its inter-

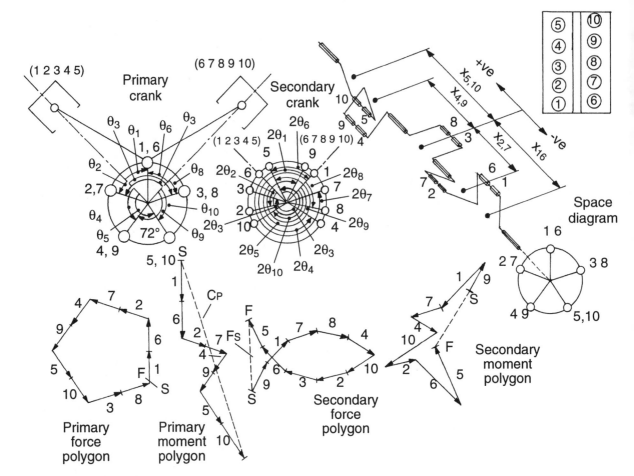

Fig. 3.53 90° vee ten-cylinder engine balance

connecting crankshaft configuration balance out the primary forces.

However, a method using twin countershafts can, in very special circumstances, be adopted to absorb and cancel out the unwanted vertical up-and-down shake, which results from this imbalance, without increasing the number of cylinders.

The fact that the primary inertia force balance is very rarely tackled by using countershafts and weights does not exclude a close study of this method as it is the basis of balancing primary couples and secondary forces when opposing crank-throw configurations are unable to eliminate these effects completely.

The primary inertia force balance is obtained by a pair of counterweights, each being equivalent to half the primary reciprocating inertia force mounted on twin countershafts, which are driven by a pair of meshing gear wheels so that they

rotate in opposite directions to each other (Fig. 3.57). These gears are then made to rotate at the crankshaft speed by meshing the left-hand countershaft to an input gear, which is of the same size as the countershaft gears and, which is itself, mounted on the crankshaft.

When the revolving crankshaft approaches TDC position (Fig. 3.57(a)) the twin countershaft weights are so timed as to point downwards in a BDC direction. Thus, the primary positive inertia force which is now at a maximum with an upward TDC direction will be exactly balanced by the centrifugal force produced by both counterweights.

Rotating the crankshaft 90° from TDC position (Fig. 3.57(b)) brings the piston to a mid-stroke position where the primary force has diminished to zero and the countershaft weights have also turned 90° so that they now face each other in the

124

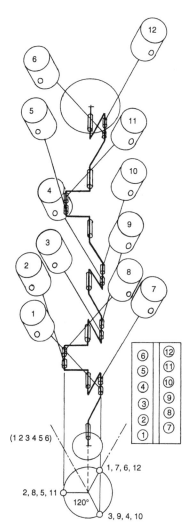

Fig. 3.54 60° vee twelve-cylinder engine crank and piston layout

horizontal plane. Accordingly, the opposing centrifugal forces produced by these counterweights now cancel out leaving the engine in a state of complete primary balance.

Rotating the crankshaft 180° from TDC position (Fig. 3.57(c)) brings the piston to BDC and the twin countershaft weights pointing upwards in a TDC direction. Correspondingly, the primary negative inertia force which now has a downward BDC direction will be neutralized by the upward direction of the centrifugal forces produced by both counterweights.

Rotating the crankshaft 270° from the TDC position (Fig. 3.57(d)) brings the piston to its mid-stroke return position where its primary in-

ertia force has been reduced to zero. Similarly, the countershaft weights will have rotated the same amount so that they are now outward facing in the horizontal plane. Consequently, the opposing centrifugal force produced by each weight is cancelled.

Finally, rotating the crankshaft 360° from TDC position (Fig. 3.57(e)) brings the piston back to its TDC position and the countershaft weights to their downward BDC direction. Accordingly, the positive primary inertia force with its upwards TDC direction is exactly balanced by the twin countershaft weights which both point in a downward BDC direction.

3.5.2 Primary couple balancing of an in-line three-cylinder engine using a single countershaft
(Fig. 3.58(a–c))

The balancing of the primary couples cannot be eliminated in a three-cylinder in-line engine by means of balance-weights attached to the crank-shaft alone. This is because, although these weights could produce balance in the vertical plane they, in turn, would introduce unbalanced horizontal couples.

Complete balancing of the primary couples in the vertical and horizontal planes, however, can be achieved using a reverse rotating balance-shaft (countershaft) in conjunction with weights attached to the crankshaft (Fig. 3.58(a–c)). The countershaft is driven at crankshaft speed in the reverse direction by a pair of meshing identical gear wheels attached to the front end of the crank and balance-shaft. Weights are attached in the form of extended crankwebs on both sides (split weights) of crankpins Nos 1 and 3, 30° before crank-throw No. 1 and 30° after crank-throw No. 3 looking from the front in a clockwise direction. Likewise, a pair of diametrically opposed weights are attached to the front and rear end of the balance countershaft. The meshing gear wheels are so timed as to position the crankshaft and countershaft weights in the vertical plane with both front end weights pointing downwards and the rear end weights pointing upwards with crankpin No. 1 positioned 30° before TDC.

Considering the crankshaft rotating at a steady speed, then, when piston No. 1 reaches TDC, pistons Nos 2 and 3 will both be 60° from either side of BDC position (Fig. 3.58(a)). Thus, the

125

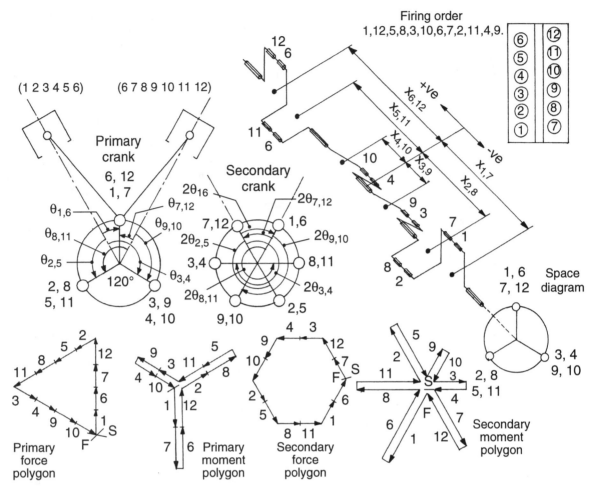

Fig. 3.55 60° vee twelve-cylinder engine balance

balance weights on the crankshaft and counter-shaft will be in planes inclined 30° after BDC and before TDC, respectively. Therefore, the upward primary maximum force of piston No. 1 will be opposed by the front split weights B_1 and weight B_2 whereas the downward primary force components of pistons Nos 2 and 3 will be equally opposed by the rear split weights B_1 and weight B_2, thus eliminating the vertical clockwise primary couple.

Rotating the crankshaft 60° from crank No. 1 TDC position brings piston No. 2 to BDC and pistons Nos 1 and 3 will both be 60° either side of the TDC position. In this position the primary maximum downward force of piston No. 2 balances the upward component forces of pistons Nos 1 and 3 and since they are offset from the central piston an equal amount on either side, there is no vertical primary couple pitching. The

balance weights of the crankshaft and counter-shaft will now be in the horizontal plane where these centrifugal forces cancel-out.

Rotating the crankshaft 120° from crank No. 1's TDC position brings piston No. 3 to TDC, and pistons Nos 1 and 2 will have both moved 60° on either side from BDC position (Fig. 3.58(b)). The balance weights of the crankshaft will now be in a plane 30° before TDC whereas the countershaft weights are in the plane 30° after TDC position. Consequently, piston No. 3's maximum upward primary force will be opposed by the downward direction of the rear balance weights B_1 and B_2, whereas the downward direction of pistons Nos 1 and 2's component forces is counterbalanced by the upward effects of the front balance weights B_1 and B_2.

Rotating the crankshaft 180° from crank No. 1's position brings piston No. 1 to BDC and pistons

126

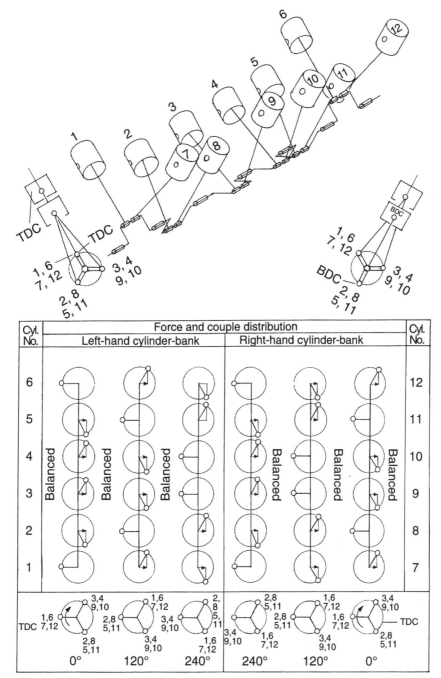

Fig. 3.56 60° vee twelve-cylinder engine primary force and couple balance

Nos 2 and 3 will now both be 60° either side of TDC position. In this position the primary maximum downward force of piston No. 1 will be opposed by the upward component forces of pistons Nos 2 and 3, thus causing an anticlockwise pitching. However, the front and rear balance weights now act in the opposite sense to cancel out the primary couple and thus eliminate any vertical pitching.

Rotating the crankshaft 240° from crank No. 1's TDC position brings piston No. 2 to TDC and pistons Nos 1 and 3 to 60° either side of BDC

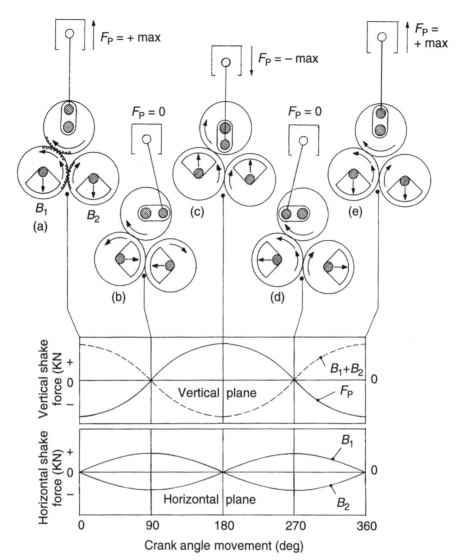

Fig. 3.57 Single-cylinder engine with twin primary force balance shafts

position (Fig. 3.58(c)). The central piston's (No. 2) upward maximum primary force will now equal the downward force components of pistons Nos 1 and 3, which are similarly displaced on either side of the central piston so that the varying rotating primary couple in the vertical plane is now zero. Correspondingly, the balance weights of the crankshaft and countershaft will be in the horizontal plane where their centrifugal forces cancel each other.

Rotating the crankshaft 300° from crank No. 1's TDC position brings piston No. 3 to BDC and pistons Nos 1 and 2 to 60° on either side of TDC position. Accordingly, the timed balance weights of the crankshaft will be in a plane 30° after TDC

and the countershaft weights will be in a position 30° before TDC. Consequently, the centrifugal forces of these weights will now oppose the downward dip at the rear and the upward lift at the front, thus neutralizing the vertical primary couple.

3.5.3 Primary couple balancing of a 60° vee four-cylinder engine using a single countershaft
(Fig. 3.59(a–d))

The unbalanced primary couple produced by the crankshaft and cylinder configuration can be dealt

128

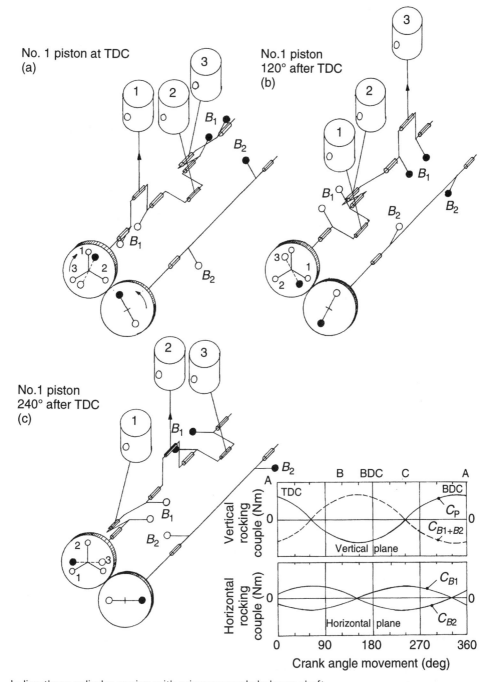

No. 1 piston at TDC
(a)

No.1 piston 120° after TDC
(b)

No.1 piston 240° after TDC
(c)

Fig. 3.58 In-line three-cylinder engine with primary couple balance shaft

with by incorporating a single countershaft driven by the crankshaft via a pair of meshing gears so that the shaft rotates at crankshaft speed but in the opposite direction (Fig. 3.59(a–d)).

Balance-weights of half the horizontal force components are attached on the common webs

between crankpins Nos 1 and 3 and on the common webs between crankpins Nos 3 and 4. Likewise, balance-weights of half the horizontal force components are mounted on the countershaft. The disposition of both crankshaft and countershaft weights are such that when piston

129

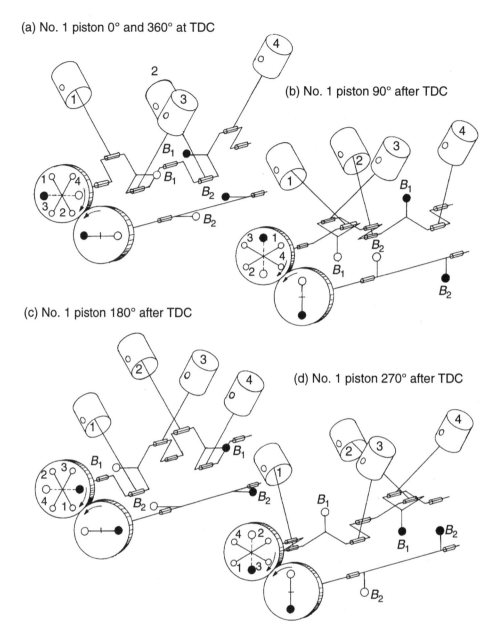

(a) No. 1 piston 0° and 360° at TDC

(b) No. 1 piston 90° after TDC

(c) No. 1 piston 180° after TDC

(d) No. 1 piston 270° after TDC

Fig. 3.59 60° vee four-cylinder engine with primary couple balance shaft

No. 1 is at TDC both sets of weights will be in the horizontal plane (Fig. 3.59(a)).

The front crankshaft and countershaft weights will then point in the same direction, whereas the rear crankshaft and countershaft weights will point in the opposite horizontal direction. Thus, these weights are so timed that they will oppose and cancel out the primary couples in the horizontal plane.

Rotation of the crankshaft 90° from piston No. 1 at TDC position (Fig. 3.59(b)) brings all four pistons to their mid-stroke positions where the piston assembly does not exert any primary inertia force. Consequently, the crankshaft front and rear balance-weights will point downwards and upwards respectively, whereas the counter-shaft front and rear weights will point upwards and downwards respectively. Therefore, since

130

there is no primary inertia force, the opposing centrifugal force of the front pair of weights and the rear pair of weights cancel, thus eliminating all primary inertia force and balance-weight imbalance.

Rotation of the crankshaft 180° from piston No. 1 at TDC position (Fig. 3.59(c)) brings the front two pistons (No. 1) to BDC and No. 3 to TDC whereas the rear two pistons (Nos 3 and 4) will now be at TDC and BDC respectively, all the pistons now exert their maximum primary inertia force and, of course, their horizontal force components. Correspondingly, the crankshaft and countershaft balance-weights have moved round so that the front pair are facing in the same direction to oppose the front horizontal force component and, likewise, the rear pair of balance-weights are together facing in the opposite direction to oppose the rear horizontal force components. Accordingly, all primary inertia couples are eliminated.

Rotation of the crankshaft 270° from piston No. 1 at TDC position (Fig. 3.59(d)) brings all four pistons to mid-stroke position where their primary inertia forces are zero. Correspondingly, the crankshaft front weights point upward and the countershaft front weights point downwards, whereas the crankshaft and countershaft weights at the rear point downwards and upwards respectively. Thus, the front crankshaft and countershaft balance-weight centrifugal forces cancel, and, likewise, the rear two balance-weights also cancel, so again all primary forces and centrifugal forces are neutralized.

A further rotation of the crankshaft 360° from piston No. 1's TDC position brings the crankshaft back to its original position as in Fig. 3.59(a), thus completing the cycle of events.

3.5.4 Secondary inertia force balancing of a single-cylinder engine using twin countershafts
(Fig. 3.60(a–i))

The secondary inertia force created by the piston and that part of the connecting-rod which is considered to reciprocate has a magnitude equal to approximately one-quarter of the primary inertia force.

With engine-cylinder swept volumes of a half a litre or more, the resultant vertical shake produced can create unwanted vibrations and harshness which is unacceptable in certain applications.

However, these secondary unbalanced forces can be balanced out by introducing a pair of counterweights (Fig. 3.60(a–i)) each being equivalent to half the secondary reciprocating inertia force, supported on twin shafts, which are driven by a pair of meshing gear wheels so that they rotate in opposite directions to each other at the same frequency as the periodic varying secondary force; that is, twice the crankshaft speed via a 2:1 chain and sprocket drive.

The balance shafts are so timed that when the piston is at either dead-centre (positive secondary forces) or at mid-stroke (negative secondary forces) then the counterweights are both pointing in the opposite direction to the piston's secondary inertia force and, therefore, exactly counteract the unbalanced reciprocating inertia force.

When the piston is at TDC (Fig. 3.60(a)) the twin counterweights both point downwards in a BDC direction, thus causing their centrifugal forces to counteract exactly the unbalanced maximum positive secondary piston inertia force, which has an upward direction towards TDC.

A 45° movement (Fig. 3.60(b)) of the crankshaft brings the piston to a quarter stroke. This moves the balance-shafts 90° in clockwise and anticlockwise directions where they are diametrically opposite each other, facing outwards, in a horizontal plane. In this position the secondary inertia force has decreased to zero and therefore the centrifugal force of each counterweight is timed to oppose and cancel out.

A 90° movement (Fig. 3.60(c)) of the crankshaft from TDC position brings the piston to the mid-stroke position where the secondary force has a maximum negative value. Correspondingly, the counterweights will now both have been rotated 180° to the TDC position where their centrifugal forces exactly balance the negative downward secondary piston inertia forces.

A 135° movement (Fig. 3.60(d)) of the crankshaft from TDC position brings the piston to a three-quarter stroke, to zero the secondary inertia force position. Correspondingly, the countershafts will have rotated 270° and will therefore point towards each other so that the centrifugal force generated by each weight will cancel out, therefore eliminating any imbalance in the horizontal plane by the counterweights.

A 180° movement (Fig. 3.60(e)) of the crankshaft from TDC position brings the piston to BDC, the secondary force is therefore at a positive maximum with its directional sense towards TDC in opposition to the primary inertia force of

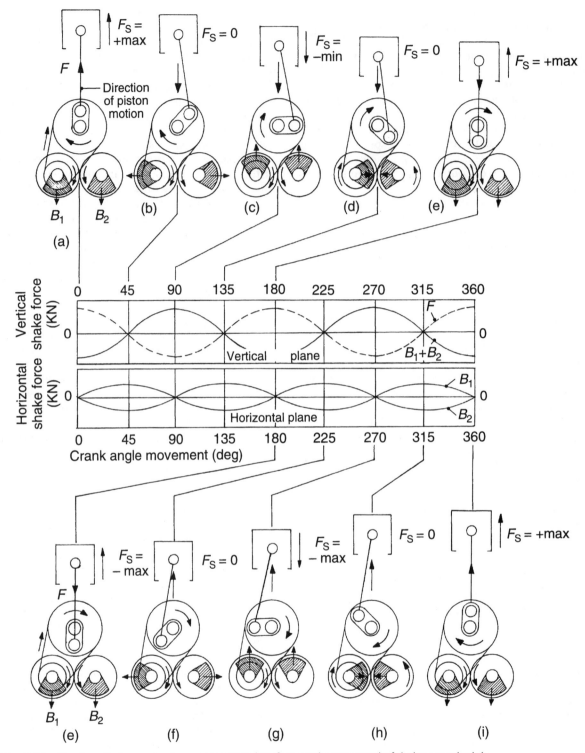

Fig. 3.60 Single-cylinder engine harmonic secondary force twin-counter shaft balancer principle

the piston. At the same time the countershafts will have turned through 360° so that both counterweight centrifugal forces will point downwards in opposition to the secondary inertia force and will thus neutralize the secondary imbalance.

Rotating the crankshaft 225° from TDC position (Fig. 3.60(f)) brings the piston to its return quarter-stroke position, where the secondary forces are again zero. Likewise, the countershaft will have revolved 450° so that the counterweights are now in a horizontal plane pointing away from each other and, accordingly, their generated centrifugal forces cancel out.

Rotating the crankshaft to the 270° position (Fig. 360(g)) brings the piston to its return mid-stroke position where its secondary inertia force creates a negative maximum in a BDC downward direction. However, the countershaft will have rotated 540° and will therefore point upwards in a TDC direction.

Rotating the crankshaft to the 315° position (Fig. 3.60(h)) brings the piston to its return three-quarter position where the secondary inertia force has diminished to zero. The countershafts revolving at twice the crankshaft speed will have rotated 630°, and therefore, their counterweights will neutralize each other since they will be facing each other in the horizontal plane.

Finally, rotating the crankshaft 360° from TDC position (Fig. 3.60(i)) brings the piston again to TDC position, while the countershafts will have rotated 720° causing the counterweights to point in a downward BDC direction. Thus, the secondary inertia force, which is at a positive maximum, points in an upward TDC direction and will accordingly be balanced by the twin counterweights pointing in the opposite direction.

3.5.5 Secondary inertia force balancing of an in-line four-cylinder engine using twin countershafts
(Fig. 3.61(a–d))

When pairs of crank-throws approach TDC or BDC with an in-line four-cylinder engine, the secondary inertia forces for all four piston and connecting-rod assemblies will be at a maximum and will simultaneously be acting upwards towards TDC. If the cylinder capacity of the engine starts to exceed two litres the resultant vertical shake at certain speeds can be prominent, which is often reflected in engine roughness and vibration.

These imbalanced secondary forces can be balanced out by incorporating twin countershafts and their weights, these being equivalent in magnitude to the secondary inertia forces of all four cylinders (Fig. 3.61(a–d)). The countershafts are mounted underneath the crankshaft and they are made to revolve in the opposite direction to each other by a pair of meshing gears driven at twice the crankshaft speed via a 2:1 chain and sprocket wheel drive.

Consider a steady rotating crankshaft in various crank-throw positions 0°, 45°, 90°, 135° and 180° as follows.

With pistons Nos 1 and 4 at TDC, pistons Nos 2 and 3 will be at BDC position (Fig. 3.61(a)), whereas the twin countershaft weights are timed to point in a downward direction. Under these conditions, the secondary inertia forces caused by the pistons at TDC act with the primary forces in an upward direction, whereas the secondary inertia forces caused by the pistons, which are at BDC, act in opposition to the downward primary forces and therefore also have an upward sense. Accordingly, the upward direction of all four secondary forces will be counteracted by the centrifugal forces of the two counterweights which, together, act in opposition.

Rotating the crankshaft 45° from cranks Nos 1 and 4's TDC position (Fig. 3.61(b)) brings pistons Nos 1 and 4 to quarter-out-stroke position and pistons Nos 2 and 3 to quarter-return-stroke position, where both pairs of pistons experience zero secondary force. Correspondingly, the countershaft weights will both have rotated 90° in a clockwise and anticlockwise direction where they are diametrically opposite each other and face outwards in the horizontal plane. Therefore, the centrifugal force of each countershaft weight opposes and cancels the other one, thereby producing a resultant zero horizontal reaction.

Rotating the crankshaft 90° from cranks Nos 1 and 4's TDC position (Fig. 3.61(c)) brings pistons Nos 1 and 4 to the mid-out-stroke position and pistons Nos 2 and 3 to the mid-return-stroke position, where both pairs of pistons exert a maximum negative secondary inertia force which has a downward direction. Correspondingly, the countershafts will both have rotated 180° in a clockwise and anticlockwise direction, thus causing both counterweights to face upwards in the vertical plane. Therefore, the centrifugal force of both counterweights acting upwards opposes and neutralizes the four secondary inertia forces with their downward direction.

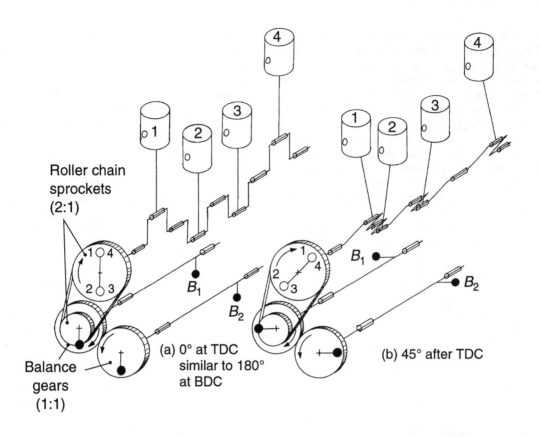

Roller chain
sprockets
(2:1)

Balance
gears
(1:1)

B_1

B_2

(a) 0° at TDC
similar to 180°
at BDC

B_1

B_2

(b) 45° after TDC

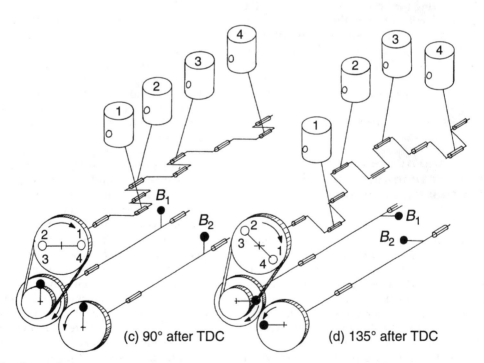

B_1

B_2

(c) 90° after TDC

B_1

B_2

(d) 135° after TDC

Fig. 3.61   In-line four-cylinder engine harmonic secondary force chain drive twin-counter shaft balancer principle

Rotating the crankshaft 135° from cranks Nos 1 and 4's TDC position (Fig. 3.61(d)) bring pistons Nos 1 and 4 to the three-quarter out-stroke position and pistons Nos 2 and 3 to the three-quarter return-stroke position, where again the secondary inertia forces of each piston assembly have decreased to zero. At the same time, the countershafts will both have rotated 270° in a clockwise and anticlockwise direction, thus causing both counterweights to be diametrically opposite and facing towards each other in the horizontal plane. Again, the centrifugal force of each counterweight will oppose and cancel the other one and so result in a zero horizontal reaction.

Rotating the crankshaft 180° from cranks Nos 1 and 4's TDC position brings pistons Nos 1 and 4 to BDC and pistons Nos 2 and 3 to the TDC position where again both pairs of pistons exert a maximum positive secondary inertia force, which has an upward direction. At the same time, the countershafts will both have rotated 360° in a clockwise and anticlockwise direction so that both countershaft weights now face downwards in the vertical plane. Accordingly, the centrifugal forces of both counterweights exactly equal and balance-out the secondary forces created by all four piston assemblies.

A similar procedure could be repeated for the remaining crank angular movements of 225°, 270°, 315° and 360° to complete the cycle of events.

3.5.6 Central gear drive transverse twin-countershaft harmonic secondary force balancer (Lanchester harmonic secondary force balancer)
(Fig. 3.62)

The first secondary force twin countershaft balancer was invented and patented in 1911 by Dr Frederick Lanchester and was known as the Lanchester harmonic analyser.

In this design, short twin countershafts are transversely mounted between two bearings underneath the middle region of the crankshaft (Fig. 3.62). The centre of each countershaft has skewed helical teeth machined around it while a pair of balance weights are positioned on either side. Each countershaft skewed gear meshes with ring gears, shrunk onto the two circular central crankwebs, and this gearing provides a 2:1 speed step-up from the crankshaft. The two ring-gear helical cut teeth are of the opposite hand so that

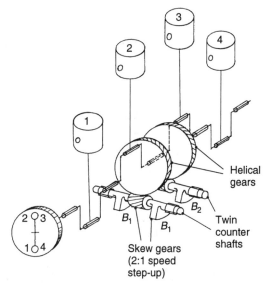

Fig. 3.62 Four-cylinder engine central gear drive transverse twin-counter shaft harmonic secondary force balancer (Lanchester harmonic balancer)

the countershafts rotate at the same speed but in the opposite direction to each other.

The basic principle of the harmonic balancer is that the countershaft weights oppose and neutralize the maximum secondary forces in the vertical plane, while the centrifugal force of each counterweight opposes and cancels the others in the horizontal plane.

3.5.7 Front-toothed belt-drive parallel twin countershaft harmonic secondary force balancer
(Figs 3.63, 3.64 and 3.65)

A belt-drive twin-countershaft secondary force balancer (Fig. 3.63) incorporates long twin countershafts with the right-hand countershaft being directly driven by the belt via a pulley, whereas the left-hand countershaft is indirectly driven by a pair of similar-sized spur gears which drive this shaft in the opposite rotational direction to the directly driven shaft. The belt pulleys are chosen to give a 2:1 speed step-up and a jockey pulley is included to provide the correct tension to the belt drive.

Thus, the balance countershafts rotate counter to each other at twice the crankshaft speed and are so phased that they counteract the positive secondary forces at either dead centre, and the negative secondary forces at mid-stroke (Fig. 3.60). This is achieved by the weights facing in the opposite direction to those secondary

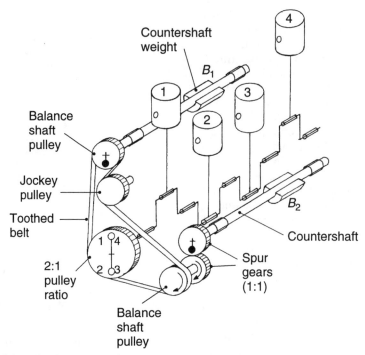

Fig. 3.63 Four-cylinder engine front-toothed belt drive parallel twin-countershaft harmonic secondary force balancer

forces when in the vertical plane, and then for both balance-weights to face either inward or outward and thereby oppose each other when they have moved to one of the two horizontal positions.

An important feature of this secondary force balancer is that the countershafts are both mounted the same distance out from the cylinder's line-of-stroke but the left-hand countershaft is mounted considerably higher than the right-hand countershaft, which is itself positioned just above the crankshaft centre (Fig. 3.63).

The object of positioning the balance countershafts at different heights is so that when the countershaft weights are in the horizontal plane, the height offset produces a balance–countershaft torque which opposes the fluctuating torque generated by all the piston secondary forces which occur at both dead centres and mid-stroke, otherwise this secondary inertia torque will promote engine roll (Fig. 3.64).

In addition to neutralizing the secondary torque fluctuations the offset balance–countershaft torque partially counteracts the explosion impulse torque produced over a wide speed and load range.

Measurements of secondary force vibration (Fig. 3.65) at the gearbox mounting rise at an increasing rate with engine speed, but when twin countershafts are incorporated these secondary force vibrations are considerably reduced, thus producing a much smoother running engine, particularly in the upper speed range.

3.6 Offset vee six and vee eight crankshaft big-end journals

3.6.1 Purpose of crank-pin offset
(Fig. 3.66)

Six-cylinder vee engines may utilize wide 90° inclined cylinder banks to reduce the engine's overall height. This configuration will result in unevenly spaced firing impulses. However, the power impulse can be made to occur evenly by dividing each big-end crank-pin in two and offsetting each half by 30°. Firing intervals will now take place at every 120° crankshaft movement, which is similar to that obtained with the commonly used six-cylinder 60° vee banked arrangement.

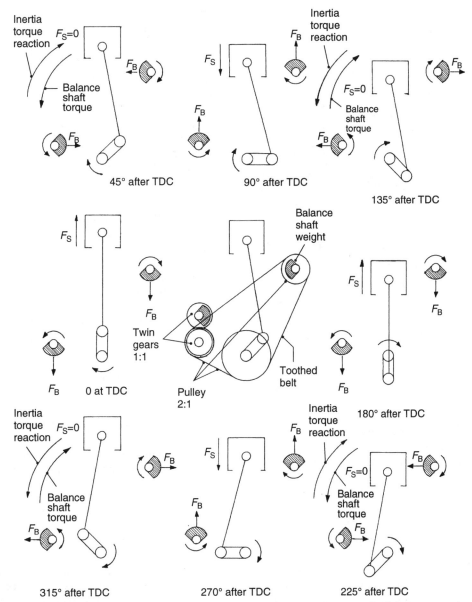

Fig. 3.64 In-line four-cylinder engine harmonic secondary force off-set twin-countershaft balancer cycle of events

A 90° six-cylinder engine permits the same machining line as in a production plant set up for a 90° vee cylinder-bank configuration to be used; thus, there are considerable cost savings by utilizing the same equipment for both engines. Conversely, vee eight-cylinder engines with narrow 60° inclined cylinder-banks to reduce engine width have unevenly spaced firing impulses, but even firing intervals can be obtained by splitting each big-end pin in half and offsetting the two halves by 30°. The firing pulses then take place at intervals of 90°, similar to those achieved with 90° vee eight-cylinder engines.

3.6.2 Cycle of piston events with offset big-end journal
(Fig. 3.66(a–d))

Consider a vee eight-cylinder engine having its cylinder banks inclined 60° to each other with the

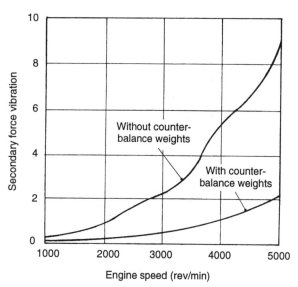

Fig. 3.65 Engine vibration caused by unbalanced secondary forces with and without twin-balance shafts

crankshaft big-end crankpins split and offset 30° to each other.

Left-hand piston crankpin at TDC in its line-of-stroke
(Fig. 3.66(a))

If the crankshaft is positioned such that the piston in the left-hand cylinder-bank is at TDC, then the right-hand cylinder-bank piston is approximately at mid-stroke with its crankpin $30° + 60° = 90°$ from the right-hand cylinder-bank TDC position.

Crankshaft movement 90° from left-hand cylinder-bank line-of-stroke
(Fig. 3.66(b))

If the crankshaft is now rotated 90° from the left-hand cylinder-bank crankpin TDC position, then the left-hand cylinder-bank piston will be approximately in its mid-stroke position with its crankpin $30° + 60° = 90°$ from the left-hand cylinder-bank line-of-stroke.

Crankshaft movement 180° from left-hand cylinder-bank line-of-stroke
(Fig. 3.66(c))

If the crankshaft is now rotated 180° from its starting position, the left-hand cylinder-bank piston will be at BDC and the right-hand cylinder-block piston will be approximately in mid-stroke with its crankpin perpendicular to the line-of-

stroke; that is, 90° from the right-hand cylinder's line-of-stroke.

Crankshaft movement 270° from left-hand cylinder-bank line-of-stroke
(Fig. 3.66(d))

If the crankshaft is now rotated 270° from its original starting position, the right-hand cylinder-bank piston will be at BDC and the left-hand cylinder-bank piston crankpin will be perpendicular to its line of stroke; that is, 90° from the left-hand cylinder line-of-stroke.

Crankshaft movement 360° from left-hand cylinder-bank line-of-stroke
(Fig. 3.66(a))

Rotating the crankshaft 360° from its starting position brings the left-hand cylinder-bank piston back to its original TDC position with the right-hand piston at approximately mid-stroke. This cycle of events for a single-throw crankshaft will therefore be continuously repeating.

3.6.3 Summary
(Figs 3.66 and 3.67)

It can be seen that having the big-end crankpin offset 30° has enabled the crankshaft angular displacement between the left-hand piston at TDC to the right-hand piston reaching TDC to be extended from 60° to 90°. Consequently if the vee twin cylinder-banks and crankshaft are adapted to accommodate eight cylinders by arranging the crank-throw in two planes—that is, in a cruciform configuration—then the firing pulse intervals of the various cylinders can be evenly spaced to occur every 90°. As a result, the regular interval between each power stroke produces a smooth power response which is comparable with the 90° vee eight-cylinder engine.

A comparison of the torque fluctuation over one cycle for a 90° vee six-cylinder engine with and without off-set big-end crank-pins is shown in Fig. 3.67. The graph illustrates the distorted engine torque due to the uneven firing intervals of 90°, 150°, 90°, 150° etc for the engine with the conventional crankshaft. This is compared with the regular torque pattern generated when an even firing interval of 120° is produced when using an offset big-end crank-pin crankshaft.

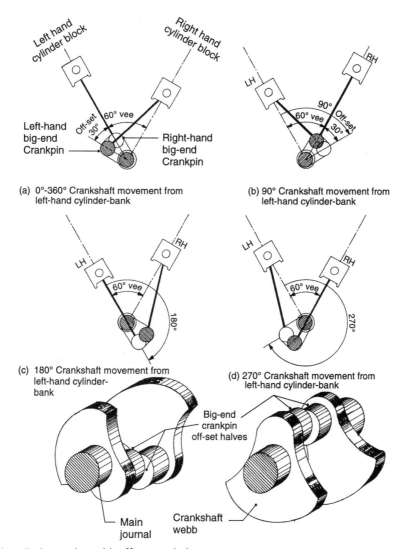

Fig. 3.66 Vee twin-cylinder engine with offset crankpin

3.7 Torsional vibration

If a torque impulse twists a shaft, it creates a strain which, when the torque ceases to act, starts to bring the shaft back to its original position. However, owing to the mass of the shaft the latter will overshoot and twist in the opposite direction, thus creating a new restoring strain. The amplitude of this angular back-and-forth deflection through the equilibrium position decays over a number of cycles due to internal friction, but its frequency of vibration is unique for the particular crankshaft's cross-section cranks, webs and counterweight configuration. Therefore, any shaft with cranked arms can be treated as an elastic shaft which will oscillate longitudinally with its own natural frequency of vibration.

3.7.1 Cylinder pressure and crank tangential torque
(Fig. 3.68)

Rotation of the crankshaft gradually increases cylinder pressure as the piston approaches TDC on its compression stroke. Ignition and combustion then cause a rapid pressure rise which peaks just after TDC, the pressure then begins to fall as the piston travels down towards BDC and the

139

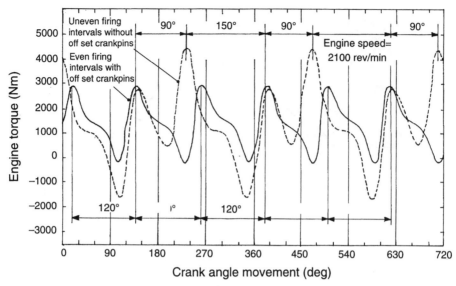

Fig. 3.67 Graph illustrating the variation in regularity of the torque pulses with and without offset big-end crankpins for a 90° six-cylinder engine

combustion gases expand (see Fig. 3.68). This pressure acting on the piston relays a force to the big-end crankpin via the connecting-rod. However, it is the tangential force which is actually transformed to do useful work for a rotating crankshaft. The resulting tangential pressure, or torque plus the turning effort due to the inertia spread over 720°, is shown in Fig. 3.68.

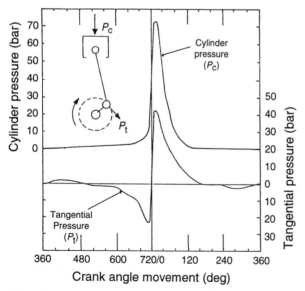

Fig. 3.68 Periodic cylinder and tangential pressure curves for a typical four-stroke single-cylinder engine

3.7.2 Harmonic (sine wave) torque curves
(Fig. 3.69)

The variable periodic tangential torque curve can be resolved into a constant mean torque and an infinite number of sine wave torque curves, known as harmonics, with orders that depend on the number of complete vibrations per engine revolution (see Fig. 3.69). Thus, the tangential crankshaft torque is made up of a large number of harmonics of varying amplitudes and frequency. The order of a harmonic is the number of sine wave (harmonic) cycles completed per crankshaft revolution. Thus, for a four-stroke cycle engine a half-order harmonic, known as *the fundamental order*, completes a period (one cycle) every second revolution; that is, half a cycle per revolution, whereas the first-order harmonic completes one period (one cycle) every revolution. Hence, for a four-stroke cycle engine the harmonic orders are as follows:

$\frac{1}{2}$-order harmonic completes a half cycle/cycles every revolution
1st-order harmonic completes one cycle/cycles every revolution
$1\frac{1}{2}$-order harmonic completes $1\frac{1}{2}$ cycle/cycles every revolution
2nd-order harmonic completes two cycle/cycles every revolution

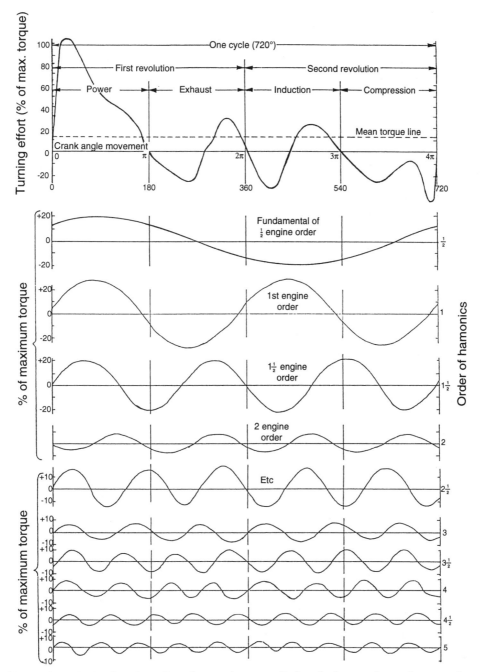

Fig. 3.69 Harmonic analysis of turning effort diagram for one cylinder of a four-stroke engine – resultant diagram of both gas pressure and inertia forces

$2\frac{1}{2}$-order harmonic completes $2\frac{1}{2}$ cycle/cycles every revolution

3rd-order harmonic completes three cycle/cycles every revolution

3.7.3 Critical speeds

When the shaft is revolving at such speed that the tangential torque frequency, or one of the harmonic sine wave frequencies (harmonic order number × crankshaft speed), synchronizes—that is, coincides with the natural frequency of vibration of the shaft—resonance occurs. Thus, the crankshaft speed at which this resonance occurs is said to be a critical speed for that particular harmonic order. If the tangential torque or harmonic impulses are applied at the same number of times per minute as the natural frequency of the elastic system, the amplitude of the vibration will increase until eventually the crankshaft is destroyed by fatigue failure or the crankshaft speed moves out of phase with the harmonic frequencies. It therefore follows for a four-stroke cycle engine that there is a critical speed for every whole and half-order harmonic.

3.7.4 Normal elastic line
(Fig. 3.70(a–c))

Crankshaft torsional vibrations can be analysed by replacing the rotating and reciprocating components of each cylinder (Fig. 3.70(a)) by single rotating masses attached to a cylindrical shaft which has an equivalent elasticity to that of the actual crankshaft so that respective masses align with the centre-line of each cylinder axis (Fig. 3.70(b)).

With a relatively heavy flywheel, the natural torsional vibration of any crankshaft system normally produces the maximum angular deflection or amplitude at the mass furthest away from the flywheel.

The relative twist between adjacent equivalent masses along the length of the crankshaft for free vibration can be derived and plotted to form a normal elastic curve over the length of the shaft (Fig. 3.70(c)). The first mass furthest removed from the flywheel which has the maximum amplitude is given a value of unity while all other masses have relative values less than unity. This information is very useful since, within the elastic working range of the material, the relative twist in the crankshaft remains in the same proportion as given, whatever the actual magnitude of vibration in the crankshaft at any one time.

It should be observed that the elastic curve intersects the shaft axis before it reaches the flywheel and then continues for a short distance below the axis. This point where the elastic curve crosses the axis is therefore a node where no torsional deflection occurs in the crankshaft.

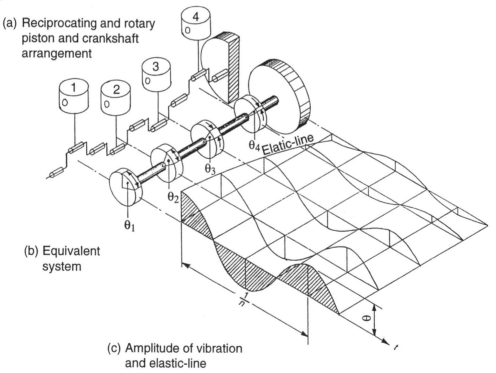

(a) Reciprocating and rotary piston and crankshaft arrangement

(b) Equivalent system

(c) Amplitude of vibration and elastic-line

Fig. 3.70 Illustration of natural torsional vibration of a four-cylinder crankshaft

3.7.5 Major and minor critical speeds

The magnitude of the vibration in a multi-cylinder engine varies greatly for different critical speeds, even taking into account the amplitude of the various harmonic torques. This is due to some harmonic torque orders assisting one another in producing large vibrations, whereas other harmonic torque orders partially cancel each other out. Hence, the important critical speeds have harmonic torque components which act cumulatively to intensify the torsional twist to the shaft in its normal mode of vibration, these critical speeds are known as the *'major'* criticals. In contrast, all other critical speeds which may exist individually, or partially overlap other harmonic orders, tend to neutralize and damp the excited oscillations. These critical speeds are therefore referred to as *'minor'* criticals.

Minor critical speeds are generally unimportant, whereas major critical speeds normally have orders which are a direct multiple of the number of torque pulses per revolution and are dangerous, and must either be avoided or must be severely damped.

3.7.6 Determination of major and minor critical speeds
(Fig. 3.71)

Vibration severity for various harmonic torque orders can be approximately determined with the knowledge that work done in the form of kinetic energy during a torsional oscillation is proportional to each order of harmonic torque T, applied at a respective cylinder axis, times the torsional twist θ taken from the elastic line ordinates in the same cylinder plane.

Hence, the kinetic energy for each cylinder harmonic torque pulse is proportional to $T\theta$

i.e. kinetic energy (KE) $\propto T\theta$

Thus, for the total number of cylinders producing phased harmonic torques, the entire kinetic energy is given by: a constant × harmonic torque × vector sum of the phased elastic line ordinates at which the torque is applied.

i.e. KE = a constant × $T\Sigma\theta$

where $\Sigma\theta$ = vector sum of the elastic line ordinates.

$T\Sigma\theta$ implies that the magnitude of the vibration

in a critical speed is directly dependent on the harmonic torque, on the shape of the normal elastic line, on the crank throw sequence and firing order.

The vector sum $\Sigma\theta$ is readily derived by means of phase and vector diagrams, as shown in Fig. 3.71.

3.7.7 Constructing the phase diagram
(Fig. 3.71)

1 For each order of critical speed a phase diagram is constructed. For the half-order critical speed, 360° represents one complete vibration which is equivalent to two crankshaft revolutions.

2 For the fundamental half-order critical speed, divide the phase circle into as many intervals as there are cylinders and, in a clockwise direction, number each successive radial arm in the firing order sequence.

3 For the 1st-order critical speed the harmonic vibration is completed in half the crank-angle movement compared with that of the half-order critical speed, so that the angular displacement between each firing interval will be doubled on the phase diagram. Thus, the 1st-order interval becomes $2 \times \frac{1}{2}$-order interval; that is, for the six-cylinder engine $2 \times 60 = 120°$.

4 For the $1\frac{1}{2}$-order critical speed, each harmonic vibration is completed in one-third of the crank-angle movement compared with that of the $1\frac{1}{2}$ order. Thus, the angular displacement between each firing interval now becomes $3 \times \frac{1}{2}$-order interval; that is, for the six-cylinder engine $3 \times 60 = 180°$.

5 For the 2nd-order critical speed, each harmonic vibration is completed in one quarter of the crank-angle movement compared with that of the $1\frac{1}{2}$ order. Thus, the angular displacement between each firing interval now becomes $4 \times \frac{1}{2}$-order interval; that is, for the six-cylinder engine $4 \times 60 = 240°$.

6 In general, the critical-speed order-number angular intervals above the fundamental $\frac{1}{2}$ order are obtained by multiplying the $\frac{1}{2}$ order number angle by 2, 3, 4, 5 and 6 for the 1st, $1\frac{1}{2}$, 2nd, $2\frac{1}{2}$ and 3rd order critical speeds respectively. After order number 3 the phase diagrams repeat themselves; that is, order numbers $3\frac{1}{2}$ and $\frac{1}{2}$, 4 and 1, $4\frac{1}{2}$ and $1\frac{1}{2}$, 5 and 2, $5\frac{1}{2}$ and $2\frac{1}{2}$ and 6 and 3 are all identical. Note also

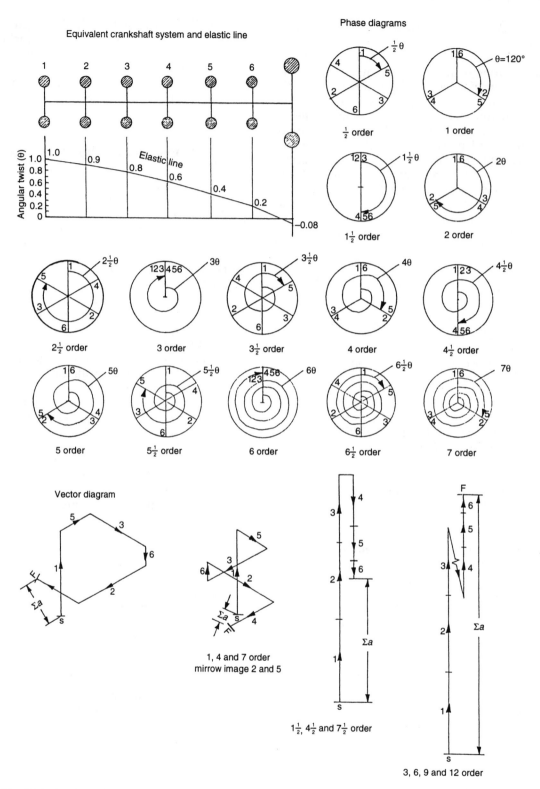

Fig. 3.71 Critical orders of torsional vibrations for in-line six-cylinder engine firing order 153624

numbers 1, 4, 7 etc have the same phase diagrams respectively.

3.7.8 Constructing the vector diagram
(Fig. 3.71)

1 Choose a suitable scale to represent the elastic curve ordinates.
2 Starting from point (S) and preceding in the firing order sequence draw to scale (1.0) a line parallel to the phase diagram radial arm '1' with an arrow pointing outwards from the centre of the phase diagram. Then, from the end of the first line, draw to scale (0.4) a line parallel to radial arm '5' with an arrow pointing outwards. Next, draw to scale (0.8) a line parallel to radial arm '3' with an arrow also pointing in an outward direction.
3 Continue with the other scaled elastic curve ordinates in turn, drawing each scaled elastic ordinate at the end of the previous one and in all cases watching the direction of the arrows, then label the end of the last scaled ordinate to be drawn (F).
4 Join points (S and F), this line represents the magnitude of the resultant of the elastic curve ordinates for that particular order number, this being a measure of the degree of additive alignment of the harmonic torques. It will be observed that the phase diagrams for orders $\frac{1}{2}$, $3\frac{1}{2}$ and $6\frac{1}{2}$ are the mirror images of orders $2\frac{1}{2}$ and $5\frac{1}{2}$ respectively. Likewise, orders 1, 4 and 7 are the mirror images of orders 2 and 5 respectively. Thus, only one vector diagram is required for each set of phase diagram configurations.

Further observations show that for critical speed orders 3, 6, 9, 12 etc the harmonic torques are all in phase and will therefore enforce the crankshaft oscillations in their normal mode of vibration. Hence, the in-phase harmonic-order critical speeds are the most potent and are consequently known as 'major' critical speeds, whereas all other critical speeds are referred to as minor criticals.

3.7.9 Inherent torsional vibration damping
(Figs 3.72 and 3.73)

When a crankshaft is subjected to repeated cycles of torsional stress reversals the stress and strain

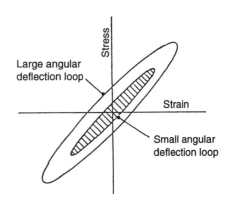

Fig. 3.72 Stress–strain hysteresis loops

do not obey Hooke's law; that is, the strain is not strictly proportional to the stresses within the elastic limit. There is, in fact, a very small lag of the strain during loading and unloading of the strain, so that a stress–strain graph encloses what is known as a hysteresis loop (Fig. 3.72) the width of which at any given stress is the strain difference between the unloading and loading curves.

The lag in strain between cycles of fluctuating stresses may cause the material at comparatively low stresses to accumulate considerable local strains, which may eventually result in shaft failure. Thus, the energy dissipated in hysteresis at a critical speed due to stress reversals in the first instance may contribute to a small amount of damping of the vibration, but eventually it may lead to the rupture of the shaft at stresses far below the elastic limit of the material.

When the stress is low, the area which represents the energy imparted by the hysteresis loop is

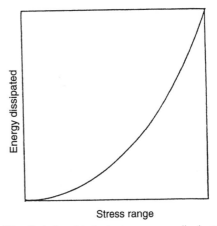

Fig. 3.73 Relationship between energy dissipated with increasing strain range

small. However, as the stress range is increased, the area increases enormously; that is, the energy dissipated within the shaft material rises rapidly (Fig. 3.73) and, if permitted to persist, will destroy the shaft even though the alternating stresses are relatively low.

Crankshaft torsional vibrations are inherently damped to a limited extent by either internal or external means; that is, within the engine or in the transmission and auxiliary drives. Internal engine damping is made up partially by energy losses due to hysteresis but more so by oil film viscous shear in the main and big-end bearings. This damping is adequate to protect the shaft during minor criticals but is not sufficient to cope with the more violent major criticals.

3.7.10 Identifying critical speeds

Critical speeds can be responsible for producing a rumbling noise caused by the relative side movements of revolving and reciprocating components such as journals and pistons pounding from side to side in their respective clearances and by backlash between meshing gear teeth. Critical speed noise may be suppressed when the engine is cold and the viscosity of the oil is high but, when the oil becomes hot, the reduced damping capacity of the lubricant permits the rebounding repetition of the bombardment of one component against another to become more violent and audible. Critical speed rumble can be identified by running the engine up to high idle speed and then cutting out the fuel supply so that the engine speed runs through its speed range. If the engine runs quietly over much of the speed range, but the noise occurs at certain defined speeds then there is a good possibility that the trouble is due to resonant torsional vibration.

3.7.11 Suppressing critical vibrations

For engines which have to operate over a relatively wide speed range it is almost certain that the crankshaft will be excited into torsional vibratory resonance at critical speeds, and if the rotary speed remains at or near one of the major criticals for any length of time, fatigue failure of the shaft could easily result.

Crankshaft rumble and, in the extreme, torsional failure can be prevented by mounting some form of vibration damper at the front end of the

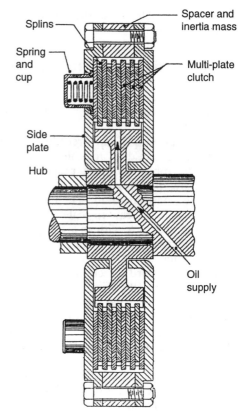

Fig. 3.74 Lanchester viscous friction vibration damper

crankshaft that is capable of absorbing and dissipating the majority of the vibratory energy.

These torsional vibration dampers can be divided into four basic kinds:

1 viscous friction damper
2 Coulomb or solid friction damper
3 viscous fluid damper
4 tuned rubber damper

Viscous friction 'Lanchester'-type vibration damper
(Fig. 3.74)
This form of damper comprises a multi-plate clutch attached between a hub and a ring mass spacer. There are two sets of friction disc plates: the drive plates which are internally splined to the hub and the driven plates externally splined to the inertia ring mass. The clutch pack is held together by side-plates which are bolted on either side to the ring mass spacer. The friction discs are lightly squeezed together by six evenly spaced coil springs. Engine oil supplied from the crankshaft

146

front main bearing submerges the friction discs in lubricant.

During crankshaft rotation the inertia disc mass and outer splined driven plates are dragged around with the drive plates and input hub. When the crankshaft accelerates the ring mass slips and holds back, whereas the ring mass maintains its forced momentum for a short time when the shaft decelerates so that slip again occurs. Consequently, when the shaft approaches a critical oscillation order, the vibratory energy generated will be dissipated by the relative slip between the many pairs of rubbing surfaces.

In theory, the wet multi-plate damper produces viscous friction. That is, the friction is independent of the spring pressure but is proportional to the area of the friction discs.

Solid friction vibration damper
(Fig. 3.75)
The inconsistency of the viscous friction multi-plate damper due to the variation in oil viscosity with the change in lubrication temperature calls for a modified friction damper to be used.

This consists of a pair of matched inertia discs separated by a number of equally spaced coil springs and mounted on a hub between two friction disc linings. The discs are dowelled so that both inertia discs are compelled to move together. The axial thrust of the coil springs presses each half-inertia disc against its respective friction-disc linings. The dry, or solid, friction generated is sufficient to clamp both halves of the inertia mass to the hub side walls, causing them to revolve with the mass all the time the crankshaft accelerates or decelerates. However, if the shaft goes through a critical speed the resultant torsional oscillation will be violent enough to cause both half-inertia discs to slip between the side walls of the hub. Thus, the excited torsional strain energy will be absorbed by the friction torque. This then is dissipated by convection and radiation to the atmosphere. This form of friction is not 'viscous' but approaches the state known as 'solid'—it is independent of the slipping velocity of the rubbing discs but is proportional to the pressure between them.

Viscous fluid vibration damper
(Figs 3.76 and 3.77)
This design of damper consists of an annular-shaped flywheel mounted on a bronze bearing lining and enclosed in a light steel casing which is rigidly attached to the crankshaft via the pulley wheel hub (Fig. 3.76). The flywheel annular chamber is completely sealed by a side cover fitted into a recess machined on the exposed side of the casing, which is then rolled over. The flywheel has a radial clearance of about 0.25 mm, and a lateral clearance of up to 0.5 mm in the chamber, and the existing space between the flywheel and casing walls is filled with a very viscous silicon fluid which changes little with temperature.

When there is very little or no vibration at the end of the crankshaft, the viscous torque drag causes the annular damper mass to rotate with the casing. If the damper casing is suddenly rotated the flywheel will move only so far, as it is dragged by the fluid, and if the direction of rotation is reversed the flywheel will overrun before it is itself reversed. Consequently, with the rapid and continual changing of direction during a torsional vibration, the fluid in the case is being sheared, and it is the resistance to this shearing action

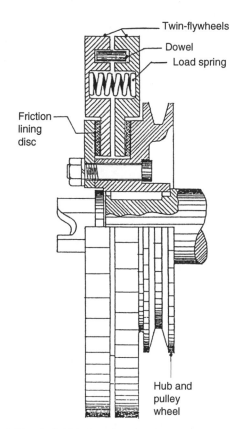

Fig. 3.75 Solid friction vibration damper

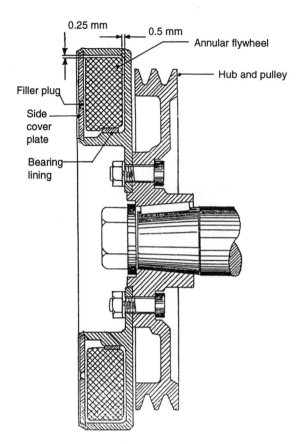

Fig. 3.76 Viscous fluid vibration damper

which forms the damping force. Thus, when the frequency and amplitude of the oscillations cause the vibration torque acting on the damper mass to exceed the viscous torque, relative slip occurs. It is this slippage between the casing and flywheel (that is, the shearing of the viscous fluid) which is responsible for the absorption of the vibratory energy which is simultaneously dissipated in the form of heat to the surroundings.

The main difference between the viscous fluid damper and the rubber damper is that the viscous damper is untuned but is very capable of reducing the amplitude and smoothing out major critical orders. Conversely, the rubber damper has a natural frequency of its own, the value of which is chosen to detune major criticals so that their vibrations occur at different speeds with much reduced amplitude (Fig. 3.77).

Tuned rubber torsional vibration damper
(Figs 3.78, 3.79, 3.80, 3.81(a–c) and 3.82)

The damper consists of a rubber band interposed between an outer inertia ring mass and a central hub, which is attached to the front end of the crankshaft. The rubber may be bonded to both members or it may be unbonded and held in position by compression. The auxiliaries can be driven directly off the pulley wheel hub (Figs 3.78, 3.79 and 3.80) or from the belt groove

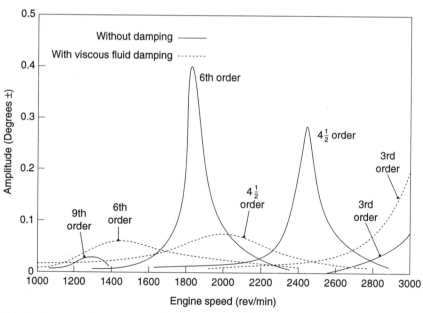

Fig. 3.77 Torsional vibration amplitude for six-cylinder diesel 6.8 litre engine with and without viscous fluid damping

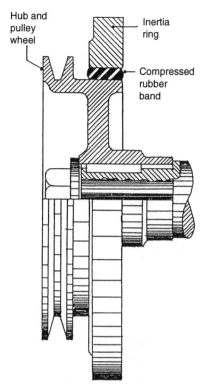

Fig. 3.78   Tuned rubber band vibration damper

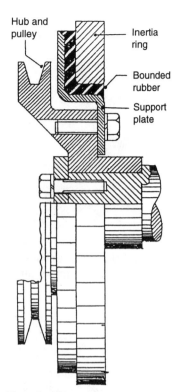

Fig. 3.79   Tuned rubber with right-angled rubber bonding vibration damper

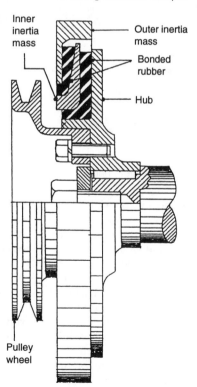

Fig. 3.80   Tuned two-piece inertia ring rubber vibration damper

149

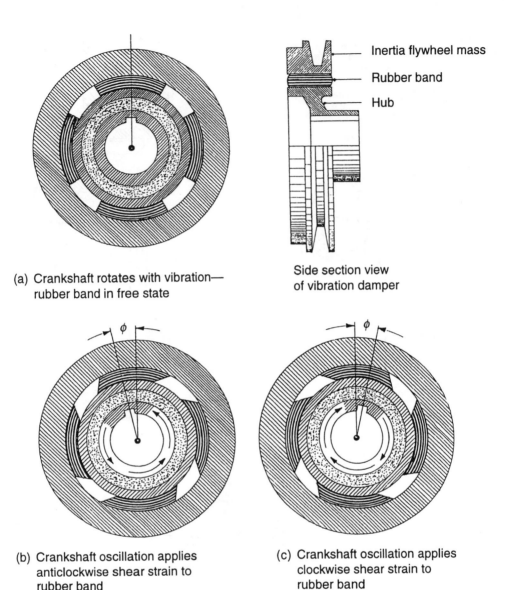

(a) Crankshaft rotates with vibration—
    rubber band in free state

Side section view
of vibration damper

Inertia flywheel mass

Rubber band

Hub

(b) Crankshaft oscillation applies
    anticlockwise shear strain to
    rubber band

(c) Crankshaft oscillation applies
    clockwise shear strain to
    rubber band

Fig. 3.81 Tuned rubber vibration damper principle

formed on the outside of the inertia ring (Fig.
3.81(a–c)).

When there is no vibration at the end of the
crankshaft the rubber strip sandwiched between
the inner and outer cast-iron vibration damper
members causes the inertia ring mass to rotate
freely with the pulley wheel hub (Fig. 3.81(a)).

If the front of the crankshaft is suddenly
accelerated clockwise (Fig. 3.81(c)) the inertia
ring mass will tend to lag behind, but is, however,
dragged around via the rubber ring which will
now be subjected to a shear-strain type of deformation, due to the reluctance of the inertia

mass to keep up with the rotary speed increase of
the crankshaft. In addition, if the direction of
oscillation is suddenly changed, the ring mass will
overrun before it is itself reversed (Fig. 3.81(b)).

Thus, with a continual reversal in the direction
of angular twist superimposed on the crankshaft's
normal rotational direction during a torsional
vibration, the ring mass, due to its inertia, will
repeatedly lag behind the crankshaft's oscillatory
movement. Consequently, the rubber will be distorted alternately in a clockwise and anticlockwise direction (Fig. 3.81(b and c)).

Hence, at a critical speed, the energy of excit-

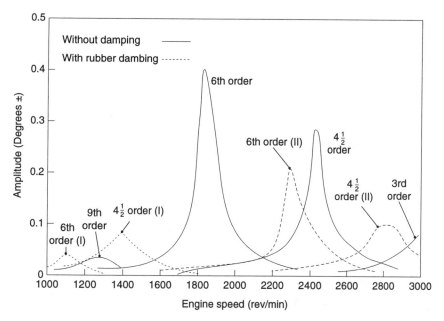

Fig. 3.82 Torsional vibration amplitude for six-cylinder diesel engine with and without tuned rubber damping

ing oscillations within the crankshaft is spent in deforming the rubber; that is, this unwanted energy is damped by transforming it to the rubber in the form of strain energy, elastic hysteresis losses and to heat rejection to the surroundings.

The rubber damper may be considered to be similar to attaching an additional mass to the crankshaft through a rubber band; the latter being equivalent to a flexible shaft of similar stiffness as the rubber. Thus, the dual stiffness of the crankshaft system produces a second frequency of vibration, which results in a frequency below and above the original frequency when no rubber vibration damper is attached.

It therefore follows that by choosing an appropriate damper mass, stiffness and damping properties, the crankshaft amplitude of vibration can be restricted to safe working limits.

Figure 3.82 shows how the rubber damper detunes a major critical vibration and simultaneously introduces a similar critical order of much reduced amplitude on either side of the undamped vibration.

Properties of rubber

Rubber stiffness decreases with high deformation while its damping ability remains constant or even decreases. Likewise, heat reduces both the stiffness and the damping ability. Rubber has the ability to store enormous amounts of energy and can release most of this energy on retraction. Thus, as a comparison, 150 times as much energy can be absorbed by rubber than for tempered spring steel of a similar mass. Rubber also has the highest internal friction (hysteresis) of all the engineering materials.

4

Combustion Chamber Design and Engine Performance

4.1 Introduction to combustion chamber design

Engine torque, power output and fuel consumption are profoundly influenced by the engine compression ratio, combustion chamber and piston crown shape, the number and size of the inlet and exhaust valves, and the position of the sparking plug.

The object of good combustion chamber design is (1) to optimize the filling and emptying of the cylinder with fresh unburnt and burnt charge respectively over the engine's operating speed range; and (2) to create the condition in the cylinder for the air and fuel to be thoroughly mixed and then excited into a highly turbulent state so that the burning of the charge will be completed in the shortest possible time.

To achieve these fundamental requirements the designer and development engineer has to be aware of the factors that contribute towards inducing the charge to enter the cylinder, to mix intimately, to burn both rapidly and smoothly and to expel the burnt gases.

An appreciation of the influencing parameters for good combustion chamber design will now be considered.

4.1.1 Induction swirl
(Fig. 4.1(a))

Swirl is the rotational flow of charge within the cylinder about its axis. Induction swirl is produced by positioning the induction port passage to one side of the cylinder axis so that the flow discharges into the cylinder tangentially (Fig. 4.1(a)).

Directed straight port
(Fig. 4.2(a))
The air or mixture charge is made to flow through a straight and parallel port passage towards the valve opening in the desired tangential direction relative to the cylinder axis. It then discharges into the cylinder tangentially towards the cylinder wall where it is deflected sideways and downwards in a spiral or whirling motion.

Deflector wall port
(Fig. 4.2(b))
The air or mixture charge is made to flow through a slightly curved and tapered port passage so that the inner side walls of the port provide a semi-circular flow path towards the valve opening. It then discharges into the cylinder with a predetermined downward spiralling motion about the cylinder axis. In the main, the charge tends to rely on the outer curved passage wall and only partially on the inner wall to obtain the directional effect, therefore the port areas are less restrictive.

Masked valve port
(Figs 4.2(c) and 4.3(a))
The air, or fuel and air mixture, is usually made to flow through a straight and parallel-port passage towards the valve opening where the valve mask forces it to discharge into the cylinder with a downward swirling motion about the cylinder

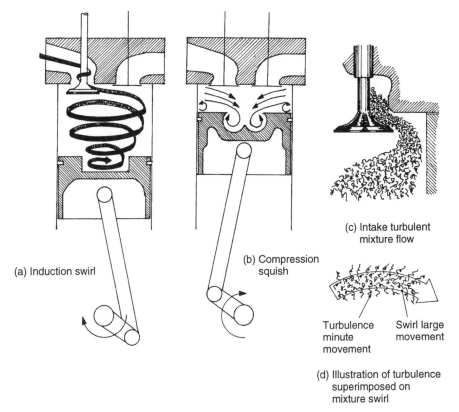

(a) Induction swirl

(b) Compression squish

(c) Intake turbulent mixture flow

Turbulence minute movement

Swirl large movement

(d) Illustration of turbulence superimposed on mixture swirl

Fig. 4.1　Terminology of air/mixture movement inside the cylinder

axis. The closing of a portion of the valve seat periphery restricts the flow discharge and therefore reduces the possible volumetric efficiency of the cylinder but with the benefits of a large amount of swirl.

Helical port
(Figs 4.2(d) and 4.36)

Swirl is generated within the port above the valve seat and about its axis before the air or mixture is discharged into the cylinder (Fig. 4.2(d)). The charge flow in the induction port is guided by the passage walls, which make it spiral around and downwards in the helix-shaped passage above the valve seat (Fig. 4.36). Therefore, when it ejects into the cylinder it will have acquired a rotational swirl motion which will continue as it is drawn into the cylinder during the piston's outward movement from TDC.

The intensity of swirl is influenced by the steepness of the port helix and the mean diameter of the spiral flow path about the valve axis.

Helical ports usually provide higher flow discharges for equivalent levels of swirl compared with directed ports because the whole periphery of the valve opening area can be fully utilized and, as a result, higher volumetric efficiencies can be obtained in the low-to-mid speed range of the engine.

Helical ports are less sensitive to their position relative to the cylinder axis since the swirl generated depends mainly on the port geometry above the valve and not how it enters the cylinder. Generally, the magnitude of swirl rises with increased valve lift.

However, helical high swirl induction ports (due to the pressure drop across the valve when creating the swirl) suffer from a loss of volumetric efficiency in the upper speed range of the order 5 to 10%.

Chamber wall deflected induction swirl
(Fig. 4.3(b))

The mixture charge entering the cylinder is made to flow between the steep semi-circular chamber wall close to one side of the inlet valve head and

153

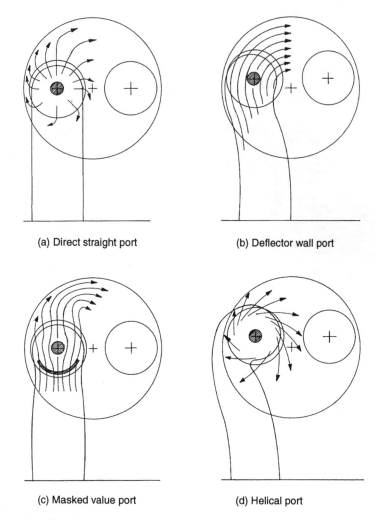

(a) Direct straight port

(b) Deflector wall port

(c) Masked value port

(d) Helical port

Fig. 4.2 Induction port classification

the slanting and widening roof portion of the chamber on the opposite sparking plug side, which curves with the plan of the cylinder in a heart or kidney shape. The nose of the steep chamber wall partially separates the inlet and exhaust valves so that the incoming charge is forced not only to move downwards but also to rotate around the wall of the cylinder. Thus, the resultant downward and circular movement of the mixture generates an expanding and then a contracting spiral swirl about the cylinder axis during both the induction and the compression strokes, respectively.

In contrast to induction, during the exhaust stroke, the burnt products of combustion are gently directed by the vertical circular chamber wall and the curved roof of the chamber so that

the least resistance is experienced during the expulsion of the exhaust gases.

4.1.2 Methods of intensifying the rate of burning

Squish
(Figs 4.1(b) and 4.4)
Squish is caused by the compression of the trapped charge formed between the parallel opposing portion of the cylinder head and piston crown (Fig. 4.1(b)) as the piston approaches TDC so that the air, or air and fuel mixture charge, is squeezed at a relative high velocity (Fig. 4.4) towards the much enlarged combustion chamber

154

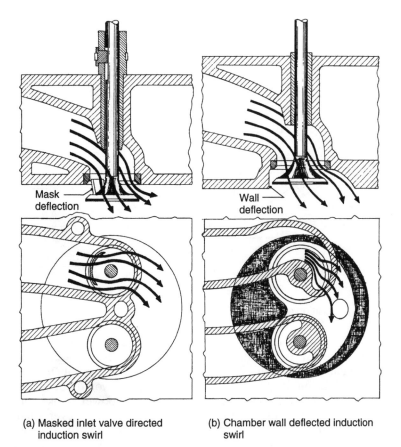

(a) Masked inlet valve directed
induction swirl

(b) Chamber wall deflected induction
swirl

Fig. 4.3 Directed induction swirl

space situated either in the piston or cylinder head.

The larger the parallel opposing portions of the cylinder head and piston crown, and the closer they are at TDC, the greater will be the flow velocity of the charge entering the main combustion chamber (Fig. 4.4).

The highly pressurized air or air–fuel mixture in the lamina region of the chamber, suddenly expanding into the larger space, considerably increases the charge turbulence, it therefore promotes fuel mixing and the rate of heat transference throughout the combustion chamber.

Quench area
(Fig. 4.4)
The quench area is defined by the parallel portions of the piston and cylinder head which almost touch each other as the piston approaches TDC. These opposing flat surfaces sandwiching a thin lamina of charge between them, have a large

surface area relative to the small volume trapped between them. Consequently, there will be a large amount of heat transferred from this thin lamina of hot charge through the metal walls. The result is a rapid cooling or quenching effect, by these parallel surfaces.

The quench area is defined as the percentage of opposing flat area relative to the piston crown area.

Turbulence
(Fig. 4.1(c and d))
Turbulence consists of randomly dispersed vortices of different size which become superimposed into the air, or air and petrol mixture, flow stream (Fig. 4.1(c)). These vortices, which are carried along with the flow stream, represent small irregular breakaways that take on a concentric spiral motion (Fig. 4.1(d)).

As the vortices whirl they will contact adjacent vortices causing viscous shear interaction. This

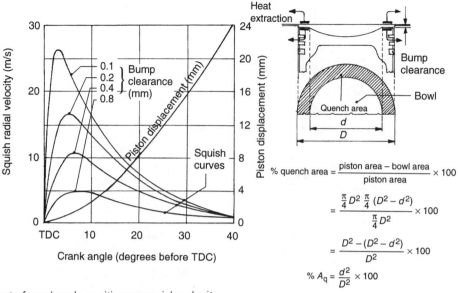

$$\% \text{ quench area} = \frac{\text{piston area} - \text{bowl area}}{\text{piston area}} \times 100$$

$$= \frac{\frac{\pi}{4} D^2 \; \frac{\pi}{4} (D^2 - d^2)}{\frac{\pi}{4} D^2} \times 100$$

$$= \frac{D^2 - (D^2 - d^2)}{D^2} \times 100$$

$$\% \, A_q = \frac{d^2}{D^2} \times 100$$

Fig. 4.4 Effect of crankangle position on squish velocity

rapidly speeds up the rate of heat transfer and fuel mixing. This is in contrast to laminar or streamline flow where heat is relatively slowly transferred by molecular diffusion, and fuel mixing has either been achieved before it enters the cylinder (as in the petrol engine) or it relies entirely on the injection of fuel spray atomizing and mixing as it penetrates the air stream (as in the diesel engine).

The amount of vortex activity, that is the formation of new vortices and the disintegration of others, increases the turbulent flow with rising engine speed.

Flame propagation
(Fig. 4.5(a, b and c))

If the induced combustible mixture moves within the cylinder with a laminar (streamline) steady motion, then when the spark produces ignition the fuel molecules which are then burning, raise—by conduction and radiation—the temperature of adjacent molecules until they also ignite. At the same time, the temperature of the molecule will rise, thereby speeding up the random movement of the gas molecules so that the collision rate amongst themselves and with adjacent boundary molecules of the unburnt gas mixture considerably increases. Accordingly, the resulting pressure increase causes an expansion of the burning molecules which greatly assists in propagating the ignition.

Unfortunately, the speed at which burning spreads to the unburnt charge by this method alone would be much too slow and therefore combustion would not be completed early enough during the power stroke for it to be an effective driving force. However, the induced air–fuel mixture contained in the cylinder will normally be in a state of turbulent swirl, its turbulent nature consisting of directed vortices and random pulsations on the velocity of the main gas stream.

1 The intensive pulsating vortices superimposed onto the flow stream distort the flame front and break it into separately burning focal points which enlarge the actual flame front surface area many times, thus increasing the rate of heat transference (Fig. 4.5(a, b and c)).
2 The very small turbulent pulsations, which are no thicker than those caused by a laminar streamline flow, do not produce any flame-front distortion, but their agitation does intensify the heat transfer and diffusion process between the boundaries of the burning flame front and the unburnt gas mixture.
3 The irregular and constantly changing flame-front caused by the intensive turbulence produces local currents which greatly speed up the heat transference process.

Typical flame propagation velocities range from something like 15 to 70 m/s. This would relate to the combustion flame velocity increasing from about 15 m/s at an idle speed of about

156

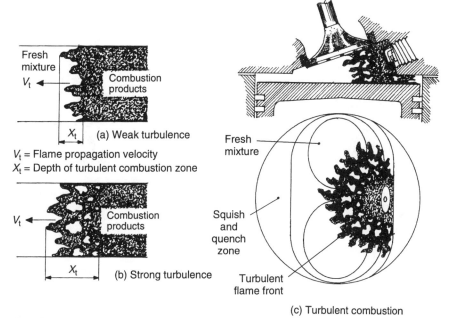

Fig. 4.5 Illustration of turbulent flame front

1000 rev/min to roughly 70 m/s at a maximum speed of 6000 rev/min.

The time interval for the completion of combustion is dependent on the intensity of turbulence, which is itself almost directly dependent on engine speed. The combustion time interval for a given intensity of turbulence will therefore correspond to one given engine speed.

Thus, if the combustion process occupies a crank angle movement of 30° at 1000 rev/min for one level of turbulence, then at 2000 rev/min with the same intensity of turbulence the crank angle interval will have doubled to 60°, an unacceptable angular movement. However, if the engine speed, and therefore turbulence, is doubled the combustion time interval will be halved, so that when the engine speed increases from 1000 rev/min to 2000 rev/min the combustion crank angle interval will remain approximately constant at its original value of 30°.

Thus, without turbulence the petrol engine would be unable to operate over the necessary speed range.

Flame propagation analogy through a turbulent flow-field
(Fig. 4.6(a, b and c))
An idealistic illustration of how the initial nucleus of a flame from a spark spreads throughout a turbulent mass can be visualized in terms of many circles representing the vortices or eddies in a turbulent flow field.

When ignition occurs the nucleus of the flame spreads with the whirling or rotating vortices in the form of a ragged burning crust from the initial sparking plug ignition site (Fig. 4.6(a)).

The speed of the flame propagation is roughly proportional to the velocity at the periphery of the vortices.

For one half-rotation of the vortices (Fig. 4.6(b)) the flow will have spread 180° about the pair of inner adjacent vortices on the right-hand side and, in turn, the contact with the second pair of inner vortices on the left-hand side will have transferred the flame to their peripheries through a 90° angular movement.

By completing the second half-revolution of the vortices (Fig. 4.6(c)) the flame will have spread all around the inner pair of vortices on the right, three-quarters of the way around the inner pair of vortices on the right, halfway around the inner-upper and inner-lower pairs of vortices and only a quarter of the way around the outer left and right-hand side vortices.

Simultaneously with the peripheral flame propagation, burning will spread in a laminar fashion and therefore at a much slower rate towards the centre of each vortex.

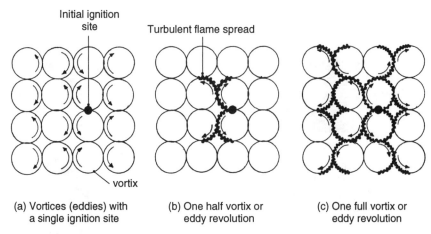

Initial ignition site

Turbulent flame spread

vortix

(a) Vortices (eddies) with a single ignition site

(b) One half vortix or eddy revolution

(c) One full vortix or eddy revolution

Fig. 4.6 Idealistic spread of turbulent combustion

In practice, the vortices are randomly arranged and will resemble a disordered network where the vortices are continuously changing their size, many will disappear while new ones will be created.

It should be visualized that the turbulent vortices are only part of the general swirl movement within the cylinder.

Swirl ratio
(Fig. 4.7)
Induction swirl can be generated by tangentially directing the air movement into the cylinder either by creating a preswirl in the induction port or by combining the tangential-directed flows with a preswirl helical port.

Cylinder air swirl is defined as the angular rotational speed about the cylinder axis. However, it is more convenient to relate this rotational speed to the crankshaft angular rotational speed, it is then known as the *swirl ratio*; that is

$$\text{Swirl ratio} = \frac{\text{air rotational speed}}{\text{crankshaft rotational speed}}$$

$$R_s = \frac{N_a}{N_c}$$

Helical ports can achieve swirl ratios of 3 to 5 at TDC with a flat piston crown. However, if a bowl in the piston chamber is used, the swirl ratio can be increased to about 15 at TDC.

The swirl ratio varies considerably throughout the engine's cycle of operation (Fig. 4.7). If the piston crown recess is very shallow, the peak in the swirl ratio as TDC is approached is small;

however, with a deep bowl chamber there is a large swirl ratio peak in the TDC region. This instant rise in the relative swirl speed just before ignition occurs, assists considerably in promoting turbulence and accelerating the initial combustion process.

4.1.3   Surface-to-volume ratio
(Fig. 4.8(a and b))

To minimize the heat losses and the formation of hydrocarbons within the combustion chamber, the chamber volume should be maximized relative to its surface area, that is, the chamber's surface area should be as small as possible relative to the volume occupied by the combustion chamber (Fig. 4.8(a)).

A very useful comparison to compare likely heat losses and the amount of hydrocarbons produced in the exhaust gases with different chamber shapes is the ratio of the combustion surface area to that of its volume; that is

$$\text{surface-to-volume ratio} =$$

$$\frac{\text{surface area of chamber}}{\text{volume of chamber}} = \frac{S}{V}$$

The primary source of hydrocarbons formed in the exhaust gases is caused by the outer layers of the mixture becoming overcooled in the region of the chamber walls. Thus, the flame cools as it approaches the chamber walls until eventually it stops burning, this therefore leaves a film of unburnt hydrocarbons on the walls which mixes with the burned charge as it is expelled through

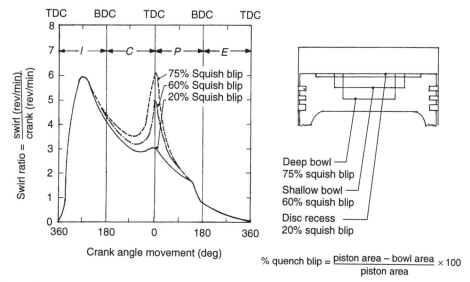

Fig. 4.7 Effect of crankangle position on swirl ratio

$$\% \text{ quench blip} = \frac{\text{piston area} - \text{bowl area}}{\text{piston area}} \times 100$$

the exhaust system. Consequently, the larger the surface-to-volume ratio, the higher will be the exhaust gas hydrocarbon concentrates (Fig. 4.8(b)).

Chamber design shape
(Fig. 4.9(a–d))

The combustion chamber with the least surface area containing a given volume is the spherical configuration. Within the limit of bore size cylinder capacity and compression ratio, the double hemispherical chamber has the lowest surface-to-volume ratio, whereas the highest is the bowl in the piston chamber (Fig. 4.9(a–d)). Generally, as the quench (squish) area is enlarged, the surface

area of the chamber increases relative to the chamber volume. Thus, chambers with large squish zones have large surface-to-volume ratios and therefore promote more hydrocarbons than do chambers with very small squish zones.

A typical comparison for a 100 mm cylinder diameter and 9:1 compression ratio engine with four different combustion chambers is shown in Fig. 4.9(a–d). It can be seen that the double hemispherical chamber has the lowest surface-to-volume ratio, whereas the bowl in the piston gives the highest value, due to the larger squish zones. In practice, most combustion chambers will include features of each basic type of chamber and will, most likely, have surface-to-volume ratios similar to the pancake chamber.

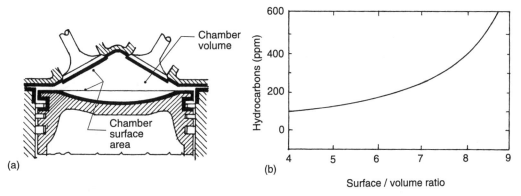

Fig. 4.8 (a) Illustration of combustion chamber surface area to volume ratio (b) Effect of surface/volume ratio on exhaust gas hydrocarbon emission

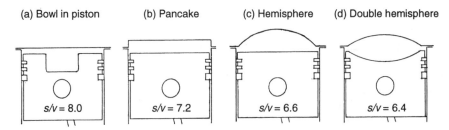

Fig. 4.9 Effect of combustion chamber shape on surface-to-volume ratio for a 100 mm diameter cylinder and a 9 : 1 compression ratio

Compression ratio
(Fig. 4.10(a))
Raising the compression ratio for a given cylinder capacity reduces the combustion chamber volume for a given cylinder diameter and piston stroke.

Therefore, the chamber surface area becomes relatively larger as the clearance volume (combustion chamber space) decreases, accordingly the chamber surface area to the chamber volume ratio increases linearly with rising compression ratio (Fig. 4.10(a)).

Cylinder size versus number of cylinders
(Figs 4.10(b and d) and 4.11)
For a given cylinder capacity, the fewer the number of cylinders and the larger their size, the lower will be the surface-to-volume ratio for the engine as a whole (Fig. 4.10(b and d)).

The larger the cylinder capacity, the greater will be the chamber volume relative to the chamber's surface area (Fig. 4.10(d)). This is because the cylinder's volume increases with the cube of its diameter whereas the surface area increases as the square of its diameter. Thus, increasing the cylinder capacity rapidly decreases the surface-to-volume ratio of the chamber. Increasing the number of cylinders decreases the size of each cylinder

and, therefore, increases the surface-to-volume ratio (Fig. 4.10(b)).

Relative variations in cylinder surface area can be made between a single-cylinder and a four-cylinder engine with both being square; that is, their stroke-to-bore ratio (s/b) is unity. The cylinder diameter for the four-cylinder engines is 86 mm and that of the single large-cylinder engine 136.6 mm.

Cylinder surface area

$$(S) = \pi DL + \frac{2\pi D^2}{4} = \pi D^2 + \frac{2\pi D^2}{4} \text{ since } L = D$$

therefore

$$S = \pi D^2 \left(1 + \tfrac{1}{2}\right) = \tfrac{3}{2}\pi D^2$$

For four cylinders

$$S_4 = 4 \times \tfrac{3}{2}\pi 86^2 = 139411.32 \text{ mm}^2$$

For a single cylinder

$$S_1 = \tfrac{3}{2}\pi 136.6^2 = 87931.1 \text{ mm}^2$$

ratio of surface areas

$$\frac{S_4}{S_1} = \frac{139411}{87931.1} = 1.58545 \simeq 1.5 : 1$$

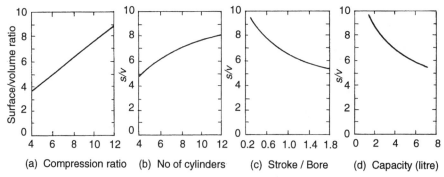

Fig. 4.10 Relationship between compression ratio, number of cylinders, stroke/bore ratio and cylinder capacity to surface-area-to-volume ratio

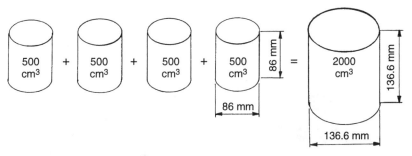

Fig. 4.11 A visual comparison of cylinder surface-area-to-volume for a single-cylinder and four-cylinder engine of similar capacity

Thus, there is 1.5 times the surface area for a four-cylinder engine as opposed to a single-cylinder engine. A visual observation (Fig. 4.11) immediately shows that there is much more surface area for the four-cylinder engine compared with the single-cylinder engine of the same cubic capacity. From these observations it can be concluded that, as the number of cylinders increases, the overall heat losses and the swept friction losses become greater, consequently this will lower the efficiency of the engine.

Stroke-to-bore ratio
(Figs 4.10(c) and 4.12(a–c))
When the stroke length and the bore diameter are equal, the engine's stroke-to-bore ratio is said to be square (Fig. 4.12(b)). However, if the stroke is smaller than the cylinder diameter the stroke-to-

bore ratio is said to be oversquare (Fig. 4.12(a)), whereas if the stroke is larger than the cylinder diameter the engine is said to be undersquare (Fig. 4.12(c)). The stroke-to-bore ratio for various engines can range from 0.6:1 to 1.4:1. Oversquare engines are more suitable for saloon car petrol engines, whereas undersquare engines are better utilized for large diesel engines.

A long stroke smaller diameter bore cylinder (Fig. 4.12(c)) more closely approaches the minimum surface area of a sphere than a short stroke large diameter cylinder of the same cylinder displacement. Thus, a long stroke-to-bore (s/b) ratio produces a smaller surface-to-volume ratio than a short stroke-to-bore ratio engine (Fig. 4.10(c)). Consequently, a long-stroke engine is preferred to minimize heat losses and the formation of hydrocarbon as opposed to short-stroke large-diameter cylinders. However, this advantage

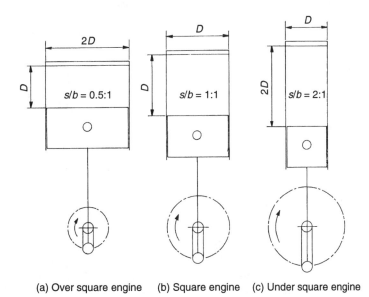

(a) Over square engine   (b) Square engine   (c) Under square engine

Fig. 4.12   Comparison of different stroke-to-bore ratio engine configurations

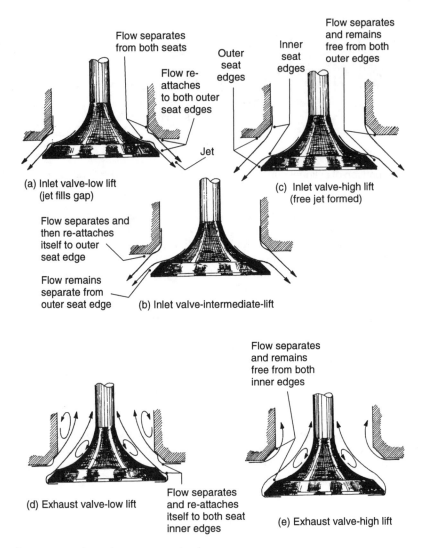

Fig. 4.13 Charge flow pattern through open poppet valve

Labels in figure:

Flow separates from both seats
Flow re-attaches to both outer seat edges
Outer seat edges
Inner seat edges
Flow separates and remains free from both outer edges
Jet
(a) Inlet valve-low lift (jet fills gap)
(c) Inlet valve-high lift (free jet formed)
Flow separates and then re-attaches itself to outer seat edge
Flow remains separate from outer seat edge
(b) Inlet valve-intermediate-lift
Flow separates and remains free from both inner edges
(d) Exhaust valve-low lift
Flow separates and re-attaches itself to both seat inner edges
(e) Exhaust valve-high lift

diminishes with very long stroke engines. A further consideration is that peak cylinder pressure tends to decrease as the stroke-to-bore ratio becomes more undersquare.

4.1.4 Poppet valve flow characteristics
(Figs 4.13(a–c) and 4.14)

The efficiency of the flow discharge from the open valve is a measure of how near the effective flow area is to the geometrical flow area. This efficiency is normally quantified by a flow discharge coefficient which is the ratio of the effective flow area to the geometric flow area, i.e.

$$\text{discharge coefficient} = \frac{\text{effective flow area}}{\text{geometric flow area}}$$

$$C_D = \frac{Ae}{\pi D_L}$$

where
Ae = effective flow areas
D = valve head diameter
L = valve lift
C_D = discharge coefficient

The inertia of an air–fuel mixture moving through an open poppet valve and its port seat prevents the charge mixture exactly following the contour of the valve port seat and the valve head seat. In fact, there will be a tendency to overshoot as it moves over abrupt surface changes; thus, the flow around such sharp edges as the open valve seats, causes the charge mixture to break away from the inner edges of both the valve-head seat and the cylinder-head port seat. The exit between the mating seats then results in the formation of a free jet. The air flow and discharge coefficient (C_D) characteristics for a typical valve and port plotted on a base of valve-lift to valve-head diameter (L/D) ratio is shown in Fig. 4.14(a). The apparatus used to measure the air flow is also shown in Fig. 4.14(b).

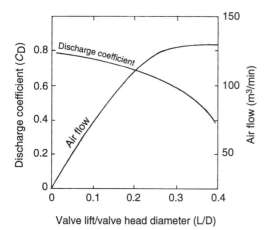

Fig. 4.14(a) Effect of valve lift on the valve discharge coefficient and air flow rate

Inlet valve flow pattern
(Figs 4.13(a–c) and 4.14(a and b))
The effectiveness of the open inlet valve flow discharge capability can be examined for three different valve lifts—small, medium and high—as follows.

1 At the lowest valve lift the flow jet separates at the inner seat corners and then re-attaches itself on both adjacent seats, thus filling the gap as it discharges into the cylinder (Fig. 4.13(a)).
2 At the intermediate range of valve lift, the flow jet separates at the inner seat corners and then re-attaches itself on the port seat side but remains free on the valve head side as it emerges between the seat gap (Fig. 4.13(b)).
3 At high valve lifts, the flow jet separates at the inner seat corners and remains free from both seats as it is expelled in a conical jet (Fig. 4.13(c)).

As can be seen, the utilization of the gap formed by the open valve seats is at its optimum somewhere between a small and medium valve lift. In general, the discharge coefficient decreases with increasing valve lift.

The flow of mixture entering the cylinder through the valve merges with the surroundings and produces zones of recirculation and pressure drop which, in response, expand the jet (Fig. 4.14(b)).

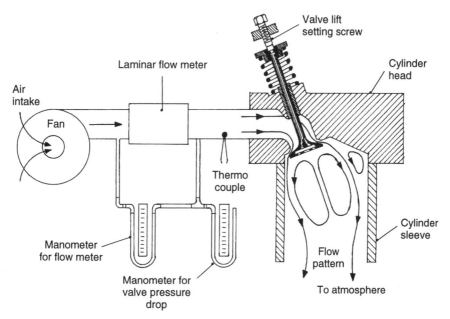

Fig. 4.14(b) Apparatus for measuring the steady air flow through open poppet valve

Exhaust valve flow pattern
(Fig. 4.13(d and e))

The exhaust gas flow from the cylinder into the exhaust port via the open poppet valve is simply the reverse of the inlet valve flow.

At low exhaust valve lift (Fig. 4.13(d)) the flow slightly separates at the outer seat corners and then re-attaches itself on both adjacent seats so that it fills the gap before it discharges into the relatively large port passage as a convergent conical jet. Consequently, the sudden expansion of the exhaust gas dissipates the kinetic energy of the gas with only the minimum being converted to prssure energy downstream of the valve port seat.

At high exhaust valve lift (Fig. 4.13(e)) the flow separates from the outer seat corners and remains free from both sides so that it establishes a jet with the centre of the valve stem as its axis. The jet is of sufficient size relative to the port cross-sectional-area to convert some of the gas's kinetic energy into pressure energy; that is, a pressure rise downstream of the valve port seat.

At valve lifts below $0.2D$, the effective area is something like 95% of the minimum geometric area of the open valve gap, whereas for valve lifts of $0.25D$ and above, the effective area is about 90% of the valve opening gap. In other words the high lift free jet is inferior to the low convergent jet.

In general, the 30° valve seat angle produces a larger effective flow area compared with the 45° valve seat angle for valve lifts up to about $0.25D$. However, for higher valve lifts, the 45° valve seat angle tends to establish a larger effective flow area.

4.1.5 Comparison between spark ignition and compression ignition engines
(Fig. 4.15)

Engine efficiency is greatly influenced by the cylinder pressure before combustion. Thus, if the compression ratio is raised the thermal efficiency will increase and vice versa.

With a petrol engine, the cylinder pressure will be at its maximum at full throttle when there is no restriction on the cylinder compression pressure just before combustion. However, as the throttle valve is steadily closed there is a corresponding increase in the cylinder and manifold depression during the induction stroke.

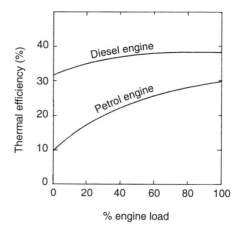

Fig. 4.15 Comparison of petrol and diesel engine thermal efficiency with varying load

This means that during the early part of the compression stroke the cylinder pressure will still be below the atmospheric pressure and so, effectively, compression will not commence until the cylinder pressure has changed from a negative depression to a positive pressure that is greater than the atmospheric pressure. In effect, positive compression may not start until the piston has travelled almost halfway along its return stroke so that the effective compression ratio and the final cylinder compression pressure before combustion commences may be halved under practically closed idling conditions. Thus, the engine's effective compression ratio will decrease as the butterfly throttle-valve is moved from the fully-open to the almost closed position, which correspondingly causes a reduction in the engine's thermal efficiency. At the same time, the increasing intake depression causes the engine to use more of its power to pull in the air–fuel mixture.

A diesel engine controls its power output by the quantity of metered fuel per injection and therefore does not need to throttle the air intake. Consequently, each cylinder receives the same amount of air per induction stroke so that the compression pressures remain roughly the same throughout the engine's speed range and, in contrast to the petrol engine, there is no power loss due to throttle restrictions as the load and speed are reduced.

Because of these basic differences in controlling engine output in both petrol and diesel engines, the diesel engine's thermal efficiency remains approximately constant for all load and speed changes, whereas the petrol engine, with its 10% thermal efficiency disadvantage at full load, deteriorates to something like a third of the diesel engine's thermal efficiency as the operating conditions move over to no-load (Fig. 4.15).

It should be appreciated that most of the engine's operating time cycle ranges between a quarter to three-quarter load and that only rarely is full load response required.

Both petrol and diesel engines are internal combustion power units which burn the air–fuel mixture in the cylinder between the cylinder-head and piston.

The basic difference between these two types of engines exists in the method of preparing the mixture for combustion and the method of igniting the combustible mixture.

With the petrol engine, a mixture of air and fuel is drawn into the cylinder during the outward moving induction stroke. On its return inward moving stroke, the mixture is compressed to something like one-ninth to one-eleventh of the unswept volume, and just before the piston reaches the end of its compression stroke a spark is used to ignite the combustible mixture.

In contrast, a diesel engine draws only fresh air into the cylinder during the induction stroke and then on its return stroke compresses this charge into about $\frac{1}{15}$ to $\frac{1}{22}$ of the unswept volume until its temperature is raised well in excess of 550°C. Just before the piston reaches the end of the compression stroke an accurately metered quantity of fuel is injected into the cylinder at pressures of 350 bar or more. The finely atomized fuel spray mixes with the hot air causing it to ignite (due to the heat) and burn rapidly.

The time for injecting the fuel over a 40° crank angle movement at 5000 rev/min is

$$\frac{60}{5000} \times \frac{40}{360} = \frac{1}{750} = 0.00133 \text{ seconds}$$

the actual effective combustion crank-angle movement may only be half of the total injection duration, that is 20°.

This is a very short interval of time and, in order that as much fuel as possible is burnt, the fuel spray must be exposed to the largest quantity of air as possible. To minimize the mixing of the fuel particles, the air is given movement so that it sweeps across the finely atomized and penetrating spray.

However, at high engine speeds, with such short injection and mixing times, the chemically correct amount of air charge will not be able to seek out, intimately contact, mix and burn all the fuel particles; therefore, very high unacceptable levels of soot and black smoke would be expelled with the exhaust gases.

Consequently, diesel engines have to operate with something like 20% excess air at full-load to come anywhere near completing the combustion process during each cycle.

This means that for the same power output, diesel engines have to have a cylinder capacity at least 20% greater than an equivalent petrol engine; for example, a 2-litre diesel engine can only be expected to develop approximately the same power as a 1.6-litre petrol engine operating under similar conditions.

Another reason why a diesel engine cannot develop the same amount of power as a petrol engine with the same cubic capacity is that, with the petrol engine, the air and fuel are already mixed before they enter the cylinder, whereas the mixing process in the diesel engine takes place only just before combustion commences. Thus, the faster the engine runs the less time there is to mix the air and fuel; in the case of the petrol engine, the mixture can be adequately prepared and burnt up to speeds of 6000 rev/min and over. However, with the diesel engine it is difficult to complete the air–fuel mixing at speeds above 3000 rev/min without using high swirl induction ports which then raise the maximum engine speed to about 4000 rev/min. For even higher engine speeds, up to around 5000 rev/min, indirect injection divided combustion chambers become essential.

4.1.6 Fundamentals of the combustion process in the spark ignition petrol engine
(Figs 4.16(a and b), 4.17(a and b), 4.18, 4.19 and 4.20)

The atomized and partially vaporized air–fuel mixture is induced into the cylinder where it mixes with the remaining hot residual exhaust gas, it also contacts the hot combustion-chamber walls, cylinder walls and piston crown. Heat is transferred from the metal surfaces to the entrapped homogeneous mixture which is uniformly distributed throughout the cylinder, with the result that the unvaporized portion of the fuel mixture entering the cylinder is vaporized before the end of the compression stroke.

The ignition temperature at which the

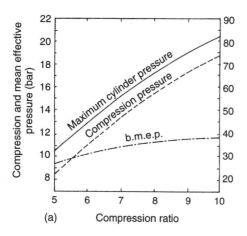

Fig. 4.16(a) Effect of compression ratio on the b.m.e.p. compression and maximum cylinder pressures

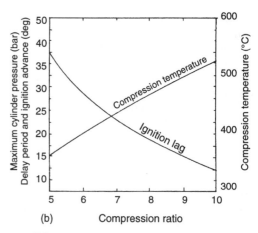

Fig. 4.16(b) Effect of compression ratio on the air temperature and ignition lag

homogeneous mixture commences to burn is known as the ignition temperature. This threshold temperature is greatly influenced by the composition of the air–fuel mixture, its dilution with the residual burned gases, the size, shape and existing temperature of the combustion chamber and, possibly the most important factor, the compression ratio which determines the compression pressure (Fig. 4.16(a)) at TDC and therefore the mean temperature of the charge (Fig. 4.16(b)). However, the ignition temperature does not rely on the cylinder compression pressure alone but more so on the heat transfer in the cylinder, which itself increases with the density of the air–fuel mixture. The effect of increasing

the compression ratio on the rise in combustion pressure is shown on a base of swept volume (Fig. 4.17(a)) and on a base of crank-angle movement (Fig. 4.17(b)). The merits and limitations of raising the compression ratio with regards to thermal efficiency and mechanical efficiency are shown in Fig. 4.18, whereas Fig. 4.19 shows the benefits of increased power and reduced specific fuel consumption with rising compression ratio.

Spark ignition petrol engines are quantity regulated—the amount of homogeneous charge of thoroughly mixed petrol and air taken into the cylinder per stroke is varied by a throttle valve to regulate the engine's speed and load conditions. This contrasts with diesel compression ignition engines, which are quality regulated—for these

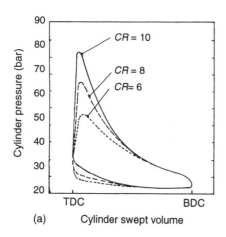

Fig. 4.17(a) Effect of compression ratio on the characteristic pressure–volume diagram for a petrol engine

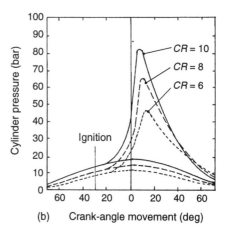

Fig. 4.17(b) Effect of compression ratio on the characteristic pressure–crank-angle diagram for a petrol engine

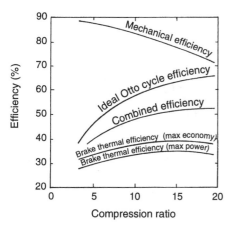

Fig. 4.18 Effect of compression ratio on an engine's thermal and mechanical efficiency

engines, the amount of air taken in per stroke is constant but the amount of fuel injected is varied to match the engine speed and load conditions. Therefore, a heterogeneous fuel–air mixture exists consisting of a fuel-rich droplet core immediately surrounded by air alone. Thus, with the diesel combustion process, for the completion of combustion, the stratified fuel droplets only have to be burnt, and this can be achieved locally at part-load, or spread out under full-load so that more air can be burnt.

Because of the nature of the homogeneous mixture in the cylinder for a spark ignition engine, the air–fuel mixture strength on the weak and rich sides of the stoichiometric ratio (14.7:1 by weight) cannot vary by more than about 20%, this will provide an air–fuel ratio operating range extending from something like 18:1 on the weak

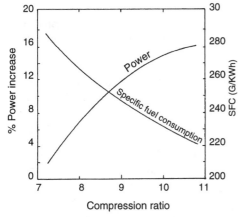

Fig. 4.19 Effect of compression on engine power and specific fuel consumption

side to very rich mixtures of 12:1. The reduced rates of burning with these extreme mixture strengths (Fig. 4.20) are only made possible for engine speeds up to 6000 rev/min by raising the engine compression ratios to something like 9:1 or, better still 10:1.

4.1.7 Compression ratio
(Figs 4.16(a and b), 4.17(a and b), 4.18 and 4.19)

Raising the engine compression ratio increases the cylinder's compression pressure from roughly 8.0 bar to 19.0 bar over a compression ratio increase from 5:1 to 10:1 respectively (Fig. 4.16(a)). Correspondingly, the maximum cylinder pressure increases from 32 bar to 82 bar and the brake mean effective pressure generated also increases from about 9.4 bar to 11.8 bar over the same compression ratio range, respectively.

The effect of higher cylinder pressure is to cause a corresponding rise in cylinder temperature from 360°C to 520°C over the same compression ratio rise. As would be expected, raising the cylinder temperature reduces the ignition delay period for one set engine speed (Fig. 4.16(b)). Thus, for an engine running in its mid-speed range, the ignition timing would be reduced from 37.5° to 12.5° before TDC if its compression ratio is increased from 5:1 to 10:1.

The steepness of the pressure rise and the peak values reached for three different compression ratios in the early part of the power-stroke can best be compared in both the pressure/volume (P–V) diagram (Fig. 4.17(a)) and the pressure/crank angle (P–θ) diagram (Fig. 4.17(b)).

The main reason for raising the engine compression ratio is due to the increased density of the air–fuel mixture at the point of ignition, so that when this energy is released it is better utilized. It therefore raises both the engine thermal efficiency and the developed power.

One of the major unwanted side effects of raising the compression ratio is that there will be a corresponding increase in cylinder pressure which, in turn, increases the piston-ring to cylinder-wall friction and the compression and expansion heat losses. Consequently, the higher compression ratio produces a reduction in the mechanical efficiency.

Subsequently, increasing the compression ratio produces an increase in thermal efficiency but at the expense of a falling mechanical efficiency (Fig. 4.18). Thus, if the efficiencies are compared

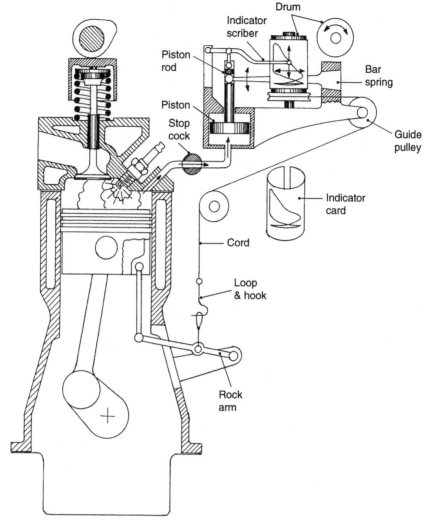

Fig. 4.20 Mechanical pressure–volume indicator (Maihak type)

on an equivalent efficiency scale, the reducing mechanical efficiency cancels out more and more of the ideal Otto cycle thermal efficiency. Hence, the resultant (combined) efficiency curve shows very little efficiency rise advantage when the compression ratio goes beyond 12:1. In theory, the maximum useful brake thermal efficiency occurs with a 16:1 compression ratio, but this would necessitate an abnormally high anti-knock octane fuel which is not readily available and would be much too expensive to refine.

The two lower themal efficiency curves (Fig. 4.18) compare the brake thermal efficiency for both maximum fuel economy and maximum generated power against a changing compression ratio.

Another way to illustrate the improvement in engine power and fuel consumption with rising compression ratio is shown in Fig. 4.19. Here, the power curve is expressed as a percentage of the maximum power developed with a 7:1 compression ratio, over the compression ratio range from 7:1 to 11:1, whereas the fuel consumption is plotted on a specific fuel consumption basis over the same compression ratio range.

One further effect of raising the engine compression ratio is that it reduces the cylinder clearance volume. This results in a lowering of the cylinder volumetric efficiency since less fresh charge will be drawn into the cylinder per stroke and, at the same time, there will be a lowering of the cylinder's residual exhaust gases.

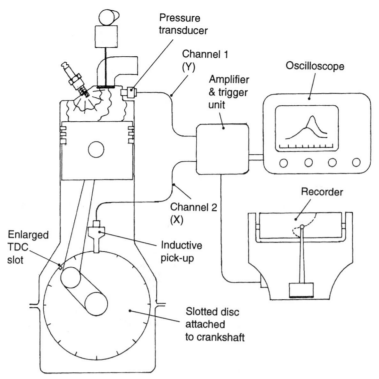

Fig. 4.21   Electronic engine indicator equipment

4.2.   Spark ignition combustion process

The characteristics of the cylinder pressure rise during the power stroke can be examined either by the cylinder pressure on a piston stroke base indicator diagram (Fig. 4.20) or by the cylinder pressure on a crank-angle-movement base indicator diagram (Fig. 4.21).

The combustion process may be considered to take place in three phases or periods:

1 the delay period;
2 the rapid pressure rise period;
3 the after burning period.

4.2.1.   Delay period (ignition and early flame development) (θ_1)

(Figs 4.22, 4.23 and 4.24)

This first phase covers the period from when a timed high-tension spark passes between the spark-plug electrode (which then ignites the air–fuel vapour surrounding the electrodes) to the time the established flame begins to release the heat energy of the burning fuel vapour fraction.

The actual end of the first period is considered as the point where the initial sign of gas expansion pressure rises above the normal compressive pressure for a given crank-angle displacement, shown on the pressure/crank-angle diagram (Fig. 4.22). This period tends to be very nearly constant in time.

The duration of this period is dependent upon the following:

1 the temperature of the flame thread passing between the electrodes of the sparking plugs (Fig. 4.23);
2 the nature of the fuel;
3 the temperature and pressure of the prepared charge;
4 the thoroughness in the mixing of the air–fuel charge;
5 the strength of the charge mixture (Fig. 4.24).

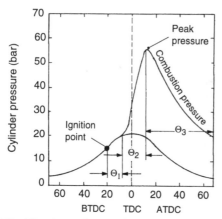

Fig. 4.22 The three phases of combustion in a spark ignition engine

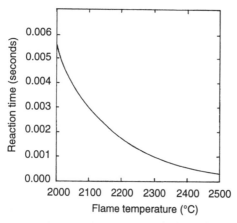

Fig. 4.23 Approximate relationship between flame temperature and rate of burning

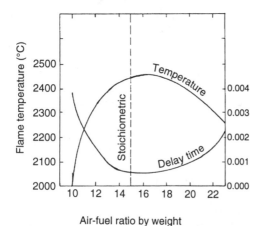

Air-fuel ratio by weight

Fig. 4.24 Approximate relationship between flame temperature delay time and mixture strength

4.2.2 Rapid pressure rise period (flame propagation) (θ_2)
(Fig. 4.22)

The second phase occupies the time between the initiation of the flame front and the beginning of the pressure rise (above the normal compression pressure) to the point in time when the rugged flame front has spread to the cylinder walls and the cylinder pressure has reached its peak valve.

Once the energy liberated from the developing flame is sufficient, it will cause the cylinder pressure to increase at a much greater rate than the normal compressive pressure as the piston approaches TDC.

The time required for the rapid pressure-rise phase depends mainly on the intensity of the turbulence or state of agitation of the mixture.

The slowest rate of burning occurs when the mixture is stagnant, whereas the rate of burning becomes faster as the turbulence in the combustion chamber increases and, since turbulence rises almost linearly with engine speed, the duration of this second period is roughly constant in terms of the crank-angle movement.

4.2.3 After burning period (termination of combustion) (θ_3)
(Fig. 4.22)

After the flame front has reached the cylinder walls there will be something like 25% of the charge which has still not completely burnt.

At this stage it becomes more difficult for the remaining oxygen in the charge to react with the petrol vapour so that the burning rate slows down, this condition is known as *after burning*.

At the same time, there will be heat liberated due to a chemical interaction caused by re-association of the combustion products throughout the expansion stroke.

Also, during this last phase, a greater proportion of the heat energy released is lost through the cylinder walls, head and piston crown and, simultaneously, the descending piston increases the clearance volume. Consequently, the cylinder pressure commences to decrease rapidly.

4.2.4 Ignition timing and engine speed
(Figs 4.25 and 4.26)

When the engine speed rises there is less time for

the cylinder to dissipate the heat of combustion per stroke and, accordingly, the time interval between the spark firing and the point of ignition when burning commences—that is, the delay period—reduces. However, in terms of crank-angle movement with rising speed the delay period increases as the square root of its speed according to $\theta_1 \propto \sqrt{N}$ where θ_1 is the delay period and N the engine speed.

With rising engine speed the intensity of turbulence and hence the rate of burning increases approximately in proportion to the engine speed. Thus, the time interval of burning from the point of ignition to peak pressure (rapid pressure rise period) reduces, while its duration, expressed in degrees of crank-angle movement, remains roughly constant.

Consequently, if the ignition timing is fixed to produce its peak pressure at about 10° after TDC at low speed then, with increasing speed, these peaks will occur progressively later in the cycle and their magnitude will also decrease (Fig. 4.25).

In contrast, if the ignition timing is advanced the same amount as the delay period increases in terms of crank-angle movement with rising engine speed, the positions at which these peak combustion pressures occur remain roughly the same, with only a very small drop in peak value due to reducing volumetric efficiency (Fig. 4.26). Thus, by automatically advancing or retarding the actual firing point with increasing or decreasing speed the optimum engine torque will be maintained.

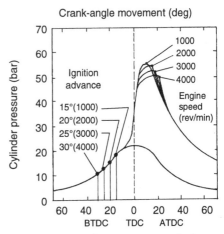

Fig. 4.26 Effect of engine speed on cylinder pressure with variable optimum ignition timing

4.3 Detonation

(Figs 4.5(c), 4.27 and 4.28(a and b))

Detonation is the creation within the cylinder of a pressure-wave travelling at such a high velocity that, through its impact against the combustion chamber walls it excites them into vibrating.

The shockwaves in the gas, which are repeatedly reflected from the walls of the combustion chamber, cause a ringing metallic knocking or high pitch 'ping' sound to be produced (Fig. 4.27). Consequently, the term 'pinking' is commonly used to describe the resulting noise caused by these abnormally high pressure-waves travelling through the previously burnt products

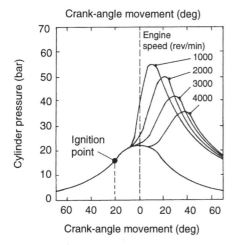

Fig. 4.25 Effect of engine speed on cylinder pressure with fixed ignition timing

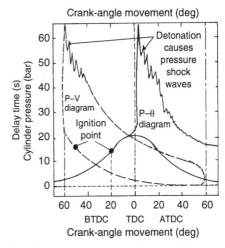

Fig. 4.27 Cylinder pressure variation when detonation occurs

171

of combustion. The waves then bombard the relatively rigid surrounding metal surfaces consisting of the cylinder-head, cylinder walls and piston crown.

To produce this knock the pressure-wave oscillation has to be within the audible fundamental frequency, and the actual tone of the noise produced will be largely influenced by the velocity of the shockwaves and the distance they travel between consecutive reflections from the enclosed chamber—that is, the size of the cylinder.

When detonation is slight it may not occur on every power stroke so that the knock may be light and intermittent whereas heavy detonation will create a continuous loud knocking noise.

Generally, the detonation pressure oscillations when superimposed on the normal pattern of combustion pressure rise and fall will show a slightly higher peak cyclic cylinder pressure.

The average vibrating frequency of detonation will be in the region of 500 cycles/second whereas the shockwaves propagating through the burnt and unburnt charge will reach velocities in the 1000 to 1200 m/s range.

During the combustion process, the atomized homogeneous fuel–air mixture is ignited by the passage of the spark between the spark-plug electrodes. At first a small nucleus of flame is created which then rapidly spreads outwards.

If there is no movement of the fresh mixture in the cylinder the nucleus of flame initiated from the spark would form an unbroken burning front travelling progressively outwards to the furthest point in the combustion chamber at a very low velocity.

However, the flame front is normally subjected to turbulence which distorts the smooth flame front into a ragged and broken one (Fig. 4.5(c)) so that the frontal area of the flame exposed to the unburnt mixture is considerably enlarged. Consequently, the rate of heat transfer from the flame to the unburnt charge by conduction, convection and radiation is greatly increased, it therefore enables and encourages the flame to travel towards the outer walls of the chamber at a very high velocity.

As the flame front moves forwards (Fig. 4.28(a and b)) it compresses ahead of it the still unburnt mixture whose temperature is raised both by compression and by radiation from the advancing flame until a point is reached when the remaining unburnt charge is subjected to such a high intensity of heat that it spontaneously ignites. The

resulting pressure-waves pass through the burnt and burning mixture at enormously high velocities. These shockwaves reflect against the combustion chamber walls and, accordingly, the walls are set into a state of vibration at an audible frequency which is heard in the form of a ringing knock.

4.3.1 Illustration of controlled and uncontrolled combustion
(Fig. 4.28(a and b))

If the speed at which the flame spreads is slow and its flame path is long (Fig. 4.28(a)) there will be more time to compress and transfer heat to the unburnt mixture ahead of the advancing flame front. Consequently, the unburnt mixture will reach its critical detonation temperature earlier during the combustion process when there is still a large proportion of mixture unburnt.

As a result, this entrapped 'end-mixture' spontaneously ignites and burns at an enormously high rate, the release of this relatively large amount of energy causes a violent pressure shockwave to travel through the combustion chamber, which thus produces detonation.

If, however, the speed at which the flame speeds through the chamber is extremely fast and its flame path is short (Fig. 4.28(b)), there will be insufficient time for the unburnt part of the mixture to become excessively compressed and overheated. Therefore, the majority of the charge will burn in a controllable manner and only the very last of the unburnt charge may actually reach the spontaneous condition which causes detonation.

Detonation can be eliminated or at least minimized by the combination of the following three methods:

1 by increasing the turbulence and the speed of flame travel thereby preventing the end-mixture overheating;
2 by reducing the flame path and increasing the surface area of the combustion-chamber in the furthest region from the sparking plug so that the 'end-mixture' is quenched;
3 by raising the octane value of the fuel.

4.3.2 Intensity of detonation

The intensity of detonation will depend mainly upon the amount of energy contained in the

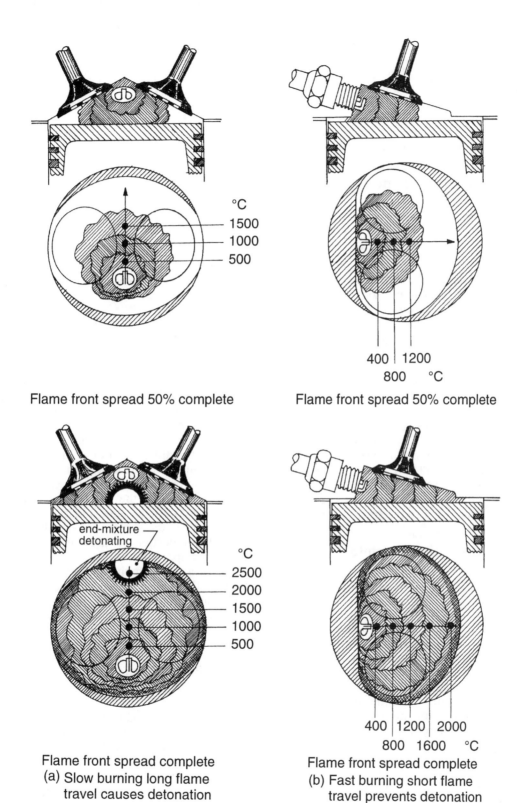

Flame front spread 50% complete

Flame front spread 50% complete

°C
1500
1000
500

400 │ 1200
800   °C

end-mixture detonating

°C
2500
2000
1500
1000
500

400 │ 1200 │ 2000
800   1600   °C

Flame front spread complete
(a) Slow burning long flame travel causes detonation

Flame front spread complete
(b) Fast burning short flame travel prevents detonation

Fig. 4.28(a and b) Illustrating controlled and uncontrolled (detonation) combustion

'end-mixture' and the rate of chemical reaction which releases it in the form of heat and a high intensity pressure-wave. Thus, the earlier in the combustion process the detonation commences, the more unburnt end-mixture will be available to intensify the detonation. As little as 5% of the total mixture charge when spontaneously ignited will be sufficient to produce a very violent shock.

4.3.3 Effects of detonation

The high detonation pressure-wave sweeping through the combustion-chamber has three basic effects:

1 to scour away the protective boundary layer of stagnant gas which always exists on the surface of the combustion-chamber so that more heat will be transmitted through the chamber wall, thereby raising the mean chamber temperature which will promote pre-ignition;

2 to scour away the oil-film which protects and lubricates the cylinder walls, thereby causing dry friction to occur between the cylinder walls and piston rings with the corresponding increase in wear;

3 the detonation shockwaves imposed on the piston may cause a vibratory load to be imparted to the gudgeon-pin, piston bosses and connecting-rod small-end bearings. Accordingly, if this is excessive it will destroy the gudgeon-pin boundary lubrication and hence cause abnormal wear to the small-end joint.

4.3.4 Factors which promote detonation

1 High compresstion ratios.
2 Low octane number fuel.
3 Long flame path from point of ignition.
4 An over-advanced ignition setting may move the cylinder peak pressure to occur at TDC thereby causing higher than normal maximum pressures to occur.
5 A slightly rich (10%) air–fuel mixture raises the combustion temperature and, correspondingly, pressure; thereby increasing the detonation tendency.
6 An inadequately cooled combustion-chamber possibly due to faulty cooling systems, slipping water pump, stuck thermostat etc.
7 Excess carburization of combustion-chamber and piston-crown.

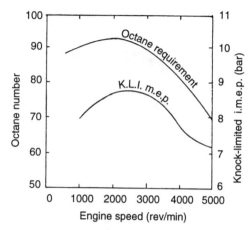

Fig. 4.29 Effect of engine speed on octane number requirements and knock-limited i.m.e.p.

4.3.5 Effects of engine speed on detonation
(*Fig. 4.29*)

At low engine speed there is more time for the flame front to preheat the end mixture, and it therefore promotes detonation, whereas at higher speeds there is less time for the end-charge to become overheated so that the end mixture will be proportionally smaller or even non-existent.

The volumetric efficiency, which is a measure of cylinder pressure, tends to decrease with engine speed so that the less severe conditions at higher speeds discourage detonation.

In contrast, the shorter time available for the heat of combustion to dissipate, increases the cylinder temperature.

However, the combined effect of shorter compression time and reduced volumetric efficiency, as opposed to increased cylinder temperature, favours reducing the detonating tendency in the cylinder with rising engine speed.

4.3.6 Preignition
(*Fig. 4.30*)

Preignition is the ignition of the homogeneous mixture in the cylinder, before the timed ignition spark occurs, caused by the local overheating of the combustible mixture. For premature ignition of any local hot-spot to occur in advance of the timed spark on the compression stroke it must attain a minimum temperature of something like 700–800°C.

The initiation of ignition and the propagation of the flame front from the heated hot-spot is similar to that produced by the spark-plug when it fires, the only difference between the hot-spot and sparking plug is their respective instant of ignition. Thus, the sparking plug provides a timed and controlled moment of ignition whereas the heated surface forming the hot-spot builds up to the ignition temperature during each compression stroke and therefore the actual instant of ignition is unpredictable.

The early ignition created by preignition extends the total time the burnt gases remain in the cylinder and therefore increases the heat transfer to the chamber walls and, as a result, the self-ignition temperature will occur earlier and earlier on each successive compression stroke. Consequently, the peak cylinder pressure (which normally occurs at its optimum position of $10°$–$15°$ after TDC) will progressively advance its position to TDC where the cylinder pressure and temperature will be maximized.

The accumulated effects of an extended combustion time and rising peak cylinder pressure and temperature, cause the self-ignition temperature to creep further and further ahead of TDC, and with it, peak cylinder pressure, which will now take place before TDC so that negative work will be done in compressing the combustion products (Fig. 4.30).

Preignition in a single-cylinder engine will result in a steady reduction in speed and power output.

If this was allowed to persist, the self-ignition temperature will commence so early in the cycle that the incoming air–fuel charge will ignite, thus causing the flame to propagate back through the induction port and manifold.

The real undesirable effects of preignition are when it occurs only in one or more cylinders in a multi-cylinder engine. Under these conditions, when the engine is driven hard, the unaffected cylinders will continue to develop their full power and speed, and so will drag the other piston or pistons, which are experiencing preignition and are producing negative work, to and fro until eventually the increased heat generated makes the preigniting cylinder's pistons and rings seize.

Thus, the danger of the majority of cylinders operating efficiently while one or more cylinders are subjected to excessive preignition is that the driver will only be aware of a loss in speed and power and therefore may try to work the engine harder to compensate for this, which only intensifies the preignition situation until seizure occurs.

Preignition is initiated by some overheated projecting part such as the sparking plug electrodes, exhaust valve head, metal corners in the combustion chamber, carbon deposit or a protruding cylinder head gasket rim, etc.

However, preignition is also caused by persistent detonating pressure shockwaves scoring away the stagnant gases which normally protect the combustion chamber walls. The resulting increased heat flow through the walls, raises the surface temperature of any protruding poorly cooled part of the chamber, and this therefore provides a focal point for preignition.

Preignition is not responsible for abnormally high cylinder pressure, but there can be a slight pressure rise above the normal due to the ignition point and, therefore, the peak pressure creeping forward to the TDC position where maximum pressure occurs.

If preignition occurs at the same time as the timed sparking plug fires, combustion will appear as normal. Therefore, if the ignition is switched-off, the engine would continue to operate at the same speed as if it were controlled by the conventional timed spark, provided the self-ignition temperature continues to occur at the same point.

4.3.7 Post-ignition or running on ignition

Post-ignition is the establishment within the combustion chamber of a hot-spot (such as the sparking plug central electrode) which supplements the timed spark ignition every time, in the compress-

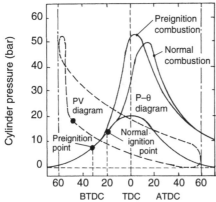

Fig. 4.30 Cylinder pressure variation when pre-ignition occurs

ion stroke cycle, the self-ignition temperature of the mixture is reached.

When the engine is at working temperature and the ignition is switched off the hot-spot will contine to be heated during the combustion phase to well above the self-ignition temperature of the mixture, after which it cools as the other three strokes' exhaust induction and compression are completed.

With the ignition switched off and the throttle closed the combustion will, at first, continue to occur every time the self-ignition temperature is reached, at roughly the same position in the compression cycle. After a time with the throttle closed, the peak temperature in the chamber will begin to fall until the hot-spot will not be able to ignite the mixture so that the engine stalls. A weak, slow-burning, mixture or retarded ignition setting can be responsible for the overheated hot-spot and thus for the occurrence of post-ignition.

4.3.8 Octane numbers
(Figs 4.31 and 4.32)

A practical measure of a fuel's resistance to knock is essential so that a comparison of commercial fuels can be made by testing the fuel in a variable compression ratio engine under controlled operating conditions.

Commercial blends of fuels are rated on an octane number scale between zero and 100. The octane number is based on two hydrocarbons which have very different knock properties enabling them to define the ends of the scale: iso-octane (C_8H_{18}) which has a very high resistance to knock and therefore is given an octane number of 100, and normal heptane (C_7H_{16}) which is very prone to knock and is therefore given a zero value. Blends of these reference fuels define the knock resistance of intermediate octane numbers, and thus a blend of 7% n-heptane and 93% iso-octane by volume has an octane number of 93.

A commercial fuel octane number is determined by measuring what blend of these primary reference fuels (iso-octane/n-heptane) matches the unknown knock property of the fuel being tested.

The knock rating of a fuel is determined by running the single-cylinder test engine under standardized operating conditions. These fixed conditions being air inlet temperature, coolant temperature, engine speed, ignition advance setting and

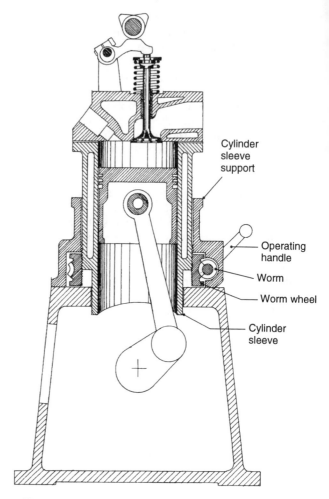

Fig. 4.31 Variable compression ratio single-cylinder research engine

Labels: Cylinder sleeve support; Operating handle; Worm; Worm wheel; Cylinder sleeve

mixture strength to give maximum knock response.

The fuel being rated is tested by increasing the compression of the test engine (Fig. 4.31) until the fuel attains a predetermined intensity of knock which is measured with a magnetostriction knock detector. Various reference fuel blends are then tested in the same way until the compression ratio of one of these reference blends exactly matches the commercial fuel in terms of similar knock intensity. This fuel is then rated by the octane percentage in the equivalent blend of reference fuel. The relationship between octane number and compression ratio is approximately as shown (Fig. 4.32).

The octane number (ON) obtained depends on the conditions of the test and the two main

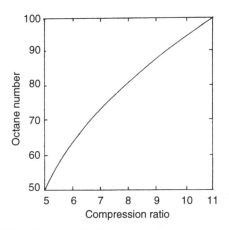

Fig. 4.32 Relationship between octane number and the highest useful compression ratio

methods in use are the Research and the Motor methods.

Research octane number (RON) method

This method measures anti-knock performance under relatively mild operating conditions. It is considered to be similar to the detonation tendency of a fuel when the engine is accelerating from low speed in top gear with a wide-open throttle under medium load. Inlet air temperature = 52°C; coolant temperature = 100°C; engine speed = 600 rev/min; ignition advance setting = 13° BTDC.

Motor octane number (MON) method

This method measures anti-knock performance under relatively severe operating conditions. It is considered to be similar to the detonation tendency of a fuel when the engine is driven at medium speed in top gear with a wide-open throttle under heavy load. Inlet air temperature = 150°C; coolant temperature = 100°C; engine speed = 900 rev/min; ignition advance setting = 19–26° BTDC.

Fuel sensitivity

The more severe operating conditions for the Motor method compared with the Research method will therefore predict a lower octane number for the MON than for RON.

This difference in octane number between the Research method and the Motor method octane numbers is known as the fuel sensitivity; thus

Fuel Sensitivity = RON – MON

Paraffinic fuels tend to have roughly the same octane numbers with either the Research or Motor method since their chemical structure is similar to that of the primary reference fuels (octane and heptane blends).

As the fuel composition deviates from the paraffinic to the napthenes aromatics or the fuel is highly cracked, the sensitivity of the fuel increases.

A comparison of the RON and MON is shown in Table 1.

Table 1 Octane number requirement for different fuel grades

Grade	RON	MON
5 star	100.0	89.0
4 star	97.0	86.0
3 star	94.0	82.0
2 star	90.0	80.0

Antiknock index

Both the Research and the Motor methods of measuring the octane number do not reflect the actual operating speed and load conditions when the vehicle is being driven on the road. Therefore, chassis dynamometer tests which predict road conditions have been devised to determine the fuel's road octane number. It has been found that the road octane number lies roughly between the Research and Motor octane ratings. It follows that the average of the two octane number rating methods, RON and MON, is a very good antiknock quality indicator which is known as the antiknock index

$$\text{Antiknock index} = \frac{\text{RON} + \text{MON}}{2}$$

4.3.9. The effects of cylinder pressure on detonation and octane number requirements
(Figs 4.29, 4.32 and 4.33)

The effects of increasing the engine's compression ratio (Fig. 4.32) is to raise the cylinder's pressure and temperature so that the end-gas becomes overheated and spontaneously burns. Therefore, fuels with higher octane numbers are necessary to suppress the detonation tendency as the compression ratio is increased.

Generally, the octane number fuel requirements for an engine, depend upon the maximum indicated mean effective pressure generated inside the cylinder, which is a measure of the volumetric efficiency over the engine's speed range. Thus, the highest knock-limited indicated mean effective pressure (Fig. 4.29) occurs roughly in the engine's mid-speed range where the volumetric efficiency and engine load are usually at a maximum, but at higher speeds the declining volumetric efficiency and, correspondingly, the reducing mean effective pressure do not demand such high octane number fuels to prevent detonation occurring.

Advancing the ignition timing so that peak pressure occurs at the engine's optimum crank angle position of about 10° after TDC with rising engine speed demands an increase in the fuel's octane number requirement to match the detonation tendency caused by the high cylinder pressures and temperature. A lower octane number fuel can be used by retarding the ignition timing slightly.

The mixture strength influences the knock-limiting indicated pressure (Fig. 4.33) because, for high engine loads and speeds, more energy has to be burnt; that is, richer mixtures which burn faster are necessary to generate the high cylinder pressures. However, for light-loads and maximum economy conditions, weak mixtures which burn slowly are used, but these tend to generate lower cylinder pressures and to overheat the end gases.

A cool intake mixture, low cylinder bore and cylinder-head wall temperatures tend to depress the temperature of combustion and therefore enable higher knock-limited mean-effective-pressures to be generated without causing detonation.

4.4 Lean or fast burn combustion chamber technology

(Figs 4.34, 4.35 and 4.36)

In recent years there has been great concern that the internal combustion engine is responsible for far too much atmospheric pollution, that it is detrimental to human health and that it is damaging the environment. Consequently, research engineers have been striving to reduce the responsible pollutants emitted from the exhaust system without sacrificing power and fuel consumption.

Pollutants are produced by the incomplete burning of the air–fuel mixture in the combustion chamber. The major pollutants emitted from the exhaust due to incomplete combustion are: unburnt hydrocarbons (HC), oxides of nitrogen (NO_x) and the highly poisonous carbon monoxide (CO). If, however, combustion is complete the only products being expelled from the exhaust would be water vapour, which is harmless, and carbon dioxide, which is an inert gas and, as such, is not directly harmful to humans.

The aim of lean burn technology has been to design engines which burn completely, or almost completely, the air–fuel mixture supplied, in an efficient manner, so that fuel consumption is reduced and the levels of the pollutants hydrocarbon (HC), oxides of nitrogen (NO_x) and carbon monoxide (CO) are within the statutory limits imposed by governments.

The conflicting characteristics of the three exhaust pollutants, the fuel consumption and developed power and how they are influenced by the mixture strength are illustrated in Figs 4.34 and 4.35.

As can be seen (Fig. 4.34) the rich and weak air–fuel mixture strength ratio for a conventional engine does not vary by more than about two numbers from either side of the chemically correct stoichiometric ratio of 14.7 parts of air to one

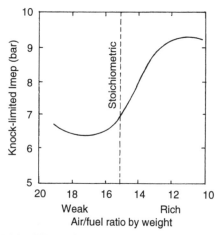

Fig. 4.33 Effect of air–fuel mixture strength on the knock-limited i.m.e.p.

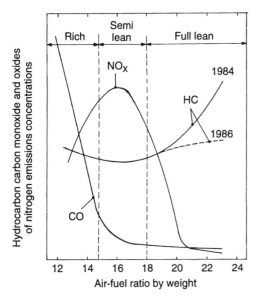

Fig. 4.34 Spark ignition engine emissions for different air–fuel ratios

part of fuel by mass; that is, from a rich 12:1 mixture to the leanest of 17:1.

Figure 4.34 shows that at an enriching mixture for full throttle maximum power, from 13.5:1 to 12.5:1, the carbon monoxide rises steeply whereas the oxides of nitrogen under similar operating conditions tend to be quickly decreasing. Conversely, under part throttle maximum economy conditions with weakening air–fuel mixture ratios, from about 15.5:1 to 16.5:1, the oxides of nitrogen have reached their worst and have peaked, whereas the unburnt hydrocarbons and carbon monoxide have decreased to their mini-

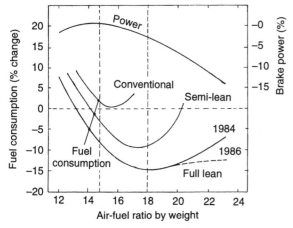

Fig. 4.35 Relationship between specific fuel consumption and the air–fuel ratio strength for a spark ignition engine

mum valves. Thus, there has always been a major conflict between the high carbon monoxide concentration with rich maximum power mixtures and the high nitrogen oxides concentration with lean part-throttle maximum economy mixtures.

This means that the engine at any one time will produce either a large amount of carbon monoxide and a modest amount of hydrocarbons or a very small amount of carbon monoxide and hydrocarbons but a large concentration of nitrogen oxide.

Generally, the air–fuel mixture ratio (Fig. 4.34) range can be divided into a rich burning zone on the left-hand side of the stoichiometric (14.7:1 to 10:1) and a weak or lean burning zone on the right-hand side of the stoichiometric, which is itself subdivided in a semi-lean zone (14.7:1 to 18:1) and a full lean zone (18:1 to 23:1).

Combustion chamber research and development has produced engines which can now operate consistently with lean air–fuel ratios of something like 18:1 and 19:1 under part throttle running conditions. This, therefore, shifts the air–fuel ratio to the decreasing portion of the nitrogen oxide curve about half-way down, whereas carbon monoxide is at a minimum and the unburnt hydrocarbons are only just on the increasing upward slope (Fig. 4.34). This has not only reduced the overall emission of the three main pollutants but it has also improved the fuel consumption when the engine is operating in the semi-lean and full-lean mixture strength region.

One of the main problems with lean burning engines is that as the mixture strength moves towards the full-lean region during transitional operating conditions, individual cylinders will experience hesitation and misfiring. This happens because it is almost impossible to produce a homogeneous mixture of exactly the same strength, identical intensity of swirl and generated turbulence, accurate and equal spark energy release and, with similar cylinder temperatures in each cylinder under the varying speed, load and throttle opening conditions.

Thus, the most important operating condition for multi-cylinder lean burning engines is that the mixture preparation, distribution and ignition energy release between cylinders is very near perfect.

With further improvements in promoting induction swirl (Fig. 4.35) and cylinder squish and with more emphasis on the precise matching of mixture strength with ignition timing, even leaner mixtures of between 19:1 to 21:1 will be capable of being reliably burnt. At the same time, with more consideration to detail for piston and piston-ring design and accurate bump clearance control, the upward trend for the hydrocarbons can be encouraged to level out.

Thus, as advancements in combustion chamber design continue, the control of combustion in the lean-burn range improves so that effective burning can extend well into the full lean-burn range, where the dominant pollutants have reached their minimum values.

The major advancement in combustion chamber design has come about by the understanding that rich mixtures burn quickly and weak mixtures burn slowly. Therefore, to obtain maximum engine power, fast-burning rich mixtures are essential to make the bulk of the pressure rise occur just after TDC where the clearance volume is at a minimum and the combustion pressure will be most effectively applied.

Conversely, with conventional combustion chambers the burning time for very lean mixtures, under part throttle, is far too long so that the rapid pressure-rise period occurs when the clearance volume is itself expanding fairly quickly. This means that the prolonged burning is spread over the least effective part of the crank-angle movement; that is, well beyond TDC.

However, lean-burn combustion chambers are able to promote large amounts of induction swirl (Fig. 4.36) which, just before ignition occurs, is excited into a highly violent and turbulent mixture but still retains some of its directional movement.

Thus, when combustion commences, the turbulent vortices within the swirling mixture break up the boundary between the flame front and the unburnt mixture into a rugged greatly enlarged frontal area, the flame therefore rapidly penetrates and propagates its way through the decreasing amounts of unburnt charge.

The initial swirl greatly helps to sweep the burning nuclei into the still unburnt charge.

By these means, the normally slow-burning rate for a weak mixture is greatly accelerated due to the vigorous activity of the mixture prior to combustion, which enables the rapid pressure rise of the second period of combustion to occur when it can be most effective; that is, just after TDC.

Thus, it can be seen that the lean-burn combustion-chamber is, in effect, a fast-burn chamber. Therefore, both terms may be used to describe a combustion chamber which, under part throttle conditions, is capable of running consistently with

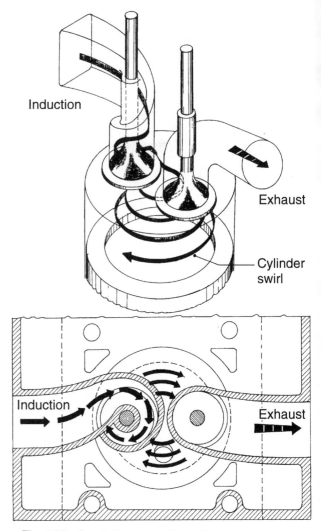

Fig. 4.36 Port induction swirl combustion chamber

extremely weak mixtures of the order 18:1 to 21:1 or even leaner.

It should be appreciated that an engine operates under part-throttle conditions for the majority of a vehicle's driving cycle and only rarely is full throttle used.

A fast combustion process compared with a moderate burn rate does produce a modest gain in engine efficiency.

A decrease in the total burn duration from 90° to 60° crank angle movement—that is, the difference between a slow and a moderate combustion burn time—results in something like a 5% decrease in the specific fuel consumption, whereas a decrease in the burning period from 60° to 30°, that is, from a moderate to fast burn rate, gives only a further 2% specific fuel consumption improvement.

4.5 A comparison of the breathing ability of various valve arrangements

(Fig. 4.37(a–f))

In general, a flat cylinder-head minimizes the valve head diameter if the valve-seat to cylinder wall and valve-seat to valve-seat spacing adequately accommodates the cooling passages within the cylinder-head.

However, by inclining both the inlet and exhaust valve stems from the vertical, in opposite directions to each other, the valve heads form an arched roof. As the included valve-stem angle is increased, the diameter of the valve-seats can also be increased, and consequently, the intake breathing and the ease of expelling the exhaust gases will be considerably improved.

A comparison between intake valve port area, which influences the intensity of cylinder filling, valve inclination to the vertical and the number of valves located in the cylinder-head, can be demonstrated by the use of the area ratios given below:

1 cylinder bore to flat cylinder-head upright single-valve seat bore area ratio A_c/A_{us}, where A_c = cylinder bore area; A_{us} = upright single-valve seat bore area;
2 tilted single-valve to upright single (flat) valve seat bore area ratio A_{ts}/A_{us}, where A_{ts} = tilted single-valve seat bore area; A_{us} = upright single-valve seat bore area;
3 tilted multi-valve to tilted single-valve seat bore area ratio A_{tm}/A_{ts}, where A_{tm} = tilted multi-valve seat bore area; A_{ts} = tilted single-valve seat bore area.

4.5.1 Calculation comparing the breathing ability of various valve arrangements
(Fig. 4.37(a–f))

Compare the breathing ability of the following valve configurations when incorporated in a cylinder-head combustion chamber that fits on top of a 80 mm diameter cylinder bore.

(a) Single inlet and exhaust valves arranged vertically (Fig. 4.37(a and b))

Flat inlet valve seat bore diameter 28 mm

Cross-sectional area of cylinder bore

$$A_{80} = \frac{\pi}{4} 80^2 = 5026.5 \, \text{mm}^2$$

Cross-sectional area of valve seat bore

$$A_{28} = \frac{\pi}{4} 28^2 = 612.75 \, \text{mm}^2$$

Ratio of cylinder bore to upright valve seat bore areas

$$\frac{A_{80}}{A_{28}} = \frac{5026.5}{612.75} = 8.203 : 1$$

(b) Single inlet and exhaust valve with a small incline of 20° from the vertical (Fig. 4.37(c))

Tilted inlet valve seat bore diameter 32 mm

Cross-sectional area of 20° inclined valve

$$A_{32} = \frac{\pi}{4} 32^2 = 804.25 \, \text{mm}^2$$

Ratio of 20° tilted valve to upright valve areas

$$\frac{A_{32}}{A_{28}} = \frac{804.25}{612.75} = 1.3125 : 1$$

this amounts to a 31.25% increase in valve seat cross-sectional area.

(c) Single inlet and exhaust valve with a large incline of 35° from the vertical (Fig. 4.37(d))

Tilted inlet valve seat bore diameter 36 mm

Cross-sectional area of 35° inclined valve

$$A_{36} = \frac{\pi}{4} 36^2 = 1017.87 \, \text{mm}^2$$

Ratio of large tilted valve to upright valve area

$$\frac{A_{36}}{A_{28}} = \frac{1017.87}{612.75} = 1.661 : 1$$

this amounts to a 66.1% increase in valve seat cross-sectional area.

(d) Twin inlet and twin exhaust valves arranged vertically (Fig. 4.37(e))

Flat inlet valve seat bore diameter 24 mm

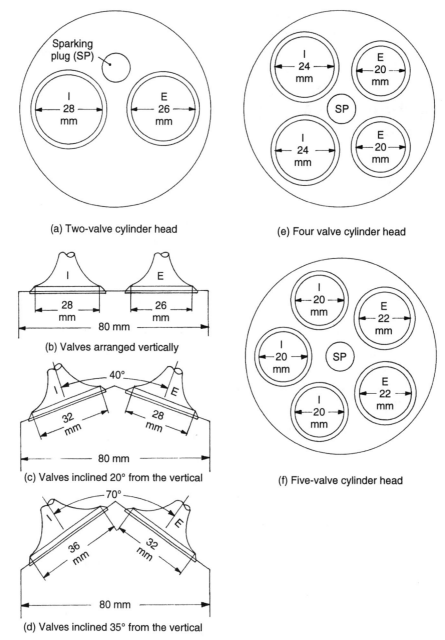

(a) Two-valve cylinder head

(b) Valves arranged vertically

(c) Valves inclined 20° from the vertical

(d) Valves inclined 35° from the vertical

(e) Four valve cylinder head

(f) Five-valve cylinder head

Fig. 4.37 Comparison of valve configuration and inclination

Cross-sectional area of twin exhaust seat bore

$$A_{24} = \frac{\pi}{4} 24^2 = 452.4 \, \text{mm}^2$$

Ratio of cylinder bore to twin valve seat bore areas

$$\frac{A_{80}}{A_{2 \times 24}} = \frac{5026.5}{2 \times 452.4} = 5.55 : 1$$

(e) Twin inlet and twin exhaust valves with a small incline of 20° from the vertical (Fig. 4.37(c and e))

Tilted inlet valve seat bore diameter 28 mm

Cross-sectional area of tilted twin inlet valve seat bore

$$A_{28} = \frac{\pi}{4} 28^2 = 612.75\,\text{mm}^2$$

Ratio of tilted twin valves to upright single-valve seat bore areas

$$\frac{A_{2 \times 28}}{A_{28}} = \frac{2 \times 612.75}{612.75} = 2.0:1$$

this amounts to a 100% increase in valve seat cross-sectional area.

Ratio of tilted twin to tilted single-valve seat bore areas

$$\frac{A_{2 \times 28}}{A_{32}} = \frac{2 \times 612.75}{804.25} = 1.524:1$$

this amounts to a 52.4% increase in valve seat cross-sectional area.

Ratio of tilted twin valve to upright twin valve seat bore

$$\frac{A_{2 \times 28}}{A_{2 \times 24}} = \frac{2 \times 612.75}{2 \times 452.4} = 1.354:1$$

this amounts to a 35.4% increase in valve seat cross-sectional area.

(f) Triple inlet and twin exhaust valves arranged vertically (Fig. 4.37(f))

Flat inlet valve seat bore diameter 20 mm

Cross-sectional area of triple inlet valve seat bore

$$A_{20} = \frac{\pi}{4} 20^2 = 314.16\,\text{mm}^2$$

Ratio of cylinder bore to valve seat bore areas

$$\frac{A_{80}}{A_{3 \times 20}} = \frac{5026.5}{3 \times 314.16} = 5.33:1$$

(g) Triple inlet and twin exhaust valves with a small incline of 20° from the vertical (Fig. 4.37(c and f))

Tilted inlet valve seat bore diameter 24 mm

Cross-sectional area of triple tilted inlet valve seat bore

$$A_{24} = \frac{\pi}{4} 24^2 = 452.4\,\text{mm}^2$$

Ratio of tilted triple valves to single upright valve seat bore

$$\frac{A_{3 \times 24}}{A_{28}} = \frac{3 \times 452.4}{612.75} = 2.215:1$$

this amounts to a 121.5% increase in valve seat cross-sectional area.

Ratio of tilted triple valves to tilted single valve seat bore

$$\frac{A_{3 \times 24}}{A_{32}} = \frac{3 \times 452.4}{804.25} = 1.6875:1$$

this amounts to a 68.75% increase in valve seat cross-sectional area.

Ratio of tilted triple valves to upright twin valve seat bore

$$\frac{A_{3 \times 24}}{A_{2 \times 24}} = \frac{3 \times 452.4}{2 \times 452.4} = 1.5:1$$

this amounts to a 50.0% increase in valve seat cross-sectional area

Ratio of tilted triple valves to triple upright valve seat bore

$$\frac{A_{3 \times 24}}{A_{3 \times 20}} = \frac{3 \times 452.4}{3 \times 314.16} = 1.44:1$$

this amounts to a 44.0% increase in valve seat cross-sectional area.

4.5.2 Summary of the breathing ability of various valve arrangements
(Fig. 3.37(a–f))

A comparison between different valve configurations in the cylinder-head to obtain the greatest intake mixture flow ability can roughly be based on the valve seat bore cross-sectional area if the port passage curvature and valve opening flow characteristics (coefficient of discharge) are ignored.

If the inlet valve seat bore cross-sectional area of the flat (upright) single inlet valve cylinder-head is taken as unity, then all other valve configurations, which enlarge the total intake valve seat bore areas for the same cylinder bore size, can be considered as an improvement.

Two-valve hemispherical combustion chamber
(Fig. 4.37(a, c and d))

The single 20° and 35° tilted inlet valves enlarge the port area and improve breathing by the same amount relative to the single flat (upright) inlet valve as follows:

a) 20° tilted single inlet valve by 31.25%
b) 35° tilted single inlet valve by 66.1%

Four-valve pentroof combustion chamber
(Fig. 4.37(c, d and e))

The twin 20° tilted inlet valves enlarge the port area and improve breathing by the same amount relative to the following layouts:

a) upright single inlet valve by 100%
b) 20° tilted single inlet valve by 52.4%
c) upright twin inlet valve by 35.4%

Five-valve pentroof combustion chamber
(Fig. 4.37(c, d and f))

The triple 20° tilted inlet valves enlarge the port area and improve breathing by the same amount relative to the following layouts:

a) upright single inlet valve by 121.5%
b) 20° tilted single inlet valve by 68.75%
c) upright twin inlet valves by 50.0%
d) upright triple inlet valves by 44.0%

4.6 General relationships between valve configurations and breathing capacity
(Fig. 4.37(a–f))

In general, it can be seen that the flat cylinder-head accommodates the smallest valve heads. However, inclining the valve stems from the vertical increases the slant length of the chamber roof, which therefore permits larger valve heads to be used. In practice, the slope of the chamber roof ranges from approximately 20° for a shallow hemispherical chamber to a limiting 35° for the deep chambers.

The alternative method of improving the breathing ability of the combustion chamber is to adopt multi-inlet and exhaust-valve cylinder-heads. Hence, by reducing the diameters of the valves and increasing their numbers, more of the combustion-chamber plan area will be utilized since less surface area between adjacent valve heads will be wasted. Thus, if two, three, four, five or six valves are incorporated in the cylinder-head, but with reducing valve-head diameters respectively, the total cross-sectional area of these valve heads increases while the exposed space around each valve rim correspondingly decreases. In other words, the chamber wall surface area becomes more effectively used with small but many valve heads as opposed to using fewer but larger valve heads. At the same time, if the valve heads are installed in a slanted roof, the port flow area is further enlarged; however, for three valves or more the arched roof is normally restricted to a 20° slope.

4.6.1 Inverted bath tub (in cylinder-head) combustion chamber (Rover and Vauxhall)
(Fig. 4.38(a–d))

With this type of in-line valve combustion chamber, the space is cast in the underside of the cylinder-head in the shape of a heart. Both valves sit in the cylinder-head recess partially separated by a protruding tongue or nose as observed in the plan view (Fig. 4.38(a)).

The flat part of the cylinder-head, which overlaps the piston from two squish regions, is a large area due to the protruding nose (tongue), and a much smaller crescent-shaped area on the opposite side of the cylinder.

The pairs of poppet valve heads are partially masked by the chamber walls on the nose side of the squish zone (Fig. 4.38(a)), whereas they are more exposed on the sparking-plug side of the chamber. This permits the induced air–fuel mixture to be reflected unrestricted in a downward spiral about the cylinder axis (Fig. 4.38(a)).

Towards the end of the compression stroke the air–fuel charge between the flat portion of the cylinder-head and the piston crown is squeezed into the combustion chamber cavity at a relatively high velocity and, in the process, it becomes highly turbulent (Fig. 4.38(b)).

Just before TDC is reached, the spark jumps between the electrodes of the sparking plug, and there is then a very small time lag in which a nucleus of flame is established 6° before TDC (Fig. 4.38(c)). In this example, the flame then propagates outwards from the ignited nucleus of flame in a circular fashion, which is shown in the

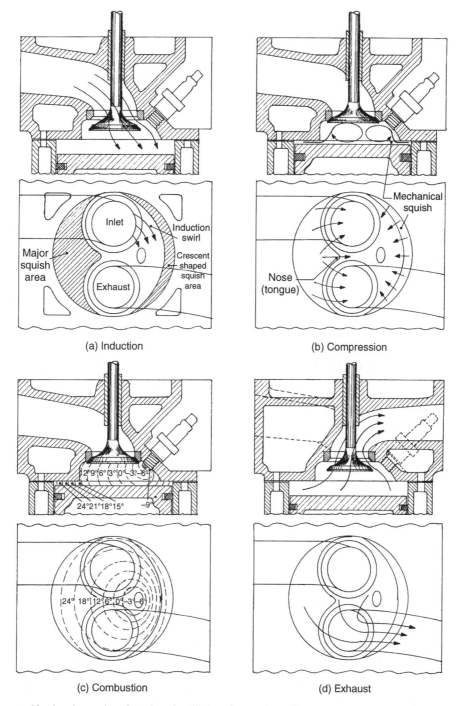

(a) Induction

(b) Compression

(c) Combustion

(d) Exhaust

Fig. 4.38 Inverted bath-tub combustion chamber illustrating cycle events

form of isocrank-angle movement loops. These expand throughout the combustion-chamber and the squish region of the cylinder until the burning flame front reaches the cylinder walls over a crank-angle movement of $6° + 24° = 30°$, 'after-burning' then completes the combustion process.

The intensity of turbulence generated by the induced swirl and squish tends to increase directly with engine speed. Consequently, the rate of burning rises roughly in the same proportion to engine speed. This means that the angular period of flame propagation from the ignition point to the extreme cylinder walls occupies approximately the same angular movement for all engine speeds.

Finally, towards the end of the expansion stroke the exhaust valve opens causing the products of combustion which are still hot and pressurized to expel themselves out through the exhaust port (Fig. 4.38(d)), mainly by flowing out on the sparking plug side of the combustion chamber.

4.6.2 Bath tub (in piston crown) combustion chamber (Rover)
(Fig. 4.39)

This combustion chamber consists of a flat cylinder-head with single-row valves facing a circular cavity cast in the piston crown. Thus, an annular squish-zone is created around the piston's rim (Fig. 4.39).

The offset downward-inclined induction port provides the incoming air–fuel mixture with a spiral motion about the cylinder axis during the piston's outward induction stroke. This induction swirl continues just as vigorously when the piston reverses its direction and commences compressing now a mostly vaporized charge.

Hence, with the piston approaching TDC, the charge is compelled to sweep around both the squish zone and the deep central bowl and, just before TDC is reached, the charge will be squeezed radially inwards entering the bowl at a relatively high velocity. This radial movement imposed on the main circular swirl produces a highly turbulent mixture charge which then sweeps past the sparking plug that is located to one side of the central bowl area.

When ignition occurs the nucleus of flame will be spread by the whirling vortices of the turbulent mixture, and the main swirl velocity of the charge so that blankets of burnt charge do not suffocate the yet unburnt vapour mixture. This enables a rapid and progressive growth and spread of the flame front which is essential if overheating of the end gases is to be avoided.

This method of maintaining rapid movement and agitation of the charge encourages the flame to travel very quickly throughout homogeneous

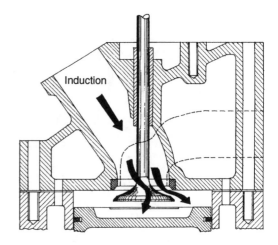

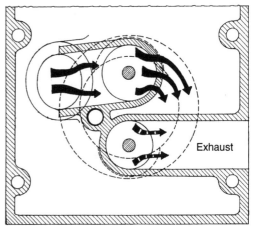

Fig. 4.39 Bath-tub combustion chamber

mixtures, that may be very weak, without hesitation and leaving local zones unburnt.

4.6.3 Wedge (in cylinder-head) combustion chamber (Ford)
(Fig. 4.40)

This chamber resembles an inclined bath tub recessed into the flat cylinder-head face. Single-row valves are tilted to accommodate the sloping roof of the chamber. The sparking plug is located on the thick side of the wedge mid-way between the valves (Fig. 4.40).

The steep end walls of the wedge chamber partially mask the air–fuel mixture flow-path from the port into the cylinder so that the charge is deflected and forced to move in a downward spiral about the cylinder axis.

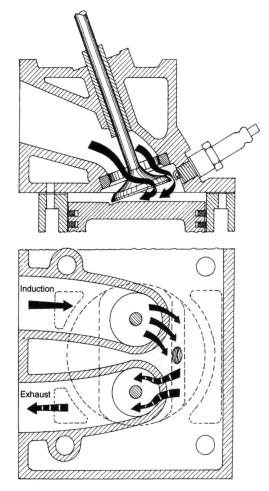

Fig. 4.40 Wedge-type combustion chamber

heated end-gases become partially quenched and are therefore prevented from detonating.

However, if the quench area is very large, not only will detonation be prevented but the remaining vapour charge may not completely burn: unwanted hydrocarbons will be produced.

4.6.4 Split level (May fireball) combustion chamber (Jaguar)
(Fig. 4.41)

This combustion chamber has the exhaust valve head deeply recessed in a circular chamber whereas the inlet valve head is sunk just below the flat face of the cylinder-head so that the valve head does not bump the piston crown when it approaches TDC (Fig. 4.41). The walls of the deep exhaust valve are steep. However, a chamber groove merges the slightly sunken inlet valve seat with the flat face of the cylinder-head. This shallow recess of the inlet valve head merges into the side wall of the deep exhaust valve chamber so that it forms a neck or passage between the valves when the piston is in the TDC region. The sparking plug intersects the steep side walls of the exhaust valve chamber, slightly to one side of the shallow passage communicating between the two valve heads.

When the piston approaches TDC the bump clearance reduces so that the mixture is squeezed with great intensity through the offset recessed passage into the deep exhaust valve chamber. This eccentric or tangential entry from the large squish zone into the deep chamber therefore produces a highly intensified squish-induced swirl about the deep exhaust chamber's axis.

At the same time, the violent inward squish due to the reducing bump clearance surrounding the exhaust chamber converts the mixture motion into a highly turbulent swirl.

At the instant of ignition, the high swirl rate and the surrounding inward squish produces a rapid and controlled propagation of the flame front from the sparking plug to the furthest point in the chamber with rich, stoichiometric and weak mixture strengths without hesitation and misfiring. It is claimed that the combustion chamber operates satisfactorily with a 97 RON octane petrol with a 16:1 compression ratio, and a 92 RON octane petrol with a 12:1 ratio, without detonation due to the elimination of most end-gases before they become overheated.

During the return compression stroke just before ignition occurs, the piston to cylinder-head squish area gap reduces to such an extent that the trapped swirling mixture is thrust violently from the thin to the thick end of the wedge chamber space.

Thus, when the electric arc spans the sparking plug electrodes, a nucleus of flame is established in the deep end of the chamber where it is protected from the very high turbulent swirl extinguishing the flame. The flame front then spreads outwards with increasing speed towards the shallow end of the wedge recess.

Once the flame front reaches the thin end of the wedge space, it enters the parallel squish-zone where the local surface-to-volume ratio is high. Consequently, these highly pressurized and over-

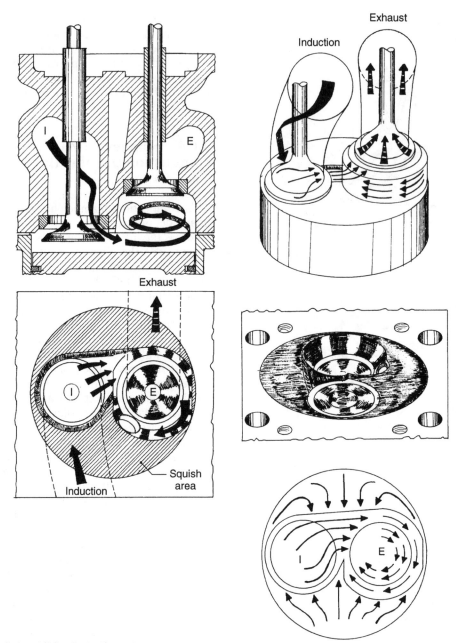

Fig. 4.41   Split level (May fireball) combustion chamber

4.6.5   Compound valve angle hemispherical combustion chamber (Ford)
(Fig. 4.42)

With this chamber the valves are inclined at 45° to each other with the plane of inclination twisted by 7° anticlockwise (looking down from above), allowing a reasonably near part-hemisphere to form the combustion chamber in spite of using rockers and a single, rather than double, camshaft arrangement (Fig. 4.42). The induction and exhaust ports are arranged in a cross-flow fashion so that the fresh charge enters the chamber on one side and is expelled on the opposite side. The

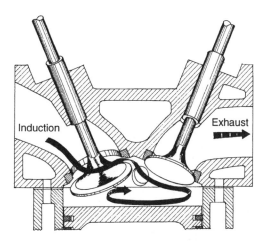

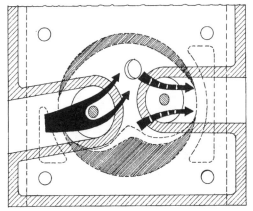

Fig. 4.42 Compound valve-angle hemispherical combustion chamber

yet it is able to spiral and flow gently around the cylinder walls as the piston moves away from TDC.

On the last part of the return stroke, the intensified swirl of the mixture is subjected to a rapid radially inward squish component which converts the slightly turbulent swirl into a highly turbulent one.

Thus, when ignition occurs just before TDC the nucleus of a flame established at the sparking plug will rapidly expand its frontal-area and, simultaneously, it will be directionally swept throughout the chamber. Consequently, this combustion phase gives very little opportunity for any of the end-gases to become over-compressed and for the temperature to reach a detonation threshold.

4.6.6 Hemispherical combustion chamber (Jaguar)
(Fig. 4.43(a and b))

By inclining both inlet and exhaust valve stems to each other, in this example by 70°, the valve heads are able to blend into the walls of the chamber without disrupting the chamber's hemispherical contour (Fig. 4.43(a)). Inclined valves are beneficial as they permit larger valve heads to be installed compared with the vertically mounted valves, as used on flat-top chambers. This feature therefore considerably improves the cylinder's filling ability, particularly at high engine speeds. Thus, for a deep hemispherical chamber, the inlet valve flow area would be increased by something like 30% or more, which accordingly, promotes high volumetric efficiencies with these chambers. A hemispherical chamber has a very low surface-to-volume ratio (S/V) so that the minimum amount of heat will be lost through the chamber walls and, consequently, a high thermal efficiency is associated with this chamber shape. Another attribute to this chamber shape is that with a curved induction port, the entering mixture is relatively free to generate a high swirl rate around the chamber towards the end of the compression stroke. A further feature of this chamber layout is the raised piston crown which contributes to a moderate squish zone around the piston's periphery.

Hence, just before TDC is reached at the end of the compression stroke, the tapered crown squeezes the mixture towards the centre so that it excites the swirling charge to become highly turbulent. Subsequently, once ignition occurs at

small inclination of the valves permits slightly larger valve heads to be incorporated in the cylinder-head chamber without causing the combustion chamber to overlap the cylinder bore. Consequently, better breathing will be achieved in the upper speed range.

This chamber resembles a shallow semi-hemisphere with a crescent squish region on the sparking plug side, and a wider nose-shaped squish region on the other side. Thus, the fuel mixture entering the cylinder will be guided by the induction port in a downward direction and, as it enters the chamber, the steep walls on the protruding nose side of the chamber mask the entry passage of the charge so that it is deflected more towards the sparking plug. As it moves beyond the chamber into the cylinder it will tend to swirl about the cylinder axis. The partially heart-shaped shallow semi-hemispherical chamber offers little resistance to the induced mixture,

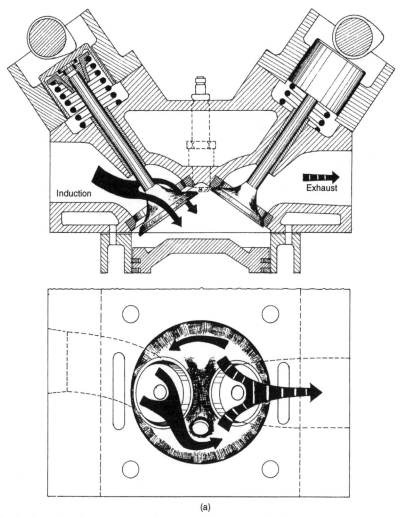

(a)

Fig. 4.43 Hemispherical combustion chamber with direct acting twin OHC

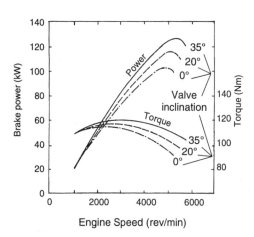

(b) Generalized effect of inclining the cylinder head valves with respect to torque and power

the offset sparking plug, a flame front will be established which spreads very quickly outwards as it is swept around the chamber.

However, some chamber designs do not rely on squish and therefore have flat piston crowns.

A comparison of torque and power (developed by an engine under similar operating conditions) with different two-valve cylinder-heads with vertical, 20° and 35° inclined valve stems, is shown in Fig. 4.43(b). These graphs demonstrate that there is very little improvement in performance in the low-to-medium speed range but the benefits become significant in the mid-to-upper speed range.

4.6.7 Triple-valve pentroof combustion chamber (Honda/Rover)
(Figs 4.44 and 4.45)

This three-valve (triple) cylinder-head employs twin inlet valves with a single exhaust valve offset to leave room for the sparking plug (Fig. 4.44). The valves are operated by individual rockers from a single overhead camshaft (Fig. 4.45). A typical inclination for both the inlet and the exhaust valves would be something like 20° from the vertical and their opposing tilt forms the pentroof-shaped combustion chamber. The sparking plug is relatively central and in the deep region of the chamber and the flat contour of the chamber where it overlaps the bore, provides the

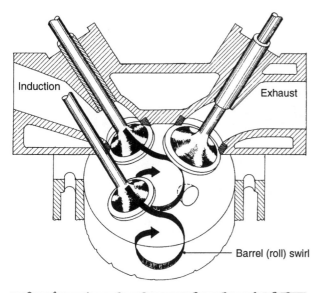

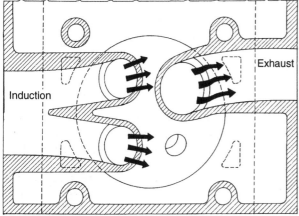

Fig. 4.44 Triple-valve pentroof combustion chamber

squish zones around the end of the chamber (Fig. 4.45). Hence, the moderately streamlined flow of the barrel (roll) swirling mixture is converted, at a very late stage on the compression stroke, into a highly turbulent one. Thus, the combination of a short flame path, wide flame-fronted area and a periphery of inward activated squish provides the essential ingredients for the fast burning of fairly lean mixtures without hesitation.

The basic benefit of this triple-valve pentroof chamber is that the port flow capability can be improved by roughly 50% compared with that for a single valve with a similar stem inclination.

One advantage of reducing the size of the valves is that the valve mass reduces so that reciprocating inertia effects are minimized.

With the overall improvements in the cylinder-head's breathing ability with twin or triple inlet valves, the necessity for the early inlet valve opening and the late exhaust valve closing is removed. Consequently, a much smaller valve overlap can be used which brings with it the benefits of far less exhaust emission under idling and light engine loads.

4.6.8 Quadruple-valve pentroof combustion chamber (Saab)
(Figs 4.46 and 4.47)

To meet the demand for more consistent torque, greater power and a reduction in fuel consumption throughout the engine's speed range, under part throttle and full throttle operating conditions, the four-valve (quadruple) cylinder-head was envisaged (Fig. 4.46). This cylinder-head incorporates twin inlet valves and twin exhaust valves with their valve ports arranged in a cross-flow fashion.

The cylinder-head intake has a single entrance which, after a short distance in, divides into two separate ports whereas the individual exhaust ports merge together to form one common exit on the side of the cylinder-head. The inclination to each other of both inlet and exhaust valves tilts the pairs of valve heads so that they resemble an arch, the chamber shape therefore takes the form of a pentroof (Fig. 4.47). Both twin intake and twin exhaust ports are symmetrically arranged on either side of the cylinder. By these means the incoming mixture passes between the cylinder-head valve seats and the underside of the valve-head seats, it is then forced to flow across the

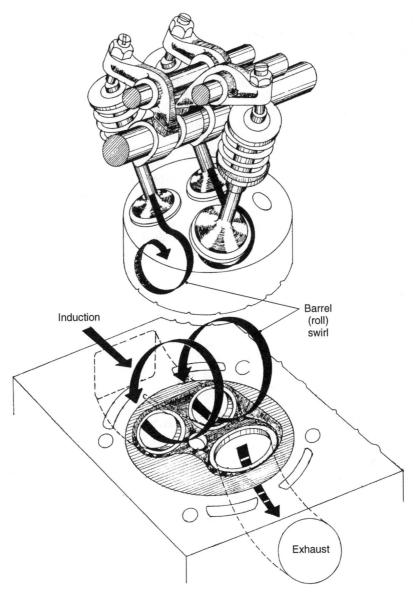

Fig. 4.45 Triple-valve pentroof combustion chamber using OHC with twin rocker shafts and direct acting rocker-arms

cylinder in a downward rolling motion at right angles to the cylinder axis. This movement can therefore be described as a transverse swirl or as a *barrel swirl*, as it is more commonly called (Figs 4.46 and 4.47).

Just before the piston reaches TDC at the end of the compression stroke the two opposing squish zones instantaneously transform the barrel swirl into turbulence. Thus, when ignition occurs the central positioned sparking plug provides a very short flame path and wide frontal area so

that a very high but controlled burning rate is achieved.

The adoption of this four-valve pentroof cylinder-head improves the intake and discharge flow to and from the cylinder by something like 50% compared with a two-valve hemispherical chamber or 100% when comparisons are made with a two-valve bath-tub chamber. Consequently, the higher overall volumetric efficiency means that the usual peak torque developed around mid-engine speed can now be maintained throughout

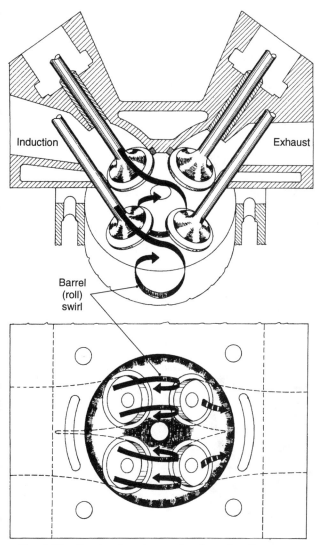

Fig. 4.46 Quadruple valve pentroof combustion chamber

inclined at 17.25° from the vertical axis, and the central one is at a very small 11.5°; through these means, lines drawn through the inlet valve stems intersect at the centre of the camshaft. On the other hand, the exhaust valves are inclined at a modest 13.75° from the vertical axis. Thus, the chosen compound valve inclination helps to form a shallow semi-hemispherical combustion chamber with a central sparking plug so that there will be a minimum outward flame travel with the widest front area in all directions. An additional feature for this chamber is the near horizontal, centre, inlet valve-head which provides a good squish zone to excite the barrel-roll movement of the mixture into a highly turbulent one, just before ignition occurs at the end of the compression stroke (Fig. 4.48). Consequently, the flame will propagate throughout the chamber at a very high speed over the shortest distance possible, with the result that a relatively high compression ratio can be used when run on 97 RON octane fuel.

The actual effective intake area around the triple inlet valve-heads is 14% larger than for an equivalent twin inlet and exhaust valve chamber.

As would be expected, the mass of each of the triple valves is considerably less compared with an equivalent single inlet-valve cylinder-head. Consequently, lighter valve springs can be used or higher engine speeds can be obtained without valve bounce occurring.

4.6.10 Sextet-valve pentroof combustion chamber (Maserati)
(Fig. 4.49(a and b))

This six-valve (sextet) cylinder-head incorporates triple inlet and triple exhaust valves arranged hexagonally to form a shallow pentroof combustion chamber (Fig. 4.49(a)). In this layout, the central inlet and exhaust valves are inclined from the vertical by only 3° whereas the two outer inlet valves are set at 11.25° and the matching outer exhaust valves are set at 10.5°. The triple inlet and the triple exhaust valves for each cylinder are operated by a pair of end-pivoted wide-spread rocker arms, these being actuated by twin camshafts; that is, one camshaft for each row of like valves.

This pentroof chamber shape with its centrally positioned sparking plug provides the optimum barrel (roll) swirl to the incoming mixture (Fig.

the upper speed range without any marked decline, in contrast to the relatively large drop-off which occurs with conventional two-valve cylinder heads.

4.6.9 Quintet-valve pentroof combustion chamber (Yamaha)
(Fig. 4.48)

This five-valve (quintet) cylinder-head utilizes three inlet and two exhaust valves placed pentagonally while their inclination is such that they form a shallow pentroof chamber (Fig. 4.48). With this design the two outer inlet valves are

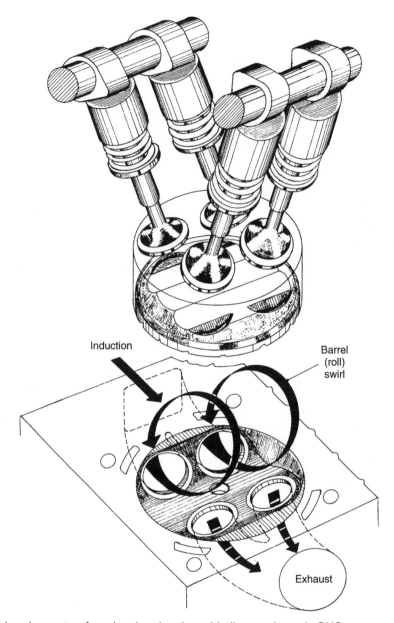

Fig. 4.47 Quadruple-valve pentroof combustion chamber with direct acting twin OHC

4.49(a)). A slight overlap of the flat portion of the cylinder-head over the cylinder bore produces a moderate amount of squish when the piston crown rim approaches the flat face of the cylinder-head. The wide frontal burning area around the centrally located sparking plug and the short and equal flame path with squish intensified regions greatly accelerates the outward flame propagation to the cylinder walls.

It is claimed that there is a 34% improvement in gas flow with the triple ports compared with the flow obtainable with an equivalent twin-valve intake valve arrangement having similar inclinations.

The benefits of multi-valve cylinder-heads, with regards to the engine torque and power developed and tested under identical conditions without ram induction charging, are compared in

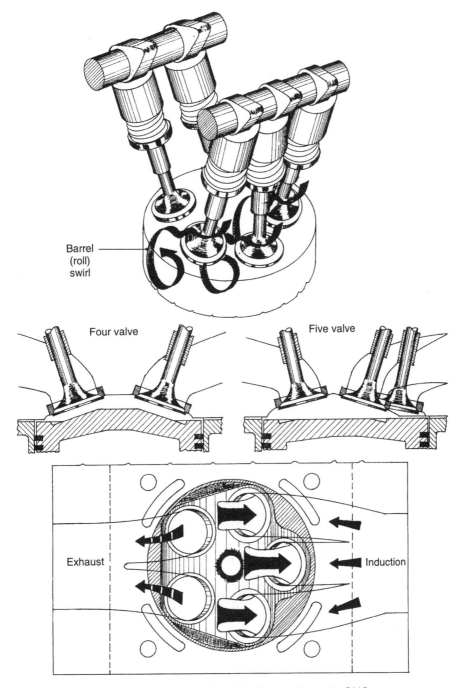

Fig. 4.48 Quintet-valve pentroof combustion chamber with direct acting twin OHC

Fig. 4.49(b)). It can be seen that the engine torque and power peak at higher speeds as the number of valves in the cylinder-head increases, but the improvement in performance in the low-to-mid speed range is relatively small. Thus, increasing the number of valves needed for emptying and filling the cylinders extends the effective speed range of the engine.

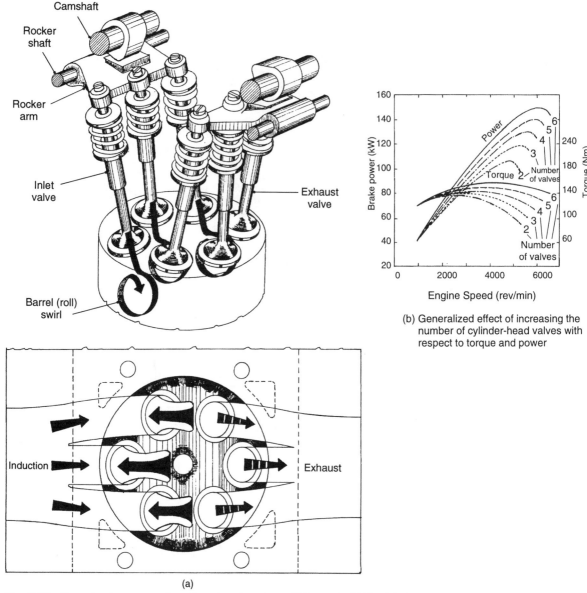

(a)

(b) Generalized effect of increasing the number of cylinder-head valves with respect to torque and power

Fig. 4.49 Sextet-valve pentroof combustion chamber utilizing twin OHC with direct-acting end-pivoted rocker arms

4.6.11 Dual ignition (twin sparking plugs) (Alfa Romeo)
(Fig. 4.50(a and b))

With a two-valve conventional cylinder-head the single sparking plug is normally located between, but to one side of, the inlet and exhaust valve (Fig. 4.50(a)). This means that the nucleus of a flame, initiated at the sparking plug, propagates only a short distance to the cylinder wall when moving away from the valves, but its flame path is much greater when it moves across the chamber in the opposite direction.

This is shown (Fig. 4.50(a)) by the flame-front crank-angle movement spread loops. Here, it can be seen that ignition commences at 12° before TDC, the flame then spreads to about the cylinder centre by the time the crankshaft has moved to 6° after TDC. The flame then spreads to its furthest point from the sparking plug in the time

196

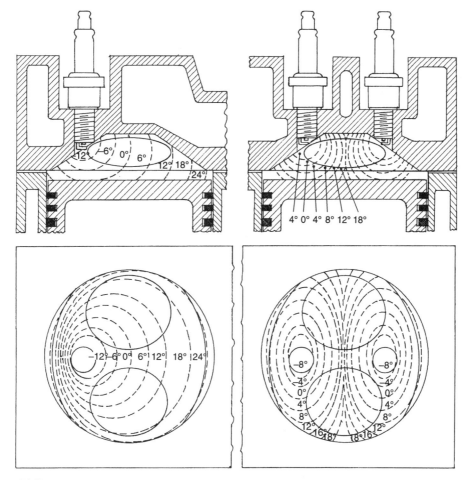

(a) Single sparking plug combustion chamber   (b) Twin sparking plug combustion chamber

Fig. 4.50   Comparison between single and twin sparking plug combustion chambers

the crankshaft has rotated to the 24° after TDC position.

Under part-throttle lean-burning conditions, the burning period would be considerably extended, this extra time causes the gases which are still unburnt to become highly compressed and over-heated. Thus, under unfavourable driving conditions spontaneous and uncontrolled combustion is produced; that is, the high pressure and temperature which the end gases are subjected to practically explode.

However, by incorporating twin sparking plugs into the cylinder-head and locating them on either side of the valves, the distance the flame or flames now travel before they reach the outer region of the cylinder wall is somewhat reduced (Fig. 4.50(b)).

The benefits of having twin sparking plugs

igniting simultaneously can be compared with the single sparking plug chamber (Fig. 4.50(a and b)) under similar operating conditions. The flame spread loops show the initial ignition timing reduced from 12° to 8° and that burning is completed when the crankshaft has moved 18° after TDC position, whereas with the single sparking plug chamber, burning is not complete until the crankshaft has moved a further 6° to the 24° after TDC position.

The completion time for flame coverage throughout the chamber makes a considerable difference to the final temperature reached by the end-gases, particularly when burning weak, slow-burning, mixtures under part throttle conditions as opposed to full-load, wide-open, throttle conditions where a rich, fast burning mixture may only take half the crank-angle burning time.

197

The merits of dual ignition as opposed to single ignition combustion chambers

a) Dual ignition under favourable conditions permits light-load part throttle mixtures to be weakened from something like 17:1 to 21:1 but there does not appear to be any benefits under full-load conditions when the mixture is on the rich side.

b) Dual ignition tends to eliminate cycle-to-cycle variations in cylinder pressure, particularly under light-load conditions, which helps to reduce rough and harsh running.

c) Dual ignition tends to permit slightly larger valve overlaps to be used without increasing exhaust emission under small throttle and idle running conditions.

d) Dual ignition tends to develop more torque in the lower speed range but this benefit becomes marginal with increasing engine speed.

e) Dual ignition tends to improve the engine's fuel consumption under part-load operating conditions.

f) Dual ignition enables the ignition advance timing to be somewhat retarded which tends to improve the engine's smoothness and response.

g) Dual ignition enables more exhaust gas recirculation (EGR) to be utilized if this technique is employed to reduce exhaust emission.

4.6.12 A summary of good combustion chamber design practice for spark ignition engines

The following are required.

1 The smallest ratio of chamber surface-area to chamber volume as possible to minimize heat losses to the cooling system.

2 The shortest flame-front travel distance as possible to minimize the combustion period.

3 The provision for quenching the mixture furthest from the sparking plug to prevent the end-gas overheating. However, it must not be excessive as this would prevent the end-gases burning and, therefore, it would cause a high level of hydrocarbons to be expelled to the exhaust.

4 The most central sparking plug position possible to minimize the flame spread path or, alternatively, twin plugs can be used to achieve the same objective.

5 The location of the sparking plug should be as close to the exhaust valve as possible to maximize the temperature of the mixture surrounding the sparking plug electrodes.

6 The incoming mixture must have adequate swirl to mix the air and fuel rapidly and intimately, but not too much as this could lead to excessive heat losses.

7 The provision for squish zones to excite the mixture into a state of turbulence just before combustion occurs.

8 The provision for adequate cooling of the exhaust valve to prevent overheating, distorting and burning occurring.

9 The provision for the incoming fresh charge to sweep past and cool the sparking plug electrodes to avoid pre-ignition under wide throttle opening.

10 The utilization of the highest possible compression ratio to maximize the engine's thermal efficiency without promoting detonation.

11 The inlet and exhaust valve sizes and numbers should be adequate to expel the exhaust gases and to fill the cylinder with the maximum mass of fresh charge in the upper speed range.

12 The degree of turbulence created should be controlled to prevent excessively high rates of burning and, correspondingly, limit very high rates of pressure rise which would cause rough and noisy running.

4.7 Fundamentals of the combustion process in diesel engines

4.7.1 Influence of compression ratio and engine speed on cylinder pressure and temperature

The power output of a diesel engine is controlled by varying the amount of fuel spray injected into a cylinder filled with compressed and heated air whereas the petrol engine is controlled by throttling the pre-mixed charge entering the cylinder.

The pressure and temperature reached at the end of the compression stroke will depend primarily upon the compression ratio intake temperature and the speed of the engine. The relationship between maximum (TDC) cylinder pressures and temperature for various compression ratios and engine speeds is shown in Tables 2

Table 4.2 Relationship between compression ratio and cylinder air compression pressure and temperature with constant air intake temperature (50°C)

Compression ratio	Compression pressure (bar)	Compression temperature (°C)
12	26	525
14	32	560
16	37	590
18	42	610
20	48	620
22	54	627
24	59	632

Table 4.3 Variation of cylinder air compression pressure and temperature with rising engine speed when using a 15.5 : 1 compression ratio

Engine speed (rev/min)	Compression pressure (bar)	Compression temperature (°C)
0	34	535
500	38	605
1000	42	680
1500	45	750
2000	48	805
2500	51	850
3000	54	885

and 3. It should be observed that injection usually commences 15° to 20° before TDC when both cylinder pressures and temperatures are much lower. As an example, a 15 : 1 compression ratio engine would have something like 600°C maximum temperature at TDC but at 15° before TDC this would only amount to 530°C.

In Table 3 it can be seen that the pressure and temperature rise in the cylinder with increased speed is largely due to the reduced time available for compressed air to escape past the piston rings and heat to be lost through the cylinder walls and head.

4.7.2 Diesel engine heterogeneous charge mixing
(Fig. 4.51)

In the diesel engine the air–fuel mixture formation is of a heterogeneous nature; that is, it is locally concentrated at various sites and is therefore unevenly distributed throughout the cylinder and combustion chamber. Injected fuel spray penetrates the highly compressed and heated air mass where it is pulverized into many very small droplets in a localized formation. The mixing of the localized spray of fuel droplets in the hot air charge causes stoichiometric (14.7 : 1 by weight) air–fuel ratio combustion zones to be established which are completely surrounded by pure air only. Thus, the overall (averaged out) air–fuel mixture ratio range may vary from a rich, full load, 20 : 1, to a weak no-load, 100 : 1, air–fuel ratio (Fig. 4.51). Most engines operate with at least 20% excess air due to the difficulty of introducing sufficient exposed oxygen to the fuel vapour in the given time available so that the combustion process can be completed before the exhaust valve opens. If the oxygen supply is partially prevented from getting to the fuel vapour early enough during the power stroke then incomplete combustion, polluted exhaust gas and dark smoke will result.

4.7.3 Diesel engine injected spray combustion process
(Figs 4.58 and 4.69)

When injection of fuel into the combustion chamber commences towards the end of the compression stroke, the quantity of fuel discharged is spread out over a predetermined period. The fuel spray enters the hot chamber but does not immediately ignite, instead it breaks up into very small droplets (Fig. 4.58) and once these liquid droplets are formed, their outer surfaces will immediately start to evaporate so there will be a liquid core surrounded with a layer of vapour. At this point it should be explained that the burning of a hydrocarbon fuel in air is purely an oxidation process. Thus, initially, heat liberated from the oxidation of the fuel vapour is less than the rate at which heat is extracted by convection and conduction, but eventually a critical temperature is reached when the rate of heat generated by oxidation exceeds the heat being dissipated by convection and conduction. As a result, the temperature rises which, in turn, speeds up the oxidation process thus further increasing the heat released until a flame site or sites (Fig. 4.69(c and d)) are established, this being known as the *ignition* and the temperature at which it occurs is called the *self-ignition temperature* of the fuel under these conditions. The heat required for further evaporation of the fuel droplets will thus

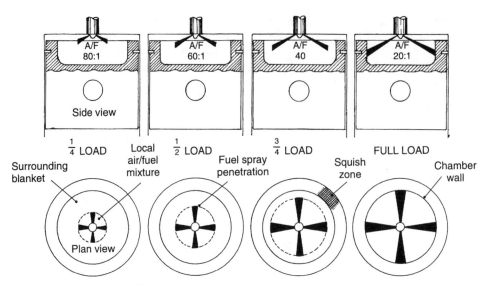

Fig. 4.51 Heterogeneous mixing of fuel and air

be provided from heat released by the oxidation process, which is referred to as combustion.

The liquid core, now surrounded by layers of heated vapour, oxidizes (burns) as fast as it can; that is, it finds fresh oxygen to keep the chemical reaction going.

Once the physical delay to convert the fuel spray into tiny droplets and the chemical reaction delay to establish ignition from the initial oxidation process are over, the rate of burning will depend upon the rate at which each burning droplet can find fresh oxygen (Fig. 4.69(a–d)) to replace the oxygen consumed as freshly formed layers of vapour oxidize. In other words, the rate of burning is largely dependent on the speed at which the droplets are moving through the air or the air is moving past the droplets.

4.7.4 Compression ratio
(Figs 4.52, 4.53, 4.54, 4.55, 4.56 and 4.57)

As the engine compression ratio is raised, the cylinder compression pressure and temperature increase; conversely, the ignition time lag between the point of injection to the instant when ignition first commences reduces. However, the cylinder pressure (Fig. 4.52) and temperature (Fig. 4.53) will also depend upon the speed of the engine since the higher the speed the shorter will be the time for air leakage and heat losses. The reduction in ignition delay period with increased compression ratio at one engine speed is shown in

Fig. 4.53. It is essential with diesel engines to have a sufficiently high compression ratio which will raise the air charge temperature just above the self-ignition temperature of the injected fuel spray.

Raising the compression ratio increases the density and turbulence of the charge, and this increases the rate of burning and, accordingly, the rate of pressure rise and the magnitude of the peak cylinder pressure reached. The characteristics of the pressure rise relative to the piston

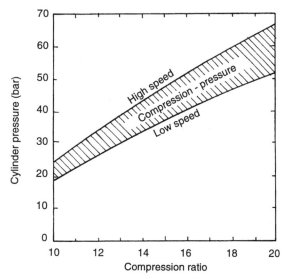

Fig. 4.52 Effect of compression ratio on the cylinder–compression pressure band between high and low engine speeds

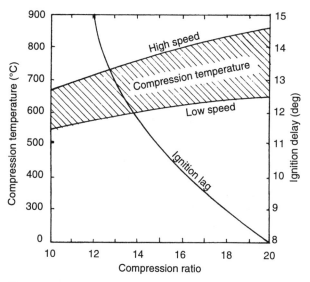

Fig. 4.53 Effect of compression ratio on the cylinder–compression temperature band between high and low engine speed

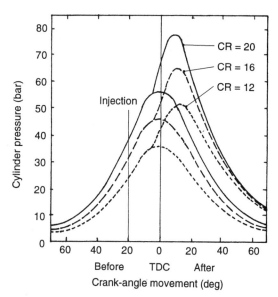

Fig. 4.55 Effect of compression ratio on the characteristic pressure–crank-angle movement diagrams for a diesel engine

stroke or crank-angle movement is illustrated in Figs 4.54 and 4.55 respectively.

The improvements in thermal efficiency and the specific fuel consumption by raising the compression ratio are clearly shown in Fig. 4.56. However, the disadvantages are that the higher cylinder pressures increase the pumping losses (induction and exhaust), friction losses (piston rings and piston) and compression and expansion

losses (heat lost to walls) as more work is done in squeezing together the trapped air charge. Thus, the result is a reduction in the mechanical efficiency with rising compression ratio (Fig. 4.57).

4.7.5 Injection spray droplet size
(Figs 4.58(a, b and c) and 4.69(a–d))

The effective rate of burning depends largely

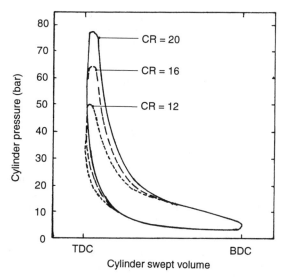

Fig. 4.54 Effect of compression ratio on the characteristic pressure–volume diagrams for a diesel engine

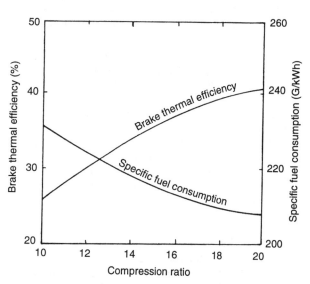

Fig. 4.56 Effect of compression ratio on the thermal efficiency and specific fuel consumption

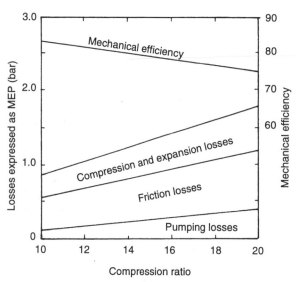

Fig. 4.57 Effect of compression ratio on the pumping, friction, compression and expansion losses and the resultant mechanical efficiency

upon the rate at which the products of combustion can be swept from the burning flame front and be replaced by fresh oxygen (Fig. 4.69(a–d)). Thus, the rate of burning depends on the relative movement of the burning droplets to the surrounding air charge.

The time taken to establish and ignite a film of vapour surrounding a liquid droplet is practically independent of the size of the droplet. However, the rate of burning and, correspondingly the pressure rise following ignition, will be dependent upon the exposed surface area of the vaporizing liquid droplets.

It would appear that the smaller the droplet size and the greater their number, the larger will be the total surface area which is subjected to ignition and therefore the greater the rate and extent of the pressure rise during the uncontrolled rapid pressure rise period.

However, if the droplets are very small they will quickly lose their momentum after being injected into the air (Fig. 4.58(a)) and therefore they will merely be dragged around with the air swirl instead of maintaining their own relative movement to the air (Fig. 4.58(c)). As a result, the air swirl will be unable to disperse the products of the burning droplets as there is no relative motion between these burning fuel droplets and the surrounding air. Consequently, the remaining unburnt fuel droplets will be suffocated by their own combustion products and starved of

any fresh oxygen, which is essential for combustion.

Conversely, if only a few large droplets were injected into the cylinder, they would be able to prolong their relative motion in the surrounding air (Fig. 4.58(b)) so that fresh oxygen could continue to be introduced to the ignited droplets, and their burning products would be equally well swept away. However, the reduced surface area of the larger but fewer number of droplets would severely slow the combustion rate.

A compromise in droplet size must therefore be made to maintain sufficient droplet size (and, therefore, momentum so that a fresh supply of air comes continuously into contact with the shrinking size of the unburnt portion of the liquid droplets) and to have available sufficient numbers of small droplets which provide an adequate surface vapour area for rapid combustion.

Droplet size, to some extent, can be controlled by the injection needle spring closing load. Generally, the greater the injector spring load, the smaller and finer will be the droplet size, whereas a light spring needle load tends to produce coarse liquid droplets.

4.8 Compression ignition combustion process

(Fig. 4.59(a))

The combustion process may be considered to take place in four phases or periods:

1 the delay period;
2 the rapid pressure rise period;
3 the mechanically controlled period;
4 the after-burning period.

4.8.1 The delay period (θ_1)
(Figs 4.59(a), 4.20 and 4.21)

The first combustion phase occupies the time between the initial discharge of fuel into the cylinder to the time where the spray droplets partially vaporize and then ignite, i.e. physical and chemical delay (Fig. 4.59(a)).

The ignition is represented by the oxidation process initially establishing a nucleus of flame, the heat liberated then raises the temperature and pressure of the air mass until the burning air charge pressure exceeds the normal compressing

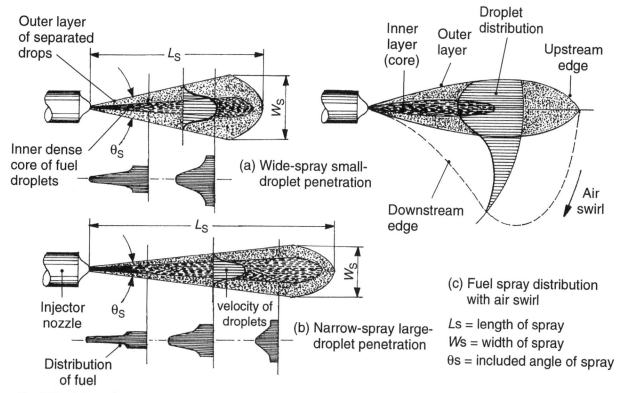

Fig. 4.58 Injected fuel spray characteristics

pressure. This point defines the end of the first phase, known as the delay period, and the start of the second rapid pressure rise period.

The delay period therefore occupies the dura-tion of the time between the beginning of injec-tion and the actual instant when combustion is first detected, usually at a crack-angle position, just before TDC.

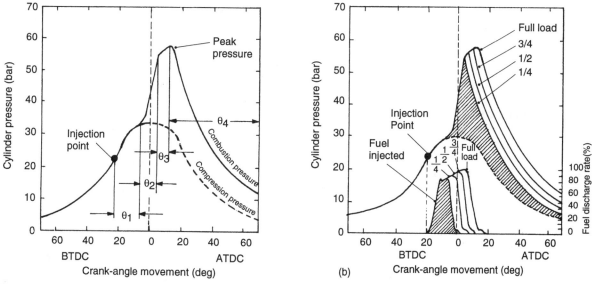

Fig. 4.59 Effect of varying the amount of fuel injected on the characteristic P–θ diagram

203

The time delay between the initial injection and ignition is influenced by the following:

1 the amount of penetration and atomization of the fuel spray;
2 the temperature and pressure of the cylinder air mass at the moment of injection;
3 the amount of pre-combustion turbulence;
4 the ability of the fuel to ignite readily, this being related to the fuel's cetane number.

4.8.2 The rapid pressure rise period (θ_2)

(Figs 4.59(a), 4.60, 4.61 and 4.62)

This is the phase of rapid combustion in which the fuel injected during the previous delay period and some of the fuel continuing to be discharged have had time and room to penetrate and spread out into the combustion chamber. Consequently, the heated liquid fuel droplets quickly become surrounded by layers of fuel vapour which burn as fast as they can find fresh oxygen to continue the combustion process.

This second period (Fig. 4.59(a)) then becomes an uncontrolled combustion phase whose rate of burning is mainly dependent on the ability of the air swirl to bring a continuous supply of fresh air to each burning droplet and to sweep away the products of combustion. It is essential that the chemical reaction is not starved of oxygen by blankets of burnt gas separating the burning vapour and the remaining fresh air charge.

Therefore, the rapid pressure rise period occupies the crank-angle movement between the first rise of cylinder pressure above the compression pressure due to the beginning of combustion, to the point where the rapid rate of pressure rise is suddenly reduced to a much lower rate of pressure increase. This slow-down in the burning rate is largely due to the increased heat burning the fuel almost immediately it is discharged from the injector nozzle, where there is very little unburnt oxygen available.

Since the rate of combustion increases with the speed of cylinder air swirl and this, in turn, increases with engine speed, the second rapid pressure rise phase remains approximately constant over the entire engine speed range.

The second period's rapid pressure rise and peak pressure is greatly influenced by the duration of the delay period. Generally, a large delay between the beginning of fuel injection and when actual ignition occurs causes a very high rate of

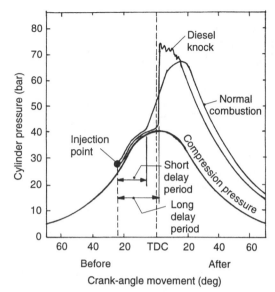

Fig. 4.60 Effect of short and long delay period on the characteristic P–θ diagram

pressure rise, whereas a small delay period results in a more gradual rate of pressure increase.

If the duration of the delay period is short, only a small amount of fuel will have entered the cylinder before ignition and burning commences (Fig. 4.60). Correspondingly, the energy released by this portion of discharged fuel at the beginning of the second period will only provide a modest rate of pressure rise during the following uncontrolled second period. It therefore has the effect of producing smooth combustion.

However, if there is a longer delay period, a considerable amount of fuel will have entered the cylinder before burning commences (Fig. 4.60). Hence, a relatively large amount of energy will be released just after the actual point of ignition so that a much more rapid and higher pressure rise occurs during the uncontrolled second period, it therefore usually results in rough running.

Some control of the second period's rapid pressure rise can be provided without sacrificing a loss in power by introducing a very small amount of fuel into the cylinder during the early part of the delay period, followed by a much higher rate of fuel discharge once combustion is established.

Provided the cylinder temperature at TDC remains very nearly the same over the engine speed range, the time delay between the commencement of injection and the point of ignition is approximately constant. However, with increased engine speed and/or load, the cylinder

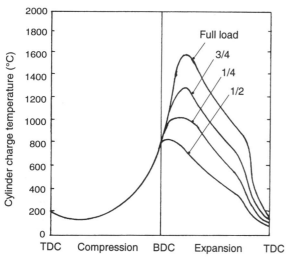

Fig. 4.61 Effect of engine load on cylinder charge temperature

pressure and temperature (Fig. 4.61) increases and therefore produces a reduction in the duration of the delay period (Fig. 4.62). The reduction of the delay period's crank-angle movement due to the increase in cylinder temperature with rising speed does not totally compensate for the increase in crank-angle duration with rising engine speed assuming a constant time duration for the delay period. Consequently, some additional fuel injection advance is necessary with the smaller wide speed range engines.

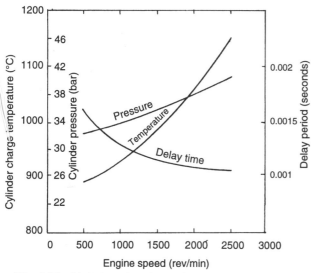

Fig. 4.62 Variation of compression pressure, temperature and delay period on base of engine speed in a direct injection–compression ignition engine

4.8.3 The mechanically controlled burning period (θ_3)
(Fig. 4.59(a and b))

This period begins at the crank-angle movement point when the very steep pressure increase has changed to a much slower pressure rise rate (Fig. 4.59(a)).

This period begins when the fuel injected in the rapid-pressure-rise second phase finds it more difficult to seek out and react with fresh oxygen in the remaining and much reduced unburnt air mass.

The crank-angle duration for fuel injection varies from a minimum, which occupies the first two phases for approximately no-load, and an extension to the end of the second phase for full-load power output.

Thus, once fuel injection crank-angle movement has reached the end of the rapid pressure rise second phase, the late droplet arrivals enter a far higher preheated charge than the initial injection droplets and, therefore, ignite almost as they issue from the injector nozzle. Any further burning of the late-injected fuel droplets will liberate more heat and again raise the cylinder pressure, but at a much reduced rate due to the burnt products of combustion in this region reducing the available oxygen and therefore slowing down the speed of burning.

This third phase is therefore mechanically controlled by the injection pump output characteristics, which can shorten or extend the duration of fuel delivery (Fig. 4.59(b)). Thus, the prolongation of this variable period is determined directly by the amount of additional fuel supplied above that needed for no-load running.

4.8.4 The after-burning period (combustion termination) (θ_4)
(Fig. 4.59(a))

The beginning of the final phase is denoted by the rapid fall-off in cylinder pressure following the end of the period during which fuel was injected (Fig. 4.59(a)).

After-burning is the fourth terminating phase in which late combustion of previously unburnt fuel or products of partial combustion is completed during the mid-expansion stroke. Ideally, there should be no after-burning, but it does take place because the mixture distribution is uneven throughout the cylinder (heterogeneous mixture).

205

Thus, in some regions the mixture will be rich and in others too weak, and at the same time, the burnt products cannot be adequately swept clear of the unburnt vapour so that there is a scarcity of oxygen causing a slow-down in the burning rate.

Consequently, not only will combustion continue until late in the cycle, but because of falling temperature some incomplete products of combustion such as carbon monoxide, aldehydes, hydrocarbons, carbon and soot will be present in the exhaust.

Subsequently, if the fuel injection period is extended too much in the third controlled combustion phase, unpleasant exhaust smoke results. Exhaust smoke is thus a result of incomplete combustion in locally overlean regions, by spray impingement and with low chamber temperatures, such as occurs when starting the engine.

Grey or black smoke is the result of incomplete combustion in locally over-rich regions in the cylinder under overload, insufficient spray penetration, or late burning caused by retarded injection. Most diesel engines are rated for full-load operation just below the point at which the exhaust smoke visibly darkens.

4.8.5 Injection timing and engine speed
(Figs 4.62, 4.63 and 4.64)

To understand the relationship between the time occupied by the delay period and the crank-angle movement over this same interval with increasing engine speed, the following factors should be appreciated.

1 The shorter time available for compression reduces the amount of gas escaping past the piston rings, enabling higher compression pressures to be attained (Fig. 4.62).
2 The shorter time available for the heat of combustion to be transferred to the cooling system raises the temperature of the compressed charge (Fig. 4.62).
3 Higher engine speed increases the intensity of the pulsating vortices which form the turbulence and, in turn, raise the rate of flame propagation throughout the combustion chamber.
4 Higher engine speed increases the injection pressure and the intensity and thinness of atomization.

The conclusion drawn from these factors is that the delay period time interval will decrease with rising engine speed (Fig. 4.62) whereas the corresponding crank-angle movement for the same time interval increases.

However, the rising cylinder pressure, temperature, turbulence and degree of atomization cannot completely compensate for the proportional increase in crank-angle movement with rising engine speed.

With indirect-injection combustion chambers (swirl and precombustion), the intensity of the four factors discussed will be such as to reduce considerably the delay period time interval at higher speeds. Thus, the corresponding crank-angle movement will only be slightly increased, whereas, with a direct-injection combustion chamber, the rise in pressure, temperature, turbulence and the amount of atomization will not be increased quite so much. Accordingly, the delay period in terms of time will not decrease quite at the same rate and, therefore, the delay period crank-angle movement will show a far larger increase with rising engine speed.

Consequently, if the engine speed range is small (500–2000 rev/min) with both direct or indirect combustion chambers, it may not be necessary to advance the point of injection (Fig. 4.63) since the rapid pressure rise and peak are only slightly retarded and reduced with rising speed. However, with large speed-range engines (700–5000 rev/min) of the direct-injection type, a

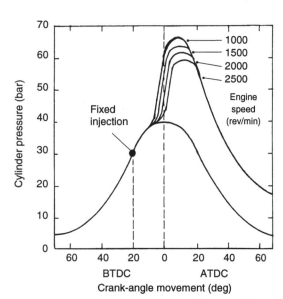

Fig. 4.63 Effect of engine speed on a P–θ diagram (narrow speed range large direct injection)

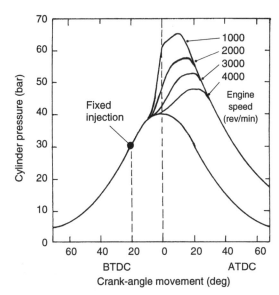

Fig. 4.64 Effect of engine speed on a P–θ diagram wide speed range small direct injection diesel

large amount of automatic injection advance becomes essential (Fig. 4.64) otherwise the peak cylinder pressure will decrease, as it occurs later in the power stroke, whereas indirect-injection types of chambers may only need a small amount of automatic advance.

4.9 Diesel knock

(Fig. 4.60)

In a diesel engine there is no pre-mixed fuel vapour and air, as in the petrol engine. Therefore, there cannot be any preignition, even if there is a hot-spot capable of creating premature ignition in the cylinder and, likewise, there is no preheating by the flame front of the unburnt mixture to cause detonation.

Diesel knock is the sound produced by the very rapid rate of pressure rise during the early part of the uncontrolled second phase of combustion. The primary cause of an excessively high pressure rise rate is due to a prolonged delay period (Fig. 4.60).

A very long ignition lag after injection causes a large proportion of the fuel discharge to enter the cylinder and to atomize before ignition and the propagation of burning actually occurs.

Accordingly, when combustion does commence a relative large amount of heat energy will be released almost immediately, this correspondingly produces the abnormally high rate of press-

ure increase, which is mainly responsible for the rough and noisy combustion process under these conditions (Fig. 4.60).

Generally, it has been found that provided the rate of pressure increase does not exceed 3 bar per degree of crank-angle movement, combustion will be relatively smooth, whereas between a 3 and 4 bar pressure rise there is a tendency to knock and, above this rate of pressure rise, diesel knock will be prominent.

An extensive delay period can be due to the following factors:

1 a low design compression ratio permitting only a marginal self-ignition temperature to be reached;
2 a low combustion pressure due to worn piston rings or badly seating valves;
3 poor fuel ignition quality; that is, a low cetane number fuel;
4 a poorly atomized fuel spray preventing early ignition to be established; that is, a worn and badly seating injector needle or blocked nozzle hole or holes;
5 an inadequate injector needle spring load producing coarse droplet formation;
6 over-advanced injection causing the fuel spray to enter the cylinder and to accumulate very early on in the compression stroke before adequate pressure and temperature has been reached;
7 a very low air intake temperature in cold wintry weather and during cold starting.

4.9.1 Cetane number (CN)
(Fig. 4.65)

The cetane number of a diesel fuel is a measure of its ignition quality. When a fuel is injected into the hot compressed air in the cylinder, it must first be raised to a temperature high enough to ignite the air–fuel mixture. This requires a certain amount of time, known as *ignition delay*. Different kinds of fuel, even though similar in appearance, may show differences in the time lag between the start of fuel injection and the commencement of ignition (Fig. 4.65). The ease with which diesel fuels ignite in terms of the time interval between injection and ignition is called ignition quality. A fuel which has a very short delay period has a good (high) ignition quality as opposed to a fuel having a prolonged delay which therefore has poor (low) ignition quality.

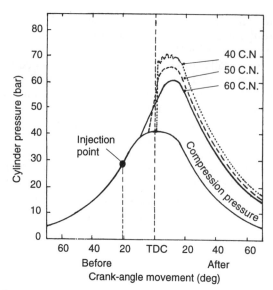

Fig. 4.65 Effect of 'octane' number on cylinder pressure rise

Ignition quality is usually determined by an engine bench test which measures the ignition time delay under standard carefully controlled conditions.

In such a test, the unknown fuel is rated on a scale between 0 and 100 against a pair of pure hydrocarbon reference fuels. Cetane ($C_{16}H_{34}$) (n-hexadecane) a straight chain paraffin which has a very high ignition quality (short delay) and does not readily knock, is assigned to the top of the scale by a cetane number of 100, whereas heptamethylnonane (HMN) which has a very low ignition quality (long delay) and readily knocks, is represented at the bottom end of the scale by a cetane number of 15. Originally, the low ignition quality reference fuel was alpha methylnaphthalane ($C_{11}H_{10}$) which was given a cetane number of zero. However, heptamethylnonane, a more stable compound but with a slightly better ignition quality (CN = 15), now replaces it.

Hence, the cetane number (CN) is shown by

cetane number (CN) = percent cetane
$\quad$ $0.15 \times$ percent heptamethylnonane

A standardized single-cylinder pre-chamber variable compression ratio engine (Fig. 4.31) is used operating under fixed conditions

inlet temperature	65.5°C
jacket temperature	100°C
speed	900 rev/min
injection timing	13° BTDC
injection pressure	103.5 bar

The engine is run on a supply of commercial fuel of unknown cetane number under standard operating conditions. With the injection timing fixed to 13° BTDC, the compression ratio is varied until combustion commences at TDC (by observing the rapid rise in cylinder pressure) thereby producing a 13° delay period of 0.0024 s at 900 rev/min.

A selection of reference fuel blends are then tested, where again the compression ratio is adjusted for each blend to obtain the standard 13° delay period. The percentage of cetane in one of the blends of reference fuels which gives exactly the same ignition delay (ignition quality) when subjected to the same compression ratio is called the cetane number of the fuel. Thus, a commercial 45 cetane fuel would have an ignition delay performance equivalent to that of a blend of 45% cetane and 55% heptamethylnonane (HMN) by volume.

The higher the cetane number, the shorter the delay period (high ignition quality) and the greater the resistance to diesel knock. Conversely, a low cetane number extends the delay period (low ignition quality) and makes the engine more prone to diesel knock.

A rough guide to cetane number requirements for different speed range engines is as follows.

Class	Speed range	Cetane number
Slow speed	100–500 rev/min	25–35
Medium speed	300–1500 rev/min	35–45
High speed	500–5000 rev/min	45–55

High octane fuels, if used in the diesel engine, would find it difficult to ignite and therefore would have a very low cetane number. Conversely, high cetane fuels, if sufficiently volatile when used in the petrol engine, would readily knock and therefore would have low octane values.

In fact, there is an approximate inverse relationship between cetane (CN) and octane (ON) number which may be given as

$$CN = 60 - \frac{ON}{2}$$

which is said to be accurate within ±5%.

4.9.2 Diesel index (DI)

It is expensive to conduct cetane number tests and therefore simpler and cheaper methods of predicting ignition quality have been devised. One such method is the diesel index.

This scale is made possible because ignition quality is sensitive to hydrocarbon compositions; that is, paraffins have high ignition quality and aromatic and naphenic compounds have low ignition quality.

The diesel index gives an indication of the ignition quality obtained from certain physical characteristics of the fuel as opposed to an actual determination in a test engine. The index is derived from the knowledge of the aniline point and the Amerian Petroleum Institute (API) gravity, which is put together in the following

Diesel index (DI)

$$= \text{Aniline point (°F)} \times \frac{\text{API gravity (deg)}}{100}$$

The aniline point of the fuel is the temperature at which equal parts of fuel and pure aniline dissolve in each other. It therefore gives an indication of the chemical composition of the fuel since the more paraffinic the fuel the higher the solution temperature. Likewise, a high API gravity reflects a low specific gravity and indicates a high paraffinic content, which corresponds to a good ignition quality.

The correlation between the diesel-index and the cetane-number is not exact and with certain fuel compositions it is not reliable but, nevertheless, it can be a useful indicator for estimating ignition quality.

4.10 Open, semi-open and divided combustion chambers for diesel engines

4.10.1 Direct-injection open quiescent-type combustion chamber
(Fig. 4.66)

With large slow and medium speed engines running up to 1500 rev/min there is sufficient time for the fuel to be injected into the cylinder and for it to be distributed and thoroughly mixed with the air charge so that combustion takes place over the most effective crank-angle movement just before and after TDC, without having to resort to induction swirl and large amounts of compression squish.

Without air swirl in the combustion-chamber there is no high hot-gas velocity, which would increase the thermal impingement on the surfaces surrounding the chamber space. Accordingly, there will be more heat available to do useful work so that higher brake mean effective pressures can be obtained where mixing of the fuel and air is achieved purely by the intensity of the spray and its ability to distribute and atomize with the surrounding air.

This is made possible by locating the injector in the centre of a four-valve cylinder-head and using an injector nozzle with something like 8 to 12 holes all equally spaced and pointing radially outwards so that they are directed towards the shallow wall of the combustion chamber.

Chambers of this design, where the air movement is almost quiescent (the air is inactive) and mixing depends nearly entirely on the discharged spray distribution and atomizing of the fuel particles, are therefore known as *quiescent open chambers* (Fig. 4.66). These chambers normally have a chamber diameter (d) to piston diameter (D) ratio of at least $d/D = 0.8$. The piston crown therefore has a flat narrow annular zone inside of which is the chamber recess, the base of the chamber from the centre to the wall has a downward dish shape which curves up and merges with the vertical wall of the chamber. The contour of the chamber is such that it conforms to the expanding spray formation so that fuel particles do not normally touch the chamber surfaces; likewise, the spray penetration should just fall short of reaching the vertical wall of the chamber.

The heat loss with this open chamber is the least compared with all other semi-open or divided combustion chambers, which is due to its very low ratio of surface area to volume, and thus its relative thermal efficiency is the highest.

For smaller high-speed engines with direct injection combustion chambers fewer injector nozzle holes have to be used since it would be almost impossible to produce the scaled-down nozzle hole sizes. Inevitably therefore, high-speed open or semi-open chambers when incorporated in high-speed diesel engines have only three to five nozzle holes. Consequently, to compensate for the higher engine speeds, smaller cylinder diameters and, therefore, fewer injector nozzle holes, it becomes essential to create air movement by adopting induction swirl, and utilizing deeper bowl-type chambers to promote air squish in order to speed up the mixing process.

Generally, open quiescent combustion chambers provide good cold starting and the lowest

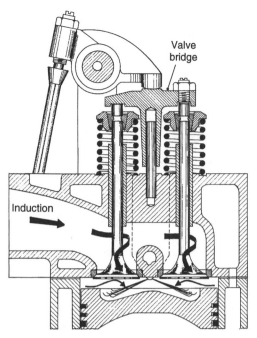

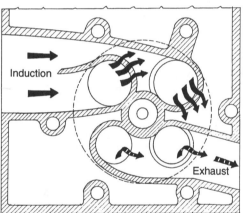

Fig. 4.66 Direct injection quiescent quadruple-valve combustion chamber

squish zone formed between the piston crown and flat cylinder-head (Fig. 4.67(a–d)).

The incoming air enters the cylinder in a tangentially and downward direction due to the valve port and seat being positioned to one side of the cylinder axis. Air is thus forced to spiral its way down and around the cylinder as it fills the space previously occupied by the outward moving piston (Fig. 4.67(a)).

At the end of the induction stroke the piston reverses its direction and then commences its compression stroke. Towards the end of the compression stroke the bump-clearance between the flat annular piston crown and the cylinder-head quickly decreases causing it to squeeze the swirling air charge inwards towards the inner chamber bowl (Fig. 4.67(b)). Thus, the air stream from all sides of the annual squish zone flows radially inwards meeting at the centre where it is then deflected downward into the bowl. At the bottom, the air disperses radially outward and then upwards to the lip of the chamber wall (Fig. 4.68(a)). At this point, the upward moving air will be met by more inwardly moving compression squish which again pushes the air towards the centre and down. This flow pattern repetition resembles a torus and therefore the air flow can be described as a toroidal movement within the combustion chamber.

The transfer of air from the annular squish zone to the inner bowl causes its rotational movement about the cylinder axis to follow an inward spiral path so that it enters the bowl in a tangential direction. Since the energy contained by the moving air is not lost as it transfers from a large to a much smaller circular path, the rotational flow of air around the chamber wall is considerably increased.

Having obtained a compact charge of swirling air towards the end of the compression stroke, fuel commences to be sprayed from the four nozzle holes into the cylinder by the centrally located injector when the crankshaft's angular position is something like 15° to 20° before TDC (Fig. 4.68(a)).

The pressure behind the discharged fuel projects it radially outwards until it strikes the chamber wall. Some of this fuel bounces off the wall while the remainder clings and spreads over the wall. The completion of the fuel injection period simply increases the amount of fuel deposited or rebounded from the chamber wall until the metered quantity of fuel per injection has been ejected (Fig. 4.67(b)).

specific fuel consumption values relative to semi-open and divided combustion chambers.

4.10.2 Direct-injection semi-open volumetric combustion chamber
(Figs 4.67(a–d), 4.68(a–e) and 4.69(a–d))

This class of semi-open direct-injection combustion chamber utilizes a two-valve cylinder-head having a semi-swirl induction port with an inclined centrally located injector. This faces an opposing slightly offset bowl in the piston combustion chamber surrounded by a large annular

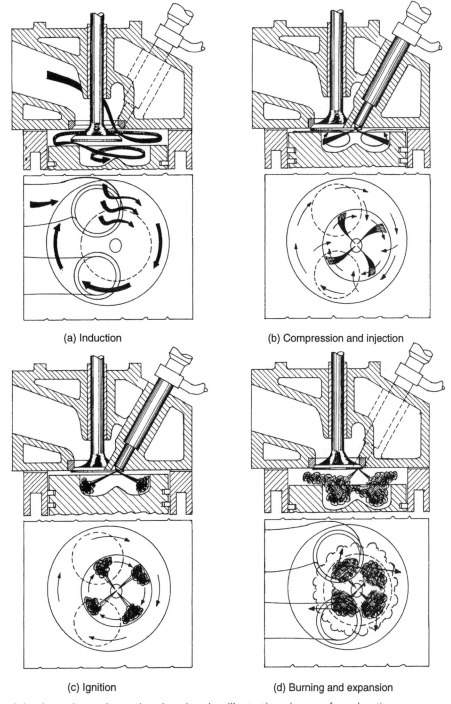

(a) Induction

(b) Compression and injection

(c) Ignition

(d) Burning and expansion

Fig. 4.67 Direct injection volumetric combustion chamber illustrating phases of combustion

The over-penetration of fuel (Fig. 4.69(b and d)) encourages the impact of the fuel against the chamber wall to spread and to apply a viscous drag effect which prevents the fuel being carried around with the air stream; consequently, the air flow passes across and through the fuel spray. Through these means, air is brought into contact with the fuel and thus considerably improves

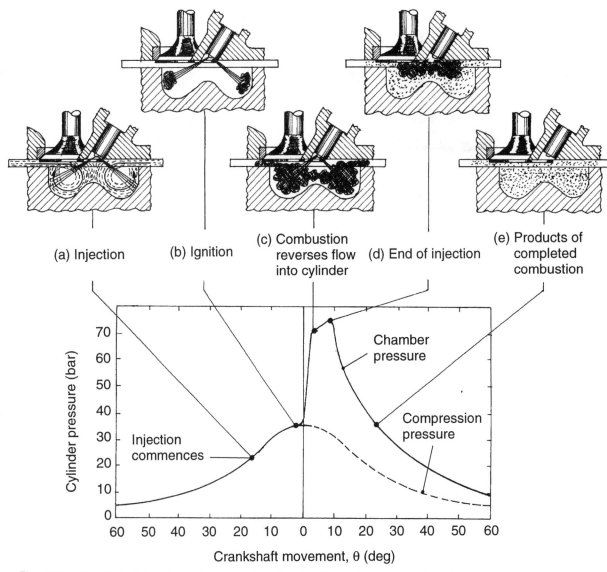

Fig. 4.68 Direct injection volumetric combustion chamber illustrating phases of combustion

the atomization and mixing of the air and fuel droplets.

However, under-penetration (Fig. 4.69(a and c))—where the fuel does not reach the chamber walls—simply allows the fuel spray to be swept around with the air swirl so that the fuel droplets become suspended in the air stream but do not actually have the opportunity to mix with the air charge.

Thus, fuel penetration and its redistribution effect due to wall interaction is essential for optimum mixing and controlled combustion.

As the centre core of the spray moves radially

outwards from the injector nozzle, its outer layers first become finely atomized and then transform into clouds of vapour. The compressed air occupying the spaces between the spray will have reached the fuel's ignition threshold temperature, and so the oxygen contained in the air in the vicinity of the fuel spray therefore reacts with the vapour, causing ignition to occur (Fig. 4.67(c)). The nuclei of flames, established randomly around the vapour clouds (Fig. 4.69(c and d)), then propagate rapidly towards the bulk of the mixture concentration near the chamber walls, the flames are then distributed and spread

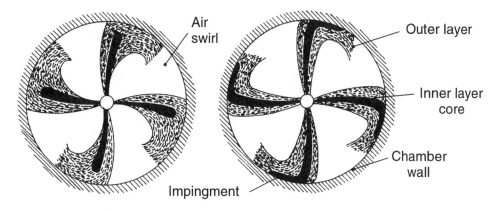

Air swirl

Outer layer

Inner layer core

Chamber wall

Impingment

(a) Spray formation without impingment   (b) Spray formation with impingement

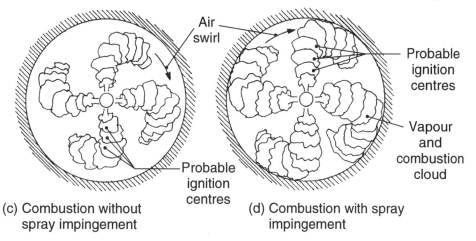

Air swirl

Probable ignition centres

Vapour and combustion cloud

Probable ignition centres

(c) Combustion without spray impingement   (d) Combustion with spray impingement

Fig. 4.69   Fuel spray development with and without combustion

throughout the bowl due to the general air movement within the chamber.

During expansion on the power stroke (Fig. 4.67(d)) the outward movement of the piston enables mixing of air and fuel to continue by the combined effect of air swirl and reversed squish (radially outward air movement from the bowl).

The rich mixture concentrated near the chamber wall is forced up and out of the chamber and is then brought into contact with the so-far untouched air located above the annular piston crown.

Finally, at something like 25° after TDC the combustion process will be about 95% complete which leaves the chamber and cylinder filled with soot and other products of combustion (Fig. 4.68(e)).

Direct-injection semi-open chambers are generally used for medium-to-large diesel engines which have cylinder capacities ranging from 0.75 to 2.5 litres per cylinder. These chambers are most suitable for operating up to 2000 rev/min for large engines; however, medium capacity engines can operate satisfactorily with smokeless exhaust up to about 3000 rev/min.

Generally, semi-open chambers have low heat losses and are good cold starters (permitting modest compression ratios as low as 12:1 to be adopted for large highly turbocharged and intercooled engines) whereas naturally aspirated medium-size engines could have an 18:1 compression ratio as an upper limit. However, the most popular compression ratio, 16:1, is normally used for naturally aspirated engines, whereas for moderate boost turbocharged engines the compression ratio would be derated to 15:1, thereby restricting excessively high peak cylinder pressures.

4.10.3 Bowl in the piston shape

Combustion chamber aspect ratio
(Figs 4.70, 4.71(a and b), 4.72 and 4.73)
The bowl in the piston semi-open combustion chamber should ideally be positioned centrally but, because the injector has to be located to one side of a two-valve cylinder-head, the chamber bowl is normally also offset so that it remains approximately concentric to the injector nozzle (Fig. 4.73). In general, an injector and chamber offset from the cylinder axis of up to 10% can be tolerated without it greatly hindering the organized air swirl or there being any loss of engine performance.

A basic comparison between chambers which are recessed in the piston head is the diameter (D) to depth (d) ratio, commonly known as the *aspect ratio*; that is, AR = D/d of the chamber (Fig. 4.70).

The extremes of chamber aspect ratios range from 5:1 for open shallow chambers to 2:1 for semi-open deep chambers.

Experience has shown that the best overall performance over a wide speed range occurs when the chamber aspect ratio lies between 2.5:1 to 3.0:1.

Deep, low, aspect-ratio chambers have shorter spray paths and therefore require less spray penetration than shallow, high aspect-ratio chambers.

Fuel injection spray penetration pressure is critical and must match the chamber aspect ratio to obtain the optimum amount of chamber wall

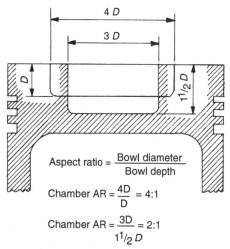

Aspect ratio = $\dfrac{\text{Bowl diameter}}{\text{Bowl depth}}$

Chamber AR = $\dfrac{4D}{D}$ = 4:1

Chamber AR = $\dfrac{3D}{1\frac{1}{2}D}$ = 2:1

Fig. 4.70 Illustration of combustion chamber piston bowl aspect ratio

wetting (Fig. 4.71(a and b)). Thus, a deep, low, aspect-ratio chamber requires less spray penetration than would a shallow, high, aspect-ratio chamber.

Multi-hole injectors (Fig. 4.71(a and b)) used with open and semi-open chambers have from three to five nozzle holes—probably the most popular number of nozzle holes for optimum performance is four—and the spray is in a cone of between 140° to 160° inclined angle.

The deeper the chamber, the higher the rotational air movement (swirl) in the chamber and the lower the maximum cylinder pressure, this being a desirable design feature. In contrast, deep chambers tend to generate slightly higher power outputs relative to shallow chambers in the engine's upper speed range.

Overswirl results in less fuel reaching the walls where the mixing is more effective, it tends to increase heat losses and can, under certain conditions, spoil the mixing through the interference of combustion products from the upstream spray.

In contrast, underswirl will restrict the rate at which the fuel spray mixes with the air stream.

However, overswirl reduces the mean effective pressure and increases the specific fuel consumption where swirl levels are considerably above the optimum compared with those in the underswirl conditions.

Isuzu square-bowl chambers
(Figs 4.71(a and b) and 4.72)

The square-bowl chamber adopted on some direct injection Isuzu engines (Fig. 4.72) has the effect of controlling swirl intensity, this is possibly due to the four-cornered chamber generating a periodic interference drag with the rotational movement of the air about the chamber's axis. It appears, therefore, that the air is excited into intensifying micro-turbulence. Consequently, the air–fuel mixing process is generally improved with relatively low swirl intensities and, correspondingly, this improves the engine's performance over the whole speed range.

Two of the major benefits with the square bowl are the much smaller deterioration of smoke and power with a retardation in the fuel injection timing, and also that substantial reductions in NO_x and noise can be obtained with only a minimum sacrifice in smoke and power.

A further refinement to the square-bowl chamber has been to add a lip to the vertical rim of the chamber which slopes at 30° to the horizontal

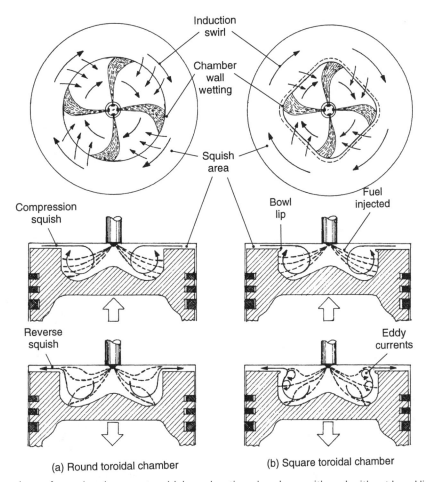

Induction swirl

Chamber wall wetting

Squish area

Compression squish

Bowl lip

Fuel injected

Reverse squish

Eddy currents

(a) Round toroidal chamber

(b) Square toroidal chamber

Fig. 4.71 Comparison of round and square toroidal combustion chambers with and without bowl lip

(Fig. 4.71(b)). This lip prevents the toroidal upward roll movement of the air ejecting fuel particles over the edge of the chamber rim into the squish zone (Fig. 4.71(a)) so that the majority of the mixing is completed and burnt inside of the piston bowl. At the same time, the lip tends to create and impose micro-turbulence within the chamber bowl.

Thus, basing performance on a constant smoke level, the brake mean effective pressure can be raised slightly at low and medium speeds; however, there is no gain in the upper speed range. Conversely, if the brake mean effective pressure and torque are not increased, then the exhaust emission can be reduced at low speed.

The square chamber engine can be started at temperatures as low as −10°C without any additional heating aid, but an unusual feature with this direct injection engine is that it incorporates a heater plug adjacent to the injector in the open chamber to facilitate easy cold engine starts at temperatures as low as −25°C.

Perkins squish lip bowl chamber (Fig. 4.73)

Perkins have developed what they call a squish lip cavity piston (Fig. 4.73) where the circular bowl has a flat base with curved side walls which taper into a large lip towards the mouth of the chamber. The mouth or opening of the chamber is approximately two-fifths of the piston diameter so that the piston crown creates a large squish zone.

Towards the end of compression, the rotational air swirl about the cylinder axis is squeezed between the cylinder-head and the annular squish area. This causes the air charge to be forced tangentially towards the centre of the chamber

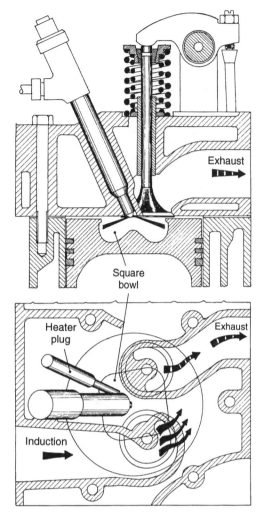

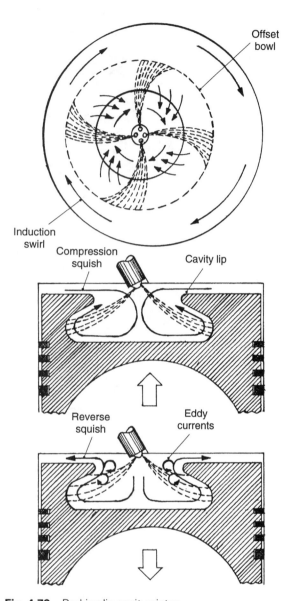

Fig. 4.72 Direct injection volumetric combustion chamber with a square-shaped bowl

Fig. 4.73 Perkins lip cavity piston

where it then takes on a downward transverse rolling motion, it then reaches the bottom and radially rolls outwards following the contour of the chamber base, walls and lip. The distorted semi-spherical shape of the chamber intensifies the rotational air swirl about the chamber axis, and the transverse rolling movement of the air speeds up and improves the mixing process due to the stratification of the fuel spray as it penetrates the vigorous and highly turbulent mass of trapped air.

The benefits of this partially enclosed chamber are a shorter and more thorough mixing phase for the fuel particles with the surrounding air, which brings about a reduction in the ignition delay period and a lower rate of pressure rise for the second combustion period. Consequently, the maximum cylinder pressures are reduced, there is less combustion noise and there is a decline in the formation of nitrous oxides and a slight improvement of the exhaust emissions as a whole. However, with the lip-type chambers, very high cyclic thermal loading subjects the piston to thermal stresses and strains which can cause the piston chamber contour to distort and crack.

4.10.4 Small direct-injection semi-open combustion chambers
(Figs 4.74 and 4.75)

The present generation of car high-speed direct-injection diesel engines are designed to reach speeds up to 4500 rev/min with low emission exhaust gas products.

With the conventional semi-open chamber the fuel is injected directly into the cylinder where it is distributed and mixed within 30° to 45° crank-angle movement at the end of compression stroke and the beginning of the power stroke.

This is achieved with the direct injection engine by promoting a moderate amount of rotational swirl and a large amount of compression squish, which combine to produce a toroidal swirl pattern within the piston bowl. The controlled movement of the air and the resulting turbulence do tend to increase roughly with engine speed; however, the time span for preparing and burning the air–fuel mixture becomes shorter and shorter until, at some point between 2500 and 3000 rev/min, the increased rate of turbulence cannot keep up with the reduced interval permitted for injection and mixing, which inevitably results in incomplete combustion and a corresponding high-level of exhaust emission (Fig. 4.74).

To overcome this inability to mix, atomize and burn in sufficient time when the piston and crankshaft are in their most effective positions, it becomes essential to increase the intensity of both

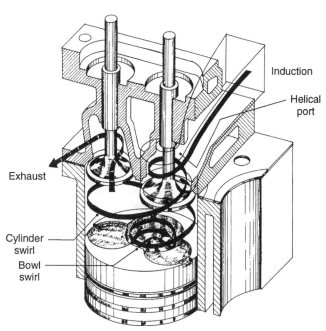

Fig. 4.75 Small diesel direct injection combustion chamber with helical induction port (Ford 2.5 lires)

the movement of air and that of the fuel spray, and at the same time to match the penetrating fuel velocity with the air swirl velocity so that the fuel particles mingle and receive the largest exposure to the oxygen in the air charge within the time available.

The extra air movement has been achieved by utilizing a helical or partial vortex form of induction port passage (Fig. 4.75) where the incoming air flow is given a helical twist or semi-vortex motion about the valve stem before it passes out between the opened valve head and its seat in a tangential direction to the cylinder axis. As a result, a high degree of air swirl is generated within the curved port passage before it is expelled into the cylinder.

To match the increase in organized air movement, the fuel injection pump camshaft has been strengthened and modified to raise the rate of injection and correspondingly to shorten the duration in which fuel is discharged into the cylinder, thus giving more time for mixing and combustion. At the same time, injection timing has been designed to be variable and precise so that it rapidly responds to the changes in operating conditions of the engine.

The level of exhaust emission generally increases with engine load and speed due to more fuel being injected with only the same quantity of

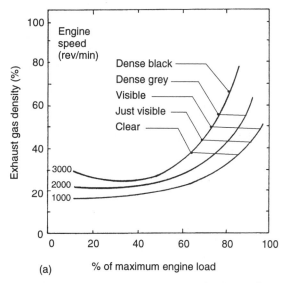

(a)

Fig. 4.74 Effect on engine speed and load on exhaust gas emission density

air charge filling the cylinders (Fig. 4.74). Thus, it becomes more difficult for all the fuel and oxygen to mix thoroughly and complete the combustion process, particularly with the decreasing combustion duration with rising engine speed.

The argument for adopting semi-open chambers as opposed to divided combustion chambers

The trend to shift from the divided indirect to the semi-open direct-injection combustion chamber is due to the lower mechanical and thermal efficiency of the divided indirect-injection combustion chamber relative to the semi-open direct-injection chamber.

The major reasons are threefold.

1 The divided chamber necessitates the forced movement of a portion of the air charge from the main cylinder chamber to the swirl or precombustion chambers, through the interconnecting restriction neck or holes, during the compression stroke and back again during the combustion stroke. Thus, a considerable amount of work is done and energy wasted in filling and emptying the air charge and the products of combustion, respectively, from the divided chamber.
2 The increased surface area of the swirl or precombustion chambers relative to their volume considerably increases the heat losses through the chamber walls so that some of the energy released by the combustion is dissipated by the cooling system via the chamber walls.
3 The increased heat losses of these divided chambers generally make it necessary to use higher compression ratios of 20 to 25:1 as opposed to 14 to 18:1 with the semi-open direct injection chambers to compensate for the otherwise lower cold start compression temperature. The effects are an increase in the pumping and friction losses in the engine.

If a semi-open direct injection engine can be designed to operate at speeds in excess of 4000 rev/min with acceptable exhaust emission but without sacrificing b.m.e.p., then the saving in the energy losses can amount to something like a 10 to 15% improvement in fuel consumption under certain driving conditions.

4.10.5 Direct-injection semi-open film (M-system) combustion chamber
(Figs 4.76(a–d) and 4.77)

The innovation and development of this film-type semi-open combustion chamber is due to Dr S. Meurer and the engine manufacturing company M.A.N. In recognition to both the inventor and the company it is known as the M-system.

This class of combustion chamber uses a two-valve cylinder-head with a high swirl or vortex type induction port with an inclined injector which is located to one side of the cylinder axis.

The combustion chamber is a spherical cavity in the piston crown with a small secondary recess to one side which aligns with the injector in the cylinder-head to provide access for the fuel spray discharge (Fig. 4.76).

Air from the high swirl generating induction port enters the cylinder where it is forced to rotate about the cylinder axis in a progressive spiral fashion as the piston moves away from the cylinder-head on its induction stroke (Fig. 4.76(a)). After the cylinder has been filled with air having a high intensity of swirl, the inlet valve closes and the air is compressed between the cylinder-head and the inwardly moving piston crown.

As the piston rapidly approaches TDC, air from the annular squish area surrounding the chamber recess is squeezed towards the centre of the chamber, it is then forced downwards and, at the bottom, outwards, where it then follows the contour of the spherical chamber wall until it again emerges at the mouth of the chamber, where further compression squish as the bump-clearance reduces, causes the transverse rolling movement to repeat itself (Fig. 4.76(b)).

The transference of air from the annular squish area to the inner chamber causes the rotational movement of the air around the cylinder to be considerably increased as it moves into the much smaller spherical chamber.

Just before the end of the compression stroke, fuel is injected into the cylinder from two nozzle holes set at acute angles to the chamber walls so that after the spray penetrates the swirling air and reaches the cylinder wall, it is not reflected but spreads over the surface in the form of a thin film 0.012–0.015 mm thick (Fig. 4.76(b)).

The swirl drag and centrifugal effect of the air on the glancing impact of the fuel against the chamber wall greatly assists the spreading of fuel

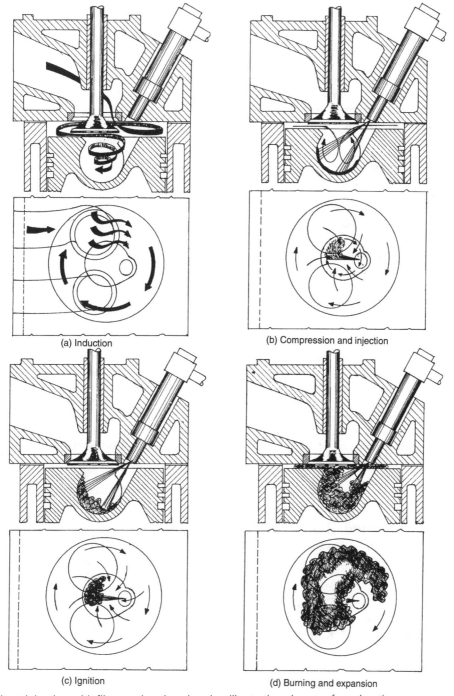

Fig. 4.76 Direct injection with film combustion chamber illustrating phases of combustion

droplets onto the chamber wall and in stratifying the liquid fuel surface into tiny particles which mix with the surrounding hot air. At the same time heat from the piston walls commences to vaporize the liquid film and any dislodged particles from the wall.

The discharge of liquid spray through the hot air charge causes the surrounding air to resist partially the jet penetration so that initial outer layers of fuel particles slow down very quickly to transform into vapour.

Immediately, this vapour commences to oxidize and to ignite (Fig. 4.76(c)); thus, something like 5 to 10% of the total quantity of fuel discharged per cycle burns in the spray stream near the injector nozzle with the minimum of delay, whereas the remainder of the spray discharge arrives shortly afterwards at the chamber walls where it spreads out to cover up to three-quarters of the chamber's hot surface area.

The vaporized fuel is carried away by the air stream and burns in the flame-front spreading from the initial ignition zone slightly beyond the injector nozzle and very nearly in the centre of the chamber.

For effective vaporization the temperature of the chamber wall must be maintained between 180 and 340°C, but if the temperature rises above this maximum, thermal decomposition and carburizing of the fuel will take place.

The energy released by the propagating combustion in the chamber bowl causes a rapid pressure rise and, simultaneously, an expansion of the burning charge (Fig. 4.76(d)).

An overall insight into the semi-open film (M-system) combustion-chamber process of combustion, from the point of injection to the end of burning, in terms of cylinder pressure variation on a base of crank-angle movement, is shown in Fig. 4.77(a–e).

4.10.6 Indirect-injection divided-chamber swirl combustion chamber (Ricardo Comet)

(Figs 4.78, 4.79 and 4.80)

This chamber was designed and developed at the research establishment of Ricardo and Company of Shoreham-by-Sea in Sussex, UK. The combustion process actually takes place in a two-stage divided-chamber system. Initially, combustion takes place in a spherical swirl chamber housed in the cylinder-head whereas the second half of the process is completed in the twin disc shaped recesses in the piston crown (Fig. 4.78).

The swirl chamber in the form of a sphere is located to one side and above the cylinder wall in the cylinder-head. The upper half of the sphere is cast directly in the cylinder-head whereas the lower half is a separate heat resisting nimonic alloy member flanged and cylindrical in shape with an upward facing semi-hemispherical chamber, it fits in a machined recess so that its underside is flush with the flat face of the cylinder-head. It is located and secured by a ball and flange while the outer cylindrical vertical wall stands away from the machined cylinder-head recess to create an isolating air gap (Fig. 4.79(a)). This chamber separator is commonly known as a heat regenerative member since it absorbs heat from combustion and dissipates it during the compression stroke.

An inclined passage through the base of the lower regenerative member forms a throat or neck between the spherical swirl chamber and twin adjacent circular cavities cast in the piston crown.

A pintle soft conical spray injector is positioned over the chamber at an acute angle to the swirl chamber whereas a cold-start heater-plug projects horizontally into the side of the chamber wall.

The lower regenerative member forms the lower half of the combustion chamber.

The delivery and expulsion of air and exhaust gases are provided by the inlet and exhaust valve ports. With the indirect-injected swirl-chamber method of combustion control, a high level of induction swirl is not so critical so that the intake port can be designed to cater more for improved breathing rather than the generation of a high intensity induction swirl. For high-speed car and van applications, an unusual triple-valve swirl-chamber manufactured by Citroën is shown in Fig. 4.78. Air is drawn tangentially into the cylinder via the twin induction port where it then moves in a circular downward path around the cylinder wall as the piston moves away from the cylinder-head (Fig. 4.79(a)).

On the return stroke the air change is compressed causing something like 40% of the air mass per induction stroke to be transferred through the throat of the regenerative member into the spherical swirl chamber (Fig. 4.79(b)).

The angle of the interlinking throat passage guides the air stream tangentially into the swirl chamber so that it is forced to follow the contour of the chamber wall in a vertical circular swirl many times during the compression stroke.

As air flows through the throat passage it absorbs heat from the hot alloy mass and from the chamber walls as it circulates around the chamber so that once combustion has been established, any fresh air entering the swirl chamber quickly

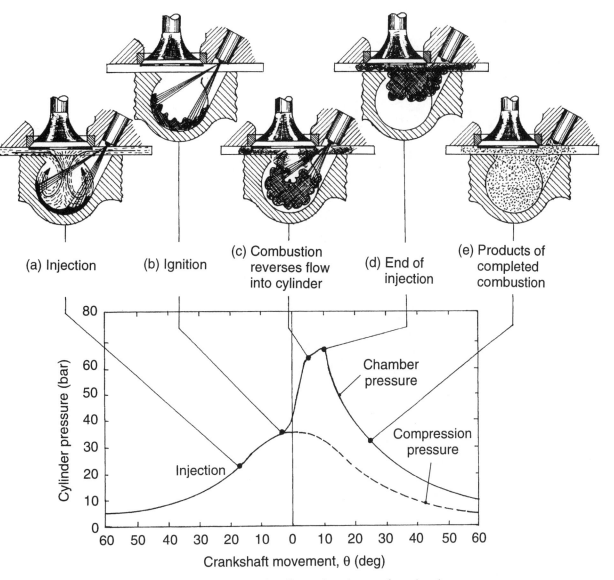

Fig. 4.77 Direct injection film combustion chamber illustrating phases of combustion

attains a temperature well above the threshold ignition temperature of the liquid fuel.

When the crankshaft is of the order of 20° to 25° before TDC, fuel is injected at an acute angle in a downstream direction to the air swirl to one side of the chamber, the spray penetrates the dense air change and impringes onto the spherical surface of the regenerative member (Fig. 4.78). Instantly the liquid fuel spreads out to form a thin film, which then vaporizes and is immediately swept around with the air stream.

The fuel vapour, oxygen and heat then com- bine to cause the oxidation reaction which is essential for ignition at random nuclei sites sur- rounding the vapour clouds within the swirl cham- ber (Fig. 4.79(c)).

Rapid flame-spread follows as unburnt vapour seeks out the oxygen in the dense but rapidly rotating air charge.

The high burning rate produces a correspond- ing rapid pressure rise in the swirl chamber. As a result, the burning charge will be blown down the throat of the regenerative member, after which it divides into two separate flame fronts as they

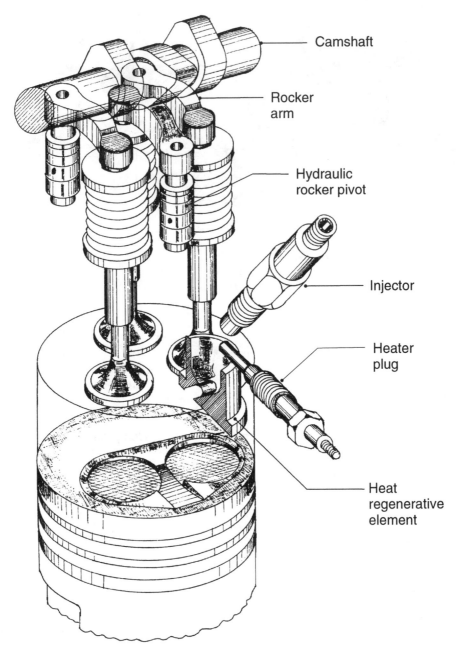

Fig. 4.78 Indirect-injection swirl combustion chamber utilizing triple-valve OHC operated with direct-acting end-pivoted rocker-arms (Citröen)

enter the twin, shallow disc-shaped recesses formed in the piston crown. The tangential entry compels the flame fronts to swirl clockwise and anticlockwise around their respective cavity walls, which gives the burning and unburnt vapour the maximum opportunity to search out the oxygen and, simultaneously, to displace the burnt products of combustion (Fig. 4.79(d)).

With further crank-angle movement, the flame-fronts now spread beyond the piston cavities into the main cylinder until all or almost all of the injected charge of liquid fuel has been consumed.

An understanding of the way the swirl chamber and main cylinder chamber pressures respond to the combustion process on a crank-angle base is shown in Fig. 4.80(a–e).

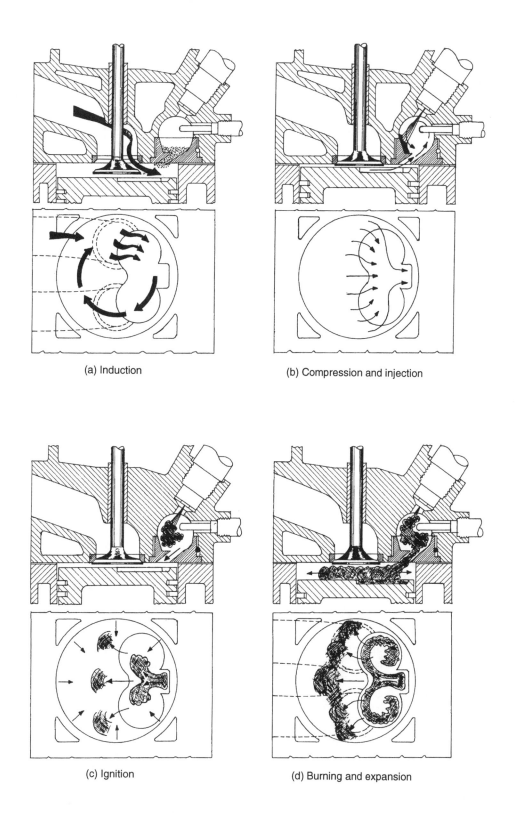

(a) Induction

(b) Compression and injection

(c) Ignition

(d) Burning and expansion

Fig. 4.79 Indirect-injection with swirl combustion chamber (Ricardo Comet) illustrating phases of combustion

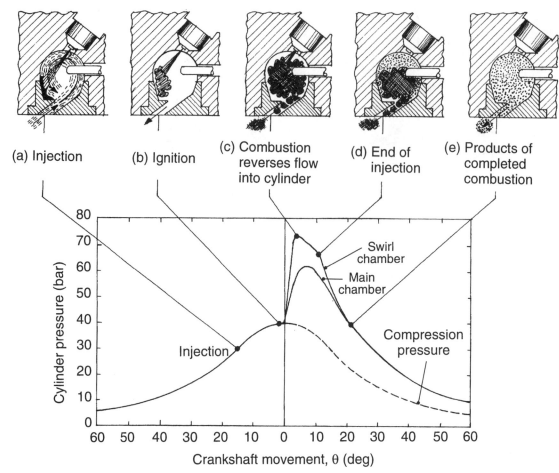

(a) Injection   **(b)** Ignition   **(c)** Combustion reverses flow into cylinder   **(d)** End of injection   **(e)** Products of completed combustion

Fig. 4.80 Indirect-injection swirl combustion chamber illustrating phases of combustion (Ricardo Comet)

4.10.7 Indirect-injection divided-chamber precombustion chamber (Mercedes Benz)
(Figs 4.81 and 4.82)

This divided-chamber two-stage combustion system incorporates a heat resisting alloy precombustion chamber mounted in the cylinder-head slightly to one side of the single inlet and exhaust valve seats (Fig. 4.81).

The precombustion chamber is a two-piece cylindrical unit consisting of a large diameter flanged section which houses the combustion chamber and a smaller diameter extended nozzle section. At the end of the enclosed nozzle are five radial holes which communicate with the main chamber, while the upper flanged end is opened up to accommodate the pintle injector. Within the

cylindrical casing is a spherical chamber with a narrow parallel passage or throat leading to the radial nozzle holes. A transverse bar with a spherical bulge in the middle (ball bar) is positioned in the lower half of the spherical chamber whereas a cold start heater plug intersects from the side of the upper half of the chamber wall.

When the inlet valve opens and the piston moves away from the cylinder-head, air enters the cylinder tangentially so that it rotates in a downward direction about the cylinder axis (Fig. 4.81(a)).

Once the piston has reached BDC it reverses its direction and commences to move inwards towards the cylinder-head until the inlet valve closes, this then completes the induction period.

As the piston approaches TDC, the air charge is compressed between the cylinder-head and the piston crown so that something like 35% to 45%

224

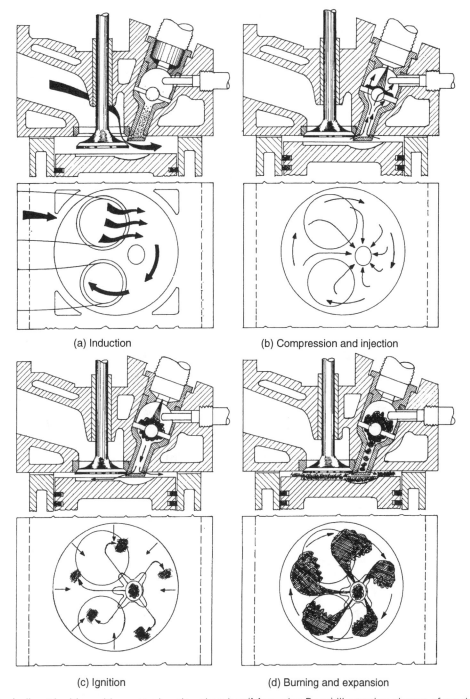

(a) Induction

(b) Compression and injection

(c) Ignition

(d) Burning and expansion

Fig. 4.81 Indirect-ignition with precombustion chamber (Mercedes Benz) illustrating phases of combustion

of the air is forced through the five nozzle holes which protrude below the flat cylinder-head (Fig. 4.81(b)).

Air will then be transferred from the cylinder into the precombustion chamber via the nozzle holes and parallel throat passage where it is excited into a vigorous and highly turbulent mass.

Towards the end of the compression stroke, fuel is discharged from the injector towards the centre of the chamber where it strikes the transverse located semi-spherical bar (Fig. 4.81(b)); however, the spray is very slightly angled so that a proportion of the spray misses the spherical bar and reaches the base of the chamber to one side of the nozzle throat. The liquid fuel now spreads out over the spherical bar and the mouth or throat of the nozzle passage. Almost immediately, the liquid film vaporizes and is torn away from the bar and chamber wall by the turbulent air movement.

Oxidation then commences causing random nuclei flame sites to form, these quickly propagate and spread throughout the hot dense air mass (Fig. 4.81(c)).

The resulting pressure rise created by the burning charge reverses the direction of air flow. The burnt and unburnt rich vapour charge is now blown down the throat of the nozzle where it then expands radially outwards through the five nozzle holes into corresponding shallow guide channels formed in the piston crown (Fig. 4.81(c)).

The thrust of combustion projects these directional jet-like flame-fronts towards the cylinder walls and, in doing so, sweeps the burnt gases and soot to one side while exposing the remaining fuel vapour to fresh oxygen (Fig. 4.81(d)).

The variation of pressure within the precombustion chamber and the main cylinder on a crank-angle movement base is shown in Fig. 4.82(a–e). This P–θ diagram illustrates the shock combustion pressure in the prechamber and the much reduced pressure generated in the main cylinder chamber.

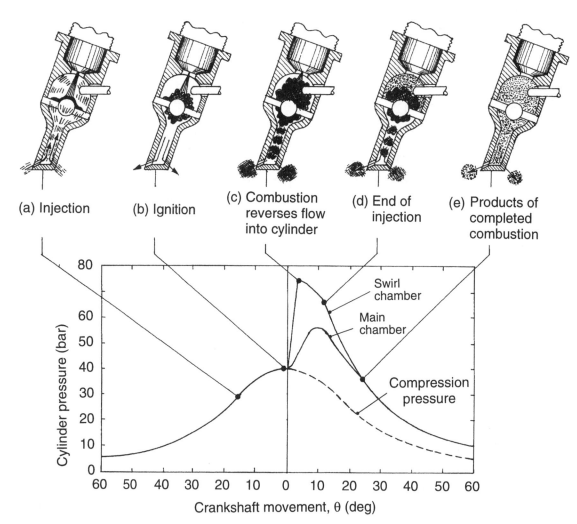

(a) Injection (b) Ignition (c) Combustion reverses flow into cylinder (d) End of injection (e) Products of completed combustion

Fig. 4.82 Indirect injection with precombustion chamber (Mercedes Benz) illustrating the combustion process

4.11 Comparison of spark-ignition and compression ignition engine load outputs, fuel consumption and exhaust compositions

(Figs 4.83, 4.84 and 4.85)

Fuel consumption loops have been plotted on a base of brake mean effective pressure (b.m.e.p.) for both petrol and diesel engines (Fig. 4.83). With the diesel engine, load and speed output is controlled entirely by varying the quantity of fuel injected into the cylinder without misfiring occurring, that is, from 0–100% of the maximum b.m.e.p. developed.

However, with the petrol engines, if there was no throttle (full throttle position) the effects of varying the mixture strength from the richest position (A) to the weakest position (E) produces a variation of b.m.e.p. (load) of only 25% of the maximum possible b.m.e.p. that is, from 75–100% of maximum b.m.e.p. (Fig. 4.83). Therefore the petrol engine's output control cannot be achieved alone by varying the mixture strength and therefore throttling the mixture coming into the cylinder becomes essential. The effects of throttling and adjusting the mixture strength are shown roughly by the $\frac{1}{4}$, $\frac{1}{2}$, $\frac{3}{4}$ and full throttle consumption loops in comparison with the single unthrottled consumption loop of the diesel engine over the engine's load range. The diesel engine is thus quality controlled and the petrol engine is quantity controlled. The different points on the consumption loops are listed below (Fig. 4.83)

A = excessively rich mixture gives slow and unstable combustion

B = maximum b.m.e.p. with something like 10–20% rich mixture

C = correct stoichiometric mixture of 14.7 : 1 by weight

D = maximum thermal effciency with something like 10–20% weak mixture (approaches ideal constant volume combustion)

E = excessively weak mixture gives slow burning and popping back through air intake

F = maximum b.m.e.p. with satisfactory clear exhaust requires mixture strength of about 18 : 1 by weight

G–H = maximum thermal effciency, minimum specific fuel consumption ranges between 50–85% of maximum b.m.e.p.

I = no-load (low speed idle) requires mixture strength 100–75 : 1 by weight

The pumping losses from the diesel engine remain reasonably constant from full load to no load whereas the pumping losses for a petrol engine progressively rises as the b.m.e.p. is reduced due to the increased throttling. This is reflected in the petrol engine by the mechanical efficiency reducing as the b.m.e.p. decreases which causes the consumption loops to take up higher and higher levels of consumption as the intake becomes more throttled (Fig. 4.83). In contrast with the diesel engine, as the load is reduced the single consumption loop only begins to rise when the b.m.e.p. drops below about half the engine's maximum b.m.e.p. (Fig. 4.83).

A petrol engine can operate effectively under steady conditions over a mixture strength ranging from 20:1 to 10:1, whereas a diesel engine which can normally only utilize 80% of its air charge with a reasonably clear exhaust, will operate from 18:1, that is 20% weak at full load to 100:1 at no load. Consequently, the petrol engine will always have a 15 to 20% higher maximum b.m.e.p. than a similar diesel engine (Figs 4.83 and 4.84).

A comparison of the composition of the dry exhaust gases for both the petrol and diesel engines is shown in Fig. 4.85. This shows that the petrol engine has considerable amounts of carbon monoxide (CO) formed as the mixture moves from the stoichiometric (14.7:1 by weight) to the

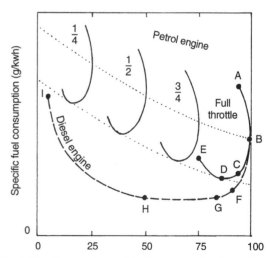

Fig. 4.83 Comparison of fuel consumption loops for petrol and diesel engines on a base of engine load (b.m.e.p.)

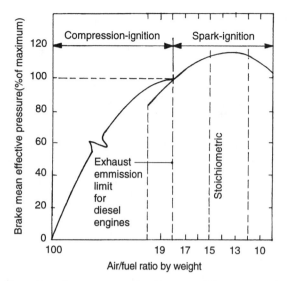

Fig. 4.84 Comparison of load (b.m.e.p.) for petrol and diesel engine on base of air–fuel ratio

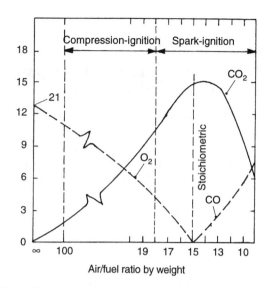

Fig. 4.85 Comparison of composition of exhaust gases for petrol and diesel engines on a base of air–fuel ratio

rich region of up to 10:1. This being within the petrol engine's mixture range when operating between idle and full power conditions.

In contrast, the diesel engine under full load never operates with mixture strengths greater than 20% rich, that is 18:1 by weight, so that no carbon-monoxide (CO) is present in the exhaust

(Fig. 4.85). A further consideration is that the carbon-dioxide (CO_2) emission produced by the diesel engine relative to the petrol engine is always much lower, particularly as the engine load is reduced, whereas the petrol engine in the stoichiometric (14.7:1) band operates with the highest level of carbon-dioxide (CO_2).

5

Induction and Exhaust Systems

5.1 Induction manifold requirements

The aims of good manifold design are as follows:

1 to provide as direct a flow as possible to each cylinder
2 to provide equal quantities of charge to each cylinder
3 to provide a uniformly distributed charge of equal mixture strength to each cylinder
4 to provide equal aspiration intervals between branch pipes
5 to provide the smallest possible induction tract diameter that will maintain adequate air velocity at low speed without impeding volumetric efficiency in the upper speed range
6 to create as little internal surface frictional resistance in each branch pipe as possible
7 to provide sufficient pre-heating to the induction manifold for cold starting and warm-up periods
8 to provide a means for drainage of the heavier liquid fraction of fuel
9 to provide a means to prevent charge flow interference between cylinders as far as possible
10 to provide a measure of ram pressure charging

5.2 Cylinder pressure under different operating conditions

5.2.1 Cylinder pressure variation during the induction period with wide open throttle
(Fig. 5.1)

During the initial part of the induction stroke the piston's outward acceleration is at a maximum. This therefore causes the flow of incoming charge to lag behind the piston movement. Consequently, the depression created by the slower moving charge being unable to keep up with the rapidly expanding cylinder space reaches a maximum of about -0.2 bar at roughly 60° after TDC (Fig. 5.1). Towards mid-stroke crank-angle movement the piston acceleration has decreased to zero, whereas its speed has reached a maximum. After this, the piston commences to decelerate, in effect permitting the charge intake to catch up and, to some extent, exceed the enlarging cylinder displacement.

Thus, the cylinder depression is steadily reduced until, at BDC, the cylinder pressure has increased again to atmospheric pressure (Fig. 5.1). Beyond BDC the late closing of the inlet valve will permit the charge, due to its inertia, to continue to enter the cylinder against the returning piston which is now on its compression stroke. Accordingly, as more mixture enters

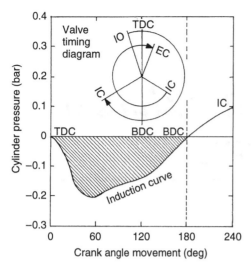

Fig. 5.1 Relationship between cylinder depression and crank-angle movement during the induction period with wide open throttle

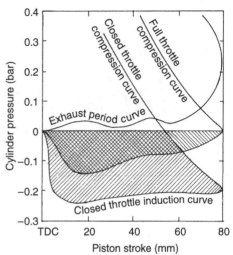

Fig. 5.2 Effect of throttle opening at constant engine speed on cylinder depression created during the induction stroke

the decreasing cylinder space the cylinder pressure tends to rise somewhat above atmospheric pressure (Fig. 5.1) reaching 0.1 bar at the point of inlet valve closure.

5.2.2 Effects of throttle opening on cylinder depression
(Fig. 5.2)

With the throttle closed (idle opening position) the restriction imposed on the charge entering the induction manifold speeds up and reduces the density of the small quantity of charge actually arriving in the cylinder. Subsequently, the depression in the cylinder during the induction stroke will be even greater, as can be seen relative to the full-throttle opening (Fig. 5.2) at similar engine speeds throughout the induction stroke. The graph also shows that, in terms of piston linear movement, the cylinder depression appears to commence very early in the induction period and continues almost horizontally right to the outermost piston position at BDC. However, on the return compression stroke the reducing cylinder volume steadily transforms the cylinder depression to atmospheric pressure at about 60° after BDC (Fig. 5.2). Only at the point when the closed throttle compression curve intersects the atmospheric line does the effective compression stroke actually commence which, in this example, implies that the effective stroke is only three quarters (60/80 mm) of the full throttle compress-

ion stroke (80 mm) which actually starts at BDC. Hence, the true compression ratio based on the effective part of the compression stroke length with closed throttle will be far less.

5.2.3 Manifold riser flow velocity
(Fig. 5.3)

A fundamental factor in induction manifold design is to choose a riser tract cross-sectional area (Fig. 5.3), which produces a minimum idling air velocity which will support and suspend the heavier fuel particles in the air stream. It has been found that an updraught manifold where the particles are drawn upwards requires a minimum charge flow velocity of 14 m/s. However, a horizontal draught and, in particular, a downdraught manifold riser readily permit the flow of mixture into the manifold branches. Hence, a larger diameter riser can be used with a horizontal draught and downdraught manifold, as lower minimum air velocities of something like 10 m/s can effectively be employed.

Conversely, the highest flow velocity which can be tolerated before there is a marked deterioration in charge density and, correspondingly, volumetric efficiency, has been estimated to be about 75 m/s.

Thus, a compromise in manifold riser diameter must be made to cope with the engine running at low speed where insufficient charge velocity may cause poor mixture distribution in the tracts, and at the top end of the engine's speed range, where

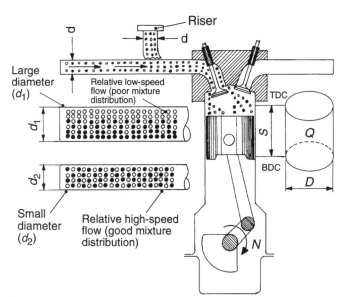

Fig. 5.3 Illustration of flow velocity effect

an excessively high flow velocity begins to reduce the engine's volumetric efficiency.

The charge flow velocity through a manifold riser during wide open throttle conditions depends on the total number of cylinder displacements, drawing from one manifold riser, engine speed, volumetric efficiency and diameter of intake riser.

Thus, for a four-stroke cycle engine

$$\text{intake velocity } (V) = \frac{nQN\eta_{v}}{2 \times 60A} \text{ (m/s)}$$

$$\text{cylinder displacement } (Q) = \frac{\pi}{4} D^{2}S \text{ (m}^{3}\text{)}$$

$$\text{cross-sectional area of tract } (A) = \frac{\pi}{4} d^{2} \text{ (m}^{2}\text{)}$$

where n = number of cylinders per riser
   Q = cylinder displacement (m³)
   N = engine speed (rev/min)
   η_{v} = volumetric efficiency (%)
   A = riser cross-sectional area (m²)
   D = cylinder diameter (m)
   d = riser diameter (m)
   S = piston stroke (m)
   V = intake velocity (m/s)

Example
A four-stroke engine has a bore and stroke of 80 mm and 82 mm respectively. If the intake riser for the induction manifold is to be 30 mm and the

engine volumetric efficiency is 85% determine the following:

a)   the minimum engine speed for a riser charge flow velocity of 10 m/s
b)   the maximum engine speed for a riser charge flow velocity of 75 m/s

a)   since $V = \dfrac{nQN\eta_{v}}{2 \times 60A}$ then $N = \dfrac{180AV}{nQ\eta_{v}}$

hence $N = \dfrac{180\pi/4(0.03)^{2} \times 10}{4 \times \pi/4(0.08)^{2} \times 0.082 \times 0.85}$

$= \dfrac{180 \times 0.03 \times 0.03 \times 10}{4 \times 0.08 \times 0.08 \times 0.082 \times 0.85}$

$= 907.9 \text{ rev/min.}$

b) $N = \dfrac{180AV}{nQ\eta_{v}}$

$= \dfrac{180 \times \pi/4(0.03)^{2} \times 75}{4 \times \pi/4(0.08)^{2} \times 0.082 \times 0.85}$

$= \dfrac{180 \times 0.03 \times 0.03 \times 75}{4 \times 0.08 \times 0.08 \times 0.082 \times 0.85}$

$= 6809 \text{ rev/min.}$

5.2.4   Mixture distribution

The fuel spray after discharging into the carburettor venturi, or into the cylinder-head inlet port in the case of a petrol injection system, consists partly of evaporated fuel, a mist of fine particles, and a considerable amount of heavier particles of fuel.

As the depression and air charge at the discharge nozzle decrease, the fuel spray becomes coarser and the proportion of heavier particles increases. The flow of air along a pipe tract from the walls to the centre varies considerably, in fact the air velocity is greatest at the centre of the passage and decreases to almost nothing on the pipe walls. Hence, fuel particles find it easier to remain in suspension if discharged towards the centre of the tract.

A manifold that provides equal charge distribution to each branch does not necessarily provide equal fuel distribution particularly when the mixture contains only partly atomized fuel droplets.

5.2.5 Effect of throttle opening on charge inertia filling at constant engine speed
(Fig. 5.4)

The effect of cylinder pressure on throttle opening at constant engine speed in relation to crank-angle movement is illustrated in Fig. 5.4.

As can be seen, with a wide-open throttle at medium engine speed, the cylinder pressure can actually reach atmospheric pressure just before BDC, it then takes on a ram charge effect with cylinder pressure rising above atmospheric press-

ure up to the inlet valve closing point at 60° after BDC (Fig. 5.4).

With two-thirds throttle the induction cylinder depression is greater during the early part of the stroke and does not reach atmospheric pressure until the crank-angle has moved 20° on the return compression stroke (Fig. 5.4). Likewise, with one-third throttle opening the induction cylinder depression is initially even greater and remains in a state of depression well beyond BDC and does not reach atmospheric pressure until the point of inlet valve closure at 60° after BDC (Fig. 5.4). This demonstrates that charge inertia filling, which can give positive pressure increases, can only be effective between half and full throttle opening.

5.2.6 Effects of engine speed with wide-open throttle on cylinder filling
(Fig. 5.5)

Tests with unrestricted wide-open throttle induction systems at varying engine speeds show that induction cylinder depression does not change to positive pressure at 1500 rev/min until the crank-angle movement reaches BDC, whereas at 3000 rev/min and 4500 rev/min the cylinder reaches atmospheric pressure earlier, at 160° and 140° crank-angle movement respectively; that is, well before the end of the induction stroke (Fig. 5.5). Beyond 180° crank-angle movement, before the inlet valve closes, the cylinder press-

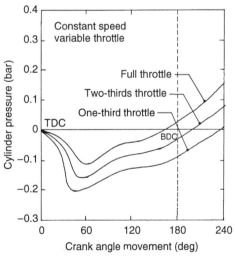

Fig. 5.4 Effect of throttle opening at constant engine speed on cylinder depression generated during the induction period crank-angle movement

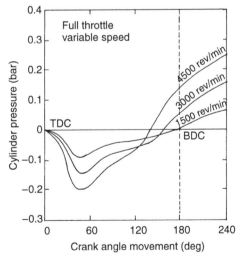

Fig. 5.5 Effect of engine speed on cylinder pressure variation during the induction stroke crank-angle movement period with a wide-open throttle

232

ure, due to the charge inertia effects, rises considerably with increased engine speed.

These results demonstrate that ram-charging increases with engine speed particularly when the inlet valve closure is delayed.

5.3 Manifold branch flow paths

5.3.1 Manifold tract cross-sections

Non-uniform flow in parallel tract
(Fig. 5.6)
The flow stream velocity distribution across the diameter of the tract takes the form of a parabola with the minimum velocity along the pipe walls rising to a maximum in the centre region. The

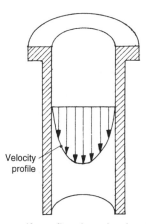

Fig. 5.6   Non-uniform flow in a circular pipe

flow velocity profile is mainly due to the large viscous drag experienced on the tract internal walls, and this flow resistance weakens towards the centre of the flow section.

Flow in diverging tract
(Fig. 5.7)
In a diverging tract, the flow is less uniform as the cross-sectional area is increased, with the charge velocity decreasing rapidly from the centre to the tract walls. The maximum charge velocity in the narrow section of the tract is high and the velocity profile decreases gently, initially from the centre, followed by a more rapid decline as the tract walls are approached. However, in the wide section of the tract the maximum velocity is much lower,

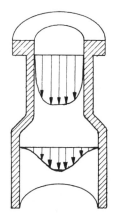

Fig. 5.7   Flow in a diverging tract

and the velocity decrease on either side of this peak velocity, is more gradual. The decrease in flow velocity from the narrow to the wide tract makes it more difficult for the suspended fuel particles to be supported by the slower moving air stream, therefore they will precipitate and be forced by the viscous drag to accumulate on the tract walls.

Flow in converging tract
(Fig. 5.8)
In a converging tract the flow velocity increases and its profile becomes more uniform as its cross-sectional area decreases. The maximum velocity in the wider cross-section intake tract is lower than for the narrow outlet tract, but in the funnel

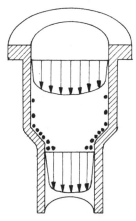

Fig. 5.8   Flow in a converging tract

or tapered transitional region, fuel particles may accumulate on the converging walls before being drawn in spurts into the high-velocity flow narrow part of the tract. In general, the mixture distribution, as the tract cross-section converges, improves, particularly downstream where the tract section again becomes parallel.

5.3.2 Internal wall surface finish
(Figs 5.9 and 5.10)

Smooth internal walls, compared with those with a rough surface finish, produce the least viscous drag on the movement of charge—whether it is pure air or a mixture of air and fuel particles—consequently marginally higher volumetric efficiency can be obtained with smooth induction tracts (Fig. 5.9).

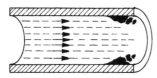

Fig. 5.9 Smooth surface finish

Slow moving air–fuel mixtures tend to precipitate fuel particles, in particular the heavier ones, onto the walls of the tract, they then tend to merge into surface films. This effect becomes more pronounced with increased throttle opening.

With rough surface finishes (Fig. 5.10) the tract

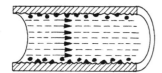

Fig. 5.10 Rough surface finish

offers considerable resistance to the flow of liquid particles suspended in the air stream near the walls, these particles then cling to the walls where they then spread and merge with each other to form films. As these films thicken they become unstable so that they are dragged back into the main air stream and immediately new fuel particles will be thrown onto the tract walls where they again accumulate before repeating the breakaway cycle. This process of build-up and breakaway of liquid fuel from the tract walls produces a continuous and effective mixing mechanism for the charge as it moves towards the inlet port.

Conversely, a smooth surface finish reduces the surface flow resistance keeping the film thickness

to a minimum, consequently there will be very little mechanical break-up or mixing of the fuel particles with the air stream until the erratically suspended fuel particles impinge on the underside of the inlet valve head.

5.3.3 Section shapes

Rectangular sections
(Fig. 5.11)
A square or rectangular section (whose width is larger than its height) provides the greatest opportunity to vaporize the precipitated fuel spreading on the roof, walls and floor of the tract. Because of the relatively larger surface area exposed with rectangular tracts for a given cross-

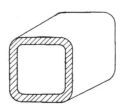

Fig. 5.11 Rectangular-sectioned tract

sectional area, the frictional drag will be somewhat greater so that marginally lower volumetric efficiencies will result. One benefit of a rectangular section is that it prevents the column of charge swirling as it moves through the tract, thus minimizing any centrifuge effect which would force the heavier liquid particles to be flung against the tract walls.

Circular sections
(Fig. 5.12)
Circular-sectioned tracts provide the minimum surface area for any cross-sectional shape. They therefore offer the least resistance to charge flow and this gives the highest volumetric efficiency and, correspondingly, the maximum mass of charge per unit time. However, the charge col-

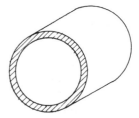

Fig. 5.12 Circular-sectioned tract

umn moving through a circular section tends to generate a longitudinal twist or whirl which causes the globules of fuel in the air stream to be thrown onto the tract walls. As a result, the distribution of mixture strength across the section may be very uneven. A disadvantage of the circular section is that the semi-circular floor of the tract provides only a relatively small surface area for evaporation of that fuel, existing in a liquid state, which has made its way to the bottom of the tract.

Segment sections
(Fig. 5.13)
This is a compromise which retains the high volumetric efficiency of the circular-sectioned tract and the large flat floor area of the rectangular-sectioned tract. It maximizes the passage vaporization ability and simultaneously minimizes

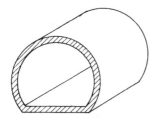

Fig. 5.13 Segment-sectioned tract

charge column swirl in the segment-shaped section. This semi-circular section provides, firstly, the least resistance to flow due to the circular portion of the tract, secondly it offers a large, flat floor area for rapid evaporation of the liquid content of the fuel, and thirdly the flat floor minimizes charge column swirl and therefore the charge retains its initial inlet density as it flows through the tract.

5.3.4 Tee junctions

Branch junction tees
The manifold tee junction divides the single input charge flow into two halves so that they have separate exits on opposite sides. The heavy fuel particles, because of their inertia, will continue to move beyond the initial entry to the left- and right-hand junction tracts until they impinge against the portion of the wall facing the riser, whereas light and well-mixed fuel particles divide

and move to the right and left within the main air stream.

Sharp-edge tee junction
(Fig. 5.14)
With the throttle open a third or more, the walls of the riser become wet; that is, a film of fuel tends to form on the riser walls. A sharp edge, or an abrupt entrance to the left- and right-hand junction, causes the riser film to be torn from the walls bringing it back into the main mixture stream so that the quality of mixture, that is the

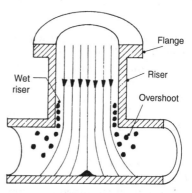

Fig. 5.14 Sharp-edge tee junction

ability of the fuel particles to be evenly divided and suspended into the air stream, is good. However, the sharp edge tee causes a sudden change in flow direction, and so it therefore tends to produce a relatively high flow resistance to the charge. Towards wide-open throttle conditions some of the heavier globules of fuel may overshoot and pile up at the bottom of the riser before this accumulated liquid fuel is gulped back into the charge stream.

Partly streamlined tee junction
(Fig. 5.15)
Curving the junction at which the riser joins the horizontal tract section increases the sectional area of the tract and therefore reduces the velocity of flow. This increases the tendency for liquid fuel particles to precipitate from the main air stream.

Liquid particles of fuel will therefore merge and collect on the tract walls just downstream of the curved bends. In the bottom of the riser, where the flow divides, there exists a sort of dead area where the mixture cannot make up its mind which way to flow and, under full throttle condi-

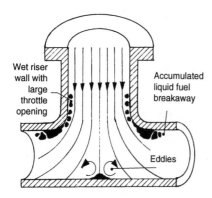

Fig. 5.15 Partly streamlined tee junction

tions, vortex filaments (eddies) will form caused by the swirling and turbulent movement of the mixture in no-man's land. Also, because of the larger tract section in this mid-position, a pressure drop exists which causes liquid particles to accumulate and breakaway continuously in this highly excited region.

Fully streamlined tee junction
(Fig. 5.16)
A fully streamlined tee junction is similar to the partly streamlined junction except that there is an inverted vee at the bottom of the riser, this being in the mid-region where the mixture flow divides in two and moves in opposite directions. This

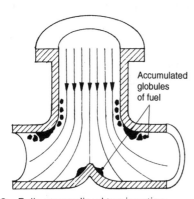

Fig. 5.16 Fully streamlined tee junction

shape ensures that the cross-sectional area at the tee junction remains constant or is even slightly reduced, this gentle division minimizes the flow path resistance. Nevertheless, there are still likely to be small amounts of accumulated liquid fuel downstream of the curved bends and on either side of the inverted vee.

5.3.5 Branch elbows

Curved elbow
(Fig. 5.17)
Charge flowing through a curved bend subjects the air and petrol particles to the effects of centrifugal force, this force being a minimum at the inner radius and a maximum at the outer radius. The radially outward direction of the centrifugal force continuously pushes the air charge towards the outer radius of the bend so that a pressure difference exists across the tract width, with the highest pressure on the outer radius bend and a relatively low pressure on the opposite inner radius region. At the same time, the air velocity in the outer radius flow path will be higher than for the inner bend flow tract

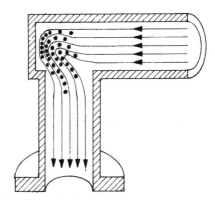

Fig. 5.17 Curved elbow

region. Thus, fuel particles will move from the outer to the inner flow path, this also causes the mixture movement to be excited and form a vortex filament region just downstream of the bend. Consequently, the low pressure and low air velocity near the inner radius curvature zone wall make it very difficult for the air to maintain its carrying capacity of the heavier fuel particles in this zone, this then causes the heavier fuel particles to be deposited on the inner bend curvature wall. The fuel therefore accumulates slightly downstream of the bend. The net result is to displace the fuel to one side of the elbow so that an unequal distribution of mixture will occur across the tract once the mixture moves beyond the bend. However, a curved bend does offer the least flow resistance, thus benefiting cylinder filling at high engine speed.

Sharp elbow
(Fig. 5.18)

In contrast to a curved bend, the sharp elbow provides an abrupt right-angle change in direction for the mixture flow. Consequently, due to the inertia of the air and fuel particles, the fresh charge column will continuously overshoot the bend so that it impacts against the blank end of the elbow. The sudden change in flow direction

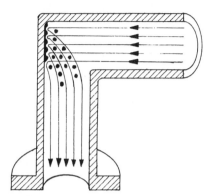

Fig. 5.18   Sharp elbow

excites the air into a state of turbulence in this region and, simultaneously, fuel particles will be dragged back into the main air stream by the high rate of flow. To some extent it can be said that the fuel droplets are pulverized by the turbulent air movement before the fuel is, in effect, bounced back into the main air stream, causing it to be thoroughly mixed with the incoming air. Mixture distribution through the tract is relatively good but this sharp bend does increase the charge flow resistance particularly at high engine speed.

Buffer end chamber elbow
(Fig. 5.19)

The short extension at each end of the main manifold tract just before the right-angle bend forms a buffer end chamber elbow. Provided that the valve ports are very close to the manifold bend the buffer chamber is able to absorb and damp out the pulsation in the inlet tract caused by the sudden opening and closing of the inlet valve. The actual volume of the buffer end chamber is important and must be tuned to respond to critical engine speeds where pulsating pressure waves may oppose and subsequently hinder the induced charge volume flow. Generally, the turbulent air movement created by trapped charge in the buffer zone improves the mixture distribution as it is projected back into the air stream.

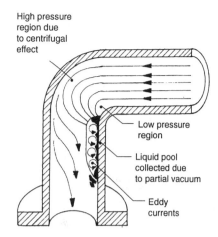

Fig. 5.19   Buffer elbow

45° angled straight elbow
(Fig. 5.20)

This elbow provides a right-angle change in flow direction by means of a 45° straight bend piece, the purpose of which is to produce a 90° directional flow change with two 45° bends so that the centrifugal force effect at each bend is low, thereby keeping the pressure difference across each

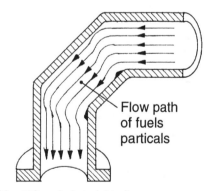

Fig. 5.20   45° angled straight-elbow

bend section to a minimum. As a result, the tendency for the liquid fuel particles to move towards the low-pressure inner walls of each bend is much reduced compared with a single curved right-angle elbow. The benefits obtained from this elbow shape therefore minimize the variation of mixture distribution across the elbow bends, and at the same time maximize the flow capacity of the tract.

Varying rectangular-sectioned elbow
(Fig. 5.21)

Right-angled elbows of this type have a wide tract flow with a low roof at the ends of the tract, whereas this gradual change in section shape to a narrow floor and a high roof at the mid-length of the elbow bend prevents a large pressure difference half-way along the tract bend. This mini-

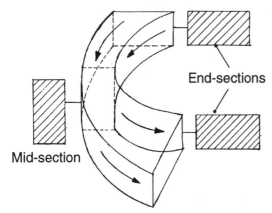

Fig. 5.21 Varying rectangular section elbow

mizes liquid fuel particles being precipitated onto the inner bend walls and yet it maintains the lowest flow resistance as the charge flows gently through the curved tract passage of the varying wall shape, but which has a constant cross-sectional area. Another benefit of this elbow section shape is that the floor area near the cylinder-head will be at a maximum, this is useful in speeding up the vaporization of any liquid fuel which might flood the tract floor when the engine is cold.

5.3.6   Branch junctions

'F' branch junction
(Fig. 5.22)

This double junction relays the incoming charge from the throttle riser to two adjacent cylinders via one curved-end elbow and an adjacent sharp junction. With this configuration there is a tendency due to the charge's inertia to overshoot the sharp junction and therefore pile into the curved-end elbow. However, when the induction period commences on the inner cylinder, charge will be drawn into the leading side of the sharp junction and a small portion of the overshoot on the trailing side of the junction will reverse its

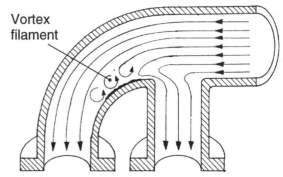

Fig. 5.22 F-branched junction

direction and will be drawn into the first branch at the trailing side of the junction. The flow resistance for the first sharp junction branch is higher than for the second curved tract and, in addition, with the overshoot charge beyond the first branch tract, the end tract provides a slightly higher filling capacity than the first branch. With the second curved branch tract there will be some vortex filament and liquid film formation at the beginning of the inner bend. Mixture distribution at the entrance of each branch is fairly even up to part throttle and light load conditions but it becomes sporadic as the throttle opens further with higher engine speeds.

'Y' branched junction
(Fig. 5.23)

This streamlined double-prong tract-type junction offers the least resistance to charge flow with a symmetrical division which produces a relatively

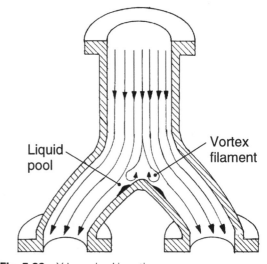

Fig. 5.23 Y-branched junction

uniform distribution of petrol liquid particles across each branch tract. Under certain open-throttle conditions there may be some vortex filament formation and a very small amount of liquid fuel clinging to the inner prong walls of the junction, and on the outer walls just beyond the parallel portion of the tract where the prong divides the branches.

5.3.7 Throttle spindle angular position relative to the tee junction branches
(Figs 5.24(a and b) and 5.25(a and b))

The butterfly-valve half-flap which opens above the spindle will pass less mixture than the half-flap side that opens below the spindle. This is because the mixture which cannot get through the upper half-flap gap (which in effect forms a sharp edge orifice) then flows to the opposite lower half-valve side of the valve where it finds it easier to wedge itself through the valve opening due to the funnel-like entry shape. Thus, if the throttle valve spindle is positioned at right angles to the tee junction branches (Fig. 5.24(a)), there will be more mixture flowing to the tee branches on the lower half-valve side than on the opposite side where the valve opening is above the spindle. From a side view (Fig. 5.24(b)) the mixture coming down from the riser to the junction branches is uniformly distributed from one side of the spindle to the other so that there is no imbalance in mixture movement which could cause a swirl in the branch tracts. If the throttle valve spindle is positioned parallel with the tee branches (Fig. 5.25(a)) the

mixture distribution will be equally divided between the branches. However, because the lower half-valve opening is nearer to the bottom of the junction than the upper half-valve side (Fig. 5.25(b)), the mixture flow from the lower half-valve remains close to the riser wall (Fig. 5.25(b)) where the mixture has begun to move away from the riser wall towards the centre of the tract. The result is an undesirable twist or swirl motion created along the tee branches (Fig. 5.25(b)).

5.3.8 Flow pattern in a four-branch inlet manifold under different operating conditions
(Fig. 5.26(a–c))

The flow pattern of an air–fuel mixture as it moves through the manifold passages (Fig. 5.26(a–c)) under different operating conditions will be described in the following.

Idle throttle conditions
(Fig. 5.26(a))
With closed idle throttle conditions the flow between the valve and riser walls is restricted so that the mixture which does pass into the riser is compelled to move at a very high velocity to keep up with the piston displacement on every induction stroke. Consequently, the manifold tracts will be subjected to a high state of depression with the result that the fuel spray discharged from the carburettor will be finely atomized. As the mixture moves through the riser and branch tracts, the fast moving air is able to keep the finely

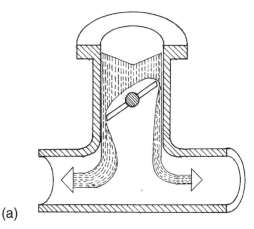

(a)

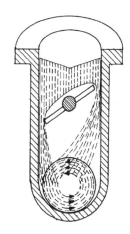

(b)

Fig. 5.24 Throttle spindle at right-angles to tee

239

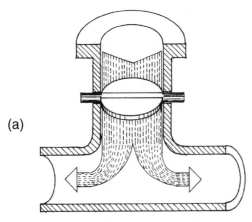

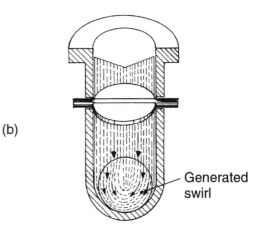

(a)

(b)

Generated swirl

Fig. 5.25 Throttle spindle parallel with tee

atomized fuel particles suspended and away from the tract walls. The overall mixture distribution under these conditions remains relatively uniform.

Part throttle conditions
(Fig. 5.26(b))

With part throttle and mid-speed conditions, one half of the butterfly throttle valve is above the spindle and the other half below. The flow of mixture between the lower half-valve edge and riser wall forms a tapered streamlined orifice, whereas the flow gap between the upper half-valve edge and riser wall forms a sharp orifice. Consequently, more mixture flows past the lower half-valve side than that of the upper half side, this therefore causes an uneven charge flow velocity across the riser tract. The effects of throttling the flow through the riser are therefore twofold.

1 The partly opened butterfly valve presents a central obstacle to the mixture moving through the riser, the mixture therefore becomes turbulent downstream of the throttle valve.
2 The partly opened butterfly valve produces an uneven flow between the two half-streams of mixture with the result that the converging mixture streams take on a swirling motion as they enter the branch tracts. The net result is that the central air stream in the branch tract tends towards a series of vortex filaments which are beneficial in suspending and thoroughly mixing the liquid fuel particles.

Full throttle conditions
(Fig. 5.26(c))

With wide-open throttle and mid-to-upper speed

operating conditions the throttle valve offers very little restriction to mixture flow from the carburettor. Under these conditions, the mixture flow velocity will only be moderately high, and consequently the manifold depression will be relatively low. The fuel spray discharged from the carburettor with the low state of depression is relatively coarse and the air movement is not totally sufficient to support the liquid fuel particles through the riser and branch tracts. It therefore follows that some of the fuel particles will be precipitated from the mixture stream onto the manifold tract walls. Under these conditions, the scrubbing action of the air stream moves closer to the tract walls and excites the liquid film and air into vortex films at the bottom of the riser and downstream of the riser tee junction as well as on the inner bends of the curved branches. These vortex liquid films grow in size and periodically breakaway from the walls and are then gulped back into the main mixture stream. However, these globules of liquid fuel do not necessarily disintegrate into very small particles as the moving column of the mixture in the tracts is drawn into the cylinder. Generally, the mixture distribution may be irregular with a wide-open throttle at only medium speeds, but at higher engine speeds the quality of the mixture (thoroughness in mixing) tends to improve.

5.4 Induction valve overlap between cylinders

(Fig. 5.27(a–d))

Induction valve overlap between cylinders occurs when the induction period of one cylinder com-

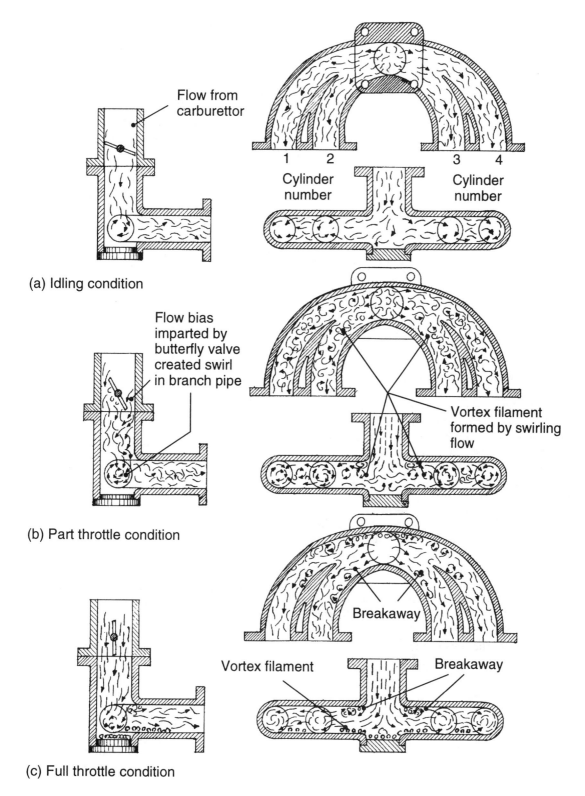

(a) Idling condition

Flow from carburettor

Cylinder number 1 2

Cylinder number 3 4

(b) Part throttle condition

Flow bias imparted by butterfly valve created swirl in branch pipe

Vortex filament formed by swirling flow

(c) Full throttle condition

Vortex filament

Breakaway

Breakaway

Fig. 5.26 Flow pattern in a four-branch inlet manifold

mences (IO) before TDC, just ahead of the induction period completion (IC) on the return compression stroke of a second cylinder. If the cylinders are joined by a single induction manifold, then the second cylinder to commence its induction period interferes with the first cylinder induction period and therefore will be responsible for the uneven filling between cylinders.

5.4.1 In-line four-cylinder engine
(Fig. 5.27(a))

Cylinder interference can be explained by considering a four-cylinder engine having a single-riser four-branch manifold. The filling of a cylinder towards the end of its induction period relies on the kinetic energy inertia effect to complete the filling process; its movement is in opposition to the returning piston at the beginning of its compression stroke. If a second cylinder sharing a common induction manifold commences its induction period slightly before TDC then the opening of its inlet valve provides an alternative and easier flow path into its cylinder due to the partial vacuum created by the outgoing exhaust gases in the TDC region just before its exhaust period terminates. Consequently, the commencement of the induction period of the second cylinder, before the first cylinder has completed its induction period, robs the first cylinder of some of the incoming mixture column at a critical time just before its filling period ends.

Induction period overlap interference for a four-cylinder engine can be overcome by having dual intake risers where the two outer cylinders are fed by one carburettor barrel and the inner adjacent cylinders are fed by a second carburettor barrel. Thus, with a firing order 1342, the induction period intervals will be such as to draw from alternative branch risers, that is from number one cylinder's outer branch and riser, then number three cylinder's inner branch and riser, followed by number four cylinder's outer branch and riser, and finally number two cylinder's inner branch and riser, these events in this order being continuous. However, twin carburettors, where one carburettor feeds number one and two cylinders and the second carburettor supplies cylinders three and four, do not solve induction overlap interference between cylinders.

5.4.2 In-line and vee six-cylinder engines
(Fig. 5.27(b and c))

Induction period overlap between cylinders on an in-line six cylinder engine amounts roughly to 120° and causes considerable interference between cylinders if a single riser manifold is employed (Fig. 5.27(b)). However, dividing the manifold in half so that a separate riser feeds adjacent groups of three cylinders completely eliminates induction period cylinder overlap interference. Thus, with a firing order of 153624, consecutive induction periods will be drawn from left and right-hand manifolds in turn. That is, cylinders 1, 2 and 3 draw from the front half-manifold whereas cylinders 4, 5 and 6 draw from the rear half-manifold. Hence, equal breathing intervals will exist between each half-manifold.

Likewise, a vee six-cylinder engine (Fig. 5.27(c)), where each cylinder-bank has its own manifold and riser, does not suffer from induction period cylinder overlap interference, so that cylinders 1, 2 and 3 draw from the left-hand manifold and cylinders 4, 5 and 6 draw from the right-hand bank manifold.

An alternative arrangement using triple manifolds and risers avoids induction period cylinder overlap, hence cylinders 1 and 5, 2 and 6, and 3 and 4 share common branches and risers, but with this layout some branch passages are compelled to be above and below each other as they cross over.

5.4.3 Vee eight-cylinder engine
(Fig. 5.27(d))

Induction period overlap between cylinders on a vee eight-cylinder engine amounts roughly to 150° and would therefore cause considerable induction period overlap interference if each cylinder-bank had its own manifold and riser with a conventional firing order such as 15486372. Therefore, to eliminate induction interference and to provide even filling intervals between manifold intakes, the induction manifold branches for cylinders numbers 1, 4, 6 and 7 merge to draw from one or a pair of primary and secondary risers, and manifold branches for cylinders numbers 2, 3, 5 and 8 merge to draw from a second single riser or from a pair of primary and secondary risers.

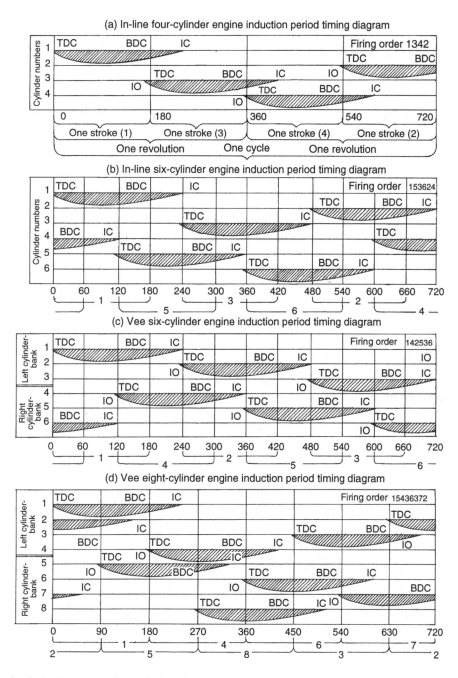

Fig. 5.27 Engine induction period linear timing diagrams

243

5.5 Flow variation and mixture distribution in manifold passages

5.5.1 Effect of branch length on individual cylinder volumetric efficiency
(Fig. 5.28)

The cylinders furthest from the manifold riser have longer tract lengths and therefore support longer columns of charge at any one time. Consequently, because of the marginally greater charge mass (and hence momentum) in the outer branches compared with the inner shorter branches, there will be a correspondingly greater ram charge effect in these outer branches. This is represented by the higher volumetric efficiency (Fig. 5.28) in the end cylinders in contrast to the inner cylinders which have slightly shorter branch tracts. However, because of the increase in flow resistance with rising engine speed and the greater amount of flow resistance with the longer tracts, the longer outer branches cause the volumetric efficiency in the cylinders they feed to peak earlier than for the inner cylinders, which are suppplied by the more direct inner branches (Fig. 5.28).

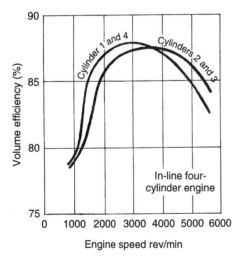

Fig. 5.28 Effect of part interference and tract length on volumetric efficiency with respect to engine speed for a four-branch inlet manifold

5.5.2 Fuel mixture distribution between cylinders
(Fig. 5.29)

The carburettor or injector supplies to the air stream a metered quantity of fuel which is partially vaporized and atomized.

The air and the pulverized fuel move through the manifold at relatively high velocities—the smaller the atomized particles the higher will be the flow velocities. Conversely, larger-sized liquid particles move at a comparatively much reduced velocity while liquid films formed on the passage walls will move erratically and relatively slowly.

With a wide-open throttle the depression in the branches will be low and the discharged fuel particles coarse whereas, as the throttle is progressively closed, the depression in the manifold increases and the discharged fuel becomes finer. However, with reduced throttle opening, the increased depression in the induction manifold will draw exhaust gases into the induction branch tracts at the end of the exhaust period and at the beginning of the induction period during valve overlap. Slightly later during the induction period this exhaust gas is pulled back into the various cylinders with the fresh incoming mixture.

Under part-throttle conditions, the residual exhaust gases at the end of the exhaust stroke and the beginning of the induction stroke will initially draw exhaust gases into the induction manifold and then back into the cylinder, the actual amount of gas moving to and fro will not always be equal, and this is mainly responsible for the poor mixture distribution during idling and part throttle opening conditions.

Mixture distribution is generally best with a wide-open throttle where the throttle valve offers the least interference with the incoming air or mixture flow stream. However, the walls of the venturi and induction manifold passage walls will tend to be wet under these particular operating conditions.

With reduced throttle opening the butterfly valve obstructs the incoming mixture stream and therefore causes an uneven quantity of mixture flow between both sides of the tilted butterfly valve plate. Therefore, the mixture distribution between induction branch tracts may vary considerably (Fig. 5.29).

As the throttle opening is reduced to idling conditions the mixture distribution between cylinders is evenly erratic due to the unpredictable amount of residual exhaust gases being pulled in

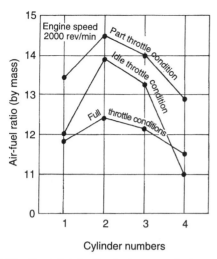

Fig. 5.29 Fuel distribution between cylinders under various conditions

and out of the induction manifold branches. This is caused by the variation in the vacuum existing in the individual branch pipes due to their flow path-length, passage curvature and shape, and how they converge to form the common manifold. The walls of the induction riser and branches tend to be fairly dry when the depression in the manifold is relatively high.

Figure 5.29 shows that the mixture strength for full throttle conditions is fairly rich to enable the engine to develop its maximum output load, whereas the mixture strength for part throttle operations is relatively weak. Generally, between these extreme operating conditions, lies the engine idle speed mixture strength which demands a sufficiently rich mixture to compensate for the exhaust gas dilution in the induction manifold. Unfortunately, the unpredictable amount of exhaust gas dilution results in a large variation of the strength of the mixture entering the individual cylinders.

The slightly richer air–fuel mixture in the outer pairs of cylinders, compared with the inner cylinders, is mostly due to the end branches collecting the inertia overshoot of the fuel particles (Fig. 5.29). This is because the mixture moves from the riser into the gallery and then outwards, and some of the mixture is then diverted to the tee junctions of the inner branches. The remaining outward moving mixture is then redirected through the end elbows to the outer branches.

The small variation in mixture strength between symmetrical pairs of branches may be put down to uneven heating of the individual branches of the manifold, causing various amounts of fuel mixture vaporization or to being due to the inclined mounting of the engine onto the chassis members.

5.6 Induction ram cylinder charging

(Fig. 5.30)

Consider a single-cylinder engine with a straight constant diameter intake tract bolted onto the cylinder-head (Fig. 5.30).

The fluctuating kinetic energy of the incoming

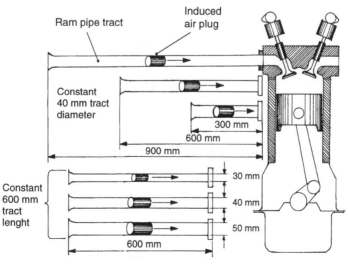

Fig. 5.30 Method of evaluating induction-ram cylinder charging

column of charge caused by the periodic opening and closing of the inlet valve can be effectively used to ram charge into the cylinder over a limited speed-range which is dependent upon the intake tract's diameter, length and throttling. Generally, the cylinder's volumetric efficiency will be at its maximum when the incoming charge is constrained to move through a smooth-walled passage having a length-to-bore diameter ratio ranging between 10:1 to 20:1.

5.6.1 Induction inertial ram cylinder charging
(Fig. 5.30)

The momentum acquired by the charge entering the cylinder during the induction period is utilized in most engines (Fig. 5.30).

At the end of the exhaust stroke and the beginning of the induction stroke the inlet valve opens and the piston commences to move away from TDC. The outward accelerating piston quickly expands the space between the cylinder-head and piston crown. Instantly, the depression generated in this rapidly enlarging space will be transmitted to the inlet port via the annular passage formed between the valve head and its seat. This drop in pressure immediately causes the column of charge in the induction tract to move as a whole towards the open inlet valve. The large cross-sectional area of the piston relative to that of the much smaller intake tract cross-sectional area, plus the acceleration of the piston, forces the column of the charge in the tract to acquire a high flow velocity.

The momentum built up by the fast moving column of charge in the intake tract is brought rapidly to a halt when the inlet valve closes against the flow. Thus, the kinetic energy generated by the fast-moving column of charge is now converted into pressure energy in the blanked-off inlet port. Consequently, the density of the trapped charge rises. It is this rise in pressure at the port which enables filling of the cylinder to continue after BDC, and for the induction period to have an early start due to the pressurized charge momentarily stored behind the inlet valve head when it opens.

The greater the momentum produced, the greater the rise of pressure, and if energy losses are very low in accelerating the flow, the inertial ram effect will be beneficial in cramming that extra mixture into the cylinder.

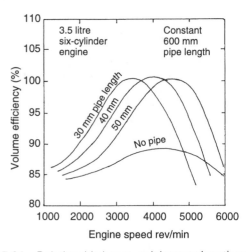

Fig. 5.31 Relationship between inlet tract length and volumetric efficiency with constant tract diameter

5.6.2 Effect of tract dimensions on volumetric efficiency characteristics
(Figs 5.31 and 5.32)

1 For a given tract diameter, the longer the tract the greater will be the surface area exposed to the air stream and the higher will be its flow resistance.

2 For a given tract length, the greater its diameter the larger will be its surface area exposed to the air stream and the higher will be its flow resistance.

3 For a given tract length, the smaller the tract diameter the higher will be the intake velocity and the lower will be the engine speed at which the ram charge pressure peaks (Fig. 5.31).

4 For a given tract diameter, the longer the tract the greater will be the charge column peak ramming effect, but the increased flow resistance causes the charge column pressure to peak at a lower engine speed (Fig. 5.32).

5 For a given tract length both small and large tract diameters produce approximately the same peak volumetric efficiency in the cylinder. However, the volumetric efficiency with the smaller diameter tract peaks much earlier than the larger diameter tract (Fig. 5.31).

6 For a given tract diameter a long passage produces, in the cylinder, a much higher peak volumetric efficiency at a relatively lower engine speed, whereas a short passage produces a much lower peak volumetric efficiency at a higher engine speed (Fig. 5.32).

7 For a given tract diameter, the longer the

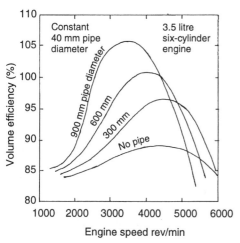

Fig. 5.32 Relationship between inlet tract diameter and volumetric efficiency with constant tract length

tract the more rapid will be the volumetric efficiency build-up with rising engine speed (Fig. 5.32).

8 For a given tract length, the smaller the pipe diameter the earlier in the engine speed range will be the rise in volumetric efficiency. The rate of volumetric efficiency increase with rising engine speed for a fixed 600 mm tract length but with different tract diameters, 30, 40 and 50 mm, appears to be similar (Fig. 5.31).

9 With no intake ram tract, induction charge flow-back through the induction port occurs owing to the late closing of the inlet valve. This flow-back continues until the inlet charge flow velocity through the short inlet port, on its own, is sufficient to produce ram pressure charging (Figs 5.31 and 5.32).

10 The longer the induction tract the greater will be the flow resistance; therefore, the more rapid will be the decrease in cylinder volumetric efficiency once it has reached its peak, and the engine's speed continues to rise. Thus, when the carburettor is bolted directly against the cylinder-head inlet port, the reduction in volumetric efficiency after it has peaked is relatively gradual compared with the 300, 600 and 900 mm tracts which show an increasing drop-off rate with rising engine speed (Fig. 5.32).

11 A small-bore tract will have a relatively high flow velocity so that it is well able to maintain a thoroughly mixed charge of air and fuel in an atomized state as it travels through the tract at low engine speeds (Fig. 5.31). In contrast, a large-bore tract will not be able to hold the liquid particles in suspension as the charge flows through the tract at low engine speed. Consequently, the heavier liquid particles may precipitate onto the walls of the tract. However, the charge velocity and the accompanying flow resistance become excessive towards the maximum engine speed, which therefore results in a rapid decrease in cylinder volumetric efficiency.

5.7 In-line four-cylinder engine induction manifolds

(Fig. 5.33(a–h))

One-piece manifolds for four-cylinder engines are usually of two or four branch arrangements with a single carburettor, either of the single or double-barrel compound type. Alternatively, some manufacturers prefer twin manifolds and carburettors which supply adjacent pairs of cylinders only.

The single carburettor manifold, be it a single or double-barrel compound type for four-cylinder engines with the usual firing order 1342, has unequal induction draw intervals between adjacent cylinders. Thus, with a firing order of 1342, the mixture is drawn first from the left-hand side main gallery (cylinder number 1) then from the right (cylinder number 3) main gallery and right (4) again, then left (2) and left (1) again, it then repeats this sequence L–R–R–L–L–R–R–L etc. This pattern of mixture draw from the main gallery of the manifold tends to produce unequal mixture distribution between branches. The reason for this being that there is less time for the main gallery to recover if adjacent cylinders in one half of the gallery draw a charge one after the other, followed by the charge being drawn from the other half of the main gallery to feed the opposite pair of cylinders in succession.

5.7.1 Two-branch induction manifold with a single carburettor
(Fig. 5.33(a))

With this arrangement, each branch feeds pairs of siamese inlet ports, the branches therefore offer the same flow resistance to each pair of cylinders since the length and bend of the branch is common to both cylinders. The limitation with siamese port arrangements is that the induction interference will occur every time adjacent cylinders are on induction, with one cylinder about to finish and the other cylinder just commencing its filling period.

5.7.2 Four-branch induction manifolds with a single carburettor
(Fig. 5.33(b–d))

This manifold has a branch to feed each of the separate inlet ports cast in the cylinder-head. However, the two outer cylinders will have a longer flow path than the inner ones. The separation of port branches as opposed to the siamese inlet ports assists in reducing the induction period interference between adjacent cylinders, but the variation of flow-path causes some unevenness in the mixture distribution between the inner and outer cylinders.

5.7.3 Four-branch log-type manifold
(Fig. 5.33(b))

With the four-branch log-type manifold, the sharp branch elbows of the end cylinders tend to bounce the overshoot fuel particles back into the main air stream of the adjacent branch passages. The quality of mixing and the mixture strength between branch passages tend to be relatively uniform over a wide range of throttle opening conditions.

5.7.4 Four-branch semi-streamlined type manifold
(Fig. 5.33(c))

With the four-branch semi-streamlined manifold the two outer branch elbows are curved to minimize the flow resistance but the two inner branches merge with the main gallery to form sharp edge tee junctions. Thus, the longer flow path for the outer cylinders is partially compensated by the streamlining of these branches so that the quantity of charge flowing into each inlet port is approximately equal.

5.7.5 Four-branch streamlined type manifold
(Fig. 5.33(d))

With the four-branch streamlined manifold, all four branches are curved, gently sweeping out from the central riser to the cylinder-head inlet ports. The streamlined junctions made between the inner and outer branch pairs as they divide from the central riser are so shaped that equal proportions of mixture are permitted to flow into the adjacent branch passages. However, the two outer branches are slightly longer and, at very high engine speeds, the outer cylinders may produce slightly higher volumetric efficiency than the inner cylinders which are fed by the shorter branch passages.

5.7.6 Four-branch 'Y'-fork type manifold
(Fig. 5.33(e))

With the four-branch 'Y'-fork manifold, a common gallery conveys charge to pairs of diverging 'Y'-branch pipes via sharp-edge elbows. The mixture is thus transferred from the gallery's central riser to the two end right-angle elbows. It then divides into two diverging branch forks which align with the cylinder-head inlet ports. The principle behind this arrangement is that similar quantities of charge exit from the end elbows, and this charge is then equally divided by the streamlined 'Y' junction so that each inlet port receives the same amount of mixture under all operating conditions. The mixing distribution, firstly between elbows and secondly between converging junctions, does not vary to any great extent. However, the short run from the 'Y' junction to the inlet ports may somewhat impair the charge column ram effects.

5.7.7 Four-branch manifold with dual barrel carburettor
(Fig. 5.33(f))

With this arrangement the inner and outer cylinders are fed separately by a pair of 'U'-shaped branches, which are themselves supplied by a dual barrel carburettor via two individual manifold risers. The middle pairs of cylinders are supplied by the inner pair of 'U' branches, which

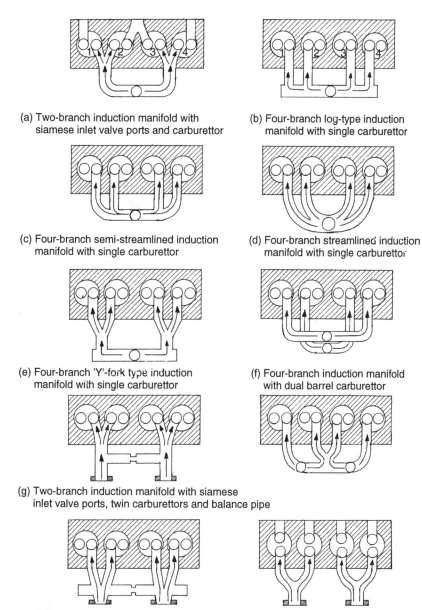

(a) Two-branch induction manifold with siamese inlet valve ports and carburettor

(b) Four-branch log-type induction manifold with single carburettor

(c) Four-branch semi-streamlined induction manifold with single carburettor

(d) Four-branch streamlined induction manifold with single carburettor

(e) Four-branch 'Y'-fork type induction manifold with single carburettor

(f) Four-branch induction manifold with dual barrel carburettor

(g) Two-branch induction manifold with siamese inlet valve ports, twin carburettors and balance pipe

(h) Two pairs of 'V' fork-type induction manifolds with twin carburettors and balance pipes

Fig. 5.33 In-line four-cylinder engine induction manifold arrangements

extend under and outwards beyond the outer cylinder branch pipes, so that the effective flow path-lengths for both inner and outer manifold branches are approximately the same.

The object of separating the inner and outer pairs of branch pipes from each other is to eliminate induction interference between adjacent cylinders, which occurs when one cylinder is about to finish its induction period and the other has just commenced. Note, with this layout both carburettor barrels are primary in action and there is no secondary barrel compound stage. Generally, equal quantities of air are supplied to all the cylinders and the mixture distribution is relatively uniform.

5.7.8 Two-branch manifold with siamese inlet ports and twin carburettors
(Fig. 5.33(g))

With this two-branch manifold, each pair of inner and outer cylinders is merged by siamese inlet valve ports which have their mixture supplied by their own carburettor via a short, straight intake tract or branch. To compensate for the irregular induction pulse, a balance-pipe is put between the two branches. When one cylinder is being robbed of charge as its inlet valve is closing, the mixture will flow through the balance-pipe from the opposite tract. The benefit of this arrangement is that the mixture strength and the quality permitted to flow to each pair of cylinders can be individually adjusted to obtain optimum performance.

5.7.9 Twin pairs of 'V' fork branched manifolds with interconnected balance pipes
(Fig. 5.33(h))

The 'V' pronged branches extend the length of the individual inlet port passages before they merge into a single flanged tract upon which the carburettor is attached. The benefits gained by having twin carburettors is that each one directly supplies the inner and outer adjacent cylinders with the prepared air and fuel mixture. The pronged division of the branches is symmetrical so that equal quantities of mixture enter each inlet port, and the length of the branches before they are joined together reduces the induction period overlap interference between adjacent inner and outer cylinders. The reduction in induction interference between adjacent cylinders is achieved because the mixture entering a branch finds it difficult to reverse its direction of flow. Thus, the beginning of the induction on one cylinder only marginally interrupts the completion of the induction period of an adjacent cylinder. An inherent problem with twin carburettors is that it is difficult to adjust the interconnecting throttle linkage so that both throttle valves are synchronized.

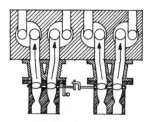

Fig. 5.34 In-line four-cylinder engine with a pair of twin-barrel carburettors

5.7.10 Twin pairs of diverging branches with a pair of twin-barrel carburettors
(Fig. 5.34)

This compact manifolding has two pairs of short 'V' fork branch pipes with twin-barrel carburettors bolted to each pair of branches. This arrangement, in effect, provides each cylinder with its own carburettor barrel thereby eliminating induction period overlap interference between cylinders. The other prominent merits for this layout are that the intake passages are direct and unrestricted so that equal filling of cylinders occurs and high cylinder volumetric efficiencies are therefore obtainable. Good atomization is maintained as the mixture flows through the branch passages, the mixture strength in each carburettor barrel can be individually adjusted. Therefore, optimum mixture distribution can be achieved in each cylinder for almost all throttle opening operating conditions.

5.8 In-line six-cylinder engine induction manifolds

5.8.1 Three-branch manifold with siamese inlet ports with a single carburettor
(Fig. 5.35(a))

This three-branch manifold feeds pairs of siamese inlet ports which divide the cylinder-head single entries into three forked adjacent inlet-port passages. The flow from each manifold branch entering the cylinder-head will be equally divided at the forked passage junction. Unfortunately, however, the flow path to the central pair of cylinders is much shorter than that of the outer pairs of cylinders. Consequently, there will be a

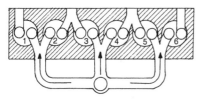

(a) Three branch induction manifold with siamesed inlet ports and single carburettor

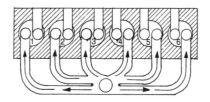

(c) Six-branch semi-streamlined induction manifold with a single carburettor

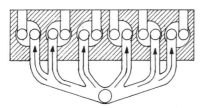

(b) Six-branch semi-streamlined induction manifold with a single carburettor

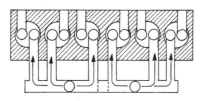

(d) Six-branch log-type induction manifold with twin carburettors

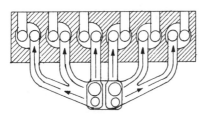

(e) Six-branch split-manifold with quadruple compound carburettor (longitudinally positioned throttle spindles)

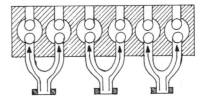

(f) Triple pairs of 'U'-fork type induction manifolds with triple carburettors

Fig. 5.35(a–f)   In-line six-cylinder engine induction manifold arrangements

greater degree of charge ramming in the outer branches compared with the central branch. In contrast, the flow resistance in the long outer branches will be higher than for the more direct central branch.

These contrasting features do not necessarily cancel each other out and under certain operating conditions may produce unequal cylinder filling and poor mixture distribution.

5.8.2   Six-branch semi-streamlined manifold with a single carburettor
(Fig. 5.35(b))

This six-branch manifold has a main gallery either side of the riser which feeds into six separate branches. The outer branches form curved elbows whereas the intermediate branches effectively form sharp tee junctions, and the inner branches form forked curved junctions relative to the central gallery region. The charge delivery to each cylinder tends to vary, as does the mixture distribution. The variation of mixture strength between cylinders is largely due to overshoot from the intermediate branches causing the accumulation of fuel droplets in the end branches. The dispersion of charge mixture from the riser to the main gallery and from there to the individual branches can be unpredictable and erratic under different throttle and load conditions. There is also the unequal tract lengths between the riser and the individual cylinder-head inlet ports which will make cylinder filling uneven at high engine speeds. This is due to the variation in branch length ramming abilities and the difference in flow resistance for the various flow paths.

5.8.3 Six-branch streamlined manifold with a single carburettor
(Fig. 5.35(c))

This six-branch manifold has individual curved branches sprouting from the central riser. Therefore, there will be three different branch lengths. The objective of this arrangement is that each branch produces a certain amount of ram charging but due to the variation in branch length the flow resistance between branches varies as does the volumetric efficiency between cylinders. The flow region between the riser and where the branches begin is very important if good mixture distribution is to be achieved at different engine speeds and loads, and under certain operating conditions the spread of mixture strength between cylinders may be large. Generally, the smooth streamlined flow-path from riser to inlet valve ports offers the least drag to the columns of charge discharging into the cylinders so that high cylinder volumetric efficiency is achieved.

5.8.4 Six or double three-branch log-type manifold with a twin carburettor
(Fig. 5.35(d))

This six-branch manifold with sharp edge tee junctions and end elbows has twin carburettors mounted along the main gallery between the inner and intermediate cylinders. Basically, with a firing order such as 153624, the cylinders draw mixture alternately from each half of the main gallery. Therefore, fuel is discharged (also alternately) from each carburettor. However, the mixture flow path to the outer inlet valve ports is greater than for the intermediate and inner cylinder-head valve ports so that the mixture distribution under certain throttle opening load and speed conditions may give a wider air–fuel ratio variation between cylinders than can be tolerated if lean mixtures are to be burnt. A further criticism with the log-type manifold is that the sharp tees and elbows offer marginally more flow resistance at high engine speeds, this therefore impedes the cylinders receiving the maximum amount of charge.

Some improvement in engine performance may be obtained by dividing the manifold in the middle (see Fig. 5.35(d) shown by the broken lines). With twin manifolds the induction pulses caused by the inlet valve's opening draws the mixture from each carburettor and half manifold alternately. However, the front and rear half-manifolds cannot be made with equal length branch tracts. If the riser is placed opposite the intermediate cylinder branches the volumetric efficiency in these two cylinders will be slightly higher than for the other four cylinders. Also, a larger amount of the fuel particles will be directed to these intermediate branches compared with the outer and inner branch tracts. A better position for the riser is between the intermediate and middle branches, as shown in Fig. 5.35(d). This is because there is no reversal of mixture flow in the manifolds before the outer cylinder's induction periods commence. On the other hand, the intermediate cylinder number 2 assists the start of the filling period of the outer cylinder number 1, and similarly the intermediate cylinder number 5 helps the beginning of the filling period of the other outer cylinder number 6. Thus, a balance is achieved to equalize the amount of flow to each branch by positioning the risers so the flow path to the outer cylinders is greater than the flow path to the other four cylinders.

5.8.5 Six-branch split-manifold with quadruple compound carburettors (longitudinally positioned throttle spindles)
(Fig. 5.35(e))

With this split three-by-three branch manifold, a quadruple compound carburettor bridges both halves of the riser platform. Each set of three branches is supplied by a small primary and a large secondary riser, the primary barrel being the furthest, and the secondary barrel on the inside therefore being the nearest to the inlet valve ports. This configuration provides the greatest distance from the primary barrel to the inlet valve ports, thereby maximizing the ram effect at the low to mid-speed range of the engine when only the primary throttles are open. At some predetermined engine speed and load combination, the secondary barrels also open and, since they are situated closer to the cylinders and the ram effect becomes less important, a more direct flow path is obtained. However, because the branch lengths are unequal, the flow resistance varies between branches and the peak ram effect for each pair of cylinders will occur at slightly different engine speeds. The mixture strength tolerance between cylinders may have to be relatively large under different throttle openings.

5.8.6 Triple pairs of 'U'-fork type manifolds with triple carburettors
(Fig. 5.35(f))

This triple manifold layout with wishbone-shaped branches provides short unrestricted and direct mixture flow to the cylinders. Consequently, equal quantities of charge will enter each cylinder so that cylinder volumetric efficiency will be at a maximum. This arrangement is thus capable of producing the maximum acceleration response time and very high power outputs in the upper speed range of the engine.

To optimize engine tuning a number of observations should be made. With an in-line six-cylinder engine, there is no overlap between the opening periods of paired cylinders. However, the induction pulses for the end pairs of cylinders occur at regular intervals whereas, the centre pairs of cylinders do have induction opening periods at regular intervals.

To compensate for the greater amount of fuel discharged into a fluctuating depression rather than a uniform one, the carburettor jet sizes for the end manifolds (which are subjected to a moderate degree of pulsating depression) are sometimes made smaller than those fitted to the central carburettor, which is exposed to a more uniform section. Thus, if care is taken in synchronizing the three throttle valves, the uniformity

and quality of the mixture induced into each cylinder is generally very good.

5.8.7 Twin pairs of three fork-type manifolds with twin-barrel carburettors
(Fig. 5.36)

This arrangement has two separate manifolds, each having three branches merging at their own common twin-barrel intake. There is no induction period overlap between cylinders with a conventional firing order such as 153624. Cylinders numbers 1, 2 and 3 receive mixture from the front twin-barrels and cylinders numbers 4, 5 and 6 are supplied by the rear twin-barrel carburettor. Cylinders 1 and 3 are fed individually by the left- and right-hand barrels of the forward carburettor

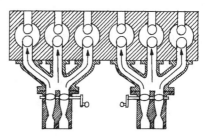

Fig. 5.36 In-line six-cylinder engine with a pair of twin-barrel carburettors

and, likewise, cylinders numbers 4 and 6 are fed individually by the rear left- and right-hand carburettor barrels, respectively. However, cylinders numbers 2 and 5 are fed by the combined mixture discharge which has not been diverted to the outer branches of each manifold. The overall result with this layout is that the quantity of the mixture entering both the middle and adjacent outer branches of each manifold are relatively similar and uniform throughout the engine's speed range, and, due to the direction of each flow path, the acceleration response lapse is minimal.

5.9 Vee four-cylinder engine induction manifolds with single and twin-barrel carburettors
(Figs 5.37 and 5.38)

A vee four-cylinder manifold occupies the space between the cylinder banks. It normally has a

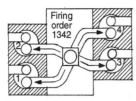

Fig. 5.37 Vee four-cylinder engine with single-barrel carburettor

centrally mounted downdraught riser (Fig. 5.37) which feeds the inward facing inlet ports on each cylinder-head through separate branch pipes. The most common firing order for this type of engine is 1342. The induction period filling pulse interval between cylinders is 180°, with the usual induction period overlap. However, the flow path distance from riser to the individual cylinder is about the same for each branch pipe and the separation of each branch reduces the induction period interference between cylinders. This is caused by one cylinder beginning its induction period before another cylinder has fully completed its filling period and has closed its inlet valve. A twin-barrel manifold (Fig. 5.38) doubles

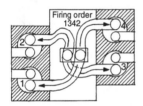

Fig. 5.38 Vee four-cylinder engine with twin-barrel carburettor

the interval between filling pulses experienced by each riser and, at the same time, eliminates induction period overlap interference. Careful design is required to keep the flow path distance and resistance from riser to port equal to the twin-barrel configuration, due to the under and overflow passage which is necessary between branches numbers 1 and 3.

5.10 Vee six-cylinder engine induction manifolds

(Fig. 5.39(a–d))

Induction manifolds for vee six-cylinder engines are normally of the two or three-riser configuration (Fig. 5.39(a–d)). These manifolds have been designed for use with the popular firing order 142536. With the twin riser layout, each feeds

three branches either from one cylinder-head (Fig. 5.39(a and b)) or, alternatively, the riser feeds two branches going to one bank with the third branch feeding an inlet port on the other cylinder-bank (Fig. 5.39(c)). If the manifold has triple risers, each riser supplies two branches—one branch for each cylinder-bank. When branch tract numbers 3 and 4 cross over branch numbers 2 and 5, respectively (Fig. 5.39(d)) care has to be taken so that each branch has the equivalent flow path. This is accomplished by a dip or hump to accommodate the cross over and under passageways if an equal mixture distribution and cylinder filling capacity is to be obtained from each set of branch tracts.

5.10.1 Equal-spacing single-plane twin-riser manifold
(Figs 5.39(a) and Fig. 5.40)

Experiments have shown that the equally spaced single-plane twin riser manifold with the direct and short branches (Fig. 5.39(a)) provides a relatively narrow (uniform) air–fuel ratio spread from cylinder to cylinder under part throttle conditions, this being 1.07 : 1 air–fuel ratio variation (7.1%), but at full throttle the air–fuel ratio distribution from cylinder to cylinder deteriorates causing the air–fuel ratio spread to widen to 3.63 : 1 (24.2%). The corresponding overall engine volumetric efficiency reaches a maximum at 2800 rev/min and remains approximately constant up to 3800 rev/min, after which there is a decrease in the engine's breathing ability (Fig. 5.40).

5.10.2 Equal spacing over and under branched twin-riser manifold
(Fig. 5.39(b))

With the equally spaced over and under branched twin riser manifold (Fig. 5.39(b)) the branches are extended someway beyond the mid-point between the cylinder-banks. Thus, the branches from one bank overlap the branches from the other bank. Consequently, the riser feeding the branches of one cylinder-bank are positioned nearer to the opposite cylinder-bank. The purpose of this is to provide a moderate amount of ram charging to take place to improve the engine's cylinder's volumetric efficiency, which peaks at 87.5% at approximately 3200 rev/min. However, the improved volumetric efficiency is

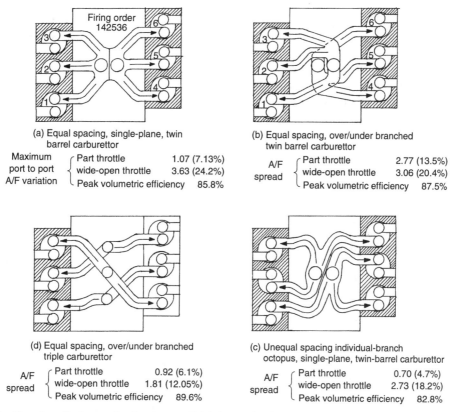

(a) Equal spacing, single-plane, twin barrel carburettor

Firing order 142536

Maximum port to port A/F variation
{ Part throttle   1.07 (7.13%)
  wide-open throttle   3.63 (24.2%)
  Peak volumetric efficiency   85.8%

(b) Equal spacing, over/under branched twin barrel carburettor

A/F spread
{ Part throttle   2.77 (13.5%)
  wide-open throttle   3.06 (20.4%)
  Peak volumetric efficiency   87.5%

(d) Equal spacing, over/under branched triple carburettor

A/F spread
{ Part throttle   0.92 (6.1%)
  wide-open throttle   1.81 (12.05%)
  Peak volumetric efficiency   89.6%

(c) Unequal spacing individual-branch octopus, single-plane, twin-barrel carburettor

A/F spread
{ Part throttle   0.70 (4.7%)
  wide-open throttle   2.73 (18.2%)
  Peak volumetric efficiency   82.8%

Fig. 5.39(a–d) Vee six-cylinder engine intake manifold alternative branch arrangements

offset by the much broader part-throttle air–fuel ratio spread from cylinder to cylinder of 2.77 : 1 (18.5%). However, there is a slight narrowing of the air–fuel ratio spread under wide-open throttle

operation, 3.06 : 1 (20.4%) compared with the manifold in Fig. 5.39(a), under similar wide-open throttle working conditions (Fig. 5.40).

5.10.3 Unequal spacing individual-branch octopus single-plane twin-riser manifold
(Fig. 5.39(c))

An alternative manifold design configuration to keep the air–fuel ratio spread to the minimum is the uneven spacing individual branch octopus single plane twin riser manifold (Fig. 5.39(c)). Here, each branch is individually curved and shaped around its respective riser so that a more uniform mixture distribution can be obtained from branch to branch, effectively under part throttle working conditions. In this case, the air–fuel ratio spread is 0.7 : 1 (4.7%), and fully opening the throttle broadens the air–fuel ratio spread to 2.73 : 1 (18.2%). However, this spread is far better than for Fig. 5.39(a) and 5.39(b). Unfortunately, the bends in the branches tend to

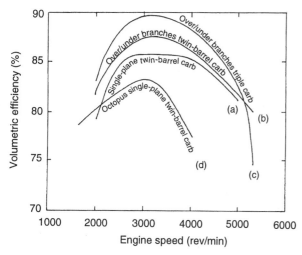

Fig. 5.40 Effect of manifold and carburettor arrangement on engine volumetric efficiency

increase the flow resistance so that a peaky and relatively low volumetric efficiency of 82.8% is achieved at 3000 rev/min (Fig. 5.40).

5.10.4 Equal spacing over and under branched triple riser manifold
(Fig. 5.39(d))

A more costly and elaborate manifold arrangement is the equal spacing over and under branched triple riser type of manifold (Fig. 5.39(d)). This layout provides a direct flow path of moderate length from the riser to the respective inlet port without experiencing induction period interference between adjacent cylinders, and it has an induction draw interval from each riser of 360°.

The result of this direct flow with no induction overlap interference, and relatively long intervals between drawing a mixture from each riser, minimizes the air–fuel ratio spread under part throttle to 0.92:1 (6.1%), and when the throttle is fully open, it widens, but only to 1.81:1 (12.05%). Similarly, the volumetric efficiency achieved in each cylinder with this branch tract configuration produces 83% at 2000 rev/min, rises to a maximum of 89.6% at 3000 rev/min, after which it decreases to 82.5% at 5000 rev/min (Fig. 5.40). Thus, from 2000 rev/min to 5000 rev/min the volumetric efficiency remains well above the other three-manifold arrangements which are being compared.

5.10.5 Manifold test conclusions

From the four manifold configurations tested (Fig. 5.39(a–d)), the equal spacing, single-plane twin-barrel arrangement (Fig. 5.40(a)) was chosen as the best compromise owing to its relatively uniform mixture distribution from cylinder to cylinder, from small to wide-open throttle operating conditions, and because it produced a comparatively high volumetric efficiency over a wide engine speed range. A final consideration, but not the least, is that the chosen manifold was the simplest and the cheapest to manufacture.

5.11 Vee eight-cylinder engine induction manifolds

5.11.1 Single twin-barrel carburettor vee eight-cylinder manifold
(Figs 5.41(a), 5.42, 5.43, 5.44)

The simplest of vee eight-cylinder manifolds has a double barrel carburettor mounted on top of a twin riser flange (Fig. 5.41(a)). Each riser passage flows through a tee junction into a gallery, and the ends of the gallery join onto branches which span the two cylinder-banks. In other words, each riser supplies a gallery which bridges one set of branches that feeds cylinders numbers 1–6 at one end and cylinders 4–7 at the opposite end. Likewise, the second riser and gallery feeds branches going to cylinders numbers 2–5 at one end and cylinders numbers 3–8 at the other end. A typical firing order for a vee eight-cylinder engine with a two-plane or cruciform crankshaft would be 15486372 and with this arrangement there would be an induction period pulse alternately, every 180° in each riser.

A moderate level of volumetric efficiency (Fig. 5.42) is obtained in the low-to-medium speed range, but there is a marked decrease in breathing as the engine speed moves beyond the mid-speed range due to venturi size resistance. This restriction is necessary to keep the air velocity at low engine speed high enough to atomize and suspend the fuel particles in the main air stream as they move towards the inlet ports. Unfortunately, at higher engine speeds the flow capacity of the twin venturi carburettors is inadequate and power performance therefore suffers.

An interesting feature of this design is that the riser feeding the lower branches on one side are actually cast into the gallery which supplies a completely different set of branch pipes; therefore, the passage space around the riser must be carefully shaped to maintain a similar flow path and resistance to that of the uninterrupted passageways.

A manifold design problem with twin or quadruple risers is that some branch passageways pass over, while others pass under a different set of branch pipes (Fig. 5.43). Consequently, the riser height for the two galleries may differ considerably. This means that the shorter of the two risers will have less distance to stabilize the mixture flow

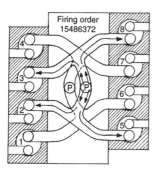

(a) Vee eight-cylinder engine with
single twin-barrel carburettor

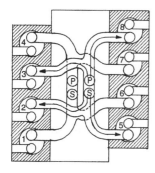

(b) Vee eight-cylinder engine with single
compound quadruple carburettor

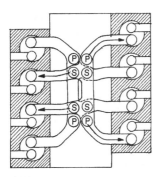

(c) Vee eight-cylinder engine with
dual quadruple carburettors

Fig. 5.41 Vee eight-cylinder engine induction
manifold arrangement

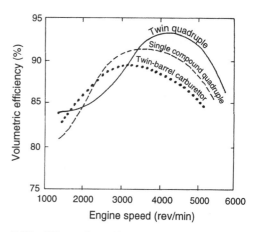

Fig. 5.42 Effect of manifold and carburettor
arrangement on engine volumetric efficiency

opening) conditions, but there is very little variation in air–fuel ratio from cylinder to cylinder
with wide throttle (50° throttle opening) operating conditions (Fig. 5.44). A further comparison
can be made (Fig. 5.44) with the mixture strength
variation between cylinders. Thus, cylinders numbers 6–7 and 2–3 operate slightly on the lean side
of the carburettor's air–fuel ratio, whereas cylinders numbers 1–4 and 5–8 operate just on the rich

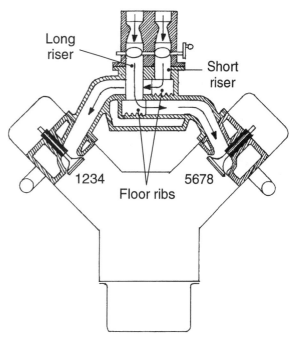

Fig. 5.43 Vee eight-cylinder engine sectional view of
an intake manifold showing long and short riser

leaving the throttle butterfly plate before it enters
its respective gallery and, as a result, the mixture
distributions between the two sets of passages will
vary. Generally, the short riser in comparison
with a long riser produces a wide spread in the
air–fuel ratio under part throttle (25° throttle

257

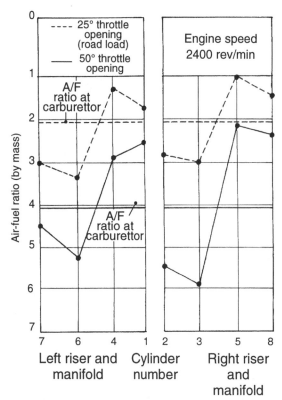

Fig. 5.44 Distribution of mixture strength between cylinders for a vee eight-cylinder engine fitted with a single twin-barrel carburettor

side of the carburettor's air–fuel ratio discharge. This variation is probably due to the easier flow path (Fig. 5.41(a)) to the diverging branches, numbers 1–4 and 5–8, as opposed to the converging branches, 2–3 and 6–7, which tend to have sharper bends where the respective galleries and branches merge.

5.11.2 Single compound quadruple-barrel carburettor vee eight-cylinder manifold
(Fig. 5.41(b))

To raise and extend the cylinder's volumetric efficiency over a wider speed range and simultaneously improve low-to-medium speed drivability a single compound quadruple carburettor is frequently adopted (Fig. 5.41(b)). This four-riser arrangement provides the high intake air–fuel velocity necessary for maintaining a high quality of air–fuel mixing as the charge moves towards the inlet ports in the low-to-mid speed range, by only opening the primary barrel throttles.

However, in the mid-to-upper speed range induction throttling caused by too much passageway restriction is overcome by opening the secondary barrel throttles, this causes the engine's volumetric efficiency to rise above that of the single twin-barrel carburettor at high engine speeds (Fig. 5.42).

5.11.3 Dual quadruple-barrel carburettor vee eight-cylinder manifold
(Figs 5.41(c) and 5.42)

If a more direct mixture distribution is sought to produce a rapid acceleration response without impairing the breathing ability of the manifold, dual quadruple carburettors are sometimes utilized (Fig. 5.41(c)). This configuration has four barrels at each end of the gallery feeding almost directly into each branch tract. If all the barrel throttles are of the primary action kind, in that there is no two-stage opening operation, then there will be very little variation in the quality of the mixture reaching the inlet valve ports compared with that initially entering the riser passageways from the carburettor venturi. However, if each carburettor is a compound of two primary and two secondary-barrel types, then there may be some flow bias under primary-barrel part-load opening operating conditions to the branches immediately under the primary risers.

Typical volumetric efficiency characteristics for a dual quadruple-carburettor manifold are shown in Fig. 5.42. Here, it can be seen that up to about 3000 rev/min the cylinder volumetric efficiency is inferior to both the single twin-barrel and single compound-quadruple carburettor manifolds, but from mid-speed onwards the dual quadruple-carburettor layout shows itself to be superior. The dip in volumetric efficiency in the low speed range of the engine is due to the secondary throttles being closed to raise the air flow velocity, thereby maintaining the quality of the mixture as it passes through the manifold passageways.

5.12 Other approaches to cylinder charging

5.12.1 Induction wave ram cylinder charging
(Fig. 5.30)

When the engine is running, a column of air

moves through the induction tract passageway from the point of entry to the inlet port and valve and then into the cylinder (Fig. 5.30). Every time the inlet valve opens, the reduction in cylinder pressure produces a negative pressure-wave which travels (at the speed of sound) through the column of air from the back of the inlet valve to the open atmospheric end of the tract. Immediately this pressure-wave pulse reaches the atmosphere, rarefaction occurs; that is, the air at the tract entrance suddenly becomes less dense and therefore creates a depression. Instantly, the surrounding air rushes in to fill this depression. As a result, a reflected positive pressure-wave is produced due to the inertia of the air, and this causes the pressure pulse to travel back to the inlet valve port. It is this first reflected pressure-wave that, if correctly timed, is responsible for ramming the air into the cylinder towards the end of the induction period. When the pressure-wave again reaches the back of the inlet valve it reverses its direction and is reflected outwards. Thus, these negative and positive pressure waves are continuously reflected backwards and forwards between the open intake and the inlet valve port but with a decaying amplitude until the inlet valve closes.

To utilize fully the pressure-wave pulse it must be timed so that its first positive pressure-wave arrives at BDC towards the end of the induction period at its peak amplitude. Therefore, it is important to know the time it takes for a pressure wave to travel through the column of air from the open inlet valve to the intake entrance, to be reflected and then to travel back again to the inlet valve.

Therefore, the time interval for the pressure wave to travel the full length of the tract and back again can be obtained as follows.

$$\text{Speed of sound} = \frac{\text{distance pulse travels}}{\text{time taken}}$$

therefore, time taken to travel the tract length and back again

$$= \frac{\substack{\text{distance pulse travels} \\ \text{(from one end to the other and back again)}}}{\text{speed of sound}}$$

$$t = \frac{2L}{1000C} \qquad (1)$$

where t = time for the pulse to travel the tract length and back again (s)

L = length of tract from open end to inlet valve head (mm)

C = speed of sound through air (approximately 330 m/s)

The crankshaft angular displacement during the same interval of time would be as follows.

Crankshaft displacement = time to travel tract length and back again × angular speed

$$\theta_t = t \times \frac{360}{60} N \text{ but } t = \frac{2L}{1000C}$$

therefore

$$\theta_t = \frac{2L}{1000C} \times 6N$$

$$= \frac{0.012LN}{C} \text{ (deg)} \qquad (2)$$

where θ_t = crankshaft angular displacement (deg)

N = engine crankshaft speed (rev/min)

Note: inertial and wave ram effects always coexist to some extent and the one which predominates at any one time depends upon the engine speed, throttle opening and the tract dimensions.

Coil spring analogy of a propagating pressure wave
(Fig. 5.45(a–e))
A pressure-wave pulse moving through a column of air can be visualized as a wave propagating through a coil spring (Fig. 5.45) which has been lightly stretched apart. If a sudden jerk is given to the left-hand end of the coil, then a number of coil loops close up momentarily at the end of impact, they then quickly travel to the other end of the spring without there being any major movement of the spring itself (Fig. 5.45(a–e)). Similarly, when an inlet valve opens, the sudden induction pulse makes a negative pressure-wave run through the column of air to the exit, it then reverses its direction and returns through the air column to the inlet valve port and head as a positive pressure-wave.

Analysing wave ram effects for various lengths of tract ($\theta_t = 60°$, $90°$ and $120°$)
(Fig. 5.46(a–c))
When the inlet valve opens at the beginning of the

259

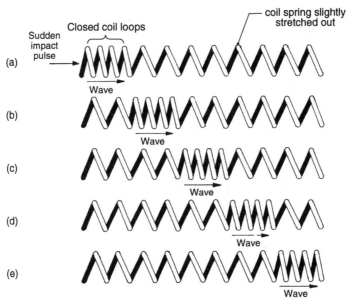

Fig. 5.45 Illustration of wave moving from one end of a spring to the other

induction period a negative pressure-wave pulse travels from the inlet valve to the tract intake, this is then reflected back as a positive pressure-wave.

The crank-angle interval, at which this primary negative pulse created at the inlet valve moves out, is denoted by θ_t. If the crank angular displacement, for the pressure wave to move out and return to the inlet valve, is 90° then the first reflected pressure wave will be first experienced at the inlet valve when the piston is in the mid-stroke region, it will then reach its peak value at BDC. At this point, the second reflected negative pressure-wave propagates from the inlet valve to the intake opening, and thus a pattern of negative and positive pressure-waves, moving from one end would continue with decreasing amplitude if time would permit.

Owing to the positive and negative half-pressure waves (broken curves) overlapping in the inlet port, the resulting pressure is shown by the full curves (Fig. 5.46(a)). The chain curve represents the pressure in the cylinder and the shaded area represents the difference in pressure between the inlet valve port and the cylinder. It is the magnitude of the pressure difference across both sides of the inlet valve head from BDC to the point at which the valve effectively closes (this is about 20° before the inlet valve timing diagram shows the valve closed (IC)) that determines the pressure-wave's effectiveness in boosting the density of the cylinder's charge.

Thus, if the induction tract length is chosen so that $\theta_t = 90°$, then it can be seen (Fig. 5.46(a)) that the cylinder pressure has risen to about atmospheric pressure at BDC and then moves to a positive pressure before the effective inlet valve closing point (EIC), i.e. effective inlet closing is reached. It can also be seen that the pressure-wave has peaked at BDC, which provides time for the charge in the inlet port to be transferred to the cylinder before the cylinder is cut off from its supply. This phasing of the resultant pressure-wave is therefore at its optimum for cramming the greatest amount of charge into the cylinder.

If a short intake tract was chosen so that $\theta_t = 60°$ (Fig. 5.46(b)) then the cylinder pressure would still be negative at BDC and would barely reach atmospheric pressure at the effective valve closure point. Due to wave overlap above and below the atmospheric pressure line and the 60° offset between the positive and negative wave peaks the resultant full-line positive pressure peak is small in magnitude and peaks at 30° before BDC. Thus, there is no pressure-wave assistance between BDC and the effective closure of the inlet valve (EIC). Instead, the resultant pressure-wave over this angular interval is negative so that the wave actually tries to extract charge from the cylinder.

Conversely, if a long intake tract was selected so that $\theta_t = 120°$ (Fig. 5.46(c)), then the resultant full-line pressure-wave in the inlet valve port

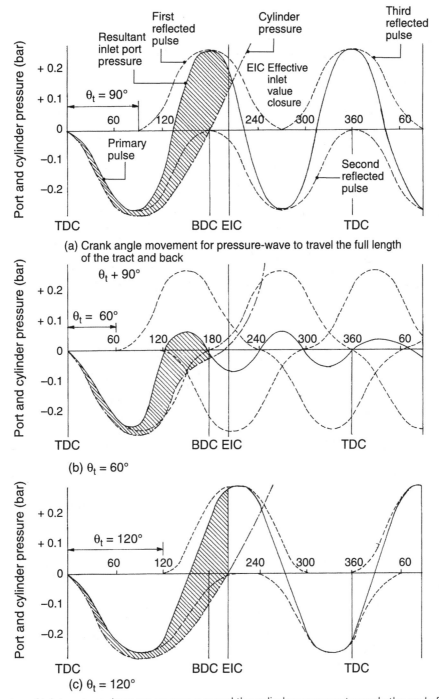

(a) Crank angle movement for pressure-wave to travel the full length of the tract and back

(b) $\theta_t = 60°$

(c) $\theta_t = 120°$

Fig. 5.46 Phasing of inlet port resultant pressure wave and the cylinder pressure towards the end of the induction period

attains a broad peak just beyond the effective inlet valve closing point. Also, the cylinder pressure is well in the negative at BDC and only reaches atmospheric pressure at the effective inlet valve closing point (EIC). Thus, it can be seen that the shaded pressure difference between the port and cylinder over the last part of the induction period from BDC to the effective point of

261

valve closure has built up relatively late to contribute much to the wave ram effect.

Experiments have shown that the optimum value for θ_t lies between 80° and 90°, which shows a close correlation with the calculated results based on organ-pipe theory. Thus, the formula $t = 0.012LN/C$ (deg) can be used to tune the induction tract length, for any engine speed, to maximize the wave ram effect.

Example

Determine the total induction length, from the point of intake entry of the tract to the head of the inlet valve, to optimize the wave ram effect at the following engine speeds: 2000, 3000 and 4000 rev/min.

Take the crank-angle movement (for the pressure-wave to travel through the tract from end to end and back again) to be $\theta_t = 85°$ and the velocity of sound in air to be $C = 330$ m/s. Thus

$$\theta_t = \frac{0.012LN}{C}$$

therefore

$$L = \frac{\theta_t C}{0.012N}$$

for an engine speed of 2000 rev/min

$$L_{2000} = \frac{85 \times 330}{0.012 \times 2000} = 1168.75 \text{ mm or } 1.168 \text{ m}$$

for an engine speed of 3000 rev/min

$$L_{3000} = \frac{85 \times 330}{0.012 \times 3000} = 779.17 \text{ mm or } 0.779 \text{ m}$$

for an engine speed of 4000 rev/min

$$L_{4000} = \frac{85 \times 330}{0.012 \times 4000} = 584.38 \text{ mm or } 0.584 \text{ m}$$

The relationship between tract length and engine speed has been plotted from the formula assuming an average value for θ_t to be 85° (Fig. 5.47). This enables a rough approximation of the tuned pipe length for any engine speed to be picked off directly from the graph.

5.12.2 Helmholtz resonator cylinder charging
(Figs 5.48, 5.49 and 5.50)

Ram induction charging will produce an improvement in volumetric efficiency at certain engine

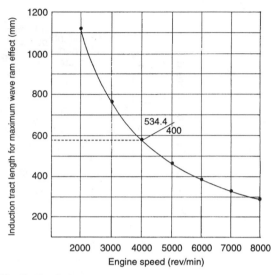

Fig. 5.47 Relationship between induction tract length for maximum wave ram effect and engine speed

speeds by means of individual cylinder induction pipes or tracts. The extra charging effects are due to the inertia and elasticity of the charge in the inlet tract and cylinder. For slow-speed heavy-duty engines, which may have a maximum engine speed of only 2000 rev/min, the individual induction tract length would be far too long, so that it would be very difficult to accommodate.

An alternative and more compact induction system known as the Helmholtz resonator tuned induction is sometimes used on large diesel engines in the single stage form or in the dual and triple stage configuration for some high performance petrol engines.

The Helmholtz resonator originally simply consisted of a spherical chamber with a pipe projecting from it (Fig. 5.48), the chamber being the equivalent of the manifold gallery and branch pipes in addition to the inlet valve ports and cylinder, whereas the pipe projecting from the chamber becomes the tuned induction tract.

The resonant frequency of this chamber and pipe is given by

$$n = \frac{c}{2\pi} \sqrt{\frac{A}{LV}}$$

where n = resonant frequency (Hz)
   c = velocity of sound in air (m/s)
   A = cross-sectional area of tuned pipe (m^2)
   L = length of tuned pipe (m)
   V = resonating volume (m^3)

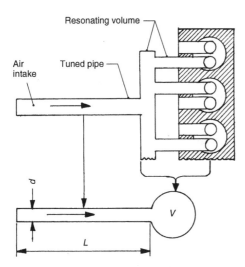

Fig. 5.48 Comparison of the Helmholtz resonator with tuned manifold system

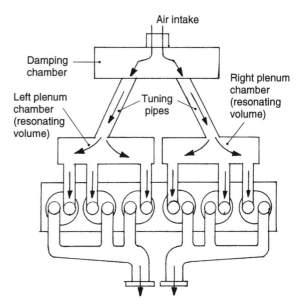

Fig. 5.49 Tuned induction system

The operating principle of the Helmholtz resonator is as follows: air filling any given sized chamber, when shaken, will oscillate to and fro at its own unique frequency—known as its natural frequency of vibration. If now the unique chamber was connected almost directly to an inlet valve port and the open-ended pipe becomes the air intake tract, then every time the inlet valve opened a negative pressure-wave pulse would disturb the air in the resonator chamber and pipe. Now, if the engine's speed is increased until the pressure-wave pulse frequency corresponds to the natural frequency of the air in the resonator chamber, then the air in the system will be excited into a state of resonance at which the amplitude of the pressure-wave moving through the incoming air stream produces a series of pressure-waves or jolts. If properly timed, these waves will bombard the cylinder with surges of compressed air towards the end of the induction period when the piston is moving against the incoming mixture and, as a result, more charge will be crammed into the cylinder before the inlet valve closes.

The trick is to choose a manifold resonator volume which resonates at an engine speed at which a boost in torque is desired (this is usually the normal peak torque for the engine) and also to experiment with the tuned pipe length to obtain optimum results. Either side of the engine speed, at which the tuned passageways resonate, the pressure-wave ram effect quickly deteriorates. In fact, the pressure-wave pulses may interfere with the normal inertial ram effects so that cylinder filling may be impeded. Thus, careful

experimental work has to be carried out to get the best compromise between inertial and wave ram effects.

In multi-cylinder engines, cylinders which have overlapping inlet valve events cannot share their manifold, but adjacent cylinders which do not have overlapping induction periods usually share a common induction manifold. It is this manifold, with the inlet port and half the cylinder displacement of one cylinder, which becomes the resonating volume. With a six-cylinder in-line engine (Fig. 5.49) cylinders numbers 1, 2 and 3 do not overlap and are therefore able to form a common manifold. Likewise, induction periods of cylinders numbers 4, 5 and 6 do not interfere with each other so they can also use a common manifold. Similarly, because there are only two resonating manifolds then only two tuning pipes are needed. These, in turn, are fed by a single damping chamber which smooths out any large intake turbulence which might occur, particularly if the engine is to be turbocharged, and also serves to separate the two tuned pipes. It should be appreciated that it is the manifold volume, along with the length and cross-sectional area of the tuned pipe, which dictates the tuning frequency of the system.

Typical volumetric efficiency characteristics are shown in Fig. 5.50, it can be seen that with the conventional manifold the volumetric efficiency rises gradually but, at about two-thirds engine speed, it then begins to decrease slowly. This

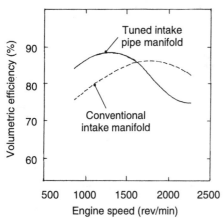

Fig. 5.50 Helmholtz resonator tuned induction system volumetric efficiency characteristics

contrasts with the tuned intake pipe manifold operating on the Helmholtz resonator principle, if it is tuned to resonate in the low-speed range to improve the back-up torque rise, which is the desired response for large slow-speed diesel engines. In this case, the engine volumetric efficiency will peak early in the engine speed range, but at higher engine speeds the resonator becomes counter productive and reduces the cylinder volumetric efficiency well below that obtained with the conventional standard manifold.

5.13 Tuned induction manifold systems

5.13.1 Dual ram induction system (Vauxhall)
(Fig. 5.51(a and b))

This air charge intake layout has a small plenum chamber (air collecting box) which acts firstly as a common throttle valve fixture point, secondly as the junction for two large-diameter feed pipes approximately 30 cm long, and thirdly it serves the purpose of separating the two tuned feed pipes to prevent interference of the resonating air columns. Each pipe supplies a plenum resonating gallery which, in turn, feeds three adjacent curved (shown straight) ram pipes of smaller diameter but of increased length (approximately 40 cm) directly to individual inlet valve ports. The separation of the front and rear three cylinder intakes enables the power unit to be treated virtually as two separate three-cylinder engines.

Thus, each half of the engine has an induction period every 240° and there is no valve overlap interference between cylinders in either the front or rear cylinder groups.

The consequences of separating the intake for the engine's cylinders into two halves is that, for each cylinder, the system resonates as a whole; that is, one of the large-diameter feed pipes with its plenum gallery and the three branch pipes together form one resonating circuit.

Operating principle

With the power valve closed (Fig. 5.51(a)) the tuned intake volume and tract length (roughly $30 + 40 = 70$ cm) is designed to resonate at an engine speed at which torque is to be maximized, in this case 3300 rev/min. It is these resonating pressure-waves within the ram pipe columns of air which are responsible for reinforcing the ram charging effect. At a higher engine speed the resonating ram boost will become ineffective at about 4000 rev/min when the interconnecting power valve, joining the two resonating plenum galleries, opens (Fig. 5.51(b)). Immediately the two plenum galleries are in communication with each other, there will be interference between the two tuned systems so that any remaining oscillating resonance will be destroyed in each of the two tuned systems so that they become a massive single volume plenum chamber. However, the volume of each individual ram pipe will, at a higher engine speed, form its own resonating system. Thus, the ram pipe dimensions can be chosen to excite resonance pressure-waves to boost the cylinder charging at an even higher engine speed—in this case 4400 rev/min. The resulting engine torque characteristics produce a relatively flat curve between about 1500 rev/min and 5000 rev/min (Fig. 5.52).

5.13.2 Dual-stage induction system with a two-valve cylinder-head
(Fig. 5.53(a and b))

This system has a plenum chamber positioned alongside the engine with a throttle valve situated in its entrance to control the air flow. Long curved ram pipes supply air from the plenum chamber to the individual inlet ports and a secondary short pipe, also supplied from the plenum chamber, intersects the long pipe further downstream. Each short high-speed pipe has its own butterfly power

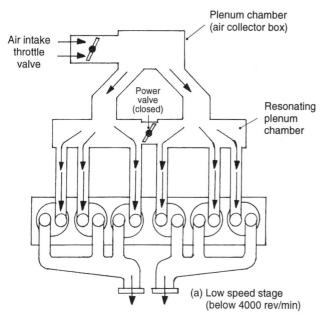

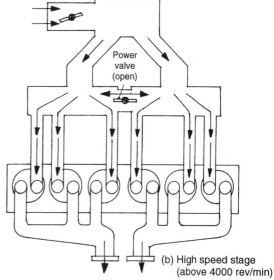

(a) Low speed stage
(below 4000 rev/min)

(b) High speed stage
(above 4000 rev/min)

Fig. 5.51 Dual ram induction system (Vauxhall)

valve to close off the secondary passageway at low engine speeds.

Operating principle

Low-to-mid engine speeds (below 3800 rev/min) (Fig. 5.53(a)). At low engine speeds the power valves are closed, and therefore air can only flow from the plenum chamber to the valve port through the tuned long ram pipes. The dimensions of the long pipes maximize the inertial pile-up of the charge in the engine's speed range

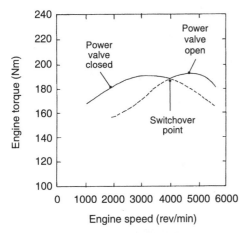

Fig. 5.52 In-line six-cylinder engine dual ram induction system torque characteristics

and also tune the reflected pressure-waves to reach their peaks just before the effective closure point of the inlet valves at some critical engine speed (3200 rev/min). On either side of this critical engine speed, the phasing of the pressure-wave relative to the closing point of the inlet valve will have shifted and thus reduced the effectiveness of the pressure-wave. Refer to section 5.12.1 and Fig. 5.46(a–c).

High engine speed (above 3800 rev/min) (Fig. 5.53(b)). As the engine speed rises beyond 3800 rev/min, the power valves open so that the inlet valve ports now draw charge from both the long and short branch pipes. Accordingly, there is now a much shorter tuned ram pipe length which requires a much higher engine speed for the first reflected pressure-wave peak to align with the inlet valve closure point. Thus, the volumetric efficiency curve will have two peaks, and hence this second stage tuned system upholds the engine's torque in the upper speed range.

5.13.3 Dual-stage induction system with four-valve cylinder-head
(Fig. 5.54(a and b))

This induction system has long curved and short straight pipes branching from the plenum cham-

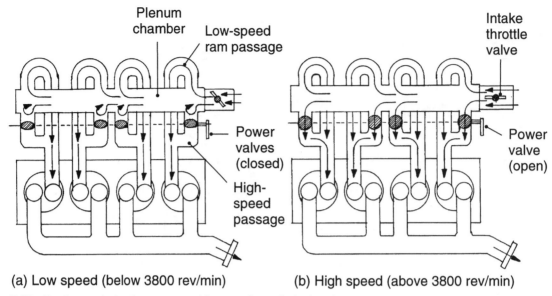

(a) Low speed (below 3800 rev/min)　　(b) High speed (above 3800 rev/min)

Fig. 5.53　Dual stage induction system with two-valve cylinder head

ber to feed each of the two inlet valve ports. The long curved branches provide the air passages for low-to-medium speed operation, whereas the short straight branches provide an additional supply of air to the cylinders at higher engine speeds, when the power valve inside these short passages is open.

Operating principle

Low-to-medium engine speed (below 3800 rev/ min) (Fig. 5.54(a)). At low engine speeds the power butterfly valves close off the short pipe

flow path. The air supply to the cylinders is therefore entirely through the long curved tract and one of the inlet valve ports. Consequently, the restricted air passage to the cylinder, and the offset position of the low speed inlet valve port in the cylinder produces a high air velocity in the long passageway, and a large amount of induction swirl as the mixture enters the cylinder. Both of these factors are essential for promoting good mixing, particularly with lean mixtures at relatively low engine speeds. The length of the long pipe is tuned to synchronize the first reflected pressure-wave peak with the effective closure

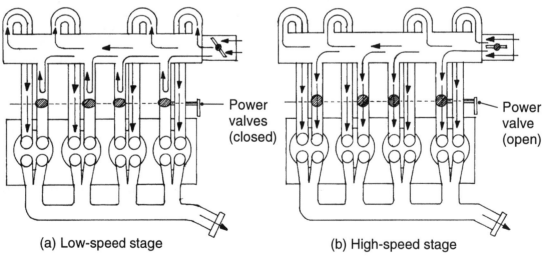

(a) Low-speed stage　　　　　　　　(b) High-speed stage

Fig. 5.54　Dual stage induction system with four-valve cylinder-head (Nissan)

266

point of the inlet valve. Thus, the long single passageway optimizes the inertial and wave ram effect and also promotes a high level of induction swirl which is the ingredient for good low-speed engine torque characteristics.

High engine speed (above 3800 rev/min) (Fig. 5.54(b)). With higher engine speeds the butterfly power valves open, thus permitting additional air to flow from the plenum chamber to the cylinder via the short branch pipes and the high-speed inlet valve ports. The opening of the second passageway improves the volumetric efficiency in the high engine speed range, and the flow through the second inlet valve port now changes the induction spiral swirl motion in the cylinder into a barrel roll motion which is more suitable for promoting effective combustion at high engine speeds and load conditions.

5.13.4 Vee six-cylinder engine two-stage ram induction system
(Fig. 5.55(a and b))

The intake plenum chamber (air box) is mounted between and along the cylinder-banks. Pipes branch out from the plenum chamber from alternate sides in a semi-spiral fashion to form loops before continuing to the inlet valve ports. Each long ram pipe has a power butterfly-valve positioned above and slightly to one side of the plenum chamber, there are three branch pipes and valves for each cylinder bank but only the right-hand power valve and ram branch pipe is shown. When these valves open they short circuit the full ram pipe flow path length so that air is diverted directly from the plenum chamber to points much further downstream from the initial entry to the long curved ram pipe.

Operating principle
Low-to-medium speed (below 4000 rev/min) (Figs 5.55(a) and 5.56). At low engine speeds, both rows of butterfly valves are closed. Under these conditions air from the plenum chamber is supplied to the extreme outer end of the looped ram pipe where the 79 cm long flow path produces a high level of inertial ram, up to about 4000 rev/min. At the same time, the tuned length produces reflected pressure-waves which peak just before

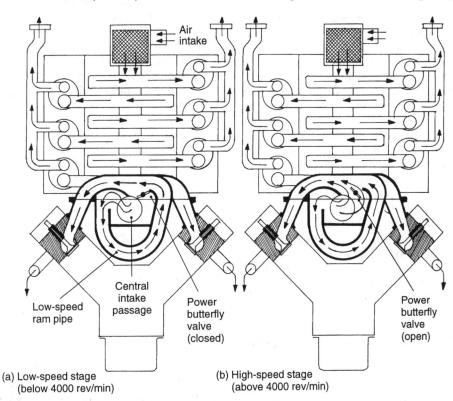

(a) Low-speed stage
(below 4000 rev/min)

(b) High-speed stage
(above 4000 rev/min)

Fig. 5.55 Vee six-cylinder engine two-stage ram induction system (Audi)

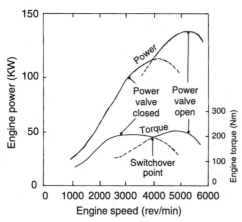

Fig. 5.56 Vee six-cylinder engine two-stage ram induction system torque and power characteristics

the inlet valve closes at roughly 3000 rev/min. At speeds below and above this critical engine speed the pressure-wave peak will be out of phase with the valve closure point so that the wave ram effect will not be prominent. However, the combined inertial and wave ram effect does produce relatively broad peak torque characteristics (Fig. 5.56).

High engine speed (above 4000 rev/min) (Figs 5.55(b) and 5.56). With engine speeds above 4000 rev/min the power butterfly valves open. Air will therefore not only enter the ram pipe at the beginning of the looped pipe but it will also be fed directly through the butterfly valve passageway, which reduces the flow path to a length of 37 cm. This reduces the distance the pressure-wave has to travel outwards from the inlet valve before it is reflected back again to the inlet port. Accordingly, there will be a new critical engine speed where the pressure-wave peak reaches the inlet valve head just before the valve closes, this condition therefore optimizes the ramming of the charge towards the end of the induction period. Simultaneously, the inertial ram build-up with the short, more direct tract, upholds the volumetric efficiency in the upper speed range, particularly between 5500 and 6000 rev/min. This results in a steep power rise as the maximum engine speed is approached (Fig. 5.56).

5.13.5 Triple-stage induction systems (Peugeot)
(Fig. 5.57(a–c))

The induction system is made up of two parallel plenum chambers positioned between the cylin-

der banks. Each plenum chamber feeds the opposite bank cylinder-head via three individual ram pipes. Thus, each cylinder bank effectively operates as two separate three-cylinder engines in each of which there are evenly spaced induction periods commencing at intervals of 240°. The air intake at one end of each plenum chamber is controlled by its own throttle valve whereas the far ends of these chambers are joined together by a long passage which can be blocked off by a pair of intermediate-speed butterfly valves situated at both entrances of the long passageway. Mid-way along the plenum chambers is a short passageway which also bridges the two chambers when the single high-speed butterfly valve, positioned inside this short passage, is opened.

Operating principle

Low engine speed (below 4000 rev/min) (Figs 5.57 and 5.58). With both intermediate and high-speed butterflies closed, the total volume of each plenum chamber with each of its three main ram pipes presents a complete resonating system which combines both an inertial and wave ram effect and which peaks the volumetric efficiency (and similarly engine torque) at some critical engine speeds, in this case 3500 rev/min. Above this speed the amplitude of resonant pressure-waves decreases so that the benefits of the tuned system decreases. Correspondingly, it will cause a proportional fall off in each cylinder's volumetric efficiency (Fig. 5.58).

Intermediate engine speed (between 4000 and 5000 rev/min) (Figs 5.57(b) and 5.58). As the engine speed exceeds 4000 rev/min the long passage joining both plenum chambers together is opened by the intermediate-speed butterfly valves, this therefore partly destroys the first-stage tuned resonating pressure-wave frequency. A new resonating system is then created further downstream which is confined almost wholly to the volume of the individual ram pipes. The second induction resonating system is tuned to be critical at around 4500 rev/min, thus the resonating pressure-waves become most violent and effective in cramming more mixture into the cylinders (Fig. 5.58).

High engine speed (above 5000 rev/min) (Figs 5.57 and 5.58). With still further rising engine speed, once 5000 rev/min has been reached, the short-passage high-speed butterfly valve opens and the

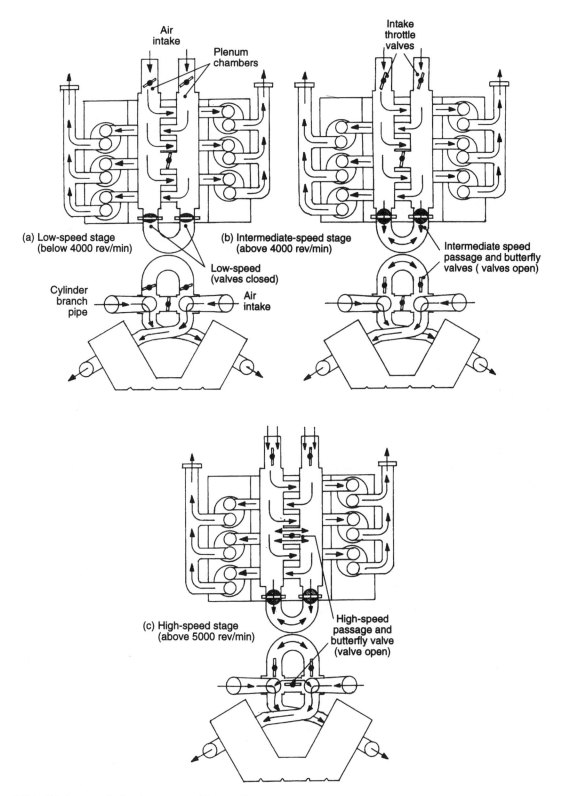

Fig. 5.57 Triple stage induction system (Peugeot)

269

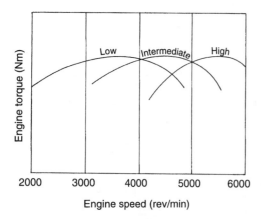

Fig. 5.58 Torque characteristics for a triple stage induction system

short and more direct passageway between the two plenum chambers now increases the interference between both cylinder-bank induction systems. Thus, the second stage resonance in the plenum chambers is completely neutralized. Consequently, induction resonance is now confined entirely to the ram pipes, which thus delay their critical resonance to an even higher engine speed of something like 5500 rev/min. Accordingly, this further boosts the ram charging effect in this narrow engine speed range (Fig. 5.58).

Control of staged valve openings. The electronic control unit has, stored in memory, the characteristic engine speed points and the corresponding load points as provided by a throttle potentiometer. This enables the solenoid-operated low and high-speed butterfly valves to adapt the induction volumes to match the operating conditions.

5.13.6 Two-valve cylinder-head with swirl control valve
(Fig. 5.59(a and b))

This unusual inlet valve port system is utilized to increase the air swirl at part throttle, light to medium-load conditions, while retaining a large flow capacity under wide-throttle high-load driving. These dual conditions are obtained by splitting the inlet valve port into two halves by a wall upstream of the inlet valve. The flow path through the outer half of the inlet valve port is open at all times and is known as the primary helical passage, whereas the inner flow-path half

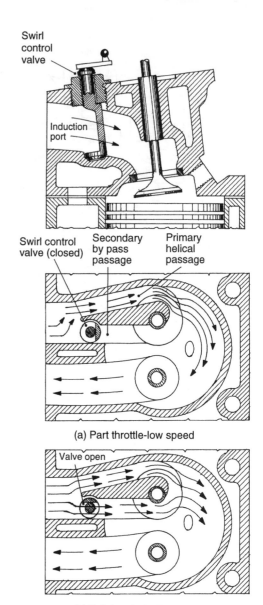

(a) Part throttle-low speed

(b) Full throttle-high speed

Fig. 5.59 Two-valve cylinder-head with swirl control valve (Toyota)

of the port has a swirl control valve at its entrance and is known as the secondary bypass passageway.

Operating principle

Part-throttle light-load conditions (Fig. 5.59(a)). When driving under part throttle operating conditions the swirl control valve closes, thus causing all the charge mixture entering the cylinder to pass through the primary helical passageway. This

270

restricted flow path speeds up the mixture movement and also directs the flow tangentially into the cylinder so that a very high rate of induction swirl is achieved. This is essential if fast burning is to be obtained with a lean mixture without the cylinder misfiring. It is claimed that mixtures as lean as 22:1 can be effectively burnt on part throttle without intermittent firing.

Wide-throttle high-load conditions (Fig. 5.59(b)). With a wide-open throttle and high engine load the swirl control valve opens so that the flow-path cross-sectional area has considerably increased. Mixture will now flow through both halves of the port thus reducing the mixture velocity while, at the same time, raising the density of the mixture entering the cylinder. Mixture flowing out from the secondary bypass passage will now partly move across the primary flow stream causing the resultant intensity of swirl in the cylinder to be reduced. These conditions at very high engine speeds are desirable for maintaining a high level of volumetric efficiency and limiting the swirl to match the combustion requirements at these high engine speeds.

5.13.7 Four-valve cylinder-head with swirl control valve
(Fig. 5.60(a and b))

This unique twin-inlet valve port induction system has a swirl control valve situated at the mouth of the secondary inlet valve port whereas the primary port is partitioned so that the outer-half of the primary inlet port is permanently open. Conversely, the inner-half of the primary inlet port is supplied from the secondary passageway, which is itself regulated by the swirl control valve at its entrance.

Operating principle

Part-throttle light-load conditions (Fig. 5.60(a)). To obtain the benefits of high air velocity in the induction port and a high swirl intensity in the cylinder, which are necessary if the engine is to burn very lean mixtures under part load conditions, the swirl control valve is closed. The mixture is therefore forced to enter the cylinder via the primary passage which forms the outside half of the low-speed inlet. This restriction is responsible for speeding up the air movement and also raising the amount of spiral swirl in the cylinder.

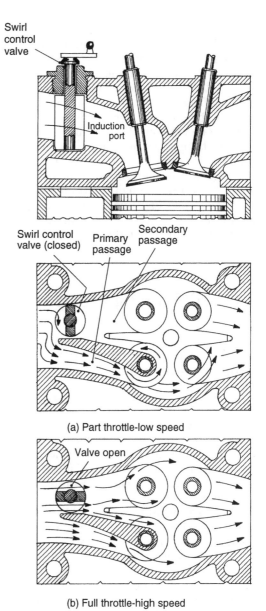

(a) Part throttle-low speed

(b) Full throttle-high speed

Fig. 5.60 Four-valve cylinder-head with swirl control valve (Toyota)

Wide-throttle high-load conditions (Fig. 5.60(b)). With increasing engine speed and load, the primary passage cannot cope with the cylinder's demands for more mixture charge, therefore at some predetermined stage the swirl control valve is made to open. The charge of air and petrol mixture is now not only permitted to flow through the high-speed inlet valve port but also through the inside half of the low-speed inlet valve port.

The considerable port cross-sectional area increase therefore reduces the port air velocity and causes the charge's density to increase, and simultaneously it will permit much larger quantities of air to enter the cylinders. At the same time the directional air swirl from the low-speed inlet port will now be counterbalanced by the high-speed inlet port flow stream. The overall result will be to convert the mixture movement in the cylinder from a basic spiral swirl into a barrel roll swirl. With this mixture movement in the cylinder, the flames can travel through the combustion chamber at relatively high speed, particularly with a rich full load mixture so that combustion can be effectively completed in the minimum time.

5.14   Exhaust gas extraction

5.14.1   Kinetic energy theory of cylinder scavenging
(Fig. 5.61(a–c))

Possibly the most important mechanism for scavenging (extracting) the residual exhaust gases from the combustion chamber at the end of the exhaust period is to utilize the kinetic energy of the outgoing exhaust gases to produce a compression wave followed by an expansion wave in which the gas pressure is reduced to a depression in the exhaust port region of the exhaust system.

The opening of the exhaust valve towards the end of the power stroke releases the products of combustion, which are under intense pressure, to the exhaust port and pipe. This sudden expulsion of high pressure gas from the cylinder to the exhaust port rapidly displaces the column of gas occupying the port and exhaust pipe passageway, thereby causing the gas to attain a high flow velocity in the primary tract. In other words, the compression-wave pulse expelled from the cylinder transfers its pressure energy to the exhaust gas column in front of it, in the form of kinetic energy (Fig. 5.61(a)). Consequently, the pressure-wave travels outwards, the leading compression side raises the pressure while the trailing expansion side reduces its pressure in the pipe (Fig. 5.61(b)). By the time the piston has moved up to TDC at the beginning of the induction stroke and the end of the exhaust stroke, the

compression wave will have reached the end of the pipe.

The speed of the pressure-wave pulse greatly exceeds the gas discharge speed through the exhaust port and pipe, caused by the upward moving piston pushing the exhaust gases out of the cylinder and into the exhaust port. Therefore, the exhaust gas on the trailing side of the expansion wave becomes less dense and this therefore causes a corresponding drop in exhaust port pressure which now becomes negative (Fig. 5.61(c)). This depression created during the valve overlap period considerably helps to draw residual exhaust gases out of the combustion chamber and into the exhaust port and, at the same time, pulls in the fresh charge from the induction port to fill this evacuated space.

If the exhaust manifold only has short branch pipes before they merge together, there will be insufficient time for the compression wave to leave behind it a depression capable of pulling out the stagnant gas so that the fresh charge arriving at the inlet port is prevented from entering the combustion chamber in the early part of the induction period. Conversely if the pipe length is very long the flow resistance may become excessive thereby creating its own back pressure, which will also slow down the scavenging and filling process.

5.14.2   Exhaust gas speed
(Fig. 5.61(a))

It is sometimes convenient to have some idea of the speed at which the exhaust gases travel through the exhaust port.

The exhaust gas speed can be calculated knowing the engine crankshaft speed in the following way:

let D = piston diameter (mm)
$\quad\quad d$ = port diameter (mm)
$\quad\quad S$ = piston stroke (mm)
$\quad\quad V_p$ = mean piston speed (m/s)
$\quad\quad V_g$ = mean gas speed (m/s)
$\quad\quad N$ = crankshaft speed (rev/min)

then mean piston speed $= \dfrac{2SN}{60 \times 1000}$ (m/s)   (1)

swept cylinder $= V_p \times \dfrac{\pi}{4} D^2$ (m³/s)   (2)

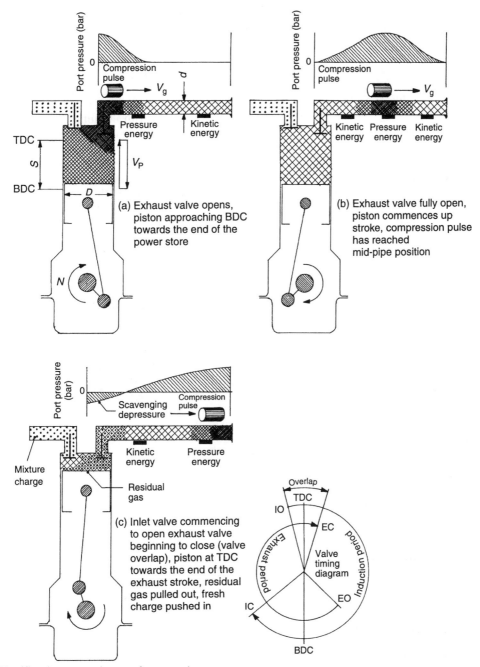

Fig. 5.61 Kinetic energy theory of scavenging

discharge gas volume per second $= V_g \dfrac{\pi}{4} d^2 \ (\text{m}^3/\text{s})$ (3)

$$V_g = V_p \frac{(\pi/4)D^2}{(\pi/4)d^2} = V_p\left(\frac{D}{d}\right)^2$$

equating (2) and (3)

therefore

$$V_g \frac{\pi}{4} d^2 = V_p \frac{\pi}{4} D^2$$

$$V_g = \frac{SN}{30\,000}\left(\frac{D}{d}\right)^2 \ (\text{m/s}) \qquad (4)$$

273

This formula only provides a very rough indication of the gas speed since it does not take into account that the exhaust valve lift is always varying.

Example

An engine has a bore diameter of 80 mm, stroke of 85 mm and exhaust port diameter of 25 mm. If the engine crankshaft speed is 6000 rev/min determine the following:

a) mean piston speed
b) mean gas speed

a) $\quad V_p = \dfrac{SN}{30\,000} = \dfrac{85 \times 6000}{30\,000} = 17 \ (\text{m/s})$

b) $\quad V_g = \dfrac{SN}{30\,000}\left(\dfrac{D}{d}\right)^2 = 17\left(\dfrac{80}{25}\right)^2$

$\qquad\quad = 17 \times 10.24 = 174 \ (\text{m/s})$

5.14.3 Reflected pressure-wave at open end of pipe
(Fig. 5.62(a–d))

The use of exhaust gas interference to improve scavenging during valve overlap necessitates an understanding of a reflected pressure-wave at a passage opening. The concept of a reflected pressure-wave can be explained as follows. Imagine a compression wave 'C$_2$' arriving at an opening (Fig. 5.62(a)) and then spreading in all directions, whereas a compression (1–1) within the pipe 'C', is prevented from spreading out by being confined within the walls of the pipe. A layer of gas at the end of the pipe is unrestricted in amplitude of oscillation because the compression has spread out and its pressure is low, whereas a layer of gas (1–1) in the rear of the compression 'C', is restricted by the compressed gas in front of it. Thus, the amplitude of vibration of a layer (2–2) at the exit will therefore be much greater and it will move out to (3–3) due to the inertia (Fig. 5.62(b)) leaving a rarefaction 'R' (depression) behind it. This depression (rarefaction) will then travel back along the pipe so that the pressure-wave is reflected as a negative pressure (rarefaction) wave. When a pressure-wave rarefaction 'R' (depression) arrives at the end of the pipe (Fig. 5.62(c)) there will be a rush of gas from all directions to the region of low pressure. Hence, a rarefaction (low density) is reflected as a compression wave 'c' from right to left (Fig. 5.62(d)).

5.14.4 Velocity of sound in a gas

The study of exhaust gas scavenging depends on being able to estimate the velocity at which sound travels through the exhaust gas, the following calculations are therefore provided.

The velocity of a sound wave in a gas is given by the formula

$$C = \sqrt{\dfrac{\gamma p}{\rho}} \ (\text{m/s})$$

where $\quad \gamma$ = ratio of molar heat capacities (for air $\gamma = 1.4$)

$\qquad\quad p$ = pressure of gas (N/m^2)

$\qquad\quad \rho$ = density of gas (kg/m^3)

$\qquad\quad C$ = velocity of sound (m/s)

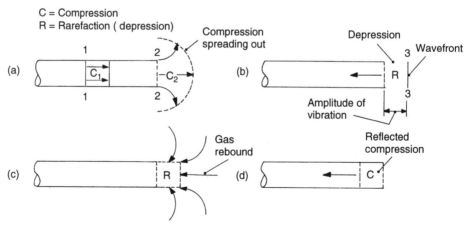

Fig. 5.62 Reflected pressure wave at open end of pipe

Example

Determine the velocity of sound through air at standard temperature and pressure (STP) taking the air density as 1.29 kg/m³ at 0°C, $\gamma = 1.4$ and $p = 101$ kN/m², then

$$C = \sqrt{\frac{\gamma p}{\rho}} = \sqrt{\frac{1.4 \times 101 \times 1000}{1.29}} = 331 \text{ (m/s)}$$

The effects of temperature on the velocity of sound in a gas can be shown in the following way

where
m = mass of gas (kg)
V = volume of gas (m³)
ρ = density of gas (kg/m³)
R = gas constant (kJ/kg K)
T = absolute temperature (K)

thus

$$m = V\rho \quad \text{or} \quad \rho = \frac{m}{V}$$

and

$$pV = mRT \quad \text{or} \quad \frac{pV}{m} = RT$$

then

$$C = \sqrt{\frac{\gamma p}{\rho}} \quad \text{but} \quad \rho = \frac{m}{V}$$

therefore

$$C = \sqrt{\frac{\gamma p}{m/V}} = \sqrt{\frac{\gamma p V}{m}} \quad \text{but} \quad \frac{pV}{m} = RT$$

therefore

$$C = \sqrt{\gamma RT} \text{ (m/s)}$$

Since γ and R are constant for a given gas

$$C = \text{constant } \sqrt{T}$$

where constant $= R$

or $C \propto \sqrt{T}$

Hence, the velocity of sound in a gas is proportional to the square root of the absolute temperature. Therefore, if C_1 is the velocity of sound at a temperature of T_1 (K) and C_2 is the velocity at a temperature of T_2 (K), then

$$\text{constant} = \frac{C_1}{\sqrt{T_1}} = \frac{C_2}{\sqrt{T_2}}$$

therefore

$$\frac{C_2}{C_1} = \sqrt{\frac{T_2}{T_1}}$$

The exhaust gases entering the exhaust port are at about 800°C but this drops to about 150° at the tail pipe. Calculations have confirmed that a good approximation for the mean temperature of the exhaust gases passing through the exhaust system would be 400°C. This temperature takes into account the water vapour in the exhaust gas, since the density of water vapour is only about 5/8 of the density of air. Hence, it causes the velocity of sound to be higher when the gas becomes moist compared with dry exhaust gas.

Example

Calculate the velocity of sound in an exhaust gas operating at a mean temperature of 400°C if the velocity of sound at standard temperature and pressure is 330 m/s, then

$$\frac{C_2}{C_1} = \sqrt{\frac{T_2}{T_1}} \quad \text{note } T_1 \text{ at STP} = 0 + 273 \text{ K}°$$

$$\frac{C_1}{330} = \sqrt{\frac{273 + 400}{273 + 0}} = \sqrt{2.4652} = 1.57$$

therefore

$$C_1 = 330 \times 1.57 = 518 \text{ (m/s)}$$

This shows that the velocity of sound is much higher at a mean temperature of 400°C compared with the velocity at standard temperature and pressure, i.e. 330 m/s.

5.14.5 Pressure-wave exhaust gas scavenging
(Fig. 5.63)

Every time the exhaust valve opens towards the end of the power stroke a compression wave is released into the exhaust port. This positive pressure-wave pulse travels to the open end of the exhaust pipe where it is expelled into the atmosphere leaving a rarefaction behind, that is, a momentary drop in density of the surrounding air at the pipe exit. The elasticity of the surrounding air will make it rebound towards the pipe exit thus causing a negative wave to be reflected all the way back to the exhaust port.

When the pulse reaches the exhaust port it will again be reflected towards the pipe outlet as a positive wave. Once again, as it reaches the open end of the pipe a wave will be reflected inwards. This cycle of events will continue indefinitely with

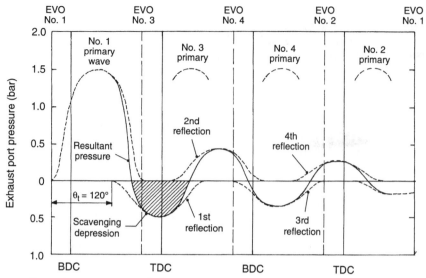

Fig. 5.63 Exhaust gas reflected pressure-wave timing

a decaying amplitude, if time permits, before the next exhaust period discharge takes place.

For best results, the exhaust pipe length should be so chosen such that a pressure-wave will travel from the exhaust valve and port to the pipe exit and back again during a crankshaft interval 'θ_t' of about 120° at a given engine speed (Fig. 5.63). This will ensure that the first reflected negative wave is at its lowest pressure when the piston has just passed TDC at the end of the exhaust period. Under these conditions the residual exhaust gas can readily be pulled out (scavenged) from the combustion chamber.

However, at lower and higher engine speeds, compared with the tuned exhaust pipe length, the first negative reflected wave will shift relative to the exhaust closure point. Thus, the depression created by the pulse in the exhaust port will not be able to extract the residual exhaust gases and induce the fresh charge to enter the combustion chamber. In fact, the positive part of the primary or secondary reflected waves may become partially aligned with the exhaust valve closure point and will therefore prevent the expulsion of the residual gases from the chamber.

5.14.6 Determination of exhaust pipe length for optimum wave scavenging
(Fig. 5.63)

To take full advantage of the pressure-wave pulse it must be timed so that the first negative reflected

pressure-wave reaches TDC towards the beginning of the induction and the end of the exhaust period at its peak negative amplitude. To obtain the correct phasing of the depression wave relative to the closure of the exhaust valve, it is essential to be able to estimate the time it takes the pressure-wave to travel through the exhaust gas column from the exhaust valve exit to the end of the exhaust pipe and for this wave to be reflected and returned to its starting point at the exhaust valve exit.

The same principles apply as for induction wave ram cylinder charging, that is, the time taken to travel the exhaust pipe length and back again is equal to the distance the pulse moves from the exhaust valve to the end of the pipe, and for it to return to its original starting point, divided by the speed that sound moves through the gas media operating under average working temperature conditions.

Let t = time for pulse to travel from the exhaust valve to the end of the pipe and back (s)

L = length of tract from exhaust valve exit to end of pipe (m)

C = speed of sound through exhaust gas (518 m/s)

N = engine crankshaft speed (rev/min)

θ_t = crankshaft angular displacement (deg)

$$\text{Thus, time } (t) = \frac{\text{distance}}{\text{velocity}} = \frac{2L}{1000C} \text{ (s)} \qquad (3)$$

Therefore, crankshaft angular displacement 'θ_t' during the same interval of time is equal to:

276

θ_t = time to travel tract length and back × angular speed

$$\theta_t = t \times \frac{360}{60}N \quad \text{but} \quad t = \frac{2L}{1000C}$$

therefore

$$\theta_t = \frac{2L}{1000C} \times 6N \text{ (deg)}$$

$$= \frac{0.012LN}{C} \text{ (deg)} \qquad (4)$$

Example

Calculate the total exhaust tract length from the exhaust valve to the exhaust pipe end to maximize the wave scavenging effect at the following engine speeds: 2000, 3000 and 4000 rev/min.

Take the crank-angle movement, for the pressure-wave to travel from the exhaust valve to the exhaust pipe exit and back again, to be $\theta_t = 120°$ and the velocity of sound through the gas to be 518 (m/s).

$$\theta_t = \frac{0.012LN}{C}$$

therefore

$$L = \frac{\theta_t C}{0.012N}$$

for an engine speed of 2000 rev/min

$$L = \frac{120 \times 518}{0.012 \times 2000} = 2560 \text{ mm or } 2.56 \text{ m}$$

for an engine speed of 3000 rev/min

$$L = \frac{120 \times 518}{0.012 \times 518} = 1726.7 \text{ mm or } 1.726 \text{ m}$$

for an engine speed of 4000 rev/min

$$L = \frac{120 \times 518}{0.012 \times 4000} = 1280 \text{ mm or } 1.28 \text{ m}$$

Note the tuned exhaust passage flow path-length is greater for wave gas scavenging compared with induction wave ram charging due to the higher speed of sound in an exhaust gas relative to that through a column of air.

5.14.7 Interference in 'Y'-branch pipes
(Fig. 5.64(a and b))

Exhaust gas compression wave interference between cylinders by utilizing an idler pipe can be beneficial in producing a negative (depression) wave in the exhaust port when the piston is in the TDC region with the exhaust valve still open.

When the exhaust valve opens (Fig. 5.64(a)) the compression wave released travels from the exhaust port 'A' to the junction 'B', the increased flow area then causes a sudden expansion of the exhaust gases. This produces a rarefaction (depression) which sends a reflected wave back to the open port, thus subjecting the exhaust valve passageway to a slight vacuum.

The original compression wave also travels around the forked junction to the blanked end of the idler pipe, and here it is reflected as a

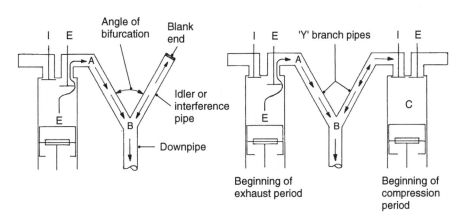

(a) Idler pipe interference   (b) 'Y'-branch pipe interference

Fig. 5.64 Paired branch pipe interference

compression wave back to the junction, its wavefront then divides with one wavefront moving back through the branch pipe to the open exhaust port, whereas the other part of the wavefront travels downstream to the downpipe exit.

The net result is that the negative pressurewave at the open exhaust port is delayed so that it occurs over the TDC valve overlap period, with the piston at approximately TDC. It thereby extracts the residual exhaust gases from the cylinder and induces the fresh charge to enter the cylinder.

Likewise, a second cylinder branch pipe with its exhaust valve closed can be considered to be the equivalent to the idler interference pipe (Fig. 5.64(b)). Therefore, similar depressions in the TDC region during valve overlap can be obtained when pairs of branch pipes such as cylinders numbers 1–4 and 2–3 merge into two downpipes, provided the correct length between the port and junction is chosen.

5.15 Exhaust manifold configurations

5.15.1 In-line four-cylinder engine exhaust manifold branch pipe gas interference
(Figs 5.65(a–d) and 5.66(c and d))

With a four-cylinder engine there is an exhaust stroke every 180° of crankshaft rotation so that, in theory, when one stroke is on the point of finishing another stroke is about to commence. However, the inlet valve opens early (10°–25°) before TDC, whereas the exhaust valve closes late (10°–25°) after TDC, this means that the inlet and exhaust valve openings overlap in the TDC region.

Similarly, there will be a certain amount of exhaust valve opening overlap between cylinders due to the late closing after TDC of the exhaust valve. Thus, when one cylinder is beginning its exhaust upstroke a second cylinder will be starting its induction downstroke with its exhaust valve initially still open (Fig. 5.65(a–d)).

This exhaust valve opening overlap between cylinders can cause the cylinder starting its exhaust period (which is expelling relatively highpressure exhaust gas out into the branch pipe) to interfere with an adjacent cylinder which is com-

ing to the end of its exhaust period and has started its induction period (Fig. 5.65(a–d)). The blowthrough charge pressure in the cylinder is low and the remaining outward flowing exhaust gas pressure in the branch pipe will also be low. Consequently, the high pressure gas flowing from the cylinder at the beginning of its exhaust stroke will not only move towards the outlet of the exhaust manifold, but will also attempt to flow down the adjacent cylinder exhaust branch pipe, thereby opposing the flow of the relatively low-pressure gas trying to escape from that cylinder.

Thus, if the branch pipe is short (Fig. 5.66(c)) reversal of the exhaust gas flow into the adjacent branch pipe may occur, thereby impeding the cylinder filling process. However, if the pipes are extended somewhat (Fig. 5.66(d)) the highvelocity column of exhaust gas will easily flow into the downpipe and will tend to overshoot the other branch pipes.

At the same time, the momentum of the column of exhaust gas moving through the uninterrupted flow path of the long branch pipe will leave behind a depression in the exhaust port. This produces an extraction affect on the residual exhaust gas towards the commencement of the induction period and the termination point of the exhaust period. A periodic scavenging process will therefore be established where the fresh charge is induced into the combustion chamber at the same time as the exhaust products are being pulled out.

5.15.2 Four-cylinder engine with three and four branch exhaust manifolds
(Figs 5.66(a and b) and 5.67)

With the four-cylinder engine, exhaust and inlet ports are situated on the same side of the cylinder-head and the central adjacent exhaust ports are siamese (Fig. 5.66(a)) so that there are only three exhaust exit ports from the cylinder-head. There are three branch pipes coming away from the cylinder-head which converge further downstream to form a short down-pipe passage which has a flanged joint exit. The four-cylinder engine has an exhaust stroke every 180° but the central pair of cylinders, 2 and 3, discharge exhaust gas at 360° intervals to each other so that exhaust gas interference in the siamese port fork area is almost minimal.

An alternative in-line four-cylinder engine inlet and exhaust port arrangement, where again both

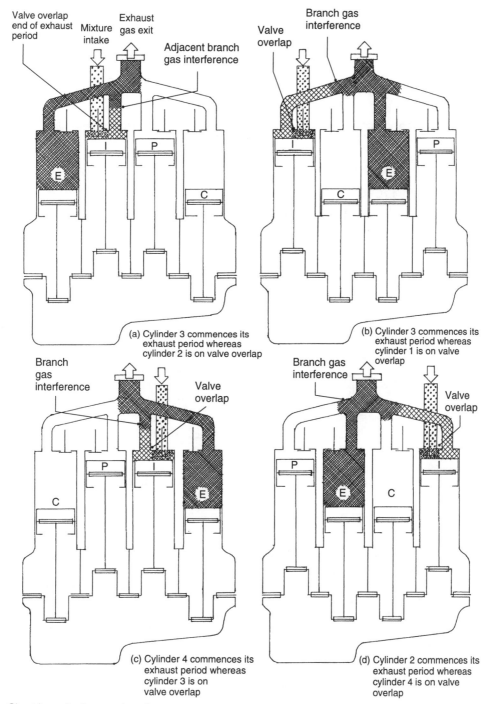

Fig. 5.65 Short branch pipe gas interference

sets of ports come out on one side of the cylinder-head, has separate inlet and exhaust ports for each cylinder (Fig. 5.66(b)). Therefore, each of the four exhaust ports has its own manifold branch pipe, and these converge downstream so that they form a short, centrally positioned, single downpassage with a flanged mouth at its outlet.

With individual exhaust ports, the gases from

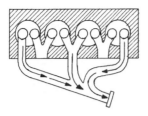 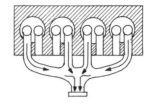

(a) Three-branch exhaust manifold with centre siamese exhaust valve ports

(b) Four-branch exhaust manifold with central downpipe

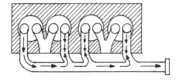 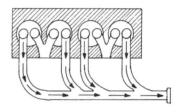

(c) Four short-branched exhaust manifold

(d) Four long-branched exhaust manifold

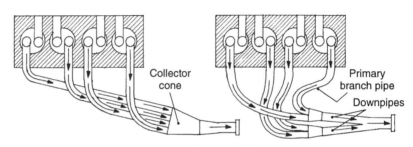

(e) Four-branch exhaust manifold arrangement for independence working

(f) Four-branch exhaust manifold with sub-divisions to obtain equal discharge interval at each junction

Fig. 5.66 In-line four-cylinder engine exhaust manifold arrangements

each cylinder will be discharged into the manifold branch pipe under identical conditions but the shortness of branches can still produce exhaust gas interference between cylinders as the gas back-pressure builds up in the down-pipe with increased engine speed. Figure 5.67(a) shows the exhaust gas pressure recorded near the flanged joint of an in-line six-cylinder engine exhaust manifold which has all the branch pipes merging into one downpipe outlet. At the lowest speed of 2000 rev/min (Fig. 5.67(a)) the pressure (shaded) is below the atmospheric line and coincides with the valve overlap at TDC—this condition being conducive to good gas scavenging. As the engine speed rises to 3000 rev/min (Fig. 5.67(b)) the pressure increases to roughly 0.4 bar but the negative pressure decreases to just 0.1 bar. With

even higher engine speeds of 4000 rev/min (Fig. 5.67(c)) the gas pressure in the manifold down-pipe remains about the same as for 3000 rev/min, but the lower pressure trough has now moved from a depression into a compression. The congested manifold will obviously make it more difficult for the residual exhaust gases in the combustion chamber to be expelled.

5.15.3 In-line four-cylinder engine exhaust manifold arrangement for independence working
(Fig. 5.66(e))

By progressively increasing the size of the branch pipe loops from the front to the rear of the

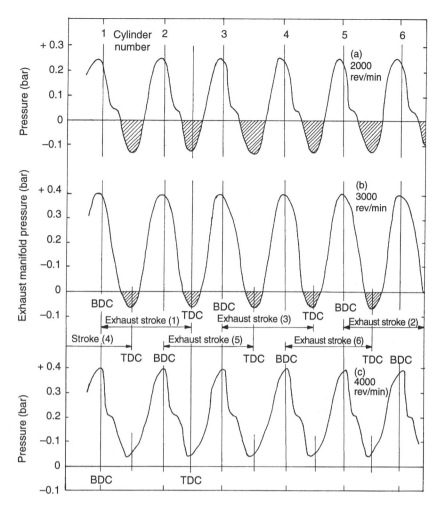

Fig. 5.67 Pressure variation at different engine speeds for a six-cylinder exhaust manifold with single outlet

engine, each branch pipe can be of equal length. The long branch pipes enable each exhaust port to discharge a column of exhaust gas independently to the other pipes for a relatively greater time period before all the pipes converge at the collector cone (Fig. 5.66(e)). If these tuned branch lengths are correctly chosen each pipe will reflect back a pressure-wave at the cone junction which is phased to become negative in the exhaust port when the piston reaches TDC. This greatly assists the expulsion of the residual exhaust gases and the early beginning of the induction period.

5.15.4 In-line four-cylinder engine four-branch exhaust manifold with pipe subdivisions tuned for interference working
(Fig. 5.66(f))

With the pairing of branch pipes 1–4 and 2–3 this separates the other cylinders so that equal exhaust intervals are obtained for each downpipe, the two downpipes then join together further downstream to form a single outlet pipe (Fig. 5.66(f)).

The opening of the exhaust valve in one cylinder releases a compression wave to the junction of one pair of primary branch pipes, the increased passage space at this point causes an expansion of the gases which instantly reflects a rarefaction (low density) wave back to the exhaust port. At the same time, the compression wave front travels further downstream to the downpipe junction where, again, the exhaust gases expand in the enlarged flow-path space, hence sending back a second reflected rarefaction wave to the exhaust port.

The fundamental (first) compression wave also travels round the primary branch pipe junction to the second branch pipe, thereby sending a wave-front to the closed exhaust port. The resulting reflected compression wave then returns to the junction where it divides one wavefront, moves downstream of the downpipe, and another travels upstream back to the open exhaust port. Thus, provided that the primary branch pipes and the downpipe lengths are carefully selected, the two rarefaction reflected waves and the reflected compression wave return to the open exhaust port. This results in a depression at the open exhaust valve during the valve overlap period when the piston is somewhere near TDC at the beginning of the induction period and the completion of the exhaust period.

5.15.6 Horizontally opposed four-cylinder engine exhaust manifold
(Fig. 5.68)

The horizontally opposed four-cylinder engine normally has branch pipes 1–3 and 2–4 joined together with a downpipe connecting each pair of branch pipes to an intermediate pipe via a collector cone junction (Fig. 5.68). With a typical firing order 1432 this manifold layout provides a 360° crank angle rotation discharge interval from each downpipe.

Cylinders 3 and 4 have much longer branch pipes since the downpipes are situated to the side

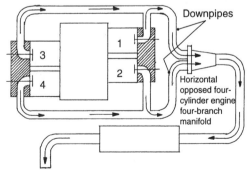

Fig. 5.68 Five-cylinder engine five-branch exhaust manifold for interference working

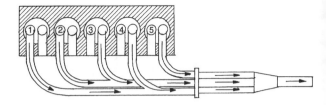

Five-cylinder engine five branch exhaust manifold for interface working

Fig. 5.69 Five-cylinder engine five-branch exhaust manifold for interference working

of cylinders 1 and 2. For normal operating conditions, this unequal-length branch pipe layout only slightly upsets the reflected negative pressure-wave timing. However, for more highly tuned engines, centrally divided branch pipes can be provided but they are more difficult to accommodate under the floor of the car.

5.15.7 In-line five-cylinder engine exhaust manifold arrangement
(Fig. 5.69)

A popular in-line five-cylinder exhaust manifold arrangement joins branch passages 1–4 and 2–3 together and keeps the branch passage 5 separate from the others (Fig. 5.69). The manifold therefore has three exit passages sharing a common flanged coupling, the gases are then transferred to a single large-bore intermediate pipe via three long downpipes of equal length. With a cylinder firing order of 12453, this manifold provides similar irregular discharge intervals for exit passages 1–4 and 2–3 of 288° and 432° crankshaft rotation, and 720° for the third passageway for cylinder number 5.

The periodic opening of the exhaust valve sends pressure-waves downstream from the exhaust ports to the intermediate pipe via the branch passages and then through the downpipes. At junctions 1–4 and 2–3 negative pressure-waves will be reflected back to their respective exhaust port and further downstream, at the downpipe junction, negative pressure-waves will again be reflected back, but this time to all the exhaust ports in turn during their respective exhaust discharge period. This results in negative pressure-waves arriving at each individual cylinder exhaust port during valve overlap; thus, it considerably assists the blow-through process, where the fresh charge chases after the residual exhaust gases

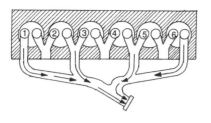

(a) Four-branch exhaust manifold with two
    siamese inner ports

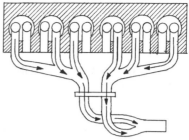

(b) Six-branch divided exhaust manifold

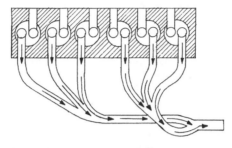

(c) Six-branch exhaust manifold to
    obtain equal discharge intervals at
    each junction and for interference
    working

Fig. 5.70 In-line six-cylinder engine exhaust-manifold arrangements

being extracted by the resultant depression created in the exhaust port.

5.15.8 In-line six-cylinder engine exhaust manifold arrangements
(Fig. 5.70(a–c))

With the in-line six-cylinder engine, cylinders numbers 2–3 and 4–5 exhaust ports can be siamese, so that only four exhaust ports and branch pipes are necessary to dispose of the exhaust gases from all the cylinders (Fig. 5.70(a)). These branch pipes then converge to form one short exit passage so that there will be an exhaust discharge from this single outlet every 120° crankshaft rotation. Hence, the exhaust stroke between cylinders overlaps, so that with increased engine speed there will be a certain amount of undesirable gas interference between cylinder discharges, and a progressive build-up of gas pressure in the congested exit pipe will result (Fig. 5.67(a–c)).

However, if the exhaust manifold is divided into two halves (Fig. 5.70(b)) so that cylinders 1–2–3 feed into one exit passage and cylinders 4–5–6 feed into another, then with a firing order such as 153624, there will be an equal exhaust discharge interval from each outlet passage every 240° crankshaft rotation. Thus, there will be no exhaust period overlap. Subsequently, there is no exhaust gas interference in either of the divided manifolds and, by the time the gas has travelled to where both downpipes discharge into one common intermediate pipe, any exhaust gas interference this far downstream will be at a minimum.

For high performance it may justify having long branch pipes of equal length, in groups of three, merge at each downpipe, the pair of downpipes themselves then converge further downstream to form one intermediate pipe (Fig. 5.70(c)). The junction where the three pipes merge into one suddenly increases the passage cross-sectional area so that the pressure-waves released when the exhaust valves open will be reflected at this enlarged passage volume. The pressure-waves travel upstream to the other branch pipes, to the closed exhaust ports, where they are again reflected. The return journey takes the wave around the forked junction, back to the exhaust port, so that it arrives with a negative wave when the piston is at TDC with the inlet valve just open and the exhaust valve is beginning to close. Thus, the interference of these pressure-waves can be used to improve the extraction of the residual exhaust gases from the cylinder towards the end of the exhaust period.

5.15.9 Vee six-cylinder engine exhaust manifold arrangement
(Fig. 5.71(a and b))

Each cylinder bank of a vee six-cylinder engine can be treated as a separate three-cylinder engine, but not sharing a common crankshaft with the adjacent cylinder bank. Thus, provided that a firing order such as 143625 is adopted, the exhaust periods will be discharged alternately from each cylinder bank, which provides a 240° crank-angle interval between exhaust expulsions from each cylinder bank exhaust manifold.

Short branch passages of equal length may have their flow path curve rearwards, with each branch merging with its adjacent one, so that they form a single trailing discharge passage with a flanged joint at its exit (Fig. 5.71(a)).

If high performance is the criteria, then separate, long, curved branch pipes of equal length, which have loops of increasing size as the branches approach the rear of the engine, can be effectively used (Fig. 5.71(b)). These branch pipes converge at a collector cone to form a single downpipe which may lead directly to two separate silencers or they may merge into a single intermediate pipe before feeding into a common silencer.

The long uninterrupted flow path encourages both kinetic energy scavenging and pressure-wave exhaust scavenging. Thus, the pressure-wave released when the exhaust valve opens sends a pressure-wave down each pipe to the collector cone. This enlarged passage junction is then responsible for reflecting a negative wave back to the open exhaust valve so that the waves arrive in the exhaust port during valve overlap with the piston having completed its exhaust stroke. Thus, the greater the resultant negativeness of the pressure in the exhaust port, the greater will be the follow through of the fresh charge chasing the outgoing, but reluctant, residual exhaust gases.

5.15.10 Vee eight-cylinder engine exhaust manifold arrangements
(Fig. 5.72(a and b))

Vee eight-cylinder engines can be considered as a pair of four-cylinder engines joined together by a common crankshaft and crankcase. Simple ex-

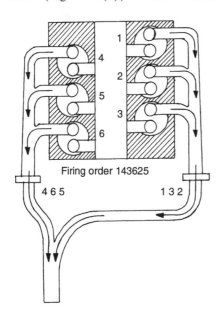

(a) Three-branch exhaust manifold with trailing downpipe for longitudinal mounted engine

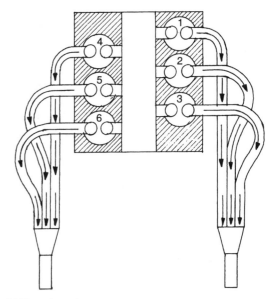

(b) Three-branch exhaust manifold layout for independence working

Fig. 5.71 Vee six-cylinder engine exhaust manifold arrangements

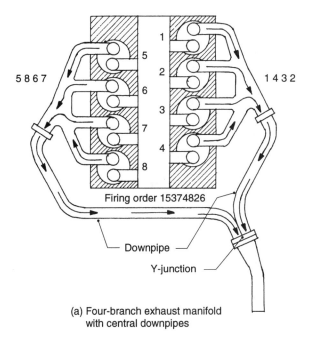

5 8 6 7

1 4 3 2

Firing order 15374826

Downpipe

Y-junction

(a) Four-branch exhaust manifold with central downpipes

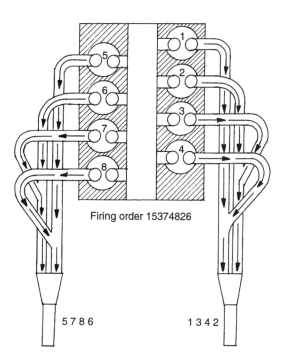

Firing order 15374826

5 7 8 6

1 3 4 2

(b) Four-branch exhaust manifold adapted for interference working used with single-plane crankshaft

Fig. 5.72 Vee eight-cylinder engine exhaust manifold arrangements

haust manifolds can be arranged with short branch passages which merge into a central discharge passage (Fig. 5.72(a)). The exhaust gases then flow through their respective downpipes, which may converge centrally or to one side of the engine to form a single intermediate pipe leading onto the exhaust silencer. A vee eight-cylinder engine has an exhaust stroke every 90° of crankshaft movement. However, each downpipe discharges exhaust gas at intervals of 180°.

With a firing order such as 15486372 (popular with the two-plane crankshaft) there will be an exhaust discharge from the left-hand cylinder-bank exhaust port in the sequence 1–4–3–2–1 with correspondingly irregular intervals of 180°, 270°, 180° and 90° crankshaft rotation, respectively. Similarly, the exhaust gas discharge from the right-hand cylinder bank will be in the order 5–8–6–7 corresponding to intervals of 180°, 90°, 180° and 270° crankshaft movement, respectively. Thus, it can be seen that there is some unevenness in exhaust port discharge intervals.

By adopting a single-plane crankshaft and choosing a firing order such as 15374826, each cylinder bank discharges exhaust gas at equal intervals of 180° crankshaft rotation in a sequence of 5786 and 1342 for the left- and right-hand cylinder banks respectively (Fig. 5.72(b)). This discharge sequence is similar to an in-line four-cylinder engine, and therefore, to extend the crank-angle interval to 360° before the expelled exhaust gases from individual ports merge together, the outer branch pipes (1–4) are paired. Likewise, the inner branch pipes (2–3) are paired for the right-hand branches. Similarly, the left-hand branch pipes (5–8) and (6–7) are paired. To equalize the lengths of each branch pipe, the pipes are individually looped so that those near the rear of the engine have much larger loops than those at the front.

The pairing of the branch pipes enables the exhaust discharge pulse pressure-wave to enter the adjacent pipe, and travel through to the closed exhaust valve port, which reflects it back to the forked junction. The pressure-wave then divides, with one wave continuing to travel downstream to the collector-cone while the other is reflected and returns to the original open exhaust port. If the lengths of the branch pipes are correctly tuned, the pulse arrives in the form of a negative wave towards the end of the exhaust period to pull out the residual exhaust gases from the combustion chamber.

For larger high-performance engines, a two-

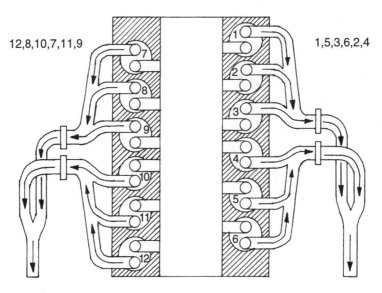

12,8,10,7,11,9   1,5,3,6,2,4

Firing order 1,12,5,8,3,10,6,7,11,4,9

Fig. 5.73   Vee twelve-cylinder engine exhaust manifold arrangement

plane crankshaft provides better balance than a single-plane crankshaft for the reciprocating components and is therefore preferred.

5.15.11   Vee 12-cylinder engine manifold arrangement
(Fig. 5.73)

This double-cylinder banked-engine is, in effect, two six-cylinder engines coupled together by a single crankshaft (Fig. 5.73). A suitable firing order for this class of engine would be 1,12,5,8,3,10,6,7,2,11,4,9; therefore, the exhaust discharge sequence down the left-hand bank will be 153624 and on the right-hand bank 12,8,10,7,11,9. With a 12-cylinder engine there is an exhaust discharge every 60° of crankshaft rotation. It therefore follows that there will be an exhaust expulsion period interval between cylinder banks of 120° crankshaft movement. The manifold branch passages on each bank are divided into two groups of 1–2–3 and 4–5–6 on the left-hand bank whereas this will be 7–8–9 and 10–11–12 on the right-hand bank. This subdivision of branch passages provides equal discharge intervals of 240° crank-angle movement between the four exit passages leading to the two down-pipes.

6
Supercharging Systems

6.1 The fundamentals of supercharging

The purpose of supercharging an engine is to raise the density of the charge, be it air or air and fuel mixture, before it is delivered to the cylinders. Thus, the increased mass of charge trapped and then compressed in each cylinder during each induction and compression stroke makes more oxygen available for combustion than the conventional method of drawing the fresh charge into the cylinder. Consequently, more air and fuel mixture per cycle will be crammed into the cylinder, and this can be efficiently burnt during the combustion process to raise the engine power output to higher than would otherwise be possible. The rising supercharger delivery pressure with increasing engine speed compensates for the reduction in cylinder filling time so that the rate of power increase will continue in proportion to the rising engine speed, particularly in the upper engine speed range. This is in contrast to the naturally aspirated engine which finds it difficult, towards the limit of the engine's speed range, to induce sufficient charge into the cylinder with the decreasing time available.

A correctly matched supercharger will raise the cylinder's brake mean effective pressure (b.m.e.p.) to well above that of a naturally aspirated engine without creating excessively high peak cylinder pressures; the actual increase in the brake mean effective pressure is basically determined by the level of boost pressure the supercharged system is designed to deliver.

Large commercial vehicle diesel engines are frequently turbocharged, with the objectives of raising b.m.e.p. (and therefore torque and power output) and at the same time reducing the engine's maximum speed. The other benefits of raising the cylinder mean pressure and decreasing the engine's limiting speed is that the engine mechanical losses and noise are reduced and there is an improvement in fuel consumption with, normally, an added bonus of prolonged engine life expectancy.

The gain in raising the cylinder brake mean effective pressure (b.m.e.p.) is partly offset by having to transfer some of the engine's power to driving the supercharger as shown in Fig. 6.1. The graphs show that the power absorbed by a supercharger is dependent upon its displacement capacity per revolution and its rotational speed for a given boost pressure.

The theoretical operating cycles for both the natural aspirated and supercharged petrol engines

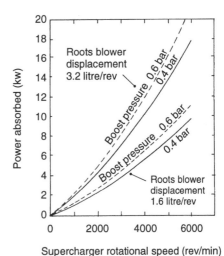

Fig. 6.1 Relationship between supercharger speed and power absorbed for two different-sized positive displacement compressors and two lost pressures

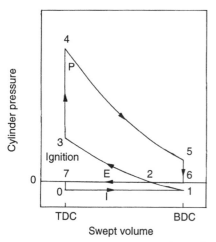

Fig. 6.2 Theoretical naturally aspirated petrol engine (constant volume) pressure–volume diagram

can be compared on pressure–volume (P–V) diagrams as shown in Figs 6.2 and 6.3 respectively. The large upper loop is a measure of the positive power developed in the cylinder while the lower loop represents the negative power needed to fill the cylinder with fresh charge.

Naturally aspirated cycle of operation
(Fig. 6.2)
The large area enclosed in the upper loop (2, 3, 4, 5 and 6) is proportional to the useful work performed by combustion on the piston whereas the small lower loop area (0, 1, 2 and 7), which is below the atmospheric line, is a measure of the work done in inducing the fresh charge into the cylinder. The four phases of the naturally aspi-

rated engine's cycle are represented as follows: induction 0 to 1, compression 1 to 3, power 3, 4 and 5 and exhaust 5, 6 and 7.

Supercharged cycle of operation
(Fig. 6.3)
With this pressurized charging system the large upper loop area (1, 2, 3 and 4) represents a measure of the work done in moving the piston to and fro so that the crankshaft rotates. Conversely, the lower small loop area (0, 1, 5 and 6) which is above the atmospheric line represents the work done in pumping the fresh charge into the cylinder. The four phases of the supercharged engine's cycle of operation are represented as follows: induction 0 to 1, compression 1 to 2, power 2, 3 and 4 and exhaust 4, 5 and 6.

Comparison of actual naturally aspirated and supercharged engine pressure volume diagrams
(Fig. 6.4)
A direct comparison of how the actual cylinder pressure varies relative to the cylinder's swept volume for both a naturally aspirated and a supercharging petrol engine, with wide-open throttle, is shown in Fig. 6.4. The two enclosed loops show that the lower compression stroke part of the curves for the supercharged engine is higher than for the naturally aspirated engine, whereas the peak cylinder pressures for the supercharged engine are only marginally higher. However, the main advantage gained by supercharging the cylinders is that the vertical distance

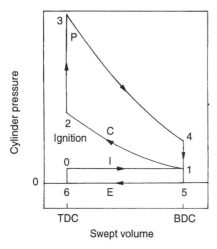

Fig. 6.3 Theoretical supercharged petrol engine (constant volume) pressure–volume diagram

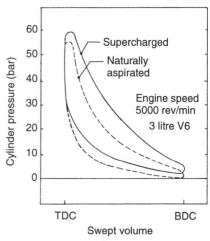

Fig. 6.4 Comparison of actual naturally aspirated and supercharged petrol engine pressure–volume diagrams

between the upper and lower curves for the supercharged engine is greater throughout the cylinder swept volume, which indicates that the mean effective pressure is greater. Finally, the loop area enclosed by the supercharged engine is much larger, the proportional difference being a measure of the increased power in the supercharged engine.

6.1.1 Boost pressure and pressure ratio

The fundamentals of supercharging are based on changes in pressure exerted on the gas being delivered to the cylinders, therefore it is worthwhile defining some of the terms normally used.

Atmospheric pressure is the pressure exerted at sea-level by the air and gas layers surrounding the Earth's surface. Atmospheric pressure at sea-level may be taken as being equivalent to $101.3 \, kN/m^2$, $760 \, mm$ Hg or 1 bar. Gauge pressure is the pressure measured above atmospheric pressure and which registers on the gauge. Absolute pressure takes into account both the atmospheric pressure, which does not register on the gauge, and the pressure above atmospheric pressure, which does register on the gauge; thus, absolute pressure = atmospheric pressure + gauge pressure.

Boost pressure refers to the gauge pressure recorded when the air or mixture supply has passed through the supercharger. A comparison of boost pressure relative to the atmospheric pressure may be expressed in terms of a pressure ratio; that is, the ratio of absolute pressure to that of the atmospheric pressure

$$\text{boost pressure} = \frac{\text{absolute pressure}}{\text{atmospheric pressure}}$$

$$= \frac{\text{boost pressure} + \text{atmospheric pressure}}{\text{atmospheric pressure}}$$

The intensity of supercharging can be broadly classified as follows.

Degree of charging	Boost pressure (bar)	Pressure ratio
Naturally aspirated	0.0 and below	1.0 and below
Low	0.0–0.5	1.0–1.5:1
Medium	0.5–1.0	1.5–2.0:1
High	1.0 and above	2.0 and above

6.1.2 The relationship between discharged air pressure temperature density and engine performance

The power developed in the cylinder is proportional to the rotational speed of the engine and the mass of charge compressed in the cylinders. Thus, assuming the engine's speed is fixed the only other way the engine power can be increased is to raise the mass of charge entering the cylinder. Thus,

$$\text{mass of charge} = \text{constant} \times \frac{\text{pressure}}{\text{temperature}}$$

$$m = \left(\frac{V}{R}\right)\frac{P}{T} \tag{1}$$

where V/R = a constant
P = absolute pressure (bar)
T = absolute temperature (°C)

Note, R is known as the universal gas constant.

Therefore, the mass of charge that can enter a cylinder is proportional to its pressure and inversely proportional to its absolute temperature. Hence, the greater the charge pressure the greater will be the mass entering the cylinder per induction stroke. Conversely, raising the temperature of the charge before it enters the cylinder reduces the mass of charge occupying the cylinder space and vice versa.

Similarly, the mass of charge that can enter a cylinder will be equal to the swept volume of the cylinder multiplied by the density of the air, or air and fuel mixture, for diesel and petrol engines respectively. That is,

$$\text{mass of charge} = \text{swept volume} \times \text{density}$$

$$m = V\rho \tag{2}$$

where V = cylinder volume (m^3)
ρ = density (kg/m^3)

This means that, as the density of the charge delivered to the cylinder increases, so the swept volume mass will increase. In contrast, reducing the charge density reduces the mass of charge filling the cylinder.

Equating equations (1) and (2)

$$m = V\rho = \frac{V}{R}\frac{P}{T}, \text{ cancelling out } V$$

then

$$\rho = \frac{1}{R}\frac{P}{T} \qquad (3)$$

where

$$\frac{I}{R} = \text{a constant}$$

Equation (3) shows that the charge density is proportional to the charge pressure and inversely proportional to its temperature. Thus, if

$$\rho_1 = \frac{P_1}{RT_1} \quad \text{and} \quad \rho_2 = \frac{P_2}{RT_2}$$

then $\rho_1 P_1 T_1$ and $\rho_2 P_2 T_2$ are the inlet and outlet pressure and temperature conditions across the compressor respectively.

$$\text{then density ratio} = \frac{\rho_2}{\rho_1} = \frac{P_2/RT_2}{P_1/RT_1}$$

$$\text{hence density ratio} = \frac{\rho_2}{\rho_1} = \frac{P_2 T_1}{P_1 T_2} \qquad (4)$$

Equation (4) enables a comparison to be made with the air density before and after compression.

Example
Air at atmospheric pressure (1 bar) and temperature of 25°C is drawn into the compressor where it is compressed to a boost pressure and temperature of 0.6 bar and 120°C respectively. Determine the output density ratio of the compressed charge and compare this with the ideal value if there were no temperature increase.

With temperature increase

$$\frac{\rho_2}{\rho_1} = \frac{P_2 T_1}{P_1 T_2} = \frac{(1+0.6)\times(273+25)}{1.0\times(273+120)}$$

$$= \frac{1.6\times 298}{1.0\times 393} = 1.213:1$$

With no temperature increase

$$\frac{\rho_2}{\rho_1} = \frac{P_2}{P_1} = \frac{1.6}{1.0} = 1.6:1$$

The considerable difference $1.6 - 1.213 = 0.387$ between the two density ratios illustrates the tremendous influence temperature has on charge density.

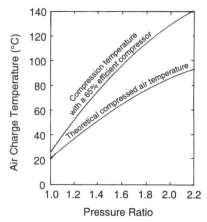

Fig. 6.5 Relationship between charge temperature and boost pressure ratio if heat is not dissipated

6.1.3 The effects of pressure ratio on air charge temperature
(Fig. 6.5)

The relationship between supercharger pressure ratio increase, and air charge temperature is shown in Fig. 6.5 where it can be seen that as the boost pressure ratio increases, so does the discharge air temperature. The lower curve shows the theoretical temperature rise with increased pressure ratio, whereas the band between the lower and upper curves is the working temperature variation likely to be encountered due to the turbulence and friction resistance generated by the compression process.

Thus, the minimum compression air temperatures with an intake temperature of 20°C for pressure ratios 1.2, 1.4, 1.6, 1.8, 2.0 and 2.2 are 35.7°C, 49.5°C, 62°C, 73.6°C, 84°C and 94°C respectively. In practice, with compressor efficiencies of the order of 60–75% and the churning, turbulence and frictional factors, the actual output air charge temperature can be considerably higher, particularly at the higher boost levels.

6.2 Sliding vane supercharger

6.2.1 Description of the components and the blower construction
(Fig. 6.6)

The three basic components of the vane type blower are the cast casing, rotor drum and vane blades (see Fig. 6.6).

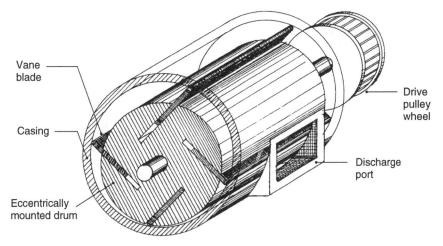

Fig. 6.6 Inclined vane-type supercharger construction

The casing is a nickel iron casting, the exterior of which is circumferentially finned to assist heat dissipation whereas the interior is bored and honed to a finish of 80 to 300 μm, which greatly helps to minimize the friction between the vanes and the casing. The normal diameter of the bore is 100 mm and the two different drum lengths manufactured are 130 mm and 240 mm. Both end covers are aluminium pressure die castings which house and support the bearings of the 19 mm diameter rotor shaft. This shaft is mounted eccentrically to the bore axis. The rotor drum is an aluminium alloy 82.5 mm diameter casting which is cast directly onto the steel drive shaft. To avoid excessive rubbing friction between the vanes and the casing, the four equally spaced slots for the vanes are tangential to a base circle, the diameter of which is about half that of the rotor body.

By locating the vanes tangentially instead of radially to the centre of the rotor, part of the centrifugal load on the vanes is transferred to the outer slot wall; therefore, there will be less load imposed between the tips of the vanes and the casing cylindrical wall, as would be the case if the vane slots were machined radially in the drum.

It is also claimed that the curved blade tips enable the compressed charge to get underneath and therefore produce an inward force which opposes the centrifugal force. Thus, at high rotating speed there is probably a near state of equilibrium where the charge inward reaction pressure acting on the blade tips largely counteracts the centrifugal effect so that the vane-tip to casing contact frictional force is relatively small. Another advantage of having tangentially positioned vanes is that the depth of the slots can be increased, which therefore provides additional overhang support for the vanes by spreading the twisting effect of the vanes over a larger surface area within the slots.

The vanes are made from laminates of linen impregnated with phenolic resin; however, Tufnol, which appears to have similar qualities, has also been used. The qualities sought are quiet running, low frictional reaction, low coefficient of thermal expansion and resistance to prolonged exposure to petrol and oil.

To prevent a pressure build-up in the slots, which would restrict the free movement of the vanes between the slot walls, two venting grooves, which are situated towards the ends, extend from the inner edge to within 6 mm of the tip.

Lubrication of the vanes and slots is achieved via a 2.25 litre oil reservoir that provides a controlled drip which is adequate for about 3200 km. The drip control is an inverse type in which the feed is inversely proportional to the depression downstream of the carburettor butterfly. Thus, the vanes receive the minimum amount of oil under part throttle conditions when the vacuum in the manifold is high; conversely, as the throttle is opened and the vacuum diminishes, an increased amount of oil will be supplied to the components subjected to frictional rubbing. To ensure that there is adequate but not excessive lubrication which might impede the vane movement or cause the sparking plugs to become heavily carburized and oiled up, the lubricator is adjusted by observing a sight glass at tick-over speed, to the correct drip rate.

A drive ratio for engine to blower of about one

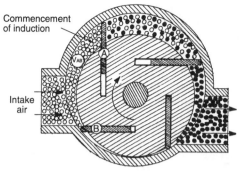

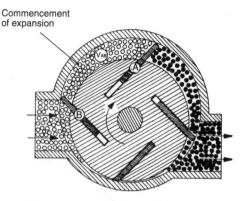

(a) Rotating drum causes charge to enter the enlarging chamber V$_{AB}$, formed between blades (A and B)

(b) Rotating drum causes trailing blade (B) to close and trap the charge in the expanding chamber V$_{AB}$

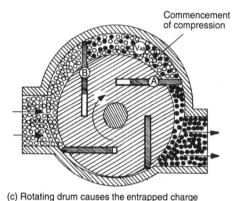

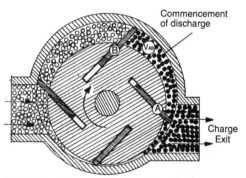

(c) Rotating drum causes the entrapped charge in chamber V$_{AB}$ to compress the charge by reducing its volume

(d) Rotating drum causes the leading blade (A) to open the chamber V$_{AB}$ to the exit port so that bade (B) pushes the compressed charge into the outlet passage

Fig. 6.7 Vane-type supercharger operational cycle

to one is chosen, low-speed engines may have this ratio increased to up to 1.5 : 1; conversely, high-speed engines will have to reduce the ratio to something of the order of 0.8 : 1 to avoid mechanical problems.

6.2.2 Operating principle
(Fig. 6.7(a–d))

The rotor drum and casing axes are offset to each other so that a crescent-shaped gap is created between both the cylindrical surfaces (Fig. 6.7(a–d)). The crescent-shaped gap formed between the casing and drum is divided into four equally spaced cells. However, as the drum rotates, the vane tips are maintained in contact with the cylindrical casing wall so that the vanes are repeatedly forced to move in and out of their respective slots. Consequently, the volume in

each cell will go through the cycle of expanding the cell space to a maximum, followed by a shrinkage of space to a minimum, for every revolution of the drum.

The inlet port in the outer casing is so positioned that the semi-crescent shaped space formed between pairs of adjacent vanes increases in volume as it moves past the inlet port region. At low rotational speed this cell space is subjected to atmospheric pressure but, with rising rotational speed, a slight vacuum is created which assists the fresh charge in entering and filling the expanding cell space (Fig. 6.7(a)).

As the trailing vane moves beyond the inlet port the cell formed between the leading and trailing vanes continues to expand so that there is an increase in vacuum in this confined space (this expansion phase does not contribute to the preparation of the discharge process but is an inherent and unwanted part of the cycle (Fig. 6.7(b)).

292

With further rotation of the drum the cell volume will expand to its maximum, it then commences to contract as the casing's internal cylindrical wall, and the external circular face of the drum forming the semi-crescent shaped cell, move closer together (Fig. 6.7(c)). This compression phase continues until the leading vane uncovers the discharge port, at this point the gap between the casing and drum cylindrical walls is still decreasing and squeezing the charge (Fig. 6.7(d)). However, the trailing vane now becomes the main driving force in displacing the compressed charge into the discharge port, against the fluctuating back-flow of charge previously accumulated and stored in the manifold, in an irreversible manner.

6.2.3 Cycle of events on a pressure–volume diagram
(Figs 6.7 and 6.8)

Consider the rotor drum rotating at a steady speed, then for one revolution of the drum, there will be a cyclic volume and pressure change in the cell, V_{AB}, formed between vanes 'A' and 'B' in the following manner.

Filling phase
(Figs 6.7(a) and 6.8)
As the drum rotates, the cell volume, V_{AB}, increases from V_1 to V_2 at atmospheric pressure P_1 until the trailing vane 'B' seals off the inlet port.

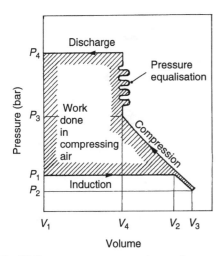

Fig. 6.8 Sliding-vane pressure–volume diagram

Expansion phase
(Figs 6.7(b) and 6.8)
The cell, V_{AB}, with its trapped mass of charge increases its volume to V_3, at which point the pressure in the cell will have decreased to P_2 (depression).

Compression phase
(Figs 6.7(c) and 6.8)
Once the cell, V_{AB}, has expanded to a volume of V_3, the space between the casing and drum commences to decrease until a volume V_2 is reached. At this point the charge begins its compression phase, the cell's volume therefore decreases to V_4, thus causing the pressure to rise to P_3.

Discharge phase
(Figs 6.7(d) and 6.8)
Once the volume in the cell, V_{AB}, is reduced to V_4 its pressure will have risen to P_3, at this point the leading vane 'A' uncovers the discharge port. The partially compressed charge is now exposed to the intake manifold's relatively large volume of stored charge which has been pressurized and is waiting its turn to enter the various cylinders. Consequently, the blower cell, V_{AB}, pressure will interact with the manifold back pressure at the same time as the charge in the cell is being pushed into the discharge port, it therefore results in an almost stabilized output pressure P_4. The discharge continues until the cell volume is again reduced to V_1, at this point the trailing vane seals off the discharge port from the cell, V_{AB}.

6.3 Semi-articulating sliding vane supercharger (Shorrock)

6.3.1 Description of the components and the blower construction
(Fig. 6.9)

The Shorrock supercharger has four semi-articulating sliding vanes slotted into an eccentrically mounted drum which is itself positioned inside a cylindrical casing.

Each vane is mounted radially to the cylindrical interior of the casing by means of two widely spaced ball races attached to a stationary carrier

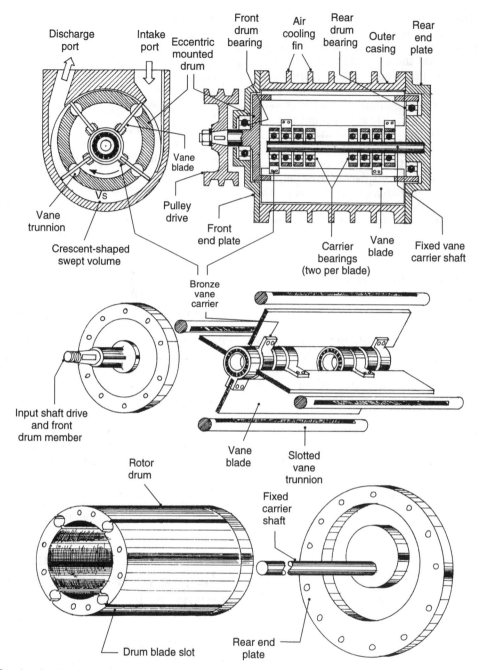

Fig. 6.9 Semi-articulated vane-type supercharger (Shorrock)

shaft, which is supported by the rear end plate centrally to the cylindrical interior of the casing.

The vanes pass through slots formed in a drum which is mounted eccentrically to the cylindrical interior of the casing by a small ball race at the front end and a much larger ball race positioned in a recess in the rear end plate.

The drum, which rotates the vanes about their central axis, is itself driven by a belt-driven pulley via the input shaft.

The offset between the drum centre and the vane carrier shaft means that the drum and vanes revolve about two separate centres.

Because of the two centre offsets, each vane is

294

supported by its own slotted trunnion shaft, and these shafts are located in circular longitudinal holes machined in the drum. Consequently, not only will the vanes slide in their slots but these slots will swivel to accommodate the small amount of articulation which must take place since the drum and vane centres of rotation do not coincide.

Thus, the four vanes are restrained radially by pairs of ball races mounted centrally to the casing on the fixed carrier shaft, whereas the drum is mounted on bearings eccentric to the cylindrical wall of the casing. Therefore, as the drum rotates, the vanes will slide and the slotted trunnions will swivel to align themselves, this therefore results in a fine vane-tip to cylindrical-wall clearance which is automatically maintained.

6.3.2 Operating principle
(Fig. 6.10(a–d))

The crescent-shaped space created by the eccentrically positioned drum relative to the internal cylindrical wall of the casing is divided into four separate cells by the equally spaced vanes which project from the drum and, almost but not quite, touch the internal walls of the casing.

The operating cycle of the semi-articulating vane-type blower can be explained by describing the four phases completed in one revolution of the drum: filling, expansion, compression and discharge.

Filling phase
(Fig. 6.10(a))
Thus, as the drum rotates, the leading vane 'A'

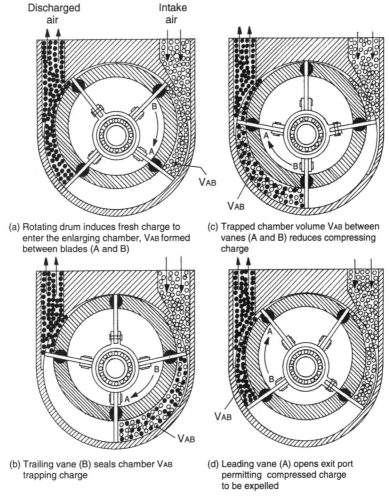

(a) Rotating drum induces fresh charge to enter the enlarging chamber, V$_{AB}$ formed between blades (A and B)

(c) Trapped chamber volume V$_{AB}$ between vanes (A and B) reduces compressing charge

(b) Trailing vane (B) seals chamber V$_{AB}$ trapping charge

(d) Leading vane (A) opens exit port permitting compressed charge to be expelled

Fig. 6.10 Semi-articulated vane-type supercharger operating cycle

moves past the inlet port, and the semi-crescent shaped volume V_{AB} between the leading vane 'A' and the trailing vane 'B' enlarges (Fig. 6.10(a)). Fresh charge at atmospheric pressure will then flow into this chamber to fill up the expanding space, which is enclosed by the vanes 'A' and 'B'.

Expansion phase
(Fig. 6.10(b))
When the trailing vane 'B' moves past the inlet port the chamber, V_{AB}, is sealed but its volume continues to expand, the charge is therefore subjected to a depression (Fig. 6.10(b)). This expansion phase with both inlet and exit passages closed is inherent and does not help to bring fresh charge into the blower or to pump the trapped charge out into the manifold.

Compression phase
(Fig. 6.10(c))
Once the chamber V_{AB} has revolved about a half revolution from the initial filling phase position, the semi-crescent shaped chamber, V_{AB}, begins to contract (Fig. 6.10(c)). Hence, the charge pressure first rises to atmospheric conditions, and then with further rotation the charge is compressed causing the pressure to increase well above atmospheric pressure.

Discharge phase
(Fig. 6.10(d))
Eventually, the leading vane 'A' uncovers the discharge port, and immediately some of the stored charge in the manifold, which is still at a higher pressure than the chamber V_{AB}, flows back until pressure equalization takes place. With further rotation of the drum, chamber V_{AB} will continue to contract and, simultaneously, the trailing vane 'B' sweeps the resultant charge forward and into the manifold via the discharge port (Fig. 6.10(d)). It therefore replaces the charge burnt and expelled from the cylinders.

This cycle of events is thus continuous and is repeated four times per revolution since the vanes divide the crescent-shaped space formed between the drum and casing walls into four equally spaced chambers.

The pressure–volume cycle of events for the semi-articulating sliding vane blower is identical to that of the simple slotted vane-type supercharger described in section 6.2.2.

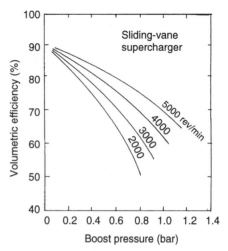

Fig. 6.11 Volumetric efficiency curves at constant drum speed relative to boost pressure

6.3.3 Vane-type supercharger performance characteristics
(Figs 6.11, 6.12 and 6.13)

The performance characteristics of a typical vane-type supercharger, with respect to the compressor's volumetric efficiency, as the drum speed and boost pressure are varied, are shown in Figs 6.11 and 6.12.

Figure 6.11 shows that for a constant drum speed there is a drop in the compressor's volumetric efficiency as the boost pressure rises and that it is more noticeable as the drum speed is reduced. Figure 6.12 shows the relationship be-

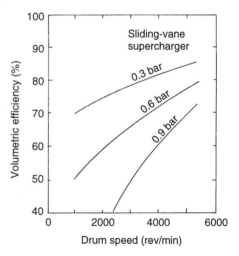

Fig. 6.12 Volumetric efficiency curves at constant boost pressure relative to drum speed

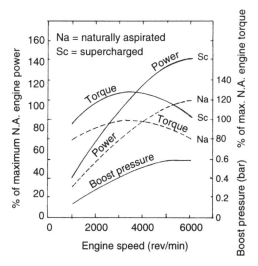

Fig. 6.13 Comparative performance curves for direct driven vane- or Root's-type supercharged petrol engine

tween volumetric efficiency and drum speed when the boost pressure is kept constant, here it can be seen that as the drum speed increases, the volumetric efficiency rises but its rise becomes less as the boost pressure is raised. The reason for the reduction in volumetric efficiency when the drum speed is reduced is due to the increased time available for the air charge to escape between the drum and vanes and casing. Similarly, there is a reduction in volumetric efficiency as the boost pressure is increased due to the increased back pressure leakage past the rotor or vanes of the compressor.

The engine torque and power characteristics of a positively driven supercharged engine are compared with the same engine in its naturally aspirated form in Fig. 6.13. The boost pressure curve shows that it rises progressively to a maximum pressure of 0.6 bar at about 5000 rev/min. This boost pressure produces an improvement in peak engine torque of roughly 30% at 3000 rev/min, which amounts to a considerable improvement in engine performance.

6.4 Roots rotating-lobe supercharger

6.4.1 Description of the blower construction
(Figs 6.14, 6.15, 6.16 and 6.17)

The Roots rotary positive displacement blower contains a pair of externally located meshing helical gears (Figs 6.14 and 6.15). Their function is to drive the pumping members, which consist of specially shaped twin contra-rotating rotors which revolve at the same speed without touching each other. These rotors may have two or three lobes made from aluminium alloy (Figs 6.15 and 6.16) each of identical design—their outer convex contour being of epicycloidal form whilst the inner concave profile is a hypocycloidal curve. This geometric form ensures that at all angular positions the high-pressure discharge space is sealed, except for a very small working clearance, from the low-pressure inlet space. This very small working clearance is maintained by the strict control of the external helical cut timing gear backlash which must not exceed 0.10 mm. The clearance between rotor lobes, when at right-angles to each other, ranges from 0.15 to 0.18 mm, whereas the rotor-lobe to casing radial clearance should be between 0.13 and 0.20 mm, while the rotor-lobe to end-casing axial clearance tolerance ranges from 0.18 to 0.20 mm. Both the radial and axial clearances are obtained by the initial installation shimming and adjustment of the support bearings.

The simple Roots-type blower tends to be noisy since the compression does not take place until the leading lobes of the rotor suddenly uncover the discharge port. Consequently, this produces pressure pulsations and turbulence, which correspondingly generate loud sound waves.

This noise can be reduced by arranging the discharge port at an angle instead of parallel to the rotor axis (Fig. 6.16) or by employing helical lobes (Fig. 6.17).

With the angled discharge port (Fig. 6.16) the straight lobes progressively uncover the leading edge of the discharge port and hence this spreads the discharge phase over a small angular movement of the rotors and thereby reduces the noise caused by the sudden interaction of the charge as it is being ejected into the manifold.

Alternatively, expensive helical lobes

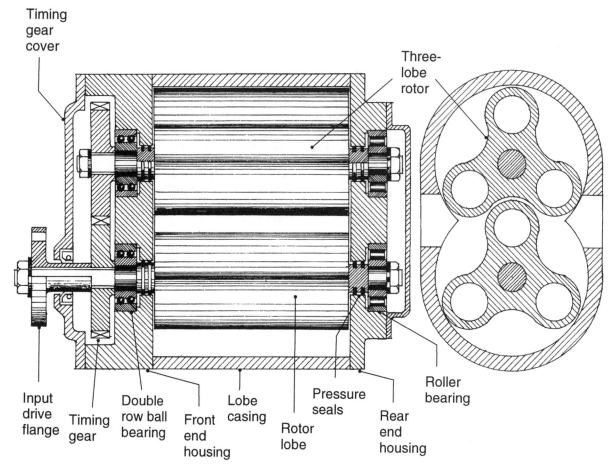

Timing gear cover

Three-lobe rotor

Input drive flange

Timing gear

Double row ball bearing

Front end housing

Lobe casing

Rotor lobe

Pressure seals

Rear end housing

Roller bearing

Fig. 6.14 Section view of a three-lobe roots-type supercharger

(Fig. 6.17) can be manufactured where the lobes spiral along their length. The two rotors have right and left-hand spiral lobes which intermesh; thus, when the rotors revolve, the trapped charge is squeezed along and between the lobes so that the discharge pressure pulsations are more gradually absorbed by the existing charge stored in the manifold. These helical formed lobes therefore operate quieter than those with straight-formed lobes.

6.4.2 Design considerations
(Figs 6.15 and 6.16)

The leakage past the rotor lobes at low rotational speeds is appreciable due to the length of the leakage path formed between the intermeshing lobes, the lobe to cylindrical-casing clearance and the axial end clearance. However, as the speed of the rotors increases, the proportion of leakage relative to the quantity of charge delivered becomes less until, at high engine speeds, it is comparatively small. This relative reduction in leakage rate with rising speed is owing to the shorter time available for the pressurized charge to escape along the leakage path created by the necessary working clearances. The leakage of air or air–fuel mixture of any appreciable magnitude is detrimental not only because of waste of power, but also because it involves the passage of high temperature charge back to the intake side of the rotors. Consequently, it increases the intake temperature of the inlet port and therefore the delivery temperature, it thus causes a reduction in the density of the delivered charge, so setting up a vicious circle.

The Roots blower can have two or three lobes per rotor, the latter reduces the tendency to leakage, but this design results in a reduced

298

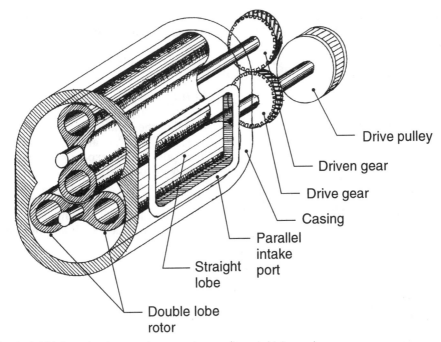

Fig. 6.15 Twin straight-lobe rotor-type roots supercharger (Lancia Volumex)

pumping space for a given diameter of rotor. The three-lobe rotor (Fig. 6.16) gives a more uniform pressure output than the two-lobe one (Fig. 6.15) as there are three air deliveries per revolution from each rotor as compared with two in the two-lobe design. In addition, as there are always two lobes practically in contact with the casing at any time with the three-lobe rotor, better sealing of the charge in its passage through the blower is obtained.

Friction losses in the Roots type blower are low as the friction amounts almost entirely to that generated in the four rotor bearings plus the losses encountered in the timing gears. This is

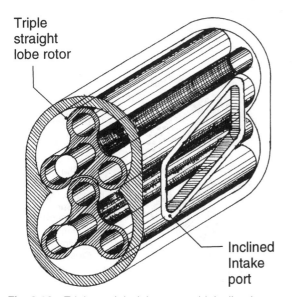

Fig. 6.16 Triple straight-lobe rotor with inclined discharge part roots supercharger

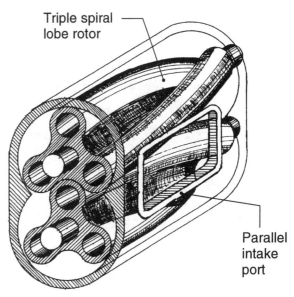

Fig. 6.17 Triple spiral-lobe rotor-type roots supercharger

because there is no contact between the two rotors or between the rotors and the casing. Thus, only the bearings and timing gears have to be lubricated whereas the rotors operate dry and should not be subjected to any form of oil spray or mist.

6.4.3 Operating principle
(Fig. 6.18(a–d))

Consider a Roots blower to be driven so that the rotors rotate in opposite directions to each other; that is, the upper rotor revolves clockwise whereas the lower one moves in an anticlockwise direction.

As each set of rotor lobes move around their

respective cylindrical casing, sealed cells are continuously being formed between adjacent pairs of leading and trailing lobes as they sweep over the cylindrical walls.

Filling phase
(Fig. 6.18(a))

Initially the space created between the intermeshing lobes 'A' and 'D' and the casing in the inlet port region expands so that fresh charge is drawn in to fill this volume. As the rotors continue to revolve, the lobe 'A' expands the space behind it and the upper cylindrical wall and, at the same time, the trailing lobe 'D' encloses the space formed between it and the leading lobe 'E'.

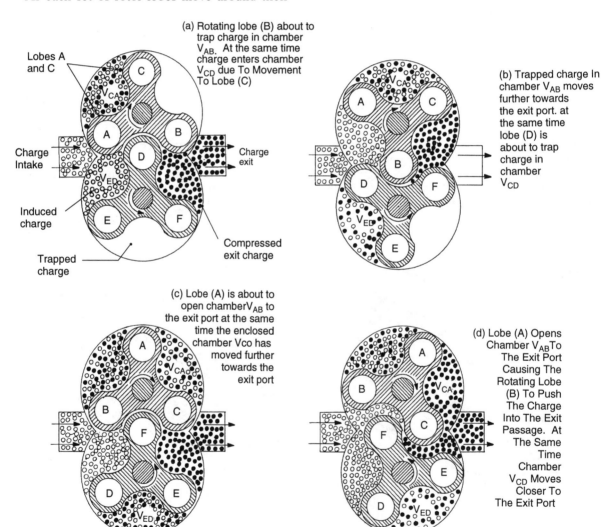

Fig. 6.18   Triple lobe roots-type supercharger

300

Displacement phase
(Fig. 6.18(b))
Further rotation of the rotors carries the trapped charge in the space between the lobes 'E' and 'D', known as cell V_{ED}, around the cylindrical casing wall until the leading lobe 'E' is about to uncover the high pressure discharge port. Likewise, the charge in cell V_{CA} sweeps around the upper cylinder wall (ahead of cell V_{ED}) until its leading lobe 'C' opens this chamber to the discharge port.

Equalization phase
(Fig. 6.18(c))
When the leading lobe 'C' sweeps past the discharge port, cell V_{CA} opens to the discharge port, immediately some of the compressed charge already in the induction manifold rushes back and mixes irreversibly with the unpressurized charge trapped in the cell V_{CA}, this rapidly increases and equalizes the pressure of the charge occupying this space.

Discharge phase
(Fig. 6.18(d))
Still further rotation of the rotors will then push both the recently arrived charge and the flow-back charge into the induction manifold via the discharge port. Thus, with the continued forward movement of the rotors, the resultant charge will reverse the flow-back, equalize the pressure fluctuation and discharge itself into the intake manifold. The magnitude of this cyclic process is therefore designed to match the engine's total cylinder-filling demands as the inlet valves periodically open and close to fulfil their induction stroke requirements.

The displaced charge entering the intake manifold occurs four times per revolution for a pair of two-lobe rotors or six times per revolution for a three-lobe rotor blower. Hence, the charge flow being forced into the manifold is a sequence of intermittent flow-back equalization and discharge even though the rotors revolve with uniform speed.

Cycle of events pressure–volume diagram
(Figs 6.18 and 6.19)
Consider the rotors rotating at a steady speed, then, for one revolution of each rotor, there will be a cyclic volume and pressure change in each cell formed between adjacent lobes. The rotation

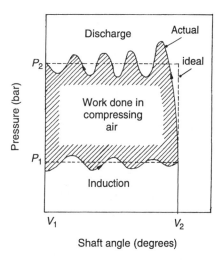

Fig. 6.19 Root's-type supercharger pressure–volume diagram

of the rotors increases the space formed between the inlet port and the intermeshing rotor lobes, and fresh charge (be it pure air or a mixture of air and petrol) will therefore be drawn in to fill the space.

Filling phase
(Fig. 6.19)
With the leading lobe at the trailing edge of the inlet port, the cell will have zero volume V_1 at atmospheric pressure P_1 since there is no cylindrical wall coverage. However, when the rotors revolve, the cell volume increases from V_1 to V_2 at which point the trailing lobe is about to seal-off and trap the charge which has been induced into the cell.

Displacement phase
(Fig. 6.19)
The trapped charge in the cell at atmospheric pressure P_1 and constant volume V_2 will then sweep around the cylindrical wall until the leading lobe uncovers the discharge port.

Pressure equalization phase
(Fig. 6.19)
Charge in the intake manifold immediately reverses and flows back into the cell, which is filled with the unpressurized fresh charge. Pressure equalization between the charge already in the cell and that which has moved back from the manifold produces a resultant output pressure P_2.

Discharge phase
(Fig. 6.19)

The resultant charge at constant pressure P_2 is then pushed into the intake manifold via the discharge port by the forward moving trailing lobe until the cell volume contracts to V_1, at which point the trailing lobe has reached the leading edge of the discharge port.

6.4.5 Electro-magnetic clutch and bypass-valve controlled Roots supercharger system (Toyota)

The need for a belt-driven supercharger clutch
(Figs 6.20 and 6.21)

The advantage of a direct belt-driven supercharger such as the Roots blower compared with an indirect exhaust gas driven turbocharger is that it supplies boost pressure in direct relation to the crankshaft speed. It therefore results in a much quicker throttle response with a corresponding improvement in acceleration.

The disadvantage of mechanically driven superchargers is that they absorb power and therefore consume more fuel compared with the naturally aspirated engine of a similar size. This increase in fuel consumption is offset by the extra torque produced when accelerating and the increased power developed in the upper speed range making it a much more flexible and responsive engine. However, an increase of torque and power is not required when driving steadily under part load conditions but, nevertheless, the supercharger still needs power to drive it proportional to the engine speed.

This problem can be partially solved by incorporating an electro-magnetic clutch between the crankshaft belt drive and the input to the supercharger, and also by an air bypass system. Hence, the blower can be engaged or disengaged as the road conditions and the driving requirements demand. The overall fuel consumption is therefore improved compared with a supercharged engine arrangement where the drive is permanently engaged.

Electro-magnetic clutch
(Fig. 6.20)

This consists of an annular-shaped solenoid-winding which is attached to the front end housing of the blower. An integral rotor and pulley-wheel is mounted on a double-row ball-bearing located in the centre of the annual shaped solenoid, the bearing is, in turn, supported by a tubular extension which forms part of the front end housing. Protruding from the centre of the tubular extension is the lobe drive shaft which has a keyed flange attached at the outer end. Surrounding the drive flange is an aligned annular armature which is held by four tangentially positioned spring drive straps. The straps are riveted at their inner end to the drive flange while the outer ends are similarly attached but to the annular armature. An annular-shaped back stop is also fixed to the drive flange on the same rivets as the inner strap attachment, and this stop is provided to control the to and fro axial movement of the armature when in service.

When the solenoid is energized by an electric current via the overload switch a powerful magnetic flux field is produced which spreads beyond the surrounding air gaps encompassing the annular armature ring. Accordingly, the intensity of the magnetic force pulls the armature hard against the rotor's parallel working face thereby completing the drive from the pulley wheel to the lobe drive-shaft. Switching off the current supply automatically collapses the magnetic force field thus permitting the spring drive strap's resilience to pull away the armature ring from the rotor, and the engine-to-supercharger drive is therefore disconnected.

Road operating conditions
(Fig. 6.21)

Under low and mid-speed operation when extra power is not needed, the computer-controlled electro-magnetic clutch disengages the Roots blower so that the rotor lobes spin purely by the small amount of air that passes between the rotor lobes to enter the manifold. At the same time, an air bypass valve controlled by a computer and operated by an actuator is made to open, thus enabling the fresh air charge to bypass the supercharger and to enter the induction manifold directly. The engine now operates under naturally aspirated conditions. It therefore provides conventional engine economy and performance. When the engine is accelerated under high engine load operating conditions, the computer switches on the electro-magnetic clutch thereby coupling the Roots blower to the crankshaft belt drive. Simultaneously, the bypass valve closes so that all the air charge entering the induction manifold is

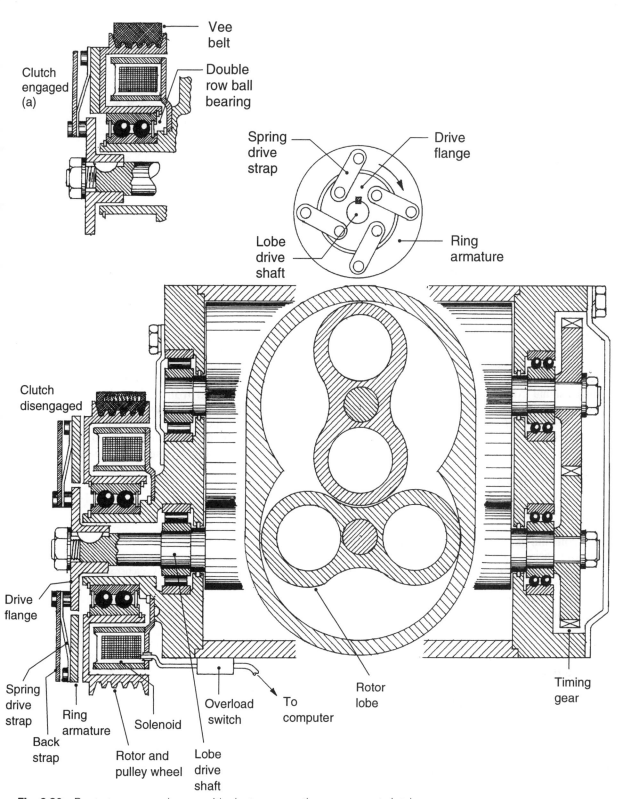

Labels in figure:

Vee belt

Clutch engaged (a)

Double row ball bearing

Spring drive strap

Drive flange

Lobe drive shaft

Ring armature

Clutch disengaged

Drive flange

Spring drive strap

Back strap

Ring armature

Solenoid

Rotor and pulley wheel

Lobe drive shaft

Overload switch

To computer

Rotor lobe

Timing gear

Fig. 6.20 Roots-type supercharger with electro-magnetic engagement clutch

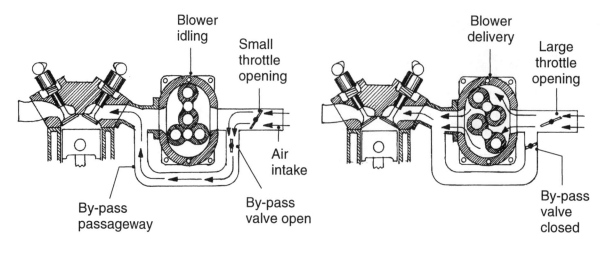

Light-load Steady Engine Speed　　**Large-load Engine Accelerating**

Fig. 6.21　Roots-type supercharger with by-pass throttle system

now pressurized. The engine's performance will now correspond to the accustomed flexibility of a postively driven supercharged engine. At high, steady engine speeds (usually above 4000 rev/min) a signal from the computer will open or close the bypass valve so that the quantity of boost air delivered (and its pressure) is continuously being regulated.

6.4.6 Roots supercharger performance characteristics
(Figs 6.22, 6.23 and 6.24)

The constant gap between the intermeshing lobes and the casing walls means that there is always a certain amount of air leakage which becomes more pronounced if either the lobe speed is reduced or the boost pressure is raised. Therefore, the volumetric efficiency will be directly influenced by the amount of lobe-to-casing wall blow-by which is related almost directly to the pressure build up in the lobe cells and the extended time the charge is swept around the casing walls; that is, the lobe rotational speed (Fig. 6.22). Investigations have been made which show that the volumetric efficiency rises as the lengths of the lobes are increased (Fig. 6.23). These graphs also show how inefficient the blower is in discharging the air out when the rotational speed is reduced below about 2000 rev/min and 1500 rev/min for the short and long lobes respectively. Another inherent limitation of the Roots

supercharger is that there are large pressure fluctuations in the induction and discharge process which are responsible for imparting a considerable amount of heat to the air charge as the air is transferred from the intake to the output side of the blower. Consequently, the adiabatic efficiency, which is a measure of the amount of input energy used to drive the blower to that which is wasted in raising the temperature of the charge, is generally low with this class of compressor (Fig. 6.24).

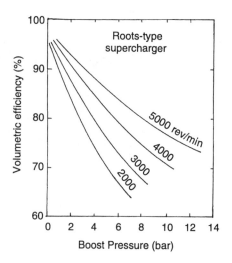

Fig. 6.22　Volumetric efficiency curves at constant rotor lobe speed relative to boost pressure

304

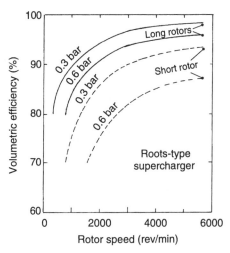

Fig. 6.23 Volumetric efficiency curves at constant boost pressure relative to rotor lobe speed

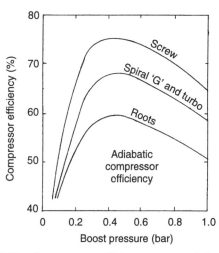

Fig. 6.24 Compressor efficiency curves relative to boost pressure for different types of superchargers

6.5 Sprintex screw supercharger (compressor)

Description of the blower construction
(Figs 6.25 and 6.26)

The Sprintex screw type supercharger (compressor) has two magnesium alloy screws of male and female form so that compression is positively contained. The lobes, or teeth as they are known, in this unit are of the helical form; that is, they are twisted along their length. The male screw has four convex-shaped lobes whereas the female screw has six concave spaces between its six lobes. The concave and convex parts of the two screws intermesh but their profile surfaces never touch each other, nor do the screw tips and ends contact

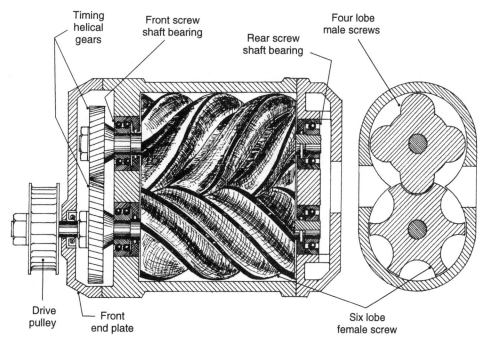

Fig. 6.25 Section view of screw-type Sprintex supercharger

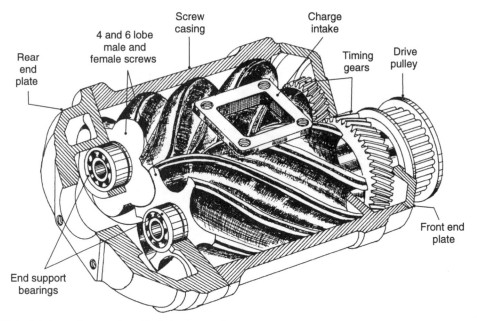

Fig. 6.26 Pictorial view of screw-type Sprintex supercharger

the cylindrical casing and end walls. Thus, the very small clearance maintained between the Teflon coated screws (50–100 µm) is achieved by two external helical cut timing gears which are meshed together in such a way that their backlash is kept to a minimum while the radial lobe (tooth)-to-casing and axial lobe-to-wall clearances (50–80 µm) are provided by the accuracy of the supporting single and double-row ball-bearings.

The female-to-male screw speed ratio is 4 : 6; that is, two to three, and it is claimed that the female screw will operate up to a maximum speed of 15 000 rev/min which corresponds to a maximum of 22 500 rev/min for the second male screw.

A toothed belt pulley drive is used to rotate the screws. To obtain a maximum input female screw speed of 15 000 rev/min would require a speed ratio for the pulleys of approximately 2.5 : 1 if the engine is to reach a maximum speed of 6000 rev/min.

6.5.1 Operating principle
(Figs 6.27(a–d) and 6.28(a–d))

Charge is drawn through the inlet port to fill the expanding space initially produced between the intermeshing lobes (threads) as the male and female screws rotate counter to each other, see Fig. 6.27(a)).

As the trailing lobes of each screw move beyond the inlet port, charge will be trapped between consecutive lobes and the cylindrical casing, these cells will therefore move around the periphery of the casing until the leading lobes of each cell arrive at the discharge port. This progressive circular movement of the trapped charge in the cells is shown from the underside of Figs 6.27(a–d) and 6.28(a–d).

The arrival of consecutive cells from both male and female screws at the discharge port, merges them, and at the same time the intermeshing lobe movement reduces the paired cell volume, see Fig. 6.27(a–b).

The intermeshing screws lie parallel to each other so that the left- and right-hand helical lobes form pairs of converging double diagonal or 'V'-shaped cells. Thus, at the same time as the trapped charge is moved in a circular direction between the cylindrical casing walls and the adjacent pairs of lobes, it is also being screwed and squeezed axially from the inlet port to the discharge port end (see Fig. 6.28(c–d)).

This sequence of events is continuously repeating itself so that eventually the induction manifold branch pipes are filled with a relatively steady pressurized column of charge, which is periodically displaced into the cylinders during each inlet valve open phase.

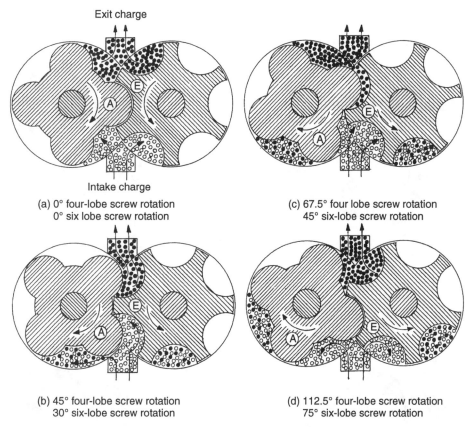

Exit charge

Intake charge

(a) 0° four-lobe screw rotation
0° six lobe screw rotation

(c) 67.5° four lobe screw rotation
45° six-lobe screw rotation

(b) 45° four-lobe screw rotation
30° six-lobe screw rotation

(d) 112.5° four-lobe screw rotation
75° six-lobe screw rotation

Fig. 6.27 Screw-type Sprintex supercharger end view of operational cycle

6.5.2 Cycle of events on a pressure–volume diagram

(Figs 6.28(a–d) and 6.29)

Consider the male and female screws rotating at a steady speed, then for one revolution of the female screw there will be six cyclic volume and pressure changes in the longitudinal cells as the intermeshing male screw completes one and a half revolutions.

Filling phase
(Fig. 6.29)
As the male and female screws revolve, the cells partially formed between each leading and trailing lobe increase in volume from V_1 to V_2 as they move around into the cylindrical casing wall region. Owing to the cell's expanding space phase, fresh charge at atmospheric pressure P_1 will enter and fill the cell.

Compression phase
(Fig. 6.29)
Once each trailing lobe seals its cell, the charge is swept around the cylindrical walls until the leading lobe uncovers the discharge port. At the same time, the male and female lobes will intermesh in such a way so that the volume formed between them, and the apex where the twin cylindrical walls meet, is progressively reduced in an axial direction from the inlet port end to the opposite discharge port end.

Thus, the charge's volume is reduced, from V_2 to V_3 with a corresponding rise in pressure from P_1 to P_2.

Discharge phase
(Fig. 6.29)
It is not until the latter part of the compression phase that any progress is made in actually pumping part of the charge being compressed into the induction manifold.

Once equalization of pressure between the

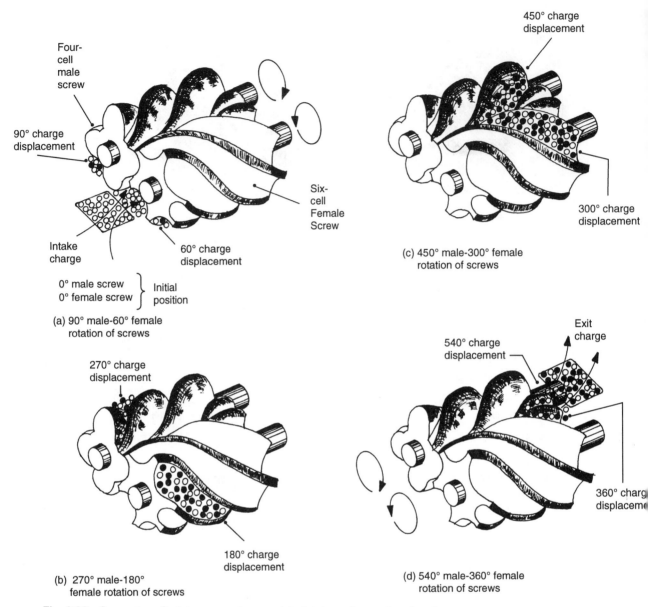

Four-cell male screw

90° charge displacement

Intake charge

60° charge displacement

Six-cell Female Screw

0° male screw ⎫ Initial
0° female screw ⎭ position

(a) 90° male-60° female
rotation of screws

270° charge displacement

180° charge displacement

(b) 270° male-180°
female rotation of screws

450° charge displacement

300° charge displacement

(c) 450° male-300° female
rotation of screws

540° charge displacement

Exit charge

360° charge displaceme

(d) 540° male-360° female
rotation of screws

Fig. 6.28 Screw-type Sprintex supercharger pictorial view of operational cycle

existing charge in the manifold and the charge being screwed into the discharge port occurs at volume V_3 and pressure P_2, then any further rotation of the screws displaces the charge positively into the manifold until the volume between the intermeshing screws and cylindrical walls in the discharge port region is reduced to V_1 again.

6.5.3 Screw-type Sprintex supercharger performance characteristics
(Figs 6.30 and 6.31)

Typical compression curves for both the four-cell male screw and the six-cell female screw are shown in Fig. 6.30. As can be seen, the male four-cell screw commences reducing the cell's volume early in the cycle whereas there is no

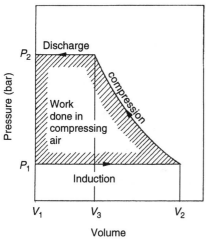

Fig. 6.29 Screw-type supercharger pressure-volume diagram

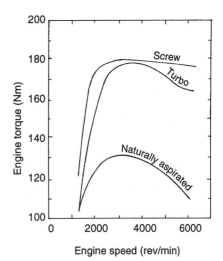

Fig. 6.31 Comparative torque characteristics relative to engine speed for screw and turbocharged superchargers

change in the female cell volume until the angular movement for the male screw is almost complete. The resultant compression for this example shows a reduction in volume of the order of 60%, but this can vary depending upon where the intake and discharge ports are situated.

The great advantage of the screw type compressor is that the air is progressively squeezed together and simultaneously displaced from the intake port to the opposite outlet end before it is discharged into the induction manifold with only a moderate rise in the compressed air temperature. As a result, these compressors consume very little power and have a relatively high adiabatic efficiency (Fig. 6.24) so that an intercooler is

needed only if the pressure ratio is likely to exceed 2:1.

The engine torque developed using both a screw supercharger and a turbocharger is shown in Fig. 6.31, and here it can be seen that the torque rise for the screw-type Sprintex supercharger commences earlier compared with the turbocharger, and that it maintains an almost flat torque peak throughout the engine's speed range.

6.6 G-Lader oscillating spiral displacer supercharger

6.6.1 Description of the components and the blower construction
(Figs 6.32 and 6.33)

The blower consists of two half-circular aluminium alloy casings, each half casing being die cast with two separate 'G'-shaped lands which protrude perpendicularly to the flat side of the casing wall (Fig. 6.32).

The spiral lands in each half casing are a mirror image of each other; however, these projected lands in each half casing do not meet, as a gap is required in the middle to accommodate a central magnesium alloy spacer disc. This displacer disc has similarly shaped spiral lands attached on either side which intermesh with the fixed spirals (Fig. 6.33).

The spiral land chambers formed between the

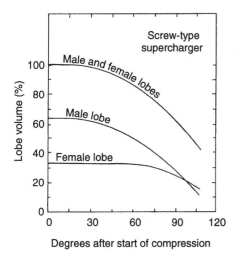

Fig. 6.30 Lobe inter-volume change relative to angular screw movement

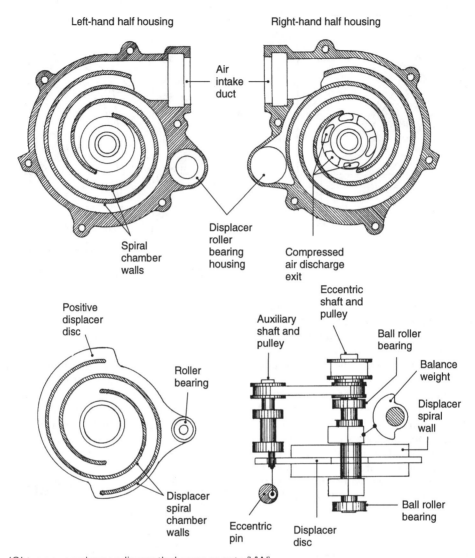

Left-hand half housing   Right-hand half housing

Air intake duct

Spiral chamber walls

Displacer roller bearing housing

Compressed air discharge exit

Positive displacer disc

Roller bearing

Eccentric shaft and pulley

Auxiliary shaft and pulley

Ball roller bearing

Balance weight

Displacer spiral wall

Displacer spiral chamber walls

Eccentric pin

Displacer disc

Ball roller bearing

Fig. 6.32 'G'-type supercharger dismantled components (VW)

fixed and moving spiral lands are sealed with sprung bronze Teflon strips which fit into recesses formed in the spiral land outer edges. The low rubbing speeds eliminate the need for oil mist or wet lubrication.

The central hub of the disc displacer is mounted on the eccentric drive shaft and additional support and directional control of the displacer movement is given to one end of the displacer by an eccentric pin attached to the auxiliary shaft. The two shafts have the same eccentricity and rotate at the same speed due to a toothed belt and pulley drive (Fig. 6.33).

The eccentric drive-shaft has integral balance weights located on either side and opposite the eccentricity, which thereby balance the cranked portion of the shaft.

The bearings of the eccentric drive-shaft are lubricated by the engine's oil circulation system, whereas the auxiliary shaft uses pre-lubricated maintenance-free bearings.

The present generation of G-Lader compressors are belt driven at 1.7:1 engine speed.

The zero-loss swept volume amounts to 566 cm^3 per revolution at a rated speed of 10350 rev/min.

A typical maximum charge–air pressure for these blowers is 0.72 bar.

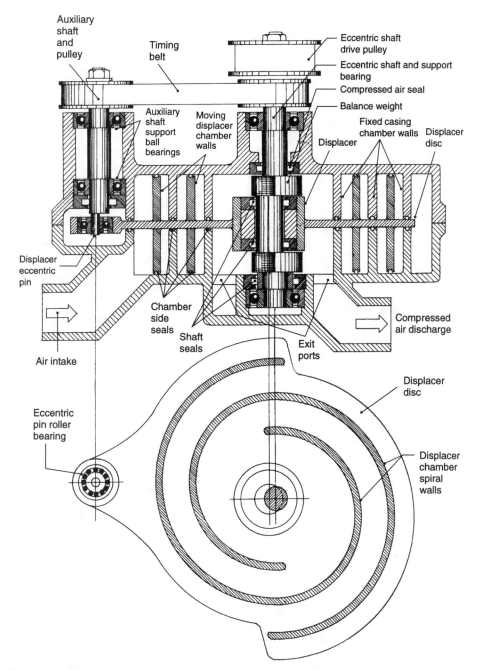

Fig. 6.33 'G'-type supercharger (VW) sectional construction detail (VW)

6.6.2 Operating principle

Filling phase
(Fig. 6.34(a and b))
Initially, the drive-shaft's eccentricity is in its highest TDC position so that the pairs of moving spirals on the displacer, interleaved between the fixed casing spirals, touch the fixed spirals at the ends and in the middle (Fig. 6.34(a)). This divides the two fixed spiral lands into two chambers open to the atmosphere: a larger outer crescent-

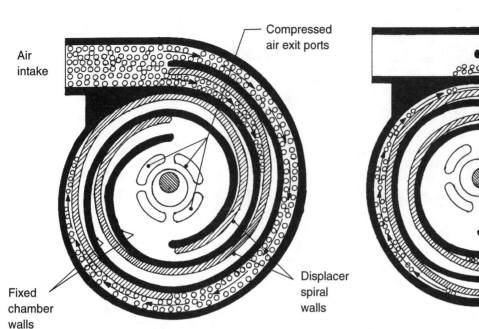

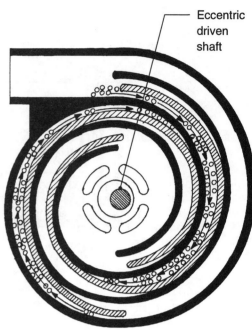

(a) Filling phase (0° and 360°) air induced
    into the open inner and outer chambers

(b) Filling and closing phase (90°) filling of
    inner and outer chambers completed,
    intake passages commences to close

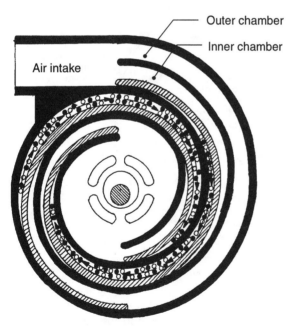

(c) Compression phase (180°) volume reduction
    of inner and outer chamber squeezes
    and compresses trapped air

(d) Discharge phase (270°) volume reduction of
    inner and outer chamber continuous until
    both chambers open to the discharge ports

Fig. 6.34 'G'-type supercharger operating cycle

shaped chamber and a smaller semi-crescent profiled inner chamber.

A 90° rotation of the eccentric shaft from its
highest position, moves the expanding inner and
outer crescent-shaped chambers clockwise, which
simultaneously induces the fresh charge to fill the
newly formed chamber spaces.

Compression phase
(Fig. 6.34(c))

A 180° rotation of the eccentric shaft from its
highest position closes both the inner and outer
chambers and then commences to squeeze and
compress the trapped charges as they are swept
clockwise in the contracting crescent-shaped
chambers.

Discharge phase
(Fig. 6.34(d))

A 270° rotation of the eccentric shaft from its
highest position sweeps the inner and outer trapped charges clockwise until both chambers open
to the four central outlet ports. Thus, any further
rotation of the eccentricity squeezes the charge
out through the exit ports and brings the moving
spirals back to the original position of 0°/360° (see
Fig. 6.34(a)).

6.6.3 Cycle of events shown on a pressure–volume diagram
(Figs 6.34(a–d) and 6.35)

Consider the drive-shaft's eccentricity rotating at
a steady speed, then the displacer will wobble so
that the moving spirals will interact between the
fixed spirals to form chambers which fill, compress and discharge the air–fuel mixture into the
induction manifold.

Filling phase
(Figs 6.34(a and b) and 6.35)

With the eccentricity of the drive shaft in the
vertical position, both the inner and outer chambers will be at their minimum volume V_1; however, as the eccentricity rotates clockwise, the
chamber spaces formed between the fixed and
moving spirals expand until a maximum volume
of V_2 is reached. At the same time fresh charge at
very nearly atmospheric pressure P_1 enters and
fills both the inner and outer crescent-shaped
chambers.

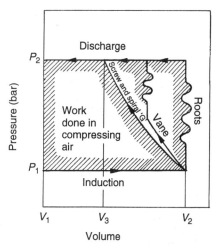

Fig. 6.35 Comparative pressure–volume diagrams
for different types of superchargers

Compression phase
(Figs 6.34(c) and 6.35)

With further rotation of the eccentric shaft the
moving spirals close the inlet passages to both the
inner and outer chambers. This is followed by the
crescent-shaped chambers moving clockwise and
simultaneously reducing in volume from V_2 to V_3
as the trapped charge is compressed from atmospheric pressure P_1 to a much higher pressure P_2,
at which point the chambers open to the exit
ports.

Discharge phase
(Figs 6.34(d) and 6.35)

With further rotation of the eccentric shaft both
chambers move around until the moving spirals
open the inner and outer chambers to the exit
ports. At the same time these chambers will
continue to contract so that the charge in each
chamber will be progressively squeezed into the
exit port against the flow-back from the manifold.
The resultant pressure between the existing
charge and the newly arrived charge in the exit
port remains approximately constant at P_2 until
the volume in both chambers is reduced again to
V_1.

6.6.4 'G'-Lader type blower with bypass intake system
(Fig. 6.36)

Under low speed and part load operating conditions the output from a 'G'-Lader type of blower
will be much greater than the engine can consume

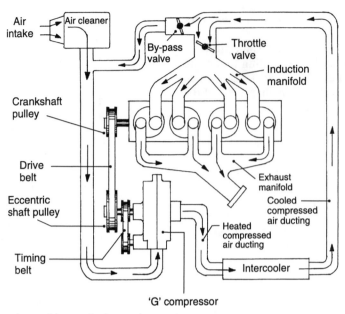

Fig. 6.36 'G'-Lader supercharged four-cylinder engine system

and pass through to the exhaust, therefore the pressure would rise against the throttle butterfly valve on the intake side. The effects of this unwanted pressure build-up against the throttle valve would be for the blower to absorb more power and for the engine to use more fuel than would be the case if the excess delivery of air could be diverted back to the intake side of the blower. An additional benefit of bypassing the air charge back to the intake side of the blower is to relieve the load on the throttle butterfly valve spindle so that the throttle will respond instantly if the driver so commands it.

Thus, to overcome the poor matching of the blower's output to that demanded by the engine at idle and low engine speeds, a bypass valve and passage, which return the discharge air from the blower back to the filtered side of the air intake, are incorporated (Fig. 6.36). This bypass valve, which is interlinked with the throttle valve is made to open at low-speed part-throttle conditions, but above some predetermined throttle opening, the bypass valve closes, so that the pressurized air charge from the blower is now pumped via the intercooler and induction manifold into the engine cylinders.

6.6.5 'G'-Lader oscillating spiral displacer supercharger performance characteristics
(Figs 6.24, 6.35, 6.37 and 6.38)

The compression curve for the oscillating spiral displacer-type supercharger is almost adiabatic and is similar to the screw-type Sprintex supercharger. Consequently, these compressors absorb very little power during the compression process (Fig. 6.35) compared with the Roots or sliding-vane type compressors. Note that the area of the enclosed pressure–volume diagram represents the amount of energy necessary to compress the air charge for the four basic types of positive-driven compressors described. As can be seen (Fig. 6.35) the screw and the oscillating spiral displacer-type compressors consume the least energy, whereas the Roots lobe-type of blower requires the most power to drive and, between these extremes, is the sliding-vane (including the semi-articulating) supercharger. The power losses encountered with rising rotational speed compared with a typical Roots blower is shown in Fig. 6.37; thus, at 4000 rev/min the Roots blower consumed 51% of the maximum speed power but the 'G'-Lader only needs 35% of the maximum power to drive it. However, there is a decrease in the volumetric efficiency with rising oscillating speed (Fig. 6.38) for a given boost pressure, at present the max-

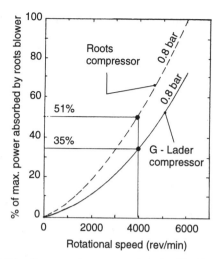

Fig. 6.37 Comparative power absorbed by Root's- and 'G'-Lader-type superchargers relative to rotational speed

imum boost pressure is limited to just under 0.8 bar.

The adiabatic compressor efficiency rises rapidly with an increase in boost pressure, reaching a peak of about 68%, and then gradually decreases (Fig. 6.24).

It is claimed that there is very little temperature rise during the compression process so that with the low amount of power absorbed by the compressor and the relatively cool delivery of the boost charge, it makes this type of supercharger very efficient and extremely suitable for boosting small- to medium-sized car petrol engines.

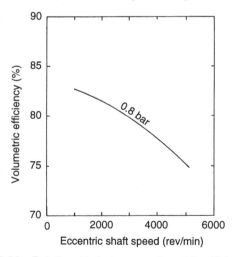

Fig. 6.38 Relationship between volumetric efficiency and eccentric shaft speed at constant boost pressure for a 'G'-Lader supercharger

6.7 Turbochargers

6.7.1 Introduction

A typical petrol engine may harness up to 30% of the energy contained in the fuel supplied to do useful work under optimum conditions but the remining 70% of this energy is lost in the following way:

7% heat energy to friction, pumping and dynamic movement
9% heat energy to surrounding air
16% heat energy to engine's coolant system
38% heat energy to outgoing exhaust gases

Thus, the vast majority of energy, for design reasons, is allowed to escape to the atmosphere through the exhaust system.

A turbocharger utilizes a portion of the energy contained in the exhaust gas—when it is released by the opening of the exhaust valve towards the end of the power stroke (something like 50° before BDC)—to drive a turbine wheel which simultaneously propels a centrifugal compressor wheel.

The turbocharger relies solely on extracting up to a third of the wasted energy passing out from the engine's cylinders to impart power to the turbine wheel and compressor wheel assembly. However, this does produce a penalty by increasing the manifold's back-pressure and so making it more difficult for each successive burnt charge to be expelled from the cylinders. It therefore impedes the clearing process in the cylinders during the exhaust strokes.

The ideal available energy which can be used to drive the turbocharger comes from the blow-down energy transfer which takes place when the exhaust valve opens and the gas expands down to atmospheric pressure (Fig. 6.39). This blow-down energy is represented by the loop area 4, 5 and 6 whereas the boost pressure energy used to fill the cylinder is represented by the rectangular area 0, 1, 6 and 7.

Turbocharged engines produce higher cylinder volumetric efficiencies compared with the normally aspirated induction systems. Therefore, there will be higher peak cylinder pressures which increase the mechanical loading of the engine components and could cause detonation in petrol engines. Therefore, it is usual to reduce the engine's compression ratio by a factor of one or

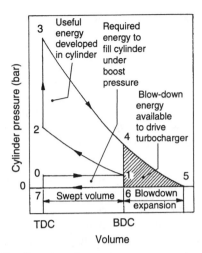

Fig. 6.39 Petrol engine cycle pressure–volume diagram showing available exhaust gas energy

two. Thus, a compression ratio of 10:1 for a normally aspirated engine would be derated to 9:1 for a low boost pressure or even reduced to 8:1 if a medium to high boost pressure is to be introduced. Similarly, for a direct injection diesel engine which might normally have a compression ratio of 16:1, when lightly turbocharged the compression ratio may be lowered to 15:1 and, if much higher supercharged inlet pressures are to be used, the compression ratio may have to be brought down to something like 14:1.

The compression of the charge entering the cells of the impellor depends upon the centrifugal force effect which increases with the square of the rotational speed of the impellor wheel. Consequently, under light load and low engine speed conditions the energy released with the exhaust gases will be relatively small and is therefore insufficient to drive the turbine assembly at very high speeds. Correspondingly, there will be very little extra boost pressure to make any marked improvement to the engine's torque and power output in the low-speed range of the engine. Thus, in effect, the turbocharged engine will operate with almost no boost pressure and with a reduced compression ratio compared with the equivalent naturally aspirated engine. Hence, in the very low speed range, the turbocharged engine may have torque and power outputs and fuel consumption values which are inferior to the unsupercharged engine.

Another inherent undesirable characteristic of turbochargers is that when the engine is suddenly accelerated there will be a small time delay before the extra energy discharged into the turbine hous-

ing volute can speed up the turbine wheel. Thus, during this transition period, there will be very little improvement in the cylinder filling process, and hence the rise in cylinder brake mean effective pressure will be rather sluggish.

6.7.2 Altitude compensation
(Fig. 6.40)

Engine power outputs are tested and rated at sea-level where the atmospheric air is most dense; however, as a vehicle climbs, its altitude is increased and the air becomes thinner, that is, less dense. The consequence is a decrease in volumetric efficiency as less air will be drawn into the cylinders per cycle, with a corresponding reduction in engine power since power is directly related to the actual mass of charge burnt in the cylinder's every power stroke. A naturally aspirated engine will have its power output reduced by approximately 13% if it is operated approximately 1000 m above sea-level. Supercharging the cylinders enables the engine's rated power above sea-level to be maintained or even exceeded.

With a turbocharged engine there will still be some power loss with the engine operating at high altitudes, but the loss will be far less than if the engine breathing depended only on natural aspiration. As can be seen (Fig. 6.40) at 1000 m the power loss is only 8% compared with the naturally aspirated engine where the power decrease is roughly 13%. The reason for the turbocharger's ability to compensate by raising its boost pressure is that the turbine speed increases

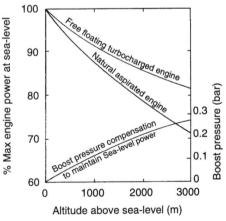

Fig. 6.40 Effect of altitude on rated engine power at sea level for both naturally-aspirated and turbocharged engines

directly with any increase in pressure difference between the exhaust gas entering the turbine and its exit pressure, which is the ambient air pressure. Therefore, as the altitude increases the air becomes thinner and the pressure drops, but the pressure in the exhaust manifold which impinges onto the turbine wheel remains substantially the same. The result is that the pressure differential across the turbine increases and therefore raises the turbine assembly's speed and, correspondingly, the compressor's boost pressure.

The necessary boost pressure required to maintain the engine's sea-level power, the loss of sea-level power rating, and the turbocharger's ability to compensate partially for the decrease in air density with increasing altitude driving, is compared in Fig. 6.40.

6.7.3   Description of turbocharger
(Fig. 6.41)

A turbocharger comprises three major components, an exhaust-gas driven turbine and housing, a centrifugal compressor wheel and housing and an interconnecting support spindle mounted on a pair of fully floating plain bearings, which are themselves encased in a central bearing housing made from nickel cast iron (Fig. 6.41).

Compressor
(Figs 6.41 and 6.42)
The impellor compressor wheel (Fig. 6.42) is an aluminium alloy casting which takes the form of a disc mounted on a hub with radial blades projecting from one side. This causes the air surrounding the compressor wheel to be divided up into a number of cells (something like 12 blades). The hub of the compressor-wheel onto which the blades are attached is so curved that air enters the cells formed between adjacent pairs of blades axially from the centre. The enclosed air is then divided by the passageway formed between the hub and the compressor-wheel housing internal wall so that the flow path moves through a right-angle causing the air to be expelled radially from the cells. Once the air reaches the periphery of the compressor-wheel it then passes to the parallel diffuser (gap) formed between the bearing housing and the compressor-wheel housing (Fig. 6.41). From the diffuser gap the air flows into a circular volute-shaped collector, which is a constant expansion passage from some starting point to its exit.

Turbine
(Figs 6.41 and 6.42)
The exhaust gas temperature at the inlet to the turbine wheel, under light-load to full-load high-speed operating conditions, may range between 600°C to 900°C. Consequently, the turbine is usually made from a high-temperature heat-resistant nickel-based alloy such as 'Inconel'. Ceramic materials such as sintered silicon nitride are being developed as an alternative to nickel based alloys, and these materials have the advantage of weighing less than half that of suitable metallic materials. These ceramic materials have a higher specific strength and a much lower coefficient of expansion than their counterparts, they are also capable of operating under the same full-load exhaust gas temperature conditions at similar maximum rotational speeds as the nickel-based alloys.

The turbine wheel (Fig. 6.42) takes the form of a hub supporting a disc at one end and a number of radial blades which project axially and radially from both the hub and disc respectively. The outer edges of the blades are curved backwards to trap the impinging exhaust gases.

Exhaust gases from the manifold enter the spherical graphite cast-iron turbine housing flange entrance, the gases then flow around either a single or twin volute passageway surrounding the turbine wheel (Fig. 6.41). The gases are then forced tangentially inwards from the throat of the turbine housing so that they impinge onto the blade faces. The flow path then directs the gases gradually through a right-angle so that they come out axially from the centre of the turbine hub, and they are then expelled into the exhaust pipe system.

In most turbocharger designs the turbine and spindle are joined together by some sort of welding process such as inert gas welding, resistance welding, or electron beam welding. The general trend these days is to attach the turbine to the spindle by the friction welding solid phase technique (Fig. 6.42). The turbine and spindle are brought together under load, with one part revolving against the other so that frictional heat is generated at the interface. When the joint area is sufficiently plastic as a result of the increase in temperature the rotation is stopped and the end force increased to forge and consolidate the metallic bonds. The surface films and inclusions that might interfere with the formation of these bonds are broken up by friction and removed from the weld area in a radial direction owing to

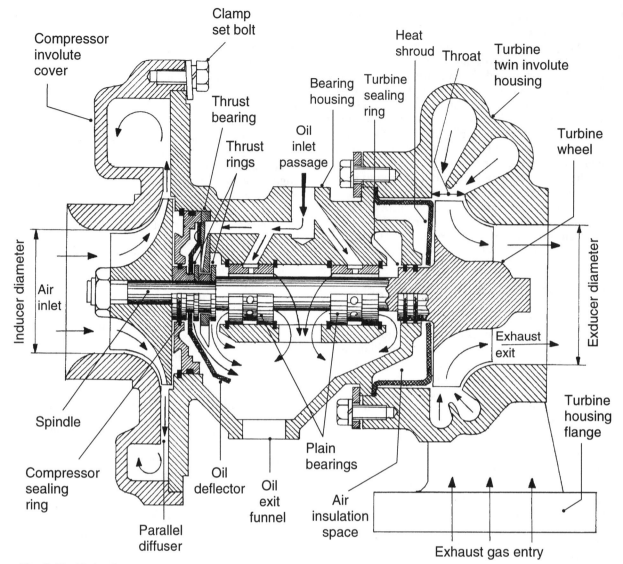

Fig. 6.41 Turbocharger construction

marked plastic deformation on the surfaces. The burrs surrounding the joint are then ground away leaving a high-quality joint.

Spindle and bearing assembly
(Figs 6.41 and 6.42)
The turbine wheel and steel spindle can be welded together in a vacuum by an electron beam. This joining process produces a neat joint in which the heat flow areas are held to a minimum so that there is very little distortion and machining is kept to the minimum. The medium carbon steel spindle is induction hardened where

the bearings contact the spindle. The hollow space between the turbine wheel and the spindle (Fig. 6.42 sectioned-view) prevents heat being transferred through the centre of the spindle. Thus, heat is carried away along the outer section of the spindle, which can readily be cooled by the lubricating oil. The spindle is supported on a pair of free-floating phosphorus bronze plain bearings and around the outside of each bearing shell are six radial holes and, depending upon design, there may be a circumferential groove machined to distribute the lubrication oil. The rotation of the spindle assembly in conjunction with the oil supply from the engine's lubrication system

318

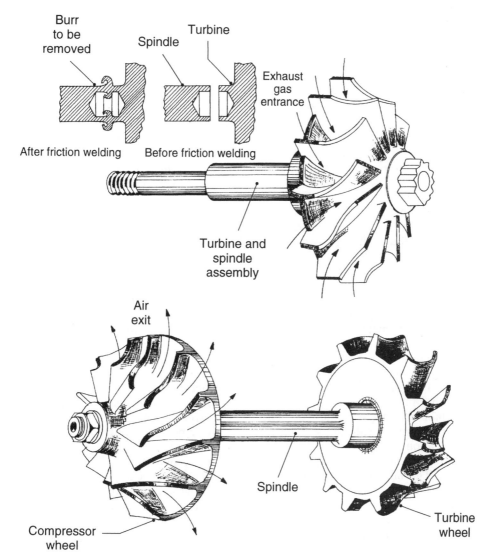

Fig. 6.42 Compressor and turbine wheel assembly

causes the viscous drag to rotate these bearings at approximately one-third of the spindle's rotational speed.

The large-diameter shouldered section next to the turbine-wheel is grooved to house two piston rings and a similar single piston ring is positioned on a grooved collar at the opposite end of the spindle next to the compressor wheel (Fig. 6.41). These piston-type rings remain stationary when the spindle rotates, their function being to prevent exhaust gas or compressed air entering the bearing housing chamber. Oil should be prevented from reaching the piston rings as any film, foam or splash entering the seal region will leak

out. The spindle bearing at the turbine wheel end is separated from the large diameter shouldered section of the spindle by a cast-in-oil drain slot cavity so that oil, spurting from the end of the bearing, spills into this cavity and then drains down to the oil exit funnel. In contrast, at the compressor wheel end, an oil deflector, fixed to the thrust bearing, protects the piston ring seal from contamination by shielding the bearing oil spray from the piston rings, the oil is then permitted to drain down to the bearing housing exit.

During operation, the exhaust gas impinging onto the turbine-wheel, and the compressed air reaction on the compressor wheel under varying speed and load conditions, produces a certain amount of end thrust as the spindle will want to move axially first in one direction and then the other.

Provision for end float or thrust control is provided by a thrust spacer (Fig. 6.41) which has a deep central groove. This thrust spacer is sandwiched between the stepped spindle shoulder and the piston-ring spacer. A phosphorus bronze thrust bearing plate, which is attached to the bearing housing, slips into the central groove. Consequently, the walls on each side of the thrust-plate become thrust-rings, whilst the reduced diameter portion between them forms a spacer sleeve, its width being critical to control any end movement of the compressor-wheel and the turbine wheel.

To minimize the heat transference from the turbine-wheel exhaust gas flow path to the bearing housing, a space is created between the turbine-wheel and the bearing housing. This air space is enclosed by a stainless-steel heat shroud pressing, shaped in the form of a cup, which is located immediately behind the turbine-wheel. This relatively large air gap provides an effective heat barrier, and therefore insulates the bearing housing from the hot turbine assembly.

Both radial plain bearings and the axial end thrust bearing are supplied with ample oil from the engine's lubrication system via drillings made in the bearing housing and in the bearings themselves (Fig. 6.41). The oil supply has two major functions: firstly, to lubricate the bearings so that a hydrodynamic oil film can be established so that, in effect, the shaft and bearings are floating on oil; and secondly, to remove excess heat from the bearing assembly. Thus, it is just as important to be able to return the circulating oil to the engine's sump as it is to flood the bearings in the first place with lubricant.

In some turbochargers, the bearing housing incorporates a liquid coolant jacket through which coolant from the engine's cooling system is made to circulate via a pair of flexible inlet and outlet pipes.

6.7.4 The operating principles of compressor and turbine

The operating principle of the compressor
(Fig. 6.43)
With the spindle assembly rotating, the air cells formed between adjacent blades sweep the en-

trapped air around the compressor housing curved wall: the air mass is therefore subjected to centrifugal force.

This force produces a radial outward motion to the air, with its velocity and, to some extent, its pressure becoming greater the further the air moves out from the centre of rotation (Fig. 6.43). The air thus moves through the diverging passages of the cells to the periphery where it is flung out with a high velocity. More air will, at the same time, be drawn into the inducer due to the forward curved blades at the entrance, and this tends to generate a slight depression. Hence, it encourages a continuous supply of fresh charge to enter the eye of the impellor.

Once leaving the outer rim of the impellor the tangential air movement relative to the impellor has its maximum kinetic energy, but, as it is pressure energy that is required, the air is expanded in the parallel diffuser so that its velocity sharply falls while, simultaneously, its pressure rises. In other words, the kinetic energy at the entrance to the parallel diffuser is partially converted into pressure energy by the time it arrives at the outer edge of the parallel annular-shaped passageway.

The air then leaving the diffuser is progressively collected in the volute, from some starting point where the circular passageway is at its smallest, to its exit where the passage is at its largest cross-section. The volute therefore prevents the air discharged from the diffuser becoming congested and, at the same time, it continues the diffusion process further; that is, the air movement is slowed down even more whereas its pressure still rises.

The operating principle of the turbine
(Fig. 6.43)
Exhaust gas from the engine's cylinders is expelled via the exhaust manifold into the turbine volute circular decreasing cross-section passageway, at a very high velocity, where it is directed tangentially inwards through the throat of the turbine housing. The released gas kinetic energy impinges on the turbine-blades, thereby imparting energy to the turbine-wheel as it passes through the cells formed between adjacent blades with a corresponding decrease in both gas velocity and pressure (Fig. 6.43). The exhaust gas with its rapidly decreasing energy moves radially inwards and, at the same time, its flow path moves through a right-angle so that it passes axially along the hub before leaving the turbine housing.

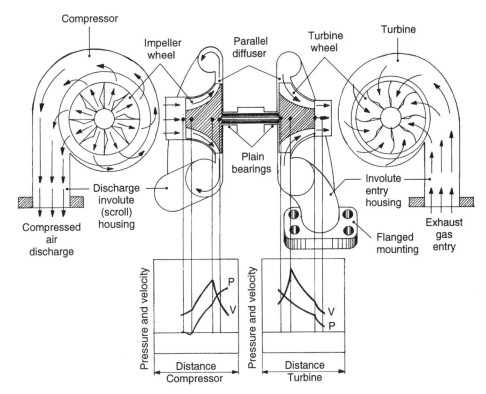

Fig. 6.43 Turbocharger principle

The expansion of the gas ejected from the turbine-wheel then produces a sudden drop in its velocity and pressure as it enters the silencer pipe system. Turbine speed and boost pressure are largely dependent upon the amount of energy contained in the hot, highly mobile exhaust gases and on the rate of energy transference from the gas to the turbine-blades. Thus, at idle speed very little fuel is supplied to the engine, and therefore the energy content in the outgoing exhaust gas will also be very low whereas, with increased engine speed and load conditions, considerably more fuel is consumed by the engine, which in turn releases proportionally more energy to the escaping exhaust gases. Hence, at light load and low speed, the turbine assembly speed can be something like 30000 to 50000 rev/min, whereas at high speed and high load operating conditions the spindle and wheel assembly can revolve at speeds up to 120000 to 150000 rev/min, depending upon design and application.

6.7.5 Compressor impellor and housing design

Compressor and housing arrangements
Closed or shrouded impellor with scroll diffuser (Fig. 6.44(a)). The impellor may be closed or shrouded; that is, the impellor is cast so that the cells or channels are completely enclosed. This construction eliminates direct leakage as the induced air is flung radially outwards in the cells. However, it is difficult to cast radial cells so that they curve backwards and also provide an axial angled inlet at the eye of the impellor. Other important disadvantages which must be considered are that the mass of the shroud is supported by the blades, such that at high rotational speeds the blades are subjected to severe centrifugal stresses. In addition, the shroud which is away from the central hub raises the impellor wheel's moment of inertia and thus impedes its ability to accelerate or decelerate rapidly. This design has been succeeded by the open-cell type impellor.

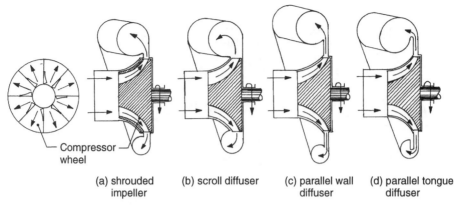

Compressor
wheel

(a) shrouded
impeller

(b) scroll diffuser

(c) parallel wall
diffuser

(d) parallel tongue
diffuser

Fig. 6.44 Vaneless diffuser compressors

Open impellor with scroll diffuser (Fig. 6.44(b)). With open impeller and scroll diffuser, the impeller is cast with blades forming the walls of the cells, these blades can be shaped so as to provide the best inducement for the incoming air and the radial flow path can be curved backwards to optimize the flow discharge at high rotational speeds. However, there is a clearance between the outer edges of the blades and the internal curved walls of the housing which encloses the rotating cells. This gap, small as it is, will be responsible for leakage losses under high boost pressure operating conditions. With this arrangement, the kinetic energy of the air at the blade tips is converted into pressure energy by directly entering into the relatively large scroll volume. In other words, the air flung out at the rim of the impellor enters the scroll where it is diffused by the relatively large mass of air already occupying the circular passageway. Thus, the intermingling of the discharged air causes its velocity to decrease and its pressure to increase, which goes someway towards producing the desired flow conditions for the charge.

Open impellor with parallel wall diffuser (Fig. 6.44(c)). If a more positive method of converting the kinetic energy to pressure energy is required, a parallel annular space between the impeller and the volute or scroll will enlarge the circular passage from the entrance at the impellor rim to where it merges with the discharge volute. Thus, as the air moves outwards in a semi-spiral and radial direction the air will expand, thus causing its speed to reduce while its pressure rises.

Open impellor with parallel tongue diffuser (Fig. 6.44(d)). To reduce the maximum diameter of the volute or scroll housing, the circular volute passageway can be cast to one side of the parallel diffuser with a much reduced diameter. Looking at the sectional view of the compressor, the wall between the diffuser and volute resembles a tongue; hence its name—parallel tongue diffuser. This, in effect, produces a right-angled flow-path similar to the normal parallel wall diffuser design but the overall dimensions of the housing will be more compact.

Compressor diffusers
(Figs 6.45 and 6.46)
The object of a compressor diffuser is to convert the air's kinetic energy produced at the rim of the impellor, to pressure energy by expanding it so that its velocity falls, thereby causing its pressure to rise.

The are three types of diffusers:

a) the scroll diffuser
b) the vane ring diffuser
c) the vaneless parallel wall diffuser

The scroll-type diffuser (Fig. 6.44(b)). The scroll or volute is the circular passageway of varying cross-sectional area which directly surrounds the impellor rim. Thus, when the air flowing through the impellor radial cells reaches the periphery of the blades it will have attained a very high velocity. The air will then be flung in a spiral flow path directly into the scroll, and the much larger circular passageway then expands the air so that it reduces speed and, at the same time, raises pressure. This form of diffuser does convert

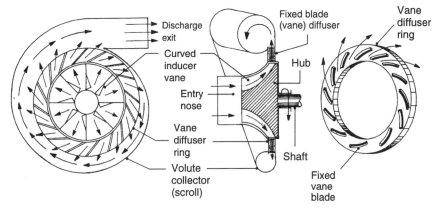

Fig. 6.45 Vane-diffuser compressor

velocity energy into pressure energy but at a relatively low rate, therefore more effective methods of diffusing the air charge are normally selected.

Vane-ring type diffuser (Fig. 6.45). This diffuser consists of an annular ring with vanes positioned tangentially around one side. The diffuser ring with its diverging multi-passageways joins the impellor cell periphery outlets to the circular, variable cross-section, volute collector. The vanes are so positioned that they guide the air discharge from the impellor rim to the volute in a tangential direction through passages of increasing cross-section. Thus, since the energy contained by the air cannot be destroyed, the effect of the expanding passages will be to slow down the air movement; therefore, if energy is to be retained the air pressure will rise.

Vaneless parallel-wall type diffuser (Fig. 6.46). Vaneless diffusers are parallel annular-shaped passageways which connect the impellor cell rim exits to the circular snail-shell shaped volute outer passageway.

The air enters the diffuser at radius R_1, through a relatively small cross-sectional area A_1, and is discharged at radius R_2 through a proportionally larger cross-section A_2. Thus, by similar triangles

$$\frac{A_1}{R_1} = \frac{A_2}{R_2}$$

therefore

$$A_1 = A_2 \frac{R_1}{R_2}$$

Hence, if R_1 is half that of R_2, then A_1 will be half the cross-sectional area of A_2 and vice versa.

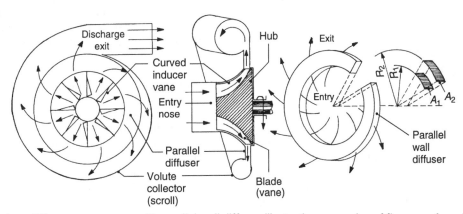

Fig. 6.46 Vaneless diffuser compressor with parallel wall diffuser illustrating expansion of flow area from inlet to exit

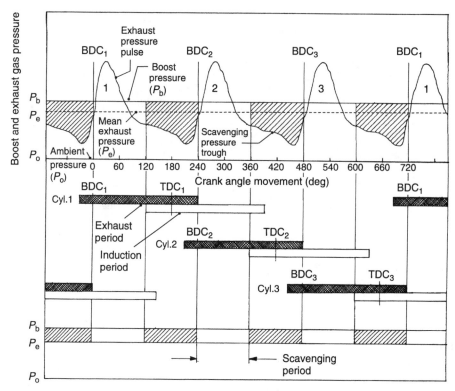

Fig. 6.47 Exhaust gas pressure variation in activated six-cylinder turbocharged engine manifold

Accordingly, the air leaving the impellor and passing through the parallel diffuser will reduce its speed in proportion to the increase in the annular passageway cross-sectional area. In contrast, the air discharge pressure rises.

Parallel-wall diffusers have a broad operating speed range over which a moderate compressor efficiency is maintained, whereas the vane ring diffuser operates the compressor at a fairly high efficiency but over a much narrower speed range.

6.7.6 Exhaust gas control and turbine housing design

Pulsed exhaust discharge
(Fig. 6.47)
It is important for effective cylinder scavenging that pulsed exhaust gas energy is introduced to the turbine wheel, in contrast to a damped steady flow of exhaust gas. With a four-cylinder engine, single-exhaust manifold, this is possible as there is an exhaust discharge every 180° so that there is very little exhaust gas interference between cylinders.

However, with more than four cylinders, exhaust gas will discharge at shorter intervals than the 180°; that is, for 5, 6 and 8-cylinder engines the intervals between exhaust discharges will be 144°, 120° and 90° respectively. To overcome exhaust gas interference in the manifold, manifolds are sub-divided so that, in the case of an in-line six-cylinder engine, cylinders 1, 2 and 3 are grouped together and, similarly, cylinders 4, 5 and 6 are grouped together so that there is now an extensive exhaust interval between sub-divided manifolds of 240°. The exhaust discharge from each half-manifold is then fed to the turbine-wheel through two separate passageways. If the exhaust gas in the branch pipes is permitted to discharge in the form of a pulse (Fig. 6.47) the initial blow-down from the open exhaust valve port will produce a rapid pressure rise until it peaks. The exhaust pressure then quickly decreases to a minimum value before the next cylinder, sharing the same common manifold gallery, discharges another lot of exhaust gas. This cycle of events will therefore be continuously repeating. By reducing or even eliminating intercylinder exhaust gas interference, by sub-dividing

324

the manifold if need be, the exhaust pressure in the manifold will fall towards the end of the exhaust stroke to a value below the mean compressor pressure (Fig. 6.47). Thus, during the valve overlap near the end of the exhaust period and at the beginning of the inlet period, a positive pressure difference will exist between the cylinder intake and the cylinder exit, which will cause a blow-through of fresh charge from the intake manifold to the exhaust manifold. If there is sufficient pressure difference the fresh charge will rush into the cylinder and push out the residual exhaust gases still remaining in the unswept combustion chamber space. The effectiveness of this scavenging action will also depend on the engine speed and the actual valve opening area during the time of valve overlap.

Divided turbine housing passageways

The turbine housing for a four-cylinder engine normally has a single volute circular passageway into which the four merged branch pipes discharge their individual exhaust gas pulses at intervals of 180° through a 360° throat entrance to the turbine wheel.

When there are more than four cylinders, better turbine response is obtained by dividing the exhaust manifold into two halves. The discharge from each half manifold is then fed to the turbine wheel through two separate passageways formed in the turbine housing. Alternating exhaust gas pulses from each group of branches discharge at relatively prolonged intervals through the throat of the turbine housing in a pulsed jet stream against the turbine-blades. This pulse impingement of the exhaust gas from individual branch pipes is more effective than the resultant steady discharge from several cylinders.

There are two kinds of divided turbine housings, the double-flow 180° divided turbine housing and the twin-flow axially divided turbine housing.

Double flow 180° divided turbine housing
(Fig. 6.48)

In the double-flow 180° circular passageway, the throats are separated, and each passage feeds only one half of the turbine-wheel circumference. With this method of exhaust gas delivery, the pulsed gas will improve the turbine speed response at very low engine speeds. However, with this arrangement the gas does have a tendency to reverse its flow due to the centrifugal force exceeding the inward pulse thrust when there are

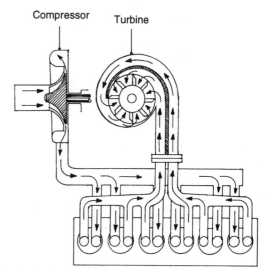

Fig. 6.48 Double-flow 180° divided turbo-housing

only low pressure pulses in the housing passageways.

Twin flow axially divided turbine housing
(Fig. 6.49)

In the twin-flow axially divided turbine housing each of the two volute passages commences feeding into a common throat which completely surrounds the circumference of the turbine wheel. With this configuration, exhaust gas is pulsed inwards alternating from each volute passage through the 360° throat which discharges directly

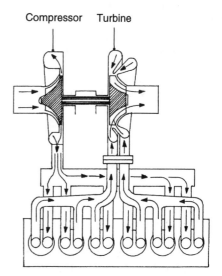

Fig. 6.49 Twin-flow (twin-scroll) axially divided turbine housing

325

onto the turbine-wheel. This method of discharging the gas onto the turbine-wheel is relatively effective in preventing the gas pulses reversing when the pressure in the volutes is low. This design is therefore suitable for engines where high torque is desired at low engine speeds.

6.7.7 Transitional response time (turbocharger lag)
(Figs 6.50, 6.51 and 6.52)

Transient response time is dependent upon the inertia of the rotating parts and the efficient projection of the exhaust gas onto the turbine blades.

Immediately the engine throttle is opened there will be an increased flow of mixture entering the cylinders with a corresponding exit of exhaust gas, which is directed onto the turbine-blades causing the wheel assembly to accelerate rapidly.

The time taken for the turbine and compressor assembly to attain the maximum operating speed is dominated by the overall efficiency of the turbocharger and the polar moment of inertia of the rotating assembly.

The polar moment of inertia I is the reluctance of the rotating body to change speed, which may be represented by

$$I = mk^2 \ (\text{kg m}^2) \tag{5}$$

where k is the radius of gyration in metres, m is the mass of the rotating assembly in kg and I is the polar moment of inertia. In this context the radius of gyration is the distance from the rotating axis to a point where all the mass may be assumed to be concentrated.

The torque T required to accelerate the rotating body is given by the following

$$T = I\alpha \ (\text{N m}) \tag{6}$$

thus

$$\alpha = \frac{T}{I} \ (\text{rad/s}^2) \tag{7}$$

where I is the polar moment of inertia (kg m^2) and α is the angular acceleration of the shaft in rad/s^2.

The acceleration of the turbine assembly is thus inversely proportional to the rotating inertia so that halving the polar moment of inertia will double the acceleration and vice versa.

Example
Compare the moment of inertia for two different-sized turbine-wheels from the following data.
a) For a large turbine—diameter 75 mm, mass 1.28 kg and radius of gyration of 20 mm.
b) For a small turbine—diameter 50 mm, mass 1.0 kg and radius of gyration of 16 mm.

a) $I_{75} = mk^2 = 1.28(20 \times 10^{-3})^2$
$\qquad = 1.28 \times 400 \times 10^{-6}$
$\qquad = 512 \times 10^{-6} \ (\text{kg m}^2)$
b) $I_{50} = mk^2 = 1.0(16 \times 10^{-3})^2$
$\qquad = 1.0(16 \times 10^{-3})^2$
$\qquad = 1.0 \times 256 \times 10^{-6}$
$\qquad = 256 \times 10^6 \ \text{kgm}^2$

Relative polar moment of inertia =

$$= \frac{512 \times 10^{-6}}{256 \times 10^{-6}} = 2 : 1$$

This comparison illustrates that only a small reduction in turbine diameter considerably reduces the polar moment of inertia and, in this example, reducing the turbine diameter by one third, halves the polar moment of inertia for the turbine. Thus, using two small turbochargers instead of one large unit for a V-cylinder banked engine goes someway towards reducing transient response time (turbocharger lag).

Turbocharger lag will depend to some extent upon the excess torque available from the turbine-wheel over that required to drive the compressor with the air flow and boost pressure existing at that instant. Therefore, a small turbine volute housing attached directly to a short and small diameter passageway (minimum volume) manifold is desirable as this will provide an undamped exhaust gas pulse directly and effectively to the turbine blades, thereby producing the least time lag.

The rotational speed characteristics of the turbine and compressor assembly relative to the engine speed for wide-open throttle operation and for normal road load driving conditions are shown in Fig. 6.50. It can be seen that with wide-open throttle (maximum acceleration) the extra exhaust gas energy released to the turbine housing produces a much earlier rise in turbo speed than for road load throttle operating conditions where the throttle opening will be, say, between a third and two-thirds less. Note the kink at about 3000 rev/min on the wide-open throttle curve and then the reduced rate of turbo speed increase as the wastegate begins to open and bypasses a portion of the exhaust gas from the

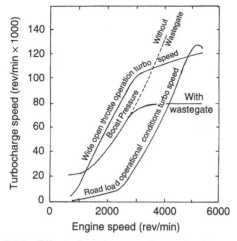

Fig. 6.50 Effect of engine speed on turbocharger spin speed and boost-pressure for wide-open throttle and road load operating conditions

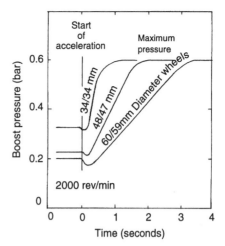

Fig. 6.51 Effect of turbine and compressor wheel size on acceleration response time to reach the designed maximum boost pressure

turbine blades. In the case of the road load throttle opening curve, the turbo speed does not commence to rise until the engine's speed has increased to about 2500 rev/min, it then rises continuously and only bends over as the engine speed approaches its maximum, indicating that the wastegate only is then required to open.

The maximum turbocharger rotational speeds obtainable are related to the turbine and compressor wheel sizes—small diameter wheels are able and do operate effectively at much higher rotational speeds than do large diameter wheels.

Likewise, small turbine and compressor wheel combinations are capable of much higher acceleration rates due to their low inertia compared with larger wheels. Consequently, small-diameter wheels can reach their maximum boost pressure before the wastegate opens much earlier than larger wheels. These facts are demonstrated in Fig. 6.51. Here, three different-sized turbochargers have been tested from a starting engine speed of 2000 rev/min and their boost pressure-rise against accelerating time has been recorded. As can be seen, the large turbocharger takes between 3.0–3.5 seconds to reach full boost pressure from 2000 rev/min, the medium-sized turbo takes only 1.5–2.0 seconds, whereas the small turbo only needs 0.5–1.0 seconds to attain its maximum boost pressure. These large, medium and small turbine and compressor wheel diameters have maximum rotational speeds respectively as follows: 60/59 mm–150 000 rev/min, 48/47–180 000 rev/min, and 34/34–270 000 rev/min.

However, the reduced transitional response time with the smaller-sized wheels is partly offset by the reduction in efficiency caused by the proportional increase of leakage past the wheel as its diameter decreases, and also due to the increase in exhaust gas back pressure imposed by the reduced flow path created between the turbine blades.

Improvements in low-speed boost pressure by using a smaller and faster turbocharger can raise the torque and power developed in the engine's lower speed range and, at the same time, can reduce the specific fuel consumption (see Fig. 6.52).

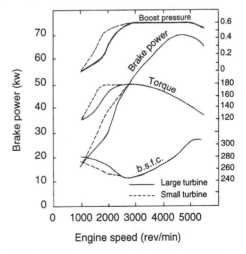

Fig. 6.52 Effect of reducing diameter of turbine-wheel on engine low-speed performance

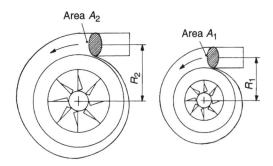

Fig. 6.53 Comparative A/R ratios for vaneless turbine housings

6.7.8 *A/R* ratio
(Fig. 6.53)

The speed and acceleration of the turbine and compressor wheel assembly is influenced by a number of factors, but one of the most critical and important controlling parameters is the *A/R* ratio.

The *A/R* ratio (Fig. 6.53) is the smallest cross-sectional area (CSA) of the intake passages in the turbine housing before the flow path spreads around the circumferential throat leading to the turbine-wheel divided by the distance from the centre of the turbine-wheel to the centroid of area '*A*' i.e.

$$A/R \text{ ratio} = \frac{\text{smallest CSA of passage leading to volute}}{\text{distance between CSA centroid and centre of shaft}}$$

Some turbocharger manufacturers choose not to use the *A/R* ratio as a design parameter and instead only quote the intake passage cross-sectional area at its smallest point just before the gas enters the volute surrounding the turbine wheel.

A large *A/R* ratio reduces the turbine spin speed for a given exhaust gas flow, conversely a small *A/R* ratio raises the turbine-wheel spin speed for a similar exhaust gas delivery. *A/R* ratio values tend to range between 0.3 and 1.0.

A large intake passage radius '*R*' will slow down the turbine-wheel just like a large intake passage cross-sectional area '*A*'. A small *A/R* ratio will speed up the turbine-wheel for a given engine speed and throttle opening, whereas a large *A/R* ratio will slow it down under the same operating conditions.

For example, if a turbine housing has an *A/R* ratio of 0.7 and an earlier boost is demanded, then a smaller turbine housing *A/R* ratio of something like 0.6 or 0.5 should be tried. However, if a slightly later boost is required then a turbine housing with a larger *A/R* ratio, say 0.8 or 0.9, should be fitted.

6.7.9 Carburettor location

The carburettor may be located either upstream on the intake side of the compressor, in which case the air is drawn (sucked) through the carburettor venturi, or it may be positioned downstream on the outlet side of the compressor, and here the air is forced (blown) through the carburettor venturi. The merits and limitation of both arrangements are compared as follows.

Carburettor positioned upstream of compressor (suck through)
(Fig. 6.54(a))

Advantages

a) The carburettor operates at ambient pressure and therefore can still maintain the standard fuel pump system
b) The carburettor operates under normal temperature conditions
c) The carburettor does not require any modification except a matching of the jet sizes to cope with the greater volume of air flow
d) The carburettor tuning is easier than for downstream mounted carburettors
e) Charge mixture distribution under most driving conditions is generally good

Disadvantages

a) The turbocharger must have a special seal to avoid drawing oil into the compressor housing during part throttle high vacuum operating conditions
b) The longer intake flow path and the vacuum before the compressor may cause the fuel to condense on the cool manifold walls, which can upset the mixture distribution, particularly at low operating speed conditions
c) Under certain operating conditions the throttle response may be impaired due to fuel wetting of the extended intake passage walls
d) Segregation of liquid fuel from the air charge under transient driving conditions can occur in

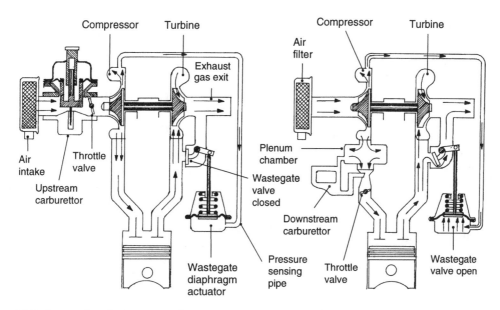

(a) Carburettor installed upstream of compressor

(b) Carburettor installed downstream of compressor

Fig. 6.54 Turbocharged engine layout with carburettor position either upstream or downstream of the compressor

the compressor volute housing, which may cause the air–fuel mixture ratio to be unstable

e) The cooling effect of the liquid fuel is transferred to the air stream when it is still cool and not when it has been compressed and heated, where it would provide some degree of intercooling

f) The greater pressure differential across the compressor, with the carburettor's venturi ahead of the compressor, will increase the temperature of the compressed charge

Carburettor positioned downstream of compressor (blow through)
(Fig. 6.54(b))

Advantages

a) The danger of high vacuum levels in the compressor housing under part throttle conditions is greatly reduced so that the need for special seals on the compressor end of the spindle is avoided

b) The carburettor can remain in the same position as for the standard naturally aspirated engine layout

c) Discharging liquid fuel on the heated output side of the compressor provides good air–fuel mixture distribution and fuel vaporization

under cold-start and warm-up operating conditions

d) Discharging liquid fuel into the compressed and heated air charge provides a certain amount of intercooling of the charge before it enters the cylinder

e) With the carburettor on the downstream side of the compressor the pressure difference across it will be smaller, which will marginally lower the output temperature of the discharged mixture

Disadvantages

a) The carburettor will be subjected to the pressurized air charge and will therefore have to be sealed against the atmosphere during operating conditions

b) The fuel supply system will be more complex since it must be able to cope with the severe fluctuation in float chamber pressure

c) The air–fuel mixture ratio adjustment may find it difficult to keep up with the constantly changing air density on the output side of the compressor

d) The carburettor will be subjected to the heat of the compressed charge as it flows through the venturi. Therefore, means must be provided to cool the carburettor assembly.

329

6.8 Boost pressure control

6.8.1 The need for exhaust gas bypass valve (wastegate) control
(Figs 6.54 and 6.57)

The turbocharger has to be prevented from over-speeding and overheating as this can have two disastrous consequences: firstly, excessively high compressor and turbine-wheel rotational speeds, when subjected to high operating exhaust gas temperatures, can very quickly destroy the re-volving components; and secondly, excessively high boost pressure will produce a correspond-ingly high cylinder pressure and temperature over a period of time, which can do considerable damage to the various rotating and reciprocating components of the engine and, in the case of a petrol engine, will certainly promote cylinder detonation during acceleration conditions.

To safeguard the turbocharger from overspeed-ing and overheating, a portion of the exhaust gas expelled from the cylinders under high engine-load and/or speed operating conditions is deliber-ately made to bypass the turbine housing and instead flow directly to the exhaust pipe. Under extreme operating conditions, something like 30% to 40% of the exhaust gas can be diverted away from the turbine throat with the effect that the turbine will not increase its speed and the output boost pressure will remain approximately constant with any further rise in engine speed.

The exhaust gas bypass passage opening is controlled by a wastegate in the form of either a poppet-type valve (Fig. 6.57) or a swinging-flap type valve (Fig. 6.54). Both types of wastegate valves are normally operated by a diaphragm actuator controlled by either the boost pressure from the volute impellor housing or by the ex-haust manifold gas pressure.

With the poppet-type valve wastegate the long stem of the valve is connected directly to the diaphragm actuator, this stem is usually enclosed in a finned housing to improve the heat dissipa-tion from the valve and actuator assembly. Con-versely, the swinging-flap type wastegate is oper-ated by a short external lever which is linked to the diaphragm actuator by a long push-rod, so that the actuator is practically insulated from the exhaust gas heat.

The wastegate and bypass passageways for small turbochargers can be integral with the tur-bine-wheel housing or, for the larger turbochar-gers, the wastegate unit and the bypass passages can be mounted separately away from the turbine wheel housing.

6.8.2 Turbine-wheel size and wastegate control
(Figs 6.55 and 6.56)

The boost pressure characteristics of a centrifugal compressor show that there is no noticeable pressure rise until the engine speed has risen to approximately a quarter of its maximum speed (Fig. 6.55). With a further increase in engine speed, if a large turbine-wheel is used, the boost pressure will rise steadily and then, as maximum engine speed is approached, the rise in boost pressure will be at a much reduced rate until the predetermined safe maximum boost pressure coincides with the maximum engine speed. However, if a small turbine is used, the rise in boost pressure will tend to commence at about the same engine speed, but its rise will be much steeper, and it will increase to a much higher pressure due to the lighter and smaller turbine-wheel's ability to increase its speed in a shorter time and for the wheel to reach a much higher spin speed.

Without a wastegate to limit the exhaust gas energy going to the turbine housing, the small turbine-wheel spin-speed would rise beyond the safe maximum burst speed of the rotating assem-bly and therefore could not be utilized. If, how-ever, a portion of the exhaust gas bypasses the turbine housing when the desired boost pressure is reached then the small turbine-wheel will de-velop a positive and potent boost pressure much earlier than a large turbine-wheel (Fig. 6.56). Thus, once the desired boost pressure has been obtained the wastegate is made to open, and the bypass passage will therefore divert a sufficient quantity of the discharged exhaust gas from the turbine housing so that the boost pressure re-mains fairly constant throughout the upper speed range of the engine.

6.8.3 Boost-pressure controlled wastegate
(Figs 6.54(a and b) and 6.57(a and b))

With the boost pressure sensing method of con-trol, a rubber pipe—connected between the com-

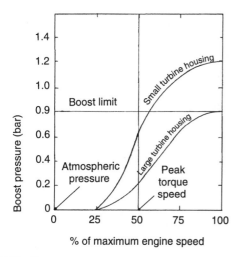

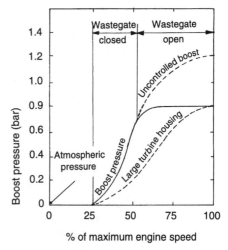

Fig. 6.55 Boost pressure wastegate control characteristics

Fig. 6.56 Wastegate and blow-off valve control turbocharged engine system

pressor volute housing and the wastegate actuator—relays a pressure signal to the actuator diaphragm. Under normal steady driving conditions the wastegate valve is closed (Figs 6.54(a) and 6.57(a)), however, when this pressure reaches the predetermined value (say 1.6 bar) the charge pressure acting on the diaphragm's cross-sectional-area will be sufficient to force back the return-spring. If a poppet-type wastegate valve is used (Fig. 6.57(b)) it will be pushed into the open position permitting exhaust gas to bypass the turbine housing. Alternatively, the swing-flap

type wastegate valve opens (Fig. 6.54(b)) due to the push-rod thrust twisting the external lever attached to the wastegate pivot.

Wastegate assembly air cooling system (Fig. 6.58)

With an integral turbine housing and poppet-valve type wastegate there is a tendency for the poppet-valve and diaphragm actuator assembly to overheat. One way of overcoming this problem is to circulate a portion of the boost air supplied to

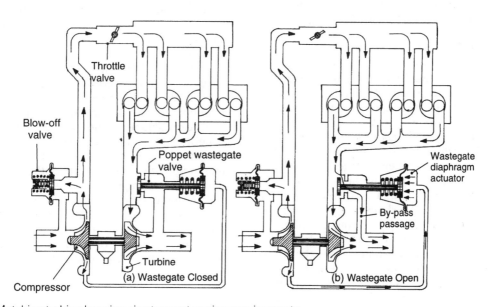

Fig. 6.57 Matching turbine housing size to meet engine requirements

331

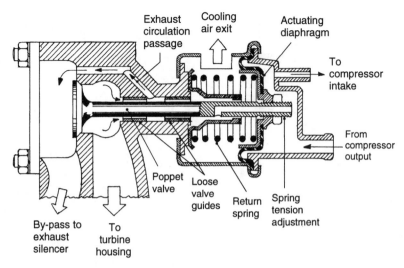

Fig. 6.58 Poppet-valve type integral wastegate with cooling air circulation and boost pressure actuated

the wastegate actuator diaphragm chamber back to the intake side of the compressor. At the same time some of the compressed air charge enters the central drilling in the valve-stem and comes out on one side in a funnel-shaped chamber surrounding the outer exposed end of the valve-guide (Fig. 6.58). Compressed air enters the relatively large clearance space formed between the valve-stem and guide so that it provides a cooling influence to the valve-stem assembly, the bulk of the air, however, will flow from the mouth of the funnel chamber into the outer spring chamber where it then exhausts into the atmosphere.

Between the split valve bush guide is a passageway drilling leading to the wastegate gas bypass exit (Fig. 6.58). Thus, exhaust gas surrounding the underside of the poppet-valve head will flow through the relatively large clearance made between the valve-stem and the inner valve-guide bush, where it meets the much cooler boost air flowing through from the opposite end. The merged gas and air is then expelled through the exhaust circulating drillings to the gas bypass exit passageway which will be at a much lower pressure. By these means exhaust gas is prevented from getting through to the spring chamber via the clearance space formed between the valve-guide bush and the valve-stem so that the whole unit is kept relatively cool. With the loose fitting valve-stem and guide assembly there is very little tendency for the valve to stick while in service.

6.8.4 Exhaust back-pressure and vacuum controlled wastegate
(Figs 6.59, 6.60 and 6.61(a–c))

This method of opening and closing the wastegate provides a steep rise in boost pressure to approximately 1.8 bar at 2500 rev/min followed by a steady decline of boost down to roughly 1.6 bar at 5000 rev/min, after which the charge pressure remains very nearly constant (Fig. 6.61).

The initial wastegate opening under mid-speed part throttle conditions is obtained by a steel pipe relaying gas pressure from the exhaust manifold to the working diaphragm chamber side of the wastegate actuator (Fig. 6.59).

Towards wide-open throttle mid-to-high engine speed operating conditions, a rubber pipe connects the inlet suction side of the compressor impellor to the protection diaphragm of the wastegate actuator, and this conveys a vacuum, generated at high turbocharger spin speeds, from the inlet side of the impellor to the protection diaphragm chamber on the other side of the working diaphragm, to assist the exhaust gas thrust on the working diaphragm in order to open the wastegate poppet-valve even further. The reducing boost pressure with rising engine speed provides a means of minimizing the pinking tendency within the engine's cylinders as its speed under boost pressure continues to increase.

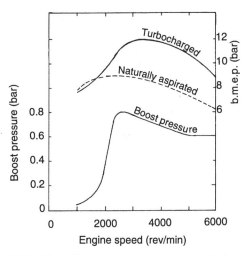

Fig. 6.59 The effects of turbocharging on the brake mean effective pressure characteristics at different engine speeds

Stages in wastegate operation
(Fig. 6.61(a–c))
During idle and low part throttle engine speeds the turbocharger spin speed may range between 5000–20 000 rev/min, these compressor wheel speeds are much too low in producing any pressure changes and, therefore, the fresh charge will be drawn into the induction manifold via the spaces between the impellor blades just like a naturally aspirated engine.

With increased engine speed and slightly wider part throttle operation, the increased exhaust gas energy released onto the turbine-wheel rapidly raises its spin speed to something like 30 000–50 000 rev/min. The higher compressor spin speed now commences to supply pressure, thus increasing the engine cylinder pressure and, accordingly, the engine's torque and power output (Fig. 6.61(a)).

As the engine speed approaches 2500 rev/min the boost pressure rises to about 1.8 bar, and at the same time the exhaust gas back-pressure thrust onto the working diaphragm will be sufficient to pull the poppet-valve wastegate (Fig. 6.61(b)) partially open, thereby allowing a portion of the exhaust manifold gas to bypass the turbine-wheel housing. This will regulate the amount of exhaust gas energy reaching the turbine so that the turbine-wheel maximum spin speed will automatically be restricted to roughly 100 000–120 000 rev/min and, correspondingly, the boost pressure will be limited to something of the order of 1.8 bar.

With still higher engine speeds and wider throttle openings the vacuum on the inlet side of the compressor wheel progressively rises so that the actuator will be subjected to a high exhaust gas back-pressure on the working diaphragm side and a relatively high vacuum on the opposite protection diaphragm side (Fig. 6.61(c)). The additive effect of both gas pressure and compress-

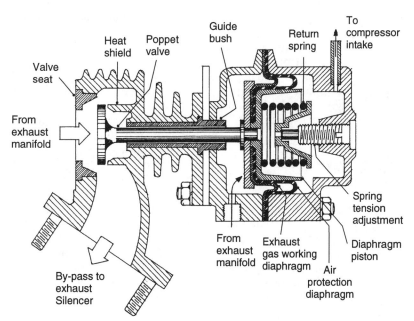

Fig. 6.60 Poppet-valve type detached wastegate unit exhaust pressure/air intake pressure actuated

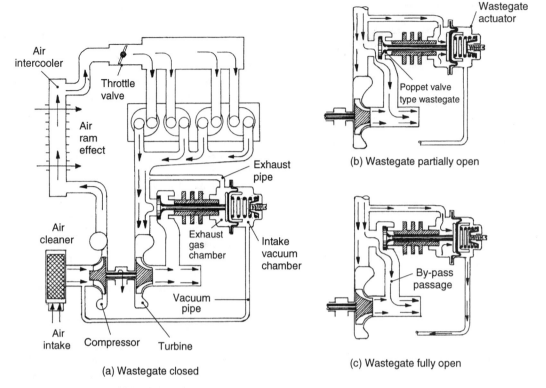

Air
intercooler

Throttle
valve

Air
ram
effect

Air
cleaner

Exhaust
pipe

Exhaust
gas
chamber

Intake
vacuum
chamber

Air
intake

Compressor

Turbine

Vacuum
pipe

(a) Wastegate closed

Wastegate
actuator

Poppet valve
type wastegate

(b) Wastegate partially open

By-pass
passage

(c) Wastegate fully open

Fig. 6.61 Exhaust back-pressure and vacuum controlled wastegate system

or inlet vacuum will steadily move the poppet-valve fully away from its seat until it butts against the annular heat shield which protects the valve neck from overheating.

The consequence of this additional opening of the poppet-valve is to reduce the turbine spin speed and, subsequently, the boost pressure will also decrease from its peak value of 1.8 bar to 1.6 bar at an engine speed of about 5000 rev/min (Fig. 6.60). Beyond this engine speed, the boost pressure is then seen to remain constant.

6.8.5 Compressor blow-off valve
(Figs 6.62 and 6.63(a and b))

Boost pressure can be controlled by blowing off either surplus exhaust gas from the turbine inlet through a wastegate valve or surplus air from the compressor delivery via a blow-off valve.

Blowing off surplus air from the compressor discharge results in higher turbocharger speeds than the exhaust wastegate method. This is because the compressed air delivery load is reduced, but there will be very little change in the amount

of gas energy passing through the turbine-wheel; consequently, the excess energy input to the turbine will raise the rotor assembly spin speed to a higher value. Since a portion of the compressed air delivery is discharged back into the atmosphere, and energy has been spent in driving the turbine-wheel, it follows that there will be some reduction in the engine's thermal efficiency during this period when the compressed air is blowing off.

Thus, because of the turbocharger's relatively low overall efficiency during the compressor discharge blow-off period (which may be prolonged under certain driving conditions) the wastegate method of diverting exhaust gas away from the turbine-wheel has been universally adopted.

However, the blow-off valve's simplicity has encouraged some engine manufacturers to incorporate this form of pressure-relief valve between the compressor and the inlet manifold (Fig. 6.62) as a secondary means of limiting boost pressure in the event of its build-up rate exceeding the wastegate's ability to divert sufficient gas energy from the turbine-wheel.

334

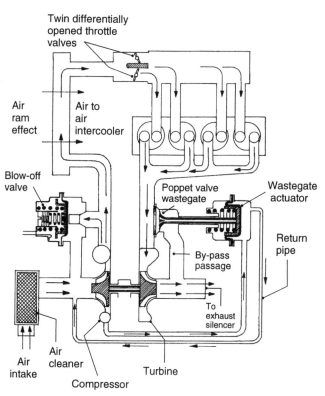

Fig. 6.62 Wastegate and blow-off valve boost pressure controlled

Stages in wastegate and blow-off valve operation

(Figs 6.62 and 6.63(a and b))

With increased engine speed and load the boost pressure will continue to rise until the predetermined maximum discharge pressure, say 0.7 bar is reached, at which point the wastegate will commence to open while the blow-off valve remains closed (Fig. 6.63(a)). However, under large engine loads or at high speed and large power outputs the wastegate may not be totally able to control the boost pressure. Therefore, under such conditions, with a further rise in charge pressure of only 0.1 bar, that is, to a boost pressure of 0.8 bar, the blow-off diaphragm valve is pushed open by the pressurized charge. Subsequently, a portion of the surplus air delivery will now be bypassed back to the intake side of the compressor (Fig. 6.63(b)). The result of firstly reducing turbine power and secondly recycling some of the air charge back to the inlet side of the compressor provides a reliable approach in controlling the delivery boost pressure under all operating conditions. If the warning contacts of the warning light closes (Fig. 6.63(b)) the light will be illuminated, indicating to the driver that excessively high boost pressure is being generated and that possibly the wastegate has become stuck in the closed position.

6.9 Turbocharged engine systems

6.9.1 Bypass priority valve turbocharged engine system
(Fig. 6.64(a and b))

With the turbocharged engine system there is a passage which bypasses the compressor housing and connects the carburettor outlet to the output side of the compressor (Fig. 6.64(a and b)). The bypass passage is divided by a priority-valve which may be interconnected to the carburettor throttle valve or it may be a flexible flap-valve. When an interlinked priority valve is employed the valve remains open while the carburettor throttle opening is less than one-third fully open, the engine is then naturally aspirated (Fig. 6.64(a)). Above this throttle setting, however, the interlinkage closes the priority valve (Fig. 6.64(b)). Therefore, all the mixture entering the turbo compressor comes out as a pressurized charge mixture. Alternatively, if a flexible flap priority-valve is used with a small throttle opening, the high vacuum on the engine-side of the priority valve pulls the flap-valve open (Fig. 6.64(a)). Thus, the mixture from the carburettor is able to bypass the turbocharger and therefore goes directly to the induction manifold. Under these operating conditions, the engine is naturally aspirated. With increased engine speed the compressor boost pressure will apply a back pressure to the flap-valve region until eventually it closes the valve (Fig. 6.64(b)). Thus, with any further rise in engine load and speed all the mixture from the carburettor passes through the compressor, hence causing the charge to be compressed before it is delivered to the cylinders.

Thus, the priority-valve overcomes the disadvantage of the high inlet flow resistance at low engine output and operates as a naturally aspirated engine until the turbine and compressor spin-speed has built up sufficiently to deliver a positive boost pressure. The result is an almost spontaneous throttle response, with the fuel eco-

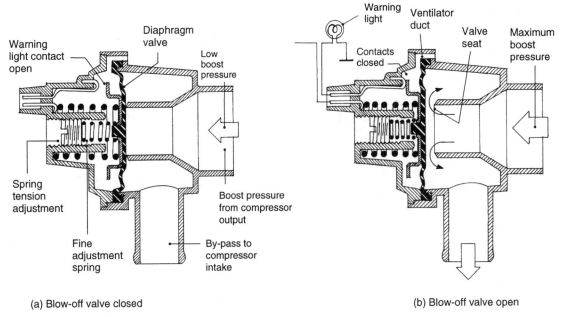

Fig. 6.63   Diaphragm-type blow-off valve

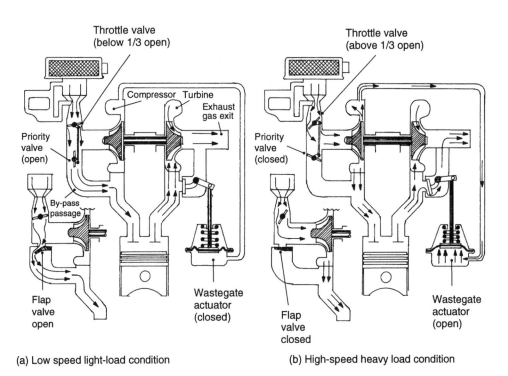

Fig. 6.64   Turbocharged engine layout with part throttle by-pass priority valve and passage

nomy at the low engine-speed of a naturally aspirated engine.

6.9.2 Automatic performance controlled (APC) turbocharged engine systems
(Fig 6.65(a and b))

The performance of the turbocharger can be better matched to the engine requirements if, every time there is a tendency for the engine to operate with excess cylinder pressure or for detonation to occur, then the turbocharger spin speed is automatically reduced by diverting some of the exhaust gas from the turbine housing.

The problem with the conventional wastegate boost pressure or exhaust back pressure control is that the opening of the wastegate is totally determined by the predetermined spring stiffness of the actuator's return-spring and the opposing pressure acting on the actuator diaphragm. This simple method of control is not sensitive to a combination of factors such as engine speed, cylinder pressure or combustion roughness.

However, a more sensitive and accurate method of regulating the turbine spin speed, and therefore boost pressure, is to use a wastegate actuator with a relatively weak return-spring and then to control the delivery of boost pressure from the compressor to the actuator diaphragm via a wastegate solenoid control valve unit (Fig. 6.65(a and b)).

An electronic (microprocessor) control unit forms an important part of the automatic performance control system: its function is to receive inputs from the knock sensor pressure transmitter as well as the ignition pulse frequency as a measure of the engine speed. This information is then processed, an output signal is then passed to the solenoid control valve which will either open or close off the passage connecting the compressor to the wastegate actuator.

The actual engine speed and load combinations at which the wastegate opens will therefore depend upon the microprocessor instantly responding to input signals, such as those supplied by the pressure transmitter (which measures boost pressure) and signals from the piezometric knock sensor (which detects any incipient pinking). The outcome is an instant lowering in boost pressure under severe and abnormal operating conditions, which therefore protects the engine against any

possible damage. An ignition retardation device may be installed to reduce the ignition advance whenever there is a tendency for the engine cylinder combustion process to become rough.

This boost control system enables the engine to use fuel, with the usual octane rating between RON 91 and 98 without reducing the engine's compression ratio by more than one, i.e. say from 9.5:1 to, say, 8.5:1.

Operating conditions
Moderate-speed light-load conditions (Fig. 6.65(a)). When the engine is operating under moderate speed and part throttle opening the wastegate solenoid control valve is made to open due to the energized solenoid but, with higher engine speeds and large throttle opening or under rough combustion (detonation), the microprocessor will cut off the electrical signal to the solenoid so that the wastegate control valve closes and therefore returns the boost pressure back to the intake side of the compressor, so that all the exhaust gas is directed towards the turbine, which has to provide the maximum boost relative to the turbocharger spin speed.

High-speed heavy-load conditions (Fig. 6.65(b)). With higher engine speeds and large throttle opening or under rough combustion (detonation) the microprocessor will cut off the electrical signal to the solenoid so that the wastegate control valve closes. Accordingly, the compressed air from the compressor will be conveyed to the wastegate actuator diaphragm via the control valve. This, therefore, pushes open the wastegate and thereby reduces the amount of exhaust gas energy reaching the turbine housing.

6.9.3 Dual-stage twin-volute turbine housing turbocharged engine system
(Fig. 6.66(a and b))

The twin scroll turbine housing has two volute passages surrounding the turbine-wheel which are separated by an integrally cast wall. Both of these passages feed into a common throat which is completely exposed to the periphery of the turbine-wheel. One of the twin exhaust gas inlet passages to the volute is equipped with a trapdoor like flap-valve (Fig. 6.66(a and b)). Thus, the turbine housing intake passage flow path has two cross-sectional areas, A_1 and A_2. However,

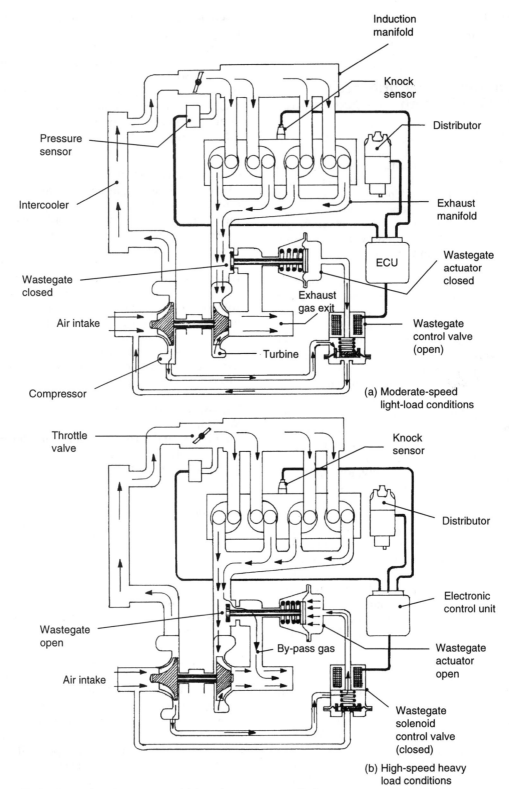

Fig. 6.65 Turbocharged engine system with knock sensor controlled wastegate

the gas can either go through the primary passage A_1 to the outer volute when the flap-valve is closed or through both the primary and secondary passages A_1 and A_2 respectively when the flap-valve in the secondary passage is open.

In both cases, the distance R between the turbine shaft centre and the centre of areas A_1 and A_2 is similar. Consequently, the turbine housing has two A/R ratios. That is, A_1/R ratio equal to $0.4:1$ for the primary passage and $(A_1 + A_2)/R$ equal to $1.0:1$ for the combined primary and secondary passageways.

At low engine speeds (Fig. 6.66(a)) exhaust gas flows through the primary passage filling the outer volute and then discharges itself onto the turbine-wheel via the circumferential throat. The small cross-sectional area of the primary passage has a throttling effect which accelerates gas speed, and the volute and throat shape directs gas towards the turbine at roughly right-angles to the blades. This maximizes the thrust imparted to the turbine-wheel so that it rapidly increases the spin speed of the turbine and compressor assembly.

Once the engine has reached something like 2500 rev/min the change-over flap-valve opens (Fig. 6.66(b)). This now increases the flow-path cross-sectional area, and consequently the exhaust gas speed is reduced which will cause the gas to strike the turbine-blades at a more obtuse angle. With the slower gas speed the turbine will progressively slow down to a speed which matches the new flow conditions.

The small turbine housing primary passage provides good low engine speed response and minimizes turbo lag during transitional operating conditions, whereas the large turbine housing passages (primary and secondary) provide a measure of self-regulating speed and boost pressure control and, at the same time, reduce exhaust back-pressure, thereby improving cylinder filling and thermal efficiency.

The change-over flap-valve is operated by a diaphragm actuator which is, in turn, controlled by a change-over solenoid control valve and an electronic (microprocessor) control unit.

Operating conditions
(Fig. 6.66(a and b))
When the engine's speed and boost-pressure signal inputs to the microprocessor reach a predetermined set of combinations, an output signal to the change-over solenoid-controlled valve will open the valve so that boost is conveyed to the di-

aphragm actuator. This then forces open the change-over flap-valve (Fig. 6.66(b)). Accordingly, the flow path to the turbine will now be via both the primary and secondary passageways leading to the turbine's circumferential throat. If the engine speed and boost-pressure drop below the designed change-over from large to small intake cross-sectional area (Fig. 6.66(a)), then the change-over solenoid valve will be energized. This opens the control-valve so that the boost pressure will by bypassed back to the intake side of the compressor. It thus releases the diaphragm actuator so that the change-over flap-valve closes again. The turbine and compressor-wheel assembly will then be able to build up rapidly the boost pressure in the low-speed range or during a transition period where the engine is being accelerated.

Under excessive boost pressure or rough combustion the wastegate solenoid control-valve closes, thereby permitting boost pressure to push the actuator diaphragm against the return-spring until the wastegate flap-valve opens and redirects some of the exhaust gas away from the turbine housing.

6.9.4 Variable nozzle area single volute turbine housing turbocharger engine system
(Fig. 6.67(a and b))

The narrowing flow path leading to the circular volute passageway surrounding the rim of the turbine wheel is known as the nozzle and has a variable cross-sectional area provided by a curved tapering flap hinged at one end (Fig. 6.67(a and b)). The impellor wheel for the compressor is 53 mm in diameter, and is driven by a 49.9 mm diameter turbine wheel. The angular movement of the flap between the smallest to the largest nozzle cross-sectional area, amounts to 27°. With the flap nozzle closed the smallest cross-sectional area A in the nozzle region is 313 mm², whereas with the flap fully open it is enlarged to 858 mm². Thus, taking the distance between the centroid of the nozzle cross-sectional area and the centre of the turbine shaft as R, then the A/R ratio for the smallest nozzle is 0.21 going up to 0.77 for the largest opening.

The variable nozzle flap is operated by a diaphragm actuator, which is itself controlled by a variable flap control valve (pressure control mod-

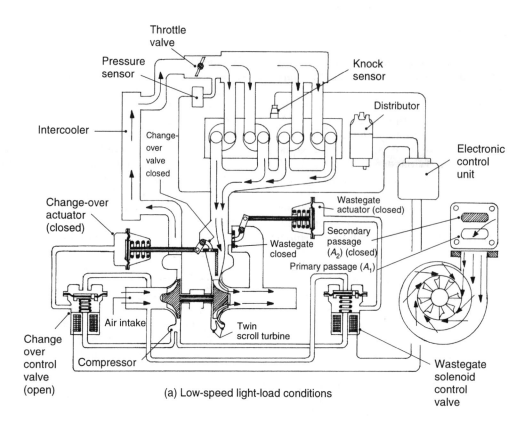

(a) Low-speed light-load conditions

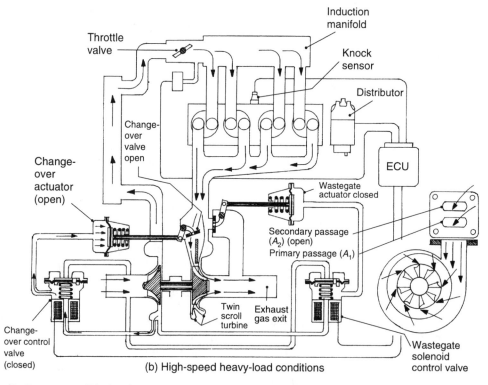

(b) High-speed heavy-load conditions

Fig. 6.66 Dual-stage scroll (volute) area turbocharged engine system

340

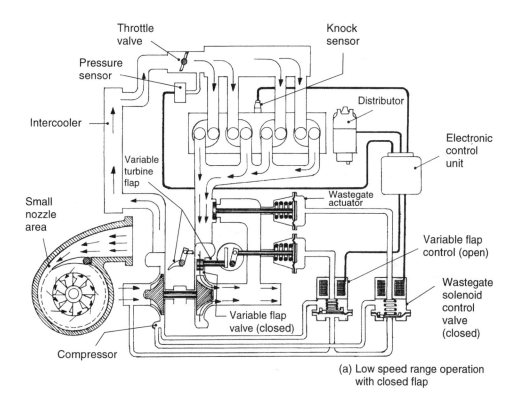

(a) Low speed range operation with closed flap

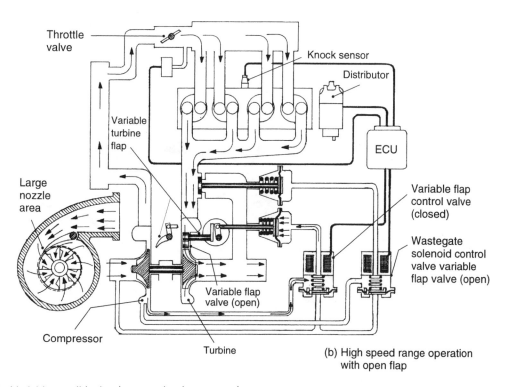

(b) High speed range operation with open flap

Fig. 6.67   Variable scroll (volute) area turbocharger engine system

341

ulator) which permits boost pressure to be transmitted to the diaphragm actuator to enlarge the nozzle area to allow the air pressure to escape from the actuator to reduce the flow path.

Engine speed, load and boost pressure are continuously being monitored by the computer which then calculates the optimum nozzle area required to match the engine's demands. The electronic control unit then immediately transmits the signal to the pressure control modulator to operate the nozzle flap diaphragm actuator accordingly.

Thus, with a small nozzle setting (Fig. 6.67(a)) the exhaust gas flow will be accelerated and will therefore raise the pulsed gas pressure acting on the turbine-blades so that the turbine spins faster at low engine speed and light load conditions. With increased engine speed (Fig. 6.67(b)) the priority is for a decrease in pulse gas pressure since the turbine is already working in the effective upper speed range, which produces the boost pressure, and instead the exhaust gas flow resistance should be minimized.

This is achieved by enlarging the nozzle flow area which therefore slows down the gas speed and makes it easier for the gas to escape, the result being a marked improvement in cylinder charge filling.

The tilting of the nozzle flap between the small and large passageway cross-sectional area is a stepless and a continuous operation which is dictated by the information received by the electronic control unit (microprocessor).

Sudden over-boost and incipient pinking is accommodated by the wastegate, which automatically opens thereby reducing the amount of exhaust gas energy reaching the turbine-wheel.

6.9.5 Variable geometry multi-nozzle turbine housing turbocharger
(Figs 6.68(a and b) and 6.69)

The variable-geometry multi-nozzle turbine has the usual circular volute passage but the surrounding flow path to the turbine-wheel is first made to pass through a ring of nozzle vanes mounted on pivot spindles which are positioned parallel to the turbine axis (Fig. 6.68(a and b)). These vanes interlink to a nozzle control ring via nozzle bell-crank levers so that they are all synchronized to open and close together through an angular movement of about 30°.

Under light-load and low engine-speed conditions, the nozzle vanes are tilted to reduce the flow-path cross-sectional area so that the gas velocity is increased (Fig. 6.68(a)). At the same time, the exhaust gas is directed through 360°, almost at right-angles, onto the outer periphery of the turbine-blades in a semi-jet fashion where the

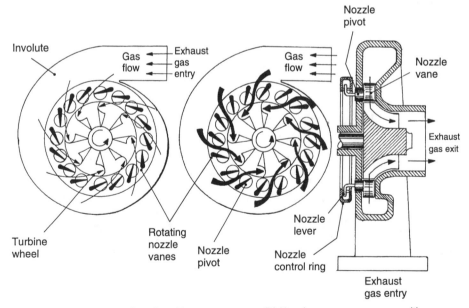

(a) Nozzle vanes near closed position   (b) Nozzle vanes near open position

Fig. 6.68   Variable nozzle geometry turbocharger

gas can impose the maximum effective thrust to the turbine-wheel. The result is that the power developed by the turbine will be sufficient to provide high compressor and turbine spin speeds at very low engine speeds.

With greater engine load and higher speeds the nozzle-vanes are proportionally twisted to enlarge the nozzle flow areas (Fig. 6.68(b)), thereby reducing the gas speed and, at the same time, spreading out the gas discharge impinging against the turbine-blades. Consequently, the angle of impact and the force of contact will not be at its optimum. There is a self-regulating turbine upper spin-speed, particularly as boost pressure approaches the designed upper limit.

The ability to change rapidly the angle of the nozzle-vanes produces a very quick acceleration response which is not generally possible with the conventional turbine housing in which the gas from the volute discharges directly onto the turbine-blades.

The other major benefit of being able to open the nozzle flow-path at high engine outputs is to reduce considerably the exhaust-gas back-pressure so that the cylinders can be cleared and filled more effectively.

The control of the rotating nozzle-vanes can be actuated by boost pressure exhaust back-pressure or intake depression. Whichever source of energy is used it will be programmed by a microprocessor, which signals the engine requirement to a solenoid-controlled pressure modulator.

A comparison of boost-pressure characteristics of a vaneless nozzle and variable-geometry multinozzle turbine-turbocharged engines, with different angle settings, is shown in Fig. 6.69.

6.9.6 Turbocharger and turbocharged engine performance characteristics

Turbocharged engine performance characteristics
(Figs 6.70, 6.71 and 6.72)
Turbocharging diesel engines can reduce the specific fuel consumption from about 3 to 14% in the engine's speed range. The reduction in fuel consumption becomes more marked as the engine's load is reduced, as can be seen from the family of constant load (b.m.e.p.) curves ranging from $\frac{1}{4}$, $\frac{1}{2}$, $\frac{3}{4}$ and full engine load (Fig. 6.70). However, at full load below 1400 rev/min and $\frac{3}{4}$ load below 1000 rev/min, the specific fuel consumption is

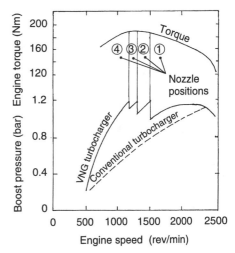

Fig. 6.69 Boost pressure characteristics of a vaneless and variable geometry multi-nozzle turbocharged diesel engine

inferior to that of the naturally aspirated engine. Thus, the improvement in fuel consumption becomes more effective as the engine load is reduced.

With the turbocharged engine, the level of exhaust smoke emission is considerably reduced with increasing engine speed, as excess air is supplied to the cylinders, which is in contrast to the naturally aspirated engine (Fig. 6.70). In the upper speed range the naturally aspirated engine finds it difficult to clear and fill the cylinders with sufficient quantities of fresh air, it therefore results in a rapid rise in the level of exhaust smoke as the engine approaches maximum speed.

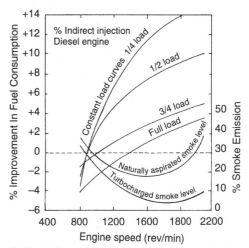

Fig. 6.70 Effect of engine speed on level of exhaust smoke and full consumption for various engine loads

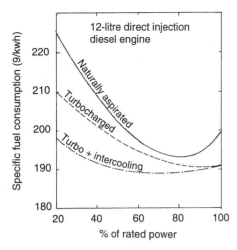

Fig. 6.71 Effect of varying the percentage of power on engine-specific fuel consumption for a naturally aspirated, turbocharged and turbocharged and intercooled state of tune

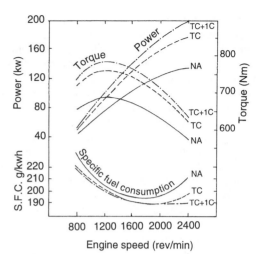

Fig. 6.72 Comparative performance for an engine operating under the following conditions: naturally-aspirated turbocharged and turbochaged and intercooled

Frictional losses rise rapidly with an increase in engine speed but do not rise in direct proportion to the engine load output. Thus, if the engine brake mean effective pressure (b.m.e.p.), and thus engine torque, is made to peak at lower engine speed where the mechanical losses are least, then there will be an improvement in the specific fuel consumption if the engine operates under these conditions. Figure 6.71 shows the benefits in specific fuel consumption compared with the naturally aspirated engine of an uncooled and intercooled turbocharged engine over the engine's operating power range. The graphs show that far less fuel is needed as the engine power is reduced when turbocharged, and particularly if the compressed and heated charge is intercooled. Furthermore, if the power developed exceeds 80% of its rated power, the naturally aspirated engine showed an upturn in fuel consumption due, possibly, to breathing difficulties, but the specific fuel consumption for both the uncooled and cooled turbocharged engine just continues to level off. The general performance characteristics of engine torque, power and specific fuel consumption against engine speed are shown in Fig. 6.72 for three different stages of engine tune: (1) naturally aspirated; (2) turbocharged; and (3) turbocharged and intercooled. The specific fuel consumption curves indicate that there is very little difference between uncooled and cooled charging on either side of the 1400 to 1800 rev/min speed band but the difference is more significant towards maximum speed.

Turbocharged petrol engines generally have reduced compression ratios to accommodate the high cylinder pressures and, under load, the ignition timing is automatically retarded to prevent detonation taking place, while at full load a rich mixture is necessary. Consequently, the turbocharged petrol engine's efficiency may not equal that of an equivalent sized naturally aspirated petrol engine although the engine's torque and power will be far superior.

6.10   Turbocharger fault diagnosis

6.10.1   Engine starting and stopping procedure

Starting
As soon as the engine fires and is revolving by its own power, remove one's foot from the accelerator pedal for about five seconds to permit the engine's lubricating pump to circulate oil to the turbocharger shaft and bearings before the turbine and compressor assembly becomes operational. If insufficient time is permitted for the oil to establish hydrodynamic lubrication conditions between the shaft and bearings then only boundary lubrication conditions prevail during the starting phase and the consequences will be a rapid wear rate.

Stopping

After completing a journey, particularly if the engine has been driven hard just before the vehicle stops, the engine should be allowed to idle for a short period until the turbine and compressor assembly have had time to reduce their speed to no-load rotational conditions. Hence, when the engine stops and the oil supply is cut off to the turbocharger, there will still be sufficient residual oil coverage between the shaft and bearings during the time it takes the rotating assembly to come to rest. If this precaution is not taken when stopping the engine, then it is highly likely that only boundary lubrication conditions will exist during the spin down time needed to bring the turbine and compressor assembly to a standstill.

One method used to ensure that there is adequate stopping time for the rotating assembly to slow down is to incorporate a time delay into the ignition switch circuit so that when the engine is switched off the engine will continue to run for a predetermined period before automatically cutting out.

6.10.2 Turbocharger service considerations

Faults

Symptoms which could indicate turbocharger malfunction are:

1 loss of power
2 excessive exhaust smoke
3 high fuel consumption
4 overheating
5 high exhaust temperature
6 oil leakage from turbocharger

Oil quality

Oil should be filtered below 15–20 μm. Only recommended oil specified by the manufacturer should be used. If inferior oil is used carbon deposits can be formed at the turbine end of the turbocharger causing excess wear of the sealing rings.

Oil supply

The minimum oil pressure when the engine is in load is 2.0 bar, however the pressure should not exceed 4.0 bar as this can force the oil past the seals into the back of the turbine or compressor wheels. Under idle conditions the pressure should not fall below 0.7 bar. When the engine is started the oil should reach the turbocharger inlet oil connection within 3 to 4 seconds of the engine firing.

Joints and connections

1 Gas leakage at the manifold or turbine inlet could cause loss of turbine speed and power accompanied with excessive exhaust smoke.
2 Air leakage at the turbocharger compressor outlet connection and the engine induction manifold joints could cause a loss of charge pressure, power and excessive exhaust smoke.

Oil leakage and restriction

1 A blocked air filter will produce a depression in the compressor wheel-chamber. This can cause oil to be drawn through the seals into the space behind the compressive-wheel with the result that it can clog up the compressor.
2 A damaged or badly fitted supply pipe connection will starve the turbine bearing assembly of oil causing excessive wear.
3 A blocked or kinked or otherwise damaged oil return drain-pipe may cause a build up of oil in the bearing housing.
4 High crank case pressure due to piston blow-by or clogged crankcase ventilation will restrict the flow of oil from the turbocharger bearing housing.

Air filter restriction

Excessive air cleaner restriction results in a shortage of air needed for combustion, it therefore reduces charge pressure, reduces power and increases fuel consumption. A restriction can produce flooding of oil through the seals (due to suction), which will eventually clog up the compressor-wheel, and if the restriction is serious, the engine can even overheat. The air-cleaner pressure-drop should not exceed 40 mm Hg or 0.05 bar.

Exhaust back-pressure

Excessive exhaust back-pressure will restrict turbine speed response, causing overheating, loss of power, and high fuel consumption. Excessive back-pressure can also force the gas through the seal and between the bearing and shaft; thus, contamination will carburize the oil and clog the turbocharger components. Exhaust gas back-pressure should not exceed 40 mm Hg or 0.05 bar.

Boost pressure

Boost pressure should be checked every 50 000 km. The maximum boost pressure when the engine is running at maximum speed at full-load will vary according to the type of turbo-charger and power unit. A typical boost pressure reading would be 0.8 ± 0.1 bar at 2100 rev/min for a large diesel engine.

Crankcase pressure

Crankcase pressure build-up should be checked as this indicates leakage of combustion gas from the cylinders, which will then contaminate the oil and shorten the service life of the various components. Crankcase pressure should not exceed 8 mm Hg or 0.01 bar.

Compressor housing checks

(Charts 1, 2 and 3 on p. 363)
Remove the inlet direct to the turbocharger periodically and check for the following.

1 Dirt and dust build-up on the impellor or in the housing, and for any signs of contact between the impellor and the housing. Excessive accumulation of dirt indicates either a leak in the ducting or a faulty air filter.
2 Observe if there is oil entering the compressor housing after the turbocharger has been operated for some time under load conditions. If there is, refer to chart 1.
3 Observe if there is oil in the inlet or outlet ducts or dripping from either housing. If there is refer to chart 2.
4 Spin the compressor-wheel by hand observing how freely it revolves. If the shaft tends to drag refer to chart 3.
5 Check for any unusual turbocharger vibration and noises when the engine is operating at rated output.

A bearing which is failing will produce a shrill whine over and above normal turbine whine. Usually, noises may result if there is improper clearance between the turbine-wheel and its housing.

6.10.3 Common turbocharging system faults

The most common turbocharger system faults come under the following headings:

1 noise
2 excessive smoke in exhaust gas
3 loss of power output
4 oil leakage from the turbocharger unit
5 damaged turbocharger malfunctioning

Noise
The probable causes are:

1 gas/air leakage
2 gas/air restriction
3 damaged turbocharger

Excessive smoke in exhaust gas and loss of power output
The probable causes can be similar for high levels of smoke and loss of power:

1 gas/air leakage
2 gas/air restriction
3 malfunctioning fuel pump and/or injectors
4 incorrect injection pump setting
5 dirt or soot deposits on compressor or turbine-wheel
6 oil leakage from engine
7 damaged turbocharger

Oil leakage
The probable causes are:

1 gas/air leakage
2 gas/air restriction
3 blocked crankcase ventilation
4 blocked oil return in the turbocharger
5 excessive idle
6 engine blow-by
7 damaged turbocharger

Damaged turbocharger
Damage to the turbocharger is usually due to worn bearings so that the turbine and/or compressor wheel fouls the housing.

An inspection can be made by removing the inlet connection and the exhaust pipe from the turbocharger unit. The rotating assembly should then be turned by hand to check whether it revolves easily or if a grating or scraping noise is experienced. If the bearings or seals are damaged then the cause must be traced.

Damaged compressor wheels can be caused by foreign particles bypassing the air filter so that they enter the compressor-wheel cells and then jam themselves between the blades and housing. They either severely damage the blade edges or bring the rotor assembly to a standstill.

Damaged turbine-wheels can be due to solid metallic objects being expelled from the engine cylinders and passing through the turbine-wheel passageways. These objects usually consist of engine components such as broken valves, pistons, piston rings and injector nozzles etc.

6.11 Intercooling

6.11.1 The need for intercooling
(Figs 6.73, 6.74, 6.75 and 6.76)

The density of air charge, that is, the mass per unit volume entering the cylinder, is of vital importance since squeezing more mass into the cylinder increases the power generated in the cylinder. With supercharged engines the air change entering the cylinder will be above the normal atmospheric density of air so that a comparison must be made between the actual density of charge in the cylinder to the air density under normal temperature and pressure (NTP) conditions (in which the temperature is taken as 16°C and the pressure as $101.3 \, \text{kN/m}^2$ (1 bar)). The comparison used to relate different air densities in the cylinder to a known air density under normal temperature and pressure conditions is the air charge density ratio.

air charge density ratio =

$$A/R \text{ ratio} = \frac{\begin{array}{c}\text{density of charge in}\\\text{cylinder under operating conditions}\end{array}}{\begin{array}{c}\text{density of charge in}\\\text{cylinder under NTP conditions}\end{array}}$$

therefore air density ratio $= \dfrac{\rho}{\rho_{\text{NTP}}}$

The relationship between pressure ratio and air charge density ratio if the air temperature is held constant at three different temperatures is shown in Fig. 6.73. The graphs show that as the boost pressure ratio increases, the air density ratio increases likewise, however, the more the air is intercooled and its temperature reduced, the

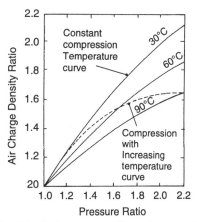

Fig. 6.73 Effect of pressure ratio on increase in charge density if the charge temperature is held constant or is allowed to rise

greater will be the rise in air charge density. Thus, if the air temperature is maintained at 30°C, the density at a pressure ratio of 2.2:1 will be about 2.1:1, whereas if the air temperature is kept constant at 90°C the air density ratio only rises to around 1.64:1. Well-designed intercoolers can hold the compressed air temperature to about 60°C.

The broken curve shows how the air density ratio increases if there is no intercooling. Here, it can be seen that as the boost pressure ratio increases beyond about 1.6:1 there is very little useful increase in the air density ratio for a considerable further increase in the pressure ratio.

When including the compressor efficiency it can be shown (Fig. 6.74), on a number of constant

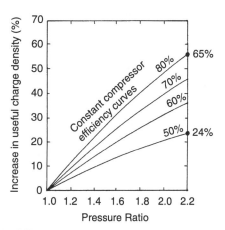

Fig. 6.74 Effect of pressure ratio on increase in useful charge density for various compressor efficiencies

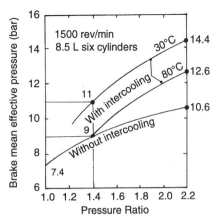

Fig. 6.75 Relationship between brake mean effective pressure and charge pressure ratio subjected to various temperature conditions

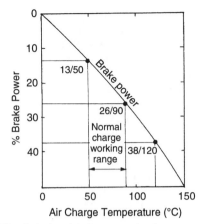

Fig. 6.76 Relationship between brake power and charge and temperature

efficiency curves, that as the pressure ratio increases, the useful charge density as a percentage increases, but the increase is greatly influenced by the efficiency of the compressor. Thus, for a 50% efficient compressor at a pressure ratio of 2.2:1 the increase in air density amounts to about 24%, whereas for an 80% efficient compressor the density rise is as much as 55% for the same pressure ratio increases.

The effects of boost pressure ratio on the 'brake mean effective pressure' (b.m.e.p.) developed in the cylinder are clearly illustrated in Fig. 6.75. The lower curve shows that if there is no intercooling, so that the compressed air temperature is allowed to rise uncontrolled then, as the boost pressure ratio increases from the naturally aspirated condition to 2.2:1, the b.m.e.p. also rises from 7.4 bar to 10.6 bar respectively. Note the intercooler has very little effect on the b.m.e.p. below a pressure ratio of around 1.4:1.

However, if the air charge is intercooled by an air-to-liquid intercooler, so that the temperature is maintained at about 80°C, then there is a marked increase in b.m.e.p. for a boost pressure ratio of 1.4:1 to 2.2:1 which raises the b.m.e.p. from 9 bar to 12.6 bar. Even better results can be obtained if an air–air intercooler is used, where the compressed air can be cooled down to 30°C, as the graph shows. Here, with a 1.4:1 pressure ratio, the b.m.e.p. rises to 11.0 bar, and as the pressure ratio reaches 2.2:1 the b.m.e.p. will be as high as 14.4 bar.

A very convincing case for the employment of an intercooler as one of the stages of improving engine performance is shown in Fig. 6.76. The graph shows that as the air–charge output temperature increases from zero to 150°C the percentage of engine power decreases by 50%. If the air discharge temperature can be reduced to 50°C, the power lost amounts to only 13% while if the temperature increases to 80°C the power wasted will now be about 26%. With an uncooled supercharging system, where the temperature may reach 120°C, the power lost may be as high as 38%.

6.11.2 Intercooler effectiveness
(Fig. 6.77)

The cooling of the delivery charge after it has been compressed contributes considerably to the recovery of the charge's density ratio. The benefit of an intercooler to reduce the charge's temperature and thereby raise its density ratio is shown in Fig. 6.77. However, the ability to increase the density of the compressed charge for a given pressure ratio by cooling the heated charge is dependent upon the effectiveness of the intercooler. Intercooler effectiveness is defined as follows:

intercooler effectiveness =

$$\frac{\text{actual heat transfer}}{\text{maximum possible heat transfer}}$$

$$\varepsilon = \frac{T_2 - T_3}{T_2 - T_1}$$

where T_1 = coolant temperature
T_2 = charge output temperature from the compressor
T_3 = charge output temperature from the intercooler

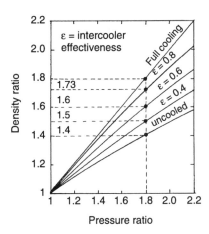

Fig. 6.77 Effects of intercooler effectiveness on air density with different pressure ratios

Example 1

With an aged and scaled heat exchanger matrix, the charge air temperature leaving the compressor is 125°C and the coolant temperature is 85°C. If the charge temperature on the output side of the intercooler is lowered to 105°C determine the intercooler effectiveness.

$$\varepsilon = \frac{T_2 - T_3}{T_2 - T_1} = \frac{125 - 105}{125 - 85} = \frac{20}{40} = 0.5$$

Example 2

An efficient heat exchanger should be able to drop the air charge temperature from 120°C to 90°C when the cooling media is at 80°C. What is the intercooler effectiveness?

$$\varepsilon = \frac{T_2 - T_3}{T_2 - T_1} = \frac{120 - 90}{120 - 80} = \frac{30}{40} = 0.75$$

The benefit of improving the intercooler effectiveness (ε) is illustrated in Fig. 6.77, the curves show that, with a pressure ratio of 1.8:1, the uncooled output charge has a density ratio of 1.4:1 whereas, with increasing intercooler effectiveness of 0.4, 0.6 and 0.8, the density ratios rise to 1.5:1, 1.6:1 and 1.73:1, respectively.

6.11.3 Benefits and limitations to intercooling

Reducing the intake temperature to the cylinder produces a corresponding reduction in exhaust temperature; thus, if the intercooler lowers the charge temperature entering the cylinder from say 120°C to 40°C, then there will be a roughly similar fall in exhaust gas temperature. This can be significant. Hence, if the full load exhaust gas temperature is approximately 750°C uncooled, then a 80°C reduction in intake temperature will reduce the exhaust gas temperature to 670°C which can be very important in prolonging exhaust valve life.

With the overall operating temperature of the engine reduced, the combustion chamber pressure for a given brake mean effective pressure will also be lower; consequently, there will be a similar reduction in the thermal stresses imposed on the engine components.

With a perfect heat exchanger the temperature of the charge will be reduced to that of the cooling media without any reduction in air pressure. In practice, there is always some resistance to charge movement through the intercooler flow path matrix. At the same time, the heat transfer from the heated input charge cannot be brought down to the same temperature as the coolant due to thermal resistance to heat flow through the metal walls of the tubes from the hot to the cold surface areas.

6.11.4 Summary of the advantages of intercooling

Supercharger intercoolers provide a means of reducing the charge inlet air or air–fuel mixture temperature between the compressor outlet and the engine's inlet ports. This achieves several objectives.

1 It keeps the cylinder-head temperature low even under heavy load conditions, thus reducing thermal stresses, and therefore it prolongs the life of the engine's components.
2 It increases the mass of charge that can be crammed into each cylinder during each induction stroke thereby increasing the engine power.
3 It reduces the oxides of nitrogen (NO_x) emission due to the lower combustion temperature.
4 It reduces diesel engine black smoke emission at low engine speeds and high loads due to the reduction in the charge temperature.
5 It raises the knock limit for petrol engines and therefore permits a higher mean effective pressure.

Test results have shown that for every 10°C reduction in the compressed charge temperature there is an increase in power output of roughly

3%. It has, however, been found that intercooling can only be justified if the charge can be cooled by a minimum of 20°C, corresponding to a power increase of around 6%.

6.12 Intercooling systems

The purpose of supercharging an engine is to increase the density of the charge entering the cylinder, thereby maximizing the oxygen content involved in the combustion process. The supercharger performs the task of increasing the charge density by taking in the fresh charge, reducing its volume and then discharging it. This is achieved by compressing the charge. Unfortunately, however, it raises its temperature which has the opposite result, as it tends to expand the air or air–fuel mixture and force the molecules further apart. Hence, in effect, it reduces the number of molecules packed into each unit volume of charge being continuously conveyed from the compressor outlet to the cylinders via the induction passageways. In other words, an increase in temperature counteracts the cramming together of the charge molecules and therefore reduces the rate of density increase caused by a corresponding rise in pressure ratio.

To overcome these difficulties, the charge leaving the compressor can be passed through many small but long passageways which are surrounded by a cooling media, which is itself well below the compressed output charge temperature. Consequently, heat will be transferred from the hot moving charge through the metal passageway channel walls to the outside cooling media, be it air or liquid. Thus, by the time the charge arrives at the cylinder entrance its mean temperature will be substantially reduced and thus, the function of the intercooler is to transfer heat from the compressed charge to another source. This component is therefore also known as a heat exchanger.

There are two basic approaches to intercooler heat exchanger design, these are:

1 air-to-air heat exchanger
2 air-to-liquid heat exchanger

6.12.1 Air-to-air intercoolers
(Figs 6.78, 6.79, and 6.80)

The exchange of heat is achieved by passing the compressed hot charge through many vertically mounted elongated flat cross-sectional tubes, whereas the cooling atmospheric air flows between and across the tubes (Fig. 6.78). Heat is dissipated from the hot charge to the air stream by conduction through the copper or aluminium alloy walls of the tubes and then by convection current and radiation to the atmosphere. To increase the efficiency of the heat-exchanger, the external surface area of the tubes is greatly increased by attaching fins made from corrugated copper sheet between the vertical columns of the tubes. Atmospheric air will either be drawn by the engine's cooling system fan, or by the ram effect caused by the vehicle's movement, through the spaces formed between adjacent tubes. Thus, the air on the cold side of the tube matrix will be forced to enter the small but relatively long triangular passageways created by the zigzag pattern of the copper sheet fins and the vertical columns of the tubes. Heat will be effectively transferred from the metal fins to the air stream as it continuously scrubs the walls of the triangular channels on its way from the front of the matrix to the rear before it is returned to the surrounding atmosphere.

With a typical air inlet temperature of about 25°C the pressurized output charge temperature from the compressor may reach 120°C or even more, if it then passes through the air-to-air intercooler the charge's temperature can be brought down to approximately 40°C to 60°C before it enters the cylinders.

The coolant air stream may be supplied by two methods.

1 By placing the tubular matrix in a frontal position so that it is subjected to the ram effect as the vehicle moves forwards (Fig. 6.79). This arrangement can reduce the hot compressed charge to a temperature between 50°C and 60°C, but it is greatly dependent upon the speed of the vehicle and the direction of any air flow.

2 By placing the heat exchanger matrix immediately in front of the engine cooling system's radiator, so that the engine's fan determines how much air is drawn through the finned spacing between the tubes (Fig. 6.80). This more stable layout is used on commercial vehicles where there is more room, it is capable of bringing the heated charge's temperature down to a consistent 40°C.

The major advantage of the air-to-air intercooler compared with the air-to-liquid intercooler

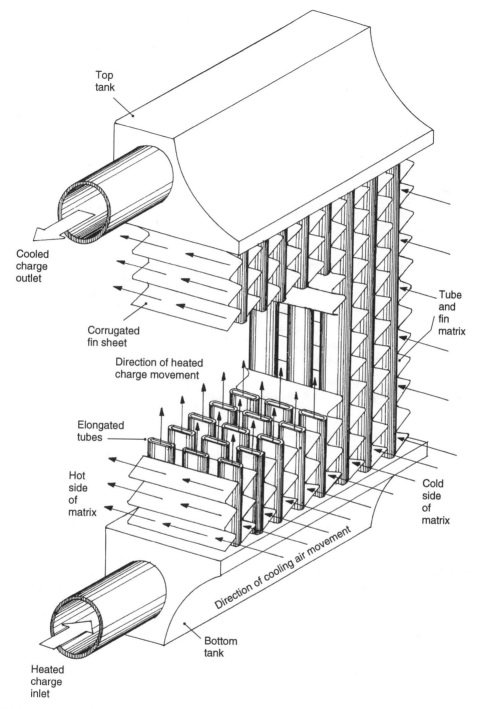

Top tank

Cooled charge outlet

Corrugated fin sheet

Direction of heated charge movement

Elongated tubes

Hot side of matrix

Tube and fin matrix

Cold side of matrix

Direction of cooling air movement

Bottom tank

Heated charge inlet

Fig. 6.78 Air-to-air intercooler

is that it can bring the charge's output temperature down to around 40°C to 50°C, whereas the air-to-liquid method of cooling can only reduce the charge's delivery temperature to a value of about 5°C to 10°C above that of the engine's cooling temperature, i.e. 80°C to 90°C.

351

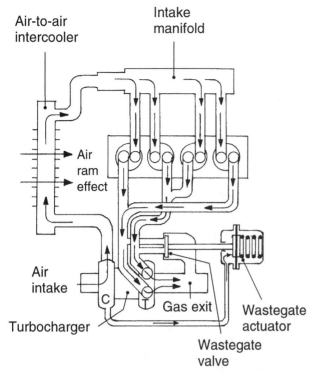

Air-to-air intercooler

Intake manifold

Air ram effect

Air intake

Turbocharger

C

Gas exit

Wastegate actuator

Wastegate valve

Fig. 6.79 In-line four-cylinder engine with double-entry turbocharger, wastegate and air-to-air intercooler

6.12.2 Air-to-liquid intercooler
(Figs 6.81 and 6.82)

Heat exchangers of this type have a number of horizontally mounted circular tubes, through which coolant liquid passes, while the compressed and heated air charge flows around and across them (Fig. 6.81). Heat transfer takes place between the heated air and cooling liquid by means of conduction through the copper nickel alloy walls of the tubes and by convection current as the coolant is forceably circulated from and to the engine's cooling system. To speed up the heat transfer process many closely spaced copper cooling fins are stacked perpendicular to the axes of the tubes. The hot pressurized charge therefore flows between pairs of parallel thin copper sheets, thus causing the heated air to skim across and over the flat and relatively very large surface areas. Heat is thus readily transferred to the coolant liquid via conduction through the fins and tubes and then by convection current as the liquid coolant flows along the internal walls of the tubes.

The compact surface area of the tube and fin matrix is capable of dissipating heat so that the hot air charge temperature, which may be as high as 120°C–150°C, can be reduced to something like 85°C–90°C with an engine coolant jacket temperature of around 80°C. Thus, cool liquid coolant is pumped from the water pump outlet to the intercooler tube matrix where heat is extracted from the hot charge. It is then passed back to the engine radiator header tank to be cooled in the usual way.

A typical air-to-liquid intercooled turbocharged engine layout is shown in Fig. 6.82, the intercooler tube matrix is positioned parallel to the engine and forms part of the intake manifold plenum chamber. This very compact arrangement is commonly used for large diesel engine applications.

6.12.3 Intercooler arrangements for vee six- and eight-cylinder engines
(Figs 6.83, 6.84 and 6.85)

For a vee six-cylinder petrol engine, each cylinder bank can be treated as a separate engine having its own turbocharger, wastegate and an air-to-air intercooler positioned at the front and to each side of the engine, with a balance pipe joining the two intake systems together (Fig. 6.83).

With heavy-duty vee eight-cylinder diesel engines having a single turbocharger, an air-to-liquid intercooler casing can be conveniently located between the cylinder-banks (Fig. 6.84) with its twin core horizontally mounted cooling tubes facing each cylinder-bank. Thus, compressed heated air charge from the single turbocharge outlet is divided into two flow paths as it winds its way around the spaced cooling tubes before passing into the cylinder-head inlet ports.

If twin turbochargers are employed, then a pair of parallel air-to-liquid intercoolers can be installed in the trough formed between the cylinder-banks (Fig. 6.85(a)). Each intercooler operates independently from the other so that there is a separate inlet passage to each intercooler casing, the air is then forced to flow across and between the tubes before entering the cylinder via the inlet ports.

Observe that the slow-speed heavy-duty diesel engines do not generally need wastegate relief to control boost charge pressure.

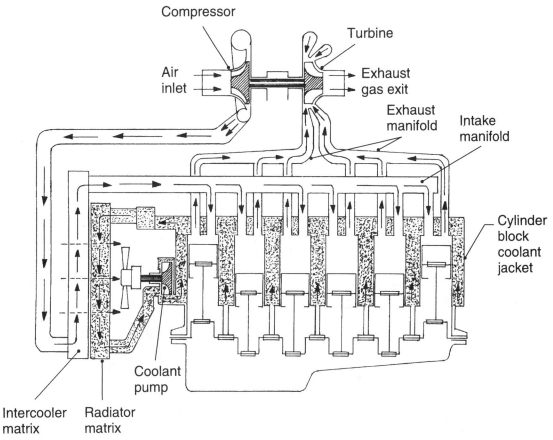

Fig. 6.80 Turbocharged six-cylinder engine with air-to-air intercooler

6.13 Turbocompound engine (Scania)

(Figs 6.86 and 6.87)

The turbocompound engine uses a conventional engine which utilizes a pair of turbine wheels. These units are connected in series to the exhaust gas discharge (Fig. 6.86).

The first-stage compressor-turbine drives the centrifugal compressor in the usual manner through the energy imparted to the turbine-wheel by the exhaust gas, so that it can compress and deliver high density air charge via an air-to-air intercooler to the engine cylinders.

The second-stage power-turbine receives the remaining heat energy from the exhaust gas and converts it into mechanical rotational power which is transferred to the crankshaft via a viscous coupling and a 25:1 reduction gear train. The

object of the large gear reduction is to reduce the power-turbine's maximum rotational speed of about 60 000 rev/min to some value slightly above the governed maximum engine speed. The viscous coupling cushions the power-turbine to crankshaft drive and absorbs the inherent vibrations created in the system.

A measure of the energy transformed from the exhaust gas to the compressor turbine-wheel is the amount the exhaust gas temperature drop on its way through the turbine blades. In its simplest form, the thermal energy absorbed by the turbine wheel can be given by the formula

$$Q = mC(T_2 - T_1) \tag{8}$$

where Q = rate of energy transfer (J/min)
       m = gas flow rate (kg/min)
       C = specific heat capacity
       T_1 = temperature at turbine inlet (°C)
       T_2 = temperature at turbine outlet (°C)

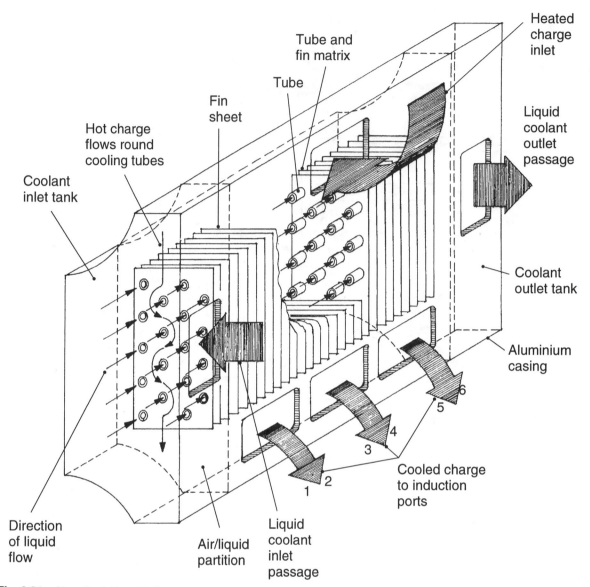

Fig. 6.81 Air-to-liquid intercooler

In the first turbine stage there is a temperature reduction of 100°C to 150°C from an input temperature of between 650°C–700°C to an output temperature of roughly 550°C. The second turbine stage is the extra energy extracted from the gas, this again produces something like a 110°C drop in temperature from 550°C to 440°C. The energy released in the second turbine stage would normally be lost to the atmosphere but the power-turbine converts this energy to rotational power, which is then transmitted back to the engine crankshaft.

With this system, the turbocharger can be matched to maximize engine torque in the low-speed range, whereas the power-turbine is matched to reach maximum rotational speed and power in the engine's upper-speed range.

It is claimed that the power turbine increases the rated power at 1900 rev/min from 380 kW to 400 kW and the maximum torque from 1660 Nm to 1750 Nm (Fig. 6.87). At the same time, the best specific fuel consumption improves from 191 g/kWh to 186 g/kWh.

Thus, the engine's maximum power and torque

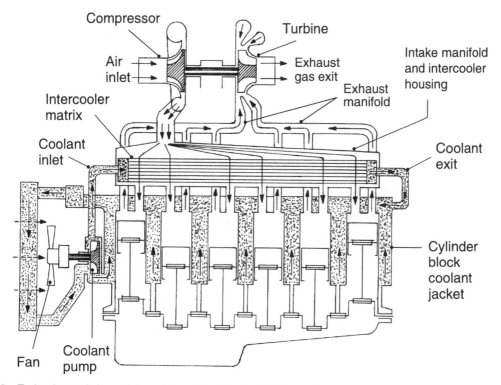

Fig. 6.82 Turbocharged six-cylinder engine with air-to-liquid intercooler

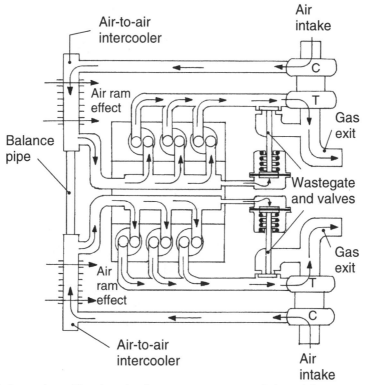

Fig. 6.83 Vee six-cylinder engine with twin turbocharger, wastegates and air-to-air intercoolers

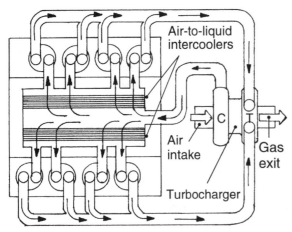

Fig. 6.84 Vee eight-cylinder engine with single turbocharger and air-to-liquid intercooler

is raised by 20 kW and 90 Nm respectively, whereas the best specific fuel consumption is improved by 5 g/kWh which is substantial.

The manufacturers state that this compound system increases the engine's overall efficiency from 44% to 46% and reduces the emission performance, particularly the NO_x levels.

6.14 Comprex pressure wave supercharger

Brown Boveri of Baden, Switzerland, have investigated the development potential, in the 1960s, of the pressure wave supercharger (original patent 1928) and named it Comprex because it was meant to compress and expand gases at the same time.

Comprex pressure-wave chargers can operate with pressure ratios of up to 2.8 : 1 although they are generally restricted to around 2.0 : 1, which makes them particularly suitable for road vehicle diesel engine applications, although petrol engines can also achieve direct performance benefits due to the charger's boost characteristics.

6.14.1 Design and construction
(Figs 6.88 and 6.89)

The principle component in the Comprex blower is the heat resistant iron–nickel alloy investment-casting cylindrical rotor, which has two rows of radial vanes separated by an intermediate shroud. The spacings between the vanes form straight axial channel cells which are open at both ends (Fig. 6.88). The inner and outer vanes are staggered from one another by half a pitch, and it is claimed that this double row staggered cell configuration considerably reduces the narrow band whistle by interference action, which was experienced with the original single-row vane-type rotor. The rotor has an overhanging mounting on its shaft which is carried in self-lubricating ball bearings (Fig. 6.89). An extension of the shaft beyond the bearings supports the pulley for the 'V'-belt drive, which revolves the rotor at a

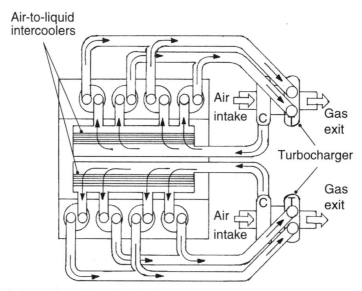

Fig. 6.85 Vee eight-cylinder engine with twin turbochargers and air-to-liquid intercooler

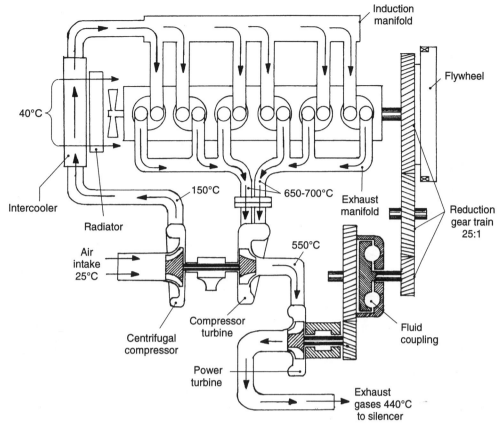

Fig. 6.86 Turbocompound engine (Scania)

multiple of normally three or four times the engine speed, although a multiple of six has been used for an agricultural tractor diesel engine.

The rotor revolves within a cylindrical fabricated steel casing. Both ends of the rotor are enclosed (Fig. 6.89). The overhang end of the rotor has a ductile-iron casting which forms the hot end wall to the rotor cells, this is known as the exhaust gas casing. Within this exhaust gas casing are two ports, an inlet gas port which receives exhaust gas from the engine manifold and a much wider exit gas port to extend the discharge period as the exhaust gases are expelled into the silencing system. In contrast, the aluminium casing supporting the rotor at the opposite end forms the wall to the cold end of the rotor cells, in which are cast both the air-charge wide-angled intake port and the narrower exit port.

6.14.2 Comprex pressure wave supercharger cycle of operation
(Figs 6.90 and 6.91(a–d))

The energy for supercharging is transmitted by pressure-waves from the exhaust gas to the incoming fresh air. This energy exchange takes place in the cell channels of the cylindrical rotor at the speed of sound (Fig. 6.90). Both ends of these cell channels are open. However, as the rotor revolves, the ends of the cells align periodically, with the intake and exit air charge ports at one end, and the exhaust gas inlet and exit ports at the opposite end. These cells therefore form the communicating passageways between the fresh charge and the outgoing exhaust gases.

For optimum results, the length of the rotor channel cells, the port timing and the rotor speed all have to be matched to one another and then to the engine.

The cycle of operation (Fig. 6.91(a–d)) may be conveniently divided into four phases:

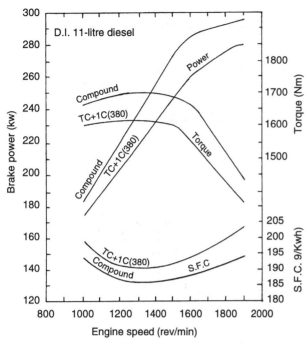

Fig. 6.87 Comparative performance of an engine when it is turbocharged and intercooled as opposed to it being turbocompounded

1 cell charge compression
2 cylinder charge filling
3 cell exhaust gas expulsion
4 cell charge filling

Cell charge compression phase
(Fig. 6.91(a))

Exhaust gas will continuously be discharged from the exhaust manifold to the exhaust gas inlet port of the compressor where it enters the momentarily aligned channel cells of the revolving rotor. The sudden introduction of the exhaust gas's kinetic energy to the existing fresh air charge in the rotor channels generates a pressure-wave which moves at sonic speed through the column of air charge to the other end of the rotor.

Cylinder charge filling phase
(Fig. 6.91(b))

The incoming pressure-waves produced by the exhaust gas, simultaneously accelerate and compress the fresh air charge trapped in the revolving channel cells of the rotor. Almost immediately, the far end of the pressurized cells will move in sequence into alignment with the charge exit port in the air charge casing thereby causing the fresh air charge to be driven out through the charge exit port into the engine's intake manifold.

Cell exhaust gas expulsion
(Fig. 6.91(c))

The angular movement of the rotor during the cylinder filling phase is relatively short, so that before the exhaust gas in the channels can flow out through the air charge port, the relevant

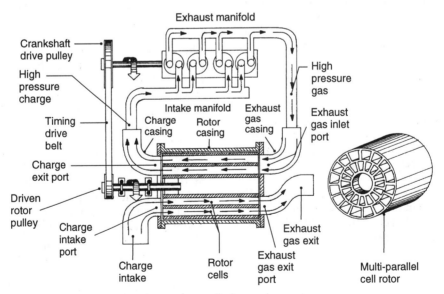

Fig. 6.88 Comprex pressure-wave supercharger four-cylinder engine system

358

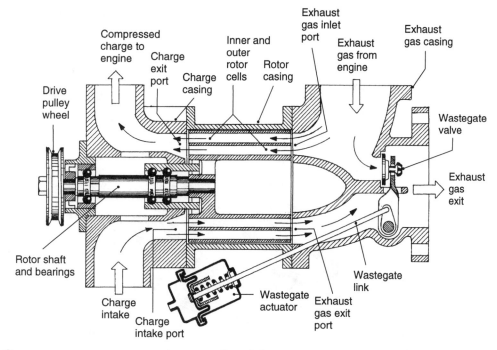

Fig. 6.89 Comprex pressure-wave supercharger sectioned view

Labels in figure:
Compressed charge to engine · Charge exit port · Charge casing · Inner and outer rotor cells · Rotor casing · Exhaust gas inlet port · Exhaust gas from engine · Exhaust gas casing · Drive pulley wheel · Wastegate valve · Exhaust gas exit · Rotor shaft and bearings · Charge intake · Charge intake port · Wastegate actuator · Exhaust gas exit port · Wastegate link

energized rotor channels will have moved beyond both the exhaust gas inlet port and the air charge exit port. Thus, exhaust gas will be prevented from entering or leaving these cells. However, with further rotation of the rotor, the cells containing the trapped exhaust gas will align with the exhaust gas exit port in the exhaust gas casing. Thus, the exhaust gas which is still highly energized will now be permitted to expand and flow out into the vehicle's exhaust system.

Cell charge filling phase
(Fig. 6.91(d))
The rapid movement of the outgoing exhaust gas from the active rotor cell creates a vacuum in the channel spaces it originally occupied, and shortly afterwards the revolving rotor aligns the channels with the charge intake-port leading edge. Consequently, the fresh air charge at atmospheric pressure waiting at the charge intake port will now rush into the lower pressure channel cells to endeavour to catch up with the fast escaping exhaust gases. This results in the fresh air charge flowing straight through the exposed rotor channels, where it then enters the exhaust gas exit port, it therefore provides a useful scavenging and cooling effect.

With further rotation of the rotor, these cells filled with the freshly charged air, will again align with the exhaust gas inlet port to repeat the supercharging cycle of events.

6.14.3 Unmatched engine speed and low load compensation
(Fig. 6.91(a–d))

The basic Comprex blower can only operate over a narrow performance speed and load band. This is because the pressure-wave travel time and peripheral speed have to be matched, but this must be related to engine load and speed operating conditions. Once the match has been made, the supercharger is only capable of operating at something like 15% above or below the matching speed conditions and, therefore, the blower requires a bypass valve to control part-load operation, which makes it unsuitable for road vehicle applications. This major problem was resolved by introducing recesses or pockets in both end casings (Fig. 6.91(a)). Consequently, these pockets alter the end flow boundary conditions so that additional pressure-waves will be generated which provide a regulating influence on the response to changes of engine speed and load.

359

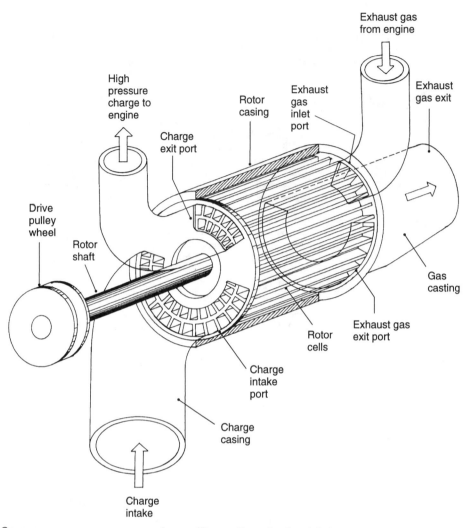

Fig. 6.90 Comprex pressure-wave supercharger (Brown Boveri): pictorial view

The air charge casing has two pockets, a compression pocket and an expansion pocket, whereas the exhaust gas casing has only a single gas pocket (Fig. 6.91(a)). The inclusion of these pockets modifies the flow path between the rotor cells and port passageways. At low peripheral speeds, the compression pocket enables the charging process to continue so that a relatively flat pressure curve will be obtained as the speed is varied. Thus, with rotational speeds as low as one-third of its maximum speed there will still be a modest rise in the air density. In contrast, the expansion pockets and the gas pockets ensure efficient scavenging from idle to maximum rotational speed.

Wastegate control
(Figs 6.89 and 6.92)

The pressure ratio increase can be controlled by the conventional wastegate method, as used with turbocharged systems (Fig. 6.89) where exhaust gas is bypassed directly from the engine's exhaust manifold when the air charge boost pressure reaches the desired maximum (Fig. 6.92). This is achieved by conveying a measure of the boost pressure via a rubber pipe to the wastegate diaphragm actuator, it then 'cracks open' at the predetermined diaphragm spring setting. A portion of the exhaust gas will then be diverted from the rotor channel cells and, instead, escapes through the open wastegate passage directly to the exhaust silencing system. High-torque back-

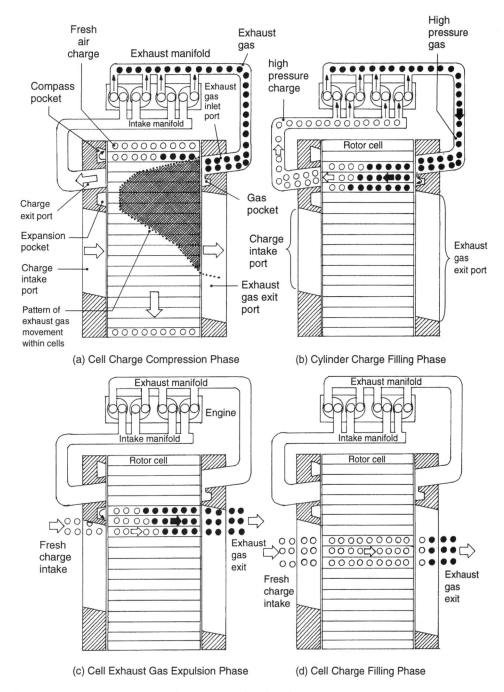

Fig. 6.91 Comprex pressure-wave supercharger operational cycle

up characteristics can be obtained by progressively reducing the pressure ratio once it has reached its peak by means of a pressure modulator, which dictates the degree of air-charge pressure arriving at the actuator.

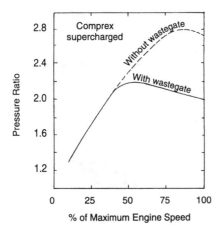

Fig. 6.92 Effect of engine speed on comprex supercharger boost pressure characteristics without and with wastegate control

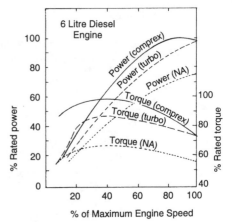

Fig. 6.93 Comparative performance for an engine operating naturally aspirated, turbocharged or with a comprex supercharger

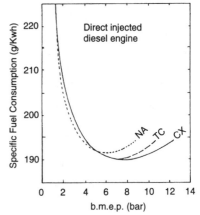

Fig. 6.94 Effect of engine load (b.m.e.p.) on specific fuel consumption for naturally-aspirated turbocharged and comprex superchargers

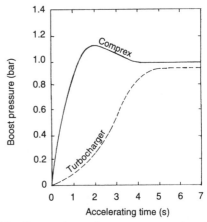

Fig. 6.95 Comparison of comprex supercharging and turbocharging accelerating boost pressure response times

6.14.4 Comprex supercharger performance characteristics
(Figs 6.93, 6.94 and 6.95)

The power consumption needed to drive the Comprex pressure-wave blower is very low, being of the order 2% at low-to-half load and down to about 1% of full-load. These losses, which are negligible, are due to bearing friction and windage since no mechanical work is done in compressing the air-charge, as is the case for positively driven mechanical superchargers such as the sliding vane, Roots, screw and oscillating spiral displacer types.

The peak adiabatic compressor efficiency for the Comprex charger is about 90% due to the internal cooling action by scavenging air, while the overall efficiency of the blower under optimal operating conditions is claimed to be 56%.

The Comprex blower specific fuel consumption for steady-state operation is not quite as good as it is for the turbocharger, but this is the opposite situation under actual road operating conditions. The better road operating fuel consumption of the Comprex charger compared with the turbo-charger is achieved by the higher torque developed at lower engine speed (Fig. 6.93) and the much better acceleration response in the low-to-mid speed range of the engine.

The specific fuel consumption for the Comprex charger (Fig. 6.94) improves with increased engine load (b.m.e.p.) compared with both the naturally aspirated and turbocharged engines;

362

however, in the light-load region, the naturally aspirated engine shows marginally better specific fuel consumption.

Generally, full-throttle pick-up from idle speed with the Comprex charger is not so good as the positive displacement charger, but it is better than the turbocharged engine, which has to accelerate the turbine and compressor assembly before it gets any boost (Fig. 6.95).

At present, the cost of manufacturing the Comprex pressure-wave type supercharger is nearly twice that of the turbocharger, which makes this class of supercharger a long-term economic proposition when weighing up the pros and cons.

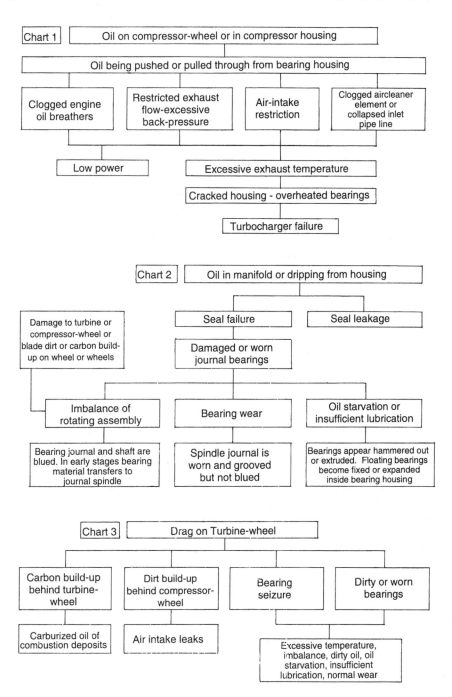

7

Carburetted Fuel Systems

7.1 Pollution derived from products of combustion

The main pollutant in urban air is carbon monoxide (CO) which is a colourless odourless gas of about the same density as air. Carbon monoxide is a poisonous gas which, when inhaled, replaces the oxygen in the blood stream so that the body's metabolism cannot function correctly. Small amounts of carbon monoxide concentrations, when breathed in, slow down physical and mental activity and produces headaches, while large concentrations will kill. Hydro-carbons—derived from unburnt fuel emitted by exhausts, engine crankcase fumes and vapour escaping from the carburettor—are also harmful to health. Oxides of nitrogen and other obnoxious substances are produced in very small quantities and, in certain environments, can cause pollution; while prolonged exposure is dangerous to health.

If the air–fuel mixture is too rich there is insufficient air for complete combustion and some of the fuel will not be burnt or at least only partly burnt. Since hydrogen has a greater affinity for oxygen, the hydrogen will take all the oxygen it needs leaving the carbon with a deficiency of oxygen. As a result of the shortage of oxygen a percentage of the carbon will be converted to carbon monoxide as well as carbon dioxide, and, with very rich mixtures, particles of pure unburnt carbon may be expelled from the exhaust as black smoke.

Incomplete combustion due to partial oxidation of the hydro-carbon fuel also produces other products such as acelylene and aldehyde. These products, when expelled from the exhaust, leave an unpleasant smell and are particularly notice-able during engine warm-up when a rich mixture is provided.

Conversely, if the mixture is made too weak it is unlikely that the atomized liquid fuel will be thoroughly mixed throughout the combustion chamber so that slow burning, incomplete combustion and misfiring may result.

A further characteristic of weak mixtures is that the excess oxygen (which has not taken part in the combustion process) at very high temperature is able to combine with some of the nitrogen that constitutes about three-quarters of the air, to form oxides of nitrogen such as nitrogen peroxide (NO_2). The amount of the nitrogen peroxide produced will increase as the mixture weakens until it peaks at just over an air–fuel ratio of 15.5:1, beyond this point the temperature of combustion begins to fall below that necessary for the formation of nitrogen peroxide so that with further reduction in mixture strength the amount of nitrogen peroxide progressively decreases.

7.1.1 Mixture strength and combustion product characteristics
(Fig. 7.1)

The chemically correct air–fuel ratio by mass for complete combustion is known as stoichiometric ratio. From Fig. 7.1 it can be seen how the three main exhaust pollutant products vary from different air–fuel ratios operating on either side of the stoichiometric ratio from a very rich mixture (11:1) to a very lean mixture (18:1).

The amount of carbon monoxide produced in the exhaust is about 8% for an 11:1 air–fuel ratio, but this percentage steadily decreases to zero as

364

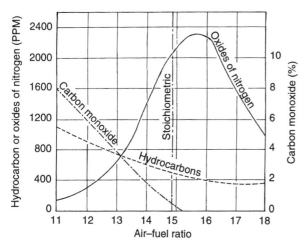

Fig. 7.1 Effects of mixture strength on exhaust composition of a petrol engine

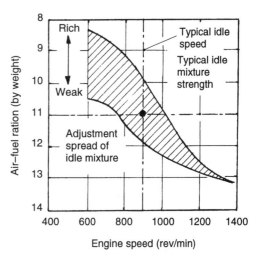

Fig. 7.2 Effects of slow running on mixture strength

the mixture strength is reduced to just beyond the stoichiometric ratio (on the lean side).

Hydro-carbon produced in the exhaust gases amounts to about 1100 parts per million (PPM) with a rich 11:1 air–fuel ratio and, as the mixture strength approaches the stoichiometric ratio, it progressively falls to around 500 PPM. A further weakening of the mixture to 18:1 air–fuel ratio only reduces the hydro-carbon content to approximately 350 PPM.

Oxides of nitrogen products formed during combustion are very low at 100 PPM with a rich air–fuel ratio of 11:1. As the mixture strength approaches the stoichiometric ratio it rises fairly rapidly to 2000 PPM, and a further reduction of the mixture strength to 15:1 peaks the oxides of nitrogen to something like 2300 PPM, weakening the mixture beyond this point rapidly reduces it until, at an 18:1 air–fuel ratio, it is 1000 PPM.

7.1.2 Mixture strength and performance characteristics
(Figs 7.2, 7.3 and 7.4)

A mixture of air and fuel which is homogeneously mixed (uniformly distributed throughout the combustion chamber) and which has just sufficient oxygen for complete combustion of the fuel is known as the stoichiometric mixture ratio. For typical petrols this ratio is about 14.8 to 1, although it does vary slightly with the composition of the blended fuel.

An engine operates over a wide range of speed,

load and temperature conditions which are not necessarily satisfied by the stoichiometric mixture—which is only a theoretical ratio based on ideal burning conditions.

During idle running the scavenging of the exhaust gases from the cylinders is poor and the atomizing of the mixture subjected to the high depression in the induction manifold is incomplete, so that a very rich mixture of about 10:1 to 12:1 is necessary (Fig. 7.2). With some modern lean-burn engines, much leaner mixtures of the order of 12 to 16:1 can be burnt while the engine is at idle speed.

Under full load conditions only a slightly enriched air–fuel mixture of about 12:1 to 13:1 is necessary, so that all the oxygen in the air–fuel mixture may be rapidly and completely burnt and transformed into an effective cylinder pressure before the exhaust valve opens (Fig. 7.3).

Conversely, under light load and medium engine speeds where maximum power is not required and economy is desired, weak mixtures from 16:1 to 18:1 may be burnt in well-designed combustion chambers without noticeable signs of engine misfiring, provided the slower rate of burning is compensated for by an additional ignition advance (Fig. 7.3).

Under cold starting cranking conditions the incoming air stream velocity is very low and is insufficient to support the heavier induced liquid fuel droplets, which are only partly atomized, so that the majority of the mixture is used up in wetting the induction pipe, ports and cylinder walls. Excess fuel is thus needed to make up for

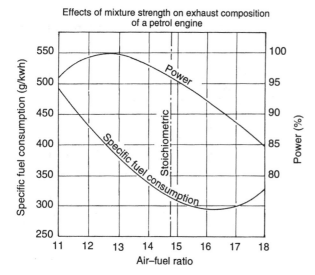

Fig. 7.3 Effects of mixture strength on power and specific fuel consumption of a petrol engine

the very small proportion of atomized mixture entering the cylinder, and in this country air–fuel ratios as rich as 6:1, and in some cases even richer 4:1 ratios are not uncommon for cold starting (see Fig. 7.4).

7.2 Carburation

Carburation is the process of supplying the correct air–fuel mixture ratio in an atomized form to the engine cylinders under the following operating conditions.

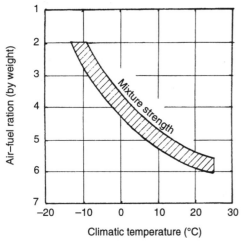

Fig. 7.4 Effects of cold starting on mixture strength

1 Cold starting
2 Slow running and progression
3 Full load, both low and high speed
4 Part load and cruising economy
5 Acceleration and deceleration

The simple single-jet carburettor cannot satisfy all these requirements. Consequently, both the constant-choke variable-depression and the variable-choke constant-depression carburettors have been modified so that the mixture strength is adjusted to match the changing engine demands automatically. An in-depth survey of the various approaches to carburation design, to meet the contrasting environmental, climatic, terrain, engine capability and drivers' expectations, will now be explored.

7.2.1 Multi-venturi choke tube
(Fig. 7.5)

The purpose of a multi-venturi, in contrast to a single one, is to increase the air velocity, and hence the depression in the region of the discharge spout, at low engine speed without seriously affecting the air flow capacity of the carburettor at high speed. The high air velocity at low engine speed increases the amount of airborne fuel flow at low speed and, more important still, permits greater low-speed pulsation of the fuel. In addition, the annular blanket of air provided by the annulus between the two venturi assists in keeping the wet fuel off the walls of the induction tract for a greater distance, so that more of the liquid particles of fuel reach the hot areas of the inlet ports directly.

7.2.2 Compensated idle system

This system comprises two separate idling circuits a basic supply circuit and an additional (auxiliary) circuit, both of which merge into one common outlet in the choke tube below the throttle valve.

Basic idle system
(Fig. 7.6)
The basic idle system is designed to supply a suitable fuel mixture to the engine to produce an exhaust gas composition within the specified CO

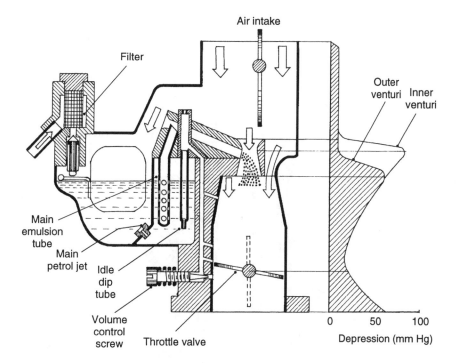

Fig. 7.5 Double venturi depression characteristics

content. The CO content is regulated by the mixture control screw (CO control) which is initially adjusted by the manufacturer and sealed in position.

Fuel flows through the main jet to the idle petrol jet, it is drawn up the tube where it meets and mixes with the idle air supply to form the basic mixture emulsion. The basic mixture then passes through the passages leading to the mixture control screw (CO control), and the metered mixture is then drawn out of the discharge duct into the choke tube on the engine side of the throttle.

Auxiliary idle system
(Fig. 7.6)
To ensure that the CO setting does not alter during idling adjustment, which is necessary to compensate for the variation of frictional resistance caused by the extreme of either a new tight engine or one that has bedded in over a period of time, an auxiliary idle system which is adjustable supplies additional mixture to that of the basic idle mixture (Fig. 7.6).

Fuel flows through the main jet to the auxiliary petrol jet, it is drawn up the tube and mixes with the incoming air bled from the auxiliary air jet to

form an additional emulsion mixture. It then passes down the additional mixture emulsion tube where it is further bled from ducts in the side of the venturi. The resultant weak emulsion then transfers to the slow-running control screw (bypass), the metered weak mixture then flows along a horizontal passage to meet up with the basic idle mixture. Finally, the combined basic and additional mixture is discharged into the choke tube on the engine side of the throttle. For idle speed regulation, only the slow running control screw should be reset.

Idle shut-off valve
(Fig. 7.6)
Engine stopping and starting is controlled entirely by the ignition switch, which interrupts the electrical primary coil circuit so that no high-tension spark plug voltage is generated. Unfortunately, the heat from the combustion chamber and valves is sufficient to ignite the incoming fuel mixture so that the engine runs on, this situation is sometimes known as 'dieselling'.

A solenoid-operated idle shut-off valve prevents the engine running on. When the ignition is switched off the solenoid is de-energized permitting the plunger shut-off valve to block the idle

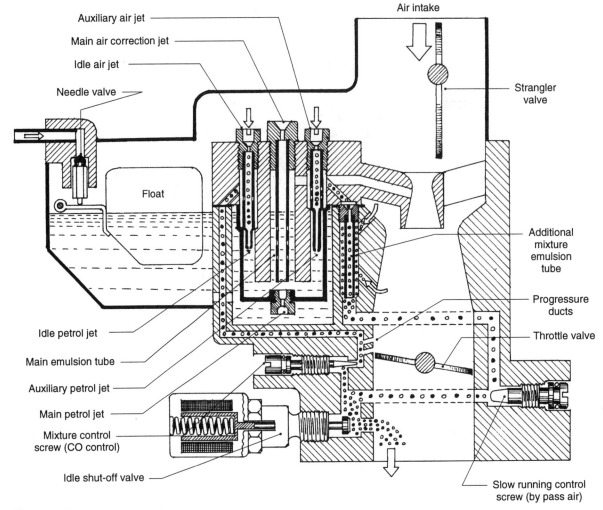

Fig. 7.6 By-pass idle system

discharge duct, to stop the fuel mixture supply to the choke tube. Conversely, switching the ignition on energizes the solenoid whch pulls back the plunger valve allowing the idle system to function again.

7.2.3 Air bleed and capacity well mixture correction, with air bleed economy device

Air bleed and capacity well mixture correction
(Fig.7.7)
Petrol enters the inlet pipe and passes through the centre of the needle seat where the flow is con-

trolled by the needle and the float. As the petrol level rises, the float lifts and, by means of the lever arm, closes the needle on its seating when the correct level has been attained. While the engine is running, petrol is drawn from the float chamber lowering the float and permitting more fuel to flow through the needle seat. If the petrol supply exceeds the engine demands the fuel level will rise and, in so doing, will push the needle against its seat, this action therefore automatically keeps the fuel level approximately constant under all operating conditions.

The capacity well, air bleed well, inclined discharge spout, main and compensation jets are all housed in a single casting known as the emulsion block (Fig. 7.7). Petrol in the float chamber surrounds this block and also enters the internal

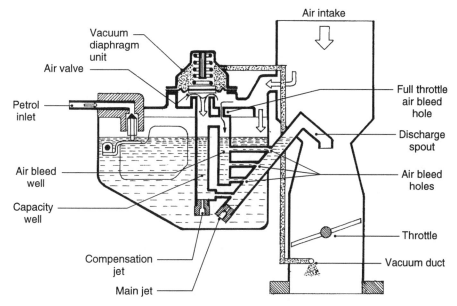

Fig. 7.7 Air bleed and capacity well mixture correction with air bleed economy devise

passage up to the level that exists in the float chamber, by way of the two submerged jets.

With the engine running and a progressively opening of the throttle, the depression in the venturi and discharge-beak region increases. This causes fuel fed from the main and compensating jets to be drawn out of the discharge spout and beak, into the venturi throat, where it is atomized by the fast-moving air stream. As the demands of the engine draw more fuel out than the main and compensation jets can supply, the fuel level in both wells falls and progressively uncovers the air bleed hole passages. Air thus enters into the discharge spout as these holes are exposed, and so weakens the mixture. This air and petrol therefore forms a well-mixed emulsion prior to being forced out of the spout into the venturi where it is atomized by the fast moving air stream. A further fall in fuel level empties the capacity well so that air will now enter the passage joining the capacity well to the discharge spout, it thereby relieves the suction on the capacity jet and further weakens the overall mixture strength.

Air bleed economy device
(Fig. 7.7 and 7.8)
The purpose of the economy device is to permit practically unlimited air bleeding (which weakens the mixture) in the discharge spout under part throttle cruising conditions, and to restrict the amount of air bleed (which enriches the mixture) when the throttle is wide open and the engine is pulling hard under load.

Under part throttle cruising conditions the manifold depression is high. This depression is imposed on the spring loaded side of the diaphragm, thereby lifting the valve from its seat. Air from the intake will now be admitted from both the full throttle air bleed hole and through the air valve into both the capacity and air bleed wells. As the fuel level drops with increased engine demands, unrestricted air bleeding through the air bleed passages occurs so that the overall mixture strength is weakened (Fig. 7.7).

Towards wide throttle maximum power engine operating conditions, the depression in the manifold reduces so that the diaphragm spring compressive force overcomes the force caused by the vacuum and closes the air valve. The air supply to the wells will now be restricted to the full throttle air bleed hole only. Consequently, the degree of air bleeding reduces and causes more fuel to be drawn from the discharge spout to meet the richer mixture demands of the engine under these conditions (Fig. 7.8).

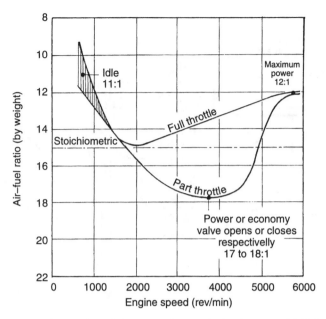

Fig. 7.8 Effects of mixture strength over the engine speed range for both part and full-throttle operating conditions

7.2.4 Vacuum operated power jet
(Figs 7.8, 7.9 and 7.10)

The purpose of a power jet is to enable a moderate-sized main jet to control mixture correction between idle and near maximum speeds with part throttle, and to allow a richer mixture to be supplied by the power jet when the throttle is wide open under full load conditions (Fig. 7.8). These devices may be either diaphragm or piston actuated.

With small throttle openings the depression in the manifold and carburettor on the engine side of the throttle is high. This depression is imposed by way of the vacuum duct on either the diaphragm (Fig. 7.9) or the piston (Fig. 7.10), it overcomes the compression of the return spring and so raises the diaphragm or piston stem clear of the extended valve stem, thus keeping the power jet closed.

Upon opening the throttle under full engine load conditions however, the depression in the throttle region of the carburettor is reduced. This allows the extended diaphragm or piston stem to push downwards and so force the power valve off its seating. A further quantity of petrol is admitted over and above that supplied by the main jet to the discharge emulsion tube. Here it mixes with the metered air passing through the air correction jet before it is expelled with the main

jet supply out of the discharge beak into the venturi.

7.2.5 Natural economy stage
(Fig. 7.11)

A pulsating flow in the inlet tract causes a given jet size to discharge more fuel than does a steady flow. The increased amount of fuel discharged with a pulsating manifold pressure is probably due to the push-and-release or jerk effect created between the constant atmospheric air pressure acting on the petrol head in the float chamber and the fluctuating depression experienced at the venturi. When the throttle is closed by as little as twenty degrees or so from the fully open position, the butterfly disc restriction in the choke tube is sufficient to damp the pressure pulsations reaching the venturi, so that fuel flow is reduced. This condition is known as the natural economy stage and is most apparent in engines where full throttle pulsations are greatest, for example four-cylinder engines with twin carburettors and six-cylinder engines utilizing triple carburettors. Thus, under part throttle cruising conditions pulsating depressions are sufficiently damped on the air intake side of the throttle as to reduce the amount of fuel discharged into the venturi. However, with the throttle about 2/3 or more open, the manifold induction pulsations are able

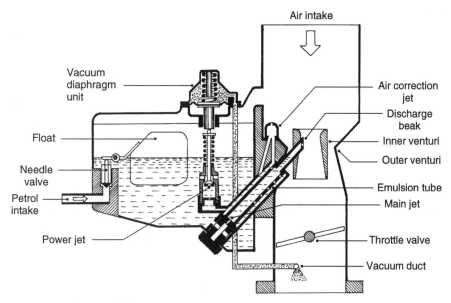

Fig. 7.9 Vacuum operated power jet

to reach the venturi region with very little interference so that more fuel will be drawn out from the discharge spout (Fig. 7.11).

On single-carburettor six and eight-cylinder engines, where induction depression pulses considerably overlap in the induction manifold, this economy stage is almost non-existent so that either an economy air valve or a power petrol valve is necessary.

7.2.6 Acceleration pumps
(Fig. 7.12)

The purpose of the accelerator pump is to provide additional fuel in an atomized form to enter the air stream when the throttle is opened quickly, as for example when a car is being accelerated. This extra fuel is necessary because when the throttle is suddenly opened, the passage at the throttle

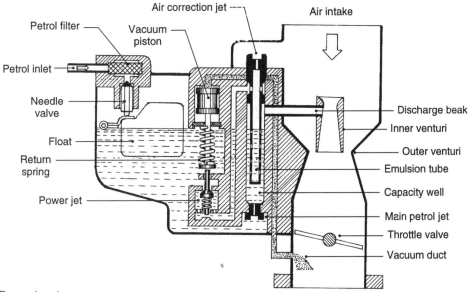

Fig. 7.10 Power jet piston type vacuum operated

371

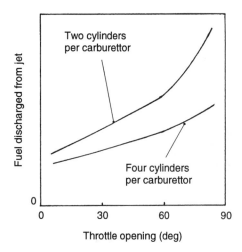

Fig. 7.11 Effects of throttle opening on the fuel discharged from jet

becomes greater and the air passing through responds very rapidly and speeds up, conversely the petrol being relatively heavy cannot increase its speed in the same time so that it flows at a slower rate than the air. As a result, the mixture becomes very weak and the power temporarily drops. Therefore, to compensate for the sluggish petrol response compared with that of the air, petrol is positively injected into the main air stream entering the venturi (Fig. 7.12).

The accelerator pump may be of the direct action type where the operating linkage responds to the slightest accelerator pedal movement, or it may be the more usual delayed action arrangement in which any sudden acceleration pedal movement is relayed through a compression spring to the pump diaphragm or piston (Fig. 7.12).

Piston-type cam-operated acceleration pump (Fig. 7.13)

When the throttle is closed the piston return spring pushes the piston to the top of its stroke. Petrol from the float chamber will then enter and fill up the piston cylinder through the open-return inlet valve.

As the throttle is opened, the accelerator cam will rotate and push away the relay lever, which thus rocks about its pivot. This forces the relay plunger down and simultaneously (due to the sluggishness of the compressed fuel to be displaced) will compress the delay spring. As this pressurized fuel is steadily pushed through the non-return discharge ball-valve and jet out into the venturi, in the form of a spray, the stiff delay spring will expand forcing down the piston against the opposing upthrust of the weak return spring. Continued expansion of the delay spring, to the extent of the total downward movement of the relay plunger, ensures a follow-up of the pump

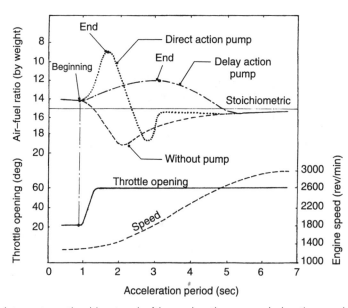

Fig. 7.12 Effects of mixture strength without and with acceleration pump during the acceleration period

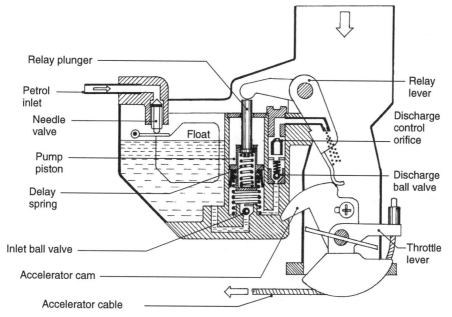

Fig. 7.13 Accelerator pump piston-type cam-operated

operation after the throttle movement accelerating the engine has ceased (Fig. 7.13).

Note that the spring-loaded discharge ball valve will only permit fuel to be drawn out from the discharge jet, when there is sufficient depression in the region of the discharge jet, this condition might occur with a wide open throttle at very high engine speeds and will thus strengthen the mixture under these conditions.

The purpose of the discharge control orifice is to allow fuel to flow from the discharge valve to the jet by way of the outer conical nose of the control valve under steady acceleration conditions. If, on the other hand, the accelerator pedal is opened sharply, the discharge control orifice will jerk upwards and so restrict the fuel flow through the centre of the orifice only. It thereby prevents over enriching and prolongs the acceleration discharge time.

Diaphragm-type lever-and-cam operated acceleration pump
(Fig. 7.14)

When the throttle is closed, the lower position of the cam profile contacts the rocking-lever. This relieves the load on the disphragm return-spring which therefore expands and pushes the diaphragm outwards. A depression will be created in the enlarged diaphragm chamber housing the

return-spring, and the inlet non-return ball-valve will be drawn away from its seat while the economy ball valve will be closed by its spring, and the discharge ball-valve also closes due to the weight of its ball. Petrol will then be admitted from the float chamber through the inlet valve to fill up the diaphragm chamber housing the spring.

Upon the throttle being opened, the cam profile pushes against its roller follower and so pivots the rocking-lever, this compresses the delay-spring, the inlet-valve will then drop onto its seat and close while both the economy and discharge valves will be forced to open. Petrol will thus be displaced through both outlet-valves to the discharge jet, where it is then expelled in the form of a spray into the venturi air stream. The amount and time taken to discharge this fuel will depend upon the jet size and the delay-spring stiffness. Thus, since the delay-spring is stronger than the diaphragm return-spring, it will progressively expand the diaphram against the resistance of the weaker return-spring, and it is this action which enables the accelerator pump discharge time to be extended (Fig. 7.12). During the pumping period, a proportion of the pressurized fuel, along with any trapped vapour, will be pushed back into the float chamber through the bypass jet, the result being that it damps out, to some degree, the acceleration pump response when the throttle is only gently opened.

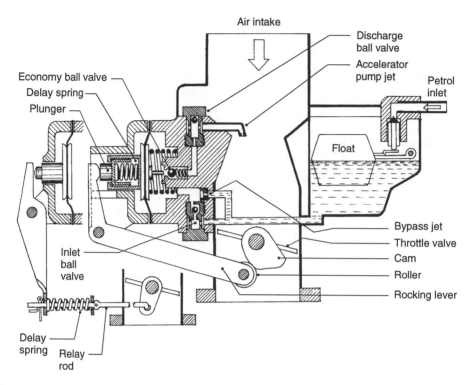

Fig. 7.14 Accelerator pump diaphragm type lever and cam-operated

As the throttle approaches the fully open position, the stem protruding out from the diaphragm will dislodge the economy ball-valve so that petrol can be drawn out from the discharge jet if a sufficient vacuum exists in the upper venturi region. This action provides, under wide-throttle heavy-load conditions, a continuous discharge of fuel in addition to that normally supplied from the venturi discharge beak. In other words, it has a similar function to an economy or power jet device. Note that the diaphragm stem and economy ball valve is only adopted for certain engine applications.

7.2.7 Cold starting
(Fig. 7.15)

To enable an engine to be started in cold weather, considerably more petrol must be supplied to the venturi tube compared with normal running requirements. Under these conditions, the air–fuel ratio could be as rich as 7 or 4 to 1. This additional fuel is necessary to compensate for the large proportion of petrol which, when drawn into the

venturi mixing chamber is not able to atomize and therefore will not remain suspended in the cold slow moving air stream when the engine is being cranked (Fig. 7.15). This unprepared, poorly mixed, petrol thus spreads around the cold metal

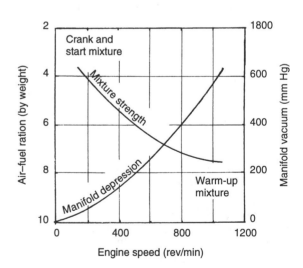

Fig. 7.15 Effects of cranking and slow running speeds at manifold vacuum and mixture strength

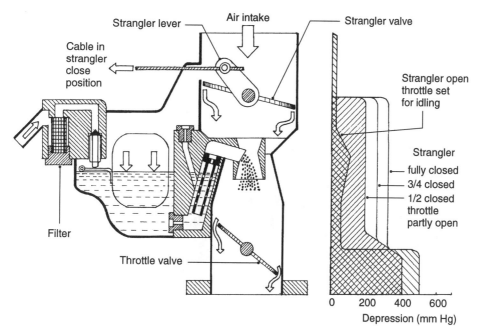

Fig. 7.16 Manual operated choke (strangler)

walls, collects together and gravitates to the flow of the induction manifold and ports, it is then gulped into the cylinders in the form of large liquid droplets quite unsuitable for efficient burning.

Improved atomization of petrol and mixing with the air stream will result if the air intake is restricted so that not only will the air stream velocity increase, but the resultant rise in depression in the mixing chamber forces more petrol, in the form of a spray, to be discharged.

Manually controlled strangler cold start choke
(Fig. 7.16)
This simple device consists of a butterfly-type flap (plate) valve supported by a spindle and mounted centrally on the intake side (upstream) of the venturi. When a cold engine is to be started, the strangler spindle lever is rotated until the valve is closed by pulling a cable which is spanned between the carburettor and dashboard.

A large depression on the engine side of the strangler will then draw petrol from the discharge spout, and the relatively high velocity of the passing air stream will stretch and break up the majority of the liquid petrol into very small particles as it moves through the venturi tube, induction manifold and ports.

As the engine warms the ports, manifold and venturi receive heat. This pre-heats both the fresh air entering the venturi tube and the petrol transferred from the float chamber into the venturi. As the engine warms to its normal working temperature the strangler valve should be progressively opened by pushing in the dashboard cable knob.

Semi-automatic choke with offset strangler spindle
(Fig. 7.17)
The effectiveness of the simple manual strangler valve choke may be improved by slightly offsetting the valve plate spindle to one side of the centre line of the intake tube. With this device the operating cable is connected indirectly to the strangler by way of a relay lever, pull-off arm and link-rod. When the cable is pulled out from the dashboard, the relay lever will be rotated anti-clockwise which closes the strangler valve and partially opens the throttle valve.

To start the engine from cold the strangler is closed and the engine is cranked, once the engine starts, the rapid rise in engine speed will create an excessively high vacuum on the engine side of the strangler valve. The pressure difference between the atmospheric side and the vacuum side of the strangler plate will apply a downward force on both sides of the spindle plate but, since there is a

375

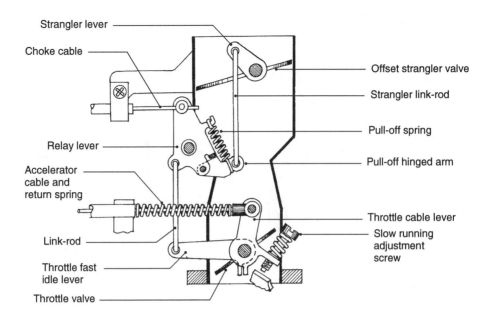

Strangler lever

Choke cable

Relay lever

Accelerator
cable and
return spring

Link-rod

Throttle fast
idle lever

Throttle valve

Offset strangler valve

Strangler link-rod

Pull-off spring

Pull-off hinged arm

Throttle cable lever

Slow running
adjustment
screw

Fig. 7.17   Cold start choke devices

larger area of plate on the left-hand side, this side will be pulled down and the other smaller plate side will move upwards. In other words, the resultant force applied to the two half-strangler plates produces an anticlockwise torque about the spindle. This turning effect will be opposed by the stiffness of the pull-off spring holding the relay-lever and hinged arm together. Eventually, the twisting effect on the spindle will stretch the pull-off spring so that the pull-off hinged arm will pivot about the relay-lever, thus permitting the strangler to open. An increased quantity of air will now enter the venturi to relieve the vacuum partially and increase the mass of air entering the cylinders so that the engine does not choke or stall.

Semi-automatic choke with offset strangler spindle and pull-down diaphragm
(Fig. 7.18)

This choke device incorporates an offset strangler spindle which is interconnected by way of a spring-loaded link-rod to a fast-idle cam and cable lever. When the choke cable knob is pulled out, the fast-idle cam and cable lever rotates about its pivot, this pushes up the spring-loaded link-rod so that the strangler lever closes the valve. At the same time the fast-idle cam pushes out the throt-

tle fast-idle lever so that the throttle valve is partially opened.

As the car is being driven from standstill, the carburettor throttle will be progressively opened and this increases the demand for air flow through the intake tube. Consequently, the velocity of air, passing on both sides of the strangler between the restricted valve plate gap and venturi tube wall will increase to such an extent as to raise the depression underneath the strangler until, due to the offset spindle, there will be sufficient force to twist the strangler anticlockwise against the stiffness of the push-down spring, which is compressed. The outcome will be that an increased amount of air will enter the carburettor system to meet the engine's requirements, enabling it to accelerate without hesitation.

In addition to the automatic strangler opening by the offset spindle, a diaphragm unit is also included, this device is sensitive to the sudden rise in venturi depression when the engine speed rises from cranking to fast-idle conditions. The diaphragm is connected to a bell-crank lever which relays its movement to the strangler lever when a high vacuum is created on the engine side of the throttle. Once the engine fires and speeds up to a fast idle, the vacuum created will act on the spring side of the diaphragm and so draw it to the right; simultaneously, the push-off spring will be com-

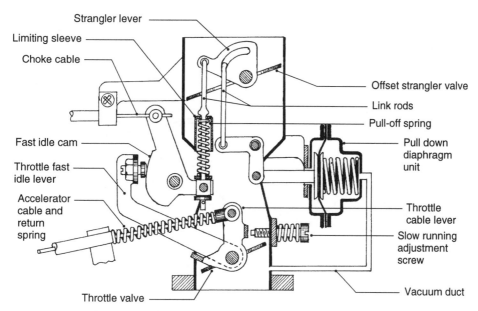

Fig. 7.18 Semi-automatic choke with off-set strangler spindle and pull-down diaphragm

pressed, enabling the strangler valve to rotate anticlockwise and open. Immediately this occurs, the excessively large depression will be relieved and so prevent the engine from becoming choked and stalling.

Note that the offset spindle and vacuum unit action are no substitute for the eventual return of the cable to the strangler fully open position.

Automatic choke
(Fig. 7.19(a and b))

Fully automatic chokes are designed to sense and respond to four different operating conditions:

1 the increased frictional drag of the engine when cold;
2 the relatively slow change in the engine's coolant temperature between cold starting and normal operating conditions;
3 the rapid rise in manifold vacuum as the engine speed changes from cranking to fast idling speed when the engine fires;
4 the mean increase of air velocity past the strangler butterfly valve when the vehicle is accelerated from a standstill with the engine cold.

These cold start carburettor operating conditions are satisfied in the following way.

1 A fast cold idle, to prevent the engine stalling when cold, is provided by a cam attached to the relay-lever and a rocking-lever with an adjustment screw which contacts the cam (Fig. 7.19a). When starting a cold engine, the accelerator pedal must be fully pressed and released, this lifts the adjustment screw away from the cam profile and permits the bimetallic spring tension to rotate the cam to the cold-start fast-idle position. As the engine warms up the bimetallic spring will progressively rotate the cam to the hot position allowing the adjustment screw on the rocker lever to move to a lower part of the cam profile. Consequently, the throttle valve will be able to close proportionally until eventually the hot-idle position is reached (Fig. 7.19(b)).

2 The angular position of the strangler spindle is controlled by a bimetallic spiral spring mounted on an anchor part at the centre with the outer end of the spring linked to a pivoting relay lever arm. A second arm forming part of the bimetallic spring relay-lever is connected indirectly to the strangler lever by means of the relay link-rod. During operating conditions the bimetallic spiral leaf spring tends to straighten out as it absorbs heat from the coolant system or, conversely, curls when it cools. Subsequently, as the engine warms up, the spring

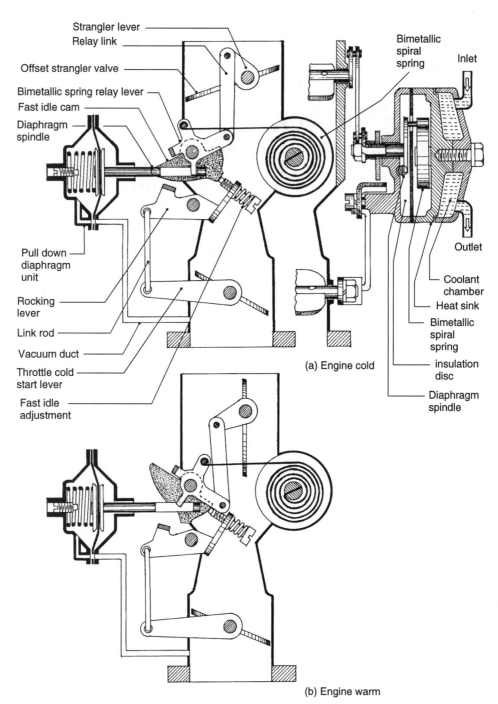

Strangler lever

Relay link

Offset strangler valve

Bimetallic spring relay lever

Fast idle cam

Diaphragm spindle

Bimetallic spiral spring

Inlet

Pull down diaphragm unit

Rocking lever

Link rod

Vacuum duct

Throttle cold start lever

Fast idle adjustment

Outlet

(a) Engine cold

Coolant chamber

Heat sink

Bimetallic spiral spring

insulation disc

Diaphragm spindle

(b) Engine warm

Fig. 7.19 Constant choke weber-type automatic choke

will unwrap itself and so pull the relay-lever about its pivot in a clockwise direction. This movement is relayed to the strangler lever and hence rotates the strangler spindle anticlockwise to open the strangler-valve (Fig. 7.19(b)).

3 Initially, when the engine is cranked from cold and it fires, there will be rapid changes in speed from something of the order of 150 rev/min to 1000 rev/min and, as a result, an excessively high vacuum would be created in the venturi

region which would consequently flood and choke the engine. To prevent the discharge spout supplying too much petrol, the depression is released by means of a pull-down diaphragm unit which has a vacuum duct intersecting the venturi on the engine side of the throttle. The sudden rise in vacuum will be experienced in the vacuum diaphragm chamber, and will overcome the spring tension and draw the diaphragm and spindle into the chamber (to the left). The result will be that the relay lever is given a clockwise rotation, and the strangler spindle an anticlockwise movement, this partially opens the strangler valve and so relieves the depression in the venturi mixing chamber (Fig. 7.19(a and b)).

4 With a cold engine and high engine speeds the air velocity past the tips of the offset spindle valve plate applies an opposing turning force on both sides of the spindle, and since the resultant twisting moment will be greater on the larger valve plate side, the strangler-valve will tend to rotate towards the open position against the tension of the bimetallic spiral spring. Additional air will then be able to enter the venturi to meet the engine's demands without engine hesitation while accelerating (Fig. 7.19(a and b)).

Disc-type progressive fast idle cold starting device
(Fig. 7.20)

This type of cold start device relies on the closed throttle valve itself creating the high vacuum necessary to draw excess petrol for cold starting.

The mechanism comprises a starter petrol and air jet, capacity-well, emulsion chamber, control-disc valve, suction-disc valve and control spindle.

To start a cold engine the operating lever is pulled backwards so that the disc-control valve orifices align with the petrol supply duct and the emulsion discharge duct. When the engine is cranked with the throttle closed the high depression on the engine side of the throttle will be conveyed through the emulsion chamber to the capacity-well. Petrol will then be drawn through the supply-duct disc orifice to the emulsion chamber where it mixes with air coming through the external air duct and the air passage just below the venturi. The air and petrol emulsion is then drawn through the discharge duct into the main air flow below the throttle. The excess amount of air which bypasses the throttle valve increases the engine speed to a fast idle condition.

Once the engine starts the increased depression will empty the starter capacity-well very rapidly

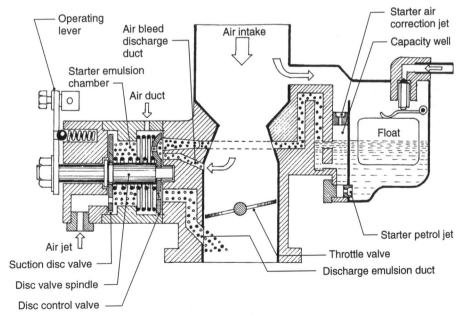

Fig. 7.20 Solex disc-type progressive fast idle cold starting device

with the result that air will enter the supply ducts in the capacity-well to bleed the petrol supply coming from the starter petrol jet, this results in a reduction of petrol passing through to the emulsion chamber. An additional mixture-weakening stage is introduced by the suction-disc valve which will be drawn inwards against the inner spring if the depression should become excessive. Thus, it permits air to pass around the edge of the disc to release some of the depression established in the emulsion chamber.

As the throttle is opened under acceleration conditions, the depression on the engine side of the throttle will move up to the venturi region so that additional fuel in an emulsion form will then be drawn from the air bleed discharge duct into the venturi-tube.

During the warm-up period the disc control valve should be progressively rotated, thus exposing smaller disc discharge orifices to reduce the quantity of emulsion passing to the engine. Eventually the disc will completely block-off the petrol duct and emulsion discharge ducts.

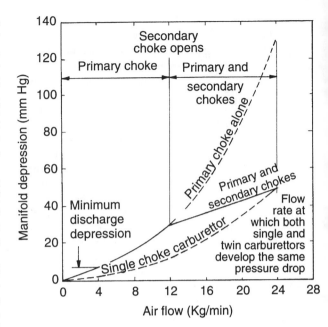

Fig. 7.21 Compound barrel carburettor flow characteristics

7.2.8 Justification for a compound twin barrel carburettor
(Fig. 7.21 and 7.22)

This type of carburettor has two fixed venturi barrels which operate in two stages, the first stage functions only through the primary barrel while the second stage uses both the primary and secondary barrels.

The objective of the primary barrel stage is to restrict the air intake passage at low engine speed outputs thereby creating a relatively high air charge speed; fuel entering the venturi will therefore be discharged into a fast moving air stream to become highly atomized and, consequently, will become thoroughly mixed on its way through to the intake ports. Over the primary operating stage the fuel droplets will be broken up into very tiny particles light enough to remain suspended in the rapidly flowing air stream. Thus, fuel wetting of the induction passage walls and the accumulation of globules of liquid fuel which, at random, tear themselves away from the metal surfaces and enter the cylinders in an erratic form, are prevented. The improvement in mixture quality and the ability to maintain it in this condition on its way into the engine's cylinder, greatly improves

low speed engine response flexibility, torque and fuel economy.

With increased engine speed and load, the air speed and, thus, the pressure drop in the primary venturi will become excessive with the result that the engine's volumetric efficiency will begin to decrease. At this stage, the secondary barrel throttle valve will therefore be made to open

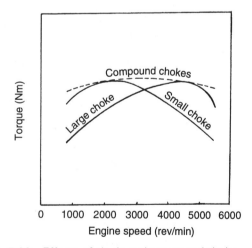

Fig. 7.22 Effects of single and compound choke tubes on engine torque over its speed range

380

progressively to give increased air flow. The primary barrel pressure drop is now slightly reduced as some of the engine's air intake is bypassed through the secondary barrel, thus producing a small pressure drop in the secondary venturi so that fuel will also be drawn from this barrel. As the engine's speed and power continues to rise, the overall depression in the two barrels will increase but at a much reduced rate (Fig. 7.21).

Thus, the compound two-barrel carburettor overcomes the limitations of a single-barrel venturi which has difficulties in meeting the engine's demands at both low and high speed and load conditions (Fig. 7.22). Thus, at low engine speeds and load conditions, a high air velocity is required to mix thoroughly the air and fuel charge if good torque and fuel economy is to be obtained. Unfortunately, at high engine speed and power operating conditions a large-bore barrel is essential to reduce the air reistance and so raise the engine's volumetric efficiency if optimum power is to be achieved. Therefore, if a large barrel was to be used at the lower engine speed range, the quality of charge mixing would be poor so that the engine's response and performance would also be inferior. It therefore follows that a twin-barrel carburettor, which can adjust to both extremes of engine speed and power range, is highly desirable.

Auxiliary-throttle operated secondary-stage barrel
(Fig. 7.23(a–c))
This class of compound two-barrel fixed-choke carburettor has a primary and secondary-stage barrel, with each having its own throttle valve. Both the spindles which support these valves are, in this particular example, interconnected by meshing quadrant gears and are synchronized to open and close together. An auxiliary throttle valve is also incorporated inside the secondary barrel, this valve being situated with its spindle slightly to one side and above the secondary throttle, and a counterweight on an external arm attached to the auxiliary throttle spindle, holds the valve in the normally closed position (Fig. 7.23(a)).

Should the accelerator pedal be fully pressed to open both throttles at low speed, the auxiliary throttle will remain closed, so that in all accelerations, and low-speed running under load, only the primary barrel operates. This ensures that sufficient depression exists in the primary venturi to

discharge a highly atomized charge mixture into a rapidly moving air stream, thus fulfilling the requirements for good low-speed torque and fuel economy (Fig. 7.23(b)).

When the air flow demands exceed those provided by the primary barrel, the manifold depression will sufficiently rise to some predetermined value to apply an anticlockwise turning moment to the offset auxiliary throttle spindle, this eventually lifts the counterweight and arm and so opens the auxiliary valve. The secondary barrel stage now comes into action. Thus, in addition to the primary barrel charge supply, the secondary venturi discharge will provide the extra charge mixture necessary at high engine speed and load (Fig. 7.23(c)).

Differential linkage operated secondary stage throttle
(Fig. 7.24(a–c))
The interconnecting linkage between the primary and secondary throttle valves takes the form of a relatively long spring-loaded primary relay-lever with a protruding peg at one end which fits into an elongated slot formed in the much shorter secondary throttle-lever.

The primary throttle spindle is operated by the throttle-lever which is clamped to it, this spindle also acts as a pivot post for the primary relay-lever which does not move until the throttle spindle and lever have rotated to about 2/3 full-throttle position (60°) (Fig. 7.24(a)). Further primary throttle-spindle movement will convey rotation to the relay-lever due to the right-angled lug on the throttle-lever contacting the edge of the relay-lever (Fig. 7.24(b)). Motion from the primary relay-lever to the secondary throttle-lever is transferred by means of the peg and its slot. Thus, when the relay-lever begins to rotate its peg pushes down on the elongated slot formed in the secondary throttle-lever and therefore also rotates the secondary throttle (Fig. 7.24(c)).

An important feature, necessary with the throttle interconnecting mechanism, is that the secondary throttle opening action must be rapid; that is, the secondary throttle must move from closed to fully open in the same time as the primary throttle completes its remaining 1/3 throttle opening movement, so that both throttles reach the fully-open position simultaneously. This differential in throttle opening speeds is achieved by the leverage ratio between the primary relay-lever and the secondary throttle-lever. The actual

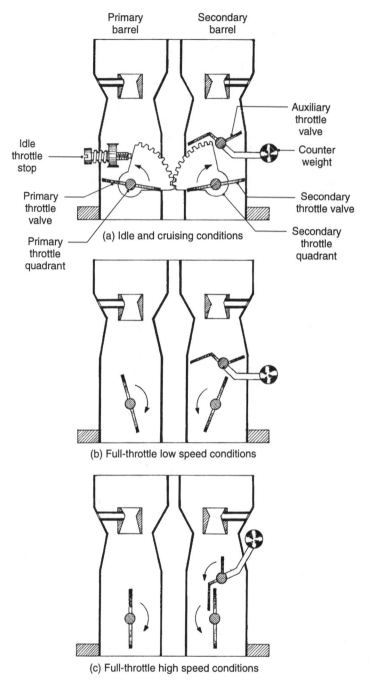

Primary barrel **Secondary barrel**

Auxiliary throttle valve

Counter weight

Secondary throttle valve

Secondary throttle quadrant

Idle throttle stop

Primary throttle valve

Primary throttle quadrant

(a) Idle and cruising conditions

(b) Full-throttle low speed conditions

(c) Full-throttle high speed conditions

Fig. 7.23 Bendix Stromberg compound barrel auxilliary throttle operated carburettor

leverage ratio obtained between the two levers changes slightly due to the contact point between the two levers varying as their relative angular positions alter. Throughout the secondary-stage throttle movement the leverage will range between 2.5:1 to 3:1.

Vacuum diaphragm operated secondary stage throttle
(Fig. 7.25)

With this two-barrel fixed-venturi carburettor there are two independently operated throttle valves, one for each barrel. The primary-stage

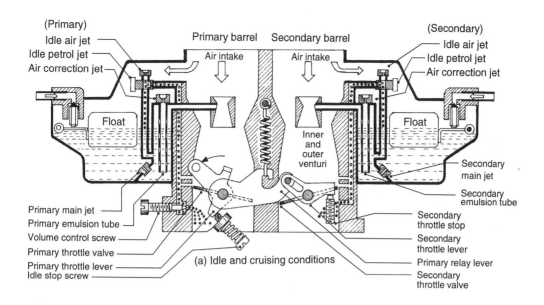

(Primary)

Idle air jet
Idle petrol jet
Air correction jet

Primary barrel   Secondary barrel

Air intake   Air intake

(Secondary)

Idle air jet
Idle petrol jet
Air correction jet

Float

Inner
and
outer
venturi

Float

Secondary
main jet

Secondary
emulsion tube

Primary main jet
Primary emulsion tube
Volume control screw
Primary throttle valve
Primary throttle lever
Idle stop screw

(a) Idle and cruising conditions

Secondary
throttle stop

Secondary
throttle lever

Primary relay lever

Secondary
throttle valve

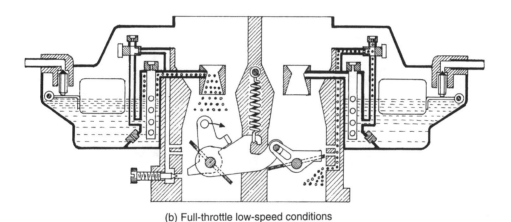

(b) Full-throttle low-speed conditions

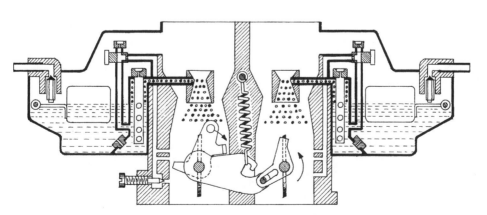

(c) Full-throttle high-speed conditions

Fig. 7.24   Weber compound barrel differentially operated carburettor

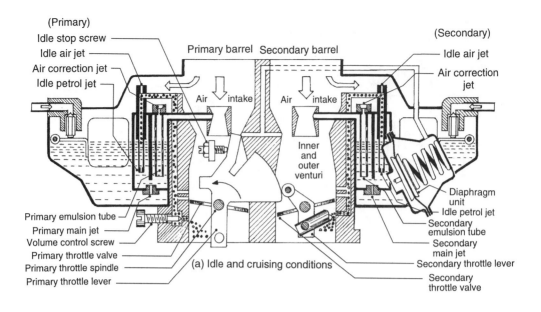

(Primary)
Idle stop screw
Idle air jet
Air correction jet
Idle petrol jet

Primary barrel   Secondary barrel

(Secondary)
Idle air jet
Air correction jet

Air intake   Air intake

Inner and outer venturi

Diaphragm unit
Idle petrol jet
Secondary emulsion tube
Secondary main jet
Secondary throttle lever
Secondary throttle valve

Primary emulsion tube
Primary main jet
Volume control screw
Primary throttle valve
Primary throttle spindle
Primary throttle lever

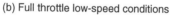
(a) Idle and cruising conditions

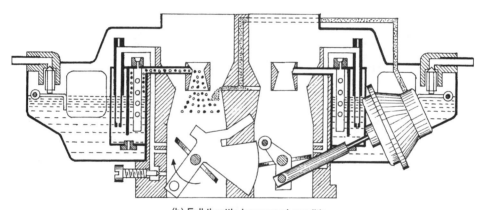

(b) Full throttle low-speed conditions

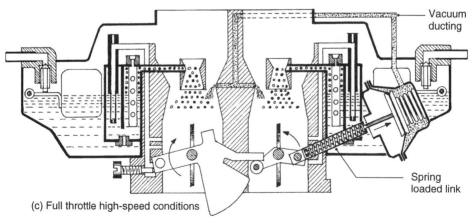

Vacuum ducting

Spring loaded link

(c) Full throttle high-speed conditions

Fig. 7.25   Solex compound barrel vacuum operated carburettor

384

barrel throttle-valve is operated directly by the accelerator pedal linkage which is connected to the primary throttle lever, and part of this lever has a sector-plate which contacts a roller on the secondary throttle-lever. This sector acts as an interlock and prevents the secondary throttle from opening until the primary throttle valve has itself opened to at least 2/3 to 3/4 of its full amount (Fig. 7.25(a)). At this point the secondary-throttle lever-roller clears the sector profile, enabling the secondary throttle to open or close independently of that of the primary first-stage barrel and valve. This secondary stage is then controlled by a vacuum diaphragm and opens or closes according to the depression in the venturi of both barrels.

For cruising and part-load low-speed conditions up to about 2/3 full throttle, only the primary-barrel stage will operate, this ensures that, for these operating conditions there will be an adequate degree of depression in the venturi to produce a metered highly-atomized mixture supply entering the fast-moving air-stream (Fig. 7.25(b)).

Upon the primary-throttle valve-spindle rotating more than about 60° to 70° of its maximum angular movement, the interlocking mechanism of the secondary throttle valve is released, which allows it to open according to the engine's demands when approaching full-load conditions (Fig. 7.25(c)).

The secondary-barrel stage will be actuated by the depression created in both the primary and secondary venturis, this being communicated to the diaphragm spring chamber which then pulls back the diaphragm against the return-spring in proportion to the degree of depression sensed, and consequently opening or closing the secondary throttle-valve. Thus, the opening of the secondary throttle-valve automatically reduces the restriction of the air charge passing through the barrels so that at full-load high-speed conditions the engine's volumetric efficiency, and therefore possible power, will be at a maximum.

The spring-loaded diaphragm link-rod prevents the secondary throttle-lever and roller applying a heavy load on the primary throttle sector-plate, if the diaphragm is subjected to high depressions when the interlock mechanism stops the secondary throttle-valve opening.

Summary and comparison of secondary stage throttle control

The secondary-barrel butterfly can be controlled by the following methods.

1 Auxiliary throttle operated

The pressure difference created across the auxiliary butterfly valve with the offset spindle produces a torque which counteracts the balance (counter) weight and opens the secondary barrel at some predetermined air-velocity in the primary barrel. There is a tendency for the auxiliary offset spindle valve opening to be erratic and not necessarily progressive.

2 Differential-lever linkage operated

A differentially operated mechanical throttle linkage, which does not actuate the secondary butterfly until about 2/3 full opening of the primary throttle valve, and a quick release link-arm, then rapidly open the secondary butterfly valve. This mechanism operates purely on the degree of throttle opening and is insensitive to engine speed and load conditions.

3 Diaphragm operated

The degree of vacuum sensed at the primary stage venturi operates a diaphragm actuator which is linked up to the secondary butterfly-valve and tuned to operate at some predetermined depression in the primary-barrel. An interlocking device prevents the secondary throttle beginning to open before a definite primary throttle opening has been attained. This arrangement enables the secondary throttle-valve opening to be proportional to the engine speed and load demands.

7.2.9 SU constant vacuum variable choke (HIF) carburettor

Normal operating conditions
(Fig. 7.26)

The full metering is accomplished by a tapered needle which is raised or lowered in a jet to alter the effective annular jet orifice, and hence fuel flow. The needle projects from underneath the flat face of the cylindrical air-valve, which alters the choke area as it is raised or lowered. The upper part of the air-valve is enlarged to form a piston which fits into the lower open end of the vacuum chamber. A spindle situated in the centre of the air-valve guides the assembly into the

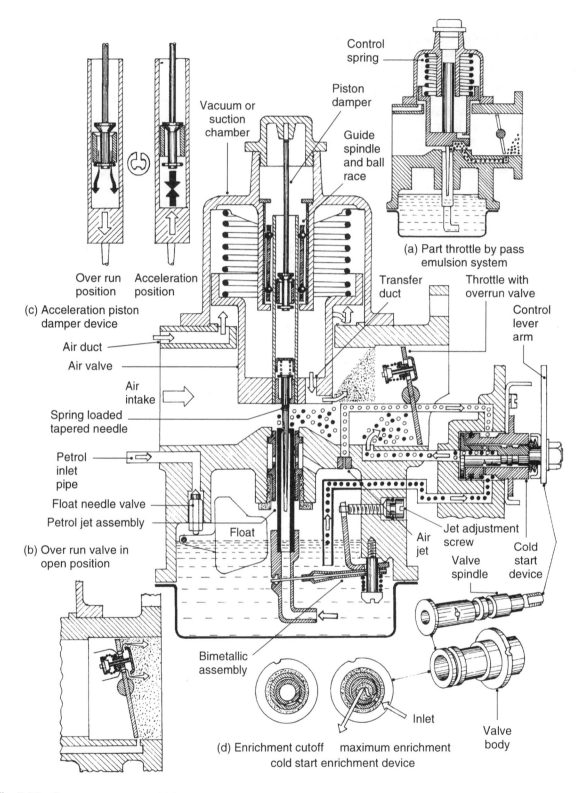

Control
spring

Vacuum or
suction
chamber

Piston
damper

Guide
spindle
and ball
race

(a) Part throttle by pass
emulsion system

Over run
position

Acceleration
position

(c) Acceleration piston
damper device

Transfer
duct

Throttle with
overrun valve

Control
lever
arm

Air duct

Air valve

Air
intake

Spring loaded
tapered needle

Petrol
inlet
pipe

Float needle valve

Petrol jet assembly

Float

Jet adjustment
screw

Air
jet

Cold
start
device

Valve
spindle

(b) Over run valve in
open position

Bimetallic
assembly

Inlet

Valve
body

(d) Enrichment cutoff   maximum enrichment
cold start enrichment device

Fig. 7.26 Constant vacuum variable choke SU carburettor (type HIF)

386

cylindrical vacuum chamber. To improve the accuracy and time response of the air-valve vertical movement with very small changes in engine demands, the friction between the air-valve spindle and guide is sometimes reduced by installing a ball-race between the two sliding surfaces.

When the engine is running, the effect of the depression above the piston in the upper chamber, and the atmospheric pressure underneath, is to raise the air-valve and piston assembly against its own weight and the stiffness of the return spring. Since the downward load is almost constant, a constant depression is needed to keep the air-valve stationary in any raised position. The amount the air-valve lifts depends on the flow rate of air which passes through the mixing chamber, this being controlled by the throttle opening position and engine speed.

Mixture adjustment and fuel temperature compensation
(Fig. 7.26)
Initial adjustment of the jet height and hence mixture strength can be made by altering the tilt of the right-angled lever which is attached to a spring-loaded retaining screw and a bimetallic strip which extends to the petrol jet. To change the jet height, the horizontal jet adjustment screw is screwed inwards to lower the jet and enrich the mixture, and outwards to raise the jet and weaken the mixture.

To compensate for the variation in fuel viscosity with changing temperature and the reluctance of fuel to flow through an orifice as its viscosity rises, a bimetallic strip submerged in the fuel senses a temperature change and alters the effective jet size accordingly.

When the fuel temperature rises, the bimetallic strip curls upwards and pushes the jet further into the tapered needle. Conversely, if the fuel becomes cooler, the strip bends downwards and lowers the jet to increase the annular jet orifice.

Part throttle bypass emulsion system
(Fig. 7.26(a))
This is a passageway which bypasses the mixing chamber, it spans the distance between the feed duct at the jet bridge and a discharge duct at the throttle butterfly edge.

With a small throttle opening, the bypass passage delivers a quantity of mixture in a well-emulsified condition from the jet to a high depression point near the edge of the throttle. Since

the bypass passageway is much smaller than the mixing chamber bore, the mixture velocity through this passage will be much greater and therefore the air–fuel mixing will be that much more thorough.

Overrun valve
(Fig. 7.26(b))
The engine operating conditions frequently change from those of drive, when the throttle is opened and the engine propels the vehicle, to those of overrun, when the throttle is closed and the potential and kinetic energy of the vehicle and wheels provide the power to drive the engine.

Under overrun working conditions, the closed throttle will create a very high depression on the engine side of the throttle and in the induction manifold. Consequently, the effective compression ratio will be low, burning will be slow and erratic, and the exhaust products will contain high values of hydrocarbon.

To improve the burning process so that more of the fuel is doing useful work and less is passed through to the exhaust as incomplete combustion products, a spring-loaded plate-valve is incorporated in the throttle butterfly disc.

When the engine is operating in overrun conditions, the manifold depression at some predetermined value will force open the spring-loaded plate-valve to emit an additional quantity of correct air–fuel mixture. The increased supply of air–fuel mixture will reduce the manifold depression with the result that the denser and better prepared mixture charge will improve combustion, and hence less unburnt products will be passed through to the exhaust.

Hydraulic damper (acceleration enrichment device)
(Fig. 7.26(c))
Due to the reluctance of fuel deposits entering the mixing chamber to respond as quickly as air to the increased demands of the engine, when the throttle is snapped open, a temporary enrichment of the overall mixture strength is necessary.

The hydraulic damper device provides the means to enrich the mixture strength when the throttle is opened rapidly but it does not interfere with the normal air-valve lift or fall as the mixing chamber depression changes with respect to steady throttle opening.

The damper valve is mounted on the lower end of a long stem inside the hollow guide spindle of the air-valve and is submerged in a light (SAE 20) oil. The damper consists of a vertically positioned loose fitting sleeve, its underside resting on a spring clip attached to the stem, while its upper end is chamfered so that it matches a conical seat formed on the central support stem.

When the throttle is rapidly opened, the sudden rise in depression in both mixing chamber and air valve upper chamber tends to jerk up the air valve assembly. Simultaneously, the viscous drag of the oil in the hollow spindle will lift the sleeve and press it against its seat, and so the oil is thus temporarily trapped beneath the damper so that it prevents any further upward movement of the air valve. For this brief period a temporary increase in the depression over the jet orifice is achieved, and more fuel will therefore be drawn to enrich the resultant mixture strength.

As the engine's change in speed steadies, the depression in the upper air-valve vacuum chamber will also stabilize and there will be a slight leakage of oil between the sleeve and its spindle bore. Consequently, any oil pressure created underneath the sleeve damper will now be released enabling the sleeve to drop down onto the spring clip. Oil will now move freely through the annular space made between the sleeve and its seat so that the air-valve vertical movement can again react to small changes in engine demands.

Cold start device
(Fig. 7.26(d))

This device takes the form of a rotary-valve consisting of a cylindrical valve body, which has an annular groove in the middle region with a single radial hole drilled in its side. Fitted inside the valve body is a spindle which has an axial hole bored half-way along from one end, while at the other end a control-lever is bolted. A double taper notched radial hole intersects the axial hole in the spindle. The whole assembly of the valve body and spindle is positioned in a larger hole made in the side of the floor chamber.

For cold starting the choke knob situated on the instrument panel is pulled out, the interconnecting cable rotates the control lever and spindle to a position where the radial hole for both spindle and valve body are aligned. When the engine is cranked a high depression is created in the mixing chamber formed between the jet bridge and throttle valve, and this depression is conveyed to the axial hole in the control spindle where it then passes to the annular groove on the

outside of the valve body. Here it divides and draws fuel from the dip tube and atmospheric air from the float chamber by way of the air jet. The emulsified mixture is then drawn into the hollow spindle along the discharge passage duct and out into the mixing chamber.

As the engine warms up, the choke knob can be pushed back steadily, this rotates the control lever and spindle so that the notched hole passageway becomes progressively smaller and thus restricts the quantity of air and petrol emulsion trying to enter the mixing chamber.

7.2.10 Zenith-Stromberg constant depression (CD) emission control carburettor

Temperature compensation
(Fig. 7.27)

Due to the temperature changes which occur with varying climatic and operating conditions, the fuel temperature and therefore its viscosity will vary. Since the fuel flow-rate is sensitive to viscosity, the amount of fuel discharged from the needle and jet orifice will also vary. This then causes minor undesirable changes to the mixture strength.

Corrections to carburettor and fuel temperature variations can be achieved by using a bimetallic spring temperature compensator unit. This device permits a controlled amount of air intake to bypass the jet bridge and pass directly into the mixing chamber. The additional air passing into the mixing chamber slightly reduces the chamber's vacuum which, in turn, lowers the vacuum above the diaphragm with the result that the air-valve moves to a lower position. Thus, the effective jet orifice is reduced so that less fuel is withdrawn (Fig. 7.27).

The quantity of air bypassing the air-valve bridge is controlled by a tapered-plug and a bimetallic strip. When the carburettor temperature rises, the bimetallic strip curls back, permitting more air to pass through the orifice made by the tapered-plug positioned in the centre of the air passage. Alternatively, if the temperature drops, the bimetallic strip bends and pushes the tapered-plug further into the passage, thus reducing the quantity of air passing through. Increasing the amount of air bypassing the air-valve bridge reduces the depression in the mixing chamber. This lowers the air-valve so that the tapered-

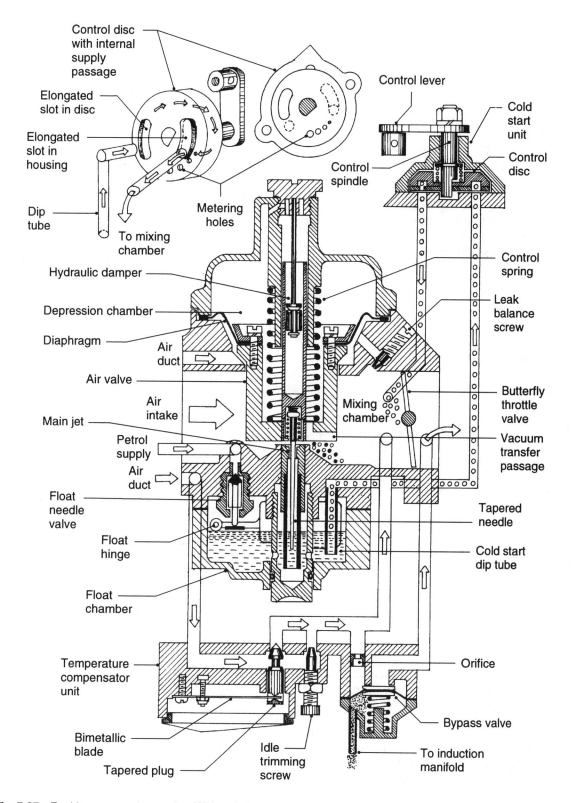

Fig. 7.27 Zenith constant depression (CD) emission control carburettor

389

needle moves further into its jet and thus restricts the fuel discharge. Conversely, reducing the quantity of air bypassing the bridge increases the mixing chamber depression, which thereby raises the air-valve and needle so that more fuel will be discharged. Through these means, very small variations in mixture-strength can be made to compensate for the differing flow resistance which occurs over the working temperature range.

To compensate for the differing mixture strengths required for a new tight engine, and one that has bedded in, an idle trimming screw is provided. With a new engine, more air bleeding is necessary in addition to setting the throttle-valve idling higher. Initially, the trimming screw is set to bypass the maximum amount of air to the mixing chamber, but as the engine wears in and runs freely, the trimming screw should be screwed in over a period of time until eventually the screw seats and prevents any more air entering by this route.

Leak balance screw
(Fig. 7.27)
The flow-rate of both air and fuel must be matched for different carburettor applications. One method is deliberately to leak atmospheric air from the underside of the diaphragm into the mixing chamber by way of a manufacturing tolerance compensation or leak balance screw. This air bleed will then lower the mixing-chamber depression. Since the mixing-chamber depression for a given throttle opening determines the amount of air-valve lift, adjustment of the leak balance screw by the manufacturers will enable the correct needle-to-jet setting to be obtained. This adjustment should not be altered. Therefore, the screw is sealed with a plug which should not be tampered with in service.

Idling
(Fig. 7.27)
The main jet orifice provides the idling fuel supply. The amount metered is determined by the positioning of the jet in the bridge and the needle in the base of the air-valve, which is set by the manufacturer during assembly. The needle is spring-loaded and tilts to one side so that it is permanently in contact with one side of the jet. Thus, for a given needle profile the fuel flow will be consistent. The throttle stop screw regulates the idle speed by limiting the amount the throttle valve closes.

Bypass valve
(Fig. 7.27)
Very high manifold depressions (550 mm Hg or more) will cause high rates of hydrocarbon and CO emission to be expelled from the exhaust system. A throttle bypass valve is incorporated to provide an additional air path from the air intake duct to the engine side of the throttle whenever the manifold depression becomes excessive.

The bypass valve forms the centre portion of a diaphragm which is subjected to manifold depression on the return-spring side. When the depression is low, the diaphragm return-spring forces the bypass valve against its seat. However, under light-load high-speed conditions, the high manifold vacuum created will be sensed in the diaphragm-chamber and, at some predetermined setting, will draw the diaphragm and valve open. Air will then bypass the throttle valve to relieve partially the manifold depression.

Cold starting
(Fig. 7.27)
For cold starting purposes, an extra rich mixture is necessary to compensate for the poor atomization of the fuel droplets owing to the low temperature.

To start the engine from cold, a choke control cable on the instrument panel is pulled fully out, this rotates the control lever, spindle and disc-valve and, at the same time, slightly opens the throttle by means of a fast-idle cam interconnecting linkage. The disc-valve has an elongated hole cut out on one side, and on the other side there is a series of different-sized drilled holes. In the cold-start position, the elongated slot in the disc aligns with the dip-tube passage duct, while all the drilled holes on the other side of the disc align with the mixing-chamber passage elongated-hole, formed in the disc housing. Fuel is therefore drawn from the dip-tube, enters the annular passage space in the disc-valve and comes out the other side where it is then discharged into the mixing-chamber.

When the choke control-cable is pushed gradually in, the disc-valve rotates so that fewer holes in the disc match up with the mixing-chamber passage, the mixture will thereby be progressively weakened until there are no more holes supplying fuel to the mixing-chamber.

7.2.11 Ford constant depression (VV) carburettor

Sonic idling system
(Fig. 7.28)

Whenever the throttle valve is closed, such as when the engine is idling or on the overrun, the induction expansion strokes of the engine create a very large suction in the region of the throttle.

Upon the manifold pressure dropping to half the atmospheric pressure, the air-flow between the almost-closed throttle valve-disc and barrel is raised to the speed of sound (332 m/s). Any further increase in pressure differential does not raise the air flow-rate, and the throttle-to-barrel passage gap under these conditions is then said to be choked.

Fuel is drawn from the main jet pick-up tube, metered through the idle jet where it then meets the air bled through the air jet. The rich air–fuel mixture is then drawn through internal passages down to the volume control screw which provides an adjustable restriction path into the bypass gallery.

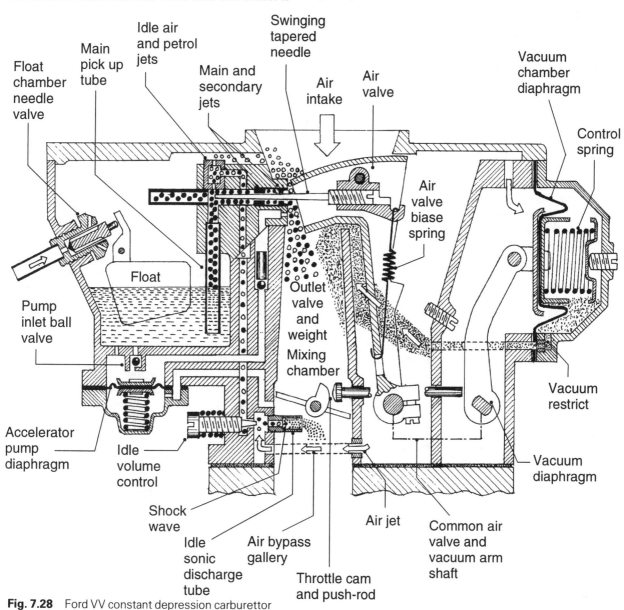

Fig. 7.28 Ford VV constant depression carburettor

Air is drawn past the gap made between the air-valve and venturi wall adjacent to the needle and main jet, it then divides into two streams. One stream passes directly downwards between the barrel wall and throttle disc and enters the manifold, while the other stream passes into the chamber housing the lower part of the air-valve, through a large air jet, into the idle bypass gallery to merge with the volume control screw orifice mixture discharging into the enlarged part of the gallery at the far end. Through these means the majority of the mixture (70%) will enter the sonic tube, and only a small proportion (30%) will pass directly down the venturi to the throttle in the normal manner.

The emulsion weakened mixture now enters the sonic discharge tube and accelerates to sonic speeds, owing to the large pressure difference between the one in the bypass gallery, and the much lower pressure established at the mouth of the sonic tube on the engine side of the throttle. At the exit of the tube, shockwaves are created which break the fuel droplets into a very fine mist. This is then discharged into the centre of the main stream to the manifold. The very high quality of air and fuel mixing enables an engine to idle with a 16:1 air–fuel ratio instead of the more usual value of between 9:1 and 11:1, which would be expected from a conventional idle bypass system.

Main mixture control
(Fig. 7.28)

This type of constant depression carburettor features a swing arm air-valve with a tapered needle attached, this fits into a horizontal primary and secondary jet orifice. The valve opens or closes proportionally to the engine's demand, created by the forces resisting the vehicle's motion and the degree of throttle opening. A diaphragm subjected to a venturi mixing-chamber depression is linked by an arm to the air-valve pivot shaft, and a high or low depression acting on the diaphragm in opposition to the control-spring will accordingly open or close the air-valve. Since the free end of the tapered-needle rests in the jet and the other end is attached to the air-valve, any opening or closing of the valve will increase or decrease the effective jet orifice. Through these means the discharged metered fuel matches the quantity of air flowing through the venturi and so provides the correct air–fuel mixture strength over the main operating speed range.

Acceleration pump
(Fig. 7.28)

This is a diaphragm-operated pump controlled by the degree of depression on the engine side of the throttle. When the throttle is closed, the high depression below the throttle will be sufficiently felt on the underside of the diaphragm to pull it down against the resistance of the return-spring. Petrol will now pass through the non-return ball-valve to fill up the space above the diaphragm. Upon opening the throttle, the depression beneath the diaphragm now drops, permitting the return-spring to push up the diaphragm. Immediately, the inlet ball-valve closes and the outlet ball-valve opens, and fuel will then be discharged in the form of a spray, into the venturi to supplement prolonged and heavy acceleration. A back bleed duct allows any excess fuel vapour to return to the float chamber when prolonged idling causes the carburettor temperature to rise and the fuel above the diaphragm to become overheated.

Automatic choke
(Fig. 7.29)

The fuel supply to the automatic choke device is drawn from the pick-up tube, fed along passages to the tapered needle-valve in the choke housing, and then metered fuel flows along a passage towards the control-spindle, through a space made by the cut-back of the gasket squeezed between both carburettor housings, to enter the mixing-chamber formed in the mixture-control spindle.

An extra supply of air is also required and this enters the spindle mixing-chamber from the side of the venturi wall just below the air-valve, at the same time metered fuel is drawn into this chamber, and the resultant emulsified mixture now passes through a radial hole drilled through the side wall of the hollow-spindle which, under cold starting conditions, aligns with a vertical discharge passage which feeds into the barrel just below the throttle-valve.

A bimetallic spiral spring exposed to coolant temperature is connected through link arms to both the control-spindle valve and the tapered needle.

When the engine is cold the bimetallic-spring contracts and rotates the control spindle anti-clockwise. Consequently, the tapered needle will be pulled out of its jet and the mixture-valve will be fully opened.

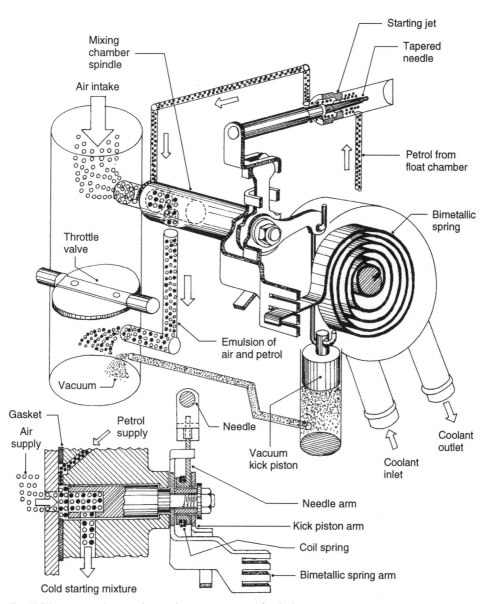

Fig. 7.29 Ford VV constant depression carburettor automatic choke

As the engine warms up, the bimetallic-spring tends to unwind and, in so doing, rotates the control-spindle clockwise. In turn, this action progressively pushes the tapered-needle into its jet, reducing the fuel supply and, at the same time, it misaligns the radial hole in the mixture-control valve-spindle with the discharge passage, thus reducing the effective passage size until eventually it will completely close.

An additional feature of this cold-start device is a vacuum pull-down piston which is also con-nected by means of a rod and arm to the control-spindle. The underside of this piston is subjected to manifold depression by means of a passage intersecting the barrel on the engine side of the throttle.

When the engine initially starts, or is operating at light-load or cruise conditions, the depression below the throttle rises steeply, this is immedi-ately sensed underneath the piston so that the piston is pulled down against the resisting tension of the bimetallic coil-spring. Consequently, the rotation of the spindle closes both the needle-valve and the mixture control-valve and therefore prevents the engine over-choking. This reduces the overall mixture strength so that fuel economy improves and emission levels during engine warm-up are reduced.

7.2.12 Constant depression carburettor incorporating idle and primary fixed jet system

Idle system
(Fig. 7.30)
With the throttle closed, virtually all the induction vacuum will be on the engine side of the butterfly valve and there will be insufficient depression in the mixing chamber, formed between the air valve and throttle, to draw fuel either from the primary or secondary jet systems. Therefore, an idle system which bypasses the mixing chamber and feeds directly into the high vacuum on the engine side of the throttle is used. Fuel from the float chamber is drawn through the primary-jet to the idle-jet where it meets air passing through the idle air-jet. The emulsified mixture then passes along passages to the volume-control screw, which regulates the quantity of mixture entering the barrel on the manifold side of the throttle.

Primary system
(Fig. 7.30)
The vertical movement of the air-valve is controlled by the pressure difference between the top and bottom of the air-valve piston. Atmospheric pressure acts underneath the piston and a variable depression provided by the mixing chamber fills the space above it. When the throttle is partially opened a proportion of the vacuum craated in the induction manifold will be transferred from the engine side of the throttle to the mixing-chamber and also to the vacuum-chamber, by way of the transfer port in the base of the air-valve. The air-valve will be raised to a height determined by the upthrust on the piston caused by the pressure difference above and below it, and the opposing air-valve assembly weight and control-spring resistance. The greater the depression in the vacuum-chamber, the more the air-valve will open. Additional air will now enter the mixing-chamber on its way into the manifold and cylinder and, at the same time, petrol mixture will be drawn from the primary main jet into the emul-

sion-tube where it meets air from the primary air-jet. The emulsified mixture is then discharged into the air-stream passing through the mixing-chamber.

Secondary system
(Fig. 7.30)
The needle profile for the secondary system is divided approximately into two halves: an upper parallel half and a lower tapered section. While the parallel portion of the needle is in the jet orifice no fuel mixture will be drawn from the jet. Once the air-valve has lifted the needle at least half-way out under full-throttle, high-load and speed conditions, then the tapered portion of the needle aligns with the jet orifice. It now permits the fuel mixture from the secondary jet to be discharged into the mixing-chamber, in addition to that which is already being supplied by the primary system.

Accelerator pump
(Fig. 7.30)
Unlike most other constant depression carburettors, a damper cannot be fitted because the secondary system does not function until the air-valve is at least half open. To provide the extra-rich mixture necessary for acceleration, a diaphragm-type acceleration pump is incorporated to replace the normal damper action. It is operated by a mechanical linkage interconnected to the throttle-spindle.

Upon the throttle-valve closing, the diaphragm return-spring pushes up the diaphragm to increase the chamber space housing the return-spring. Petrol will then be drawn through the inlet-valve to fill up the enlarged return-spring chamber. When the throttle is depressed the interconnecting linkage push-rod pushes down the diaphragm, and this squeezes the petrol underneath, closing the inlet-valve and forcing petrol through both the outlet-valve and accelerator pump-jet to be ejected in the form of a spray into the intake air stream.

Power valve
(Fig. 7.30)
A power valve is utilized to provide lean mixtures under light-load conditions and still have adequate fuel for full-load low-speed operations, such as prolonged hill climbing and full-load high-speed running.

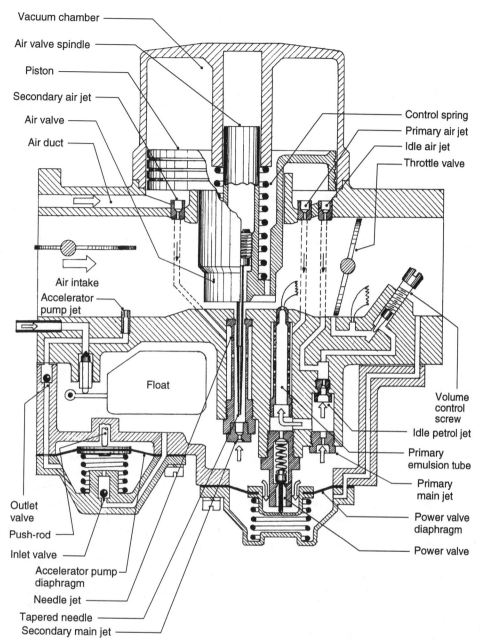

Vacuum chamber

Air valve spindle

Piston

Secondary air jet

Air valve

Air duct

Control spring

Primary air jet

Idle air jet

Throttle valve

Air intake

Accelerator pump jet

Float

Volume control screw

Idle petrol jet

Primary emulsion tube

Primary main jet

Power valve diaphragm

Power valve

Outlet valve

Push-rod

Inlet valve

Accelerator pump diaphragm

Needle jet

Tapered needle

Secondary main jet

Fig. 7.30 Constant depression carburettor incorporating idle and primary fixed jet system

Under cruising conditions, while there is a high depression in the manifold, the power-valve diaphragm will be drawn down. This allows the power-valve return-spring to close the valve allowing the engine to operate on fuel supplied from the primary main-jet only.

Upon acceleration and hill climbing, while there is a low depression in the manifold, the diaphragm return-spring pushes the diaphragm up and this opens the power-valve. This allows fuel to pass through the power-valve directly to the primary-emulsion tube and is in addition to the fuel supplied from the primary main-jet.

8

Petrol Injection Systems

8.1 An introduction to fuel injection

The function of a fuel injection system is to monitor the engine's operating variables, to transfer this information to a metering control, then to discharge and atomize the fuel into the incoming air stream. The position where the fuel is injected into the air charge considerably influences the performance of the engine and will now be classified and compared.

8.1.1 Direct injection
(Fig. 8.1)

With this layout the fuel injectors are positioned in the cylinder-head so that fuel is directly discharged into each combustion chamber (Fig. 8.1). It is essential with this arrangement that injection is timed to occur about 60° after TDC on the induction stroke. Because of the shorter time period for the fuel spray to mix with the incoming air charge, increased air turbulence is necessary. To compensate for the shorter permitted time for injection, atomizing and mixing, the injection pressure needs to be higher than for indirect injection. More overlap of exhaust and inlet valves can be utilized compared with other carburetted or injected systems, so that incoming fresh air can assist in sweeping out any remaining exhaust gases from the combustion chambers. The injector nozzle and valve has to be designed to withstand the high operating pressures and temperatures of the combustion chamber, this means that a more robust and costly injector unit is required. The position and the direction in

which the injectors point are of the utmost importance to obtain optimum performance, and no one position will be ideal for all operating conditions. Generally, direct-injection air and fuel mixing is more thorough in large cylinders than in small ones because fuel droplet sizes do not scale down as the mixing space becomes smaller. All condensation and wetting of the induction manifold and ports is eliminated but some spray may condense on the piston crown and cylinder walls.

8.1.2 Indirect injection
(Fig. 8.1)

Fuel is injected into the air stream prior to entering the combustion chamber. Fuel spray may be delivered from a single-point injection (SPI) source, which is usually just upstream from the throttle (air intake side of the throttle), or it may be supplied from a multi-point injection (MPI) source, where the injectors are positioned in each induction manifold branch pipe just in front of the inlet port (Fig. 8.1).

Indirect injection can be discharged at relatively low pressure (2–6 bar) and need not be synchronized to the engine's induction cycle. Fuel can be discharged simultaneously to each induction pipe where it is mixed and stored until the inlet valve opens.

Because indirect injection does not need to be timed, it requires only low discharge pressures and the injectors are not exposed to combustion, the complexity of the operating mechanisms can be greatly reduced, which considerably lowers costs.

The single-point injection system has the same air and fuel mixing and distribution problems as a carburettor layout but without the venturi restric-

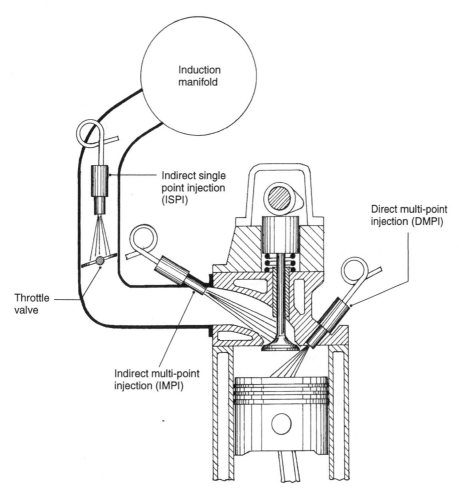

Fig. 8.1 Classification of the three-principal injector positions

tion so that higher engine volumetric efficiencies are obtained. High injection pressures, compared with the carburettor discharge method of fuel delivery, speed up and improve the atomization of the liquid spray.

In contrast to the single-point injection method the multi-point injection layout has no fuel distribution difficulties since each injector discharges directly into its own induction port and the mixture then has only to move a short distance before it enters the cylinder. Since the induction manifold deals mainly with only induced air, the branch pipes can be enlarged and extended to maximize the ram effect of the incoming air charge.

A major feature with petrol injection is that there is separate air and fuel metering and that fuel metering is precise under all engine operating conditions.

8.1.3 Injection considerations

With indirect injection arrangements there are two methods of discharging the fuel into the air stream.

1. **Continuous injection**
With this form of injection the injector nozzle and valve are permanently open while the engine is operating and the amount of fuel discharged in the form of a spray is controlled by either varying the metering orifice or the fuel discharge pressure, or a combination of both of these possible variables.

2. **Intermittent or pulsed injection**
Fuel is delivered from the injector in spray form at regular intervals with a constant fuel discharge pressure and the amount of fuel discharged is

controlled by the time period the injector nozzle valve is open.

3. Comparison of pulsed or continuous injection

For fuel to be supplied continuously to a typical engine operating over a 10:1 speed range between idling and maximum speed, i.e. 750 and 7500 rev/min respectively, the fuel flow rate has to vary by a factor of 50:1 by volume. This requires either a very large fuel pressure range of about 2500:1 with injectors having fixed orifices, or a variable area orifice injector or metering control in the region of 50:1, or a combination of both variables—pressure and orifice area.

Conversely, pulsed injection where the injector nozzle valve is opened for only a short time every engine cycle requires a much smaller pulse range of approximately 5:1 during normal engine operation, this range only being increased significantly for cold starting where the control accuracy requirement is much reduced.

Timed injection

This is where the start of fuel delivery for each cylinder occurs at the same angular point in the engine cycle, this can be anything from 60° to 90° after TDC on the induction stroke.

Non-timed injection

In contrast to timed injection, this is where all the injectors are programmed to discharge their spray at the same time, therefore each cylinder piston will be on a different part of the engine cycle.

8.1.4 Injection equipment controlling parameters

The petrol injection system must be able to sense the changes to the influencing parameters, pass this information to a central co-ordinating mechanism, which then integrates these individual signals and interprets the fuel requirements. The resultant needs are then transmitted by either mechanical, hydraulic or electrical means to the pumping and metering devices, which then supply the correct quantity of fuel to the induction for the particular operating conditions.

The controlling parameters to sense are as follows.

1 Engine speed (determines frequency and amount of fuel delivered).

2 Amount of inlet air flow (a measure of engine load).
3 Throttle position (correction for sluggish movement of fuel droplets during speed transition conditions).
4 Air temperature (correction for air density variation).
5 Coolant temperature (correction for poor atomization and wall wetting).
6 Altitude (correction for air density change with height).
7 Cranking speed (allowance for low air velocity and poor air–fuel mixing).
8 Exhaust oxygen content (correction for emission control).
9 Battery voltage (correction for supply voltage to control unit and injectors).

8.1.5 Comparison of petrol injection and carburetted fuel supply systems

Merits of petrol injection

1 Due to the absence of the venturi there is the minimum of air restriction so that higher engine volumetric efficiencies can be obtained with the corresponding improvement in power and torque.
2 Hot spots for pre-heating the cold air and fuel mixture are eliminated so that denser air enters the cylinder when the engine has reached normal operating conditions.
3 Since the manifold branch pipes are not greatly concerned with mixture preparation they can be designed to utilize the inertia of the air charge to increase the engine's volumetric efficiency. (This does not apply for single point injection.)
4 Acceleration response is better due to direct spray discharge into each inlet port.
5 Atomization of fuel droplets is generally improved over normal speed and load driving conditions.
6 There is a possibility of using greater inlet and exhaust valve overlap without poor idling, loss of fuel or increased exhaust pollution.
7 The monitoring of engine operating parameters enables accurate matching of air and fuel requirements under normal speed and load conditions which improves engine performance, fuel consumption and reduces exhaust pollution.
8 Fuel injection equipment is precise in meter-

ing injected fuel spray into the intake ports over the complete engine speed, load and temperature operating range.

9 With multi-point injection there will be precise fuel distribution between engine cylinders even under full load conditions.

10 Multi-point injection does not require time for fuel transportation in the intake manifold and there is no manifold wall wetting.

11 Fuel surge is eliminated with fuel injection when cornering fast or due to heavy braking.

12 Fuel injection systems, be they either single-point or multi-point injection, are particularly adaptable and suitable for supercharged engines.

Limitations of petrol injection

1 High initial cost of equipment, and replacement parts are also expensive.

2 Generally increased care and attention is needed and there are more servicing problems likely to arise.

3 Special servicing equipment is necessary to diagnose fuel injection system faults and failures.

4 Considerably more mechanical and electrical knowledge is necessary to diagnose and rectify fuel equipment faults.

5 Injection equipment may be elaborately complicated, delicate to handle and impossible to service.

6 More electrical and mechanical wearing components to go wrong.

7 Increased mechanical and hydraulic noise due to pumping and metering of the fuel.

8 Very careful filtration is needed due to the fine working tolerances of the metering and discharging components.

9 Power is necessary, be it electrical or mechanical, to drive the fuel pressure pump or injection discharge devices.

10 More fuel injection equipment and pipe plumbing are required, which may be awkwardly placed and can be bulky.

8.2 K-jetronic fuel injection system (Bosch)

(Fig. 8.2)

The K-jetronic fuel injection system is a driveless mechanical system in which fuel is continuously

metered in proportion to the quantity of air induced into the engine's cylinders, 'K' stands for the German word for continuously.

This fuel injection system (Fig. 8.2) may be considered in three parts:

1 air flow measurement
2 fuel supply
3 metering and injection of fuel

1 The air flow sensor measures the throttle-controlled quantity of air drawn into the engine.

2 Pressurized fuel is provided by an electric motor driven roller-type pump which delivers fuel through an accumulator and filter to the mixture control distributor unit, and a pressure regulator maintains the fuel entering the mixture control unit at constant pressure.

3 The amount of fuel discharged into the air stream is related to the measured air flow signalled to the mixture control unit, whose function is to meter the corresponding quantity of fuel before it is transferred to the injectors.

The individual components of the complete injection system will now be considered in detail.

8.2.1 Electric fuel pump
(Figs 8.3 and 8.4(a and b))

The fuel pump which is of the roller-cell type is driven by a permanent magnet electric motor (Fig. 8.3). The pump is in the form of a rotor disc, eccentrically mounted within the pump casing, five slots are formed around the disc circumference to locate the steel rollers. Whenever the pump is rotated, these rollers are pushed outwards by centrifugal force, they thereby divide the space between the circumference of the rotor disc and outer casing into a number of sealed cells (Fig. 8.4).

The intake fuel passes through a passage to the side of the rotor disc casing which has the largest radial clearance formed by the offset-positioned drive shaft (Fig. 8.4(a)). Rotation of the rotor disc will then trap fuel within a cell formed by the casing rotor and two adjacent rollers. As the rotor continues to turn, the fuel cell becomes smaller (Fig. 8.4(b)) thereby pressurizing the fuel until the cell opens to the discharge passage, it then flows out under pressure, through the electric motor (Fig. 8.3) between the armature and poles, passing through the non-return valve to the accu-

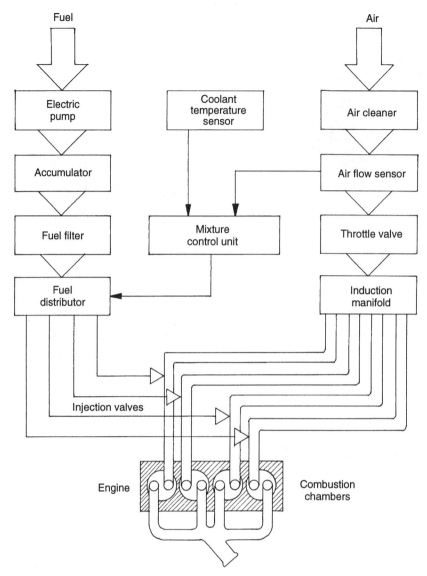

Fig. 8.2 Mechanical fuel injection system (Bosch K-jetronic)

mulator. An excess pressure valve in the pump casing bypasses fuel back to the intake side of the pump if the generated pump pressure should become abnormally high.

8.2.2 Fuel accumulator
(Fig. 8.3)

This accumulator performs two functions, firstly it maintains the fuel line pressure when the engine is switched off for some time so that the injection pump system responds immediately when the engine is cranked for restarting, secondly it quietens the noise created by the roller cell pump charging and discharging action.

The accumulator is made in two halves separated by a diaphragm: one half is shaped to house the spring while the other half forms the storage fuel chamber (Fig. 8.3).

When the fuel pump is operating, fuel pressure will compress and push back the diaphragm against its stops, and the enlarged fuel chamber remains in this position while the engine is running. A flap valve allows the accumulator chamber to fill up rapidly but restricts the outflow

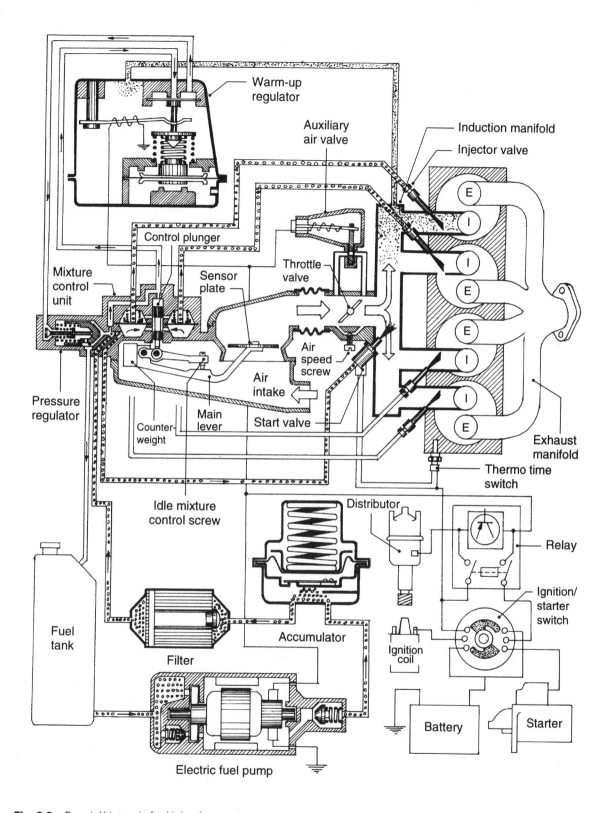

Fig. 8.3 Bosch K-jetronic fuel injection system

401

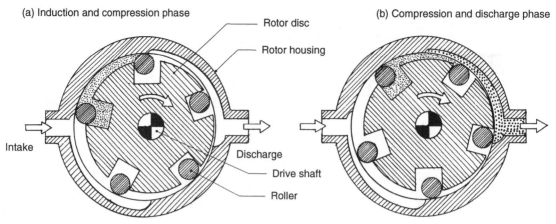

Fig. 8.4 Roller-cell pump

through a small orifice to a much lower rate. Upon the engine being stopped and restarted, fuel stored in the chamber will automatically supplement fuel being discharged into the induction system until the pump itself provides the source of fuel supply.

8.2.3 Fuel filter
(Fig. 8.3)

It is necessary to install a very fine filter in the primary pipe line to prevent particles of dirt suspended in the fuel clogging the close tolerances of the metering, distribution and injection parts of the system.

The canister filter contains a pleated-paper element plus a lint-of-fluff filter and a strainer holds in the filtering media on the output side (Fig. 8.3). It is important that the filter unit is positioned the right way round on the output side of the accumulator if filtering is to be effective.

8.2.4 Primary pressure regulator and push-up valve
(Fig. 8.3)

Primary pressure regulator
(Fig. 8.5)
This is a regulating valve which maintains the output delivery pressure (primary pressure), generated by the electrically driven pump, to be approximately constant at about 5 bar.

The regulator is a spring-loaded plunger-operated valve and when the output of the pump

exceeds the demands of the fuel system, the plunger is pushed back, thus permitting fuel to return to the fuel tank (Fig. 8.5). The amount the return port is opened will be matched by the demands of the engine and the pump's delivery pressure. The opening of the return port is progressive and is balanced by the spring thrust and the delivery pressure at any one time. If the engine is switched-off the fuel pump also stops delivery, and pressure drops below the injection valve opening pressure. Subsequently, the regulator valve closes.

Push-up valve
(Fig. 8.5)
If the engine is switched off the control pressure circuit may leak fuel back to the fuel tank by way of the warm-up regulator diaphragm valve return. To prevent control pressure circuit leakage, a non-return push-up valve is installed at the opposite end of the pressure regulator plunger valve (Fig. 8.5).

When the engine is running and the pressure regulator plunger is partially open, contact will be made between the inside of the plunger valve and the push-up valve stem, this opens the valve permitting excess fuel from the warm-up regulator to return to the fuel tank. Upon the engine being switched off the pump stops and the regulator plunger shifts over to the closed position also allowing the push-up valve to close and thus positively sealing the control pressure circuit.

402

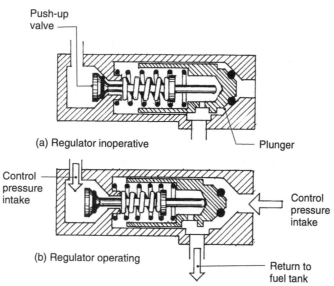

Fig. 8.5 Primary pressure regulator

8.2.5 Fuel injection valve

(Fig. 8.6(a and b))

Fuel injection valves are situated at the entrance of each cylinder induction port. The valves are insulated in holders to prevent fuel vapour bubbles forming in the fuel lines due to the heat of the engine, which would make starting difficult.

The injector valves (Fig. 8.6) do not meter fuel. Their main function is to open at a certain injection line pressure, normally about 3.3 bar, and to discharge fuel in the form of a spray into their

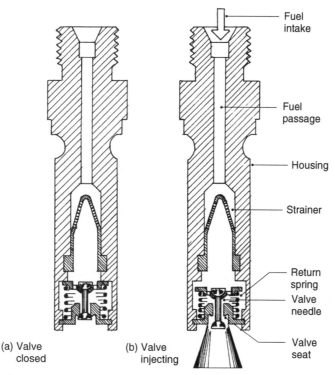

Fig. 8.6 Fuel injection valve

respective induction ports. The valves are made to open by the pressure of the fuel acting on the exposed projected annular area formed between the valve seat and the central needle. This, then, produces a thrust which exceeds the return spring compressive force. Once the needle valve opens, it oscillates at a high frequency (1500 cycles/s) (audible chatter noise) and the discharged fuel becomes highly atomized due to the needle valve's dynamic motion. It is claimed that atomization is effective under these conditions for both large and small quantities of metered fuel.

Upon the engine stopping, the line pressure drops slightly and the needle valve then closes tightly on its seat so that no dribble takes place.

8.2.6 Air flow sensor
(Figs 8.7(a and b) and 8.8)

The amount of air entering the engine can be sensed and measured by a sensor-plate operating on the suspended-body principle.

This form of sensor comprises a disc air-plate at one end of a pivoting control-lever while the other end of the lever supports a counterweight so that, in a free state, the lever will remain horizontal (Fig. 8.7). The air sensor-plate is housed in a specially shaped funnel forming part of the air intake system, ahead of the throttle valve, at the entrance to the induction manifold.

When the engine is running, air will flow through the funnel on its way to the engine, and the greater the quantity of air flow (controlled by the throttle opening) the more the sensor-plate will be lifted up by the air stream. Since the vertical movement of the sensor-plate is a direct measure of the air flow, this movement is relayed to a control plunger which meters the quantity of fuel required to match the air flow necessary for each engine load and speed condition. If the engine should back-fire into the induction manifold, the shock waves are able to swing back the sensor-plate to an enlarged section of the funnel to provide pressure relief. When the engine is stationary a leaf spring on the underside of the sensor-plate supports and holds the control lever in its zero position.

A funnel having a single conical profile will provide, over the whole of the air sensor-plate range of travel, a uniform control plunger movement and therefore a constant air–fuel ratio.

The sensitivity of the sensor-plate can be altered by changing the funnel profile (Fig. 8.8), a small conical angle produces a large change in sensor-plate movement for a given change in air flow, and more fuel is therefore metered and the mixture is richer. Conversely, a large conical angle provides very little sensor-plate movement for a given change in air flow, and hence only a small quantity of fuel is metered resulting in a lean mixture.

To provide different mixture strengths for idle, part load and full load operating conditions, the air-flow sensor-plate vertical movement can be made to move different amounts for similar changes of air flow at different positions of the sensor-plate travel.

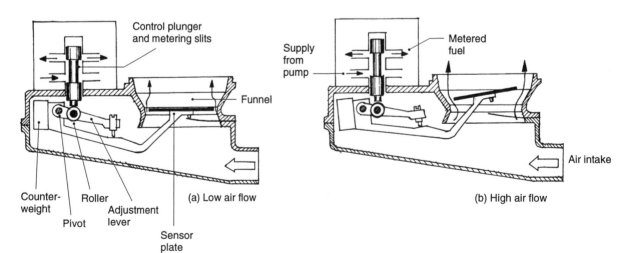

Fig. 8.7 Air flow sensor

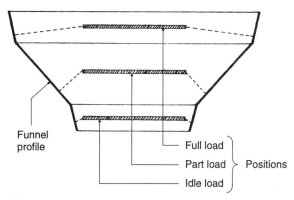

Funnel
profile

Full load ⎫
Part load ⎬ Positions
Idle load ⎭

Fig. 8.8 Air flow sensor funnel profile

This is achieved by having the early part of the sensor-plate travel correspond to the idling rich-mixture condition with a small conical angle funnel profile, whereas the middle travel range of the sensor-plate operates mostly under a part-load lean-mixture condition and requires a large conical angle for the funnel profile. Towards the end of the sensor-plate travel a small conical angle profile is used to produce full-load rich-mixture conditions.

8.2.7 Fuel distributor plunger and barrel operation
(Figs 8.9(a–c) and 8.10(a–c))

The plunger and barrel meter the quantity of fuel

delivered to individual cylinders in proportion to the air flow sensor-plate lift.

The control plunger is governed by two factors.

1 The air-flow drag tends to raise the sensor-plate, and so the greater the quantity of air passing through to the induction, the more will be its lift, and vice versa.
2 The control pressure, that is the fuel pressure acting on top of the plunger, opposes the upthrust of the air sensor-plate caused by the air-stream movement.

The control plunger has a wasted central region with a control edge formed by the upper recess step, whilst spaced around the barrel are a lower and upper row of inlet ports and metering slits respectively, there being one pair for each cylinder (Fig. 8.9(a–c)).

Fuel is supplied from the pump, it flows through the accumulator and filter to the distributor mixture control unit (here it passes to each differential pressure valve lower chamber—one for each cylinder) into the barrel inlet ports around the plunger central wasted region and out of the metering slits. Fuel now enters the differential pressure-valve upper chambers where it then passes through the injector pipes to the various injectors, and finally it is discharged out of the injectors into the induction system. Accurate metering of fuel is achieved due to the narrow width of the slits, this being about 0.2 mm.

During idling conditions the plunger will be in

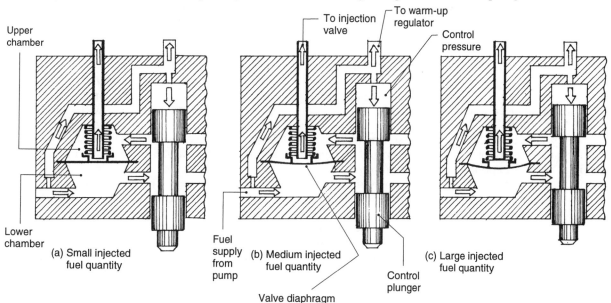

Fig. 8.9 Full distribution with differential pressure valves

405

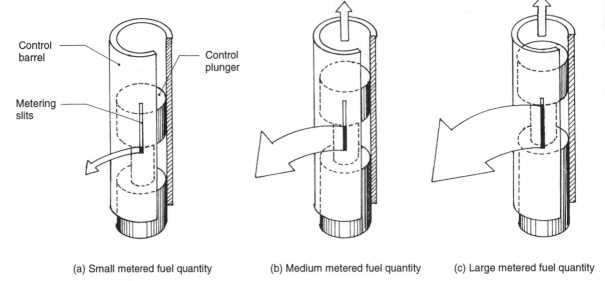

(a) Small metered fuel quantity   (b) Medium metered fuel quantity   (c) Large metered fuel quantity

Fig. 8.10 Fuel distribution output control

its lowest position so that only a small section of the slits will be open to the passage of fuel on its way to the injectors. However, with a large flow of air past the air sensor-plate, the plunger will rise accordingly and this increases the open section of the metering slits thereby permitting more fuel to flow (Fig. 8.10(a–c)).

8.2.8 Control pressure
(Fig. 8.9(a–c))

The control pressure circuit is fed from the primary pressure circuit through a restriction which, in essence, decouples the control pressure from the primary pressure. It conveys fuel to the top of the control plunger and to the warm-up regulator (Fig. 8.9(a–c)).

To counterbalance the air sensor-plate upthrust, a control pressure acting through a damping restriction applies an opposing force on top of the control plunger, it is the function of this restriction to damp out air pulsating oscillations caused by air movement.

The amount of fuel metered at any one time for a given air flow can be varied by altering the control pressure; the higher this pressure, the greater is the resistance to control plunger movement and the less fuel will be metered. Conversely, if the control pressure is low, the air sensor-plate will lift the control plunger further

for a given air-flow demand from the engine so that more fuel will pass through the metering slits.

Regulation of the control pressure is achieved by bleeding a proportion of the control pressure circuit fuel back to the fuel tank. With a cold engine, the control pressure will drop to about 0.5 bar, but with a fully warmed engine it will rise to about 3.7 bar.

8.2.9 Fuel distribution differential pressure valve
(Fig. 8.9(a–c))

Fuel flowing through the metering slits will produce a drop in fuel pressure on its way out and the greater the quantity of fuel flow, the greater will become the pressure difference across the slits. Unfortunately, this variable pressure difference across the slits would not provide a fuel quantity which depends only on the open cross-sectional area of these slits. Therefore, each metered slit output is made to pass through a differential pressure valve and this maintains a constant pressure difference between the two sides of the slit under all normal conditions (Fig. 8.9(a–c)).

There is one differential pressure valve for each metering slit, with each valve having an upper and lower chamber separated by a steel diaphragm. The lower chambers are all joined by a common ring passage which is supplied by primary pressure set at 4.5 bar. Each upper chamber is fed separately by its own metering slit and houses a spring which produces the pressure difference between both sides of the diaphragm. The upper face of the diaphragm also forms the valve seat which rests against the central discharge tube situated in the centre of the upper chamber.

When operating, the differential pressure valve exerts a pressure difference of 0.1 bar between the two chambers. The quantity of metered fuel for each slit will press down its respective diaphragm and allow fuel to discharge through the central tube towards its own injection valve. If a larger amount of metered fuel enters the upper chamber, the diaphragm deflects further, permitting more fuel to flow to its respective injection valve and this therefore maintains a pressure of 4.4 bar in the upper chamber. If a smaller amount of fuel enters the chamber, the diaphragm does not move back so much, and it therefore permits less fuel to flow to the injection valves.

8.2.10 Warm-up regulator
(Fig. 8.11(a–c))

To enable the mixture strength to be raised as the engine frictional resistance changes between cold and hot running conditions, the amount of control pressure circuit fuel leaked back to the tank is designed to be variable—this being governed by the warm-up regulator. The regulator reduces the control pressure when the engine is cold, and consequently the control plunger will be permit-

ted to lift further and increase the opening of the metering slits in proportion to that normally required for a hot running engine. The warm-up regulator is attached to the engine so that it can sense the engine temperature.

The unit consists of a spring-controlled diaphragm valve and an electrically heated bimetallic strip. There are two control springs—an outer one which rests on the casing and an inner spring which bears against a full-load diaphragm. The upper side of the diaphragm is subjected to manifold depression while the underside is exposed to atmospheric pressure (Fig. 8.11).

When the engine is cold, the bimetallic strip exerts a force which opposes the upthrust of both springs. This weakens the pressure tending to keep the diaphragm valve closed and, as a result, the control pressure fuel will force open the valve and escape back to the fuel tank so that a relatively low control pressure will be established (Fig. 8.11(a)).

Upon the engine being switched on and cranked, the bimetallic strip is electrically heated in addition to the heat it absorbs from the engine. As the bimetallic strip warms up it bends upwards, so reducing the opposing resistance to the control springs such that they now exert more upthrust onto the underside of the diaphragm valve and therefore raise the control pressure (Fig. 8.11(b)). Thus, the increased control pressure reduces the metered quantity of fuel so that the air–fuel mixture strength progressively weakens over a period of about 100 seconds. The actual mixture strength and warm-up period will

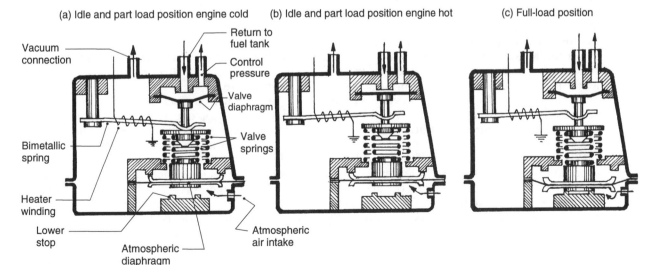

(a) Idle and part load position engine cold   (b) Idle and part load position engine hot   (c) Full-load position

Fig. 8.11 Warm-up regulator

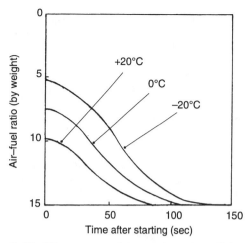

Fig. 8.12 Warm-up regulator mixture strength to time span characteristics with respect to engine coolant temperature

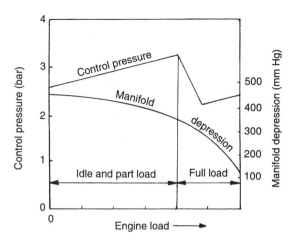

Fig. 8.13 Warm-up regulator manifold depression and control pressure characteristics with respect to engine coolant temperature

depend upon the initial coolant temperature (Fig. 8.12).

8.2.11 Full load enrichment
(Figs 8.11(c) and 8.13)

An additional feature with this warm-up regulator is the full-load diaphragm. This permits a very lean mixture to be supplied in the part-load range and, accordingly, to the manifold depression, so enriching the mixture as full-load is approached.

Full-load mixture enrichment is achieved by the full-load diaphragm sensing the change in manifold depression altering the effective stiffness of the inner control spring (Fig. 8.11(c)).

Under part-load conditions the high depression in the manifold is relayed to the top side of the full-load diaphragm and is subjected to an upthrust of atmospheric pressure. Due to these conditions, the inner spring will exert an upward load (in addition to the outer spring) against the diaphragm valve. It therefore prevents control pressure fuel bypassing back to the fuel tank so that a higher control pressure will operate (Fig. 8.13).

When the throttle is opened under full-load conditions there will be very little depression in the manifold and, correspondingly, above the full-load diaphragm, so that the diaphragm moves to its lowest position. In the lowest position the upthrust of the inner spring acting on the upper diaphragm valve will be small and therefore the leak-back to the fuel tank will increase. The control pressure will now be reduced, thus per-

mitting the control plunger to rise, and more fuel will then be able to pass through the metering slits to strengthen the mixture supply.

8.2.12 Cold-start valve

A cold engine, when being started, condenses a proportion of the inhaled fuel mixture onto the induction and cylinder walls and, in order to make up for this fuel deficiency, a separate cold-start valve is installed somewhere near the centre of the induction manifold (see Fig. 8.28 section 8.3.10).

Thermo time-switch
(Figs 8.14 and 8.15)

A thermo time-switch sets the duration of the cold start injection valve according to the engine's temperature.

This switch consists of an electrically heated bimetallic strip, one end being mounted in an insulated compound while the free end forms the moving electrical contact, the other fixed contact being earthed to the hollow brass cylindrical casing. The sealed switch unit is screwed into the engine's cylinder-head cooling system, to enable it to sense engine temperature.

When the engine is cold and the ignition is turned on, the thermo time-switch contacts are closed, and current will thus pass through both the cold start valve solenoid and the thermo time-switch winding (Fig. 8.14). Fuel will there-

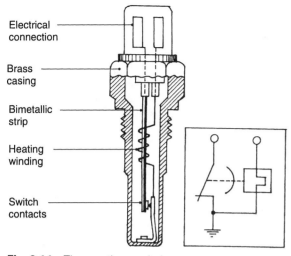

Electrical connection

Brass casing

Bimetallic strip

Heating winding

Switch contacts

Fig. 8.14 Thermo time switch

fore be injected into the manifold until the heat from the engine and winding deflects the bimetallic strip, and the contacts are then parted. This then interrupts the cold-start valve solenoid current, thereby causing the valve to close. The additional heat derived from the bimetallic strip winding enables the cold start enrichment period to be adjusted to suit the engine's requirements (Fig. 8.15).

8.2.13 Auxiliary air device
(Figs 8.16 and 8.17)

When the engine is cold, the increased frictional resistance and pumping losses are higher than if

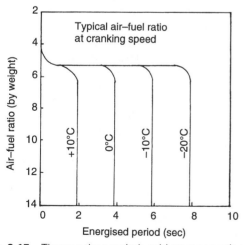

Fig. 8.15 Thermo time switch cold start operating characteristics with respect to engine coolant temperature

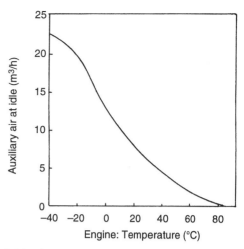

Fig. 8.16 Auxilliary air valve additional air flow requirement characteristics with respect to engine coolant temperature

the engine was operating at normal temperature. Therefore, slightly more power is needed during the idle warm-up period, which necessitates an increase in air flow and correspondingly additional fuel (Fig. 8.16).

The extra air supply is provided by the auxiliary air device which bypasses the throttle valve and, since all the air entering the manifold is measured by the air flow sensor, additional fuel will also be metered by the control-plunger accordingly.

The auxiliary air device consists of a pivoting spring-loaded disc valve and a bimetallic strip heated by an electrical winding. The unit is also warmed up by the engine (Fig. 8.17).

When the engine is cold, the bimetallic strip is straight. This permits the return-spring to pull the disc-valve open. As the bimetallic strip is heated, it bends and rotates the disc-valve progressively against the return-spring until the valve orifice is to one side of the bypass air passage, thus closing off the excess air supply. The time the valve remains open will be set by the supply current for the winding and the operating conditions of the engine.

8.3 'L'-jetronic fuel injection system (Bosch)
(Fig. 8.18)

The electronic fuel injection system is responsible for controlling the necessary amounts of air and fuel flowing to the engine cylinders to meet all the engine's operating conditions.

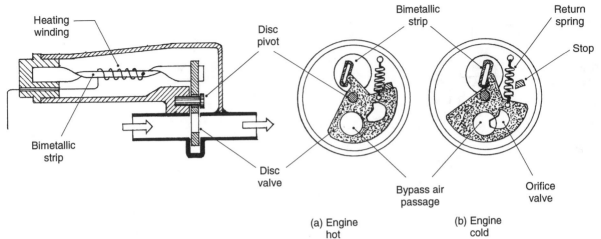

Fig. 8.17 Auxillary air valve

Clean air from the filter passes through the air flow meter, with the actual quantity of air movement being controlled by the throttle valve before it enters the induction manifold. It is then divided between the manifold branch pipes on its way into the cylinders (Fig. 8.18).

Fuel is drawn from the petrol tank to an electric-motor driven roller-type pump, the pressurized fuel then passes through a filter to a fuel distributor manifold, and here a pressure regulator valve sets the fuel injection pressure. Fuel is then directed to each electromagnetically operated injector.

To co-ordinate the metering of the fuel to match the mass flow of air at any time, an electronic control unit receives and interprets data from various operating parameters such as the air-mass rate, engine speed, coolant temperatures, air temperature, throttle opening and battery voltage. This information is then transformed into current pulses which govern the open time of the solenoid-operated injectors.

8.3.1 Electric fuel pump

Fuel pressure and circulation is produced by an electric motor driving the roller cell-type pump described in 8.2.1 (Figs 8.3 and 8.4(a and b)).

8.3.2 Pressure regulator
(Fig. 8.18)

The three possible variables in controlling the quantity of fuel injected are: the injector opening orifice area, injector pressure and injection opening time. With this fuel injection system, both injector opening orifice area and pressure are maintained constant and the variable quantity is made the opening time of the injectors.

To obtain accurate metered quantities of fuel, a diaphragm-type pressure regulator bypasses excess fuel back to the fuel tank and, in this way, maintains a fuel pressure of 2.5 or 3 bar depending upon the fuel injection system employed.

The pressure regulator is divided into two chambers by the diaphragm, a control spring chamber and a fuel overflow valve chamber (Fig. 8.18). The regulator is connected to one end of the fuel manifold and when pump pressure exceeds the pressure-regulator pressure setting, fuel will push back the diaphragm against the control spring thus opening the disc valve. Fuel will therefore be returned back to the fuel tank.

To enable the pressure difference across the injectors to remain constant with manifold depression variation, the pressure regulator control spring chamber is subjected to manifold pressure. Therefore, as the manifold pressure changes, the pressure acting on the regulator diaphragm alters correspondingly. This means that if the manifold pressure drops, as would be the case for part load conditions, the pressure acting on the regulator diaphragm in the spring chamber will also be reduced, and thus less fuel pressure will be required to open the regulator valve in order to bleed fuel back to the tank. As a result, the fuel pressure will be slightly reduced relative to atmospheric pressure, but the pressure difference across the injector, that is the injector outlet

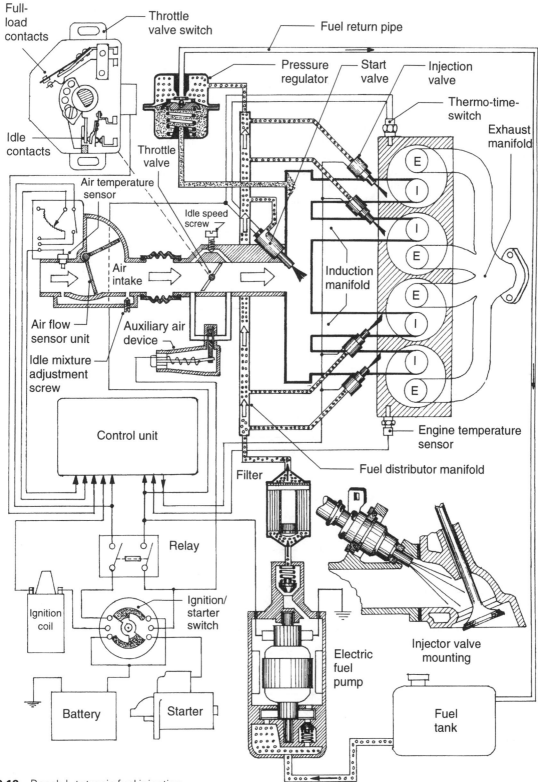

Fig. 8.18 Bosch L-tetronic fuel injection

(manifold pressure) to that of the fuel supply pressure, will remain constant.

8.3.3 Electronic control unit
(Figs 8.18, 8.19 and 8.20)

The electronic control unit consists of integrated circuits, diodes, resistances and capacitors which are arranged to form a small computer. It is the function of this control unit to collect all the signal voltages provided by the sensors on the operating conditions of the engine and so evaluate the quantity of fuel required. The input data are combined and converted into control unit current pulses, which are then transmitted to the injector valves. The pulsed time the injector solenoids are energized, and therefore the duration the injectors are open, determines the quantity of fuel injected into the induction ports.

The frequency of the injector pulses is derived from engine speed, while both the engine speed and the amount of air consumed by the engine establishes the basic time the injectors are open.

Ignition pulses from the negative side of the ignition coil determine the engine speed and hence the frequency of the injector pulses (Fig. 8.19). These ignition peak pulses are then

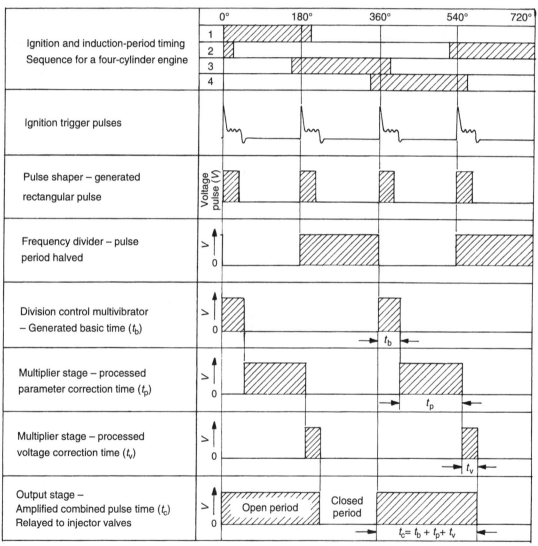

Fig. 8.19 The processing and generation phases for an injector valve opening pulse time cycle in an electronic control unit

412

fed to the pulse shaper which generates rectangular pulses from the trigger pulses. A frequency divider or splitter then halves the pulse sequence in order to provide triggering pulses for the injectors. A division control multivibrator now co-ordinates both the frequency divider inputs (a measure of engine speed) and the mass air-flow rate input to bring about a basic injection time (t_B); that is, the quantity of fuel to be injected per induction stroke without taking into account other engine operating conditions.

The basic injection time then passes to the multiplying stage where additional information parameters—such as engine coolant temperature, air temperature, engine load, throttle opening position—are received, processed and converted into a correction factor (k). This is then multiplied by the basic injection time (t_B) to produce a parameter correction time (t_p). Thus, correction time = correction factor × basic time, i.e. $t_p = kt_B$. The multiplying stage also computes a voltage correction time allowance (t_v) for the injector solenoid opening response delay which is sensitive to the battery supply voltage. Thus, the lower the voltage the longer will be the voltage correction time and vice versa. Initially, the basic time, parameter correction time and voltage correction time are transformed into a combined time pulse (t_c), this is then amplified and relayed to the injector valves. Consequently, short and long pulse durations reduce or increase fuel delivery and mixture strength, respectively, to match the engine load and speed requirements (Fig. 8.20).

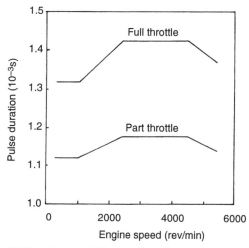

Fig. 8.20 Part and full operating injection pulse duration requirement characteristics

8.3.4 Fuel injector
(Figs 8.21(a and b) and 8.22)

The function of the injection valve (injector) is to inject fuel into the individual manifold branch pipe directly in front of the engine's respective inlet port.

The injectors are solenoid-operated valves which open and close by means of an electrical pulse signalled from the control unit. The injector-valve unit consists of a solenoid winding mounted on the rear section of the soft-iron injector body, and a fluted needle-valve (which also supports the armature plunger) is guided in the nozzle body which forms the forward section of the injector. Both sections of the valve are enclosed and held together by the valve casing.

When the current energizes the solenoid, the generated magnetic field lifts the armature and needle-valve against the helical return-spring resistance, to a position where the needle stop collar contacts a stop-plate—this limited movement having been pre-set to about 0.1 mm. The needle-valve tip is of the pintle type and, when open, forms with the nozzle body a calibrated annular orifice resulting in a hollow conical fuel spray discharge (Fig. 8.21(a and b)).

With this type of injection system both fuel pressure and the injector needle–nozzle orifice are constant. Therefore, the only other variable factor to control the amount of injected fuel is the duration of the needle-valve opening (Fig. 8.22). The quantity of fuel to be injected will depend upon a number of sensed operating parameters, such as air flow rate, engine speed, engine temperature, inlet air temperature and atmospheric pressure. Consequently, to meet all these different running conditions, the needle-valve pull-in and release times are able to range from 1.0 to 1.5 ms. A current of approximately 1.5 A is needed to actuate each injector solenoid and valve.

Upon the engine ignition being switched off, the solenoid current is interrupted and the return-spring will push the needle-valve onto its seat, this completely seals the injector from the intake port and prevents the fuel line pressure decreasing.

Each injector is mounted in a rubber moulding located in a holder, the rubber insulator prevents heat from the engine promoting fuel vapour bubbles forming in the pipe lines and it also dampens transmitted vibrations from the engine.

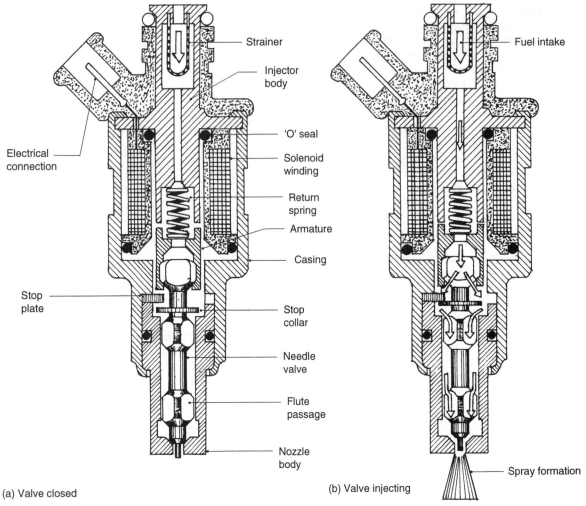

(a) Valve closed

(b) Valve injecting

Strainer

Fuel intake

Injector body

Electrical connection

'O' seal

Solenoid winding

Return spring

Armature

Casing

Stop plate

Stop collar

Needle valve

Flute passage

Nozzle body

Spray formation

Fig. 8.21 Fuel injection valve

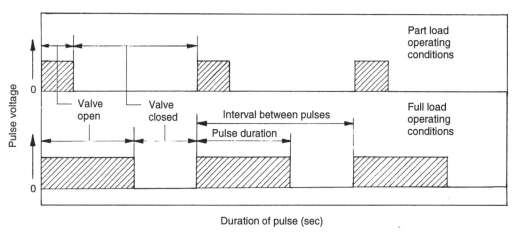

Part load operating conditions

Pulse voltage

Valve open

Valve closed

Interval between pulses

Pulse duration

Full load operating conditions

Duration of pulse (sec)

Fig. 8.22 Comparison of injection valve open and closed pulse cycle for both part and full load operating conditions at constant speed

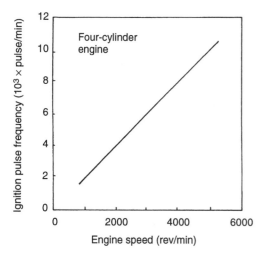

Fig. 8.23 Engine speed and ignition trigger pulse relationship

amounts of fuel mixture will accumulate in each induction branch pipe and port before being induced into the respective cylinders every time the valve opens. In order to minimize the quantity of fuel injected at any one time, the injectors are programmed to inject fuel every half-cycle: that is, every revolution of the crankshaft.

To measure the engine speed the ignition primary current impulse is signalled to the electronic control unit when a contact breaker ignition system is used; alternatively, if a breakerless ignition system is utilized a high-tension pulse from one of the high tension leads is used (Fig. 8.23). The measurement of air mass flow and engine speed enables the electronic control unit to compute the air mass per stroke and hence the fuel required per stroke.

8.3.6 Air flow meter
(Figs 8.24 and 8.25)

Engine load may be measured by sensing the amount of air drawn into the engine, this being a far more accurate method of predicting engine load than relying on the throttle opening position.

The air flow meter is of the vane type: it

8.3.5 Engine speed
(Fig. 8.23)

Fuel injection is not timed. In fact, all the injector valves open simultaneously and, therefore, when the individual engine inlet valves close, different

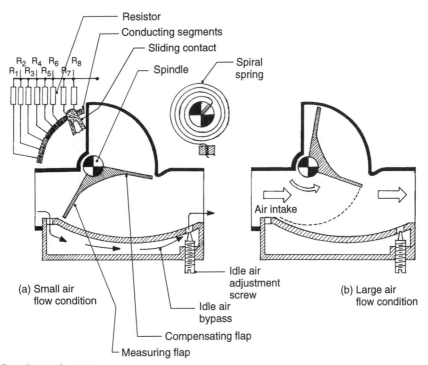

Fig. 8.24 Air flow (meter) sensor

415

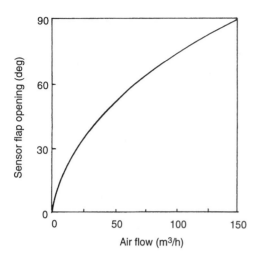

Fig. 8.25 Air flow meter sensor flap movement to rate of air flow relationship

consists of a pivoting aluminium alloy mono-casting of a pair of rectangular-shaped flaps set approximately at right angles to each other. The lower measuring flap is enclosed in a similarly shaped air tunnel while the upper compensating flap seals off a semi-circular chamber situated above the intake air tunnel. Mounted on the end of the pivoting spindle is a potentiometer which relates to the angular movement of the flaps (Fig. 8.24).

Air drawn into the engine passes through the air tunnel and, in the process, tends to drag the measuring flap open in proportion to the amount of air drawn into the engine in a given time. This flap deflection is opposed by a reaction torque provided by a spiral spring having a constant stiffness rate. The pressure difference created across the flap due to the air stream movement produces an opening force which is balanced by the spring force. Therefore the flap rotates according to the amount of air drawn into the engine (Fig. 8.25).

While the engine is running, the induction strokes create pressure pulses which are transmitted back to the intake manifold and air tunnel in the form of pressure wave oscillations. To prevent these oscillations interfering with the measuring flap function, a second compensating flap is incorporated on the same spindle. Thus, when pressure oscillations occur in the induction, pressure waves are reflected back to the air tunnel where they react against opposing faces of both the measuring and compensating flaps. As a result, both flaps tend to twist in opposite directions

and, therefore, any air intake pressure fluctuations acting on these flaps are neutralized.

Measurement of the amount of air flow is achieved by the angular movement of the sensor flaps varying the resistivity of the potentiometer and converting it into a voltage. The potentiometer resistance is so arranged as to provide a signal voltage output which is inversely proportional to the amount of air movement. This will then be sent to the control unit.

An adjustable air bypass is provided in order to set the idle air–fuel mixture ratio to suit the warmed-up engine's frictional and pumping losses and so maintain stable operating conditions.

An initial 5° movement of the measuring flap is used to deflect a very light spring which triggers the fuel pump relay.

When the air flow is below this equivalent movement the spring will switch-off the relay, and this ensures that there is the minimum of fire risk in the event of pipes fracturing during a collision.

8.3.7 Air temperature sensor
(Fig. 8.26)

The mass of air drawn into the engine depends on its density, which varies directly with its temperature. The colder the air, the denser it becomes. Therefore, for a given throttle opening, a greater mass of air will enter the cylinder as its temperature is lowered. Conversely, as the temperature of the air rises, its density decreases so that less air is drawn into the cylinder.

To allow for the change in air density as the temperature rises or falls, a thermister air temperature sensor is incorporated in the air flow meter so that the air temperature is measured. This information is then passed to the electronic control unit, a correction is then made to the injector valve opening time so that the metered fuel exactly matches the air mass flow for any given operating temperature (Fig. 8.26).

8.3.8 Throttle valve switch
(Fig. 8.18)

The throttle valve switch senses three operating conditions—idle, part-load and full-load which are translated into voltage signals to the control unit.

The throttle valve switch casing is attached to the induction manifold while the operating shaft

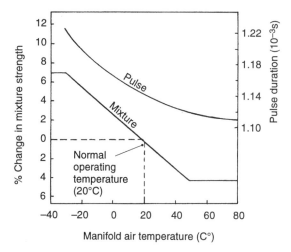

Fig. 8.26 Pulse duration and mixture strength characteristic with respect to manifold air temperature

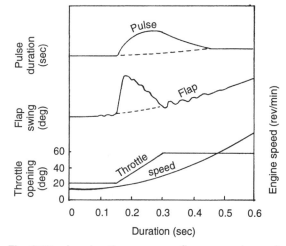

Fig. 8.27 Acceleration response flap over swing and injection pulse duration characteristics

is coupled to the throttle valve shaft, the initial and final angular movement of the operating shaft closes the idle and full-load contacts respectively—there being no contacts for the between part-load angular movement of the throttle and operating shaft.

When the throttle is released, it moves to the idle position. This automatically closes the idle contacts, signalling to the control unit that a rich air–fuel mixture is necessary under these conditions. Between idle and full-load throttle position there will be no contacts closed, this being the part-load range, and no signal is transmitted. Therefore, the control unit assumes that a lean air–fuel mixture is required. Upon the throttle approaching the fully open position, the full-load contacts close so that a signal is passed to the control unit, indicating the mixture should be strengthened.

8.3.9   Acceleration response
(Fig. 8.27)

When the engine changes from one operating speed condition to another, additional fuel mixture is necessary to compensate for the inertia lag of the heavier fuel injected into the relatively light incoming air stream, and this temporarily produces an overweak mixture.

Enrichment for acceleration is achieved by the throttle opening suddenly, so that the rapid rush of intake air through the air flow-meter will cause the air sensor-flap initially, to overswing, which

correspondingly passes an increased fuel demand signal to the electronic control unit (Fig. 8.27).

If the engine is warming up, the overswing enrichment may have to be supplemented with additional fuel, which is signalled to the electronic control unit by the speed at which the air sensor-flap is deflected.

8.3.10   Cold start system
(Figs 8.14, 8.28 and 8.29)

During engine cold-start cranking, the very low crankshaft speed of rotation does not produce sufficient air velocity to atomize and support the fuel droplets in the air stream while, at the same time, some of the fuel which has atomized will then condense in the induction ports and on the cylinder walls. To compensate for the loss of effective air–fuel mixture reaching the cylinders, a cold start valve is provided to supply extra fuel only when the engine is being cranked.

When the starter is operated and the engine is cold, current will be supplied to the solenoid winding of the start valve. The magnetic-field produced, then pulls down the armature plunger against the tension of the return spring and opens the inlet valve. Fuel now flows through the valve around the armature plunger and along the central passage to enter the side of the nozzle chamber. At this stage a high degree of fuel swirl is created within the nozzle cavity before it is discharged in the form of a highly atomized spray into the manifold (Fig. 8.28).

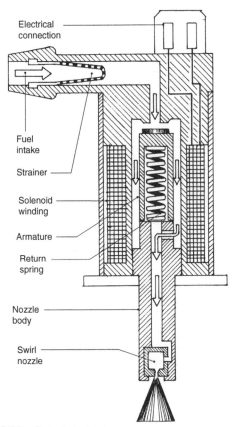

Fig. 8.28 Petrol start valve

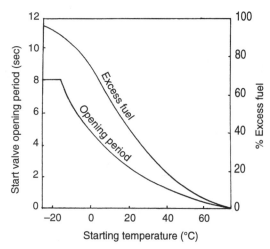

Fig. 8.29 Cold start valve opening time and excess fuel to starting temperature relationship

A thermo time-switch (Fig. 8.14) completes the electrical circuit to the cold start valve and sets a maximum time at a given temperature for which the cold-start injector is energized. At very low engine temperatures, the bimetallic strip controlled contacts are closed, but heat from the engine and energized winding very quickly deflects the bimetallic blade and so interrupts the cold-start valve electrical circuit. This switches off the cold-start valve. The cold-start valve-solenoid energized period is a maximum of 8 s at −20°C, but progressively shortens as temperature rises (Fig. 8.29).

8.3.11 Warm-up period
(Figs 8.30 and 8.31)

When cranking a cold engine, between 30% and 60% more fuel is required according to the surrounding temperature. Once the engine is operating, a much reduced additional amount of fuel to that normally supplied is still necessary to counteract mixture condensation, and this should progressively decrease as the engine warms up, until only the normal operating quantity of fuel is injected into the induction manifold.

Information on the operating temperature of the engine is obtained by the coolant temperature sensor which relays a signal voltage to the electronic control unit to enable a mixture correction to be made (Fig. 8.30).

The temperature sensor is a thermistor which consists of a hollow sealed brass tube, in which a semiconductor resistor (having a negative temperature coefficient) is enclosed (Fig. 8.31). The temperature sensor is screwed into the engine block cooling system and, as the temperature increases, the semiconductor resistance decreases. This correspondingly varies the signal

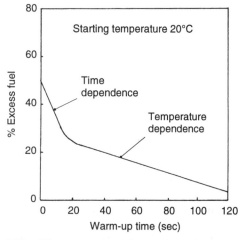

Fig. 8.30 Warm-up excess fuel to time relationship for a typical starting temperature of 20°C

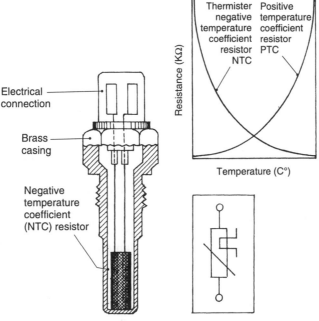

Fig. 8.31 Temperature sensor

voltage to the control unit which then progressively reduces injector opening time.

Cranking enrichment

Whenever the starter is operated at any temperature, a signal from the ignition switch to the electronic control unit increases the open time of the injectors beyond that required for normal running conditions, and the air–fuel mixture is thereby enriched.

8.3.12 Cold idle speed control
(Figs 8.16 (see p. 410) and 8.32)

Idle speed is determined by the engine developing just sufficient power to equal the frictional and pumping losses. When the engine is cold the frictional pumping losses exceed the idle power produced by the engine, so that it will tend to stall.

An auxiliary air device of the disc-valve type, with heating winding and bimetallic strip control, is incorporated (Fig. 8.17). This valve opens an air passage which bypasses the closed throttle when the engine is cold so that additional air is drawn into the engine. The increased air flow will correspondingly lift the sensor-flap further and, in so doing, will signal an increase in injector opening time so that more fuel will be supplied to

match the increased quantity of air entering the engine. Accordingly, the idling speed will increase to prevent the engine running unevenly or even stalling (Fig. 8.32). When the engine is switched on the bimetallic strip receives heat from the current passing through the heating winding, it deflects and progressively rotates the disc-valve until it eventually closes the bypass passage. The auxiliary air device is so positioned that it receives heat from the engine, and thus, when the engine is operating under normal conditions, the bi-metallic strip will prevent the disc-valve opening.

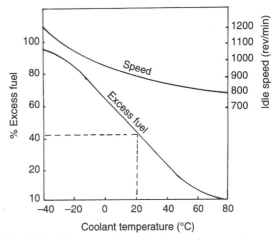

Fig. 8.32 Warm-up excess fuel and idle speed to coolant temperature relationship

419

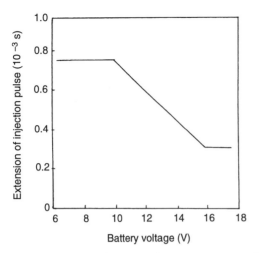

Fig. 8.33 Pulse duration extension to compensate for battery supply voltage reduction

8.3.13 Battery voltage compensation
(Fig. 8.33)

The injector performance is sensitive to the supply voltage in that the injector solenoid current affects the rise time of the valve. As the voltage decreases, due possibly to the battery condition or state of charge, the effective opening period of the injector valve is reduced so that less fuel would be discharged.

To allow for battery voltage variation, the actual voltage is signalled to the electronic control unit, and an extended pulse time is then added to the basic pulse time to compensate for the injector valve opening response delay (Fig. 8.33).

8.4 Mono-jetronic fuel injection system (Bosch)

(Fig. 8.34)

This is an electronically controlled injection system, basically designed for four-cylinder engines, in which only one electromagnetically operated injection valve intermittently injects fuel from a central point into a funnel-shaped passageway above and ahead of the butterfly throttle valve. The single-point injection arrangement simplifies the fuel system, compared with the multi-point fuel injection system where there is one solenoid-controlled injection valve situated near the inlet port of each cylinder. Being positioned above and central in the tapering passageway above the throttle valve, the fuel is injected into a high velocity region where optimum atomization and mixing is ensured. The injection solenoid is actuated twice per camshaft revolution, this being synchronized and triggered to the ignition pulses.

8.4.1 Injection valve and pressure regulator valve unit
(Fig. 8.34)

The injector unit consists of an annular-shaped solenoid surrounding a soft-iron fixed sleeve, below which is a disc-shaped armature with its dome-shaped valve head at its centre. The dome valve closes on an annular seat formed by the spherical recess made in the valve nozzle just below it. Fuel passes out of six radially arranged slanted metering holes into a deep semi-conical recess where a swirl and rebound action takes place before it is ejected in a fine conical-shaped spray. Because of the low inertia of the armature and dome valve, the pull-in and drop-out reaction time for this form of valve is well under one millisecond.

The solenoid valve assembly is mounted in the aluminium alloy valve casting, on a rectangular sectioned seal at its lowest end, and is clamped in position at its upper end with a plastic insulation cap. A nylon screen is wrapped around the solenoid valve body to filter fuel coming and going from the valve body.

A roller, centrifugally operated, electronically driven, fuel pump is installed in the fuel tank. It supplies fuel through a fuel filter to the solenoid-operated injection valve where the fuel circulates the valve body before entering radial holes to fill up the central space surrounding the dome-shaped injection valve and armature disc. The surplus fuel then passes out to the pressure regulator valve which is pushed open when the supply pressure reaches approximately 1 bar. Since the injector nozzle is positioned on the intake side of the throttle where the passageway is wide, the injector nozzle and the top side of the pressure regulator diaphragm are both subjected to the same air pressure. Consequently, the pressure regulator diaphragm maintains the differential pressure, at the injector discharge point, to be constant and independent of the injected fuel quantity.

Surplus overflow not only relieves the fuel supply pressure but it also prevents the formation of vapour locks around the solenoid-operated injection valve assembly.

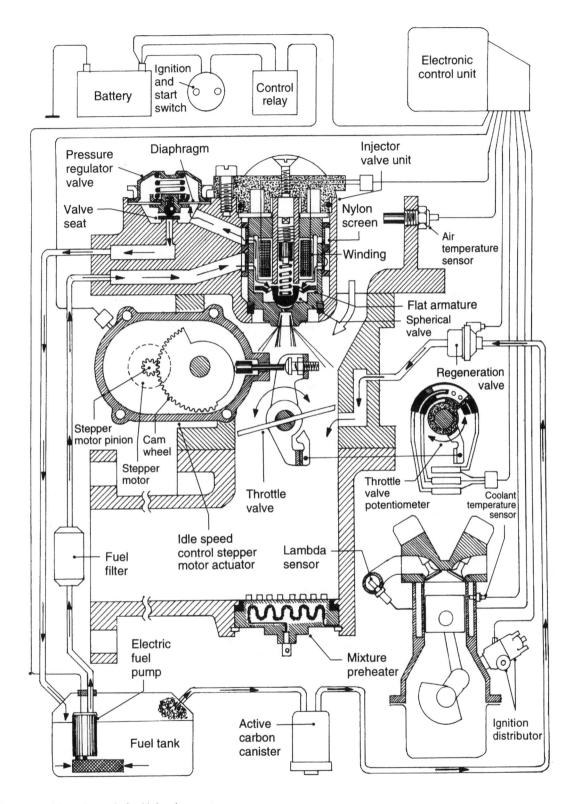

Fig. 8.34   Mono-jetronic fuel injection system

8.4.2 Throttle potentiometer
(Fig. 8.34)

The throttle potentiometer used is a variable resistor having two moving metal leaf-spring type contact brushes mounted on a spindle which sweeps over two ceramic metal (cermet) semi-circular tracks deposited on a ceramic backing. The potentiometer acts as a voltage divider. Thus, if an input voltage is applied across the ends of the resistor tracks and if the output voltage is taken between one end of the resistor track and the contact brush, then the output voltage will vary according to the distance between the brush and its starting point.

The potentiometer spindle is coupled to the throttle butterfly valve spindle so that any angular movement of the throttle valve produces an equal sweep of the brushes over their respective tracks. Since the track resistance is linear (that is, directly proportional to its length) any opening of the throttle is directly related to the change in track resistance and therefore, it proportionally changes the output voltage from the potentiometer. This is then passed on to the electronic control unit. Note that some applications of the throttle potentiometer prefer a non-linear resistor to boost the signal in the idle speed region.

The quantity of fuel required to be injected into the air intake to match the amount of air entering the cylinders under different operational conditions is calculated by the electronic control unit which receives signals from the throttle potentiometer and the ignition distributor. The throttle valve potentiometer registers the throttle valve opening angle (α), which is taken as a measure of the engine load and therefore indicates the mass of air charge needed in each cylinder per cycle, where as the engine speed (N) is sensed by the ignition distributor pulse frequency, which is a measure of the number of times the cylinder is being filled with air per minute. Thus, the overall quantity of air consumed by the cylinders at any one time, ignoring air temperature and engine temperature and any other operational variables, can be computed by knowing the factor that is constantly being monitored by the electronic control unit.

8.4.3 Closed loop stepper motor idle speed control
(Fig. 8.34)

To obtain a lower and more stabilized idle speed than would normally be achieved with the open-loop bimetallic strip operated auxiliary air device during the warm-up phase, a closed loop idle speed control system has been adopted.

This system uses input signals from the electronic control unit to energize and drive a stepper motor attached indirectly to the throttle valve spindle. The drive from the stepper motor is conveyed through a driver pinion meshing with a driven sector-gear cam-wheel, to the throttle-valve spindle-lever via a thrust pin that is actuated by the rise and fall of the cam profile, which is formed on the opposite side of the sector.

The electronic control unit supplies increments of control current to the stepper motor which are proportional to the throttle opening required to increase or decrease the air supply, depending on the amount the idle speed deviates from the optimum idle speed setting and the engine coolant temperature.

Any alteration in the throttle valve opening position is immediately sensed by the throttle potentiometer which, accordingly signals for the injection quantity to be varied to match the change in the incoming air flow. Thus, the exhaust emission during idle speed remains practically constant.

This idle speed control needs no alteration since the electronic control unit computes the variables fed into it so that the stepped signal current pulses sent to the stepper motor automatically adjust the throttle opening to compensate for engine wear.

8.4.4 Electronic control unit—processing of operating data
(Fig. 8.34)

This unit is a digital control unit which receives information from the various sensors and uses them to compute alterations for the injector opening and, thus, the quantity of fuel to be injected. The output signals are processed and delivered by means of a microcomputer, program memory, data memory and an analogue-digital converter.

The basic injection output signal is derived from a stored characteristic map which has 15 throttle opening angles and 15 engine-speed data-points. This results in $15 \times 15 = 225$ data points within which the injection duration is predicted for a stoichiometric air–fuel ratio ($\lambda = 1.0$). The appropriate injection durations for each of the 225 data points for different load and speed operational conditions have been evaluated during extensive engine trials.

Built onto the basic characteristic 15×15 map is an adapative map with 8×8 data points. If a given deviation on the basic map is exceeded, an adaptation algorithm which is based upon the evaluation of the lambda sensor signals, installs correction valves into the adaptive map. Compensation is thus provided for individual tolerances and drifts as a function of time of the engine and injection component.

8.4.5 Engine temperature sensor
(Fig. 8.34)

When the engine is started from cold, and during the warm-up period, injected fuel spray condenses onto both the cold manifold and cylinder-block walls so that only a relatively small proportion of the lighter fractions of the liquid fuel actually remains atomized to form a homogeneous mixture before and after if enters the cylinders.

Usually, with the initial cold start phase, 30% to 60% more fuel supply is necessary than is normally used when the engine is hot—the actual amount of enrichment required depends, of course, upon the ambient temperature. The first phase is time dependent and is made to last roughly 30 s. It is followed by a much reduced excess fuel phase which is engine temperature dependent.

A negative temperature coefficient semiconductor material, enclosed in a brass-sleeve type capsule, forms a variable electrical resistor whose resistance decreases with rising engine coolant temperature (Fig. 8.31). It is this variation in resistance which is used as a measure of the coolant temperature, since it changes the magnitude of the signal current passing to the electronic control unit.

8.4.6 Air temperature sensor
(Fig. 8.34)

The mass of air drawn into the cylinders is, to some extent, dependent upon its density, which is also, in turn, influenced by its temperature. Thus, to get an accurate match of fuel supply to the air consumed over the engine's speed and load range, an adjustment to the amount of fuel injected must be made to compensate for any variation in the temperature of the air passing through the manifold.

A temperature sensor projecting into the intake manifold entrance and operating on the same principle as the engine coolant temperature sensor is used to measure the temperature of the incoming air stream. This is then relayed in the form of a signal voltage to the electronic control unit.

8.4.7 Fuel mixture control
(Fig. 8.34)

Idle speed
Idle speed conditions exist when the throttle-valve lever contacts the stepper-motor thrust pin. Adjustment of the air intake stream is performed by the stepper-motor changing the throttle-valve opening during idle speed. This being immediately signalled to the electronic control unit.

Part load
Part load operating conditions are detected by the throttle-valve potentiometer angular movement. This being signalled to the electronic control unit. During this mode of running the air–fuel ratio is set to the minimum emission in the range $\lambda = 1.0$.

Full load
Full load operating conditions are sensed by the throttle valve potentiometer when its contact brush is near the end of its track. The electronic control unit characteristic map then computes the excess fuel required by extending the injection opening period.

Acceleration
Transitional speed-increased operating conditions are detected by how quickly the throttle valve potentiometer contact brush moves over its track towards the fully open position. This movement is signalled to the electronic control unit, which then prolongs the current pulse to the injection valve solenoid so that adequate fuel enrichment is provided to avoid any hesitation during the time the engine speed rises.

Overrun
Engine speed overrun can be detected by the combination of the throttle valve potentiometer

brush being in the throttle closed position and the ignition distributor speed pulse being well above idle speed. The electronic control unit computes these signals, and when the conditions are correct, cuts-off the current pulse to the injector solenoid. Since no fuel will be injected during these operating conditions, fuel consumption and exhaust emission are greatly improved.

Engine speed limiter

The electronic control unit can be programmed to interrupt the current feed pulse to the injector solenoid when the ignition distributor speed pulse reaches the designed maximum frequency, as determined by the manufacturers, to prevent the engine over-speeding.

Supply voltage correction

Voltage fluctuation, which occurs during starting, idling and rising speed operating conditions, is compensated for by the electronic control unit which computes a correction factor for any delay in the injector's opening response time.

8.5   KE-jetronic fuel injection system (Bosch)

(Fig. 8.35)

The KE-jetronic fuel injection system, where the 'E' in the 'KE' stands for electronic, is based on the mechanical K-jetronic system (Fig. 8.39), and has been provided with electronic components and an electronic mixture control unit to increase the flexibility and accuracy of matching the mixture to changing load, speed and other operating conditions.

A description of the modified components and the additional components incorporated (Fig. 8.35) are as follows:

1 air flow sensor
2 differential pressure valve
3 fuel distributor control plunger and barrel
4 electro-hydraulic pressure actuator
5 air flow sensor potentiometer
6 primary pressure regulator

Other components used in the KE-jetronic system which are common with the K-jetronic and the L-jetronic systems, and have been described under those headings, are listed as follows:

1 thermo time-sensor
2 engine coolant sensor
3 injection valve
4 cold start valve
5 auxiliary air device
6 throttle-valve switch
7 ignition distributor speed sensor
8 fuel filter
9 accumulator
10 electric fuel pump

8.5.1   Air flow sensor
(Fig. 8.35)

The quantity of air flowing into the engine's induction manifold is measured by suspending a sensor-plate at right angles to the air stream. A control lever mounted on a pivot supports the sensor-plate at one end, whereas a counterweight at the opposite end is used to balance the lever and keep it horizontal when there is no air movement. (Note: some models use a counter spring instead of the counterweight to balance the system.)

When the engine is operating, air movement through the induction manifold system controlled by the throttle opening tends to pull down (air drag) the sensor-plate in the direction of the air movement.

Since the lever contacts the control plunger on the opposite side of the pivot point to the sensor-plate, the control plunger will move in direct proportion to the movement ratio of the lever, but in the opposite direction. Accordingly, the control plunger opens or closes the barrel metering slits in direct proportion to the intensity of the downward drag (fall or lift) of the sensor-plate, which therefore causes the quantity of metered fuel to increase or decrease in a similar relationship to the sensor-plate movement.

8.5.2   Differential pressure valve
(Fig. 8.35)

The quantity of air drawn into the induction manifold is directly related to the amount the sensor-plate falls or lifts and, correspondingly, to the control plunger up and down travel.

However, with a small slit opening, there is a far greater pressure drop (resistance to fuel flow) across the metering slit than with a much larger slit opening so that, under these conditions, the

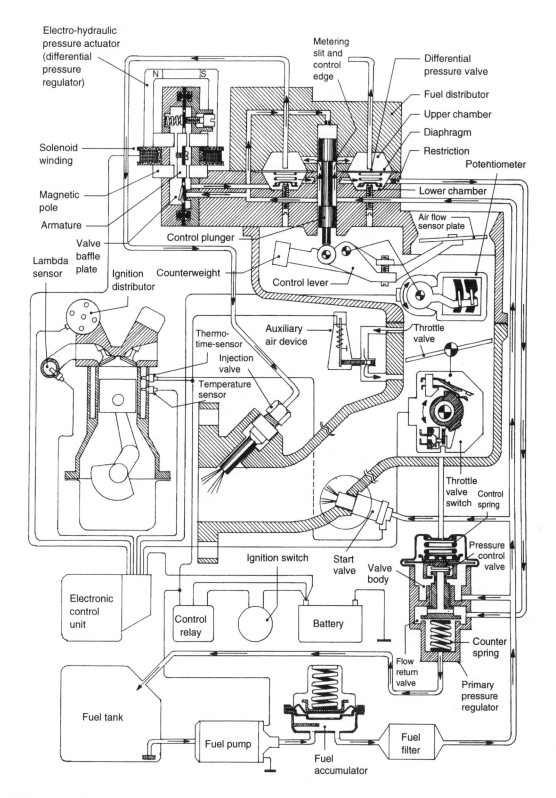

Fig. 8.35 Bosch KE-jetronic fuel injection system

425

quantity of fuel passing through the slits will not be in direct proportion to the sensor-plate downward movement.

Therefore, to maintain a constant pressure drop between both sides of the metering slits for both large and small quantities of metered fuel, so that the quantity of fuel injected directly relates to the sensor-plate lift, a spring loaded diaphragm-type differential pressure valve is incorporated on the output side of each metering slit. These differential pressure valves, which are spring loaded from underneath, keep a constant pressure difference between the upper and lower chambers of approximately 0.2 bar under nearly all operating conditions. Thus, if a large quantity of fuel flows into the upper chamber, the diaphragm deflects downwards until the set differential pressure is reached, with the fuel discharging through the central opening gap between the valve seat and the diaphragm. If the air flow demand of the engine is now reduced, the smaller slit openings will pass less fuel into the upper chambers. Consequently, the diaphragm deflection will be reduced so that the effective orifice opening of the differential valve will be smaller and proportionally less fuel will be delivered to each injector.

With this constant differential pressure control, accurate predictions of the basic injection quantity (relative to the sensor-plate movement) are achieved.

8.5.3 Fuel distributor control plunger and barrel operation
(Fig. 8.35)

Fuel metering is achieved by the up and down movement of a control-plunger inside its barrel partially uncovering and covering each of the metering slits formed radially around the barrel.

There are as many metering slits and differential valve upper and lower chambers as there are fuel injectors and engine cylinders, whereas the diagram, Fig. 8.35, shows only the cross-sectional view of two valves and chambers. The vertical lift and fall of the plunger is controlled by the position of the sensor-plate which is attached to the hinged control lever.

When the air flow-rate entering the induction manifold is low, the downward pull of the sensor-plate (and correspondingly, the control plunger upward lift) is small so that only a very tiny portion of the slits is exposed. However, if the throttle valve is opened wider and the incoming air is at a much higher rate, the air movement is able to drag down and support the sensor-plate at a lower level. Accordingly, the much larger slit openings will now permit more fuel to be metered through to the upper differential valve chamber.

To prevent the plunger overshooting or vibrating up and down, the primary fuel supply pressure is applied, through a restriction, against the plunger's crown so that any pulsations in the air stream passing through to the induction manifold are opposed and damped by the fuel trapped between the plunger and its restrictor. This drag, or resistance to the sensor-plate's upward movement, produces a constant pressure drop in the air movement across the sensor-plate, which tends to stabilize the fuel metering process. For further explanation read section 8.2.9.

8.5.4 Electro-hydraulic pressure actuator
(Figs 8.35 and 8.36(a–c))

The electro-hydraulic pressure actuator varies the pressure in the lower chambers of the differential pressure valves, and it therefore changes the amount of fuel delivered to the injection valves.

This unit is, therefore, able to regulate the pressure difference between the upper and lower differential valve chambers. Thus, it controls the effective orifice opening of the differential valves and hence the mixture strength by a signal from the electronic control unit.

Construction and design
(Figs 8.35 and 8.36(a–c))

The control unit consists of a flat armature piece suspended on a thin spring steel and a pivot baffle plate between two double magnetic poles. The outer edge of the rectangular pivot plate is clamped between the two halves of its supporting housing (Fig. 8.36(a and b)).

A portion of metal is removed from around and between the mid-section and the outer rectangular rim of the spring-steel pivot plate, near the mid-point where the inner and outer sections of the plate merge to form a narrow band. This then functions as a frictionless spring-loaded pivot point (Fig. 8.36(c)). One half of the mid-section acts as the valve face while the other half of the plate is tensioned by an adjusting screw to produce the correct preload tilt stiffness to the pivoting section of the plate.

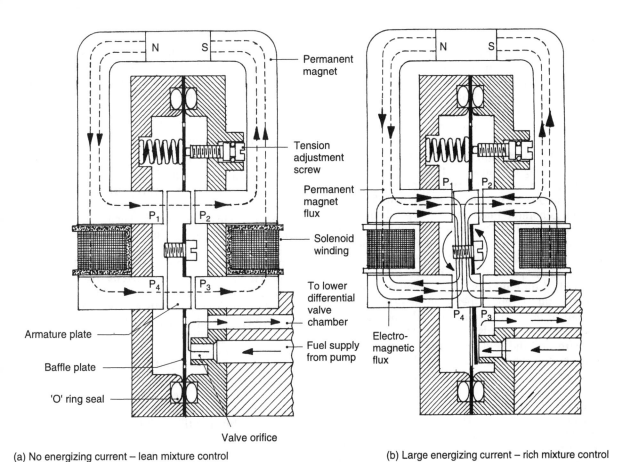

Permanent magnet

Tension adjustment screw

Permanent magnet flux

Solenoid winding

To lower differential valve chamber

Fuel supply from pump

Armature plate

Baffle plate

'O' ring seal

Valve orifice

Electro-magnetic flux

(a) No energizing current – lean mixture control

(b) Large energizing current – rich mixture control

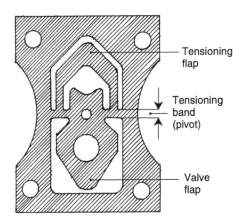

Tensioning flap

Tensioning band (pivot)

Valve flap

(c) Baffle plate with tensioning and valve flap

Fig. 8.36   Electro-hydraulic pressure actuator

Actuator operation
(Figs 8.35 and 8.36(a and b))

There are two magnetic fluxes, one from the permanent magnet (Fig. 8.36(a)) and the other from a electromagnet, both superimposed upon each other around the magnetic flow path of the pole members, air gaps and armature (Fig. 8.36(b)).

When the electromagnet is energized, the permanent magnetic flux and the solenoid magnetic flux join together to strengthen the flux at poles P_1 and P_3, whereas the permanent flux and solenoid generated magnetic flux repel each other at poles P_2 and P_4. Thus, when the current flows through the solenoid winding, diagonal poles P_1 and P_3 will have a strong attraction to the armature, whereas poles P_2 and P_4 are only weakly attracted. Consequently, the imbalance of the magnetic flux strength between air-gaps, draws one half of the armature towards the pole P_1 and the other half towards pole P_3. This twists the armature and the mid-section of the pivot plate about the tensioning band (Fig. 8.36(b)).

The permanent magnetic flux remains constant but the electromagnetic flux increases proportionally to the control current, which is fed by the electronic control unit. The flow of fuel through the pressure actuator valve orifice tends to bend the spring valve baffle away from it against the spring resistance of the twisted band of the baffle plate. Thus, the higher the control current passing to the solenoid winding, the greater will be the magnetic attraction force tending to close the spring pivot baffle plate against the fuel supply primary pressure.

Lean mixture
(Fig. 8.36(a))

When a lean mixture is required, the control current is so small that fuel will flow to the lower differential pressure valve chamber almost unrestricted. This permits a higher pressure to exist in the lower chambers, which therefore deflects the diaphragm further upwards and thus restricts the output flow from the upper chambers to their respective spray injectors.

Rich mixture
(Fig. 8.36(b))

When a rich mixture is required, the control current is increased so that the spring baffle plate twists more towards the pressure actuator orifice valve's closed position. Thus, the fuel pressure in

the lower chamber will be much more reduced so that the higher pressure in the upper chamber deflects the diaphragms downwards. This increases the differential pressure valve opening and thus permits more fuel to be transferred to the spray injectors.

Uncontrolled mixture

When no current flows to the electromagnetic solenoid winding, the baffle plate position relative to the pressure actuator valve orifice produces a differential pressure between the upper and lower chambers. This pressure is such that the fuel delivered to the spray injectors via the central differential pressure valves maintains the air–fuel mixture to something like the stoichiometric ratio. Thus, if the control current from the electronic control unit should fail, the fuel distributor assembly will operate with a single mixture setting, which is adequate for the engine still to be driven. This condition being known as the 'limp home' facility.

The pressure difference in the actuator unit can be increased from 0.4 bar at 0 mA to 1.4 bar at 120 mA, when the fuel flow is roughly 10 litres/hour. The low mass of the armature and baffle plate assembly and the spring preload of the pivot band-plate makes changes to the pressure actuator-valve opening orifice respond rapidly to any control current variation.

Overrun and engine speed limitation fuel cut-off

A directional change in the control current flow pulls the baffle plate away from the pressure actuator-valve orifice so that the pressure in the lower chamber prevents fuel in the upper chamber deflecting the diaphragm downwards. Thus, no fuel can be delivered to the spray injectors. This characteristic of the baffle plate reverse movement can be used to provide overrun fuel cut-off and engine speed limitation.

8.5.5 Air-flow sensor potentiometer
(Fig. 8.35)

This variable resistor, known as the potentiometer, has two semi-circular tracks made of a resistive film deposited on a ceramic base, over which a number of fine wires welded to a control lever brush over. The control lever is attached to one end of the sensor-plate spindle so that any movement of the sensor-plate produces a corresponding semi-circular sweep of the brushes over their respective tracks.

The resistance of the pick up is linear, whereas the main track is made non-linear by varying the track's width along a portion of its length (Fig. 8.35). The potentiometer provides the following information to the electronic control unit.

1 The relative position of the air-flow sensor-plate relative to its no-air flow static position— this being a measure of the power output of the engine.
2 The direction in which the air-flow sensor-plate is moving: that is, opening or closing, which relates to the engine increasing or decreasing speed, respectively.
3 The speed at which the air sensor-plate is pulled down with the air stream: that is, the magnitude of the engine's acceleration.

The potentiometer resistance change is made to be non-linear; therefore, the accelerating signal is designed to be at its maximum when accelerating from idling speed, but the signal then decreases at a faster rate as the sensor-plate is dragged down by the increased quantity of air entering the induction manifold as the engine's power output rises.

Thus, the potentiometer's voltage signals to the electronic control unit are then processed and, from this information, an appropriate control current is relayed to the electro-hydraulic pressure actuator. The mixture strength will therefore be adjusted to match the demands of the driver and the engine's operating conditions. Acceleration enrichment requirements are greatly influenced by the engine's temperature, as some of the atomized fuel will be attracted to the cold walls of the induction manifold instead of passing directly into the cylinders.

A rapid opening of the throttle valve, by the driver depressing the accelerator pedal, can demand a maximum factor of 1.7 times the stoichiometric air–fuel ratio for mixture enrichment, while a slow change in throttle opening may need a factor no greater than 1.1 times the stoichiometric air–fuel ratio.

The voltage signal to the electronic control unit ranges from approximately 0.2–0.3 V at idling speed, to approximately 0.7 V at full load.

8.5.6 Primary pressure regulator
(Fig. 8.35)

This diaphragm-operated pressure regulator maintains the primary pressure in the pipeline to be constant. In contrast to the K-jetronic, where the warm-up valve regulates the control pressure, for the KE-jetronic the pressure on the control plunger in its barrel is equal to the primary pressure. Pressure settings normally vary from 5.3 and 6.4 bar depending on the engine model.

Engine running
(Fig. 8.35)
Fuel supplied by the electric-motor driven pump flows into the chamber beneath the diaphragm, causing it to move up against the control spring. Simultaneously, the relatively weak counter-spring is able to push up the valve body until its shoulder contacts the casing stop. Excess pressure in the fuel supply line will then lift the diaphragm further so that the pressure control valve disc opens, and surplus fuel then flows down the centre of the valve body, out through the open return flow valve at the base of the valve body and back to the fuel tank. At the same time, fuel coming out from the distributor's lower differential pressure chambers, via the restriction, is also permitted to return to the fuel tank.

Engine switched off
(Fig. 8.35)
When the engine is turned off, the fuel pump stops pumping fuel, and therefore the diaphragm pressure regulator determines the holding pressure in the system, which is now determined by the fuel accumulator. The system's pressure drops by a certain amount through the pressure control valve, which remains open for a short period. However, the stiff control spring presses down the diaphragm until the pressure control valve closes, it then continues to exert a downward thrust compressing the weaker spring so that the valve body moves down and closes the return flow valve, thus preventing any more fuel from returning to the fuel tank. Under these conditions the stored fuel in the accumulator maintains the fuel supply line full and under pressure. It therefore prevents the formation of vapour locks and thus contributes to responsive hot starts.

8.6 LH-jetronic fuel injection system (Bosch)

(Figs 8.18 (see p. 411) and 8.37)

The LH-jetronic fuel injection system where the 'H' stands for hot-wire is based on the solenoid-controlled intermittent injection L-jetronic system (Fig. 8.18, p. 411), which has been provided with a hot-wire air flow meter in place of the swing-flap actuated potentiometer air flow meter.

With the hot-wire air-mass flow meter the mass of air entering the induction manifold is measured directly and therefore does not suffer from any of the following limitations:

1 errors due to inertia lag of moving components
2 errors due to altitude changes
3 errors due to air pulsation
4 errors due to air temperature change

All the other components of the LH-jetronic system are common to the L-jetronic system and, as such, have been explained in section 8.3.

8.6.1 Hot-wire air-mass flow meter
(Fig. 8.37(a–c))

The air-flow flap-type sensor measures the volume of air flow, but what is really required is the air mass passing to the engine cylinders. Thus, the airflap flow sensor does not take into account the change in air density which changes with altitude and the surrounding air temperature. It should be observed that air mass = volume × density. A further source of error is that the air flap is susceptible to air pulsations caused by the sudden opening and closing of the inlet valves. This problem has been overcome with the hot-wire mass flow meter (sensor) which does not have any working parts, is independent of air density changes, does not suffer from pulsation error and measures the air mass flow directly with the minimum of response time (Fig. 8.36(a)).

The hot-wire air mass flow meter operates on the constant temperature principle. The hot wire is made of 70 μm thick platinum wire and forms one of four electrical resistance arms making up a Wheatstone bridge circuit.

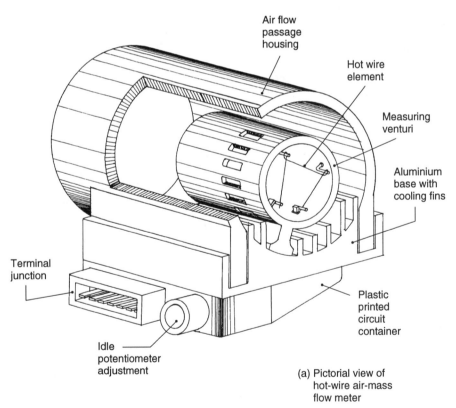

(a) Pictorial view of hot-wire air-mass flow meter

Fig. 8.37(a–c) Hot wire air mass flow meter

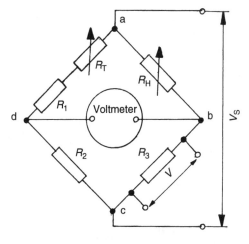

(b) Basic hot-wire
Wheatstone bridge
circuit

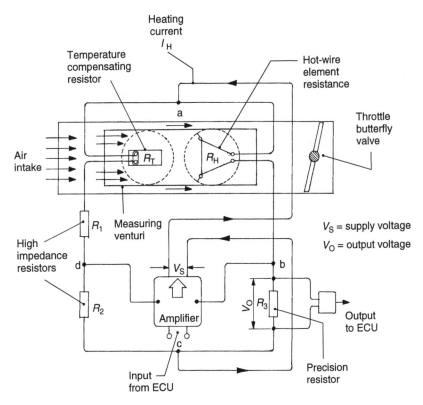

V_S = supply voltage

V_O = output voltage

(c) Circuit layout hot-wire
air-mass flow meter

The bridge consists of four resistances linked together in the form of a diamond ring so that there are four junction points (Fig. 8.37(b)). A voltmeter is connected between opposite junction points 'd' and 'b' and, when an input voltage is applied between opposite junction points 'a' and 'c', current will be divided between two parallel conducting paths, a left-hand path $R_T R_1 R_2$ and a right-hand path R_H and R_3. A current then flows through the voltmeter but if the resistance of each arm is adjusted so that their ratio relation is

$$\frac{R_H}{R_3} = \frac{R_1 + R_T}{R_2}$$

then the current through the voltmeter becomes zero and a balance is said to be obtained.

The Wheatstone bridge just described has the platinum hot-wire arm of the bridge suspended across the measuring venturi whereas the platinum temperature-compensating film resistor, supported on a blade, projects into the venturi slightly behind the hot-wire (Fig. 8.37(c)).

Current supplied by the amplifier passes through the two parallel halves of the bridge and, in so doing, heats and maintains the hot-wire at a temperature of approximately 100°C. When there is no air movement through the venturi (Fig. 8.37(c)) the bridge will be balanced. However, as the air movement commences, the hot-wire is cooled, and this decreases the hot-wire resistance in proportion to the flow rate of the air mass and consequently produces an imbalance in the bridge between junction points 'd' and 'b'. This imbalance produces a voltage between the connections feeding into the amplifier, hence causing the amplifier output to increase its voltage supply (V_s) to the bridge across junction points 'a' and 'c'. Accordingly, the current flow through the hot-wire increases, which simultaneously raises its temperature to its original value—at which point the bridge will again be in a balanced state. Between no flow and maximum air flow the heating current varies from 500 mA to 1200 mA respectively. Conversely, a drop in air-flow speed raises the hot-wire temperature and therefore reduces the current flow rate due to the increased resistance of the hot-wire.

The resulting changes in heating current are signalled to the electronic control unit in the form of an output voltage, V_o, developed across the precision resistor R_3, as this corresponds directly to changes that must be made to the amount of fuel to be injected into the induction manifold, according to the engine load. The heating current response time is very short, due to the hot-wire's low mass, so that it is able to maintain an approximately constant hot-wire temperature under normal operating conditions.

The temperature compensating resistor, R_T, has a resistance of around 500 Ω and, with the two high impedance resistors, R_1 and R_2, in series, the current flow through these three resistors is proportionally much smaller than through the hot-wire and the precision branch of the bridge.

Corrections for changes in ambient air temperature are made by inserting a temperature compensation resistor in one arm of the Wheatstone bridge and positioning it so that it projects into the incoming air stream via the measuring venturi.

To maintain a constant resistance, be corrosion resistant and to respond quickly, a platinum-film resistor is used for the temperature compensation resistor, and to obtain the required resistance to balance this arm of the bridge, a second resistor R_1 is placed in series with it.

Air temperature variations change the resistance of both the temperature compensation resistor and the hot-wire at the same time, this therefore compensates for air density changes.

During operating conditions the hot-wire will become coated with dirt and this will affect the balance of the bridge. Therefore, a command from the electronic control unit passes current through the hot-wire so that it is heated to something like 1000°C for one second each time the engine is switched off. As a result, any dirt accumulated on the hot-wire will be burnt off.

9
Ignition Systems

9.1 Electronic ignition

The ignition distributor has two basic functions: firstly, it acts as a switch to make and break the ignition coil's primary winding circuit, thus enabling the growth and collapse of the primary winding current to be controlled. Secondly, it distributes the generated high-tension voltage pulses, created in the ignition coil's secondary winding, to the individual spark plugs in the order of cylinder firing. At the same time, it optimizes the crank angle ignition point with respect to both engine speed and load.

Electronic ignition is primarily concerned with the on–off switching of the primary winding current through solid-state semiconductor devices which have largely replaced the conventional mechanical contact breaker point switching.

Electronic solid-state semiconductors are not only utilized within the ignition system, but are extensively used throughout the motor vehicle in equipment for fuel metering, engine speed governing, generator rectification and regulation, air and coolant temperature compensation, exhaust emission control, transmission control and cruise control, as well as for vehicle wiring and instrumentation, and brake skid control. There are also many more applications. Because of the importance of understanding solid-state semiconductor technology, the fundamentals of electronics will now be explained.

9.1.1 Semiconductor materials

Semiconductors, as their name implies, do not conduct as effectively as conductors; for example, metals. The electrical resistance of the semiconductor will lie somewhere between the low resistance of a metal and the high resistance of an insulator. The most important semiconductor materials are germanium and silicon.

9.1.2 Electric structure of an intrinsic semiconductor
(Fig. 9.1)

The atoms of conductors, semiconductors and insulators are held together by forces called bonds. Insulators and semiconductor atoms share their outer shell (valence) electrons with those of neighbouring atoms, and each shared pair of electrons makes a covalent bond formed between two atoms (Fig. 9.1). However, semiconductors' outer-shell (valance) electrons are not so strongly bound to their parent atoms as those of insulators and therefore can be detached when a source of electrical energy, such as a battery, is applied across the material.

9.1.3 Electron and hole charge carriers

The conductivity of semiconductors depends on the existence of two different types of charge carriers in the semiconductor material. One of these charge carriers is the free negative electron, moving from one atom to the next; while the other is the positive hole, the hole being the vacancy left when one of the electrons of the outer valency shell of the germanium or silicon atom breaks its bond and migrates to another atom.

433

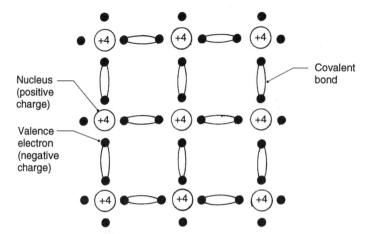

Fig. 9.1 Intrinsic (pure) silicon crystal lattice

9.1.4 Electric conduction through a semiconductor

Conduction in a semiconductor material is due to a voltage applied between the two ends of the material causing free electrons to move towards the more positive end of the semiconductor and holes to move towards the more negative end. The resultant current flowing through a semiconductor is caused by free electrons flowing in one direction and holes flowing in the opposite direction.

9.1.5 Doping of semiconductors

High purity semiconductors, such as germanium and silicon, are known as intrinsic conductors— they are very poor at conducting electric current. If traces of impurities such as arsenic, boron, etc, are added to a high-purity semiconductor material, its conductivity is considerably increased— the addition of these impurities to an intrinsic material is known as doping. The amount of doping is used to adjust the conductivity of the semiconductor to suit specific applications.

9.1.6 *N*-type semiconductor material
(Figs 9.2 and 9.3)

Pure semiconductor material can have its conductivity improved by adding other types of atoms (doping), so increasing the number of free electrons within the crystal structure.

When additive elements such as phosphorus, arsenic, antimony, and bismuth—which are pentvalent atoms (they have five valence electrons in their outer shell)—are added to a pure semiconductor material, such as silicon, they form four covalent bonds with the semiconductor atoms, while the extra fifth electron is bound only to its own outer shell (Fig. 9.2). Because this fifth electron has not been able to pair off with other electrons, it has only a very weak attachment to the additive atom and therefore can be easily dislodged from its outer shell.

These additive atoms fit into the general crystal structure of the semiconductor material and so provide the extra free electrons necessary to increase the flow of current. Since it is the fifth

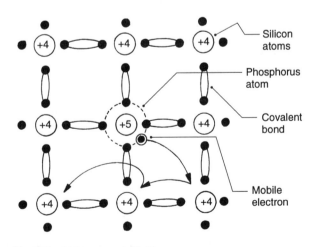

Fig. 9.2 N-type crystal lattice

434

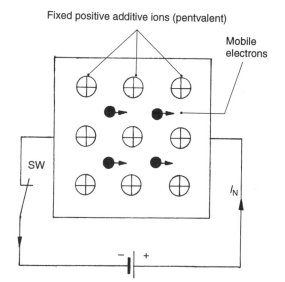

Fig. 9.3 Drift of electrons in N-type semiconductor

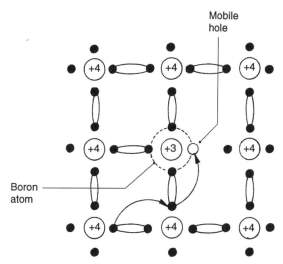

Fig. 9.4 P-type crystal lattice

additive electron which is mobile and not electrons breaking away from covalent bonds, no additional holes are created. In other words the crystal with the additive atoms is neutral so that there will be a fixed number of negative charged ions (atoms with excess electrons in the outer shell) and the same number of holes moving in the crystal. Some holes, although only a small number, may be formed by thermal ionization. Therefore, the major charge carriers will be the electrons, while the few holes become minority charge carriers. When free electrons are produced by additive atoms, these atoms are called donors and semiconductors containing such additives are said to be *N*-type materials. The electron charge carriers in an *N*-type crystal move from the negative end of the semiconductor to the opposite positive terminal (Fig. 9.3).

9.1.7 *P*-type semiconductor material
(Figs 9.4 and 9.5)

The conductivity of a semiconductor material can be improved by introducing additives which, instead of creating free electrons, produce extra holes.

Additives capable of producing extra holes in silicon are indium, aluminium, gallium and boron. These elements are all trivalent atoms (have three valence electrons in their outer shell). Therefore, the additive atoms can only form three

covalent bonds with neighbouring semiconductor atoms, it thereby causes the fourth electron of the semiconductor atom to form an incomplete bond (Fig. 9.4). This unstable gap, normally occupied by the missing electron, is known as a hole, and these positively charged holes attract and accept electrons from other neighbouring atoms, which correspondingly form their own holes and accept electrons from other atoms, and so on. Accordingly, it can be seen that conduction under these conditions is due to the continuous hole movement from one atom to the next. Since the extra holes are created by the additive atom and not by

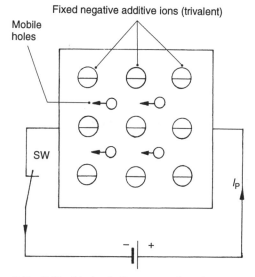

Fig. 9.5 Drift of holes in P-type semiconductor

the breaking of covalent bonds, no additional free electrons will be released. The few free electrons which do exist are due to thermal ionization. Consequently, the doped crystal is neutral, and therefore there will be fixed positive charged ions (atoms short of electrons in the outer shell) and an equal number of electrons free to move at random in the crystal. Therefore, the major charge carriers, when semiconductors are doped with trivalent atoms, are the positively charged holes and the minority charge carriers are the free electrons. When additive atoms accept electrons from the covalent bonds of a neighbouring semiconductor, they are known as acceptor atoms, and semiconductors containing these additives are said to be *P*-type materials. The hole charge carriers in a *P*-type crystal move from the positive terminal of the semiconductor through the crystal structure to the negative end (Fig. 9.5).

9.2 Semiconductor junction diodes

(Fig. 9.6)

A semiconductor junction diode is made from a single crystal having two regions—an anode and a cathode—each having an external terminal. The region with the negative applied voltage is known as the cathode and the other region, with the positive voltage potential, is the anode. Current can flow easily in one direction through the diode, passing in at the anode and coming out at the cathode (Fig. 9.6). However, when it tries to flow

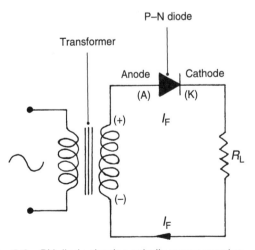

Fig. 9.6 PN diode circuit symbolic representation

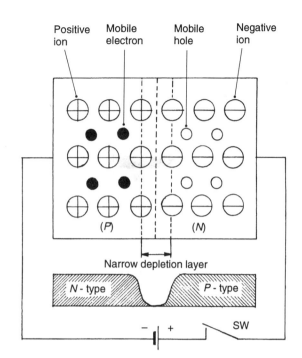

Fig. 9.7 Unbias

in the opposite direction, that is from cathode to anode, it is blocked. In other words the diode resistance is low in the conducting or forward direction but it offers a very high resistance in the opposing or reverse direction.

9.2.1 Unbiased *P–N* junction
(Fig. 9.7)

A *P–N* junction is made when *P* and *N*-type regions back on to each other in the same piece of semiconductor material (Fig. 9.7).

When the free-electron rich *N*-type region and the hole rich *P*-type region come together, they form a *P–N* junction. Free electrons will diffuse in (move through) the *P–N* junction just inside the *P*-type region where they fill holes, and similarly holes diffuse from the *P*-type region to just inside the *N*-type region capturing electrons there. This diffusion process occurs because the concentration of one type of charge carrier is large in one region and small in the other. The exchange of charge soon stops because the negative charge on the *P*-type material opposes the further flow of electrons and the positive charge on the *N*-type opposes the further flow of holes. As a result of this diffusion, the narrow region on either side of the junction becomes fairly free of majority charge carriers, subsequently it is known as the depletion layer (emptied of charge carriers) and is something like 0.001 mm in width and does not conduct—in fact, it acts as an insulator.

Due to the charge diffusion a potential difference builds up across the junction, which acts as if an imaginary battery exists at the junction. This *junction voltage*, as it is called, amounts to 0.2 V for germanium and 0.6 V for silicon semiconductor material.

9.2.2 Forward bias P–N junction
(Fig. 9.8)

If a polarity of an externally applied voltage across a *P–N* junction is such that the *P*-type material is connected to the positive terminal and the *N*-type material is joined to the negative terminal of an electromotive force (EMF) source such as a battery, the applied voltage opposes the junction voltage and the depletion layer narrows (Fig. 9.8). When the battery voltage exceeds the junction voltage the depletion layer disappears permitting holes in the *P*-type region to drift towards the junction and free electrons in the *N*-type region also to drift towards the junction.

At the same time, the majority charge carriers are able to cross the junction so that holes and electrons in the depletion layer are continuously combining; that is, electrons fill holes and holes capture electrons.

Simultaneously, electrons enter the *N*-type region from the battery's negative terminal and other electrons enter the positive terminal of the battery from the *P*-type region, thereby creating new holes which move in the opposite direction to that of the electrons in the *P*-type material. The current in the semiconductor *P–N* junction diode is therefore due to hole flow in the *P*-type region and electron flow in the *N*-type flow region and a combination of the two in the vicinity of the junction.

9.2.3 Reverse bias P–N junction
(Fig. 9.9)

When the polarity of the applied external battery voltage is reversed so that the *P*-type material is connected to the negative terminal and the *N*-type material is joined to the positive terminal of the battery, the direction of the applied voltage is such that it enlarges the junction voltage. Immediately, electrons and holes are repelled farther from the junction and the depletion layer widens (Fig. 9.9). At the same time, holes in the

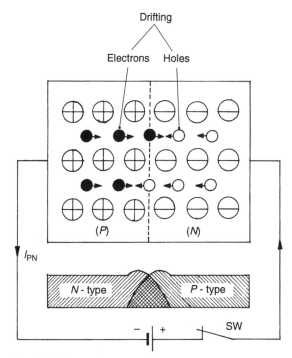

Fig. 9.8 Forward bias

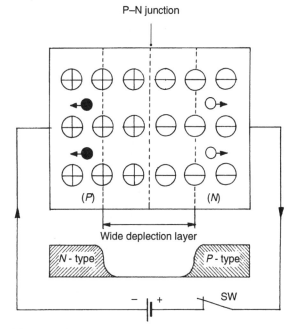

Fig. 9.9 Reverse bias

P-type region will be attracted towards the negative terminal of the battery, and free electrons in the *N*-type region will have a preference to move towards the battery's positive terminal. As a result of the major charge carriers moving away from the *P–N* junction, the depletion layer (where there are almost no charge carriers—that is holes or electrons—apart from the relatively few that are produced spontaneously by thermal agitation) becomes larger. Consequently, the enlarged depletion layer strengthens the high resistance barrier at the junction and, in other words, the junction under these conditions behaves as an insulator.

9.2.4 *P–N* junction diode characteristics
(Fig. 9.10)

An ideal diode is a device which has no resistance to current flow in the forward direction (seen as the vertical thick line in Fig. 9.10); that is, when current flows through the diode crystal from anode (A) to cathode (K). Conversely, the diode exerts a very high resistance (seen as the horizontal thick line in Fig. 9.10) when the battery polarity is reversed, so that current is prevented from moving through the diode from cathode to anode.

In practice, a small threshold voltage (V_o) applied across the diode in the forward direction is necessary to make the diode conduct (Fig. 9.10) and that when the applied voltage is reversed between the diode terminals, a very small reverse

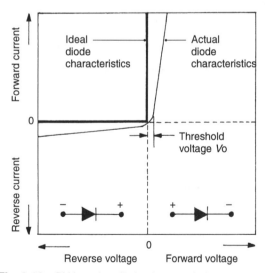

Fig. 9.10 PN junction diode characteristics

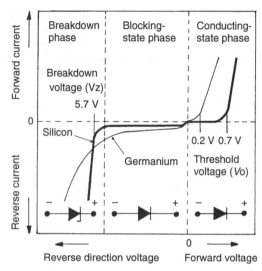

Fig. 9.11 Zener diode unidirectional reverse voltage breakdown characteristics

current will leak across the depletion layer at the *P–N* junction (see the actual diode characteristics curve, the thin line in Fig. 9.10).

9.2.5 Zener diode
(Figs 9.11 and 9.12)

The Zener diode is similar to a *P–N* junction diode but it is designed to conduct in the reverse breakdown voltage region without causing damage to itself, provided that an external current limiting resistor is connected with it in series (Fig. 9.11).

The large current, which is able to pass through the diode when the reverse voltage reaches a given value, is caused by two factors—known as the Zener and the avalanche effects.

The Zener effect occurs with reverse voltages below 5.7 V. As the reverse voltage increases to just over 5.0 V, the electric field, that is the attraction exerted between opposite charges next to the junction, is sufficient to pull electrons out of their covalent bonds holding the atoms together. Subsequently, additional pairs of holes and electrons are created (this being known as the Zener effect) and contribute to the reverse current.

The avalanche effect occurs when the reverse voltage exceeds 5.7 V. The speed at which the charge carriers move through the crystal lattice near the depletion layer becomes so fast that the carriers collide with valence electrons of the

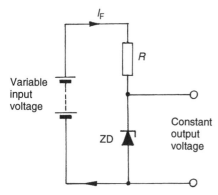

Fig. 9.12 Zener diode voltage stabilizer circuit symbolic representation

atoms within the crystal lattice and knock some of them out of their shells.

These extra newly produced holes and electrons are then, in turn, accelerated by the strong electric field and may also collide with other atoms to produce more charge carriers. As a result, a chain reaction, or avalanche, of charge carriers occurs with very little increase in voltage.

Zener diodes are used to stabilize power supplies (provide overvoltage protection) where it is important to keep voltage outputs constant (Fig. 9.12).

9.3 *N–P–N* junction transistor

9.3.1 Transistor construction
(Figs 9.13 and 9.14)

A transistor is a semiconductor device which can either be used as a current amplifier or as an electronic solid-state switch. Transistor action was discovered in 1948—the name transistor being derived from transfer–resistor.

The *N–P–N* junction transistor consists of a crystal of semiconductor material (either germanium or, more commonly, silicon) in which there are two *N–P* junctions with a common region of *P*-type material. The middle region, known as the 'base', is made from a thin slice (0.025 mm) of *P*-type material, lightly doped with boron to produce current hole charge carriers. On either side of the base region are two thicker regions heavily doped with phosphorus (the opposite *N*-type material) enabling it to produce free electrons. The region on one side of the base is known

as the emitter since it emits electron charge carriers while the region on the opposite side is called the collector as it collects electron charge carriers. The three-region semiconductor material can, in fact, be regarded as two diodes placed back to back.

9.3.2 Transistor action
(Figs 9.13, 9.14 and 9.15)

If the base–emitter (b–e) junction circuit switch is closed and the base voltage exceeds +0.6 V, in the case of a silicon semiconductor transistor material, electrons cross from the emitter *N*-type region into the base *P*-type region (Fig. 9.13). At the same time, electrons from the (b–e) negative battery terminal will replace the free electrons lost by the emitter, and so form the emitter current. In contrast, holes drift from the lightly doped *P*-type region base to the heavily doped *N*-type emitter, but since there are only a few

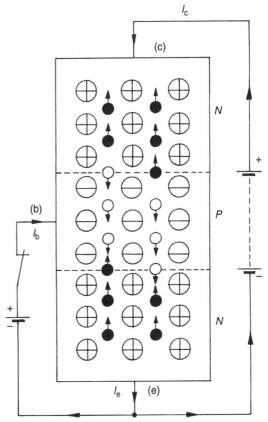

Fig. 9.13 Base-emitter junction switch on collector-emitter circuit conducting

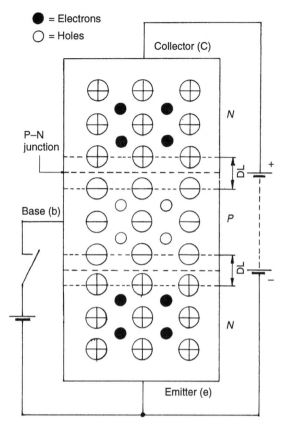

= Electrons
= Holes

Collector (C)

P–N junction

Base (b)

Emitter (e)

Fig. 9.14 Base-emitter junction switch-off collector-emitter circuit blocked

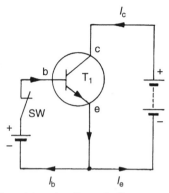

Fig. 9.15 Transistor circuit symbolic representation

so that the (c–e) circuit is able to carry relatively large amounts of current.

When the (b–e) circuit voltage is interrupted the base immediately reverts back to *P*-type material with both (c–b) and (b–e) junctions again forming non-conducting depletion layers (Fig. 9.14).

It can therefore be seen that a small external (b–e) terminal circuit voltage, when applied, can instantly convert non-conducting (blocked state) junctions (c–b) and (b–e) into low-resistant large current carrying circuits (Fig. 9.15).

9.3.3 Thyristor
(Figs 9.16, 9.17 and 9.18)

A thyristor is a semiconductor rectifier which can control the output power supplied to an electrical load with the minimum wastage of energy (Fig. 9.16). The name thyristor is an abbreviation

holes compared with the electrons flowing in the opposite direction, the electrons are the majority charge carriers in the semiconductor material.

Only a small proportion—something like 1%—of electrons entering the base are needed to fill the holes, because the base is thin and only lightly doped. Thus, the majority of free electrons entering the base will be attracted by the (c–e) circuit battery's positive terminal connected to the *N*-type material of the collector. These electrons therefore cross the base–collector junction to become the collector current in the (c–e) battery circuit.

Recombination of electrons and holes in the *P*-type base provides the base region with a negative charge tendency, which is maintained as long as the (b–e) junction continues to supply the base with positive holes. This drift of holes from the (b–e) battery circuit to the base produces a small base current which keeps the base–emitter junction forward biased. Consequently, the middle base region virtually becomes *N*-type material

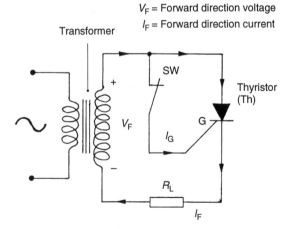

V_F = Forward direction voltage
I_F = Forward direction current

Transformer

SW

Thyristor (Th)

V_F

I_G

G

R_L

I_F

Fig. 9.16 Thyristor circuit symbolic representation

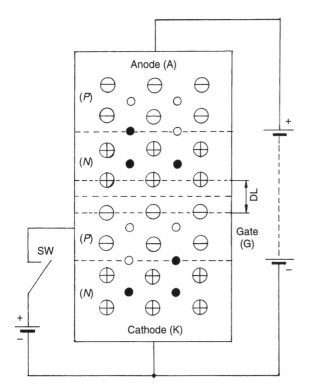

Fig. 9.17 Open gate forward bias thyristor

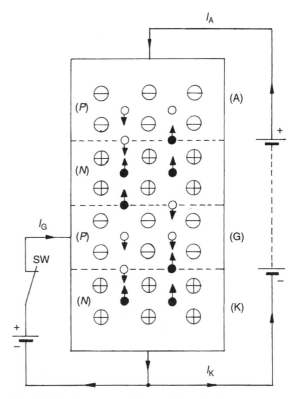

Fig. 9.18 Closed gate forward bias thyristor

of thyratron and resistor, the thyratron-tube being the predecessor of the thyristor.

The thyristor has four layers of alternative *P*-type and *N*-type silicon semiconductors forming three *P–N* junctions. The two outer layers of *P*-type and *N*-type material each have terminals and form the anode (A) and cathode (K) respectively while one of the inner regions also has a terminal and is known as the gate (G) (Figs 9.17 and 9.18). Thyristors are designated *P*-gate or *N*-gate according to which type of material is attached to the gate terminal.

When the anode is connected to the positive terminal and the cathode to the negative terminal of a power supply the outer two junctions are forward biased and therefore conduct with a very low resistance, whereas the middle junction is reverse biased so that it has a wide effective depletion layer and a very high resistance, and thus blocks (Fig. 9.17). Conversely, if the anode and cathode are connected to the negative and positive terminals respectively the middle junction becomes forward biased while the two outer junctions change to reverse biased and so block.

When a voltage is applied between the gate and the cathode in a direction which makes the gate positive and the cathode negative, current will flow between the gate and cathode junction. Immediately the depletion layer of the *P–N* middle junction is exposed to many charge carriers with the result that the depletion layer is destroyed (Fig. 9.18). Holes now move from the gate *P* region across the *P–N* junction into the *N*-type region and, conversely, electrons travel from the inner *N*-type region to the gate *P* region. In other words, the thyristor is conducting the principal current (I_p), i.e. I_A and I_K.

The characteristic of a thyristor is that it continues to conduct when the gate voltage is removed and stops only if the principal supply voltage is switched-off or reversed, such as occurs with an alternating power supply or the anode current falls below a given holding current.

9.4 Inductive semiconductor ignition with induction-type pulse generator (Bosch)

9.4.1 Induction-type pulse generator
(Figs 9.19, 9.20 and 9.21)

The pulse generator is used to generate an alternating voltage, which is used instead of contact breaker points for controlling the make and break of the current build-up in the primary winding of the ignition coil.

The components of this generator comprise the permanent magnet teeth, core and induction winding—which makes up the stator members—while the trigger rotor (reluctor) and teeth made from soft magnetic steel are mounted on the distributor shaft and thus form the rotating member (Fig. 9.19). There are as many trigger and stator teeth as there are engine cylinders. A sectioned view of the complete distributor including both the centrifugal and the vacuum advance is shown in Fig. 9.20.

A minimum air gap of about 0.5 mm exists between the trigger rotor and stator teeth, when they are opposite each other. In this position, the magnetic flux established by the permanent magnet stator teeth will provide the maximum linkage with the trigger rotor teeth. When the trigger rotor rotates, the two sets of teeth may be aligned, or move towards or away from each other. Correspondingly, the magnetic flux will vary in strength as the distance between the teeth changes and this induces an alternating voltage into the induction stator winding.

Thus, as the rotating teeth approach the stator teeth the magnetic flux intensifies, reaching a maximum just before the teeth are directly opposite each other (Fig. 9.21(b)). With further rotation, the distance between the teeth increases and the generated voltage switches polarity, at this point ignition occurs (point t_i).

9.4.2 Voltage stabilizer
(Figs 9.21 and 9.22)

The supply voltage from the battery must be closely regulated so that the generated input pulse can be accurately reshaped and extended to provide the desired dwell angle. This is achieved with the voltage stabilizer circuit using resistor R_1, capacitors C_1 and C_2 and a zener diode ZD_1 (Fig. 9.22).

The battery voltage fed to the trigger unit is reduced as current passes through resistor R_1, to provide the operating supplied voltage. If the supply voltage should exceed a predetermined value, the zener diode automatically switches on and conducts, and this increased current flow causes an additional voltage drop across R_1 so that the supply voltage does not rise but actually remains constant (Fig. 9.21(c)).

To prevent brief voltage peaks from the battery supply and charging system entering the trigger

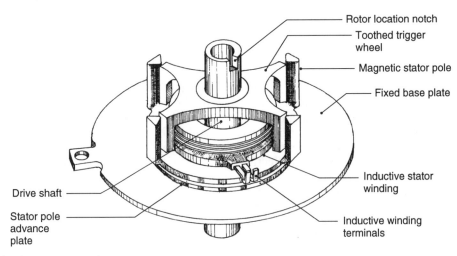

Fig. 9.19 Inductive generator trigger

442

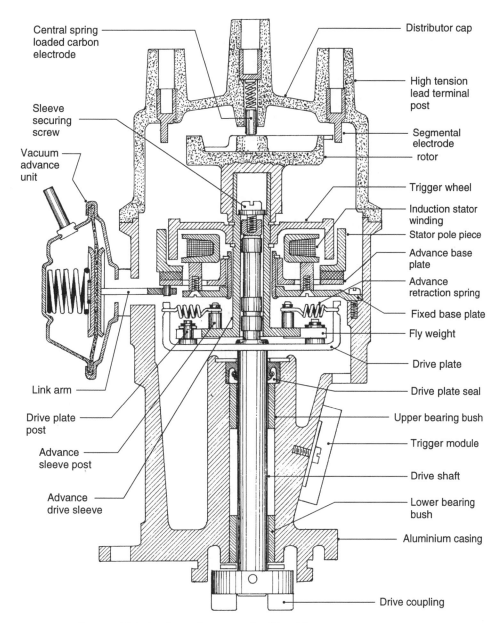

Fig. 9.20 Ignition distributor with induction pulse generator (Bosch)

Central spring loaded carbon electrode

Sleeve securing screw

Vacuum advance unit

Link arm

Drive plate post

Advance sleeve post

Advance drive sleeve

Distributor cap

High tension lead terminal post

Segmental electrode rotor

Trigger wheel

Induction stator winding

Stator pole piece

Advance base plate

Advance retraction spring

Fixed base plate

Fly weight

Drive plate

Drive plate seal

Upper bearing bush

Trigger module

Drive shaft

Lower bearing bush

Aluminium casing

Drive coupling

unit, a capacitor C_1 is shunted across the two input terminals, its function being to absorb rapid voltage rises and to discharge itself as the input voltage decreases.

A small supply voltage ripple, originating from the generator, is smoothed out by the capacitor C_2.

9.4.3 Pulse shaping stage
(Figs 9.21 and 9.22)

It is the function of this stage to convert the control alternating voltage of the pulse generator into rectified rectangular current pulses (Fig. 9.21(d)). This pulse shaping stage is performed by the threshold switch circuit known as the Schmidt trigger.

The induction generator pulse is electrically

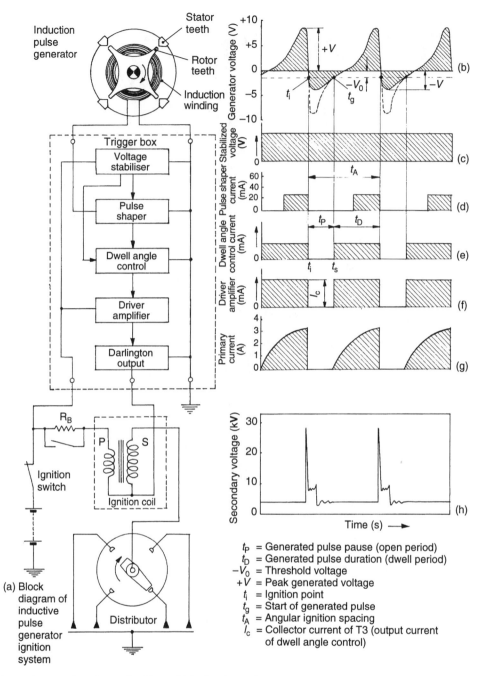

Fig. 9.21 Inductive semiconductor ignition with induction pulse generator

t_P = Generated pulse pause (open period)
t_D = Generated pulse duration (dwell period)
$-V_0$ = Threshold voltage
$+V$ = Peak generated voltage
t_i = Ignition point
t_g = Start of generated pulse
t_A = Angular ignition spacing
I_c = Collector current of T3 (output current of dwell angle control)

loaded only on the negative phase of the alternating pulse, and therefore the positive phase voltage amplitude is much larger than that of the loaded negative half-cycle (Fig. 9.21(b)).

The shaped circuit is triggered by a threshold voltage $(-V_o)$ (pre-set minimum voltage) on the

negative side of the alternating voltage pulse caused by the polarity of diode D_4. The pulse generator output feeds the base (b) of transistor T_1 by way of diode D_4 and resistor R_4 and also the emitter (e) of T_1 and T_2 through resistor R_6 (Fig. 9.22). As soon as the alternating voltage on

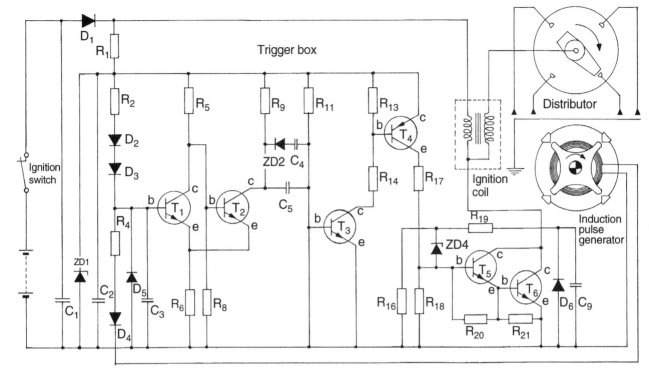

Fig. 9.22 Induction semiconductor ignition with induction pulse generator circuit layout

the negative half-cycle rises to the threshold value $(-V_o)$, the transistor T_1 is switched on and the emitter–collector path conducts. At the same time, transistor T_2 stops conducting; that is, it switches to the blocked state. Consequently, the output of the pulse-shaping circuit does not conduct for a period (t_p). Transistor T_2 remains switched-off until the alternating control voltage drops below the threshold voltage $(-V_o)$. At this point, the base–emitter junction circuit of T_1 switches off so that its collector–emitter path stops conducting (blocks). The base (b) of T_2 now becomes positive by way of R_5, thus switching on the base–emitter junction circuit. Immediately, the collector–emitter path circuit of T_2 commences to conduct (point t_s).

This cycle of events for T_1 and T_2 alternately conducting and blocking is continuously repeated.

To compensate for variations in temperature, two series diodes D_2 and D_3 are included in the shaper circuit.

9.4.4 Dwell angle control
(Figs 9.21 and 9.22)

Dwell angle refers to the angular period when the contact breaker points of a conventional distributor remain closed, during which time the primary winding of the ignition coil is energized. The dwell angle control automatically advances the point at which the dwell period begins with rising engine speed, and since the switching-off is fixed (the point of ignition), the period when current is permitted to flow in the primary winding circuit is prolonged (Fig. 9.21(e)).

Dwell duration control is achieved with a resistor–capacitor network that charges and discharges a capacitor C_5 by way of resistors R_9 and R_{11}, while the switching stage of the circuit is controlled by transistors T_2 and T_3 (Fig. 9.22).

Starting with the transistor T_2 in the blocked state (point t_s), the capacitor C_5 will become charged to a capacitor voltage of almost 12 V at low engine speed by way of R_9 and the base–emitter junction of transistor T_3.

At point t_i (ignition), the trigger transistor T_2 of the pulse shaper switches on, the capacitor C_5

445

discharges through R_{11} and the collector–emitter path circuit of the transistor T_2. During this discharge period, T_3 will not conduct since its base (b) is negative with respect to earth. The discharge phase continues until that instant when the polarity of capacitor C_5 changes from negative to positive (point t_i), transistor T_3 will then immediately commence to conduct via the base–emitter junction. The capacity now charges in the opposite direction until transistor T_2 switches off (blocks) (point t_s). Beyond this point, the charge path again switches to R_9 and the base–emitter junction of T_3 so that the charge potential is reversed and the cycle of events is repeated.

Dwell duration control is derived from the polarity charge of the capacitor switching on the collector current (I_c) of transistor T_3. This simultaneously switches on T_4, T_5 and T_6. The energizing of the ignition coil primary winding is provided by the power transistor T_6 collector–emitter path circuit current flow.

With rising engine speed the charging time of the capacitor is insufficient to charge it up fully to the battery voltage, therefore capacitor C_5 will begin the discharge correspondingly earlier (point t_i) resulting in the capacitor polarity change point also occurring earlier. Thus, since the polarity change point switches on T_3, the beginning of the dwell period will be advanced and, consequently, the duration for the ignition coil's primary current flow extended.

9.4.5 Driver and Darlington output stage
(Figs 9.21 and 9.22)

It is the purpose of this stage to enlarge the dwell angle control current pulse (Fig. 9.21(f)), so that it can act as an input signal to the output power transistor whose function is to regulate the energizing and interruption of the ignition coil's primary winding current supply.

A rectangular current pulse forming the collector current (I_c) of transistor T_3 provides the output of the dwell angle control. This signal feeds into the base (b) of the driver transistor T_4 which then switches on the collector–emitter circuit, the amplified current flowing in this circuit now forms the input driver to the Darlington output stage (Fig. 9.22).

The current pulse flows into the base (b) of the transistor T_5, the amplified collector–emitter junction current is then fed into the base (b) of

the power transistor T_6. The collector–emitter path of T_6 now functions as the battery and primary winding circuit switch so that when it conducts, the coil winding will be energized (Fig. 9.21(g)). Note that when the base current to T_5 ceases, T_6 blocks, breaking the ignition coil primary circuit so that a high tension voltage is immediately produced in the secondary spark-plug circuit (Fig. 9.21(h)).

9.5 Inductive semiconductor ignition with Hall-type generator (Bosch)

9.5.1 The Hall effect
(Fig. 9.23(a and b))

When a current (I_s) passes through a metal conductor layer free of any magnetic flux, the electron charge will be fairly uniformly distributed throughout its crystal lattice structure (Fig. 9.23(a)). If a magnetic field (B) is applied at right angles to a supply current (I_s) moving through a conductor, then an electromagnetic force (F_L) (Lorentz force) acts on each electron perpendicular to the direction of both the magnetic field and current flow. As a result, this electromagnetic force deflects the electron charge to one side (Fig. 9.23(b)). Subsequently, there will be an excess number of electrons on one (top) surface, while the other (bottom) surface will be depleted of electrons and, because of this, a potential difference (V_H) will exist between the two opposing surfaces.

The deflection of electrons due to the magnetic field crossing the current flow is known as the Hall effect, and the voltage created between the two end surfaces is called the Hall voltage. This was discovered by the American, Hall, in 1879.

The Hall effect is very small in metals (several millivolts) but is considerably increased when applied to semiconductors such as germanium and silicon.

The Hall voltage may be given by the following

$$\text{Hall voltage } V_H = \frac{KI_s B}{t}$$

where K = the Hall coefficient
I_s = supply current
B = magnetic flux density
t = conductor layer thickness

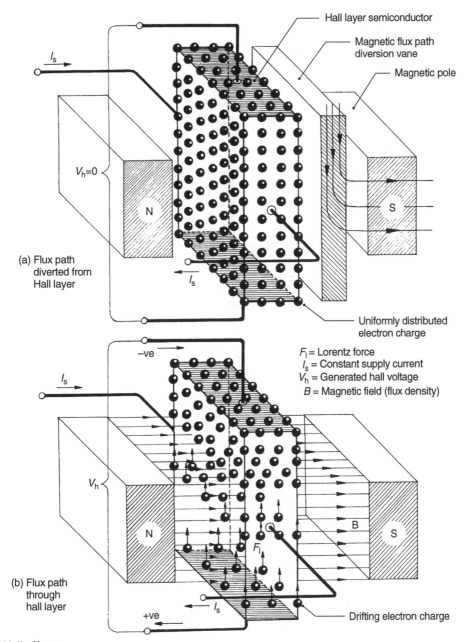

Hall layer semiconductor

Magnetic flux path diversion vane

Magnetic pole

I_s

$V_h=0$

N

S

(a) Flux path diverted from Hall layer

I_s

Uniformly distributed electron charge

F_l = Lorentz force
I_s = Constant supply current
V_h = Generated hall voltage
B = Magnetic field (flux density)

$-ve$

I_s

V_h

N

B

S

F_l

(b) Flux path through hall layer

$+ve$

I_s

Drifting electron charge

Fig. 9.23 Hall effect

9.5.2 Hall generator stage
(Figs 9.24(a and b), 9.25 and 9.26)

The Hall generator consists of two members, one being stationary while the other rotates (Fig. 9.24(a and b)).

The stationary member is divided into two halves which are separated by an air gap. Posi-tioned on one side of this air gap is a permanent magnet with a half magnetic flux conductor, and situated on the opposite side facing the south pole of the magnet is a semiconductor strip known as the Hall layer. Backing onto the Hall layer is the Hall integrated circuit (Hall IC). Performing the task of both an electronic switch and a pulse shaper, this very tiny circuit is mounted on a ceramic substrate. Set a little further back from

447

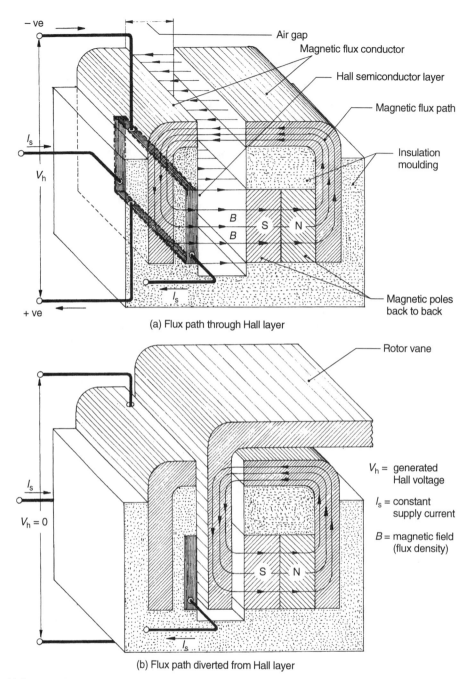

Fig. 9.24 Hall generator

the Hall IC is the second half magnetic flux conductor. The magnetic flux originating from the magnet therefore follows a path through the two half conductors and the air gap.

The rotating component known as a trigger wheel rotor is driven by the ignition distributor drive shaft, it is made in the form of a steel circular disc with downward bent lugs, known as vanes, which act as magnetic-flux path diverters when they enter the air gap (Fig. 9.25). There are as many vanes as there are engine cylinders and the width of these vanes determines the dwell-

448

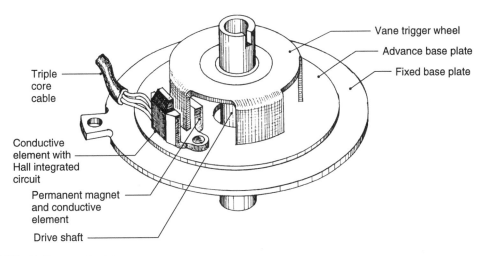

Fig. 9.25 Hall generator trigger

angle needed for primary current build-up. Since the angular distance between vanes is fixed, the dwell-angle remains constant at all times.

When the vanes are clear of the air gap, the magnetic-flux is permitted to saturate the Hall layer, causing the Hall voltage to rise to its maximum value (Fig. 9.24(a)). Conversely, each time the trigger wheel vane passes through the Hall generator air gap, the Hall layer is screened causing the flux path to divert from the Hall layer and instead complete its circuit through the conducting vane (Fig. 9.24(b)). While the vane obstructs the air gap the flux density through the Hall layer decays almost to zero except for a small residual flux density caused by stray lines of force. Accordingly, the Hall voltage will reduce to a minimum.

An exploded view of a Hall-type generator distributor is shown in Fig. 9.26. It can be seen that the Hall integrated circuit and conductive elements are small compared with the distributor case itself.

9.5.3 Hall integrated circuit (amplifier pulse shaper and output signal)
(Figs 9.27(a–h) and 9.28)

The Hall integrated circuit (Fig. 9.28) first of all utilizes transistor T_1 to amplify the very small trapezium-shaped Hall voltage pulse (Fig. 9.27(c)). A pulse-shaper current using T_2 and T_3 then converts these pulses into rectangular shaped ones (Fig. 9.27(d)), a further amplifica-

tion occurs through T_4 which functions as an output signal stage.

As the trigger wheel rotates, so that the vane moves out of alignment with the air gap, the flux-density will rise until a cut-in threshold B_1 (point of ignition) is reached (Fig. 9.27(b)). At this point, the Hall voltage feeding base (b) of transistor T_1 is sufficient to switch on its collector–emitter (c–e) circuit (Fig. 9.28). Immediately, T_1 conducts, and an amplified current flows to the base (b) of T_2 switching on its (c–e) circuit.

At the same time, T_2 switches to the conducting state, and T_3 and T_4 switch to the blocked state because the current previously flowing through R_1 and R_6, which fed the base (b) of T_3, now finds a lower resistance path through the (c–e) circuit of T_2 and resistor R_8 (Fig. 9.28). Note that, to switch T_4 on, a base current is necessary, and this can be supplied by the (c–e) circuit of T_3 when it conducts. Thus, when T_1 and T_2 switch on, T_3 and T_4 switch off, so that the generated pulse voltage coming from the output signal drops to a value below 0.5 V.

As the trigger wheel continues to rotate, one of the vanes enters the air gap and this screens the Hall layer so that the flux abruptly decreases down to the cut-out threshold B_2, and T_1 and T_2 cease to conduct (Fig. 9.27(b)). Consequently, current is now available for the base (b) of T_3 via R_7 and R_6 and for base (b) of T_4 via R_9 and the (c–e) circuit of T_4 and R_{10} (Fig. 9.28). Accordingly, T_3 and T_4 switch to the 'on' state. At this point, the generated voltage V_G is allowed to rise rapidly to a value of a few volts (Fig. 9.27(e)).

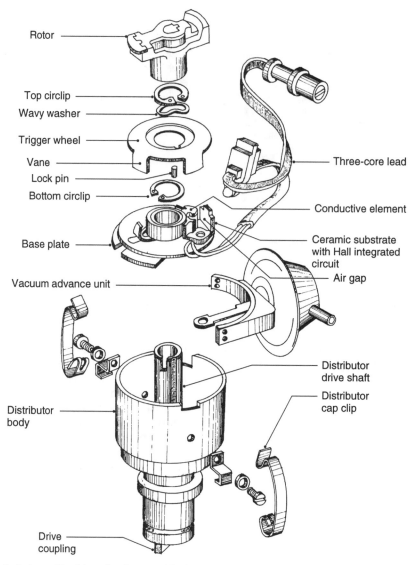

Rotor

Top circlip

Wavy washer

Trigger wheel

Vane

Lock pin

Bottom circlip

Base plate

Vacuum advance unit

Distributor body

Drive coupling

Three-core lead

Conductive element

Ceramic substrate with Hall integrated circuit

Air gap

Distributor drive shaft

Distributor cap clip

Fig. 9.26 Exploded view of ignition distributor with Hall generator trigger

9.5.4 Trigger box
(Figs 9.27 and 9.28)

The rectangular current pulse in the integrated circuit is used to trigger the driver in the trigger box transistor T_5, and the current from the driver transistor is used to drive the power output stage transistors T_6 and T_7 (Fig. 9.28). The Darlington circuit output stage consists of an amplifier transistor T_6 and an output power transistor T_7 which switches on and off the ignition coil primary current, both of these transistors are coupled together as one.

As the vane moves through the air gap the magnetic flux path is diverted from the Hall layer, and this switches off the Hall integrated circuit (Hall IC) output signal current. The trigger box, which includes the driver and Darlington stages, then switches on the output transistor which abruptly supplies current to the ignition coil primary winding (Fig. 9.27(g)).

When the vane moves beyond the air gap, the Hall voltage rises and switches on the signal current, immediately the Darlington output transistor switch interrupts the ignition coil primary current. The instantaneous collapse of primary

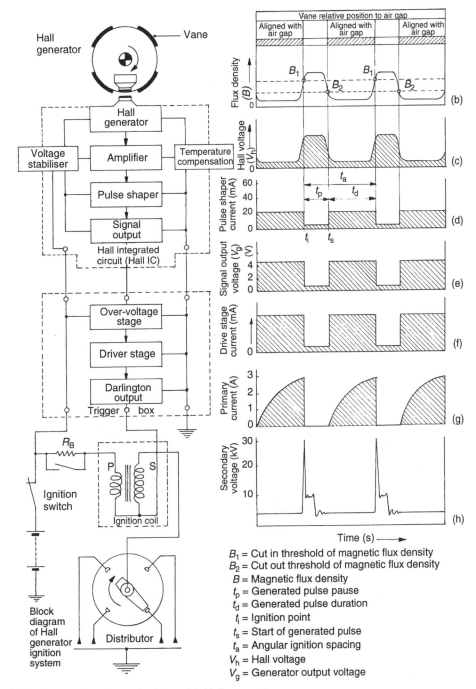

Fig. 9.27 Inductive semiconductor ignition with Hall generator

current induces a very high voltage pulse into the secondary winding of the ignition coil (Fig. 9.27(h)).

As soon as the vane aligns with the air-gap, the integrated circuit output transistor T_4 switches off. This permits the base (b) of the trigger box transistor, T_5, to become positive due to the voltage division, and the (c–e) circuit of T_5 to conduct (Fig. 9.28). There is now a voltage drop across R_4, the (c–e) circuit of T_5 and R_6 so that

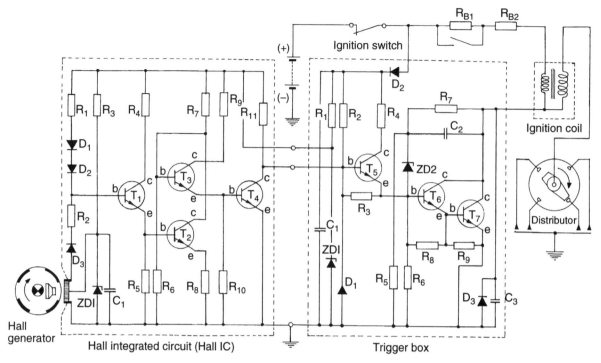

Fig. 9.28 Inductive semiconductor ignition with Hall generator circuit layout

the base (b) of T_6 goes positive and the voltage between the (b–e) junction due to R_8 switches on T_6. The conducting (c–e) path of T_6 simultaneously applies a positive potential to base (b) of the output power transistor T_7, so that its (c–e) circuit also switches on, and the current now flowing in the primary winding current enables the magnetic flux to build-up within the ignition coil (Fig. 9.27(g)).

At the point when the vane just moves out of the air gap, the Hall voltage rises and the generator output transistor T_4 is switched on. The consequence of the (c–e) circuit of T_4 conducting is that the trigger box driver transistor T_5 switches off; this being caused by its base (b) going to a negative potential (Fig. 9.28). Correspondingly, the bases (b) of both the Darlington stage transistors T_6 and T_7 will also now become negative. Accordingly, the power transistor T_7 abruptly cuts-off the ignition coil primary current, and at this point (t_i) ignition occurs (Fig. 9.27(g and h)). To prevent excessively high primary self-induced voltages reaching and damaging the output transistors T_6 and T_7, a zener diode ZD_2 is included in the output stage.

9.6 Inductive semiconductor ignition with optical trigger circuit (Lumenition)

9.6.1 Optical pulse stage
(Fig. 9.29)

Optical trigger units are assembled inside the distributor cap and replace the contact-breaker points. There are two basic parts—the rotating chopper segmented disc (shaped like a propeller with as many blades as there are cylinders) which is attached to the distributor drive cam, and a stationary member light-source, directed onto a photo-cell (Fig. 9.29).

The radiation source is a gallium arsenide cell operating at a constant level by means of a zener-diode (ZD) stabilizer, and the photo-cell is a silicon phototransistor directly coupled to a second transistor; both transistors forming what is known as a *Darlington amplifier*.

The gallium arsenide lamp (cell) provides radiation rays in the near infra-red region of the

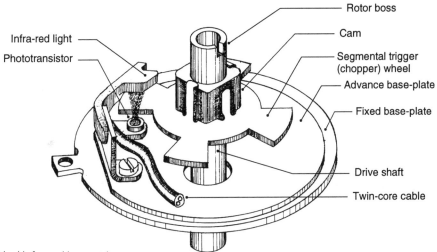

Fig. 9.29 Optical infra-red beam trigger

spectrum (wavelength 0.94×10^{-6} m), it has a nearly hemispherical lens, which focuses the beam rays at the chopping point to a width of about 1.25 mm, and switching always takes place within half this width regardless of dirt accumulation. The phototransistor is usually moulded in transparent plastic with the top layer providing protection and simultaneously acting as a lens so that radiation rays focus towards the semiconductor base junctions.

If a radiation beam falls upon a semiconductor material, valance electrons are released and holes will be created. As a result, a base current flows in the *P*-type region which switches on the transistor's collector–emitter circuit.

When the engine rotates, the chopper disc periodically interrupts (chops) and prevents radiation rays reaching the photo-cell. Consequently, its base-current ceases, switching the phototransistor into a blocked state. The width and spacing of the chopper blades are such that they provide an accurate on–off triggering of the phototransistor with a constant 66% dwell period.

9.6.2 Signal amplification and output stage
(Fig. 9.30)

When the ignition switch is closed and light falls on the phototransistor (PT), new holes and free electrons will be created in the semiconductor crystal so that current will flow in its base and be of sufficient magnitude to switch on its collector–

emitter (c–e) circuit (Fig. 9.30). Therefore, a positive bias voltage will be applied to the base of transistor T_1 via resistor R_3 and the PT (c–e) circuit. Immediately, the Darlington amplifier transistor T_1 conducts, causing current to flow from the positive battery terminal through R_3, the T_1 (c–e) circuit and R_4, to the battery's negative terminal. Voltage will thus be applied to the base–emitter (b–e) circuit of T_2 causing its (c–e) circuit also to conduct. The base–emitter voltage supplies potential so that the T_3 (c–e) current path goes into a blocked state. Current is now permitted to flow via the voltage divider's resistors, R_6 and R_7, into the base of the power transistor T_4 causing the transistor (c–e) circuit to move into a conducting state. Instantly, the current will flow in the primary winding of the ignition coil.

As the engine rotates, the chopping disc, synchronized to the engine timing requirements, will cut off the light from the phototransistor (PT) and instantly its base current ceases to flow, thus causing both the phototransistor (PT) and the Darlington amplifier transistor T_1 to revert to their blocked states (Fig. 9.30). The base of T_2 now only has a negative potential through R_4 so that it also moves into a non-conducting blocked state. Consequently, instead of current flowing through its lowest resistance path, via R_5 and the T_2 (c–e) circuit, to the negative ground it is now directed to the base of T_3, switching on its (c–e) circuit conducting path. This current will be diverted from the base of the power transistor T_4 to the lowest resistance path via the (c–e) circuit of T_3 to the negative ground potential. As a result,

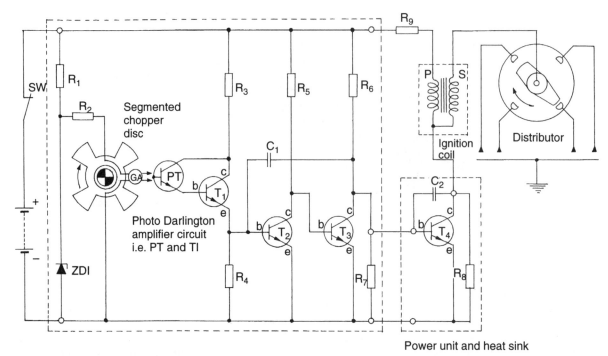

Fig. 9.30 Inductive semiconductor ignition with optical trigger circuit layout (Lumenition)

T_4 switches to a blocked state, thus interrupting the primary winding current flow. The rapid collapse of current in the winding induces a high tension voltage into the secondary winding and spark-plug circuit.

9.7 Capacitor discharge ignition system (Bosch)

(Fig. 9.31)

The capacitor discharge method of creating a high tension voltage spark across the spark plug electrodes is achieved by storing the electrical supply energy in a charging capacitor (Fig. 9.31(a)). When ignition is timed to occur, a thyristor power switch conducts and completes the capacitor–primary winding circuit of the ignition coil. Instantly, the capacitor will discharge through the primary winding. The sudden flow of current in the primary winding induces a very high voltage in the secondary winding and, since the spark-plug forms part of the secondary winding circuit, this voltage pulse will be dissipated at the plug air-gap in the form of a spark. Unfortunately, the duration of the spark, which is about 0.1 ms, is

extremely short—so this type of ignition system is only suitable for certain high performance engines.

9.7.1 Pulse shaping stage
(Figs 9.31(a, b and c) and 9.32)

When the generator rotor teeth (Fig. 9.31(a)) are approximately half-way between the stator teeth, the generated alternating voltage switches from a positive to a negative potential (Fig. 9.31(b)). As soon as this negative voltage exceeds a threshold (predetermined) value $(-V_o)$ at point (t_s) the diode D_1 conducts, this allows the negative voltage pulse to reach the base (b) of the transistor T_1 (note the diode D_1 blocks all positive voltage pulses) (Fig. 9.32). Immediately, T_1 conducts and thus permits current to flow via R_3, the c–e circuit of T_1 and R_4; simultaneously transistor T_2 is in the blocked state. During the T_1 conducting and T_2 blocked phase, the pulse shaper output current remains zero for a period (t_p) (Fig. 9.31(c)).

The switching state continues until the alternating control voltage becomes less negative than the threshold voltage $(-V_o)$ at point (t_i), the switching state of the two transistors now changes, that

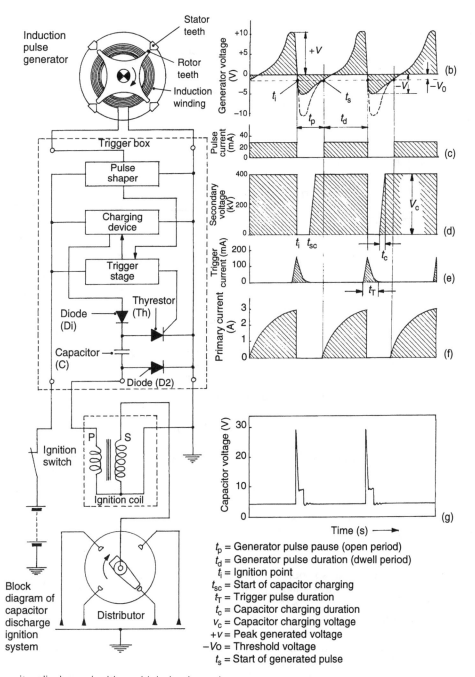

Fig. 9.31 Capacitor discharge ignition with induction pulse generator

t_p = Generator pulse pause (open period)
t_d = Generator pulse duration (dwell period)
t_i = Ignition point
t_{sc} = Start of capacitor charging
t_T = Trigger pulse duration
t_c = Capacitor charging duration
v_c = Capacitor charging voltage
$+v$ = Peak generated voltage
$-V_0$ = Threshold voltage
t_s = Start of generated pulse

is, T_1 blocks and the base (b) of T_2 becomes positive, by way of R_3, so that it conducts. The pulse shaper output during this phase supplies an output current pulse for a period t_D.

9.7.2 Charging state
(Figs 9.31(b and d) and 9.32).

To increase the energy stored in the discharge capacitor, a charging transformer first steps up the 12 V battery potential to something like 400 V

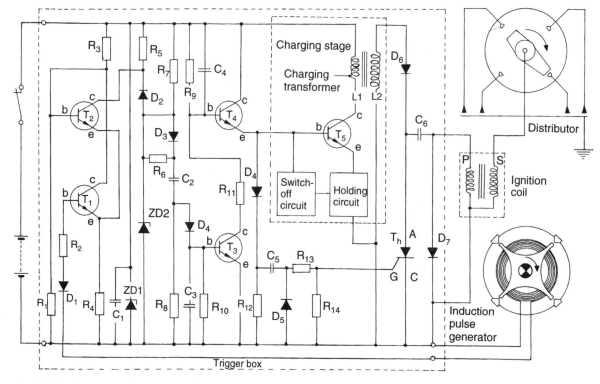

Fig. 9.32 Capacitor discharge ignition with induction pulse generator circuit layout

and a diode D_6 is included in the circuit to prevent the electric charge flowing back into the charging transformer while the capacitor remains charged (Fig. 9.32).

Ignition occurs when the rotor teeth of the pulse generator pass the rotor teeth—at this point the generated voltage polarity changes from positive to negative (Fig. 9.31(b)). At the instant of ignition, the trigger stage circuit sends a current pulse to the thyristor via the gate terminal (G), the thyristor switches on and so permits the storage capacitor to discharge through the primary winding of the ignition coil (Fig. 9.32). Simultaneously, as the ignition coil primary winding is being energized, the transistor T_3 conducts and current begins to flow in the primary winding (L_1) of the charging transformer. It is the function of the holding circuit to keep the transistor T_3 in the conducting (on) state until the current has risen to some predetermined level, when this value has been reached the switch-off circuit causes the transistor T_3 to switch to the non-conducting (blocked) state.

The current build-up in the primary winding of L_1 during the conducting state of T_3 produces the induced charging current pulse in the secondary winding L_2 of the charging transformer. Consequently, the storage capacitor is charged up to a peak value of almost 400 V by way of the circuit diodes D_6 and D_7 (Fig. 9.31(d)).

The stored energy (E) in the capacitor C_6 when the charging voltage is V_c may be given by

$$E = \tfrac{1}{2}CV^2$$

where $C = 1 \times 10^{-6}$ Farads and $V = 400$ V, therefore

$$E = \tfrac{1}{2} \times 10^{-6} \times 400^2 = 80 \times 10^{-3} = 80 \text{ mJ}.$$

9.7.3 Trigger stage
(Figs 9.31(e, f and g) and 9.32)

The trigger stage provides a small current pulse timed to open and close periodically the ignition coil primary winding circuit by means of the thyristor power switch.

When the pulse shaper output transistor T_2 switches to the blocked state, current flowing from R_5 to ground is interrupted and it is at this point (t_i) that ignition occurs (Fig. 9.32). Transistor T_3 now switches on so that its (c–e) circuit conducts and so changes the base (b) of T_4 from a positive to a negative potential. Immediately, its (c–e) circuit turns to the on state and, as a result, links the positive supply terminal via diode D_4 to the trigger capacitor C_5. The capacitor charging current is thus completed through resistors R_{13} and R_{14}, and the ground return path to the negative supply terminal. The current flowing in the circuit T_4, D_4, C_5, R_{13} and R_{14} is known as the control charging current. The magnitude of this current is designed to produce a voltage drop across resistor R_{14} equal to the trigger voltage necessary to switch on the thyristor. Therefore, when the transistor T_4 conducts, the trigger voltage across R_{14} causes a current pulse (Fig. 9.31(e)) to trigger the thyristor (Th) into a conducting state. At the instant the thyristor switches to the 'on' state (this being the ignition point t_i) the charged capacitor C_6 is permitted to discharge by way of the ignition coil primary winding (Fig. 9.31(f)). This surge of current through the primary winding induces the high voltage pulse into the secondary winding (Fig. 9.31(g)) and, consequently, produces the spark across the electrodes of the spark plug. A little while later, the pulse shaper control current switches on T_2 and so diverts current flowing through resistor R_5 directly to ground, consequently T_3 and T_4 convert to the blocked state and the trigger capacitor C_5 discharges through resistor R_{12} and diode D_5.

9.8   Electronic spark advance

(Fig. 9.33(a and b))

If engine speed and crank angle position are both to be accurately predicted so that the ignition setting can be constantly varied to match the changing engine operating conditions, a more precise method to control the distributor trigger wheel is required. An alternative way of triggering the ignition using the flywheel starter ring teeth or a toothed disc attached to the clutch pressure plate is preferred.

One such arrangement (Fig. 9.33) utilizes a flywheel trigger disc having 34 teeth spaced at 10° intervals with two double spaces 180° apart, which correspond to TDC for numbers 1 and 4 pistons and numbers 2 and 3 pistons, respectively.

When the crankshaft rotates, the trigger disc teeth pass across the sensor-probe and, in so doing, generate a series of signal pulses which are relayed to the electronic control unit. The control unit picks up the missing pulse every 180° so that the TDC position is accurately established. Every 10° of crankshaft rotation is represented by a pulse which, along with the time interval between pulses, allows the control unit to determine accurately the angular engine position and speed.

Engine load is measured by changes in induction manifold pressure, this being sensed through a pipe connecting the manifold to a pressure transducer within the control unit. Additional data obtained by other sensor units, such as air intake temperature, coolant temperature and cylinder pressure waves (only with a turbo charger), may also be signalled to the computerized control unit.

All of the varying individual inputs are controlled accordingly, and compared against a range of ignition characteristics stored in the control unit's computer memory, thereby enabling a particular ignition advance or retard to be selected for optimum engine performance.

As a result, the electronic control unit sets the ignition for a particular crank angle position and triggers the ignition coil primary circuit, thus inducing a high-tension voltage to the secondary circuit and spark plug, in the normal timing sequence.

9.9   Comparison of ignition systems

For an ignition system to be acceptable it must be moderately priced, reliable and its performance must be adequate to meet all the demands imposed on it by various operating conditions. An ignition system must be able to fulfil the following requirements.

1   It should consume the minimum of power and convert it efficiently to a high-energy spark across the spark-plug electrode gap.
2   It should have an adequate reserve of secondary voltage and ignition energy over the entire operating speed range of the engine.
3   It should have a spark duration which is sufficient to establish burning of the air–fuel mixture under all operating conditions.
4   It should have the ability to produce an ignition spark when a shunt (short) is established over the spark-plug electrode insulator surface, due possibly to carbon, oil or lead deposits, liquid fuel or water condensation.

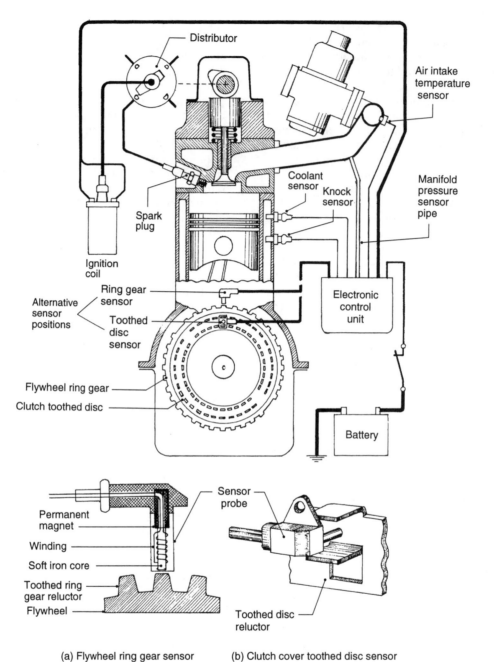

(a) Flywheel ring gear sensor   **(b)** Clutch cover toothed disc sensor

Fig. 9.33 Electronic controlled ignition advance

5 It should have a service life at least equal to that of the engine and the maintenance required is of a minimum.

The characteristics of the various ignition systems are considered in the following order.

1 Inductive ignition with contact breaker switching
2 Inductive ignition with solid-state switching
3 Capacitor discharge ignition

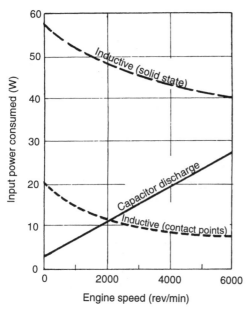

Fig. 9.34 Comparison of power consumption over normal speed range with different ignition systems

9.9.1 Inductive ignition with contact breaker switching
(Figs 9.34, 9.35, 9.36 and 9.37)

The conventional inductive coil ignition system, when operated by mechanical contact breaker points, has a decreasing power consumption with rising engine speed (Fig. 9.34) which might range from 20 W down to 10 W. An effective spark frequency range is achieved between 2000 to 18 000 sparks per minute, which is equivalent to an engine speed range for a four, six and eight-cylinder engine at 1000 to 9000 rev/min, 660 to 600 rev/min and 500 to 4500 rev/min, respectively. Correspondingly, the secondary circuit available voltage may decrease from something like 25 kV at low speeds down to around 10 kV at high engine speeds (Fig. 9.35). At low engine speeds, the spark duration is about 1.5 ms, decreasing to between 0.4 and 1.0 ms at very high engine speeds. The contact breaker points are heavily loaded by the peak primary circuit current which may normally vary between 4 to 5 A. Due to the fact that the secondary voltage rise, when the primary current is interrupted, is relatively slow compared with other ignition systems (Fig. 9.36), leakage of secondary voltage to earth is likely in damp weather, under dirty operating conditions or if the spark-plugs are fouled (Fig. 9.37). An inherent fault with mechanical contacts is that the

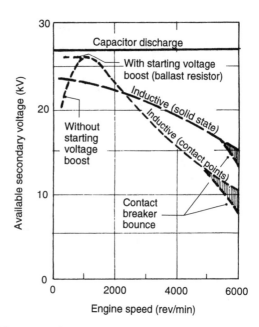

Fig. 9.35 Comparison of available secondary voltage over normal speed range with different ignition systems

low rate of points opening at idling speed may cause arcing at the points and, at high speeds, the contacts may have a tendency to bounce and therefore not precisely follow the cam opening and closing contour.

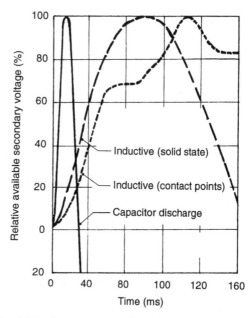

Fig. 9.36 Comparison of duration and rate of secondary voltage rise for different ignition systems

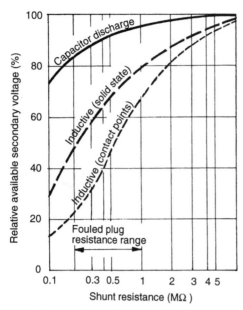

Fig. 9.37 Comparison of available secondary voltage with various shunt resistances with different ignition systems

9.9.2 Inductive ignition with solid-state switching
(Figs 9.34, 9.35, 9.36 and 9.37)

With a contact-breaker triggered solid-state ignition system, the contacts act only as a timing frequency control signal for determining the point of ignition, whereas the make and break power switching of the primary winding current flow is provided by a power transistor. In other words, this type of ignition system functions just like a relay switch where a small current signal, provided by the contact points, controls the switching on and off of a relatively large current flow, in this case the primary winding current. The benefits of this system are that the current flowing between the contact points is very small, in fact may be less than 0.5 A, but the full primary current interruption of between 4 and 8 A with these designs is absorbed by the solid-state transistor. Another inherent feature of solid-state switching is that it is much more rapid than for any mechanical-type contact points, and consequently higher secondary voltages will be generated. As a result of the reduced current loading between contacts, the contacts' lives are considerably extended and cleaning of contacts may require intervals three times longer than the conventional breaker-point system, where the contacts conduct the full amount of primary circuit current.

With transistor-assisted switching, the effective spark frequency range can be extended to something like 1000 to 24 000 sparks per minute, which in terms of engine speed would correspond to a four, six and eight-cylinder engine at 500 to 12 000 rev/min, 330 to 8000 rev/min and 250 to 6000 rev/min, respectively. Because a larger primary current is employed, the power consumed will be greater, but this again drops off with increased engine speed, typical power consumption figures may vary from initially 60 W down to around 40 W at the top end of the speed range (Fig. 9.34). The available secondary voltage may be anything from 25 kV at low engine speeds, decreasing to around 15 kV at maximum engine speeds (Fig. 9.35). Contact-breaker bounce can still occur at very high engine speeds but this can be controlled somewhat easier than with the conventional contact-breaker system.

When breakerless solid-state switching ignition systems are adopted, the effective spark frequency upper range will be extended to about 30 000 sparks per minute—this being equivalent for a four-cylinder engine at 15 000 rev/min. This is far beyond the normal range of an engine since there are no mechanical inertia problems, as there would be with breaker points.

Solid-state ignition systems generally have ignition coils with much smaller inductance; that is, fewer primary winding turns with thicker wiring. Therefore, the interruption of the primary current and the collapse of the magnetic-flux will be much more abrupt. Since the generation of any voltage is directly proportional to the rate at which the lines of magnetic-flux are cut, higher secondary voltages can be obtained and the rate of this voltage rise will be much faster (Fig. 9.36) so that it is more difficult for electric shunts to occur (Fig. 9.37) due to fouled spark-plug electrodes, compared with the conventional ignition system.

Finally, breakerless solid-state switching is not subjected to mechanical wear or contact bounce, as compared with any of the other contact breaker systems. Therefore, it does not require servicing nor does it suffer from intermittent primary current interruption at high engine speeds.

9.9.3 Capacitor discharge ignition

Capacitor discharge ignition systems provide ignition sparking which is effective over the entire speed range of any engine, right from cranking speed to well beyond the maximum speed of even higher performance engines. There is a linear power consumption increase with engine speed from about 3 W at stall to around 28 W at maximum engine speed (Fig. 9.34). As can be seen, the power characteristics are very different to the inductive ignition system where the power consumed is highest at low speed and then decreases with increasing speed. The efficiency of this system is such that the available secondary voltage is higher than for any of the inductive systems and it remains approximately constant over the whole speed range (Fig. 9.35) at something like 27 kV, the sparks thus produced have a relatively large and desirable energy content. Because the discharge of the storage capacitor is so rapid, the voltage rise is extremely fast (Fig. 9.36) making the secondary system (that is, the distributor cap), leads and spark plugs insensitive to electrical shunts (shorts) (Fig. 9.37). Thus, even when the spark plug electrodes and insulator have conducting deposits over their surfaces, the secondary voltage surge is so rapid that sparking is not hindered.

An inherent limitation with the capacitor discharge ignition is the duration of the spark, which is extremely short 0.1 to 0.2 ms, which might make it unsuitable for certain types of combustion chamber designs. Conversely, the large surplus in secondary voltage permits much wider spark-plug electrode gaps to be used, compared with other ignition systems and, as a result, partially compensates for the short time available to ignite the air–fuel mixture. Again, as with the breakerless solid-state switch, there are no moving components. Therefore, all wear is eliminated and no service is required over the entire life span.

9.10 Ballast resistor ignition system

(Fig. 9.38(a, b and c))

9.10.1 Non-ballast resistor ignition circuit operating conditions

With a conventional non-ballast resistor ignition system, the ignition coil is supplied directly from the ignition switch to the primary winding switch terminal, and when the starter cranks the engine, particularly in the cold weather, the increased electrical current load reduces the normal 12 V supply to something like 9 V. Under these condi-

tions an ignition coil designed to operate with a 12 V supply would have a greatly reduced secondary HT voltage, just when the highest voltage of the coil would be desirable to ionize the cold, damp spark-plug electrode air gap. It can therefore be seen that the supply voltage to the ignition system will be minimal when it is most needed, during starting; and a maximum when it is least required, under normal running conditions.

9.10.2 Ballast resistor ignition circuit
(Fig. 9.38(a))

This problem of low coil HT voltage output under starting conditions has been largely overcome by utilizing a low voltage ignition coil (7 V), a ballast resistor connected in series between the ignition switch and the coil's primary winding switch terminal, and a bypass lead joining up the starter solenoid and the ignition coil's switch terminal. While starting, the ballast resistor lead is bypassed allowing the full battery voltage to be applied to the coil. During normal running, only the ballast resistor is in circuit. It therefore reduces the supply voltage to the nominal working voltage of the coil.

9.10.3 Starting ignition circuit
(Fig. 9.38(b))

When the ignition switch is turned fully to the start position, current flows from the battery, through the ignition switch to the solenoid winding, and to the ignition coil through the ballast resistor and lead. Immediately, the solenoid winding is energized, the contacts close and current flows through the bypass lead directly to the switch terminal of the coil and, in effect, shorts out the ballast resistor. Thus, full battery voltage is applied to the coil which, being designed to operate at a much lower voltage, will provide a surplus overboost output of the coil during this starting phase.

9.10.4 Normal running ignition circuit
(Fig. 9.38(c))

When the engine starts, the ignition key is released and the spring loaded ignition switch moves back to the 'ignition on' position. This interrupts the solenoid winding current so that the

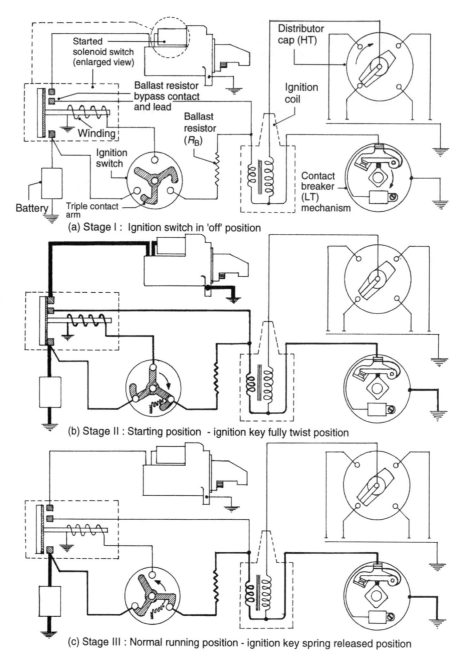

(a) Stage I : Ignition switch in 'off' position

(b) Stage II : Starting position - ignition key fully twist position

(c) Stage III : Normal running position - ignition key spring released position

Fig. 9.38 Ballasted ignition system

solenoid contacts move apart. As a result, the bypass lead is open-circuited and the only current supply is now through the ballast resistor which therefore reduces the coil's voltage to its normal safe working value. If the full battery voltage is allowed to be applied continuously to a 7 V coil it would eventually overheat, break down the wind-

ing insulation, short some of the winding turns together, and burn out. When exchanging the ignition coil, only fit an equivalent voltage coil in its place.

The ballast resistor may take the form of a separate resistance winding or the ignition-switch to coil lead may actually be made from resistance

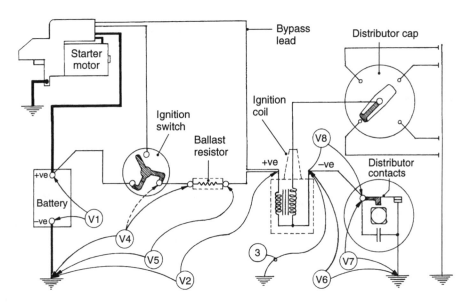

Fig. 9.39 Ignition system primary circuit voltage tests

wire. A typical resistance to limit the battery voltage to the coil's normal operating voltage of say 7 V would be 1.5 Ω.

9.11   Ignition system circuit: tests and inspection

9.11.1   Ignition primary circuit test
(Fig. 9.39)

1 *Battery voltage test*
(Fig. 9.39)
Rotate the engine until the distributor contacts close so that current flows through the primary winding of the coil. Attach voltmeter leads between both battery terminals (V1) and read the battery voltage, which should normally be between 12 and 12.5 V. Low battery voltage indicates a low state of battery charge, sulphated or shortened battery plates, or a charging system malfunctioning.

2 *Coil switch terminal (SW) voltage test*
(Fig. 9.39)
With distributor contacts still closed, connect a voltmeter between the switch terminal on the coil and a good earth (V2). Read the voltage which should be within 0.5 V of the battery voltage with an unballasted resistor ignition system, but should

drop down to 5–8 V when a ballasted resistor is included in the supply circuit.

If the voltage is low in the unballasted-resistor primary circuit, a high resistance exists. Therefore, check all terminal connections and the condition of the switch-to-coil lead wire.

With a ballasted-resistor circuit, a reading below 5 V indicates a high resistance in the primary circuit wiring or a ballasted resistor unit or ballasted resistance wire. A voltage reading above 8 V indicates a short-circuited ballast resistor or a non-ballast type coil fitted.

3 *Ballasted resistor lead bypass voltage test*
(Fig. 9.39)
Attach a shunt lead between the coil's distributor terminal and earth (3) and connect a voltmeter between the switch terminal on the coil and earth (V2). Crank the engine with the starter motor by turning the ignition switch. A rise in the voltmeter reading should be immediately observed due to the ballasted resistor having been short-circuited.

If there is a slight decrease in voltage, suspect a faulty solenoid switch or the bypass ballast lead connection, which connects the ballast resistor to the coil.

4 *Ballast resistor supply side voltage test*
(Fig. 9.39)
With the contacts still closed, connect the voltmeter between the supply side of the ballast

resistor and a good earth (V4) or, in the case where the ballast resistor is formed by the resistance cable, connect the voltmeter between the ignition switch end of the resistance cable and the earth (V5). The voltmeter should read the nominal battery voltage and, if this is low, check the supply lead from the ignition switch and its connection to the ignition switch terminal.

5 *Ballast resistor output side voltage test*
(Fig. 9.39)
Again, with distributor contacts closed, connect the voltmeter between the output side of the ballast resistor going to the ignition coil and a good earth (V5). The voltage should read between 5 and 8 V. If there is no reading suspect a faulty ballast resistor.

6 *Coil contact breaker terminal (CB) voltage test*
(Fig. 9.39)
With contacts open, this time connect the voltmeter between the contact breaker coil terminal and a good earth (V6). Switch on the ignition switch and observe the voltmeter, which should read the nominal battery voltage. A zero reading indicates an open circuit primary winding or a short circuit on the lead from the coil to the distributor or within the distributor.

Repeat test 5 with the coil-to-distributor lead disconnected. If a reading is now shown, suspect a fault within the distributor.

7 *Contact breaker voltage drop test*
(Fig. 9.39)
Connect the voltmeter between the disributor contact breaker terminal and a good earth (V7). Rotate the engine until the contacts close. Switch on the ignition and observe the voltmeter reading. With the contacts closed the voltage drop across the contact points should not exceed 0.5 V. A voltage reading above 0.5 V indicates dirty or burnt contacts or contacts not closing properly.

8 *Coil-to-distributor lead voltage drop test*
(Fig. 9.39)
With contacts closed, connect the voltmeter between the coil contact-breaker terminal and the distributor terminal (V8). Read off the voltage which should not exceed 0.1 V with a normal primary stall current flow. If the voltage drop is above 0.1 V check the lead and terminals at either end for poor continuity.

9.11.2 Ignition secondary circuit visual inspection
(Fig. 9.40)

1 *High tension cable inspection*
(Fig. 9.40)

   a) Check the condition of the coil-to-distributor central cable (1) and the individual distributor to spark-plug cable (2) by flexing

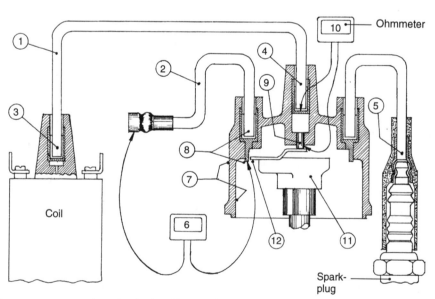

Fig. 9.40 Ignition system secondary circuit visual inspection

short lengths of cable at a time and sensing any change in stiffness or deterioration.

b) Check the condition of the HT cable end contacts at both coil socket (3) and distributor socket electrodes (4). Observe if the metal ends are clean and free of corrosion and that the conducting wire or carbon fibre strands are firmly secured to them. If ends are excessively corroded or damaged, replace the cable.

c) Check the spark-plug cable connectors (5) for tightness on plug terminals, corrosion, the condition of the insulation cover and the effective conductor-to-connector continuity.

d) Measure the ohm resistance of the conducting cable using an ohmmeter with its lead connected between the HT cable ends (6). If the resistance is excessively high replace the cable. Allowance must be made for radio suppressed cables which may have a resistance up to $30\,000\,\Omega$ per cable.

2 *Distributor cap inspection*
(Fig. 9.40)

a) Check inner and outer cap surfaces (7) for carbon runs or hairline cracks, then thoroughly clean inside and out for all dirt and oily deposits with a rag moistened with petrol.

b) Check the condition of the distributor cap, outer and inner parts (8) of the spark-plug cable electrodes, and scrape away deposits which may have built up over the internal segmental electrodes. If spark erosion has worn the segments excessively, replace the cap.

c) Check the condition of the distributor cap internal central carbon electrode brush (9) for wear. Measure the electrical continuity between the outside socket and inside carbon brush (10) with an ohmmeter connected between the two sides. Allowance should be made to the resistance if a radio suppressor is built in between the socket and brush.

3 *Distributor rotor arm inspection*
(Fig. 9.40)

Check the rotor arm (11) for drive-shaft fit, for loose electrode plate, defective leaf-spring brush conductor (if fitted), poor insulation, and tracking or cracking. Examine the electrode sickle plate (12) for erosion wear and replace if necessary. Wipe clean the rotor on the shaft. An excessive rotor-to-segmental electrode gap will show up on an oscilloscope, giving a high voltage reading.

9.12 Spark-plugs

9.12.1 Spark-plug operating heat range
(Fig. 9.41(g and h))

A spark-plug heat range is a measure of the plug's ability to transfer heat from the central electrode and insulator nose to the cylinder-head and cooling system.

If the heat absorbed by the plug's central electrode and insulator nose exceeds the capability of the plug to dissipate this heat in the same time, then the plug will overheat and the central electrode temperature will rise above its safe operating limit of about 900 to 950°C (Fig. 9.41(g and h)). Above the plug's upper working temperature-limit, the central electrode will glow and ignite the air–fuel mixture before the timed spark actually occurs. This condition is known as auto-ignition as it automatically starts the combustion process independently of the controlled ignition spark. The danger of this occurring is in the fact that it may take place relatively early on in the compression stroke. Consequently, the pressure generated in the particular cylinder suffering from auto-ignition will oppose the upward movement of the piston. Excessive mechanical stresses will be produced in the reciprocating and rotating components and an abnormal rise in the cylinder temperature would, if allowed to continue, damage the engine.

In contrast to the high working temperature region of the plug, if the plug's ability to transfer heat away from the central electrode and insulator tip exceeds that of the input heat from combustion, over the same time span, then the plug's central electrode and insulator nose would operate at such a low temperature as to permit the formation of carbon deposits around the central nose of the plug.

This critical lower temperature region is usually between 350°C and 400°C and, at temperatures below this, carbon or oil deposits will foul the insulation, creating conducting shunts to the inside of the metal casing of the plug (Fig. 9.41(g

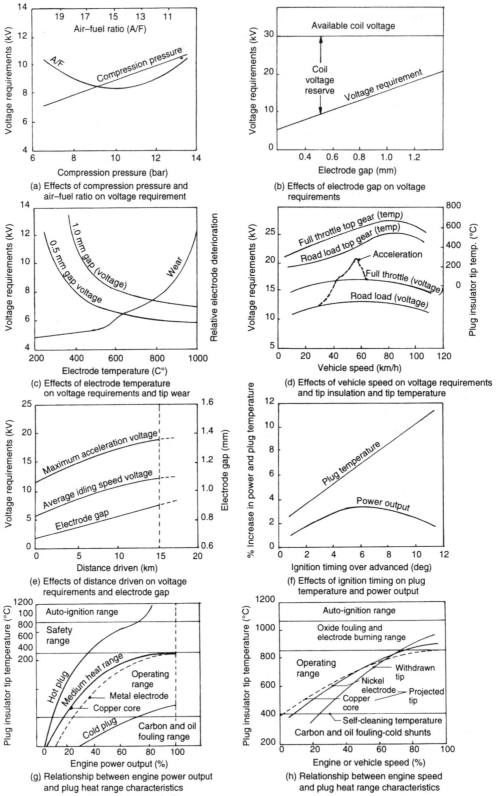

Fig. 9.41 Spark plug operating characteristics

and h)). Consequently, if deposits are permitted to form, a proportion of the ignition spark energy will bypass the plug gap so that there will be insufficient energy left to ionize the electrode with the result that misfiring will result. Establishing a heat balance between the plug's input and output heat flow, so that the plug's temperature remains just in excess of 400°C, provides a self-cleaning action on both the surfaces of the electrodes and insulator.

Good spark-plug design tries to match the heat flowing from the plug to the heat flowing into it, caused by combustion under all working conditions, so that the plug operates below the upper temperature limit at full load, but never drops below the lower limit when idling or running under light-load conditions.

9.12.2   Heat distribution
(Fig. 9.42)

When the fuel mixture is ignited in the gap formed between the electrodes, heat from the combustion process is dissipated throughout the combustion chamber walls, the valve heads and piston crown and, of course, the spark-plugs. The heat then passes to the cooling system. The actual temperature of the burning gas may reach values of between 2000 to 2500°C.

Roughly 80% of the heat absorbed by the exposed end of the plug is extracted through the plug by conduction, and the other 20% is removed by convection and radiation through the induced fresh mixture charge sweeping through the combustion chamber during the induction stroke (Fig. 9.42). About two-thirds of the heat absorbed by the plug due to the combustion process is transferred through the threaded sleeve and shoulder (seat) portion of the plug casing, while the rest flows from the insulator nose and central electrode to the main body of the plug.

A typical breakdown of the total heat distribution at the exposed end of the plug would be: 67% to the threaded casing, 8% via the earthed electrode, 21% by way of the insulator and only 4% along the central electrode (Fig. 9.42). Dissipation of heat from the plug is achieved mainly by conduction: 34% from the plug seat and 30% by way of the threaded region through the casing.

The remainder of the heat is expelled by convection and radiation in the following proportions: 10% from the upper metal casing, 4% from the ribbed insulator and 2% via the end terminal (Fig. 9.42). The actual plug heat absorption and dissipation distribution will vary somewhat under different operating conditions; for example, the location of the plug in the cylinder-head and the kind of plug lead terminal covering used.

9.12.3   Insulator nose length
(Fig. 9.43)

To meet the different demands on the spark-plug due to such variables as cylinder size, compression ratio, speed range, mixture strength, performance expectation and operating conditions, spark-plugs are manufactured with various insulator nose lengths (Fig. 9.43). Plugs which are used in high-output hot running engines, for instance, require a relatively short heat path from the central nose tip to the inner metal casing to dissipate the heat rapidly, while low-output engines which tend to run cooler, require longer heat flow paths to enable the plug to operate well within the self-cleaning temperature region.

Plugs which are to operate in high specific output engines need to dissipate more heat and therefore have shorter nose lengths—these are known as cold or hard plugs—whereas plugs used for low specific output engines need to transfer less heat in a given time, and therefore have longer heat flow path noses—these are therefore called hot or soft plugs.

Cold plugs tend to operate efficiently only under high power conditions; conversely, hot plugs are effective in the low-power output region, while a medium heat range plug would be well matched over a moderate power output range (as shown in Fig. 9.41(g)).

9.12.4   Spark-plug firing voltage

A certain minimum voltage is necessary to make the spark jump the electrode air gap, the actual magnitude of the voltage required will depend upon the following factors:

a) compression pressure
b) mixture strength
c) electrode gap
d) electrode tip temperature

a) Compression pressure (Fig. 9.41(a))
As the compression ratio is raised, the air density between the plug electrodes is in-

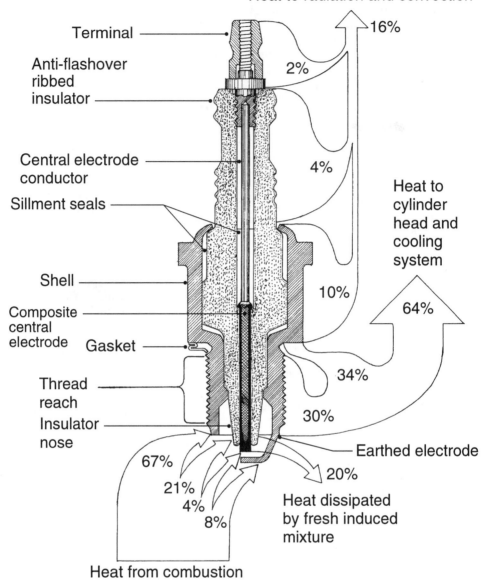

Fig. 9.42 Spark plug heat absorption and dissipation

creased so that a further voltage is necessary for a spark to bridge the air gap.

b) Mixture strength (Fig. 9.41(a))
A stoichiometric air–fuel ratio requires a minimum voltage to create a spark between the electrode as the mixture is just damp enough to improve its electrical conductivity, whereas a weak mixture is too dry and therefore needs a larger voltage applied between the electrodes. A rich mixture produces a degree of tracking over the electrode insulator surface

so that a higher voltage is also necessary to produce the spark.

c) Electrode gap (Fig. 9.41(b))
The voltage required to ionize the electrode gap is directly proportional to the gap width; thus, a large gap demands considerably more voltage to generate the spark. Figure 9.41(b) shows that, as the gap is widened, the reserve of maximum coil voltage to that needed to fire decreases.

d) Electrode tip temperature (Fig. 9.41(c))

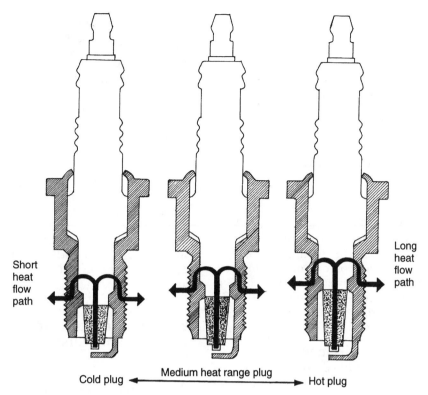

Fig. 9.43 Spark plug heat range controlled by heat flow path length

As the temperature of the central electrode tip rises, the voltage necessary to produce the spark decreases rapidly between 200 and 600°C. Above this temperature, the voltage will still fall but at a much slower rate. The actual voltage requirements with respect to tip temperature will depend to some extent on the electrode gap (as shown in Fig. 9.41(c)).

9.12.5 Projected nose (turbo action) plug
(Fig. 9.41(h))

A method of extending the effective speed and load range of a plug is to project the nose of the central electrode, and insulator nose, out from the plug casing recess. One benefit of the protruding nose is that it shortens the flame path within the combustion chamber and thereby produces a higher combustion propagation rate, resulting in greater combustion efficiency.

At low engine, no-load, speed conditions the longer projected nose with its extended heat flow path operates at a higher temperature than with a standard, slightly recessed, nose. This implies a greater degree of protection against the formation of deposits on the insulator surface. This being particularly suitable for driving in 'stop–go' traffic situations.

At the upper end of the speed range, the exposed plug nose is scavenged and cooled by the fresh air–fuel charge sweeping across the combustion chamber as it is drawn into the cylinder. This lowers the central electrode operating temperature and so raises the upper auto-ignition free-operation range of the plug.

9.12.6 Copper-cored central electrode
(Fig. 9.41(g and h))

The heat range of a plug can be widened by using a copper-cored central electrode with its ability to transfer rapidly heat absorbed at the tip, along its length, to the plug body and hence the cylinder cooling system. A nickel alloy jacket around the copper core, with its very good anti-corrosion and erosion properties, protects it from the aggressive combustion environment. With a combination of a long insulator nose and a composite copper-nickel alloy central electrode, these plugs are capable of operating at sufficiently high temperatures at light loads, to provide the necessary self-cleaning action. Yet, at full load, they are able to dissipate the additional heat without the electrode tip exceeding the upper safe working temperature.

9.12.7 Electrode deterioration versus temperature
(Fig. 9.41(c))

Spark erosion up to 580°C is largely independent of electrode temperature. Above this temperature, chemical corrosion is prominent, and is caused by the sulphur content of the fuel or deposits from fuel additives. At even higher temperatures, in the region of 870°C to 1000°C, oxidation of the electrode will dominate.

9.12.8 Voltage requirements with changing speed and load conditions
(Fig. 9.41(d))

The plug voltage requirements to produce the electrode spark over a wide speed range follow a typical engine torque characteristic curve. The voltage demand will also depend, to a great extent, on the throttle opening since this influences cylinder compression pressure. It can be seen in Fig. 9.41(d) that the voltage initially rises with increased speed, due to the engine's valve timing improving the cylinder filling and correspondingly raising the cylinder pressure, whereas towards the upper speed limit the voltage demands decrease because cylinder breathing becomes more difficult so that the induced mixture charge is less dense. Additional plug voltage is necessary when the throttle is snapped open to accelerate the vehicle, the extra mixture charge then entering the cylinder briefly raises the cylinder pressure so that more voltage must be available at the plug electrodes.

9.12.9 Electrode tip temperature variation with changing speed and load conditions
(Fig. 9.41(d))

If a vehicle's speed is steadily increased without

the driver changing gear, then the power developed by each cylinder also rises proportionally. Ideally, the rate of heat absorbed and dissipated by the central electrode and insulator should match the engine's power output, but the extra demands on the heat transference cannot keep up with the rate of heat release caused by the cyclic combustion. Therefore, the electrode tip temperature rises with vehicle speed accordingly. As the engine approaches maximum speed there will be a fall in volumetric efficiency and, accordingly, a slight reduction in power which is reflected in a decrease in the relative amount of heat absorbed at the electrode tip; correspondingly, its temperature is reduced. Note that the actual temperature level throughout the speed range is considerably influenced by the degree of throttle opening—that is, the engine loading.

9.12.10 Electrode wear and voltage requirements versus plug operating life
(Fig. 9.41(e))

As the distance driven on one set of spark-plugs increases, both electrodes wear. This is partially caused by the spark discharge eroding the portion of the electrodes facing the air-gap, and also the effects of corrosion when the plug is operating in its upper temperature range.

Spark-plugs should be cleaned and electrodes filed and regapped at about 7500 km (5000 miles) and, after 15 000 km (10 000 miles), replaced with a new set. A typical normal wear rate would be 0.015 mm per 1000 km and the central electrode shape will change from a flat sharp tip, this being ideal for releasing electrons, to a rounded profile which finds it more difficult to emit electrons.

A new plug with a 0.6 mm electrode gap operating under light load conditions would probably require a firing voltage of 8 to 10 kV. Half-way through its life at 7500 km (5000 miles) the gap will have widened to 0.6 + 0.015 × 7.5 = 0.7125 mm and the voltage then needed to bridge the spark across the enlarged gap with the rounded central electrode will rise to something like 10 to 12 kV. Figure 9.41(e) shows the increased voltage demands for both idling and acceleration conditions with respect to plug life in terms of distance driven.

9.12.11 Power output and plug tip temperatures versus ignition advance
(Fig. 9.41(f))

Advancing the point of ignition so that it occurs before TDC, increases the peak cylinder combustion pressure. It therefore raises the amount of heat transferred from the gases to the central electrode and insulator. Consequently, the plug tip temperature rises roughly in direct proportion to the degree of advance. However, the gain in power output becomes an optimum at about 6° before TDC for a typical engine and any further ignition advance actually reduces the power developed in the cylinders. This decrease in relative power is due to combustion gas pressure occurring so early that it opposes the upward movement of the piston towards the end of the compression stroke. Therefore, it can be seen that over advancing the point of ignition increases plug temperature with no improvement in engine power beyond a certain advance, but the engine now becomes detonation prone.

9.12.12 Plug central electrode polarity

A spark jumping the air gap between the plug electrodes consists of a stream of free electrons flowing from the negative-biased electrode (cathode) to the positive-biased electrode (anode).

If enough heat is applied to one of these electrodes, free electrons will be released to the surroundings, and these electrons do not tend to move in any particular direction unless urged by an electric field. If, now, the ignition coil's secondary voltage potential is applied between the two electrodes so that the hottest central electrode is made negative and the cooler earthed electrode becomes positive, then the released electrons will readily travel from the central electrode (cathode), which has an excess of electrons, to the earthed electrode (anode) which has a deficiency of electrons.

Electron emission due to a very hot electrode assists the ignition coil HT voltage in establishing the spark between the electrodes, whereas if the polarities of the electrodes were reversed so that the central electrode now had a positive potential; the electron flow would try to oppose the formation of a spark and a much higher voltage would then be required to produce the spark.

9.12.13 Spark-plug tightness
(Fig. 9.44(a and b))

The plug seat, be it tapered or of the flat type, conducts the majority of the heat absorbed at the tip and exposed metal casing through the plug to cylinder-head joint, to the cylinder-head cooling system. Therefore, the seat joint tightness is essential for good heat dissipation.

Spark-plugs should not be over tightened otherwise the plug metal casing may become distorted, causing the central electrode insulator to break its seal and become loose. Combustion gas may then escape through the plug with the result that it overheats.

An under-tightened plug may work itself loose and cause combustion gas to escape between the plug and cylinder-head plug hole threads to the atmosphere, again this will result in overheating and rapid deterioration of the electrode tips.

It is best to torque the plug to a definite degree of tightness as shown in the table but if this is not possible, then use the following technique.

Plug tightening torques

Type of seat joint	Cylinder-head	
	Steel	Aluminium
	Units of torque (Nm)	Units of torque (Nm)
Flat seat plug		
m14 × 1.25	20–40	20–30
m18 × 1.5	30–45	20–35
Conical seat plug		
m14 × 1.25	15–25	10–20
m18 × 0.15	20–30	15–23

as illustrated in Fig. 9.44(a and b).

First of all, screw in the plug as far as it will go by hand and then apply the box spanner or socket with its tommy bar to the following extent.

a) **Flat seat plug with gasket** (Fig. 9.44(a))

With a new plug, turn the plug in the threaded hole by hand until it has reached the end of its thread, then rotate the plug a further 90° with the plug spanner.

When tightening a used plug, again twist the plug by hand until the gasket is touching both seats (that is, it will go no further) then apply the plug spanner and rotate the plug a further 30°.

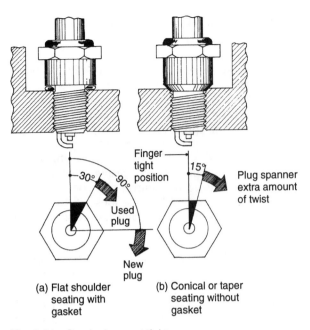

Finger tight position

−30°

90°

Used plug

New plug

15°

Plug spanner extra amount of twist

(a) Flat shoulder
   seating with
   gasket

(b) Conical or taper
   seating without
   gasket

Fig. 9.44 Spark plug seat tightness

b) **Tapered seat plug without gaskets** (Fig. 9.44(b))

With new and used plugs, turn the plug in the threaded hole by hand until it has reached the end of its thread (that is, it will go no further), then apply the plug spanner and rotate the plug a further 15°.

10

Engine Testing Equipment

10.1 Fault finding equipment

To enable the condition of an engine, and its state of tune, to be assessed, special equipment has been designed which, when used correctly, can locate and provide an accurate and rapid diagnosis of any faults which may have developed. Faults which may possibly be due to natural wear, bad driving practice or lack of servicing. Some of the more important fault finding test procedures will now be explained.

10.2 Vacuum gauge test

10.2.1 Engine functions partially as a vacuum pump
(Fig. 10.1(a))

During every fourth stroke of the engine, the inlet valve opens and the exhaust valve closes while the piston moves outwards and away from the cylinder-head. In effect, therefore, the engine becomes a vacuum pump. Thus, when the piston sweeps down its stroke from TDC to BDC the cylinder volume enlarges and the air–fuel mixture molecules induced into the cylinder are stretched apart as they endeavour to occupy the expanding space. As a result, the continuous movement and the impact of the trapped molecules against the confined cylinder walls (this being known as cylinder pressure) are far less than for the more densely populated air molecules in the atmosphere trying to enter the cylinder. The amount of vacuum created within the cylinder is therefore a measure of the induced substance's pressure,

relative to that produced outside the cylinder in the atmosphere. Vacuum is therefore measured as a pressure and, since the cylinder pressure under these conditions will be less than that of the atmosphere, it is referred to as a depression or negative pressure. The construction of a vacuum gauge is shown in Fig. 10.1(a), a full description of how it operates is given in the section covering the cylinder compression gauge test (10.5).

10.2.2 Units of pressure

The pressure of gas is a measure of the force exerted per unit area, and the unit of pressure may be in kN/m^2. It has been established that the pressure of the atmosphere at sea level is approximately $101\ kN/m^2$.

Sometimes it is more convenient to express pressure relative to atmospheric pressure, and this unit of pressure is a ratio of atmospheric pressure to some other pressure value and is known as the bar. For small vacuum pressures, the bar is subdivided into decimal fractions, i.e. 0.1, 0.2, 0.3 bar etc. Atmospheric pressure may also be defined as that pressure which supports a column of mercury 760 mm high (760 mm Hg) this being the more usual way of presenting depressions, thus $101\ kN/m^2 = 1\ bar = 760\ mm\ Hg$.

10.2.3 Engine tuning effect
(Fig. 10.1(b))

For a particular operating condition and for ignition to occur at the most optimum position in the TDC region at the end of the compression stroke, an engine's power depends upon the complete-

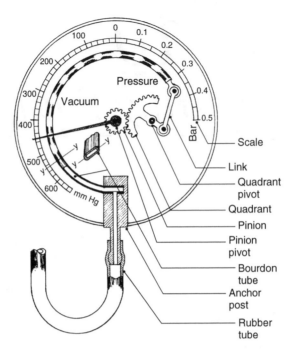

(a) Combined vacuum and pressure gauge

Fig. 10.1 Vacuum gauge engine test

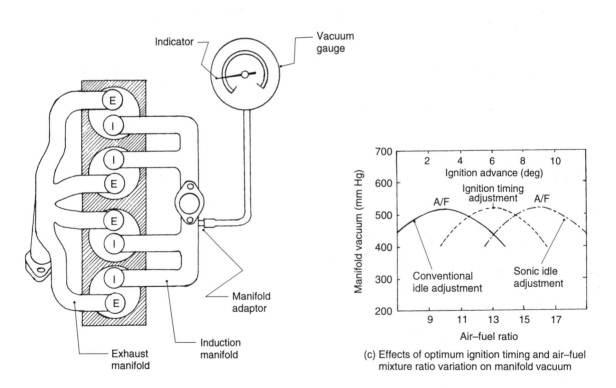

(b) Vacuum gauge testing layout

(c) Effects of optimum ignition timing and air–fuel mixture ratio variation on manifold vacuum

ness of combustion, which is controlled to a great extent by the correct air–fuel mixture ratio. An idling engine develops just sufficient power to overcome frictional and pumping losses and the small surplus power determines the idling speed of the engine.

Consider now an idling engine with a vacuum gauge connected to the induction manifold (Fig. 10.1(b)) if tuning raises the engine speed without altering the throttle valve opening, there will be an increased number of suction strokes in the same time period, consequently the average vacuum in the manifold will be raised. Conversely, if there is a drop of engine speed owing to poor tuning, the reduction in the number of suction strokes per unit time will reduce the amount of vacuum generated in the induction manifold. A change in manifold vacuum will therefore be a direct measure of the effectiveness of combustion during the tuning of the air–fuel mixture and ignition point.

10.2.4 Mechanical engine condition

The amount of vacuum generated in the induction manifold is a measure of the gas sealing tightness of the piston and valve assemblies and the effectiveness of combustion within the cylinder in producing power.

When testing a single-cylinder engine the induction vacuum will fluctuate over the operating four-stroke cycle—there being a suction stroke every 720° of crankshaft movement. A four, six and eight-cylinder engine will have a suction stroke every 180°, 120° and 90° respectively, so that the induction pulsations will be steadily reduced as the number of cylinders increases. The uniformity of each successive vacuum pulsation in the induction manifold should be equal (if there is an air leak somewhere in the induction system, the vacuum gauge needle pulsation would still be equal but the average vacuum reading would be lower).

If, however, any one or more of the cylinders has a leaky valve, sticking or broken piston rings, or a blown cylinder head gasket, then the effect of this leak would be apparent only when the suction of that particular cylinder (or those cylinders) is being measured, the result would be a momentary drop in vacuum shown by an irregular gauge reading.

10.2.5 Cranking vacuum test

Connect a jumper lead between the ignition coils 'CB' terminal and earth, then operate the engine as a vacuum pump by completely closing the throttle valve and cranking the engine on the starter motor for about 15 s. The degree of suction generated in the cylinder and manifold will be a measure of the mechanical condition of the engine. An engine in good condition will produce a high vacuum (350 mm Hg or more) owing to effective sealing between the cylinder walls and pistons, and the air tightness of the valves in the cylinder-head. Low readings would indicate induction leakage such as a manifold gasket blown, or generally poor valve sealing, or worn piston rings and cylinder bores.

10.2.6 Idle vacuum test

Run the engine at idle speed and observe the vacuum gauge reading. An engine in good condition should give a steady needle reading with values within the following range.

Four cylinders   460–510 mm Hg
Six cylinders   480–540 mm Hg
Eight cylinders 510–560 mm Hg

If there is an intermittent drop of vacuum of 75–100 mm Hg below the normal steady reading for an engine in good condition, then suspect one or more valves leaking.

10.2.7 Idle mixture setting
(Fig. 10.1(c))

With the engine at operating temperature set the idling speed to that recommended. Adjust the slow running mixture control screw to give the highest steady reading on the vacuum gauge (Fig. 10.1(c)). With double and triple carburettor installations, disconnect the inter-connecting linkage and adjust each carburettor separately until the best reading is obtained.

10.2.8 Idle dynamic ignition setting
(Fig. 10.1(c))

Operate the engine at normal working temperature and idling speed, and advance or retard the distributor to give the highest steady vacuum gauge reading (Fig. 10.1(c)). It is advisable to compare this setting with the recommended stroboscope timing light datum marks and if these differ considerably from the vacuum gauge setting investigate further.

10.2.9 Cylinder balance test

This test compares the performance of each cylinder by removing each spark-plug lead in turn, with the engine idling, and carefully observing the drop in vacuum reading. Each cylinder should give an equal fall in vacuum which may be from 15 to 50 mm Hg. A cylinder with only a small drop-off in vacuum indicates that it is not contributing as much power as the other cylinders. Further investigations should be made with a cylinder compression test to determine if the fault is in the cylinder or is due either to the fuel supply or ignition system malfunctioning.

10.2.10 Leaky piston ring test

Run the engine at idling speed and then quickly open the throttle to its fully open position, allowing the engine to reach its highest safe speed. Close the throttle rapidly and observe the vacuum gauge reading. This should increase immediately to 125 mm Hg or more, above the normal idle vacuum before settling down to its original tickover value. If this is the case, the engine is in good condition. Only a small rise in vacuum indicates loss of compression, which may be due to worn bores, rings and pistons or a combination of all three.

10.2.11 Weak or broken valve spring test

Operate the engine at idling speed and then steadily increase its speed to its highest safe value. At the same time, observe the vacuum gauge pointer. If there is a vacuum fluctuation of between 250 and 550 mm Hg, with the frequency of the fluctuation increasing as the speed rises, faulty valve springs may be suspect.

10.3 Mechanical fuel pump testing

(Fig. 10.2(a and b))

A combined vacuum–pressure gauge can be used for testing the efficiency of the mechanical fuel pumps for both petrol and diesel engines by adopting the following procedure.

10.3.1 Dry test with gauge

(Fig. 10.2(a and b))

A suspected faulty fuel pump should be given a dry (without fuel) gauge test before dismantling and then after reassembly.

1 *Suction*

(Fig. 10.2(a))
1 Connect the gauge rubber pipe to the inlet connection on the pump.
2 Operate the rocker-lever or push-rod through three full strokes.
3 Observe the minimum gauge reading on the vacuum scale and the vacuum drop rate.

Pump type	Minimum vacuum	Vacuum drop rate
Low pressure	150 mm Hg	50 mm Hg in 15 s
High pressure	225 mm Hg	75 mm Hg in 5 s

2 *Delivery*

(Fig. 10.2(b))
1 Connect the gauge to the outlet pipe connection on the pump.
2 Operate the rocker-lever or the push-rod through two full strokes.
3 Observe the minimum gauge reading on the pressure scale and the pressure drop rate.

Pump type	Minimum pressure	Pressure drop rate
Low pressure	0.2 bar	0.04 bar in 15 s
High pressure	0.42 bar	0.14 bar in 5 s

10.3.2 Dry test without gauge

An indication of the condition of the pump can be obtained without a gauge, if one is not available, by adopting the following procedure.

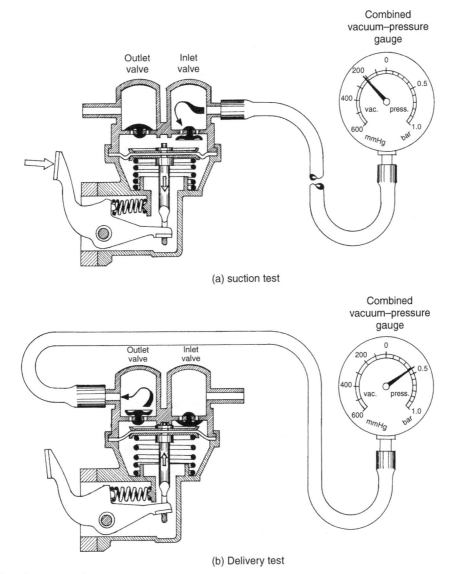

Fig. 10.2 Petrol pump testing

1 *Suction*

1 Hold a finger over the inlet pipe connection of the pump and operate the rocker or push-rod through three full strokes.
2 Release the finger and listen to the noise caused by air rushing into the suction chamber. There should be a distinct suction sound each time the diaphragm is released.

2 *Delivery*

1 Hold a finger over the outlet pipe connection and depress the rocker-arm or push-rod through two full strokes.
2 Note a feeling of pressure over the pipe outlet which should hold for at least 15 s.

10.4 Cylinder power balance

(Fig. 10.3)

An understanding of why an engine will run steady under no-load speed conditions at relatively low speed provides an insight into the cylinder power balance check.

For an engine to accelerate and increase its speed and power produced, it must first develop sufficient power to overcome all the frictional and pumping losses which are incurred—power derived in excess of this minimum is then available to propel the vehicle. The total resistance opposing the rotation of the engine itself without any external loading may be summarized as follows.

1 Journal resistance torque necessary to rotate the crankshaft, camshaft, distributor, driveshaft and jackshaft, if used.
2 Reciprocating resistance caused by the gas pressure producing piston side thrust and increasing the radial outward thrust of the piston rings against the cylinder walls.
3 Resistance torque produced by the oil pump, water pump, petrol pump, generator, power steering hydraulic pump, if incorporated, etc.
4 Resistance torque necessary to drive the timing gear train and that needed to actuate the valve mechanism against the return springs and opposing inertia forces acting on the poppet valves.
5 Periodic energy imparted to the flywheel which is necessary to accelerate the engine.
6 Energy expended to drag the fresh air–fuel mixture through the induction ports into the cylinders.
7 Energy expended in compressing the air–fuel mixture charge on compression stroke.
8 Energy expended to sweep the spent exhaust gases from the cylinders out into the atmosphere.

10.4.1 Principle of cylinder power balance test

Idling speed stability is achieved by adjusting the butterfly throttle valve opening so that there is just sufficient air–fuel mixture entering the cylinders to develop enough power to overcome all the friction and pumping losses and to provide a small surplus of power so that any selected running speed can be maintained.

If one of the engine's cylinders is prevented from producing a power stroke by interrupting the secondary voltage to the spark plug, there will be less surplus power developed above that necessary to keep the engine rotating against the opposing friction and pumping resistances of all the cylinders. Consequently, the engine's speed drops to a new state of balance to match the reduced power now available. If the inoperative cylinder originally produced its full proportion of power then, when the cylinder's power is cut out, there will be a marked reduction in engine speed. Conversely, if this particular cylinder had not been contributing its full useful power, the speed reduction with this cylinder power removed will only cause a marginal decrease in speed. It can thus be appreciated that by cutting-out the power of each cylinder in turn, and observing the corresponding change in speed, an indication is provided of which cylinders are operating effectively and which cylinders are producing less power than they should. In the latter case, further investigation will be necessary.

10.4.2 Automatic contact breaker shunt device
(Fig. 10.3)

To simplify the shorting-out of each cylinder while making a cylinder power balance check, most engine analysers provide an electronic impulse facility which short-circuits the contact breaker points when they open momentarily, to prevent any selected spark plug from firing. One such device (Fig. 10.3) uses a power output transistor (T) and when the contact open voltage is supplied to the base (b), this switches on the collector (C)–emitter (e) circuit. Immediately, the contact breaker points are shorted, so preventing the induced secondary voltage from occurring. To use this instrument, switch the control to the 'correct number' of cylinders and then turn the selector control knob to stop each plug in turn from firing.

10.4.3 Interpreting power balance results

When observing the speed reduction with each individual cylinder cut-out in turn, the expected fall in speed with all cylinders in good condition will produce an 80 to 100 rev/min drop for a four-cylinder engine, 60 to 80 rev/min drop for a six-cylinder engine and 35 to 45 rev/min drop for an eight-cylinder engine. If the difference between the largest and smallest speed reduction readings exceeds 25% of the largest drop in

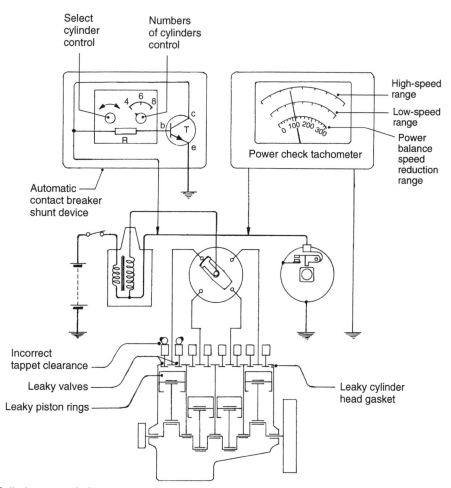

Select cylinder control

Numbers of cylinders control

High-speed range

Low-speed range

Power balance speed reduction range

Power check tachometer

0 100 200 300

Automatic contact breaker shunt device

Incorrect tappet clearance

Leaky valves

Leaky piston rings

Leaky cylinder head gasket

Fig. 10.3 Cylinder power balance test

speed, the following likely faults should be investigated:

1 tappets incorrectly set
2 air leak in induction system
3 multi-carburettors unbalanced or individual petrol injectors at fault
4 sticking piston rings or burnt valves
5 cylinder-head gasket leaking

For example, if the readings for a four-cylinder engine were as follows: 110, 60, 75, 100 then $110 - 60 = 50$ rev/min; as this is greater than 25%, check the engine for the listed likely faults.

10.5 Cylinder compression test

The purpose of the cylinder compression test is to measure the maximum individual cylinder air pressure generated by the piston on its compression stroke during the period the engine is cranked by the starter motor and, from the results, determine the mechanical condition of the engine.

10.5.1 Operating principle of pressure gauge
(Fig. 10.4(a))

The central feature of the test is the pressure gauge itself which comprises an oval cross-sectioned tube bent into a semi-circle (Fig. 10.4(a)). One end of this so-called Bourdon tube is fixed onto the anchor post while the other end is free to flex and is attached indirectly to the toothed quadrant via a link arm.

When air pressure is introduced into the bourdon tube its curvature tends to straighten out in relationship to the applied pressure. Accordingly,

479

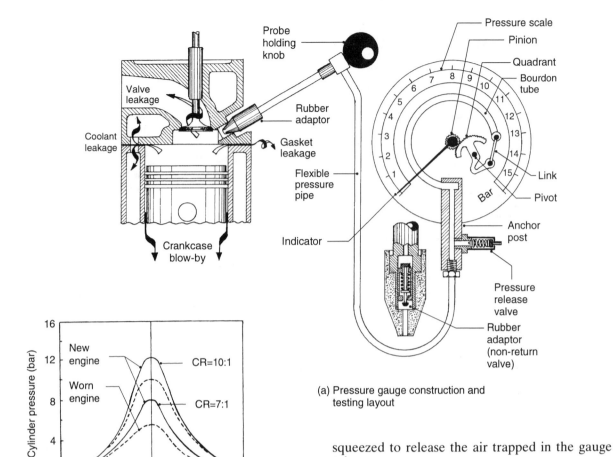

(a) Pressure gauge construction and testing layout

(b) Effect of compression ratio and state of wear on cylinder compression pressure

Fig. 10.4 Cylinder pressure test

squeezed to release the air trapped in the gauge tube.

10.5.2 Theory of the cylinder pressure test
(Fig. 10.4(a))

To take cylinder pressure readings the gauge adaptor is inserted in the spark-plug or injector hole. When the crankshaft is cranked by the starter motor there will be a compression stroke every fourth piston stroke or second revolution. The trapped air will be forced up into the cylinder-head combustion chamber space where it will seek out any available opening (see Fig. 10.4(a)). Since the interior of the pressure gauge is initially at atmospheric pressure, the non-return valve will open, allowing the compressed air to enter. This registers on the pressure gauge scale by the needle pointer movement. Immediately the inner gauge pressure is equal to that in the cylinder, the non-return valve will snap shut. The trapped air in the gauge will then continue to indicate the pressure of the air as the piston's cycle of opera-

the link arm partially twists the toothed quadrant and, consequently, the meshed pinion supporting the indicator pointer pivots on its axis. The amount the indicator rotates will depend upon the quadrant-to-pinion gear ratio and the magnitude of the air pressure being measured: a calibrated scale provides a direct pressure reading.

The gauge has a non-return valve which prevents the air forced into the tube from escaping unless the valve is manually opened by pushing a matchstick or something similar against the valve pin. With some gauges, a separate pressure release valve is produced—this being simply

tion is repeated. During the next cycle, a greater pressure will have to be reached to open the non-return valve and so allow more air to enter the gauge. This situation will continue until succeeding strokes fail to raise the gauge pressure. The accumulated air trapped in the gauge will then give the maximum pressure reached in the cylinder.

10.5.3 Preparing the engine for the cylinder compression test

Run the engine until normal warm up conditions have been reached then switch the ignition switch off. Pull off the spark-plug leads, loosen all spark-plugs about one turn and then reconnect the leads to the plugs. Alternatively, slacken off all injector clamp nuts. Restart the engine and accelerate to approximately 1200 rev/min, then reduce the speed to idling and stop the engine. This procedure is adopted to remove any carbon that may be sealing the spark-plug or injector screw joint, and which otherwise may become lodged under a valve causing low compression readings. Before removing plugs or injectors use a compressed air blast to clear dirt from the plug or injector and jam open the carburettor throttle valve, if appropriate, to remove any air intake restriction.

10.5.4 Procedure for dry cylinder compression test

Use either the screw-type adaptor, which should be screwed into the spark-plug hole, or the rubber nose adaptor which must be pressed against the plug or injector hole. Apply the gauge adaptor to the first cylinder plug or injector, then crank the engine by means of the starter motor and count the number of impulses of the gauge needle before it attains its maximum value. Record the reading then release the air trapped in the gauge. Repeat the procedure for all remaining cylinders, operating the starter motor to give the same number of gauge pulses for each cylinder and record the readings. If the gauge readings are low, carry out the wet cylinder compression test.

10.5.5 Procedure for wet cylinder compression test

Using an oil can, inject a small quantity of heavy oil into the cylinder being tested through the plug or injector hole. The oil will then spread and settle around the piston rings and cylinder wall. Thus, for a short period the viscosity of the lubricant will contribute to the sealing and gas tightness of the piston assembly while it is reciprocating within the cylinder. Repeat the cylinder compression test for each cylinder and record and compare results.

If there is an increase in cylinder pressure when testing wet, this indicates an improvement in the sealing of the rings so that worn walls and piston rings, or even broken rings, may be suspected. If there is no change in cylinder pressure, this means that air is escaping from above the piston, either by passing badly seated or burnt exhaust and inlet valves, or there is leakage through a blown cylinder-head gasket.

10.5.6 Results and conclusions

Cylinder compression pressure readings will depend to a great extent upon the following factors

1. compression ratio—the higher this is, the greater will be the cylinder pressure (Fig. 10.4(b));
2. the amount of inlet valve closing lag after BDC (40–70°)—the greater the lag, the lower the effective cylinder pressure;
3. the starter motor cranking speed (200–350 rev/min)—the higher this is, the less time there is for cylinder leakage and therefore the greater the cylinder compression pressure;
4. the mechanical condition of the engine components such as the cylinder, piston, rings, and poppet valves—the greater the wear, the lower the compression pressure (Fig. 10.4(b)).

A typical guide to cylinder compression pressure over a range of compression ratios under normal cranking conditions is tabulated below:

Compression ratio	6	7	8	9	10	11
Cylinder pressure (bar)	6–8	7–10	9–11	10–12	11–13	12–14

When checking cylinder pressures, compare the results with recommended data. Individual cylinder readings should not vary by more than one bar between any cylinder. If pressure readings between adjacent cylinders are low, suspect that the cylinder-head gasket has blown, while if the coolant in the radiator bubbles when testing, this may also indicate a leaking cylinder-head gasket. An exact cause of low or uneven compression test results can best be diagnosed with the aid of the cylinder leakage test.

10.6   Cylinder leakage test

(Fig. 10.5(a, b and c))

With the cylinder leakage test, each cylinder is brought to TDC in the position where the piston ring belt is in the greatest cylinder-wall wear zone. Compressed air, set to a reference pressure, is then permitted to enter the space above the piston through the spark-plug hole (Fig. 10.5(a)). The drop in air pressure, calibrated in percentage efficiency, which is caused by a leakage somewhere between the cylinder walls, piston and cylinder-head can then be accurately read, recorded and compared with other cylinder observations. In combination with a visual and aural inspection during the test, an accurate fault diagnosis can be predicted.

10.6.1   Preparing an engine for the leakage test

Before making the leakage test, run the engine until normal operating temperature is reached and then stop the engine. Withdraw all the spark-plugs and jam open the throttle valve to prevent air intake restriction. Remove the air cleaner, crankcase fill cap, radiator cap and disconnect one end of the positive crankcase ventilation system from the engine.

10.6.2   Determination of piston TDC on compression stroke

(Fig. 10.5(a))

The TDC position on the compression stroke can easily be found by attaching the whistle provided with the equipment to the cylinder-head adaptor, and screwing it into the plug hole (Fig. 10.5(a)). A 12 V test lamp should then be connected between the ignition coil's distributor LT lead terminal and earth. Then, the engine should be slowly cranked on the starter by jabbing the ignition switch to the start position. As the piston

approaches TDC on the compression stroke the whistle will blow with diminishing volume and, at the instant the distributor contact-points open or the electronic ignition blocks just before TDC, the test lamp will light—this being approximately the innermost piston position. To secure the crankshaft in this position engage top gear and apply the handbrake.

10.6.3   Procedure for leakage test

(Fig. 10.5(b and c))

A pressurized air supply (5–14 bar) is connected to the pressure gauge inlet coupling, and the regulating control knob is then adjusted until the indicator pointer moves to the set 100% no leakage position (100% efficiency). This is possible because the shut-off valve in the outlet coupling (chuck) automatically closes when it is disconnected from the hose joining it to the spark-plug hole (Fig. 10.5(b)).

A spark-plug adaptor is then screwed into the spark-plug hole of the cylinder being analysed, and the two ends of the hose joining the pressure gauge meter outlet to the cylinder adaptor are then coupled. As a result of the meter's outlet male and female coupling being engaged, the chuck's shut-off valve will open, thus permitting compressed air to flow into the enclosed space above the piston (Fig. 10.5(c)). This pressurized air will then seek out and escape through any possible exit caused by cylinder-wall, piston skirt and piston-ring wear, or poor or burnt exhaust and inlet valve seats, or a blown cylinder-head gasket or cracked cylinder block/liner, piston crown or cylinder-head. Any loss of air will immediately be registered as a drop-off in leak proof efficiency.

After the test, disconnect the hose coupling (chuck) from the meter and cylinder-head plug adaptor, and remove the adaptor from the cylinder-head. Attach the adaptor and whistle to the next cylinder to be tested and rotate the crankshaft again until the whistle sounds and the test lamp lights in the TDC region. Detach the whistle and couple the connecting hose between the meter and cylinder-head adaptor. Finally, observe and record the pressure drop in percentage efficiency. Repeat this procedure for all cylinders.

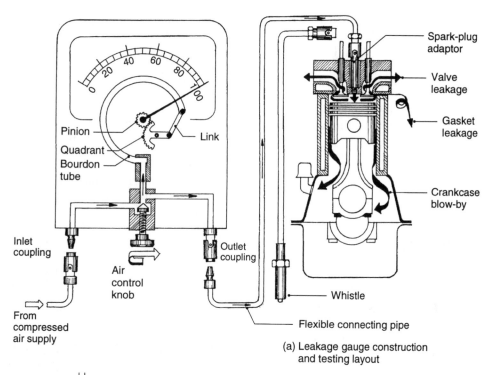

Pinion

Quadrant

Bourdon tube

Link

Inlet coupling

Air control knob

Outlet coupling

From compressed air supply

Spark-plug adaptor

Valve leakage

Gasket leakage

Crankcase blow-by

Whistle

Flexible connecting pipe

(a) Leakage gauge construction and testing layout

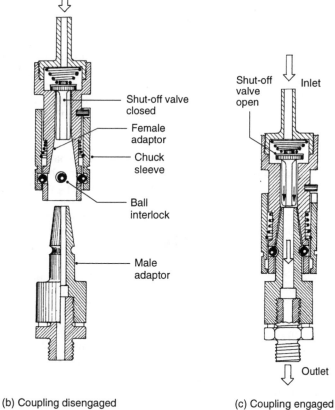

Shut-off valve closed

Female adaptor

Chuck sleeve

Ball interlock

Male adaptor

Shut-off valve open

Inlet

Outlet

(b) Coupling disengaged

(c) Coupling engaged

Fig. 10.5   Cylinder leakage test

483

10.6.4 Results and conclusions

Individual cylinder leakage should not vary by more than 5% and an engine in good condition would normally have leakage efficiencies better than 80%.

A general guide to an engine's mechanical condition, based on engine percentage efficient leakage readings, is as follows:

100–90%	very good
90–80%	good
80–70%	fair
below 70%	poor

10.6.5 Identifying engine faults

The source of a leak can be determined by listening and observing the following:

1 listening at the air intake—air escaping here indicates a badly seating or burnt inlet valve.
2 listening at the oil filler tube—air escaping here indicates worn piston rings, or a worn or scored cylinder wall.
3 listening at the exhaust pipe—air escaping here indicates a badly seating or burnt exhaust valve.
4 visually seeing air bubbles in the radiator or coolant reservoir indicates either a leaky gasket or a cracked cylinder-block or head.

10.7 Ignition timing

Ignition timing is the setting of the point when the spark-plug fires, which occurs when the contact breaker points first open or, with electronic systems, the initial interruption of the primary winding current flow, relative to the crankshaft angular position, when No. 1 cylinder is approaching the end of its compression stroke.

1 Basic ignition timing
The initial setting of the point of ignition relative to the crankshaft angular position is referred to as 'basic ignition timing' which may take two forms—basic timing with TDC setting marks and basic timing with ignition setting marks.

2 Basic timing TDC setting mark
These timing marks are designed to align when No. 1 cylinder's crank-throw angular position has reached the TDC position.

3 Basic timing ignition setting marks
These marks are designed to align when No. 1 cylinder's crank-throw angular position has reached the point where the spark is intended to occur, which is normally a few degrees before top dead centre.

4 Static and dynamic ignition timing
The crankshaft angular position when the point of ignition occurs can either be set statically when the engine is stationary, or dynamically when the engine is idling. Dynamic timing has the obvious merit over static ignition timing in that the angular point at which the spark takes place is set under operating conditions; this method of timing thereby provides the exact and actual timing requirements for the working engine, whereas the static timing method setting is useful for initially starting the engine, but does not take into consideration drive shaft gear backlash and distributor automatic advance mechanism interference.

10.7.1 Ignition timing procedure using the stroboscope light method

Power timing flash delay
(Fig. 10.6(a))

The purpose of the delay circuit control is to enable the operator to measure the angular distance that the pulley or fly-wheel has moved from the basic timing position in crankshaft degrees at different engine speeds. If the operator increases the engine speed to some specified value, the flash will appear to occur at an earlier point in its relationship to the crankshaft, due to the distributor automatic advance mechanism, and the mark on the rotating member will be seen to move against the direction of rotation. By adjusting the delay control, the time between the firing of No. 1 plug (Fig. 10.6(a)) or any other reference plug, and the flash on the power timing light can be increased. Therefore, the pulley or fly-wheel timing mark will be seen to move in the same direction as the engine rotates.

The delay between the HT voltage pulse and the flash can be extended by adjustment, until the timing marks appear to move back and align with the basic ignition timing mark position, obtained when the engine was at idle speed. At the same time as the flash delay period is altered so that it appears to occur at the basic ignition setting, the delay circuit converts this time period into crank-angle movement and records this on the advance

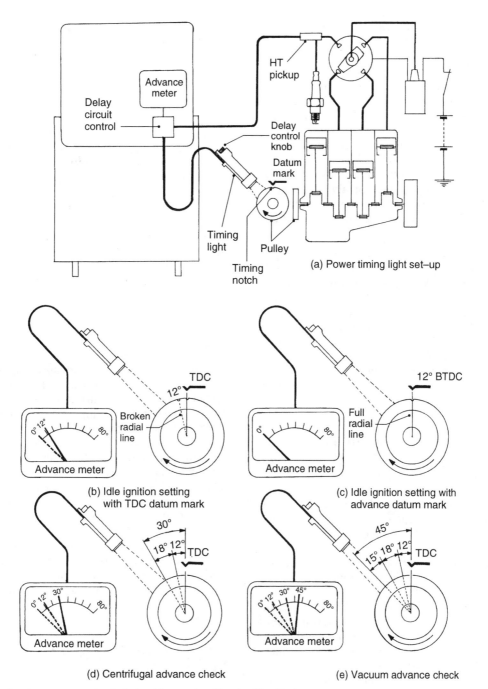

Fig. 10.6 Stroboscope method of checking and setting ignition timing

(a) Power timing light set–up

(b) Idle ignition setting with TDC datum mark

(c) Idle ignition setting with advance datum mark

(d) Centrifugal advance check

(e) Vacuum advance check

meter. Note that at various engine speeds one flash delay period will correspond to a different crank-angle movement; for instance, a small flash delay may be equal to 5° crank-angle movement at 1000 rev/min which will correspond to 10° crank-angle displacement at 2000 rev/min.

10.7.2 Basic timing TDC setting mark
(Fig. 10.6(b))

1 Remove the vacuum advance pipe to prevent interference with the basic timing (in some cases this pipe should not be disconnected

485

while checking basic timing, in which case refer to tuning data) and run the engine at normal idling speed.

2   Adjust the synchronizer control knob on the timing-light or advance meter unit until the indicator needle points to the desired basic ignition setting degree reading on the meter scale, for example 12°.

3   Slacken the distributor body clamp, focus the timing light on the fixed TDC timing marks, and then partially rotate the body either way until the moving pulley or flywheel datum mark (notch) appears to align with the stationary TDC mark. Then retighten the distributor body clamp.

What has happened in fact is that the spark-plug fires at its basic ignition setting BTDC (shown by the 12° BTDC broken radial line on the pulley in this example), but the electronic delay circuit control which received the HT voltage pulse has been adjusted to delay the lamp flashing until the pulley notch (or flywheel) reaches the fixed TDC mark on the casing, at which point the trigger signal fires the timing light flash tube (now shown by the full radial line aligning with the TDC fixed-mark position).

10.7.3   Basic timing ignition setting mark
(Fig. 10.6(c))

1   Run the engine at normal idle speed with the vacuum advance pipe either connected or disconnected, as recommended by the manufacturer.

2   Set the synchronizer control knob on the timing light or advance the meter indicator needle to read zero degrees.

3   Slacken the distributor body clamp, focus the timing light on the fixed ignition setting timing mark, and twist the distributor body either way until the rotating pulley or flywheel mark (notch) appears to align with the stationary datum mark. Finally, retighten the distributor body clamp.

With the advance meter set to read zero, the electronic delay control eliminates the time interval between the HT voltage pulse and the timing light flash so that the flash actually occurs at the point of ignition.

10.7.4   Centrifugal advance timing inspection
(Fig. 10.6(d))

Increase the engine speed from idling to a specified test speed, obtained from tuning data, and observe the timing mark moving out of alignment. Adjust the synchronizer control knob to align timing marks and observe the advance meter reading. The final meter reading comprises both the timing and the centrifugal advance. Therefore, subtract the basic timing from the total which gives the centrifugal advance, i.e. $30° - 12° = 18°$.

If the measured centrifugal advance is outside the allowable tolerance, check the distributor advance mechanism for freedom of movement and the advance return springs for correct action and stiffness.

Vacuum advance timing inspection
(Fig. 10.6(e))

Reconnect the vacuum advance pipes and increase the engine speed from idling to a specified speed (from tuning data) and observe the timing marks moving out of alignment. Adjust the synchronizer control knob to align timing marks and observe the advance meter reading. The final reading comprises the basic timing, centrifugal advance and vacuum advance. Therefore, the vacuum advance can be derived by subtracting both the basic timing and centrifugal advance from the total reading, i.e. $45° - (18° + 12°) = 15°$.

If the measured vacuum advance does not conform to specification, check the vacuum pipe or connectors for leakage, a seized base-plate, a faulty vacuum unit diaphragm or inadequate vacuum supply due to poor engine conditions.

10.8   Exhaust gas CO and HC analyser

Exhaust CO and HC content as a measure of combustion efficiency
The products of combustion are a means of determining how efficient the burning process in the combustion chamber has been. This, in turn, is greatly influenced by the accuracy of fuel metering, the thoroughness of mixing the air and

petrol, and the quantity and quality of the charge which is actually induced into the cylinder under different operating conditions.

The exhaust gas analyser may work either on the principle of the combustion Wheatstone-bridge burning method or the infra-red absorption method of measuring just the carbon monoxide (CO) or both the carbon monoxide and hydrocarbon (HC) products in the exhaust system. Carbon monoxide is measured in volume percentage (%) while hydrocarbon is measured as N-hexane in parts per million (PPM).

10.8.1 Principle of the carbon monoxide combustion Wheatstone bridge gas analyser
(Fig. 10.7)

This type of exhaust gas analyser continuously feeds a small calibrated proportion of the exhaust gas through a combustion chamber (gas sample cell) situated within the analyser (Fig. 10.7). A heating filament then ignites and burns any combustible gas from the continuous sample supply that has not previously been able to react with sufficient oxygen in the engine's cylinders to have been fully burnt. The increased heat produced by the combustion is then used as a measure of the amount of unburnt carbon monoxide passing out of the engine.

To meet this objective, an electrically driven pump draws gas from the exhaust system's tail pipe, through a water trap, a filter and calibrated jet. The gas mixture is then pumped through the sample cell containing an electrically heated platinum filament, which burns any remaining combustible gas in the exhaust gas sample.

Measurement of the heat from combustion is obtained by using a Wheatstone bridge electrical circuit that has two parallel conducting paths, one

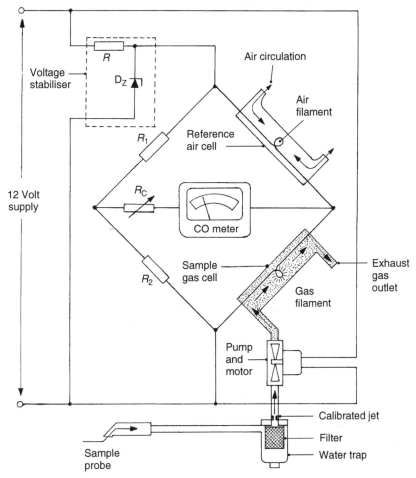

Fig. 10.7 Combustion type carbon-monoxide analyser

being via resistors, R_1 and R_2 while the other is through the air and gas heating filaments. When the resistance in the resistors and heating filaments are all equal, the current will be equally divided between the two paths so that no current will flow between the meter junctions.

Now, if the resistance of the gas filament is increased or decreased by the heat of combustion raising the filament temperature by more or less, depending on the amount of CO present in the exhaust sample, the current carried by the resistor path R_1 and R_2 will also vary in proportion. This will then create a difference in voltage across the meter between the two junctions which, in turn, deflects the CO needle indicator in proportion to the ohmic change in the gas cell filament. Now, the air filament, which is similar to the gas filament, is surrounded by air only, so that the analyser actually compares the ultimate temperature, and thereby the resistance of the gas filament in the path of the exhaust gases, with an air filament which remains at approximately its original warm-up temperature and resistance.

A voltage stabilizer is provided to compensate for any supply voltage variation that would upset the calibration of the meter. The calibration resistor R_c is used to adjust the zero setting of the meter when initially setting up the analyser.

10.8.2 Principle of the carbon monoxide infra-red exhaust gas analyser
(Fig. 10.8)

With this type of carbon monoxide (CO) analyser, use is made of the fact that pure gases and known compounds absorb infra-red radiation at certain wavelengths.

The analyser consists of a balanced pair of radiation emitter sources (Fig. 10.8). Infra-red radiation from the emitter source is produced by electrically heating a filament made from a high-resistance brittle element compound of powder sintered oxides or a rod of silicon carbide. At a temperature in the range of 1100–1800°C, depending on the filament material, the incandescent filaments emit radiation of the desired intensity covering the whole infra-red spectrum.

Two identical cylindrical tube cells are used, one for exhaust gas sampling while the other provides a reference. These tubes provide the twin radiation beam paths from the infra-red emitter source to the detector. The sample exhaust gas absorbs and so blocks some of the infra-red radiation energy being directed towards the detector, corresponding to the wavelength spectrum of the carbon monoxide (CO). The amount of energy absorbed will be directly proportional to the quantity of carbon monoxide contained within the exhaust gas.

The detector comprises two chambers filled with an infra-red absorbing gas (CO)—making it sensitive to radiation of a certain wavelength only—and separated by a thin metal diaphragm. Absorption of energy by the gas in the detector chamber occurs only at wavelengths which correspond to the spectrum of the gas. If the amount of radiating energy in the sample and in the reference beams coming from the emitted sources is equal, then the temperature and pressure in the two adjacent chambers are also equal and the diaphragm remains in the central position.

When exhaust gas is drawn through the sample cell its CO content absorbs infra-red energy from the beam, which otherwise would have been transmitted directly to the detector gas chamber. Therefore, the energy in the detector gas chamber, on the sample's side, diminishes so that a differential pressure will exist between the two chambers and the higher pressure in the reference gas chamber will deflect the diaphragm towards the sample chamber in proportion to the amount of energy the CO gas from the exhaust gas sample has absorbed; that is, relative to the percentage of CO contained within the exhaust gas sample. To eliminate long term effects, such as changes in ambient temperature, a small leak is provided between the two detector chambers, but the leak must not be large enough to reduce the pressure changes which occur.

Connected to the diaphragm is a mirror which reflects a light beam directed upon it to a photocell, the amount of light reflected will thus be in direct proportion to the difference in gas chamber expansion between the two compartment chambers and hence to the amount of infra-red radiation energy the sample CO has prevented from reaching the detector chamber. The electrical signal from the photocell is then amplified and finally passed on to the microammeter indicator, calibrated in percentage of carbon monoxide (CO). An interruption disc control is provided so that the emitter sources can be blocked for initial setting of the meter.

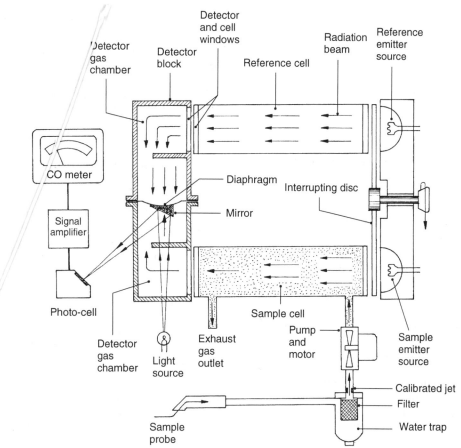

Fig. 10.8 Infra-red type carbon-monoxide analyser

10.8.3 What information do the CO and HC readings provide?

Carbon monoxide is a by-product of fuel that has only been partially burnt due to lack of oxygen—if an excess of oxygen had been available the products of combustion would have all been burnt to carbon dioxide (CO_2)—therefore, an over-rich mixture will produce a high CO content. Conversely, the amount of HC present in the exhaust gas represents petrol, which forms the part of the charge mixture in the combustion chamber that has not taken part in the burning process. Therefore, a high level of HC emission could be due to low compression or cylinder misfiring with the result that combustion is incomplete.

Samples of exhaust gas are picked up by a probe inserted in the tail pipe and are passed to the exhaust gas analyser by a flexible hose. The analyser, be it just a CO meter, or CO and HC meter, analyses the gas sample, and then converts the results to an electrical current signal feeding into the respective meter.

Relying on CO readings alone may not show up certain faults, but cross-checking with an HC reading can sometimes assist in identifying a defect. As an example, an engine's exhaust CO content is within the specification, but when a high HC meter reading is obtained, then a lean mixture could be suspected causing one or more of the cylinders to misfire.

10.8.4 Reasons for excessive emission

High CO readings may be due to the following:

a) rich mixture setting
b) incorrect idle speed
c) clogged air cleaner
d) defective cold start choke device

High HC readings may be due to the following:

a) incorrect ignition timing
b) ignition system misfiring due to malfunctioning
c) excessively rich or lean mixture setting
d) engine in a poor condition—such as low compression, leaky gaskets, defective valves, guides or lifters, excessive piston blow-by.

10.8.5 CO and HC exhaust content test procedure

1 Switch on the analyser and permit it to warm up for at least 5 minutes, then zero both meter needle indicators.
2 Allow the engine to attain its normal operating temperature and attach the sample probe well into the tail pipe.
3 Connect the tachometer to the engine's distributor.
4 Increase the engine speed to 2500 rev/min and maintain it for at least 30 seconds then take both CO and HC readings.
5 Reduce the speed to idling for 10–20 seconds to enable readings to stabilize and then observe CO and HC readings.
6 If necessary, adjust the mixture and idle speed screws to give the correct percentage CO exhaust content at a specified idle speed.
7 If emission readings still exceed the specifications, check the fuel supply system, ignition system and engine condition for the cause of the high readings.

If CO and HC emissions for the vehicle are unknown then use the following figures as a guide.

Emission controlled vehicles
CO and HC should not exceed 3% and 400 PPM respectively.

Older uncontrolled vehicles
CO and HC should not exceed 6% and 900 PPM respectively.

At present, most vehicle manufacturers quote maximum CO emission levels, at idling speed, of around 1.5%. Note that, frequently, manufacturers do not provide HC content data nor CO content above idle.

10.8.6 Accelerator pump operation

To overcome the inertia lag of fuel reaching the cylinder when the throttle is rapidly opened and there is a momentary loss of manifold vacuum, the accelerator pump sprays additional fuel into the choke tube. This should correspond with a rise in CO content.

10.8.7 Accelerator pump test procedure

1 From idling speed, quickly open and release the throttle. The engine should increase speed without hesitation and the CO meter should show an increase in CO content.
2 Check and correct the accelerator pump linkage if the CO reading decreases before increasing or does not increase at all.

10.9 Oscilloscope engine analyser

Engine diagnosis using the cathode-ray oscilloscope will be better understood if the basic principle of the instrument is initially explained.

10.9.1 Cathode ray oscilloscope: principle of operation
(Fig. 10.9)

The cathode-ray tube consists of a glass container exhausted of air; and about half its length is parallel and of small diameter, while the other half is conical with a slight convex-shaped end forming the screen (Fig. 10.9).

At the narrow tube end a heating filament (H) heats a short cylindrical tube known as the cathode (C). Partially enclosing the cathode is a grid (G) resembling a cup with a hole in its bottom, and this grid has a variable negative bias whose function is to control the electron emission from the cathode, thereby varying the brilliancy of the electrons projected onto the fluorescent screen (S). Further along the glass tube are anodes A_1, A_2 and A_3. These positive biased anodes are at a very high potential relative to the cathode so that the attracted electrons passing through the grid are accelerated at a very high rate and thus form a continuous electron beam

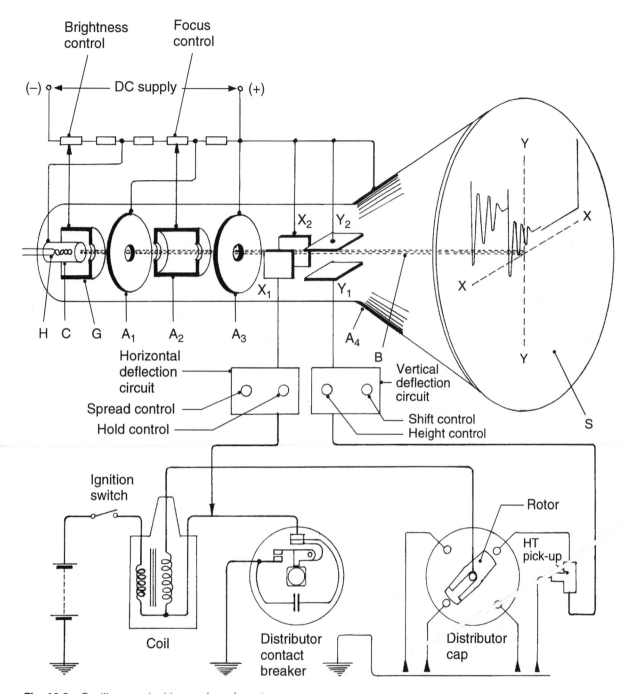

Fig. 10.9 Oscilloscope ignition analyser layout

(B). A fourth anode (A_4) consists of a coating of carbon inside the tube, its function being to help guide the electron beam towards the screen. Note that since the electron stream is completely surrounded by the anodes, the electrons will not tend to pull to one side but are projected directly onto the screen. To obtain a sharp focus the second anode (A_2) has a variable positive potential which can be independently adjusted as required. The glass screen is coated with a layer of phosphor,

491

which glows when bombarded by electrons, and this essential process is known as fluorescence. After the electrons strike the screen they are drawn away by the positive biased coating of carbon forming the fourth anode (A_4) around the mouth of the cathode-ray tube.

The electron beam is concentrated into a small spot on the screen and, for the spot to trace out a two-dimensional graph of vertical voltage and horizontal time base, a means must be provided to deflect the beam up and down and across so that a predicted pattern is produced. Two small deflection plates X_1 and X_2 are placed one on either side of the beam. If they are connected to a voltage supply, the negative biased plate will repel the beam while the positive plate will attract the electrons. Therefore, if an alternating potential is applied to these plates the beam will sweep from left to right and back again, thus producing a timed base-line across the screen. A second pair of plates situated above and below the beam, Y_1 and Y_2 respectively, is connected indirectly to either the primary or secondary circuit of the ignition system so that when a negative voltage pulse is received by the plates the beam is deflected upwards, or downwards when it is positive, and as a result a vertical trace is obtained.

The classical graphical pattern traced by the electron beam spot on the screen will thus be produced by the combination of the horizontal plates monitoring the charge in the primary circuit voltage when the contact breaker points open and close, and the vertical plates simultaneously signalling the magnitude of either the primary or secondary ignition system circuit voltage pulses.

10.9.2 Oscilloscope screen trace patterns and their interpretation

The three methods of presenting the ignition voltage process cycle of each cylinder on the oscilloscope screen, to make comparisons and diagnose faults, are the superimposed, parade and the Raster display patterns.

1 Superimposed primary and secondary patterns
These pictures show the ignition system performance of each cylinder stacked one behind the other so that if every cylinder is operating correctly the screen pattern will be seen as a single image, while a fault in one cylinder or more will result in a ghost image appearing in addition to

the normal trace. Superimposed patterns enable ignition faults to be readily identified at a glance.

2 Parade or display primary and secondary patterns
These patterns project the ignition system performance of individual cylinders side by side, as on parade, so as to display all the cylinder traces together on one screen. To enable all the trace patterns to appear on one screen, the width of each pattern has been distorted and squeezed in so that the firing line appears to predominate at the expense of both the horizontal spark line and oscillation phases, which are therefore not shown so clearly. Since the object of the Parade pattern is to compare each cylinder firing line, the lack of detail for the other parts of the trace is relatively unimportant.

3 Raster primary and secondary patterns
The so-called Raster display is similar to the superimposed one except that the cylinders are placed one above the other, rather than on top of each other. This enables the operator to assess which one (or more) of the cylinder pattern traces is faulty.

10.9.3 Primary pattern
(Fig. 10.10)

The primary pattern is obtained by connecting the oscilloscope leads between ground and the negative contact breaker lead terminal on the coil, as shown in Fig. 10.10.

To interpret the primary pattern it is best to consider the five phases (Fig. 10.10) of the trace as follows.

1 Capacitor oscillations (Fig. 10.10)
When the points open, the primary current collapses and its inertia charges the capacitor instead of jumping the contact point gap. Once the capacitor's charge has reached its peak, the direction of the current flow reverses so that the capacitor now discharges into the primary winding. There now follows a to and fro movement of current (known as an oscillation) between the series connected capacitor and the primary winding, with a diminishing voltage until all the energy initially stored in the capacitor is dissipated in heat within the winding. An effective capacitor and primary winding current should show an almost vertical sharp first oscillation without any indication of delay, which would be seen more as

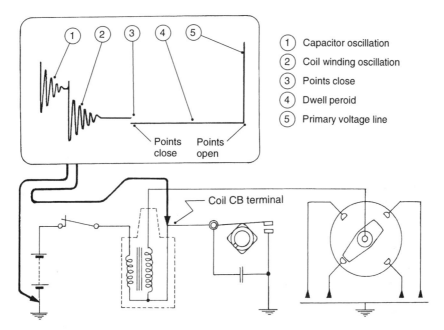

Fig. 10.10 Normal primary pattern

a sloping trace. The trace oscillation pattern should also appear as a conical or trumpet shape of diminishing voltage.

2 Coil winding oscillations (Fig. 10.10)
During the time the high-tension voltage impulse in the secondary circuit produces the spark between the plug electrodes, energy will be dissipated until eventually the arc is extinguished. At the instant this spark ceases there will still be several hundred volts in the secondary winding, which immediately discharge into the primary winding via the connection between the two. Again, the primary winding–capacitor series circuit will produce a charge and discharge sequence of oscillation of diminishing voltage amplitude. An ignition system in good order should show a pattern with four or more oscillations or peaks.

3 Points close (Fig. 10.10)
After the second phase of oscillations has dispersed to a horizontal straight line there is a break in the trace accompanied by a step. This point along the trace presents the moment when the contacts close, it should appear as a clean break with no arcing above the top line or below the bottom line.

4 Dwell period (Fig. 10.10)
The length of this horizontal 12 V base-line repre-

sents the dwell period (angle) during which the contacts are closed and current is permitted to energize the primary winding circuit. A dwell degree scale is usually situated beneath and parallel to this dwell base line to enable distributor dwell to be accurately measured and, if need be, adjusted.

5 Primary voltage line (Fig. 10.10)
At the end of the dwell base-line a vertical trace appears. This represents the induced primary circuit voltage which occurs at the instant the contact points open, due to the sudden collapse of current flow in the primary winding circuit. A peak potential of 200 to 300 V should be observed at this stage.

10.9.4 Secondary pattern
(Fig. 10.11)

The secondary pattern can be observed by connecting the oscilloscope leads between ground and the central coil high-tension terminal, as shown in Fig. 10.11.

A description of the secondary pattern is best understood by dividing the trace into five phases (Fig. 10.11).

1 Spark line (Fig. 10.11)
Once the air gap is ionized current flows between

493

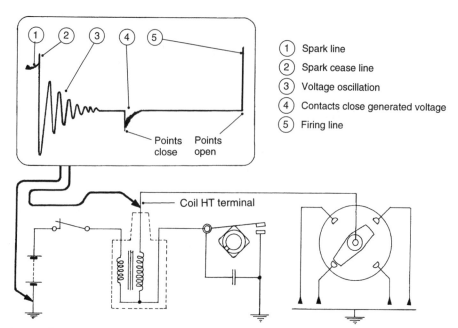

Fig. 10.11 Normal secondary pattern

the plug electrodes, thus creating a closed secondary circuit. Immediately the secondary circuit is loaded and the voltage drops to the horizontal spark line on the left-hand side, which represents the actual coil's voltage output during the time the plug fires.

2 Spark cease line (Fig. 10.11)
When the spark ceases, the secondary winding goes into open circuit, and instantly the coil voltage rises momentarily due to the electrical load having been removed. However, since there is insufficient energy in the secondary circuit to jump the air gap the voltage surge collapses.

3 Voltage oscillation (Fig. 10.11)
The collapse of current flow in the secondary winding when the spark ceases causes a shrinkage of the existing magnetic flux. Instantly the lines of force cut through both windings so that a fresh but reduced voltage surge is created in the coil. Consequently, the capacitor will be charged and discharged producing a voltage swing in the secondary winding that is represented by the decaying oscillations.

4 Contacts close generated voltage (Fig. 10.11)
Shortly after the secondary voltage oscillations have died away, the contact points close, causing current to flow into the primary winding circuit.

The relatively rapid rise of primary winding current creates a growth in the magnetic-flux in both windings. Subsequently, a voltage is generated due to the expanding lines of force, unlike that due to the collapse of current. This voltage therefore acts in the opposite sense, and it is represented by an oscillation beneath the horizontal base line. Note that there is no step down on the secondary trace, as in the case of the primary pattern, as it is no longer connected directly to the 12 V supply. Also, observe that the maximum voltage produced due to an expanding magnetic-field (growing current) is much smaller than that generated with a shrinking magnetic-flux (collapsing current).

5 Firing line (Fig. 10.11)
On the right-hand side of the screen is a vertical firing (voltage) line that represents the voltage rise necessary, at the instant the contacts open, for the plug electrodes to break down the air gap initially and thus establish an arc (spark).

10.9.5 Primary circuit faults
(Fig. 10.12(a–j))

Primary circuit faults can easily be identified by using the superimposed pattern trace.

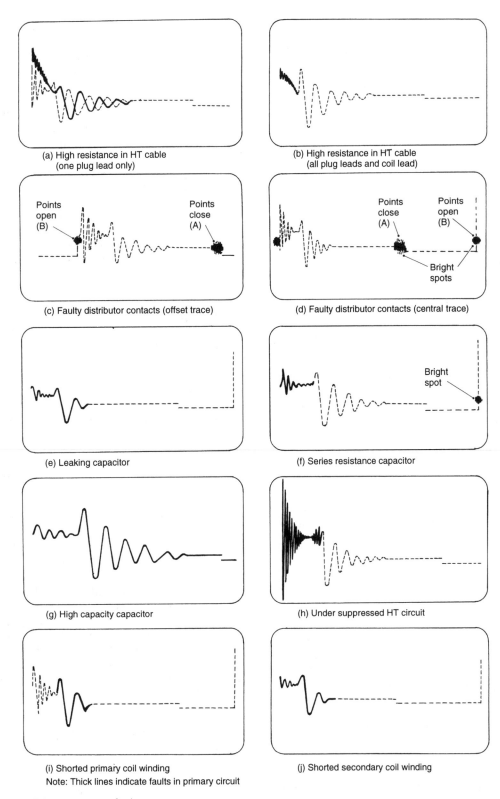

(a) High resistance in HT cable
(one plug lead only)

(b) High resistance in HT cable
(all plug leads and coil lead)

(c) Faulty distributor contacts (offset trace)

(d) Faulty distributor contacts (central trace)

(e) Leaking capacitor

(f) Series resistance capacitor

(g) High capacity capacitor

(h) Under suppressed HT circuit

(i) Shorted primary coil winding
Note: Thick lines indicate faults in primary circuit

(j) Shorted secondary coil winding

Fig. 10.12 Primary pattern faults

High resistance in HT cable (one plug lead only) (Fig. 10.12(a))

If the first oscillation trace droops with only a very small waveform, suspect a high resistance in one plug lead. This could be due to an over-suppressed plug lead or a sooted spark-plug and may cause engine misfiring. Inspect the condition and resistance of the plug lead and plug. Clean the plug or replace the plug lead as necessary.

High resistance in HT cable (all plug leads and coil lead) (Fig. 10.12(b))

If the superimposed primary pattern shows a single downward sloping capacitor trace with a very restricted waveform, suspect excessive resistance in the central HT lead or in all individual plug leads. This could be due to a small fracture, an over-suppressed central lead, or all the plug leads being over-suppressed. Measure the resistance of leads and replace where necessary.

Faulty distributor contacts (offset trace and central trace) (Fig. 10.12(c and d))

If a bright spot appears on the primary trace when the contacts close (point A) or contacts open (point B) suspect arcing at the contact breaker points caused by the following. For point (A), a weak contact return spring, a dry distributor pivot, contacts loosely riveted, ballast resistor faulty or incorrectly wired or burnt contact points. For point (B), voltage regulator setting too high, loose contact points, ballast resistor faulty or incorrectly wired, burnt contact points or faulty capacitor. Note, the offset trace is not available on all models of oscilloscope equipment.

Leaking capacitor (Fig. 10.12(e))

If both oscillation traces are reduced in height and length, suspect a leaking capacitor. Thus, when the contact points open, a proportion of the primary winding current will continue to flow between the capacitor leaking plates with the result that the coil voltage output is reduced. Suspect a capacitor-insulated terminal and, if necessary, replace the capacitor.

Series resistance capacitor (Fig. 10.12(f))

If the first oscillation trace is reduced in height while the second oscillation is unaffected and there is a bright spot on the voltage line, suspect a series resistance in the capacitor. The spot indicates points arcing when the points open. This consequently delays the collapse of magnetic-flux

in the windings and so prevents the coil from reaching its maximum possible output. Check the insulation and each connection of the capacitor and replace the capacitor if necessary.

High capacity capacitor (Fig. 10.12(g))

If the first oscillation phase is stretched out and reduced in height while the second oscillation has an extra large extended waveform, suspect the total capacity of the capacitor is too large for the contact breaker point–primary winding circuit. Causes of high capacitor capacitance may be due to the car radio suppressor capacitor being incorrectly connected to the contact breaker coil terminal instead of the ignition switch terminal. In some cars an FM filter may be connected intentionally to the contact breaker coil terminal which should not be confused with the normal ignition suppressor.

Under-suppressed HT circuit (Fig. 10.12(h))

If the first oscillation phase trace is larger than usual, suspect that the coil secondary circuit has no suppression. Note that some cars utilize a single-filament suppressed cable which gives no radio or television interference yet produces a large first capacity oscillation. Check the resistance of the central lead and the individual plug leads.

Shorted primary coil winding (Fig. 10.12(i))

If the second primary winding oscillation trace has a reduced number of wave cycles, suspect a shorted primary coil winding. As a result of shorted primary-winding turns, the output of the coil will be reduced. Measure the resistance of the primary winding and, if necessary, replace the coil.

Shorted secondary coil winding (Fig. 10.12(j))

If both first and second oscillation phases have reduced numbers of wave-cycles, suspect a shorted secondary coil winding. The consequence of shorted secondary-winding turns is to reduce the voltage step-up within the coil. Measure the resistance of the secondary winding and replace the coil if necessary.

10.9.6 Secondary circuit faults
(Fig. 10.13(a–j))

Secondary circuit ignition faults can be easily identified by using the superimposed pattern trace.

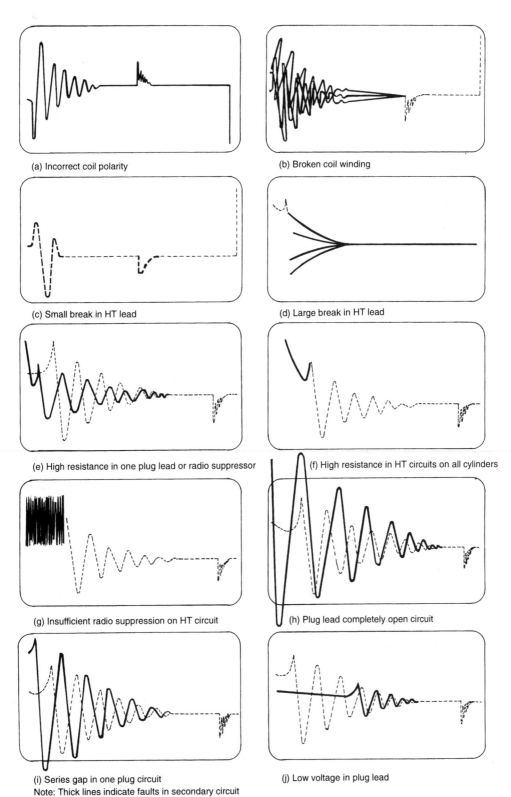

(a) Incorrect coil polarity

(b) Broken coil winding

(c) Small break in HT lead

(d) Large break in HT lead

(e) High resistance in one plug lead or radio suppressor

(f) High resistance in HT circuits on all cylinders

(g) Insufficient radio suppression on HT circuit

(h) Plug lead completely open circuit

(i) Series gap in one plug circuit
Note: Thick lines indicate faults in secondary circuit

(j) Low voltage in plug lead

Fig. 10.13   Secondary pattern faults

Incorrect coil polarity (Fig. 10.13(a))
If the coil ignition switch and distributor contact breaker terminals have been accidentally reversed, the secondary pattern will show an inverted trace. Correct the terminal connections and, if the fault still exists, check the type of ignition coil being used.

Broken coil winding (Fig. 10.13(b))
If the secondary spark line and voltage oscillation trace tend to swing vertically about the contact breaker closing point, a broken coil winding may be suspected. Check the resistance of the main HT lead and then substitute a known good ignition coil for the old one.

Small break in HT lead (Fig. 10.13(c))
If the two oscillation phases of the secondary trace appear with broken lines or are dotted, then a small break in the HT lead should be suspected. Repair or replace the cable as required.

Large break in HT lead (Fig. 10.13(d))
If oscillation phase waves are missing from the secondary trace, suspect a large break in the HT connection between the coil tower and the distributor. Check if the HT coil or distributor terminals are loose or corroded. Tighten and clean the terminal connections and, if required, change the cable.

High resistance in one plug lead or radio suppressor (Fig. 10.13(e))
If a spark line trace slopes excessively, suspect a high resistance in one of the spark-plug circuits, due possibly to a tracking distributor cap, sooted spark-plug, over-suppressed plug circuit or corroded HT terminals. Check the distributor cap, spark-plug and lead, test the plug lead for high resistance using an ohmmeter. Clean or replace the components as necessary.

High resistance in HT circuit on all cylinders (Fig. 10.13(f))
If a spark line trace droops excessively, suspect a high resistance in the central HT lead or over-suppressed individual plug leads. Over suppression will occur when radio suppressors are inserted in resistance-type plug leads. This may lead to engine misfire underload. The sloping spark line may result from an excessive voltage drop which occurs as the current flows between spark-plug electrodes. Inspect the resistance of the main

HT lead and each individual plug lead with an ohmmeter.

Insufficient radio suppression on HT circuit or cracked plug insulator (Fig. 10.13(g))
If a spark line appears as a thick fluctuating but ragged trace, suspect insufficient radio suppression, while a higher than normal but similar spark line indicates a cracked spark-plug insulator. The lack of suppression may be due to the replacement of a suppressed carbon-impregnated textile core cable with low-resistance copper-wire leads. When copper leads are used, fit spark-plug suppressors to each plug lead to reduce the radio and television interference. Remove and check each spark-plug, and replace the faulty plug if necessary.

Plug lead completely open circuit (Fig. 10.13(h))
If there is no spark line and the oscillation waves are abnormally large suspect an open circuit somewhere in the plug lead or spark-plug. Inspect the plug lead and plug, and replace as necessary.

Series gap in one plug circuit (Fig. 10.13(i))
If there is a high short spark line, suspect a break in the plug lead, the connection in the distributor cap or a faulty plug suppressor. Remove and inspect the plug, check the plug lead for breaks in insulation.

Low voltage in the plug lead (Fig. 10.13(j))
If a spark line of the secondary trace is prolonged and low, with only a few oscillations of small amplitude, suspect a low voltage in the plug lead due possibly to a shorted plug lead, closed plug gap or oiled plug. Remove and inspect the plug lead and plug, clean and test the plug, refit or replace as necessary.

10.9.7 High-tension (kV) parade pattern
(Fig. 10.14(a–e))

Voltage output from the coil, distributor cap and spark-plugs can best be observed with the secondary circuit parade pattern.

Plug kV check (Fig. 10.14(a))
With the engine running at a fast idle the HT voltage parade voltage tower should show a firing voltage between 8–14 kV with a maximum 3 kV variation between plugs. If all plugs read over

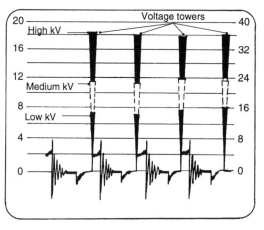

(a) Plug kV check

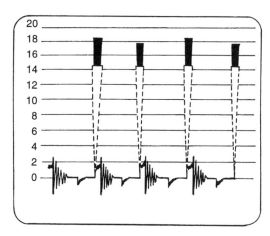

(b) Acceleration check

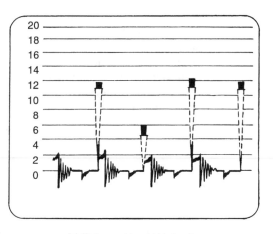

(c) Rotor and lead kV check

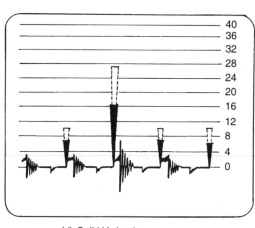

(d) Coil kV check

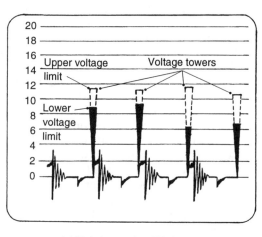

(e) Unbalance plug kV check

Fig. 10.14 High tension (kV) parade pattern

14 kV, suspect a weak mixture, worn or over-gapped spark-plugs, a wide rotor gap, a rounded central plug electrode, or a series break in the plug circuit. If all plugs read under 8 kV suspect a rich mixture, narrowed plug gaps, incorrect ignition timing or fouled plugs.

Acceleration load check (Fig. 10.14(b))
With the engine idling, snap the throttle open and observe the HT voltage parade. Voltage towers should not rise more than 3 kV or exceed 16 kV. If the firing voltage is excessive, suspect over-gapped or rounded plug electrodes, the car-burettor accelerator pump or an inoperative damper.

Rotor and lead kV check (Fig. 10.14(c))
With the engine idling, earth each plug lead in turn and observe the firing voltage tower readings, which should fall below 5 kV. If the firing voltage towers do not fall below 5 kV suspect a worn rotor arm, worn or corroded distributor cap segmental electrodes, a faulty central carbon brush in the cap, faulty plug leads or a faulty main HT lead, or the wrong distributor cap has been fitted.

Coil kV check (Fig. 10.14(d))
With the engine idling, remove each plug lead in turn and hold away from earth while observing the firing voltage tower readings, which should increase to at least 24 kV depending upon the type of coil fitted. If firing voltage towers do not rise above the recommended kV (i.e. 24 kV) suspect the coil tower or distributor cap tracking, an internal short-circuit in the coil, an incorrect-type coil has been fitted, the ballast resistor is fitted to a 12 V coil or check the coil (SW) terminal voltage, dwell angle, distributor contact-breaker voltage drop and capacitor.

Unbalanced plug kV check (Fig. 10.14(e))
With the engine running at a fast idle, the HT parade voltage tower firing voltage should not vary by more than 3 kV between cylinders. If individual cylinder firing voltages do exceed 3 kV between plugs, suspect different mixture strength in each cylinder, low cylinder compression or faulty plug leads.

10.10   Distributor dwell-angle

(Fig. 10.15(a–f))

The distributor dwell-angle is the angular movement (in degrees) through which the distributor cam rotates while the contacts remain closed. The time available for the primary current to magnetize the iron-core is dependent upon the dwell duration. At low engine speeds there is adequate time for growth of the magnetic-field within the primary-winding, but as the dwell-angle is constant, the dwell period reduces directly with engine speed until eventually there is insufficient time for the primary-winding to be energized by the flow of current, with the result that coil output voltage falls.

If the contact points are set very wide, they open earlier, advancing the ignition timing, and close later, thereby reducing the angle of dwell (Fig. 10.15(b)). Conversely, a narrow contact-breaker gap-setting delays the opening, so retarding the ignition timing, while point closure occurs earlier. This condition therefore increases the angle of dwell but may increase contact breaker arcing (Fig. 10.15(b)).

10.10.1   Dwell-angle adjustment procedure
(Fig. 10.15(a))

1 Start the engine and idle between 800–1000 rev/min. Adjust the primary pattern using the horizontal spreader so that the first oscillation wave just touches the boundary line on the left, and the firing-line aligns with the corresponding boundary line on the right (see Fig. 10.15(a)) and observe the contact closure point which indicates the dwell angle.
2 Remove the distributor cap and rotor arm. Crank the engine using the starter-motor, with the ignition switch on, and observe the contact closure point (dwell angle) on the oscilloscope or dwell meter (Fig. 10.15(f)) taking into account the difference in dwell-angle between running and cranking speeds.
3 Slacken the contact breaker locking screw and adjust the contact gap while the engine is cranked by the starter-motor, until the correct dwell-angle is obtained. Retighten the contact locking screw and recheck the dwell-angle.

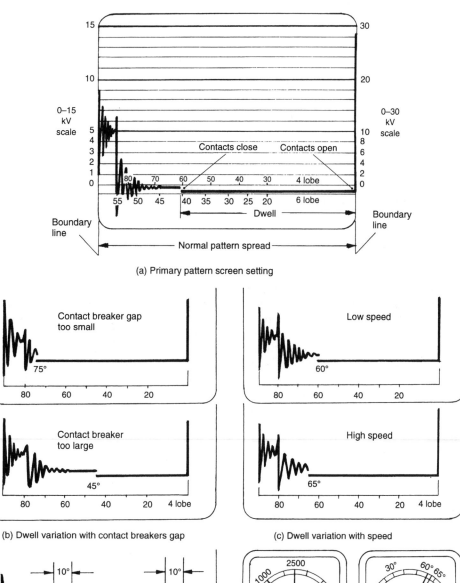

(a) Primary pattern screen setting

(b) Dwell variation with contact breakers gap

Contact breaker gap too small

75°

Contact breaker too large

45°

(c) Dwell variation with speed

Low speed

60°

High speed

65°

(d) Dwell overlap

10°   10°

65° 55°

65° 55°

Alternative display showing overlap

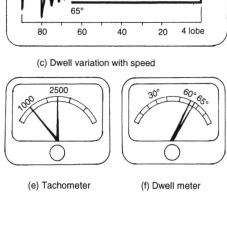

(e) Tachometer

(f) Dwell meter

Fig. 10.15   Distributor dwell angle

501

Replace the rotor and distributor cap, start the engine and recheck the dwell-angle.

Note that any engine speed variation necessitates re-adjustment of the primary pattern horizontal hold, because the sweep occurs in a definite time period which is synchronized with engine speed. A change of speed therefore alters the relationship between the oscilloscope and the engine's speed so that the sweep time must be re-synchronized.

With some types of distributor the dwell-angle can be adjusted externally. Therefore, the dwell can be altered while the engine is idling.

10.10.2 Dwell variation with speed
(Fig. 10.15(c))

1 Switch the tachometer to the low-speed range and run the engine at a fast idle (1000 rev/min). Adjust the pattern width to align with both boundary lines and observe the dwell-angle.
2 Switch the tachometer to the high-speed range and increase the engine speed to 2000 rev/min. Re-adjust the pattern width to fit between the boundary lines and read off the dwell-angle.
3 Compare the 1000 and 2000 rev/min (Fig. 10.15(e)) and the dwell-angle readings on the oscilloscope or dwell meter (Fig. 10.15(f)). These should not vary more than three degrees. A large dwell-angle variation may be due to a loose distributor base plate, a worn contact-breaker base-plate or worn distributor and bushes. Recondition or replace the distributor as necessary. Note that dwell-angle variations of 10° or more may be intentionally designed to occur with certain electronic ignition systems, which have constant current amplifiers.

10.10.3 Dwell overlap
(Fig. 10.15(d))

The extent of dwell overlap on the primary pattern between the step of the upper and lower horizontal trace-ends, at the points closed position, is a measure of individual cam-lobe variation and of individual cylinder ignition timing differences. To measure if there is any dwell overlap the oscilloscope must be switched to the superimposed all-cylinder primary pattern.

Run the engine at a fast idle (1000 rev/min) and adjust the width of the pattern using the horizontal spreader to fill exactly the space between the boundary lines and observe the amount of dwell overlap. This should not exceed 3°. Excessive dwell overlap may be caused by a worn camshaft drive, a bent distributor shaft, an eccentric distributor cam or an unevenly worn cam.

11

Diesel In-line Fuel Injection Pump Systems

11.1 Introduction to in-line fuel injection pumps

(Fig. 11.1)

The in-line fuel injection pump has a pumping element for every cylinder. These pumping elements consist of plungers, which reciprocates in the barrels, the top of each barrel being enclosed by its own delivery valve assembly. Pumping elements are mounted vertically in a straight line, side by side in an aluminium-alloy mono housing or a steel upper body and aluminium alloy low cam housing. The lower half of the pump housing supports and encloses a horizontally positioned cam shaft, which has as many cam lobes as there are pumping elements. The cam lobe profile converts the angular movement of the camshaft into a linear lift and fall plunger motion by way of the roller cam follower and plunger return spring. The individual lobe profiles are dispersed around the camshaft so that a timed upward plunger movement is obtained at angular intervals which correspond to the engine's firing order sequence.

11.1.1 In-line injection pump filling and pumping cycle of operation
(Fig. 11.2(a–c))

1 Pressure chamber filling phase (Fig. 11.2(a))
Towards the end of the plunger downward stroke both the feed and feed/spill ports are uncovered. Fuel surrounding the upper portion of the barrel at lift pump pressure will immediately enter the pressure chamber through the ports to fill up the

space between the plunger and the underside of the delivery body.

2 Commencement of fuel delivery (injection) phase (Fig. 11.2(b))
After the plunger has reached its outer dead centre and commences its upward stroke, it begins to sweep over the port aperture. Initially it pushes back a small amount of fuel through the ports until the top edge of the plunger cuts off the space above the plunger from the fuel gallery and ports. The continuing plunger upward movement pressurizes the trapped fuel until the delivery valve is forced open. Any further plunger lift relays the increase in pressure through the residual fuel in the pipeline to the injector nozzle, needle lift, fuel gallery. The pressure build-up underneath the needle pressure shoulder eventually lifts the needle off its seat. This causes fuel to be discharged from the nozzle sac into the combustion chamber via the annular-formed spray hole or through several small holes distributed around the nozzle nose.

3 End of fuel delivery (injection ceases) phase (Fig. 11.2(c))
The continuing plunger rise forces fuel through the delivery valve until the edge of the plunger helix uncovers the feed/spill port. Instantly, the fuel pressure in the barrel collapses as fuel escapes down the vertical slot and partly around the circumferential groove and out through the feed/spill port. Simultaneously, as the fuel pressure decreases, the delivery valve spring load exceeds the delivery valve opening pressure, causing the delivery valve to snap shut. The

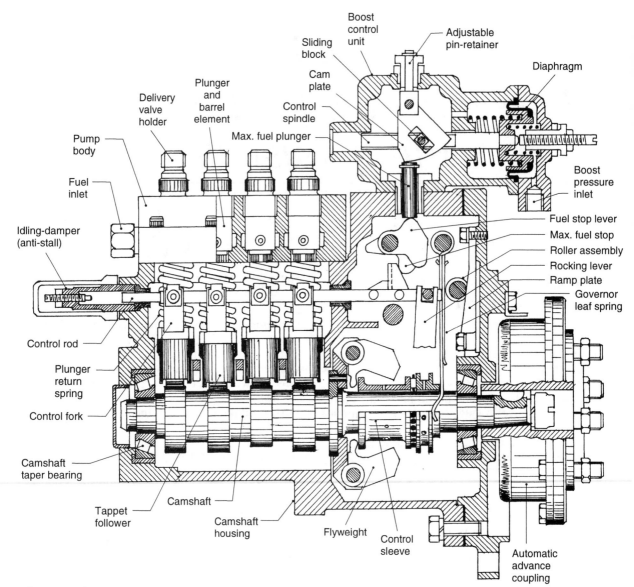

Fig. 11.1 CAV Minimec injection pump sectioned side view

sudden drop in fuel pressure in the barrel causes a corresponding drop in pipeline pressure so that the injector needle valve also snaps down on its seat, it therefore prevents any more fuel discharging from the injector nozzle spray holes.

11.1.2 Fuel delivery output control
(Fig. 11.3(a–c))

The plunger stroke is controlled by the cam lobe profile and is always constant, but the part of it which actually pumps is variable. The point on the upward travel of the plunger at which the spill occurs can be altered by twisting the plunger relative to the barrel. This enables the position of the plunger helical spill groove to be varied relative to the fixed barrel spill port, it thereby increases or decreases the effective pumping stroke of the plunger.

To reduce delivery, say to half load (Fig. 11.3(b)), the helical groove must be made to align with the spill port earlier in the plunger upstroke by rotating the plunger θ_h degrees. In this posi-

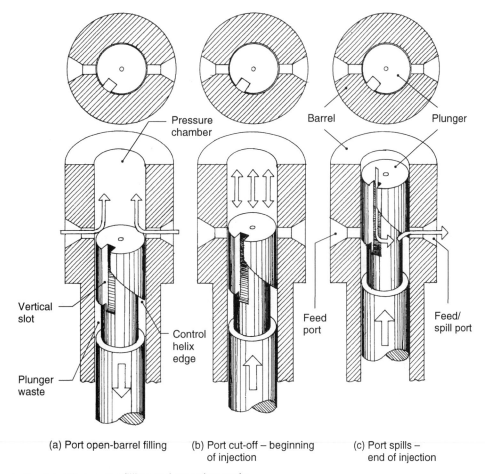

Pressure chamber

Barrel

Plunger

Vertical slot

Control helix edge

Feed port

Feed/ spill port

Plunger waste

(a) Port open-barrel filling

(b) Port cut-off – beginning of injection

(c) Port spills – end of injection

Fig. 11.2 In-line injection pump filling and pumping cycle

tion, the spill port is uncovered about mid-way along the plunger helix edge. The plunger lift before the fuel escapes from the spill port is small, and therefore the amount of fuel injected will also be small.

To increase delivery, say to full load (Fig. 11.3(c)), the helix groove must be aligned with the spill port later in the plunger stroke by rotating the plunger θ_f degrees so that the effective plunger stroke will be lengthened before spill occurs.

The plunger can be positioned in the no-delivery shut-off position (Fig. 11.3(a)) by rotating it until the helical groove uncovers the spill port in the barrel for the entire stroke, so that at no time can the fuel be compressed and displaced through the delivery valve.

11.1.3 Injection advance with increasing engine load

For some indirect pre-combustion chambers, the temperature of the compressed air in the chamber may vary considerably—from something like 400°C to 750°C from low to full engine load. It is therefore desirable to match the injection timing to the change in engine load, which is a measure of chamber air charge temperature. Thus, under light engine load, the peak chamber temperature is reached later in the compression stroke and therefore combustion is optimized if injection is set back slightly. Conversely, with increased engine load, more heat energy is released so that the peak temperature will be higher and it will occur earlier in the compression stroke. Accordingly, injection timing can be brought forward to improve the effectiveness of the combustion process.

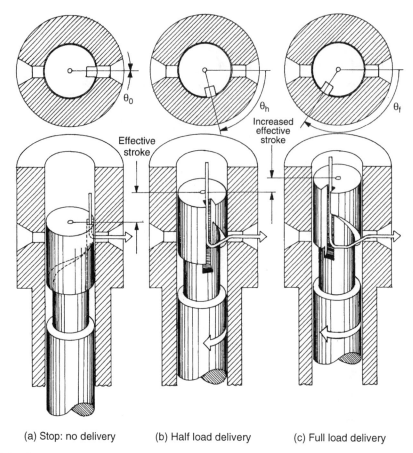

(a) Stop: no delivery   (b) Half load delivery   (c) Full load delivery

Fig. 11.3 In-line injection pump output delivery control

Load advance upper helix edge plunger action

(Fig. 11.4(a and b))

Injection advance with increasing engine load can be obtained by machining an upper helix above and over the lower spill control helix (Fig. 11.4(a)).

a) **Low-load plunger action** (Fig. 11.4(a))
When the plunger is rotated to the low-load fuel delivery position, the plunger's top crown edge on its upstroke rises beyond the spill port until the upper helix edge cuts off the spill port. Fuel above the plunger is now trapped, and thus fuel injection commences. Injection continues until the lower helix edge uncovers the spill port and injection ceases. The active working area of the plunger is the vertical distance between the cut-off edge of the upper helix and the spill edge of the lower helix.

b) **High-load plunger action** (Figs 11.4(b) and 11.5(a and b))
Rotating the plunger to a higher load fuel delivery position aligns the spill port with the upper helix, closer to the crown of the plunger. Thus, on the plunger's upward stroke, the upper helix cuts off the spill port earlier. Injection immediately commences and displaces fuel above the plunger until the low helix edge exposes the spill port. Instantly, injection ceases as the fuel pressure in the barrel collapses. The nearer the upper helix edge is to the crown of the plunger at the point of cut-off, the earlier injection begins. Conversely, the further away the plunger crown is from the upper helix edge when it cuts off the spill port, the later will be the point of injection.

If the plunger has no upper helix then the beginning of injection will be constant for any

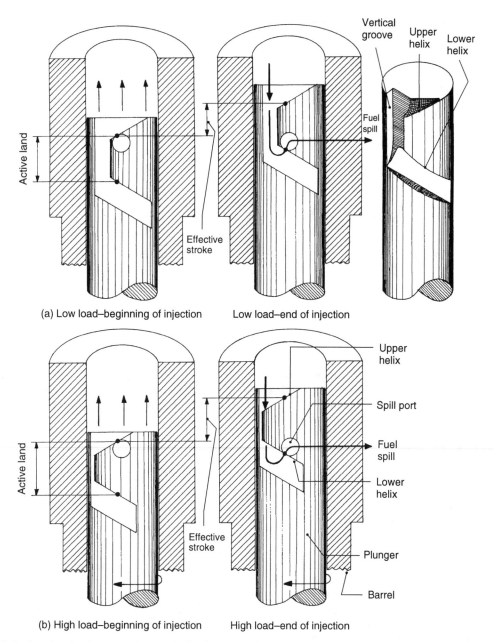

(a) Low load–beginning of injection   Low load–end of injection

(b) High load–beginning of injection   High load–end of injection

Fig. 11.4 Load reduction retard upper helix plunger action

fuel delivery output position, whereas the end of injection occurs later as the plunger is rotated towards the full-load fuel delivery (Fig. 11.5(a)).

With an upper helix the beginning of injection occurs earlier as the plunger is rotated towards the full-load fuel delivery position. However, the end of injection takes place later as the distance between the upper and lower helix portion, which aligns with the spill port, increases (Fig. 11.5(b)).

11.1.4 Cold start retarding groove
(Fig. 11.6(a and b))

To improve diesel engine startability, some engines respond better if the injection timing is

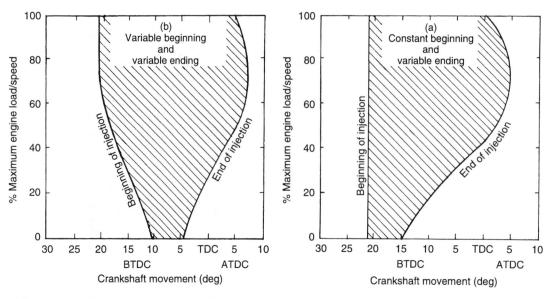

Fig. 11.5 Relationship between engine load/speed and the beginning and ending of injection for single and double helix plunger

retarded relative to the starting point of injection. One method of retarding the cold starting point of ignition is by grinding a starting groove on top of the plunger on one side. Thus, when starting the engine from cold, the control-rod opening is moved to the excess fuel position. This causes the plunger to rotate so that the starting grooves align with the spill ports (Fig. 11.6(a)). When the plunger moves upwards, injection does not occur when the plunger's top edge reaches the same height as the spill port. Instead, injection is delayed until the plunger's start groove cuts off the spill port. Further upward plunger lift displaces the fuel into the pipeline. It therefore causes a corresponding discharge from the injector nozzle holes. Once the engine fires and its speed increases sufficiently, the governor moves the control rod from the excess fuel opening to the normal operating position, which is determined by the accelerator pedal setting. Accordingly, the plungers rotate back to the normal position where part of the helix edge registers with the spill ports (Fig. 11.6(b)). Injection now commences when the top edge of each plunger closes its respective spill port.

The start groove retards the beginning of injection by something like 5° to 10° crankshaft movement.

An alternative method is to cut a slot across the crown of the plunger at a definite angle to the helix so that when the slot aligns with the spill port in the excess fuel position, no injection takes place until the plunger has moved up sufficiently for the slot to be above the top edge of the spill port (see lower plunger on right hand side).

11.1.5 Conversion of angular plunger motion to linear control-rod movement

With rack and sector gear control
(Fig. 11.7)
This device consists of a slotted control sleeve and a hollow-toothed sector split gear which is clamped around the upper end of the control sleeve (Fig. 11.7). The slotted control sleeve slips over the barrel which protrudes downwards into the pump plunger and spring inspection chamber. Each toothed sector gear meshes with the teeth cut along the control-rod, which is located longitudinally at the rear of the sector gears. The bottom of the sleeve is slotted to accommodate the plunger control lugs, which form part of the plunger at its lower end.

Rotation of the camshaft produces a vertical up and down movement of the plunger, the plunger lugs being forced to slide in the sleeve slots. At the same time, the control sleeve is permitted to rotate when the meshing control rack is pushed to and fro to increase or decrease the fuel delivery to the injectors.

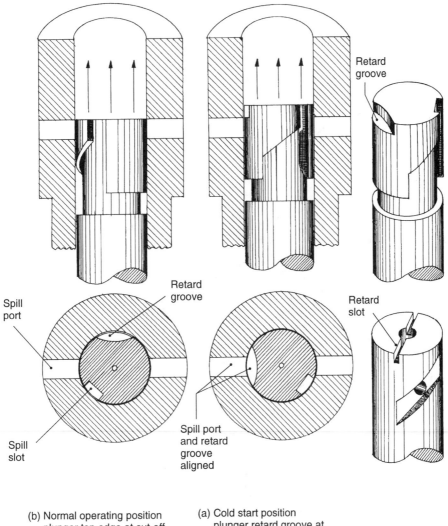

Fig. 11.6 Cold start retard groove plunger action

With sleeve arm and peg ball control
(Fig. 11.8)
A variation in fuel delivery is obtained through a control sleeve which is slipped over the lower part of the barrel. The interior of the sleeve is semi-circular with parallel flats machined in its lower end to accommodate and guide the plunger control lugs (Fig. 11.8). A control arm with either a peg (Fig. 11.8) or ball (Fig. 11.22) is pressed over the top end of the sleeve. Synchronization of each plunger helix edge angular position relative to the corresponding barrel spill port is achieved by slotting each control arm peg (Fig. 11.8) into a groove machined in the top face of each adjacent control-rod clamp. Thus, when the control-rod is moved to and fro, all the control-arms and sleeves swivel about their respective barrel an equal amount.

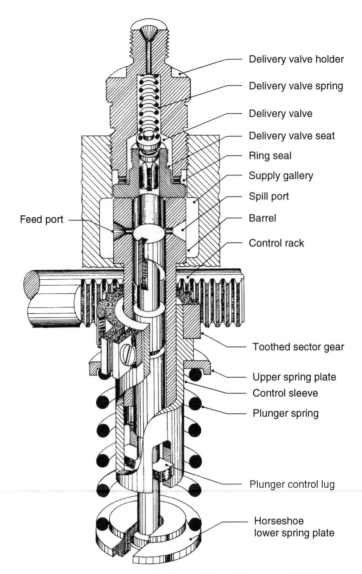

Delivery valve holder

Delivery valve spring

Delivery valve

Delivery valve seat

Ring seal

Supply gallery

Spill port

Feed port

Barrel

Control rack

Toothed sector gear

Upper spring plate

Control sleeve

Plunger spring

Plunger control lug

Horseshoe
lower spring plate

Fig. 11.7   Injection pump pumping element assembly (Bosch Type PE..A and CAV Type AA)

11.1.6   Relationship between control-rod opening, fuel delivery, pump speed and plunger–barrel wear
(Figs 11.9 and 11.10)

Fuel output from an in-line injection pump with a constant plunger stroke varies directly with the control-rod opening (CRO) (measured in mm) and upon the diameter of the pumping plunger (Fig. 11.9).

The plunger and barrel have a lapped working clearance of between 3 to 5 μm. They are therefore a matched pair and cannot be interchanged.

Wear, which may take the form of vertical abrasive scores running between the crown of the plunger and the helix control edge, reduces the discharge from the pumping elements to the high pressure pipes for a given control-rod opening. Fuel delivery with worn plungers is much more prominent at low speed than at high speed because the leakage time factor is relatively greater (Fig. 11.10). Therefore, at low speed, a much larger proportion of the theoretical discharge has time to escape down the plunger/barrel walls. Consequently, at high pumping speed, the amount of fuel which is able to leak between the

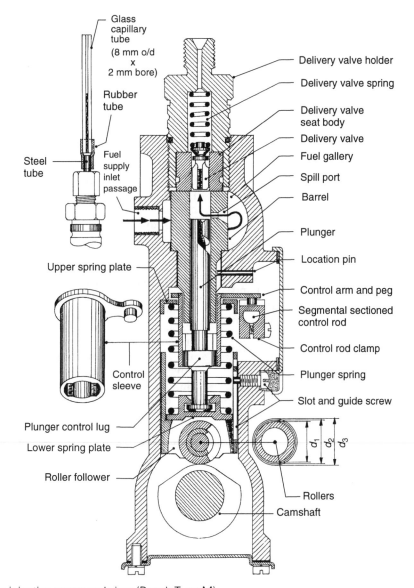

Fig. 11.8   In-line injection pump end view (Bosch Type M)

plunger and barrel walls in the much reduced time is quite small. Wear is therefore shown up when the pump is being calibrated, as it becomes more difficult for the pumping elements to discharge the specified output for a given speed and control-rod opening.

11.2   Delivery valve characteristics

11.2.1   The need for an anti-dribble delivery valve
(Figs 11.11, 11.12, 11.13 and 11.14)

The delivery valve functions as a non-return valve which permits the injection pump plunger's upward travel to displace fuel from the barrel into the high pressure pipeline and out of the injector

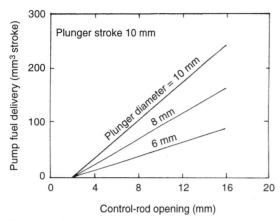

Fig. 11.9 Effect of control-rod-opening on fuel delivery output

nozzle holes. Fuel continues to enter the pipeline until the plunger helix uncovers the spill port and, at this point, the pressure in the barrel's pressure chamber collapses. The sudden reduction in delivery pressure causes the delivery valve to close so that fuel in the pipeline is prevented from returning to the plunger and barrel. A residual pipeline pressure will be established due to the existing fuel trapped in the pipe. Consequently, there will be very little contraction of the fuel when the next injection cycle takes place. However, the sudden closing of the delivery-valve does not produce an instantaneous reduction in the pipeline pressure, which is holding open the injector needle valve. Consequently, a rapid closing of the injector needle-valve will be prevented. Thus, the needle-valve will hover in a partially open position for a very brief time interval, and therefore fuel continues to discharge from the

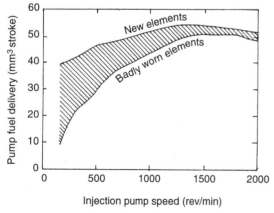

Fig. 11.10 Effect of plunger wear on fuel delivery output

nozzle holes in the form of a dribble until the needle-valve spring downward load exceeds the hydraulic upthrust.

The usual method of making the needle valve snap down onto its seat immediately after the delivery valve closes is to expand the pipeline internal passageway volume between the delivery valve and the injector needle valve fuel gallery at the same instant. The sudden enlargement of the trapped fuel in the pipe passageway produces a corresponding rapid reduction in pipeline pressure. Accordingly, the needle valve spring without the opposition of the hydraulic upthrust abruptly closes the needle valve, thus eliminating dribble at the nozzle holes.

11.2.2 Fluted delivery valve with unloading collar
(Fig. 11.11 and 11.15)

Instantaneous pressure reduction in the high-pressure pipeline is obtained by machining an unloading collar or piston between the conical valve and its guide flutes. Consequently, when plunger spill occurs, the collapse of barrel pressure enables the delivery valve spring to push down the valve until the lower edge of the unloading collar separates the fuel pipeline above the delivery valve from the barrel chamber underneath the valve. The remainder of the downward movement of the delivery valve sweeps the unloading collar further down its guide bore. The effect of the unloading collar swept volume enlarging the pipeline passageway is to produce an instantaneous drop in the fuel pressure. As a result, the needle valve spring downward force considerably exceeds the opposing hydraulic upthrust so that the needle valve is virtually free to snap onto its seat. A clean fuel cut-off is therefore achieved, which is essential if a relatively dry nozzle nose is sought at the end of each injection discharge. The greater the unloading swept volume relative to the total pipeline passageway volume, the more rapid will be the reduction in pipeline pressure and the more positive will be the instant of valve closure.

To maximize the effectiveness of the unloading collar, the pipeline passageway length and bore diameter, the delivery valve holder internal space, and the injector fuel gallery should be kept to a minimum for a given unloading swept volume. In other words, the ratio of total passageway volume between the delivery valve and injec-

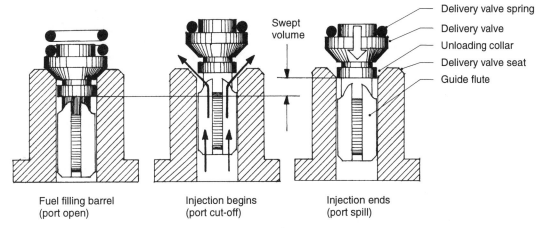

Fuel filling barrel
(port open)

Injection begins
(port cut-off)

Injection ends
(port spill)

Delivery valve spring
Delivery valve
Unloading collar
Delivery valve seat
Guide flute

Swept volume

Fig. 11.11   Fluted delivery valve with unloading collar

tor needle to the unloading swept volume is critical if the rate of fuel pressure drop in the pipeline is to be rapid enough to instigate a sharp closure of the needle valve.

The fuel delivery versus injection pump speed characteristics are shown in Fig. 11.15. The reference curve being the 'without bypass passage'.

11.2.3   Cylindrical delivery valve with unloading action
(Fig. 11.12 and 11.15)

This valve has a conical seat and a cylindrical guide which is permitted to slide in the bore of the valve seat body. On the upstroke, the plunger pressurizes the fuel causing it to push up the delivery valve against the opposing spring load

until the cross-holes project above the valve seat. Fuel now transfers from the barrel pressure chamber to the high pressure pipeline and to the underside of the injector needle valve fuel gallery. With rising pipeline pressure the needle valve lifts off its seat thereby permitting fuel to be discharged from the nozzle spray holes. Towards the end of the plunger travel the plunger helix edge uncovers the spill port, thus causing the pressure in the barrel pressure chamber to collapse. Instantly, the delivery valve spring forces back the delivery valve, thus closing the cross-holes and cutting off the high pressure pipe from the barrel. Further downward movement of the delivery valve expands the pipeline passageway volume up to the point where the valve seats.

Accordingly, the rapid enlargement of the pipeline volume creates a corresponding rapid

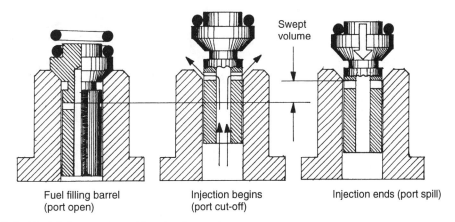

Swept volume

Fuel filling barrel
(port open)

Injection begins
(port cut-off)

Injection ends (port spill)

Fig. 11.12   Cylindrical delivery valve with unloading action

513

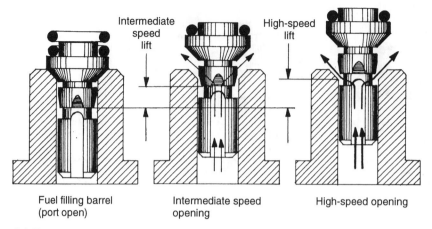

Fig. 11.13 Fluted delivery valve with by-pass chamfers (variable retraction volume)

reduction in the hydraulic pressure beneath the needle valve with the result that the needle valve snaps closed.

The fuel delivery versus injection pump speed characteristics are shown in Fig. 11.15 the reference curve being the 'without bypass passage', which is identical to the fluted delivery valve with unloading collar.

11.2.4 Fluted delivery valve with bypass chamfer (variable retraction volume)
(Fig. 11.13 and 11.15)

This valve is similar to the standard fluted delivery valve but four chamfers are ground around the lower end of the unloading collar. During low speed operation the pressure generated in the barrel above the plunger on its upstroke will lift

the valve until the chamfered passageways are just large enough to pass the fuel to the high-pressure pipeline. Therefore, when the plunger helix edge uncovers the spill port, the drop in pressure under the delivery valve makes the unloading collar sweep downwards. However, this movement, to the point where the valve seats, is relatively small under these conditions.

With increased injection-pump speed the pressure above the plunger builds up so that the delivery valve is progressively lifted higher and higher during each pumping cycle to cope with the increased fuel flow rate. Subsequently, the unloading volume produced by the down sweep of the unloading collar becomes larger as the delivery valve return-stroke lengthens.

The fuel delivery versus injection pump speed characteristics are shown in Fig. 11.15, the reference curve being the 'with bypass chamfer'.

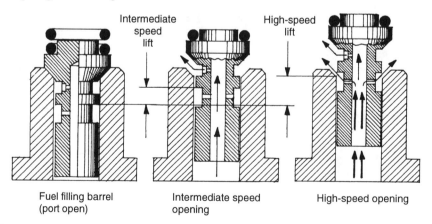

Fig. 11.14 Cylindrical delivery valve by-pass hole

514

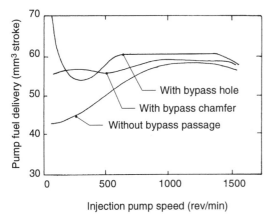

Fig. 11.15 Effect of delivery valve design on fuel delivery output

11.2.5 Cylindrical delivery valve with bypass hole
(Fig. 11.14 and 11.15)

At very low injection pump speeds the bypass hole size is able to transfer the displaced fuel from the barrel to the pipeline so that the valve lifts just sufficiently to expose the upper groove, thus permitting fuel to pass to the high-pressure pipeline. The unloading volume created by the return sweep of the cylindrical guide portion of the valve will therefore be relatively small. However, with rising pump speed the bypass passage cannot cope with the increased fuel delivery so that the delivery valve will be pushed up further until the unloading collar clears the valve seat bore. Thus, the extended upward delivery valve movement produces a corresponding return sweep of the unloading collar inside its guide bore when port spill occurs. As a result, full unloading of the pipeline pressure takes place.

The fuel delivery versus injection pump speed characteristics are shown in Fig. 11.15, the reference curve being the 'with bypass hole'.

11.2.6 Engine volumetric efficiency and fuel delivery characteristics
(Fig. 11.16)

The volumetric efficiency curve for a naturally aspirated diesel engine is initially low due to the late closing of the inlet valve. It then rises to a maximum roughly a third of the way along its speed range, and then commences to decrease as breathing becomes more difficult (Fig. 11.16).

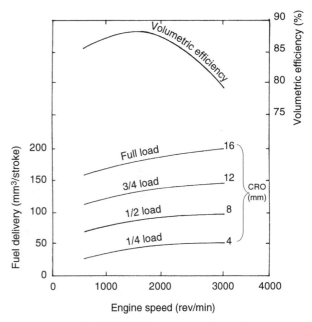

Fig. 11.16 Relationship between volumetric efficiency and fuel delivery characteristics over the engines speed range

Conversely, a fuel injection pump tends to produce an increase in fuel discharge per stroke with rising speed (Fig. 11.16). This is basically because, at low pump speed, there is more time for fuel leakage between the barrel and plunger to take place. In addition, the spill-port flow-back when plunger cut-off occurs becomes greater.

It therefore follows that the air–fuel ratio will tend to become richer at high speed instead of maintaining a more desirable constant value. Consequently, there will be an excess of fuel delivered in the upper speed range of the engine.

To obtain a satisfactory air–fuel ratio match throughout the engine's speed range, the injection pump output characteristics can be changed slightly to produce a progressively decreasing fuel delivery per plunger stroke, as the speed increases, by modifying the design of the delivery valves.

The net fuel delivery from the injector nozzle holes is obtained by subtracting the delivery valve unloading volume from the plunger discharge into the pipeline. Therefore, a diminishing fuel delivery with increasing speed can be achieved by progressively enlarging the delivery valve unloading volume.

However, with turbocharged engines, an increased fuel delivery with rising speed is more suitable and, in fact, essential if fuel delivery is to

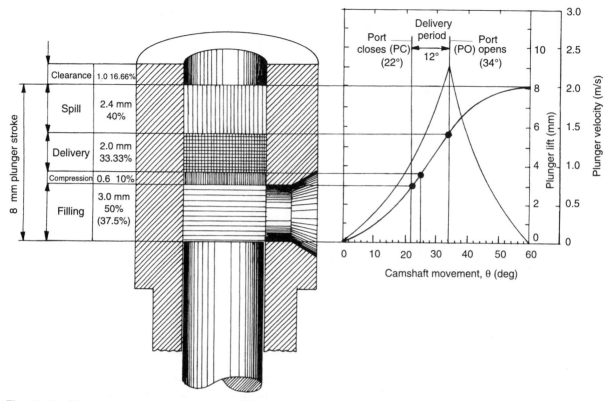

Fig. 11.17 Phases of plunger pumping stroke lift and velocity for a camshaft speed of 1000 rev/min

match the increase in volumetric efficiency in the mid and upper engine speed range.

11.3 Fuel injection analysis

11.3.1 Four phases of plunger lift
(Fig. 11.17)

The plunger's constant stroke can be divided into four phases: (1) filling; (2) compression; (3) delivery; and (4) spill. The proportions of each phase for an 8 mm plunger stroke and tangential cam operating at 1000 rev/min with full load control-rod opening are shown in Fig. 11.17.

1 **Filling phase**
Barrel filling takes place during the 3 mm plunger lift from BDC up to the point when the plunger cuts-off the feed/spill port (port closure). This filling phase occupies 37.5% of the plunger's 8 mm stroke, or 50% of the volume trapped between the top of the barrel and the top edge of the spill port.

2 **Compression phase**
Once the plunger closes the feed/spill port, the fuel trapped in the barrel is compressed so that the plunger rises about 0.6 mm before fuel begins to be discharged through the delivery valve and into the pipeline. Fuel compressibility amounts to something like 10% of the barrel's total volume between the delivery valve seat and the plunger at spill port closure.

3 **Delivery phase**
After the fuel has been compressed the plunger's upward lift displaces fuel from the barrel into the high pressure pipeline via the delivery valve for 2 mm, before its helix control edge uncovers the spill port. This 2 mm full-load plunger movement represents only one-third of the barrel volume with the spill port just closed. For 1/4, 1/2 and 3/4 load the effective plunger stroke would be 0.5 mm, 1.0 mm and 1.5 mm respectively.

4 **Spill phase**
The remaining 2.4 mm plunger lift is ineffective since the fuel in the barrel is bypassed to the fuel gallery via the open spill port. This

portion of the plunger movement is equivalent to 40% of the barrel volume when the plunger has cut-off the spill port. However, as the plunger's delivery lift is reduced by rotating the plunger to give an earlier spill, the ineffective spill movement is extended by the amount lost by the reduced delivery output.

Plunger head clearance

When the plunger reaches TDC position there will be a further 1.00 mm head clearance between the delivery valve seat and plunger crown which is equal to 16.66% of the trapped volume when the plunger cuts-off the spill port.

This clearance volume should be as small as possible to minimize the compressibility of the fuel without causing the plunger head to bump against the delivery valve seat.

11.3.2 Cam profile lift, velocity and plunger filling and pumping action
(Fig. 11.17)

During the filling phase, the plunger must be accelerated from zero velocity to a velocity high enough to give a sharp first pressure wave (Fig. 11.17). At the end of the filling phase the plunger cuts-off the spill port and, initially, the fuel which is trapped above the plunger is squeezed to roughly 90% of its original volume without any being displaced through the delivery valve. This, however, is quickly followed by the fuel delivery phase with a continuing rise in plunger velocity, as this is necessary to maintain a rising pipeline pressure when the injector needle valve opens. At the end of the delivery phase, spill occurs so that the barrel pressure collapses and, for the remainder of the plunger lift, the cam profile is designed to decelerate the plunger. It therefore causes its velocity to decrease rapidly (Fig. 11.17).

The filling, compression, delivery and spill phases take place as described but, with rising camshaft speed, the compression of the fuel in the barrel begins slightly before the plunger closes the spill port, and the collapse of the barrel pressure tends to lag a very small amount behind the spill port opening angular position. Thus, the restriction offered by the feed/spill port with rising speed tends to produce an increasing fuel delivery curve, particularly in the upper camshaft speed range.

11.3.3 Characteristic phases of operation
(Fig. 11.18)

Phase (0–1)
(Fig. 11.18)
Fuel enters the fuel port at gallery pressure, filling the space between the plunger and delivery valve. The pressure ranges from 0.3 to 1.5 bar.

Phase (1–2)
(Fig. 11.18)
The plunger moves up to cut off the feed/spill port, and the pressure begins to rise in the barrel pressure chamber before the spill port is fully closed. The trapped fuel increases its pressure until it equals the residual fuel pressure left in the high-pressure pipeline from the last cycle—this being about 30 bar. The delivery valve then opens.

Phase (2–3)
(Fig. 11.18)
The injection pump pressure will continue to rise over a cam angular displacement which sends a positive pressure wave through to the injector end of the high-pressure pipeline. There will be a short time lag before the pressure pulse reaches the underside of the needle valve. When the pressure reaches 200 bar or thereabouts there will be sufficient pressure to overcome the closing load of the injector return spring. Consequently, the needle valve commences to lift off its nozzle seat. The sudden opening of the needle valve causes an abrupt fall of fuel pressure within the injector nozzle passages (sac and fuel gallery), and during the greater part of the injection, the pressure fluctuates above and below 130 bar at the injection end of the fuel delivery system.

Phase (3–4)
(Fig. 11.18)
The plunger velocity still continues to increase so that the pressure at the pump end of the delivery valve will also continue to rise. The actual rate of pressure increase will depend upon the cam profile, cam rotational speed, nozzle spray hole sizes and opening pressure.

Phase (4–5)
(Fig. 11.18)
Eventually, the lifting of the needle valve and the

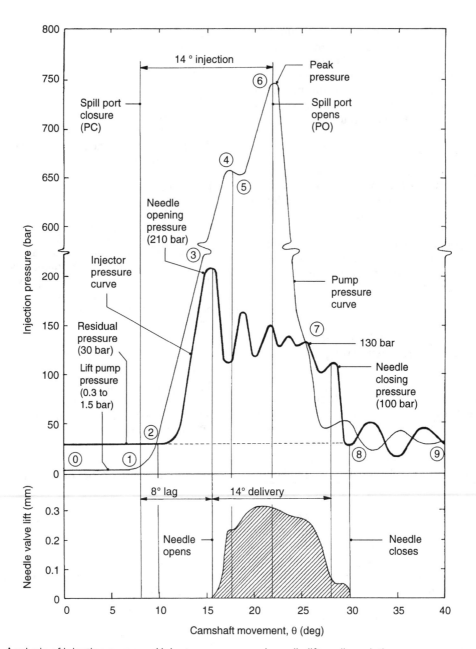

Fig. 11.18 Analysis of injection pump and injector pressure and needle lift cyclic variation

outflow of fuel from the injector exceeds the rate of fuel discharge from the barrel so that the pumping pressure tends to drop temporarily.

Phase (5–6)
(Fig. 11.18)
The hesitation in pumping pressure rise when the needle valve opens is compensated by the cam

profile raising the velocity of the plunger lift as it continues its upwards movement.

Phase (6–7)
(Fig. 11.18)
The plunger velocity and generated fuel pressure continue to rise until the plunger helix edge uncovers the spill port. Fuel spillage instantly

518

causes the collapse of barrel pressure so that the delivery valve rapidly moves towards its seat, thereby increasing the discharge passageway volume and separating the barrel from the pipeline.

Phase (7–8)
(Fig. 11.18)
There is a short time lag before the negative pressure wave reaches the injector end of the high-pressure pipeline. The injector needle valve closing pressure will be far less than the needle opening pressure. This is because there is only an annular upthrust area underneath the needle when the valve is opening, whereas a closing needle has a complete circular area for its hydraulic supporting area. Typical values for the opening and closing fuel pressures are therefore 200 bar and 100 bar respectively. Fuel stored in the region of the needle seat and nozzle sac will continue to discharge from the spray holes until its pressure energy is exhausted.

Phase (8–9)
(Fig. 11.18)
The unloading action of the delivery valve retraction collar produces a rapid pressure drop at the pump end of the pipeline and, accordingly, it sends a negative pressure wave to the injector end of the pipeline. At the instant of the needle valve closing, there will be a sudden but only slight pressure increase around the needle and nozzle passageways. Immediately, a pressure wave is produced which will travel back to the plunger end of the pipeline. These pressure waves then bound and rebound between the closed delivery valve and the closed needle valve until they are eventually damped out, they then leave a residual line pressure of around 30 bar in the pipeline.

11.3.4 Effects of control-rod opening on injection lag, the rate of delivery and the duration of delivery
(Fig. 11.19)

The angular movement between the point of spill port closure to the point when fuel commences to discharge from the injector nozzle spray holes is known as the injection lag.

At constant injection pump operating speeds, and with different control-rod openings, the injec-

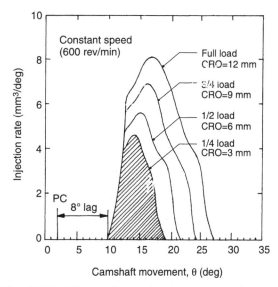

Fig. 11.19 Effects of control-rod-opening on the rate of injection and the duration and ending point of injection

tion lag and the point where injection actually commences remain constant.

However, the maximum rate of injection decreases as the fuel delivery decreases from full load to quarter load (Fig. 11.19). This is due to the plunger velocity not reaching its maximum possible value before spill, as port spill occurs earlier in the plunger stroke with a reduction in control-rod opening.

Fuel delivery output is proportional to the effective portion of the plunger's stroke. Fuel output is controlled by the plunger helix edge uncovering the spill port earlier or later during the plunger upstroke. The helix spill position is controlled by the amount the plunger is rotated which, in turn, is determined by the control-rod opening.

Thus, when the control-rod opening is set to 12 mm (full load) the plunger's effective lift before spill occurs is at a maximum. Therefore, the angular camshaft movement between the beginning of injection and when it finishes amounts to 17° camshaft displacement. As the control-rod opening is reduced, spill commences earlier in the plunger stroke, which correspondingly produces an equivalent reduction in camshaft movement over the delivery period. Thus, the delivery period amounts to 9°, 11.5°, 14° and 17° for a control-rod opening of 3, 6, 9 and 12 mm respectively.

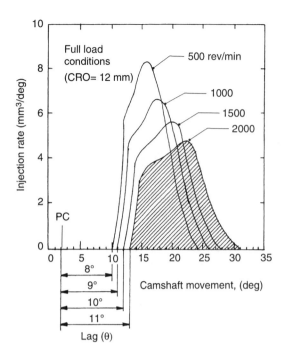

Fig. 11.20 Effects of pump speed on the rate of injection and the beginning and ending of injection

Generally, as the barrel filling time decreases with rising camshaft speed, the ability of the barrel to be fully filled decreases and, therefore, the maximum rate at which fuel is discharged from the injector nozzle also decreases (Fig. 11.20).

Because of the small spill port cross-section relative to the plunger cross-sectional area, the passageway restricts the spill flow when the plunger helix edge first exposes the upper part of the spill port. Subsequently, the throttling effect of the helix edge/spill port opening for the escaping fuel increases with rising camshaft speed and therefore prolongs the fuel period (Fig. 11.20). Fuel delivery duration with a full-load control-rod opening of 12 mm amounts to 14°, 15°, 16° and 18° for camshaft speeds of 500, 1000, 1500 and 2000 rev/min, respectively.

11.4 In-line injection pump testing

11.4.1 Phasing

Phasing is the checking and adjustment of the spill cut-off positions between each pumping element so that they take place at the correct camshaft angular displacement. In other words, phasing is setting the beginning of injection for each pumping plunger so that it occurs in the engine's firing order sequence at equal camshaft angular intervals.

11.4.2 Spill timing and phasing of Bosch and CAV Type-A injection pumps (screw tappet adjustment) procedure
(Figs 11.21 and 11.24)

1 Mount the pump (Fig. 11.21) on the testing machine (Fig. 11.24) coupling the camshaft to the machine's drive head and connect the gravity fuel supply to the pump's fuel gallery.
2 Rotate the camshaft slowly and check that each pumping element plunger has some head clearance. Head clearance is the distance between the top of the plunger and the underside of the delivery valve seat when the plunger is at top dead centre. Move and clamp the control-rack to an approximately mid-travel

11.3.5 Effects of injection pump speed on injection lag, the rate of delivery and the duration of delivery
(Fig. 11.20)

The influence of pump speed on the injection lag, delivery rate (delivery per degree camshaft rotation) and the delivery duration at full load can be examined from Fig. 11.20.

When the feed/spill port closes there is always a short time delay before fuel commences to discharge from the nozzle holes. A time lag (injection lag) is required to permit the pressure pulse initially generated in the barrel to reach the underside of the injector needle valve and then for the valve to begin its lift. This delay time can be represented as the camshaft angular lag movement between port closure and the point when fuel first begins to discharge from the nozzle holes. Injection lag, as it is known, tends to increase with rising camshaft speed and, as can be seen in Fig. 11.20, the injection lag at 500 rev/min amounts to 8°. However, with a 500 rev/min speed increase, the injection lag roughly extends by a further 1° for every 500 rev/min rise in camshaft speed.

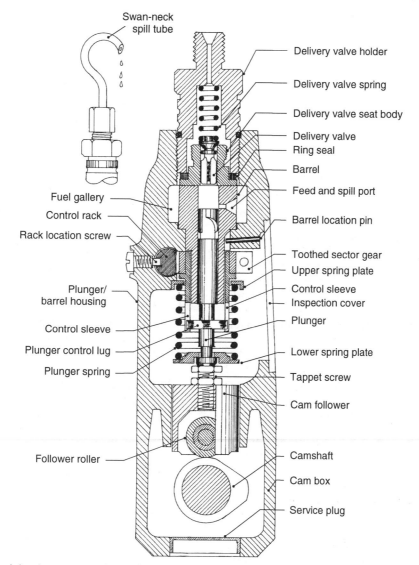

Swan-neck spill tube

Delivery valve holder

Delivery valve spring

Delivery valve seat body

Delivery valve

Ring seal

Barrel

Feed and spill port

Fuel gallery

Control rack

Rack location screw

Barrel location pin

Toothed sector gear

Upper spring plate

Control sleeve

Inspection cover

Plunger/barrel housing

Plunger

Control sleeve

Plunger control lug

Lower spring plate

Plunger spring

Tappet screw

Cam follower

Follower roller

Camshaft

Cam box

Service plug

Fig. 11.21   In-line injection pump end view (Bosch Type A)

opening position, then vent the pump gallery.

3 Turn the pump camshaft with a tommy-bar until No. 1 plunger is at TDC. Adjust the tappet screw until the head of the plunger touches the underside of the valve seat, then lower the screw a half-turn to provide the initial head clearance of 0.5 mm. (The pitch of the screw is 1.0 mm, thus half a turn is equal to 0.5 mm.) Alternatively, a dial gauge can be used to provide a more accurate reading which should be within $0.5 \text{ mm} \pm 0.15 \text{ mm}$.

4 (a) If using a swan-neck pipe for spill timing, remove No. 1 delivery valve and spring, then replace the delivery valve holder and fit the swan-neck spill pipe.

Rotate the camshaft until No. 1 plunger is at BDC. Fuel should flow freely from the swan-neck pipe. Slowly rotate the camshaft in the correct direction of rotation until the spill cut-off position is determined; that is, the point when the flow gradually diminishes and stops.

4 (b) If using a capillary glass tube do not remove the delivery valve and spring, just attach the tube to the delivery valve holder and prime the capillary tube with fuel.

When using the capillary glass tube, the spill cut-off (beginning of injection) will be indicated by the commencement of the fuel level rising in the tube.

5 Zero the phasing protractor (calibrated in degrees) at the established spill cut-off angular position and recheck again the accuracy of the spill cut-off point.

6 Remove the swan-neck pipe or the capillary tube from No. 1 delivery valve holder and replace the volume-reducer, spring and delivery valve. If the capillary tube has been used, the delivery valve assembly will not have been removed.

7 In the firing order sequence, when using the swan-neck pipe, remove the next delivery valve-holder, volume reducer, spring and delivery valve and then replace the delivery valve-holder only. The second pumping element to inject will be No. 5 for a six-cylinder engine having a firing order 153624, therefore attach the swan-neck pipe to this delivery valve-holder.

If the capillary glass tube is being used to check the spill cut-off, attach the tube to No. 5 delivery holder without removing the delivery valve assembly.

8 With the gravity fuel supply switched on, rotate the camshaft until the spill cut-off position is found on No. 5 pumping element. Check that the datum mark on the phasing protractor is at $60° \pm 0.5°$ $(90° \pm 0.5°$ for a four-cylinder engine injection pump). If it is not, the head clearance for No. 5 plunger must be altered by rotating the tappet adjustment screw.

9 The tappet screw thread has a pitch of 1 mm; that is, for one complete turn of the tappet screw it will extend or reduce its height by 1 mm. A change in tappet height of 0.1 mm will alter the phase angle by a little less than 0.5°.

Rotating the hexagonal screw head by one flat (60°) will alter the phase angle by approximately 1°; thus, half a turn, or three flats, is approximately equivalent to 0.5 mm or 3° in camshaft angular movement.

If the timing is advanced (that is, spill cut-off occurs before 60°) the head clearance must be increased, whereas if the timing is retarded (that is, spill cut-off occurs after 60° on the protractor) the head clearance must be reduced.

Thus, with a 59° phase angle, retard the

injection 1° by screwing down the tappet adjustment screw by one flat (60°). Conversely, if the phase angle reads 61.5°, advance the spill cut-off 1.5° by screwing up the tappet adjustment screw by one-and-a-half flats (90°).

10 Once the 60° phase angle is achieved, recheck the head clearance of No. 5 plunger. If the head clearance is outside the specified tolerance the head clearance of No. 1 plunger must be slightly altered. However, this means that No. 5 plunger must now be rephased.

11 Repeat procedures 7 to 10 for each pumping element in the correct firing order sequence.

12 Finally, recheck the zero datum setting of No. 1 element with the swan-neck pipe or capillary tube. If the phase protractor does not read zero the whole phasing operation must be repeated.

11.4.3 Spill timing and phasing Bosch Type-M injection pumps (roller diameter adjustment) procedure
(Fig. 11.8 and 11.24)

1 With the Bosch Type-M injection pump, the initial position on the cam profile (when the rising plunger cuts off the spill port and injection begins) can be varied by altering the size of the follower roller.

2 The manufacturer provides a range of roller diameters which are graduated in steps of 0.1 mm. A change of 0.1 mm roller diameter will alter the camshaft angular spill cut-off position by just under 0.5°.

3 Initially, No. 1 plunger head clearance is set to 0.5 mm ± 0.15 mm using a head clearance gauge. The pumping element spill cut-off timing is then determined using the swan-neck pipe spill method, at which point the phasing protractor is set to zero.

4 Phasing for all the other pumping elements is then checked and recorded relative to No. 1 element in the firing order sequence. If the spill cut-off angular intervals for any particular element or elements are outside the $\pm 0.5°$ limit, the roller or rollers, must be removed and changed for the appropriate diameter ones.

Use the relationship that a 0.1 mm roller diameter change roughly alters the phase angle by 0.5°, recheck the plunger head clearance if the roller diameters have been changed, as this clearance must still be within the prescribed tolerance.

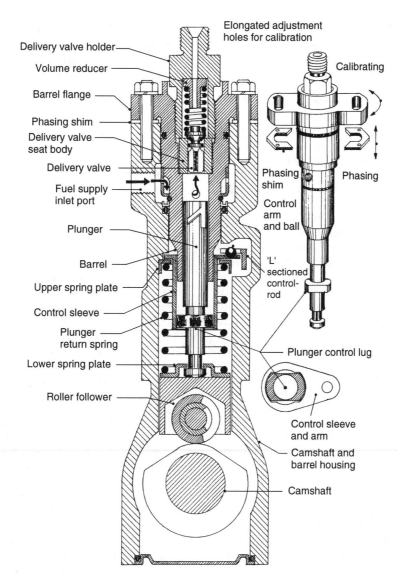

Fig. 11.22 In-line injection pump end view (Bosch Type MW)

11.4.4 Spill timing and phasing Bosch Type-MW and P injection pumps (barrel height adjustment) procedure
(Fig. 11.22 and 11.24)

1 With the Bosch Type-MW and P injection pumps, the distance between the plunger head and the lowest point of the roller follower remains constant. However, the flanged barrels can be raised or lowered by inserting different thicknesses of shim between the underside of the barrel flange and the top face of the pump housing. Consequently, raising an individual barrel relative to its plunger, retards the point of spill cut-off, whereas lowering the barrel, advances the point at which the plunger closes the feed/spill port (i.e. spill cut-off position).

2 If the injection pump has been dismantled and parts have been replaced, each plunger head clearance must be checked. First, rotate the camshaft until the plunger being checked is at TDC position. Remove both the flanged barrel horseshoe shims and then push the barrel downwards until the delivery valve seat bumps

523

against the plunger head. Measure the clearance between the barrel flange and the top face of the pump housing; the plunger head clearance will be the difference between the thickness of the shim removed and the gap formed between the lower and upper faces of the flange and pump housing respectively. Thus, if the shim thickness is 1.4 mm and the flange-to-housing gap is 0.9 mm, the head clearance will be 0.5 mm. If the head clearance is outside the recommended tolerance the shim must be changed to bring the head clearance within 0.5 mm ± 0.15 mm.

3 With the injection pump mounted and coupled to the testing machine, remove No. 1 delivery valve holder, volume reducer, spring and valve, then replace the delivery valve holder and fit the swan-neck pipe.

Using a tommy-bar attached to the camshaft/drive head coupling, rotate the camshaft until No. 1 plunger is roughly at BDC, turn on the gravity fuel supply and observe the fuel pouring out from the swan-neck pipe (Fig 11.24). Turn the camshaft slowly in the correct direction of rotation until the spill cut-off point is found; that is, the point when the fuel flow just stops. Zero the angular phasing protractor.

4 Following the engine's firing order sequence, remove the next pumping element delivery valve assembly and replace the delivery valve holder, then attach the swan-neck pipe, i.e. element No. 5 for a six-cylinder engine with a 153624 firing order.

Again, with the fuel flowing, rotate the camshaft in the correct direction of rotation until No. 5 pumping element's spill cut-off point is determined and note and record the protractor's reading.

5 Repeat this procedure for the remaining pumping elements. If the phasing angles for some of the pumping elements are outside the ±0.5° tolerance, calculate the change in shim thickness required to bring the spill cut-off within these limits. Note: if the standard cam is used, a 0.1 mm change in shim thickness will alter the point of cut-off by a little less than 0.5°. Make sure that the plunger head clearance for each plunger still remains within the specified 0.5 mm ± 0.15 mm.

11.4.5 Spill timing and phasing CAV Minimec injection pumps (tappet spacer adjustment) procedure
(Fig. 11.23(a and b) and 11.24)

1 If the pumping elements, tappets, camshaft, pump housing or barrel body have been changed then it is necessary to rephase all pumping elements.

2 Assemble the plungers with their lower spring plates in their respective barrels. Lay the pump housing on its side with the inspection cover side facing downwards. Without replacing the plunger springs, assemble the barrel body to the pump housing with the plunger arms pointing downwards and located in their respective control-rod forks. (Note: after phasing has been completed the plunger springs must be fitted by separating the barrel body from the pump housing, after which it must be reassembled.) The barrel body is then secured with four set screws inserted at each corner of the barrel body.

3 Mount the pump on the testing machine and connect the gravity fuel supply to the barrel body. Remove No. 1 element delivery valve holder, volume reducer, delivery valve spring and delivery valve from the top end of the barrel body. Attach the plunger head clearance gauge and dial indicator in place of the delivery valve holder.

4 Set the square section of the control-rod 10 mm from the supporting bush and turn the camshaft until No. 1 plunger is at the bottom of its stroke. Zero the clearance gauge and open the gravity fuel supply tap. Revolve the camshaft slowly in the correct direction of rotation until the fuel just ceases to flow out from the clearance gauge adaptor. This represents the beginning of injection from No. 1 element.

5 Observe the gauge reading, which indicates the plunger travel from BDC to the closing of the inlet port and compare the reading with the fuel setting data-sheet. If the plunger travel is within the specified tolerance, set the protractor degree-plate of the testing machine to zero (Fig. 11.24). This pumping element is then the datum setting for the remaining elements. Mark the pump flange plate in line with the coupling timing datum.

Setting No. 1 plunger to lift a definite amount before commencement of injection

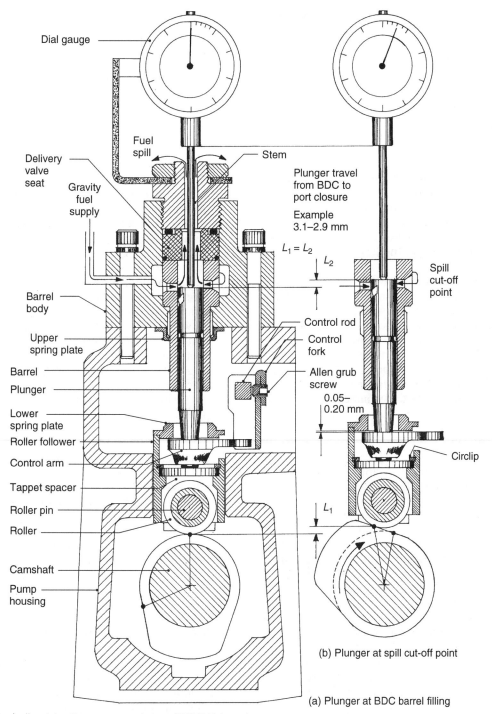

Fig. 11.23 In-line injection pump end view (CAV Minimec)

ensures that the plunger has a full effective displacement action before fuel spill occurs.

6 Adjust No. 1 tappet, if necessary, by the selection of tappet spacers which are supplied in 12 different thicknesses, graduated in steps of 0.1 mm and stamped with the numbers 1–12 with the exception of Nos 6 and 9 which have the suffix 'A'.

Table 11.1 Example for a six-cylinder injection pump phasing

	Pumping element number					
	1	2	3	4	5	6
Original spacer thickness (mm)	4.3	4.3	4.3	4.3	4.3	4.3
Correct phase angle (deg)	0	60	120	180	240	300
Observed phase angle (deg)	0	61	118.5	180.5	240	299
Error (deg)	0	+1.0	−1.5	+0.5	0	−1.0
New spacer thickness (mm)	4.3	4.5	4.5	4.4	4.3	4.1

If No. 1 setting is altered, then reset the protractor to 0° and re-mark the pump flange plate adjacent to the coupling timing datum.

7 Remove the plunger head clearance gauge and replace the delivery valve components of No. 1 element and tighten to the correct torque.

The plunger arm vertical clearance is then checked by holding the lower spring plate firmly down onto the tappet body while moving the plunger up and down. The vertical free movement between the plunger arm and the lower spring plate should be within 0.05 mm and 0.20 mm. Lower spring plates are available with 15 different thicknesses, graduated in steps of 0.01 mm and numbered from 02S–16S.

8 In the firing order sequence, remove the next delivery valve holder, volume reducer, spring and valve from the barrel body and replace with a spill pipe and adaptor (glass tube or swan-neck pipe). Turn on the gravity fuel supply. Rotate the camshaft very slowly by hand in the correct direction of rotation until fuel just ceases to flow through the spill pipe. Observe and note, on the protractor, the degrees of movement relative to the datum pointer. Switch off the fuel supply and remove the spill pipe adaptor, then replace the delivery valve assembly.

9 Repeat the process described in procedure 8 in the firing order sequence for the remaining pumping elements. This procedure therefore checks the spill cut-off angular camshaft intervals between pumping elements. It should read 120° and 240° after No. 1 element for a three-cylinder four-stroke engine, 90°, 180° and 270° for a four-cylinder four-stroke cycle engine and 60°, 120°, 180°, 240° and 300° for a six-cylinder four-stroke cycle engine.

10 If the phase angle variations exceed ±0.5°, a correction must be made by changing the tappet spacer for a different thickness. A changing tappet spacer thickness of 0.1 mm is approximately equivalent to 0.5° camshaft movement.

Make out a Table (Table 11.1) for the required and actual phase angles recorded, and insert the thickness of No. 1 tappet spacer in each column, i.e. 4.3 mm. A smaller angular movement indicates an early injection. Therefore, the replacement tappet spacer will be thinner, whereas a larger angular movement between spill cut-off means late injection so that a thicker tappet spacer must be used. For example, a 62° phase angle means that the original 4.3 mm tappet spacer must be replaced by a thinner $4.3 − (0.1 × 4) = 4.3 − 0.4 = 3.9$ mm tappet spacer. Similarly, a 119° phase angle means that the original 4.3 mm tappet spacer must be replaced by a thicker $4.3 + (0.1 × 2) = 4.5$ mm spacer.

11 Each pumping element phase angle which is outside the ±0.5° tolerance should now be adjusted by exchanging the original tappet spacer for the new calculated spacer thickness.

After changing the appropriate tappet spacers, check each plunger arm vertical float, which should be within 0.05–0.20 mm and, if necessary, change the lower spring plate to bring the clearance back to the required limits.

11.4.6 Calibration mechanisms

Calibration is the checking and adjustment of each pumping element port spill position so that each plunger and barrel delivers the same specified quantity of fuel for a given control-rod opening (CRO) (measured in mm) and camshaft speed.

Variation of fuel delivery from each pumping element is obtained by rotating either the plunger relative to the barrel or the barrel relative to the plunger, so that the plunger helix uncovers the spill port earlier or later, to reduce or increase fuel delivery respectively.

Individual pumping element calibration to a specified fuel delivery with a given control-rod opening and camshaft speed is achieved by the following methods.

Split-toothed sector gear and sleeve
(Figs 11.7 and 11.21)
By slackening the split-toothed sector gear around each control sleeve in turn, the sleeve and its corresponding plunger can be rotated clockwise or anticlockwise independently of the pumping elements, which are all meshed to the control-rack (Figs 11.7 and 11.21). Each pumping element can therefore be separately adjusted relative to the control-rack so that any movement of the rod produces similar changes in fuel delivery from individual pumping elements. After adjustment, each split-toothed sector gear is retightened.

Plunger control arm clamp
(Figs 11.8 and 11.24)
Slackening, in turn, each control arm clamp (Figs 11.8 and 11.24) relative to the control-rod permits the control sleeve and its corresponding plunger to be rotated in either direction, independently of the other pumping elements, which are interlocked (via the control sleeve-peg and control-rod slotted clamp) to the control-rod. Accordingly, the individual pumping element fuel deliveries can be adjusted by moving the slackened control-rod clamp in the left or right-hand directions, along the control-rod relative to the other plunger helix settings.

Flanged barrel or sleeve with elongated holes
(Fig. 11.22)
Slacken each pair of securing nuts (which clamp the flanged barrel or sleeve to the pumping element housing) so that the individual barrels can be rotated relative to their plungers within the limits of the elongated stud holes formed in the flanged end of the barrel (Fig. 11.22). The consequence of swivelling the barrel is to move the spill-port nearer or further from the plunger's helix, thus altering the effective stroke of the plunger. In this way the individual pumping element can be calibrated relative to the control-rod opening (CRO) setting.

11.4.7 Calibration test bench machine
(Fig. 11.24)

Injection pump plunger elements are calibrated by coupling the injection pump drive and the test bench drive-head together. While clamping the pump housing to the bench-bed top face, spacer plates may be fitted between the base of the pump and the bed top face to achieve the correct alignment of the pump camshaft to the drive coupling head. Each delivery valve holder is then connected via high pressure steel pipes to a row of test injectors mounted over an equal number of calibrated glass test tubes.

Trip cut-off solenoid valve units are mounted between each test injector and the calibrated glass tubes, the latter of which are positioned immediately underneath the units.

An electric motor drives the injection pump coupling head via a pulley and belt drive. The pulley conical side members can be moved closer or further apart by rotating the speed control hand wheel and the threaded control shaft so that the double armed control nut is screwed either to the right or left. The side thrust produced by the double armed control nut pulls one pair of conical side members apart and pushes the other pair nearer together, thereby reducing one and increasing the other pulley's effective working diameter. By these means, the speed ratio of the two pulleys can be varied, thus enabling the coupling head to be operated at various predetermined speeds.

Calibration is achieved by feeding the injection pump with test fuel fluid and driving the pump at a specified speed with the control-rod clamped to a definite opening setting, measured in mm.

The trip cut-off solenoid valve is designed to divert the test fluid away from the test tubes to the supply tank, until the trip switch toggle is flicked over, at which point, the energized solenoids open the valves, and test fluid is now discharged directly into the test tubes. After 100 shots from each plunger element of the in-line injection pump, the revolution counter electric trip automatically breaks the solenoid electricity supply. Immediately, all the valves snap close, thus preventing any more test fluid from reaching the test tubes.

Further multiples of 100 shots can be injected into the calibrated tubes as required.

Once the fluid levels have been checked, the test tubes can be turned over so that the previously drained tubes can be used immediately.

Table 11.2 Example of calibrating test plan

Test number	CRO (mm)	Speed (rev/min)	Number of shots	Fuel delivery (ml) Range	Mean	Maximum spread (ml)
1	9	600	100	7.1–8.3	7.7	0.5
2	12	900	100	12.8–14.2	13.5	0.6
3	6	200	100	0.6–1.4	1.0	0.4
4	1	1000	100	zero	–	–

If the test bench is to be operated for any length of time, then the test fluid must be cooled to maintain its viscosity within prescribed limits, otherwise the leakage between plungers and barrels may be excessive and, as a result, calibration may be inaccurate.

11.4.8 Calibration procedure
(Fig. 11.24)

Calibration is normally carried out by setting the control-rod opening and camshaft speed to something like half-load and mid-camshaft speed, i.e. CRO = 9 mm and 600 rev/min, respectively. The injection pump is then operated under these conditions while 100 shots are injected into the test tubes. Fuel discharge readings (in ml) are then taken from each pumping element test tube. The individual plunger element helix spill-ports are then adjusted (by either rotating the plunger or barrel) to be within the specified delivery tolerance.

If a comprehensive calibration check is to be made, the control-rod is then set for full-load output at a higher speed, say CRO = 12 mm and 900 rev/min, and again readings are taken from 100 injection shots: if any of the deliveries are outside the specified tolerance, further minor adjustments can then be made. This procedure is then repeated for a small control-rod opening and at a relatively low injection pump speed (idle); i.e. CRO = 6 mm at 200 rev/min.

For each set of operating conditions, the maximum fuel delivery spread must not exceed that specified in the injection pump test sheet (Table 11.2). Hence, for the pumping element set with a CRO of 9 mm running at 600 rev/min, the difference in reading between the smallest pumping element discharge and the largest discharge must not be greater than 0.5 ml. Thus, if the smallest discharge is set to 9.2 ml the largest discharge from any other element must not ex-

ceed $9.2 + 0.5 = 9.7$ ml or, if the largest delivery for an element is set to 10.2 ml, any other pumping element must not give a reading lower than $10.2 - 0.5 = 9.7$ ml.

Note: for a given plunger-to-barrel working fit and accuracy of the helix edge, the effects of wear and machining error are considerably greater for small CRO and low camshaft speeds. Consequently, the delivery spread is made relatively greater as the CRO and camshaft speed are reduced. Thus, the ratio of mean fuel delivery to fuel spread between pumping elements from Table 11.2 is

$$\frac{13.5}{0.6} = 22.5 : 1, \quad \frac{7.7}{0.5} = 15.4 : 1, \quad \frac{1.0}{0.4} = 2.5 : 1$$

for test conditions CRO/(rev/min) 12/900, 9/600 and 6/200, respectively.

A final test is made to ensure plunger delivery is reduced to zero when the CRO is moved to the specified minimum setting. From the test sheet (Table 11.2) this is CRO = 1 mm minimum at 1000 rev/min.

Many manufacturer's test plans are more basic and therefore may not involve as many test stages. The mean fuel delivery is normally not included but has been shown here to enable the ratio of mean fuel delivery to fuel spread to be obtained.

Note the first test calibration is deliberately set for roughly half the control-rod opening and mid camshaft speed, the other calibration tests are therefore more likely to be within the prescribed tolerances.

11.4.9 Balancing delivery within given tolerances

If delivery is set within tolerance at high speed and large output, but at low speed and small CRO delivery it is just outside the limit, readjustment

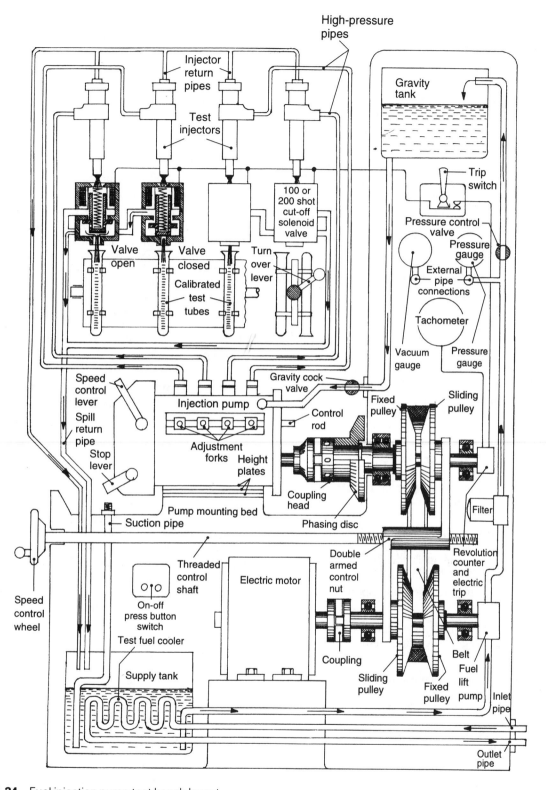

Fig. 11.24 Fuel injection pump test bench layout

of the high-speed setting nearer to one side of the tolerance will sometimes bring the lower speed small delivery within the recommended spread.

Distribution tolerance

Most manufacturers seek to maintain a maximum variation in fuel delivery between pumping elements of ±4% without causing any noticeable change in engine performance. Injection pump calibration can normally be kept to within ±2% at full load which permits a maximum variation of about ±7.5% at low idle speed, although ±5% is generally obtained. A fuel delivery variation of ±4% at part-load fuel delivery is generally acceptable to engine manufacturers.

Barrel-to-plunger leakage

The magnitude of barrel-to-plunger leakage can be measured by the variation of fuel delivery at constant control-rod opening (CRO) but at different speeds. Wear usually shows in the much reduced fuel delivery as the pumping speed is reduced (Fig. 11.10).

Helix lead (edge) tolerance

The accuracy of the helix lead can be measured by the variation of fuel delivery at constant speed but with different control-rod openings (CRO). Inaccuracy, scoring or just wear of the helix edge portion of the plunger can be shown in terms of loss in fuel delivery over the CRO working range. This is generally more pronounced for small control-rod openings.

Factors affecting calibration

Nozzle spray hole area—the larger the holes, the greater the fuel delivery.

Nozzle opening pressure—low opening pressures increase delivery.

Plunger diameter tolerance—large tolerances have a marked influence on fuel delivery under part load.

Delivery valve tightness—unloader piston clearance and valve spring stiffness considerably influence fuel delivery, particularly at low speeds.

Helix control edge (lead) variation—greatly influences the fuel delivery in the lower speed range.

Discharge high pressure tubing—the internal pipe diameter and the length of the volume of trapped

fuel controls the residual pressure and the pressure drop through the pipelines. Equal-length piping is essential to obtain consistent fuel delivery from each pumping element.

Fuel supply pressure—an increase in lift pump pressure can increase fuel delivery at high camshaft speeds. Thus, tests have shown that a rise in supply pressure from 0.35 to 1.75 bar for a given operating condition has been able to increase fuel delivery by 15%.

Fuel temperature—a rise in fuel temperature decreases its viscosity and specific gravity and, at the same time, it increases the compressibility and barrel-to-plunger leakage. The overall effect is that the fuel delivery decreases with a temperature increase. Note that when calibrating injection pumps, the test fuel temperature can undergo a considerable charge if it is not coolant controlled.

11.5 Injection pump governing

11.5.1 Comparison of petrol and diesel engine output control
(Figs 11.25 and 11.26)

The petrol engine produces a homogeneous air–fuel mixture in the cylinders, with the quantity of air, or air and petrol mixture, permitted to enter the cylinders being restricted by a throttle valve. The engine torque characteristics (the brake mean effective pressure characteristics are similar) produce a family of curves which show an increased droop with reduced throttle opening (Fig. 11.25). These curves illustrate the self-limiting speed characteristics of the spark ignition petrol engine; that is, the rapid fall off in engine torque as the throttle opening is reduced.

In contrast, the diesel engine produces a heterogeneous air–fuel mixture in the cylinders, the quantity of fuel injected into the cylinders being regulated by the injection pump control-rod opening, which determines the amount of fuel delivery metered to each cylinder. The engine torque or brake mean effective pressure characteristics produced by the diesel engine consists of a family of curves having a similar shape to the engine's volumetric efficiency curve over the engine's speed range (Fig. 11.26). These curves for each control-rod opening are stacked one above the other and, as can be seen, there are very few

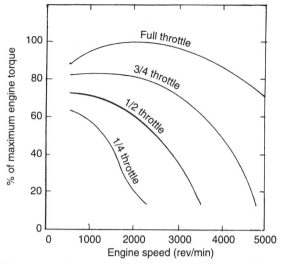

Fig. 11.25 Speed characteristics for petrol engine for different throttle openings

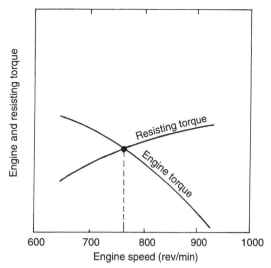

Fig. 11.27 Stable torque characteristics for a petrol engine

self-limiting speed characteristics at the top end of the engine's speed range. The fall off in engine torque for a given control-rod opening in the upper speed range is relatively small and very gradual.

11.5.2 The need to govern the diesel engine

(Figs 11.27 and 11.28)

Engine stability is the ability of the fuel system to

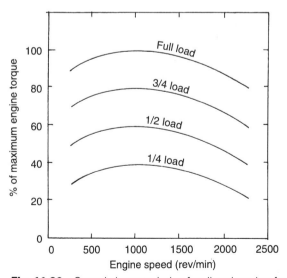

Fig. 11.26 Speed characteristics for diesel engine for different load settings

maintain the engine at a steady speed within the limits of a predetermined external load variation.

For engine stability the torque developed by the engine must equal the resisting torque caused by the engine's internal losses (friction and pumping) during idle speed or, for all other speeds, the engine's internal losses plus the external loads imposed on the crankshaft, for example the generator, hydraulic power steering pump, air compressor, propelling effort needed to drive the vehicle etc. With a petrol engine, the resisting torque curve rises with an increase in engine speed whereas the engine's output torque curve for a given throttle opening is a decreasing one (Fig. 11.27). Thus, the equilibrium speed is where the engine torque and the resisting curves cross each other. Above this equilibrium speed the resisting torque exceeds the engine torque so that the engine cannot run any faster. On the other hand, below the equilibrium speed the engine torque does exceed the resisting torque, which therefore prevents the engine speed decreasing.

Unfortunately, with a diesel engine, the developed torque tends to rise and to exceed the resisting torque caused by internal losses and external loads (Fig. 11.28). Consequently, the engine speed will continue to rise unrestricted unless the fuel delivery is proportionally decreased, thereby bringing the engine's torque back to its equilibrium value. The reduction in fuel, provided the metering is rapidly altered and regulated, will reduce the engine's torque to that of the resisting torque, thereby stabilizing the

531

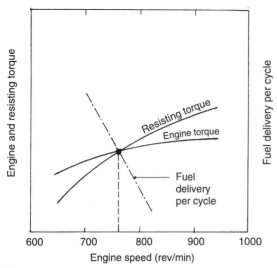

Fig. 11.28 Unstable torque characteristics for a diesel engine

engine's speed. The mechanism which senses the change in engine speed and automatically adjusts the quantity of metered fuel, to match the change in the external torque load applied to the engine, is known as the *injection pump governor*.

11.5.3 The governor's function
(Fig. 11.29)

The governor is a speed sensitive controller which regulates the engine speed within specified limits by automatically adjusting the injection pump fuel delivery to match the change in both internal and external engine loads.

The governor has two objectives.

1 To monitor engine speed and cyclic speed fluctuation and to relay these signals to the power-servo continuously.
2 To act as the power-servo mechanism, which converts the speed variation signals into the control-rod opening adjustment movement needed to counteract changes in engine speed.

The engine, injection equipment and the governor form a closed-loop system to which speed setting and external load variations are imposed (Fig. 11.29). Changes from the engine's steady-state speed conditions are signalled, in the direction of the arrows, to the governor which then activates the corrective change in fuel delivery of the injection pump equipment.

For effective speed regulation the governors must instantly sense changes in engine speed and respond rapidly by accurately altering the fuel delivery by means of the control-rod opening. With small changes in load, the engine's speed should remain uniform or change only if the external load is deliberately varied.

The amount the engine speed fluctuates due to a change in engine load is determined by how

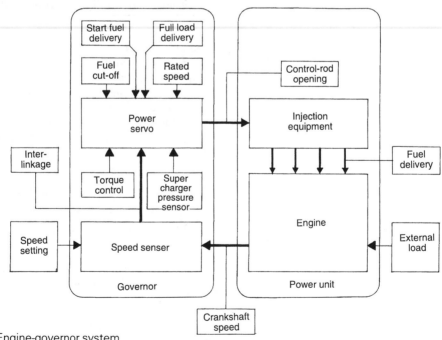

Fig. 11.29 Engine-governor system

532

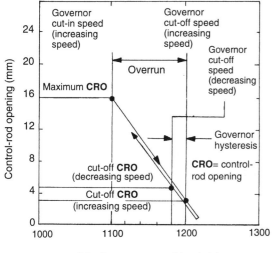

Fig. 11.30 Illustration of governor terminology

quickly and accurately the control-rod can move to its newly programmed position. Other factors are the time it takes for the engine's cylinders to react to different fuel delivery inputs, the flywheel energy absorption and the release effect created by the total rotating inertia of the crankshaft assembly, and any internal and external equipment driven directly by the engine.

There must always be some small degree of engine speed variation for the governor to detect and then to respond by repositioning the control-rod—consequently, speed uniformity cannot be perfect.

11.5.4 Governor terminology
(Fig. 11.30)

Governor cut-in
This is the injection pump speed at which the control-rod just starts to move away from the maximum fuel stop with the accelerator lever set for maximum output.

Governor cut-off
This is the injection pump speed at which the control-rod moves beyond the governor cut-in position and just reaches the fuel cut-off position, at which point fuel injection into the test tubes ceases. The accelerator lever during this check is set for maximum output.

Governor overrun
This is the difference between the governor cut-off speed where fuel delivery ceases, and the lower governor cut-in speed where the control-rod just begins to move away from the maximum control-rod opening position. In both cases the injection pump speed is increasing.

Governor overrun = governor cut-off speed
$$- \text{ governor cut-in speed}$$

Example If the governor cut-off speed is 1200 rev/min and the governor cut-in speed is 1100 rev/min, determine governor overrun

governor overrun = 1200 − 1100 = 100 rev/min

Governor hysteresis
This is the speed difference between the governor cut-off speed at which point fuel delivery ceases with increasing injection pump speed and the cut-off speed at which fuel delivery recommences with decreasing injection pump speed.

Governor hysteresis = governor cut-off speed
increasing
$$- \text{ governor cut-off speed decreasing}$$

Example If, with increasing injection pump speed, the governor cut-off occurs at 1200 rev/min whereas with decreasing speed the control-rod commences to move away from the cut-off position at 1180 rev/min, determine the governor's hysteresis

Governor hysteresis = 1200 − 1180 = 20 rev/min

Governor run-out
This is the percentage speed change when the control-rod is moved by the governor from the maximum fuel delivery position to the no-fuel delivery position. Governor run-out can be expressed by the formula

Governor run-out =

$$\frac{\text{no-fuel speed} - \text{maximum fuel speed}}{\text{maximum fuel speed}} \times 100$$

Example If the no-fuel delivery speed (fuel cut-off) is 1200 rev/min and the maximum fuel delivery speed just before the control-rod first commences to move away from the maximum control-rod opening is 1100 rev/min, determine the governor run-out

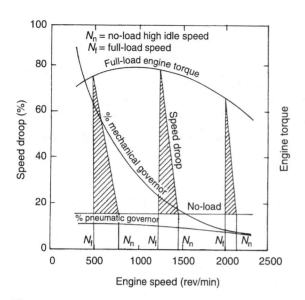

Fig. 11.31 Speed droop characteristics

$$\text{Governor run-out} = \frac{1200 - 1100}{1100} \times 100$$

$$= \frac{100 \times 100}{1100} = 9\%$$

11.5.5 Speed droop
(Fig. 11.31)

For a given maximum fuel delivery (determined by the control-rod opening) the engine will develop its full load torque characteristics within the minimum and maximum operating speed range (Fig.11.31). If the load imposed on the engine decreases without any change in the control-rod opening, then the surplus fuel energy will immediately increase the speed of the engine proportional to the load reduction. At idle speed, as the engine is warmed up from cold, the frictional and pumping losses decrease so that for a given control-rod opening the extra energy now available raises the idle speed. Conversely, if an external load is applied to the engine when it is idling, the engine speed decreases because some of the energy that was initially available for maintaining the normal idle speed is now required to overcome the external torque load.

The amount the engine speed changes, when the operating conditions change from full load to no-load, is a measure of the governor's ability to counteract the rising engine speed by moving the control-rod towards the cut-off no-load direction.

The magnitude of engine speed variation from the full load to no-load speed, expressed as a percentage, is sometimes known as *cyclic irregularity* or, more commonly, *speed droop*, due to the downward incline of the engine torque relative to the speed overrun.

Hence, speed droop may be defined by the formula

$$\frac{\text{speed}}{\text{droop}} = \frac{\text{no-load speed} - \text{full load speed}}{\text{full load speed}} \times 100$$

$$\delta_d = \frac{N_n - N_f}{N_f} \times 100$$

where N_n = no-load high idle speed
        N_f = full load speed
        δ_d = speed droop

Normally a large speed droop, in contrast to a small one, tends to stabilize speed fluctuation and the smooth operation of the engine. However, the speed droop permitted depends on the engine's application, as can be seen.

8–15% farm tractors and excavators
5–10% industrial, marine and road vehicles
2–5% generators

11.5.6 Torque control
(Figs 11.16 and 11.32)

Fuel delivery output characteristics for an injection pump should be designed to conform to the shape of the engine's volumetric efficiency curve so that the proportion of fuel delivered to that of the air mass filling the cylinders will be kept approximately constant throughout the engine's speed range.

However, the injection-pump fuel delivery characteristics are of the ever-increasing form for a given control-rod opening (Fig. 11.16), whereas the volumetric efficiency of the engine tends to increase up to about mid-speed and then decrease with further speed increases.

Torque control is the process of slightly reducing the control-rod opening once the peak volumetric efficiency has been passed so that the fuel delivery will broadly match this decrease in volumetric efficiency beyond the mid-speed point (Fig. 11.32).

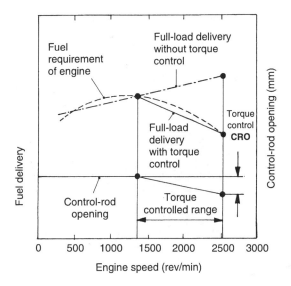

Fig. 11.32 Fuel and CRO requirements and fuel delivery characteristics with torque control

11.5.7 Maximum and minimum speed flyweight governor (Bosch Type-RQ)
(Figs 11.33(a–d), 11.34, 11.35 and 11.36)

Construction
(Fig. 11.33(a))
This two-speed governor has the governor springs located inside a pair of flyweights which are themselves mounted on a flyweight carrier hub attached to one end of the camshaft. The carrier hub is fixed to the camshaft by means of a male and female tapered fit secured by a woodruff key. Support for the hollow flyweights is provided by a pair of diametrically opposed stud-posts screwed perpendicularly into the carrier hub. The outer end of the carrier-hub has a semi-flange which supports the bell-crank lever. Flyweight outward movement is relayed to the control-rod through four pivoting bell-crank levers (two per side), a slider-pin supported on a cross-pinned slider block, a fulcrum lever and a fork-link. Normally, the inner spring plate seats on the shoulder of the spring post boss whereas the outer spring plate is held in place by a special adjustment (Fig. 11.34). The idle spring is held between the outer spring plate and the base of the flyweight hole whereas the maximum springs are held between both the outer and inner spring plates.

The accelerator lever movement is conveyed to the control-rod via the control arm and guide plunger sliding in the bore of the fulcrum lever (Fig. 11.34) which swivels about the slider block cross-pins. The slider block cross-pin which supports the fulcrum lever is mounted on the end of the slider-pin and is free to move axially with the slider-pin, but is stabilized and prevented from tilting by the guide-pin.

Idle speed operating conditions
(Fig. 11.33(b))
With the accelerator lever set against the idle speed stop the guide plunger positions itself roughly mid-way along the fulcrum-lever guide bore. The guide plunger pivot-pin becomes the fulcrum point (pivot) of the fulcrum lever, about which it tilts. The top end of the fulcrum-lever is linked to the control-rod and the bottom end is connected to the slider-pin whose function is to relay the flyweight movement via the bell-crank lever in an axial direction.

With the engine idling, the flyweight outward movement caused by the generated centrifugal force is counteracted by the outer idling spring being compressed between the base of the flyweight and the outer spring plate. If the engine load suddenly increases, the reduction in centrifugal force permits the idle spring to push the flyweights further together. As a result the bell-crank lever swivels about the carrier hub pivots, and they thus shift the slider-pin to the right. Accordingly, the fulcrum-lever swings anticlockwise about the guide plunger pivot and it therefore moves the control-rod to an increased fuel delivery position. Immediately, the engine responds with a rise in speed, and this overshoot raises the flyweight centrifugal force, compresses the outer idle spring and swings the bell-crank levers in a direction which draws the slide-pin to the left. This movement simultaneously tilts the fulcrum lever about the guide-plunger pivot in a clockwise direction, thus pulling the control-rod to a reduced fuel delivery position. Instantly, a drop in speed follows. The net result is a cyclic rise and fall in engine speed about some mean speed known as the idle speed.

The quickness at which the governor flyweights sense engine speed change and respond, by either increasing or decreasing fuel delivery, determines the amplitude of the speed fluctuation about its average value.

Intermediate speed operating conditions
(Fig. 11.33(c))
As the accelerator lever is moved to increase

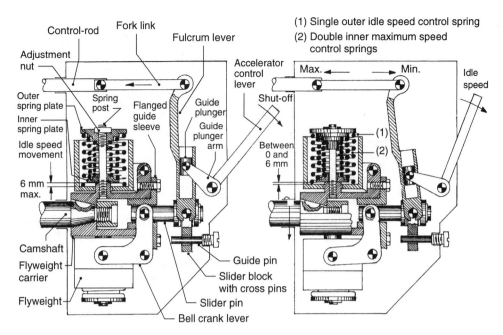

(1) Single outer idle speed control spring
(2) Double inner maximum speed control springs

(a) Governor weights engine rest (shut-off) position (b) Governor weights idle speed position

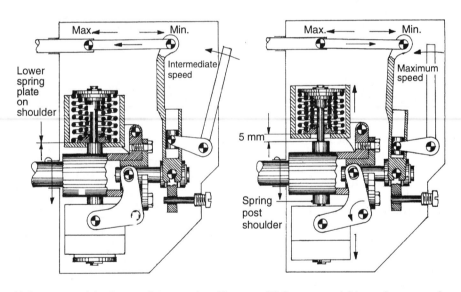

(c) Governor weights intermediate speed position (d) Governor weights maximum speed position

Fig. 11.33 Maximum and minimum speed flyweight governor (Bosch Type RQ)

output, the engine speed increases; thus causing the flyweights to swing outwards until they contact the lower swing plate. The stiffness of the outer idle spring, and now the two inner maximum speed springs, far exceeds the opposing centrifugal force created by the flyweights over

the engine's normal speed range. Consequently, the governor weights are neutralized and no further outward movement of the flyweights can take place. However, over these operating conditions the lower end of the fulcrum-lever pivots about the sliding-block cross-pins so that when

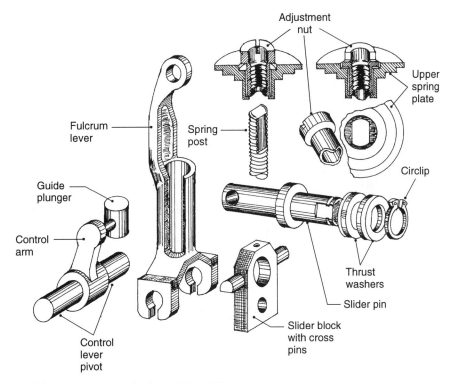

Fig. 11.34 Some of the components of a Bosch Type RQ governor

the accelerator lever is pushed forwards, it swings the guide-plunger lever so that the guide-plunger slides down the fulcrum guide bore. The fulcrum lever is therefore tilted to the left, thus causing the control-rod to move into an increased fuel delivery position.

Maximum speed operating conditions
(Fig. 11.33(d))

When the engine approaches its maximum rated speed the outward force applied by the flyweights will eventually overcome the combined stiffness of the three springs. The flyweight centrifugal force therefore compresses the springs, thus causing the lower spring plate to lift away from the shouldered boss, which is supporting the spring posts. This radial outward movement is immediately converted into an axial inward shift of the slide-pin, caused by the bell-crank lever swivelling about the flyweight-carrier pivot point. Under these conditions, the fulcrum-lever tilts clockwise about the guide-plunger pivots, thus making the control-rod move to the fuel delivery cut-off position. The governor is therefore able to override the accelerator control lever maximum

fuel delivery demands, thereby preventing the engine over-speeding.

Engine torque and control-rod opening versus speed characteristics are shown for a minimum–maximum speed governor in Figs 11.35 and 11.36.

Engine shut-off position
(Fig. 11.33(a))

The idle stop is a spring loaded plunger. However, when the accelerator-lever is pulled hard back against the idle stop, the stop spring is compressed sufficiently to permit the accelerator-lever to tilt beyond the idle speed position. With the engine running at idle speed the guide plunger and arm are able to swing the fulcrum lever clockwise so that the control-rod shifts to its shut-off position.

11.5.8 Variable speed flyweight governor (Bosch Type-EP/RSV)
(Figs 11.37(a–d), 11.38, 11.39 and 11.40)

Construction

A pair of flyweights are mounted at one end of the camshaft. These pivoting flyweights are inter-

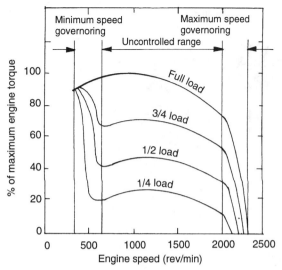

Fig. 11.35 Engine torque characteristics with a minimum–maximum speed governor

linked to the control-rack via a guide-sleeve, guide-lever, fulcrum-lever and strap-link. The guide-lever is pivoted to the case at the top end and to the sliding-pin at its lowest point, whereas the fulcrum-lever is pivoted in a slot formed in the casing at its lowest point and to the guide-lever above this point. Its upper end is attached to the strap-link which connects it to the control-rack.

A single governor spring is used to oppose and balance the flyweight centrifugal force.

Governor spring tensioning load is achieved through an accelerator-control lever and a swivel-

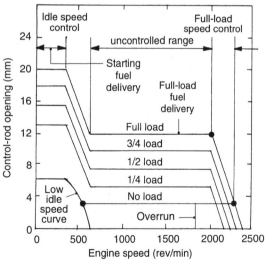

Fig. 11.36 Control-rod-opening characteristics with a minimum–maximum speed governor

lever (Fig. 11.38) sharing a common pivot and a tensioning-lever pivoted at its upper end to the casing, with its lower end free to contact the guide-sleeve sliding-pin. The governor spring is linked between the rocker-arm which forms part of the swivel-lever and the tensioning-lever.

Centrifugal force versus governor spring tension (Fig. 11.37(a))
Engine speed is increased by tilting the accelerator lever anticlockwise. This causes the rocker arm attached to the upper part of the swivel-lever to move further away from the tensioning-lever. Accordingly, the governor spring stretches and pulls the tensioning lever in the direction of the flyweights. The retraction force of the extended spring therefore shifts the sliding-pin and guide-sleeve against the opposing flyweight lever fingers. Equilibrium will be reached when the centrifugal force exerted at the flyweight fingers equals the opposing spring thrust relayed to the sliding-pin via the tensioning-lever. As a result, the fulcrum-lever will tilt and take up a position which moves the control-rack towards maximum fuel delivery, until the opposing flyweight centrifugal thrust balances the governor spring tension.

Governor response with changes in engine load (Fig. 11.37(a))
Governor control can be achieved in the following manner.

Assume the engine to be rotating at a steady speed with a given accelerator-control lever setting, without a speed sensitive governor. Then, if the load on the engine is suddenly increased or decreased, the engine will immediately respond with a fairly rapid reduction or rise in engine speed, respectively. This is because the quantity of fuel delivered to the engine remains the same and, therefore, does not increase or decrease the engine's power to compensate for changes in load imposed on the engine. However, with a flyweight governor, a decrease or increase in engine speed produces an equal variation in the injection pump camshaft speed and therefore produces an equivalent change in flyweight centrifugal force reacting on the guide-sleeve. Accordingly, the flyweight thrust opposing the governor spring thrust, which has been determined by the accelerator-lever position, is relayed to the guide-sleeve via the tensioning-lever and sliding-pin. The result is that the sliding-pin and guide-sleeve shift to a new position, this causes the fulcrum-lever to tilt so that the control-rod opening moves

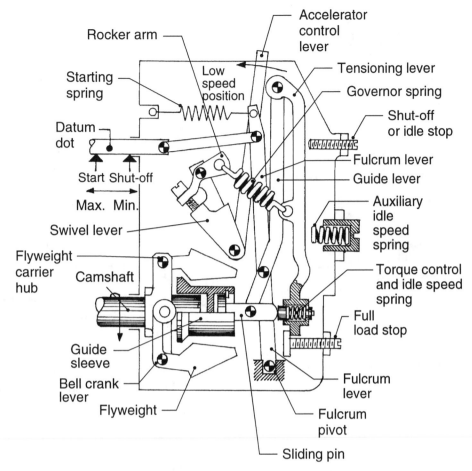

Labels on figure:

Rocker arm

Accelerator control lever

Starting spring

Low speed position

Tensioning lever

Governor spring

Shut-off or idle stop

Datum dot

Fulcrum lever

Guide lever

Start Shut-off

Max. Min.

Auxiliary idle speed spring

Swivel lever

Flyweight carrier hub

Camshaft

Torque control and idle speed spring

Full load stop

Guide sleeve

Bell crank lever

Flyweight

Fulcrum lever

Fulcrum pivot

Sliding pin

(a) Full load at low speed; start of torque control

Fig. 11.37 Variable speed flyweight governor (Bosch Type EP/RSV)

to increase fuelling if the engine speed falls and, conversely, to reduce fuel delivery if the engine speed should rise. This cycle of events is continuous and, provided the change in fuel delivery responds directly to the slightest variation in engine speed, the outcome will be to minimize the engine's overall speed fluctuation within the limits of the resultant movement relayed to the control-rack.

Engine torque and control-rod opening versus speed characteristics are shown for a variable speed governor in Figs 11.39 and 11.40.

Full-load low-speed conditions (Fig. 11.37(a))
At full-load low engine speed, the accelerator control lever has only to be tilted anticlockwise a small amount to tension the governor spring

sufficiently to oppose the flyweight centrifugal force thrust and for the tensioning-lever to tilt hard against the shoulder of the full-load stop-screw. Under these conditions the control-rack moves to the full-load fuel delivery position.

Full-load intermediate speed condition
At full-load intermediate engine speed, the accelerator control-lever is moved fully to the maximum output position to increase the governor spring tension so that the flyweights can be held together at intermediate speeds, thereby permitting the fulcrum-lever to tilt its full extent with the tensioning-lever held against the shoulder of the full-load stop-screw. The control-rack, therefore, is able to move to the full-load fuel delivery position.

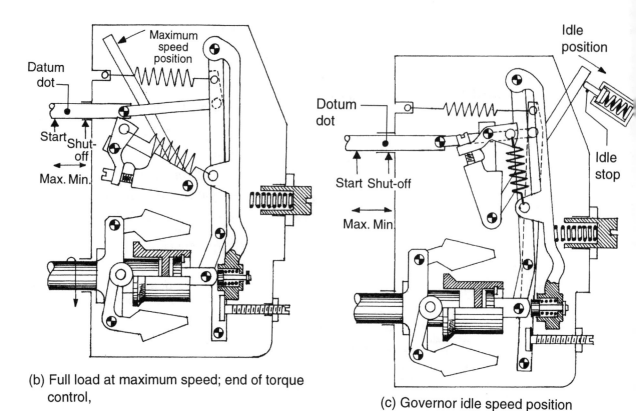

(b) Full load at maximum speed; end of torque control,

Fig. 11.37(b)

(c) Governor idle speed position

Fig. 11.37(c)

Full-load high-speed conditions (Fig. 11.37(b))
At full-load high-speed, the accelerator control lever is moved to its maximum output setting. This tensions the governor spring so that the spring thrust almost holds the flyweights together. However, at very high engine speeds the flyweight thrust, while not exceeding the governor spring thrust, is able to compress the torque control spring housed at the end of the tensioning-lever. The guide-sleeve and the sliding-pin therefore move nearer to the tensioning-lever by a small amount. As a result, the tilt of the fulcrum-lever moves the control-rack slightly back from the maximum fuel delivery position.

Maximum speed governor cut-in position
As the engine approaches the maximum rated speed with the accelerator control lever in the full-load position, the predetermined maximum governor spring stiffness now becomes insufficient to oppose and balance the flyweight centrifugal force. Accordingly, the guide-sleeve and sliding-pin shift to the right. This causes the guide-lever to swing anticlockwise about its upper

pivot and, since the fulcrum-lever is also pivoted to the guide-lever, the latter tilts clockwise at the same time. Hence, the control-rack will be pulled towards the shut-off position, thereby moving the injection pump plungers into the no-delivery position.

Idle speed position (Fig. 11.37(c))
When the accelerator-control lever is moved to the spring loaded idle-stop position, the swivel-lever and rocker-arm move to an almost vertical position so that the governor spring is in its minimum stretch position. With the very low spring stiffness, the flyweights at idling speed move outwards until the centrifugal thrust on the guide-sleeve equals the governor spring thrust exerted on the sliding-pin by the lower end of the tensioning-lever. The auxiliary idle speed spring supplements the governor spring tension, and it therefore stabilizes the idling speed and prevents the engine stalling. Under these conditions the outward movement of the guide-sleeve and sliding-pin tilts both the guide-lever and the fulcrum-lever to the right. The top end of the fulcrum-

540

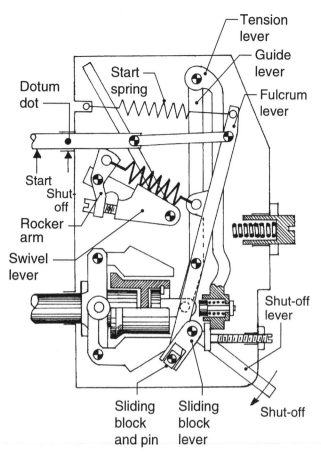

Fulcrum lever

(d) Stopping engine with shut-off device

Fig. 11.37(d)

lever therefore pulls the control-rack over to the right until the idle speed equilibrium position is obtained.

Engine starting position
When the engine is stopped the accelerator-control lever returns to the idle-speed position while the camshaft and flyweights come to rest. The starting spring stretched between the upper end of the fulcrum-lever and the casing is now free to pull the fulcrum-lever and the control-rack to the starting fuel delivery position. Thus, the engine can be started with the accelerator-control lever in the idle-speed position; that is, against the idle-stop. Once the engine fires and its speed increases, the centrifugal thrust produced by the flyweights swinging outwards opposes and exceeds the starting spring tension and will there-

fore move the fulcrum-lever and the control-rack to a reduced fuel delivery position.

Engine stopping

With accelerator control lever
When the accelerator control lever is moved fully to the shut-off position the idle stop spring is compressed so that the lever moves beyond the idle speed position. Since the accelerator-control lever and the swivel lever are attached to the same pivot shaft, the swivel-lever and rocker-arm swing around until the protruding lugs on the swivel-lever touch the guide-lever. Further rotation of the swivel-lever pushes the guide-lever and the fulcrum-lever to the right until the control-rack reaches the shut-off position; this movement being limited by the rocker arm sector face contacting the shut-off stop screw.

With separate stop lever
(Fig. 11.37(d))
To stop the engine the shut-off lever is pressed down to the shut-off position. This movement simultaneously rotates the sliding-block lever, which in turn sufficiently swings the fulcrum-lever in an anticlockwise direction to shift the control-rack to the shut-off position. The sliding-block compensates for the necessary change in distance between the shut-off lever pivot shaft and the lower pivot which is attached to the fulcrum-lever. The shut-off lever therefore overrides the accelerator-control lever which could be in any position.

11.5.9 Variable speed flyweight and leaf spring governor (CAV Minimec)
(Fig. 11.41(a–d))

Construction
(Fig. 11.41(a))
The governor consists of a pair of flyweights pivoting on a carrier rigidly attached to a flange formed on the injection pump camshaft. Fingers on the flyweights transmit the centrifugal force generated to a control-sleeve which surrounds and slides on the extended camshaft. Counteracting the flyweight thrust is a leaf-spring assembly which is hinged at the top but which contacts a grooved thrust-pad mounted on the camshaft at its lower end. Friction between flyweight control-

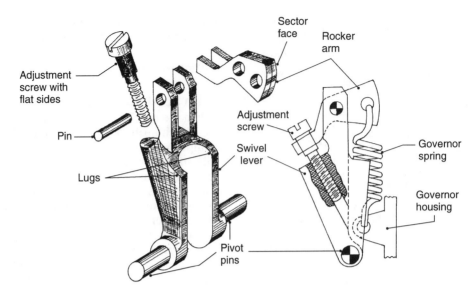

Fig. 11.38 Swivel lever assembly (Bosch Type EP/RSV governor)

sleeve and the spring thrust pad is minimized by a roller thrust race which is aligned between the two opposing members. The varying resultant thrust created by the opposing flyweights and the flat leaf-spring shifts the control-sleeve and thrust-pad to and fro, this motion is, in turn, transferred to the control-rod via the rocking-lever which is free to tilt about its pivot shaft.

A roller located between an inclined-ramp and the leaf-spring, when moved up or down the ramp, alters the effective spring stiffness. Pushing the roller down the ramp weakens the spring by

increasing its length, whereas moving the roller up the ramp shortens the spring length thus strengthening the spring.

Increasing the spring force acting on the thrust pad tends to tilt the rocker-lever clockwise and therefore increases the fuel delivery. Conversely, reducing the spring stiffness allows the flyweight thrust initially to exceed the opposing spring thrust. Consequently, the rocker-lever now pivots anticlockwise so that the control-rod is moved to a reduced fuel position.

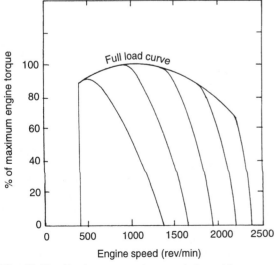

Fig. 11.39 Engine torque characteristics with a variable speed governor

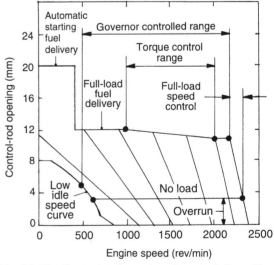

Fig. 11.40 Control-rod-opening characteristics with a variable speed governor

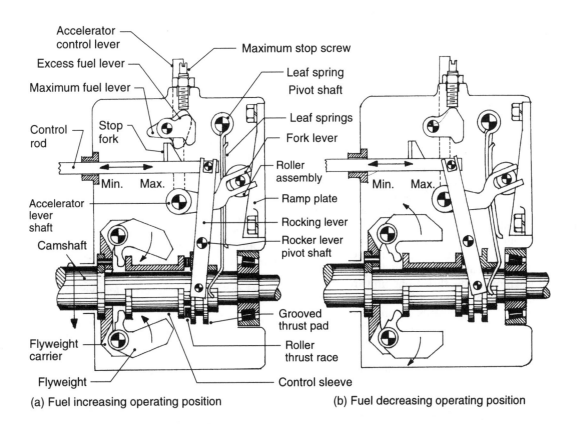

(a) Fuel increasing operating position   (b) Fuel decreasing operating position

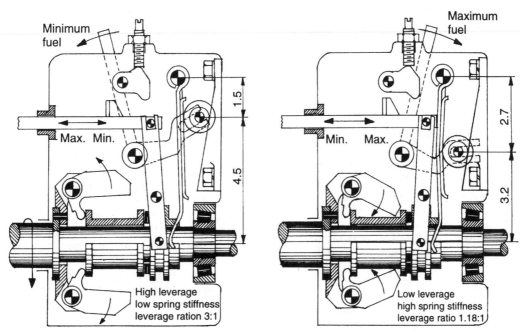

High leverage
low spring stiffness
leverage ration 3:1

Low leverage
high spring stiffness
leverage ratio 1.18:1

(c) Idle speed operating position   (d) Full load operating position

Fig. 11.41   Variable speed flyweight and leaf spring governor (CAV Minimec)

Principle of operation
(Fig. 11.41(a and b))

With the accelerator control-lever positioned so that the fork-lever moves the roller to the mid-incline position on the ramp, the centrifugal force generated by the flyweights swings them outwards until the opposing spring deflection produces an equal spring thrust. The state of balance between the centrifugal force and the spring resistance tilts the rocker-lever and therefore moves the control-rod to a fuel delivery position which matches the engine's speed and load conditions.

Should now the engine load increase, the engine speed will rapidly decrease causing a proportional reduction in flyweight centrifugal force. Accordingly, the spring thrust determined by the roller position is greater than the flyweight thrust. It therefore moves the thrust-pad inwards to a new equilibrium position. Simultaneously, the control-rod is drawn towards an increased fuel position via the tilting-rocker lever. Immediately the increased fuelling increases the engine speed back to its original value.

However, if the engine load decreases, the engine speed increases, thus also causing the flyweight centrifugal force to rise. As a result, the centrifugal force exceeds the original spring thrust and it therefore deflects the leaf-spring until the thrust-pad outward movement reaches a new equilibrium position. The interlinkage therefore tilts the rocker-lever anticlockwise and moves the control-rod to a decreased fuel delivery position. This, therefore, enables the engine speed to be restored to the initial load speed.

Idle speed operation position
(Fig. 11.41(c))

When the accelerator-lever is moved to the idle speed position, the fork-lever moves the roller down the incline towards the spring pivot shaft. Accordingly, the leaf-spring is tilted to the rear, causing the effective distance between the roller and the thrust-pad to be at its maximum, so that the stiffness of the spring will be at its minimum. Under these conditions the centrifugal force generated by the flyweights at very low speed is sufficient to oppose and deflect the governor leaf-spring. Consequently, the outward movement of the thrust-pad shifts the control-rod via the rocking-lever to the idle-speed fuel delivery position. With any speed variation caused by a sudden change in engine load, the weights immediately respond by producing a corresponding

change in the centrifugal thrust counteracting the spring thrust. Thus, the thrust-pad takes up a new equilibrium position causing a similar correction to be made by the control-rod to compensate by either reducing or increasing fuelling to the cylinders.

Full load operation position
(Fig. 11.41(d))

Rotating the accelerator control-lever fully clockwise twists the fork-lever so that the roller is pulled up the ramp incline and away from the spring's pivot shaft. This shortens the distance between the roller and thrust-pad which proportionally increases the stiffness of the spring. The downward movement of the roller also pushes the lower half of the spring to the left. Thus, the much reduced effective spring length exerts considerably more thrust to the opposing centrifugal force, and the thrust pad therefore takes up a new equilibrium position further to the left. As a result, the rocking-lever shifts the control-rod to the full fuel delivery position. Any load variation will produce a corresponding speed fluctuation which, in turn, moves the thrust-pad and control-rod to a new position; that is, a reduced fuelling position to lower the engine's speed or an increased fuel delivery setting to raise the engine's speed.

Fuel cut-off position *(Fig. 11.42(a–c))*

An external stop-lever is clamped to the outer end of the hollow stop shaft, the inner end of this shaft supports the excess fuel/stop bell-crank-lever. When the stop-lever is rotated fully clockwise the toe of the excess/stop lever butts against the stop-fork attached to the control-rod and thus pushes the control-rod to the fuel cut-off position.

Maximum fuel stop *(Fig. 11.42a)*

A bell-crank-lever mounted on the hollow shaft links the control-rod stop-fork to the tip of the maximum fuel stop-screw. When the control-rod reaches its maximum fuel opening travel, further movement is prevented by the maximum-fuel-lever interconnecting the fork-stop to the maximum-fuel-screw. Thus the maximum fuel setting can be adjusted exernally via the fuel-stop-screw.

Excess fuel stop *(Fig. 11.42c)*

The excess fuel device consists of a pair of bell-crank-levers, one limiting maximum fuel delivery and the other one affecting the control-rod excess

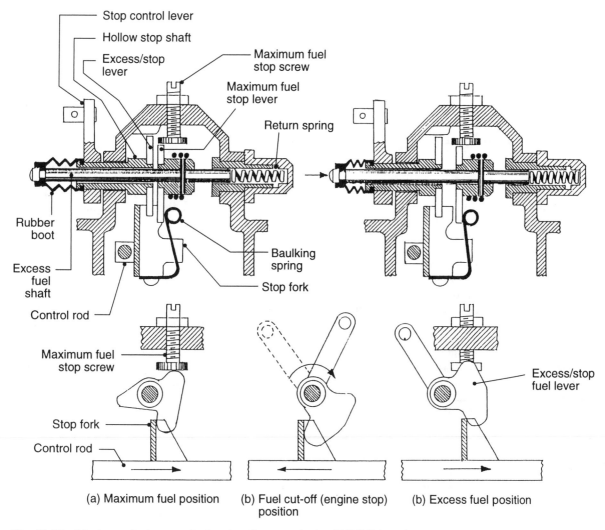

Fig. 11.42 Maximum fuel excess fuel and engine stop device (CAV Minimec)

Labels in figure:
- Stop control lever
- Hollow stop shaft
- Excess/stop lever
- Maximum fuel stop screw
- Maximum fuel stop lever
- Return spring
- Rubber boot
- Excess fuel shaft
- Control rod
- Baulking spring
- Stop fork
- Maximum fuel stop screw
- Stop fork
- Control rod
- Excess/stop fuel lever

(a) Maximum fuel position   (b) Fuel cut-off (engine stop) position   (b) Excess fuel position

fuel and the fuel cut-off position. The excess/stop-lever is mounted on the hollow stop shaft whereas the maximum fuel-stop-lever is attached to the excess fuel shaft which is itself supported inside the hollow stop shaft. During normal operating conditions the maximum fuel-stop-lever and the excess fuel shaft is pulled sideways towards the excess/stop lever by a baulking (leaf) spring which is riveted to the stop-fork which is itself mounted on the control-rod. For cold starting when additional fuel is required the excess fuel shaft is pressed inwards thereby moving the maximum fuel-stop-lever sideways out of line with the stop-fork. With the accelerator-control-lever in full-load position the governor spring will then push the control-rod via the rocking-lever beyond

the maximum fuel position until the stop-fork butts against the excess/stop-fuel-lever. Once the engine starts the governor flyweights move the control rod away from the excess fuel position so that the stop fork clears the maximum fuel stop lever. The return coil spring immediately moves the maximum fuel-stop-lever sideways and closer to the excess/stop-lever, this blocks the stop-fork movement and therefore permits the control-rod moving beyond the maximum fuel-stop-lever.

If the excess fuel shaft is held in when the engine is running the governor action draws the control-rod back from the excess fuel position until the baulking spring clears the side of the maximum fuel-stop-lever. The baulking spring therefore deflects to its free position where it aligns with the edge of the maximum fuel-stop-lever. Any further movement of the control-rod towards the excess fuel position is prevented by the circular end of the spring touching the maximum fuel-stop-lever.

11.5.10 Variable speed pneumatic governor (Bosch Type-EP/M)
(Figs 11.43(a–d) and 11.44)

Construction
(Fig. 11.43(a))

The pneumatic governor consists of a diaphragm sandwiched between the injection pump housing and the vacuum chamber. The central cylindrical extension on the pump side of the diaphragm is supported by a slotted guide-block attached lower down to the control-rod. Any tilt tendency is prevented by the guide-block slot which fits over the guide-rod.

A start/stop double-arm lever is located and pivoted just below the control-rod; when starting, its lower end contacts a full-load adjustable spring-loaded stop, whereas its upper end contacts the cylindrical extension of the diaphragm when stopping. An idle-speed screw and auxiliary spring is located to the right and in the centre of the vacuum chamber. A governor spring is positioned between the diaphragm and the vacuum-chamber casing, its extending tendency will move the injection pump contol-rod towards full-load fuel delivery.

The air intake into the induction manifold is partially throttled by a slotted butterfly valve mounted inside a venturi housing. The depression concentrated inside the small venturi is conveyed via a pipe from the small orifice situated at the throttle spindle level to the governor vacuum chamber.

The auxiliary venturi ensures that only small degrees of depression are ever experienced during idle speed and part throttle operating conditions.

The auxiliary venturi provides an escape route for exhaust gases produced if the engine should fire early, kick-back, and commence to rotate in the reverse direction. The engine must be immediately stopped to prevent damage occurring. This can be achieved by cutting off the fuel supply by means of the stop-knob. However, if the throttle has been opened the engine will race, and therefore a high gear should be selected and the clutch engaged with the handbrake on until the engine stalls.

Principle of operation
(Figs 11.43(a) and 11.44)

The control-rod opening position is dependent upon the balance between the governor spring thrust that tends to move the control-rod towards maximum fuel discharge and the auxiliary venturi depression conveyed to the governor vacuum chamber, which opposes the spring thrust and tries to move the control-rod in the opposite direction towards minimum fuel delivery. For a given throttle opening, the engine speed will increase until the auxiliary venturi depression comes within the working spring stiffness range, at which point the depression equals or even exceeds the spring thrust. This is normally between 400 and 500 mm of water column.

If, at this speed and throttle opening, the engine load either increases or decreases there will be a corresponding reduction or rise in engine speed with a proportional decrease or increase in venturi depression respectively.

If the engine load increases, the depression will decrease so that the governor spring is able to shift the control-rod slightly further towards maximum fuel delivery to compensate for the extra power needed to maintain the engine's nominal speed. Conversely, if the engine load decreases, the rise in engine speed immediately increases the depression in the venturi, and it therefore causes the diaphragm to be drawn slightly more towards the minimum fuel delivery position. Thus, the change in the depression experienced in the governor vacuum chamber quickly moves the control-rod to a new position of equilibrium. The rapid response in increasing or decreasing fuel delivery with changes in engine load is normally sufficient to keep the speed fluctuation to a minimum for one given throttle opening.

If the throttle is opened to a new setting the venturi depression momentarily decreases (Fig. 11.44), this permits the governor to increase the control-rod opening until the increase in engine speed has brought the auxiliary venturi depression back to the governor working range. When this happens, the depression opposing the governor spring commences to adjust the control-rod opening and, accordingly, the fuel delivery, in order to compensate for small changes in engine load so that engine speed variations are small. Likewise, if the throttle opening is reduced to a new setting, the venturi depression will instantly

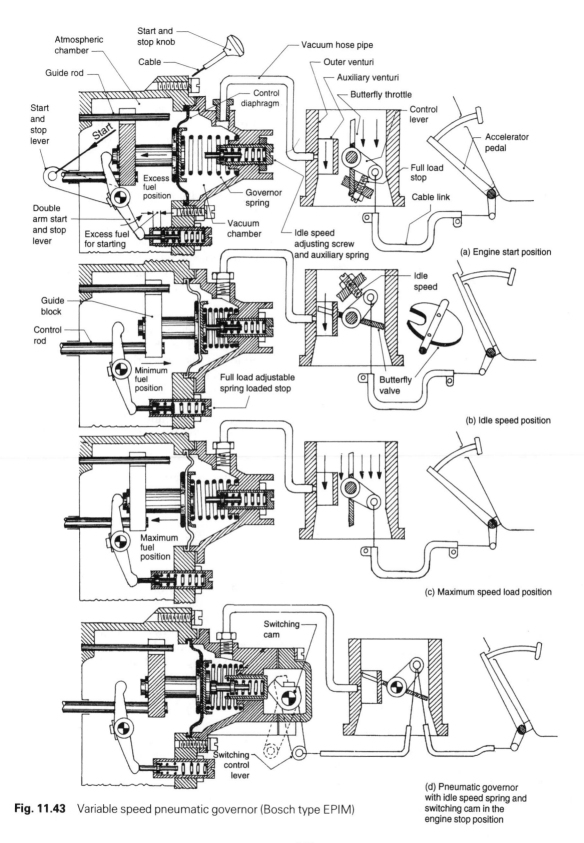

Fig. 11.43 Variable speed pneumatic governor (Bosch type EPIM)

547

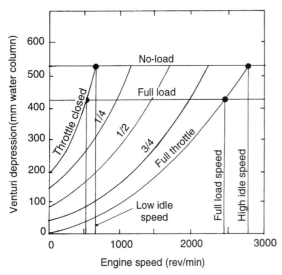

Fig. 11.44 Relationship between venturi depression throttle opening and engine speed

increase so that the control-rod is drawn back to a reduced fuel delivery position. The engine speed then decreases until the reduced engine's power requirement matches the reduction in fuel delivery, and the venturi depression again comes within its normal regulating range.

Engine starting position
(Fig. 11.43(a))

When starting the engine the start-knob cable is depressed which causes the start-and-stop lever to rotate the double-arm lever anticlockwise to a point where the lever-arm has compressed the full-load spring stop inwards. As a result, the unopposed governor spring is free to move the diaphragm and control-rod to the excess fuel delivery position.

Idle-speed position
(Fig. 11.43(b))

With the throttle closed and its lever set against the idle-speed stop the high venturi depression draws the diaphragm and control-rod towards the idle-speed spring-loaded stop. The fuel delivery is therefore set to match the nominal idle speed requirements.

If the engine load increases, its speed drops causing the venturi depression to decrease. The governor spring immediately moves the diaphragm and control-rod opening to an increased fuel delivery position, thus enabling the engine speed to recover. Conversely, if the engine load

decreases, the engine speed rises and causes the venturi depression to rise also, this pulls the diaphragm up to the idle-speed stop. If the depression becomes excessive, the diaphragm develops sufficient thrust to compress the much stiffer idle-speed spring. It therefore permits the control-rod to move to a much reduced fuel delivery position, and the idle speed fluctuation will therefore be considerably restricted and stabilized.

Maximum speed position
(Fig. 11.43(c))

As the throttle is progressively opened for a given engine speed, the venturi depression will decrease and, simultaneously, the governor spring expands until the spring thrust balances the diaphragm thrust created by the depression in the vacuum chamber.

Thus, as the engine approaches its maximum rated speed, the relatively low venturi depression is equally matched by the considerably extended and therefore weakened spring.

Hence, the slightest change in engine speed produces a corresponding change in venturi depression. Therefore, when the speed of the engine reaches its maximum rated value the slight throttling effect of the air rushing through the main and auxiliary venturi is sufficient to cause the diaphragm to pull the control-rod towards the minimum fuel delivery position. The result is a reduction in the engine's power output and a rapid fall off in speed.

Engine stop position
(Fig. 11.43(a))

The engine is stopped by pulling the stop-knob cable so that it twists the double-arm stop-lever clockwise until the upper arm contacts the diaphragm cylindrical extension. Further rotation of the stop-lever first draws the control-rod to the idle speed position, and secondly it compresses the idle-stop spring so that the control-rod is now able to move to the fuel cut-off position.

Auxiliary idle speed spring and switching cam
(Fig. 11.43(d))

The auxiliary idle-spring stop has an inner plunger and spring enclosed in a cylindrical outer plunger, which is itself located horizontally in the bore of the diaphragm chamber housing. The inner plunger stem projects into the diaphragm chamber and only butts the diaphragm when the depression in the chamber is high. The actual point of contact between the inner plunger and diaphragm is determined by the angular position of the switching-cam and the degree of depression experienced in the diaphragm chamber at any instant. When the throttle valve is partially opened the interlinking cable swivels the switching-cam so that it permits the plunger assembly to move outwards away from the diaphragm. However, when the accelerator pedal is released the throttle valve closes so that the cable twists the switching-cam lobe until it moves the plunger assembly towards the diaphragm.

When the accelerator pedal is released, the throttle valve closes and the depression in the diaphragm chamber rises, immediately the control-rod will be drawn towards the no-fuel delivery position against the opposing main governor spring thrust. Under light-load small fuel delivery conditions, the control-rod response to the slightest change in depression may not be sensitive enough to provide an instant reaction, so that the speed fluctuation may be considerable—which may even cause the engine to stall.

The object of the switching-cam is to project the spring-stop towards the diaphragm when the throttle is closed, particularly when the engine is running fast on overrun and the venturi depression exceeds the normal steady idle-speed conditions. Consequently, the additional resistance, opposing the diaphragm depression thrust, resists the control-rod surging further towards the no-load fuelling position. It therefore stabilizes the idle speed and prevents fuel cut-off during engine overrun operating conditions.

11.6 Manifold pressure compensator (boost control)

11.6.1 The need for boost control
(Fig. 11.45(a))

The engine's volumetric efficiency, when equipped with a turbocharger, shows a similar volumetric efficiency to a naturally aspirated engine at low engine speed. It then begins to rise as more exhaust gas energy becomes available to raise the turbocharger's speed and hence boost pressure. The boost pressure will eventually reach its peak so that the volumetric efficiency will then follow a similar curve to the naturally aspirated

engine—however, it is at a higher level and with less fall-off as the engine approaches maximum speed (Fig. 11.45(a)).

Fuel delivery for full-load output is determined by the maximum control-rod opening stop-setting. However, if the maximum fuel delivery is set to match the corresponding mass of air forced into the cylinders at full boost pressure then, in the lower speed range where the boost pressure practically does not exist, there would be a deficiency of air relative to the fuel delivery, with the result that smoke would be visible under these operating conditions.

To eliminate the poor match of air-mass to fuel delivery ratio over the engine's entire speed range, a manifold pressure-sensitive maximum fuel-stop is used. This device automatically adjusts the full load maximum setting of the control-rod to the varying quantity of air filling the cylinders, which is basically determined by the turbocharger output characteristics.

11.6.2 Operating principle
(Fig. 11.45(b and c))

Boost control units are variable fuel stops which limit the maximum opening of the control-rod; that is, the full-load fuel delivery according to the boost pressure entering the cylinders. This boost pressure sensing and actuating device consists of a diaphragm, one side of which is subjected to manifold pressure and the other side is exposed to atmospheric pressure and the thrust of the return spring. Attached to the diaphragm is a downward extending spindle which slides in an adjustable guide-sleeve. A set bolt with a lock nut is screwed into the lower end of this spindle and a bell-crank lever converts the vertical diaphragm assembly movement to a horizontal variable position control-rod stop. The steel pressing forming the strap-link connects the top end of the fulcrum-lever to the control-rod. It also has a vertical lug which aligns with the bell-crank lever stop. Therefore, when the fulcrum-lever is moved in the direction of maximum output, the strap-lug with the control-rod shifts towards maximum fuel delivery until it is blocked by the bell-crank lever stop.

When the turbocharger boost pressure is low (Fig. 11.45(b)) the diaphragm spring will lift up the diaphragm assembly until it contacts the upper stop set screw. Under these conditions, the control-rod will not move so far before the strap-

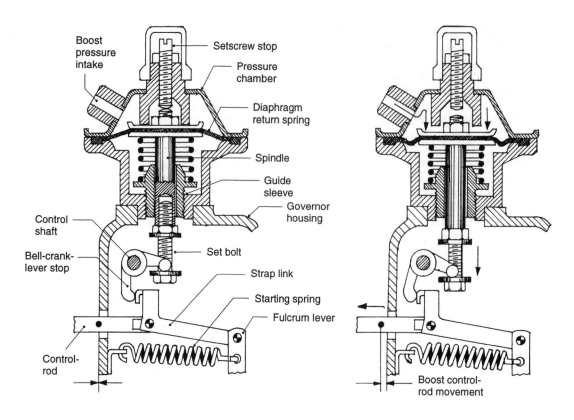

(b) Full load fuel delivery without boost control

(c) Full load fuel delivery with boost control

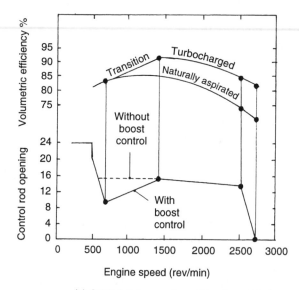

(a) Control-rod—opening setting to match the turbocharged engines volumetric efficiency characteristics

Fig. 11.45 Manifold pressure compensator (boost control) unit

550

link lug butts the lower arm of the bell-crank lever stop.

Conversely when boost pressure begins to build up, the difference in pressure across the diaphragm will be sufficient to push down the diaphragm against the spring thrust (Fig. 11.45(c)). Thus, when the accelerator-lever is moved towards maximum load output, the strap-link will move in the direction of maximum fuel delivery. The lug will then contact and rotate the bell-crank lever clockwise until the slack between the lug and vertical setbolt is taken up. The control-rod travel is therefore extended so that the maximum full-load fuel delivery is also increased to match the increased quantity of air forced into the cylinders at higher engine speeds.

The bell-crank lever is designed to move sideways when excess fuel delivery is needed for cold starting, this permits the strap-lug to move beyond the bell-crank lever stop to increase the control-rod opening.

11.6.3 Automatic speed advance device

The need for advancing the injection timing

The time it takes from the closure of the spill port to the instant fuel actually enters the combustion chamber remains roughly constant. Likewise, it takes a definite time from the point when fuel is sprayed into the combustion chamber to the instant where some of the oxidizing fuel droplets ignite. Thus, there is an injection and ignition time lag which is dependent on engine speed but is approximately constant in time.

This delay period between the point where the plunger cuts-off the spill port to the instant burning actually commences will occur over a definite number of degrees of crankshaft rotation relative to one engine speed. Hence, the angular movement of the crankshaft over this delay period will be proportional to the engine's speed; thus, if this period is equivalent to $10°$ crankshaft movement at 1000 rev/min, then at 2000 rev/min the crankshaft would rotate $20°$ in the same time span.

Consequently, if the injection timing is set at low engine speed so that the peak combustion pressure occurs at the optimum performance position after TDC (usually about $10°$), then with increasing engine speed peak cylinder pressure will progressively take place later in the power stroke where combustion becomes less effective.

The retardation of combustion relative to crankshaft angular movement can be counteracted by bringing forwards the injection timing proportional to engine speed.

It has been found that a medium speed diesel engine requires the injection timing to be advanced approximately $6°$ between engine starting and maximum speed, whereas a small high-speed diesel engine needs twice this (or more) injection advance over its speed range. Thus, it becomes essential to incorporate some means to advance the injection timing for engines which have a wide operating speed range.

Construction
(Fig. 11.46(a and b))
This unit forms a coupling between the engine injection pump drive and the injection pump camshaft. It consists of an input drive cylindrical housing and an output pump camshaft hub and cam-plate. The drive between these two input and output members is conveyed through a pair of flyweights, two pairs of drive springs, a pair of spring reaction plates, a pair of pivot posts and a pair of rollers and roller pins.

When the unit components are assembled the input cylindrical member relays its drive to the output hub and cam-plate member via the drive springs, which are mounted between the spring reaction plate supported on the pivot-posts and the driven cam-plate.

Low engine speed
(Fig. 11.46(c))
When the engine is stationary or running at idle speed, the drive springs force the flyweights with their rollers against the profiled face of the driven cam-plate. This causes the swing weights to move nearer the hub, to their innermost position. Under these operating conditions the injection timing will be fully retarded.

High engine speed
(Fig. 11.46(d))
With rising engine speed, the hinged flyweights swing outwards under the influence of the centrifugal force. This movement is opposed by the rollers which ride up and outwards on the camplate profile and, in so doing, the movement is resisted by the four drive springs which are progressively compressed. The outward movement of the rollers pushes the cam-plate and its hub

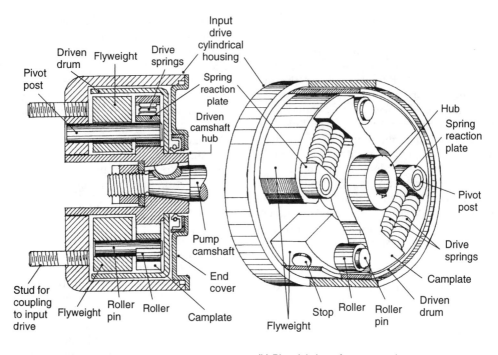

(a) Sectional view of an automatic advance device

(b) Pictorial view of an automatic advance device

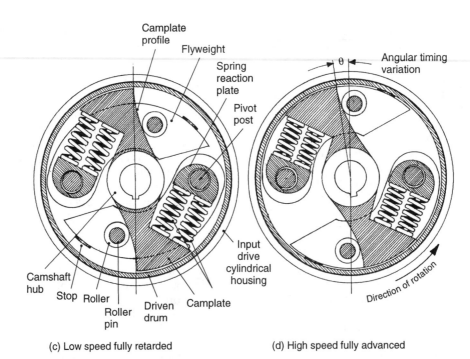

(c) Low speed fully retarded

(d) High speed fully advanced

Fig. 11.46 Automatic speed related timing device (Bosch)

552

nearer to the spring reaction plates, it therefore rotates the cam-plate relative to the input cylindrical housing in the same direction of rotation. Consequently, the point of injection is advanced in proportion to the rise in engine speed.

This automatic advance device is capable of rotating the camshaft relative to the input drive through 8° camshaft movement, i.e. 16° crankshaft movement.

The complete unit is sealed after it has been filled with oil, this lubricates the moving components and damps out vibrations transmitted from the engine. These units are now maintenance-free and are permanently lubricated.

11.7   Fuel injectors

11.7.1   Injector design features
(Fig. 11.47)

The injector can basically be considered in two parts: (1) the lower nozzle body and needle valve assembly; (2) the upper injector body which contains the valve spring and preload adjustment device and which is normally a screw adjustment sleeve and capnut or simply a spacing shim. Spring thrust (compressive load) is transferred to the needle valve via either a spindle with the high-mounted spring or with a spring plate for the low-mounted spring. Both halves of the injector unit, consisting of the upper injector body (holder) and the lower nozzle and needle valve assembly, are held rigidly together with a nozzle capnut.

Low and high spring injectors
(Figs 11.47, 11.48, 11.49 and 11.50)
Injectors may have high mounted valve springs (Fig. 11.47) which are separated from the needle valves by a long spindle. It therefore isolates the hot end of the injector from the valve spring and makes adjustment to the spring preload relatively simple. However, with maximum engine speeds for small diesel engines now reaching 5000 rev/min, which compares with something like 3500 rev/min and 2000 rev/min for medium and large-sized road transport engines respectively, particular attention must be directed to keeping the inertia for the reciprocating needle valve assembly to a minimum. The obvious solution has been to position the valve spring as near to the needle valve as possible (Figs 11.48 and 11.49).

Thus, low-mounted small-diameter valve springs have become popular as they dispense with the long stem so that spring surge is avoided and needle bounce is essentially eliminated. Thus, the opening and closing of the valve becomes more responsive and precise; that is, there is a steeper rise and fall (Fig. 11.50) of the needle with respect to crank angle movement.

Needle valve and nozzle body
(Fig. 11.51(a–e))
The nozzle body is made from a solid cylindrical bar of carbon steel which is turned to form a stepped large and small diameter body. The discharge end of the nozzle will be flat if it is of the pintle type (Fig. 11.51(a–e)) or dome-shaped in the case of the multi-hole type (Fig. 11.51(a and b)) whereas the opposite end of the nozzle, which is open, has a flat ground and lapped pressure face with an annular groove. A concentric hole is drilled and ground through the axis of the nozzle body to accommodate the needle valve. The upper portion of the cylindrical bore forms a lapped sliding fit of 2–3 μm with the large diameter portion of the needle, whereas the annular space between the needle valve and the lower half of the nozzle body bore provides the passageway for the fuel. Separating the lapped close fit and annular passageway portions of the nozzle bore is a large semi-circular fuel gallery groove. The enclosed end of the nozzle bore provides the needle valve with its tapered seat, and the small cavity between the nozzle needle seat and the discharge spray orifice (holes)—known as the sac—becomes the fuel distribution point for the spray when the needle valve opens. A fuel feed drilling connects the fuel supply passages in the injector body to the fuel gallery surrounding the tapered shoulder of the needle valve. The domed nozzle nose may contain one to ten discharge spray holes of 0.2 to 0.45 mm diameter. The number and diameter of the discharge holes depends upon the engine cylinder cubic capacity, the intensity of air swirl and the speed range of the engine. A centrally mounted injector in a four-valve cylinder-head has symmetrically drilled spray holes in the nozzle nose whereas inclined injectors fitted to two-valve cylinder-heads have asymmetrically located spray holes.

Fuel leakage between the lapped nozzle bore and needle valve stem circulates a minimal quantity of fuel which both lubricates and contributes to the cooling of the sliding pair.

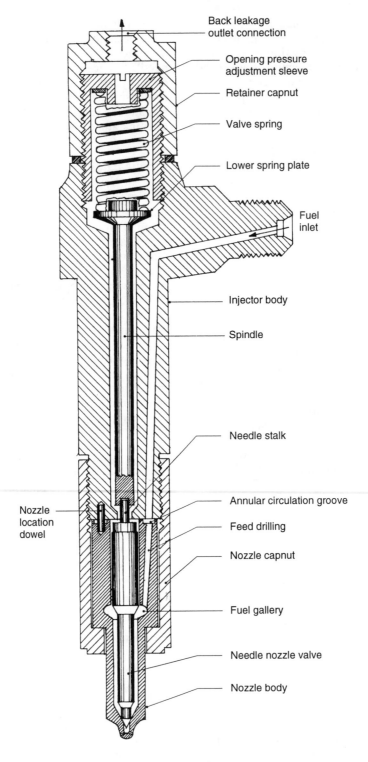

Fig. 11.47   High spring injector with multi-hole long stem nozzle

Labels in figure:
- Back leakage outlet connection
- Opening pressure adjustment sleeve
- Retainer capnut
- Valve spring
- Lower spring plate
- Fuel inlet
- Injector body
- Spindle
- Needle stalk
- Annular circulation groove
- Feed drilling
- Nozzle capnut
- Fuel gallery
- Needle nozzle valve
- Nozzle body
- Nozzle location dowel

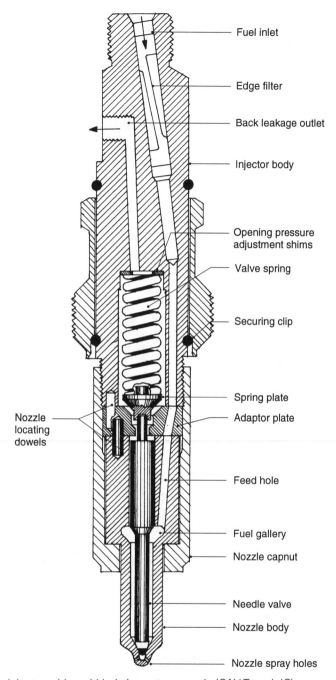

— Fuel inlet

— Edge filter

— Back leakage outlet

— Injector body

— Opening pressure adjustment shims

— Valve spring

— Securing clip

— Spring plate

— Adaptor plate

Nozzle locating dowels

— Feed hole

— Fuel gallery

— Nozzle capnut

— Needle valve

— Nozzle body

— Nozzle spray holes

Fig. 11.48 Low spring injector with multi-hole long stem nozzle (CAV Type LJC)

Adaptor plate
(Figs 11.47 and 11.48)

Originally, the injector and nozzle body lapped pressure faces were directly clamped together by the capnut (Fig. 11.47). Fuel then passed from the injector-body single supply hole to the nozzle fuel gallery via an annular groove machined in the pressure face of the nozzle and through three feed drillings joining the annular groove to the fuel gallery which surrounds the shouldered portion of

555

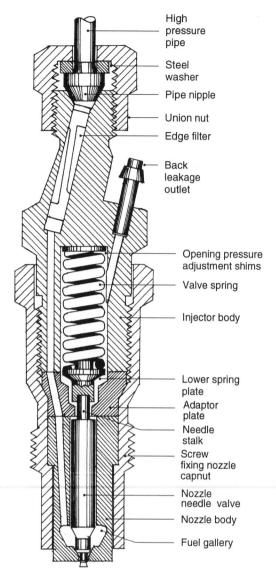

Fig. 11.49 Low spring injector with short stem pintle (Bosch and CAV)

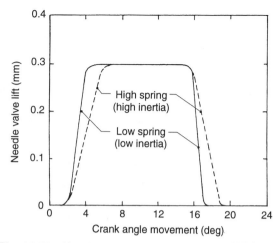

Fig. 11.50 Needle valve response to low and high inertia valve assemblies during the starting and ending of the injection open period

the needle valve. Indexing of the asymmetrical nozzle spray holes relative to the injector body is performed by a pair of solid dowels countersunk in both matting faces. However, the transfer of fuel from the injector body to the nozzle fuel gallery by the annular groove and the three feed drillings occupies a considerable amount of compressible residual fuel which can interfere with the snap opening and closing of the needle valve.

Subsequently, later injectors with low-mounted valve springs have intermediate adaptor plates with double dowels on both sides. They also have only a single hole, which is inclined to join the injector body supply-hole to a single-feed drilling made in the nozzle body leading to its fuel gallery (Fig. 11.48). With these adaptor plate injector designs there is a minimum amount of residual fuel held at any time within the injector passages. Also, the different indexing of the nozzle body for various engine applications can be accommodated by varying the position of the dowel holes in the adaptor plate.

Opening and closing phases of the needle valve
(Figs 11.47 and 11.48)

A pressure wave from the injection pump passes to the injector body; it then passes down to the nozzle annular supply passage and fuel gallery. Fuel pressure underneath that section of the needle valve, which separates the pressure tight-lapped sliding-fit portion from the reduced diameter portion, builds up on the projected annular area until it eventually overcomes the spring preload. This causes the needle valve to rise so that its conical valve tip lifts off its seat (formed in the closed end of the nozzle valve body). Immediately, the pressurized fuel trapped in the annular space formed between the needle valve and the nozzle body enters the cavity known as the sac, which is an extension of the conical valve seat in the nozzle body. The highly pressurized fuel then forces the fuel through each spray hole simultaneously so that it is discharged at very high velocity in the form of a fine spray into the

556

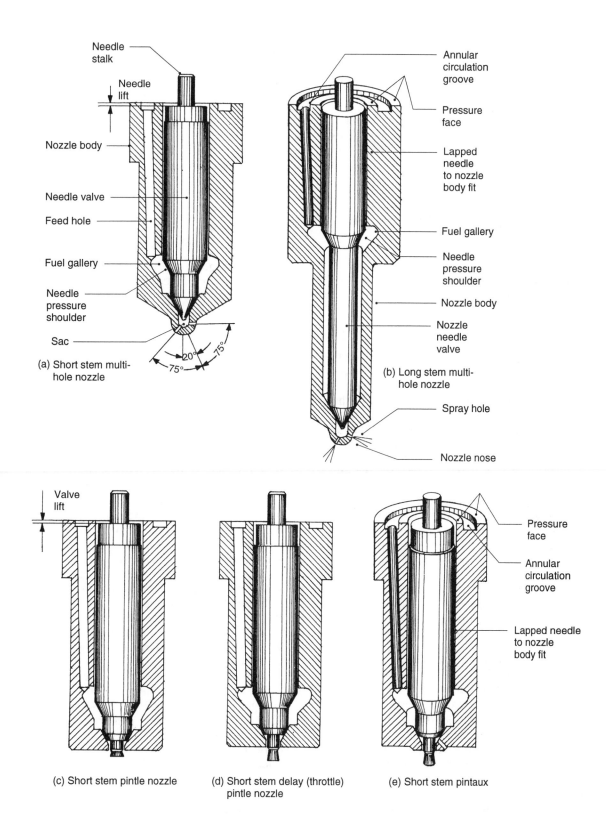

Fig. 11.51 Injector nozzle and needle assemblies

557

combustion chamber. When fuel delivery from the injection pump ceases, the collapse of fuel pressure in the pipeline and injector passageway permits the compressed valve spring to expand instantly until the needle valve tip seats. Hence, further fuel discharge into the combustion chamber is prevented. The opening and closing events of the needle valve is repeated once for every revolution or once for every second revolution for a two and four-stroke cycle engine respectively.

11.7.2 Single and multi-hole nozzle injectors

Single-hole injectors have a short stem nozzle whereas multi-hole injectors can be of the short or long-stem nozzle kind.

Short-stem single or multi-hole nozzle
(Fig. 11.51(a))
The short-stem nozzle extends the lapped sliding portion of the needle almost to the nozzle nose—the fuel gallery which surrounds the stepped portion of the needle is therefore just above the needle seat. It can be seen that the outside diameter of the nozzle body at its nose end is relatively large and therefore has to be considerably offset in the cylinder-head if the inlet and exhaust valve head diameters are to be of optimum size. These injector nozzles are only used on direct injection slow-speed large-cylinder diesel engines. They may be of the single-hole type where the injector offset in the cylinder-head permits the induction air swirl to sweep around the chamber. The air swirl meets and breaks up the downward directed fuel spray so that a thoroughly mixed charge is obtained in the region where the fuel has penetrated. To speed up the mixing process multi-hole nozzles are usually adopted, thus if there were four holes the air charge sweep would only have to be a quarter of a revolution to complete the mixing process.

Long-stem multi-hole nozzle
(Fig. 11.51(b))
The long-stem nozzle extends the reduced diameter portion of the needle valve and the nozzle body surrounding it. This enables the large diameter portion of the nozzle body, which encloses the fuel gallery surrounding the shouldered section of the needle, to be high up and well clear of the nozzle nose. Consequently, the almost pencil thin nozzle extension can be located at an inclined

angle almost centrally between the inlet and exhaust valve heads without restricting the cooling passages surrounding the valve seats. These long-stem nozzles have therefore replaced the short-stem nozzles which occupy too much space for medium and high-speed direct-injection diesel engines, where valve head size has to be maximized and it is essential to centralize the injector position in the cylinder-head.

Injector nozzle hole size and numbers
The numbers of spray holes and their size will be influenced by a number of factors as follows:

1 cylinder capacity
2 the intensity of induction swirl
3 the amount of mechanical squish
4 the engine speed range
5 the magnitude of injection pressure

Engine performance is considerably dependent upon the rate and thoroughness of fuel–air mixing within the combustion chamber.

The mixing of the air and the liquid fuel can be accomplished in two ways.

1 By a high degree of inlet port generated swirl, which spirals around the cylinder and combustion chamber before being intercepted by between three to five equally spaced, radial, penetrating, finely atomized fuel jets, which are discharged from nozzle holes at moderate injection pressure. This method is popular for medium-to-large diesel engines which have moderate maximum engine speeds of up to 4000 rev/min. Mixing with this approach relies on the high rate of air charge circulating and searching out and mixing with a small number of finely atomized fuel jets.

2 By an induction port layout, which produces a relatively low intensity of induction swirl but which is supplemented by having as many as six to ten separate spray holes equally distributed around the nozzle nose, these holes eject fuel radially outwards into the combustion chamber at high injection pressures of between 800 and 1200 bar. It is the highly penetrative fuel jets which become finely pulverized as they travel outwards and collide with the dense but relatively slow air swirl which is responsible for the stratified fuel particles seeking out and mixing with the air. Thus, with this method, less heat is produced in mixing the fuel–air charge so that high volumetric efficiency and thorough

mixing can be achieved providing the engine's maximum speed is limited. This second method of mixing the fuel and air is exceptionally effective for large diesel engines which have a maximum engine speed of no more than 2000 rev/min.

For a given cylinder capacity, the more spray holes there are, the smaller will be their diameter, the higher will be the injection pressure and the finer will be the spray formation, and vice versa, for fewer and larger diameter spray holes.

Injector inclination and spray hole geometry (Fig. 11.52)
With a two-valve cylinder-head the injector is normally inclined between 20° and 30° so that it can project out from the cylinder-head's flat surface, between the inlet and exhaust valve heads, with the smallest amount of offset to provide sufficient metal thickness to support the valve seat and ports and the injector hole boss. The inclined spray angle for the discharge holes in the nozzle nose is something of the order of 150° for optimum performance, which should be sprayed sym-metrically into the piston crown combustion chamber. However, the inclination of the injector axis means that the spray holes are formed asymmetrically in the nose of the injector nozzle. Thus, if the injector inclination is 20°, and the inclined spray angle is to be 150°, then the minimum and maximum angles of the spray holes to the injector axis will be 55° and 95° respectively.

11.7.3 Pintle needle and nozzle injectors

1 Short-stem pintle needle and nozzle (Figs 11.49, 11.51(c) and 11.53(a–c))
This injector needle and nozzle is very similar to the short-stem multi-hole nozzle but the seat end of the needle has a small pin or pintle projecting beyond its tapered seat, which protrudes through the mouth of the discharge orifice machined in the flat nose of the nozzle body (Fig. 11.51(c)). The pintle has a vee-shaped groove machined around its tip which produces a hollow parallel-sided cone spray with an angle of approximately 60° or more when the valve is fully open. The annular discharge produces a soft injection spray which

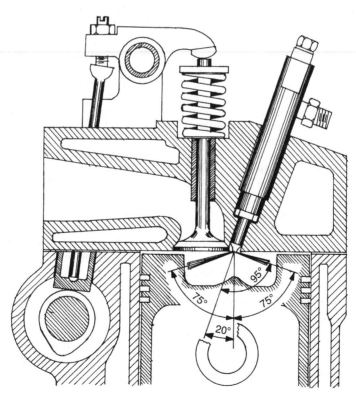

Fig. 11.52 Direct injection combustion chamber illustrating injector inclination and spray hole geometry

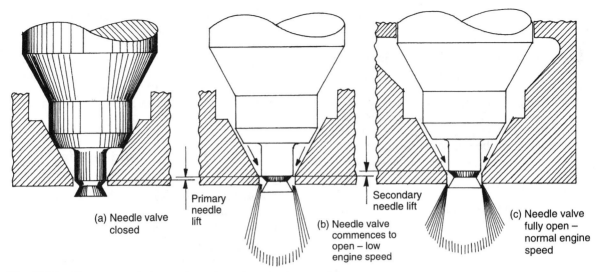

Fig. 11.53 Short-stem pintle needle and nozzle operating conditions

(a) Needle valve closed

Primary needle lift

(b) Needle valve commences to open – low engine speed

Secondary needle lift

(c) Needle valve fully open – normal engine speed

has a self-cleaning action. These injector nozzles are incorporated in indirect injection swirl (Fig. 11.53(a)) and precombustion chambers, which rely on the partial combustion in the indirect small chamber to expel the still unburnt and burning fuel spray into the main combustion chamber. The mixing process is then completed by swirling around the cylinder wall between the cylinder-head and the piston crown.

The high rate of swirl transferred into the cylinder subsequently speeds up the overall mixing and burning rate far beyond that which can be produced in a direct injection combustion chamber at high engine speed. It should be observed (Fig. 11.53(b)) that at low engine speed the needle valve lift is restricted so that only the minimum amount of fuel can be discharged in a given time, but with rising engine speed (Fig. 11.53(c)) needle valve lift increases to permit a higher rate of fuel discharge to occur.

2 Short-stem delay pintle needle and nozzle (Figs 11.49, 11.51(d) and 11.54(a–c))
This is a pintle needle and nozzle which has a longer than normal pintle (pin) (Fig. 11.51(d) and Fig. 11.54(a)). At low engine speeds when the injector pressure build-up during each operating cycle is low, the pressure reached in the nozzle fuel galley is only sufficient to lift the needle partially from its seat (Fig. 11.54(b)). Hence, some of the parallel portion of the pintle remains in the throat of the nozzle orifice so that only a limited quantity of fuel can be discharged through

the annular space formed between the pintle and orifice wall.

With rising engine speed, the needle valve lift will progressively increase due to the corresponding rise in fuel pressure until the parallel portion of the pintle clears the nozzle orifice (Fig. 11.54(c)). From here on, the conical portion of the pintle aligns with the nozzle orifice so that an increased quantity of fuel will be discharged from the annular-shaped orifice with further needle lift.

These injectors therefore provide a low rate of fuel discharge over a prolonged period at low engine speeds, which changes to an increased rate of injection with a proportional decrease in injection duration at higher engine speeds for the same fuel delivery. Injectors of this kind are particularly suitable for certain precombustion chambers, which tend to suffer from heavy diesel knock under light low-speed operating conditions. Thus, the initial low rate of injection reduces the pressure rise rate in the cylinder and therefore brings about relatively smooth combustion with minimum diesel knock.

3 Comparison of short (conventional) and long (delay or throttle) pintle flow characteristics (Figs 11.55(a and b), 11.56 and 11.57)
The relationship between needle valve lift and effective annular orifice area for both the short and long pintle is shown in Fig. 11.55(a). Initially, the annular delivery area is small and constant for a given needle lift. With the short pintle the constant first-stage lift is small before

560

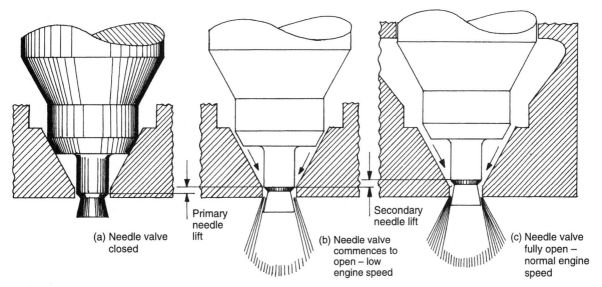

Fig. 11.54 Short-stem delayed pintle needle and nozzle operating conditions

(a) Needle valve closed

Primary needle lift

(b) Needle valve commences to open – low engine speed

Secondary needle lift

(c) Needle valve fully open – normal engine speed

the vee-grooved portion of the pintle aligns with the nozzle orifice, followed by a direct increase in annular discharge area. In contrast, the long throttle pintle has to move much further inwards before the parallel portion of the pintle clears the nozzle orifice. It thus prolongs the small constant annular orifice area lift before the main annular orifice area commences to open up with further needle lift. Note that beyond a predetermined needle lift, the pintle vee-groove begins to close-off the effective annular orifice area so that, in both cases, the area-to-lift relationship shows a decreasing orifice opening. Similar characteristics are shown for the discharge rate versus crankshaft angular movement, where it can be seen that there is almost no delivery delay with the short pintle but the long pintle extends the constant small delivery rate (delay) before there is a rapid increase in the fuel delivery rate (Fig. 11.55(b)).

Delivery characteristics can be seen in another way by relating needle lift to crankshaft angular movement for both short and long pintles

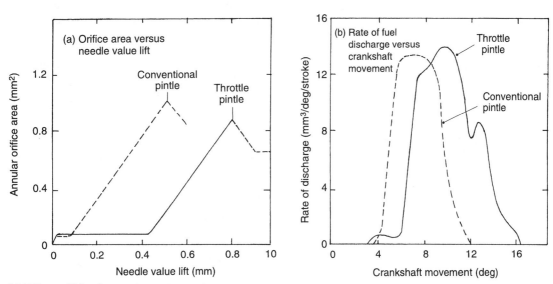

Fig. 11.55(a and b) Comparison of short (conventional) and long (throttle) pintle injectors relative to the orifice discharge area against needle valve lift and rate of discharge against crankshaft angular movement

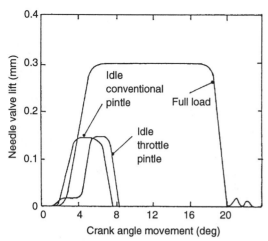

Fig. 11.56 Needle valve response to low and high inertia valve assemblies during the starting and ending of the injection open period

(Fig. 11.56). It can be seen that there is a definite restricted needle lift phase for the long pintle (delay or throttle) during idle running, whereas there is hardly any noticeable needle lift delay with the short pintle. However, for both short and long pintles at full load fuel delivery, the pressure rise rate is so fast that there is very little orifice opening delay before the needle rapidly rises to its full lift position.

A further comparison can be made between the fuel flow rate from the nozzle and needle lift (Fig. 11.57). This shows that the multi-hole nozzle provides a very high flow rate with the smallest

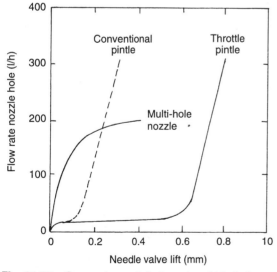

Fig. 11.57 Comparison of pintle and multi-hole flow rate for different needle lifts

of needle lift. In contrast, the short and long pintles which have initially a very small flow rate then change to a relatively high rate of fuel delivery sometime later, this being considerably delayed with the long (throttle) pintle.

4 Short-stem pintaux needle and nozzle (Figs 11.58, 11.59 and 11.60)
This development of the pintle needle and nozzle operates in two phases:

1 starting and idling speed conditions
2 normal operating conditions

Starting and idling speed phase
(Figs 11.58(b) and 11.59(b))
Under cranking and idling conditions, the pressure build-up in the nozzle fuel gallery can only partly lift the needle off its seat. Fuel now passes between the needle and nozzle seats and is then expelled partially via the restricted annular orifice formed between the pintle and nozzle hole, but mainly through the auxiliary hole. With the limited needle valve lift the predominant auxiliary spray moves against the swirl rotation in the combustion chamber, this pulverizes the fuel droplets into a finely atomized spray, some of which is directed towards the hot central region of the chamber where it readily ignites. At light loads, or when idling, the quantity of fuel delivered to the injector is relatively small so that the corresponding needle lift will also be limited. Accordingly the parallel portion of the needle still does not clear the throat of the nozzle orifice. Fuel will therefore continue to be mainly discharged from the auxiliary spray hole into the hottest part of the combustion chamber: it thereby minimizes the ignition delay period and, consequently, diesel knock.

Normal operating phase
(Figs 11.58(c) and 11.59(c))
Above idling speed running, fuel pressure in the nozzle fuel gallery during each delivery phase is sufficient to lift the needle valve fully. The parallel portion of the pintle withdraws beyond the throat of the orifice so that the bulk of each fuel delivery is now discharged between the conical end of the pintle and the parallel wall of the nozzle orifice. However, a small proportion of the fuel delivery will still be ejected from the auxiliary hole.

Thus, under normal operating conditions with full needle lift, the majority of the fuel will be

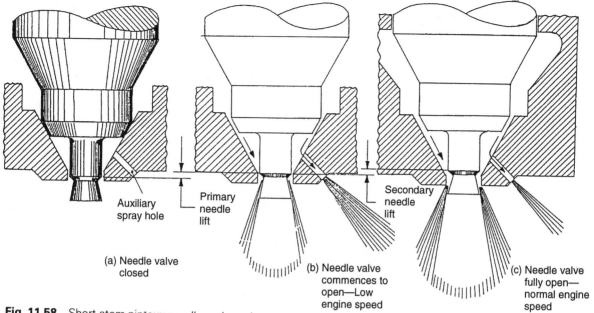

Fig. 11.58 Short-stem pintaux needle and nozzle operating conditions

(a) Needle valve closed

(b) Needle valve commences to open—Low engine speed

(c) Needle valve fully open—normal engine speed

Auxiliary spray hole

Primary needle lift

Secondary needle lift

discharged from the main pintle annular orifice in the same direction as the air swirl. The thoroughness of fuel stratification, as it is swept around the swirl chamber before it passes into the main chamber, optimizes performance throughout the engine speed and load range.

Pintaux fuel delivery versus speed characteristics
(Fig. 11.60)

The fuel flow characteristics of both the auxiliary and pintle spray holes are shown in Fig. 11.60. At cranking engine speed the auxiliary spray hole

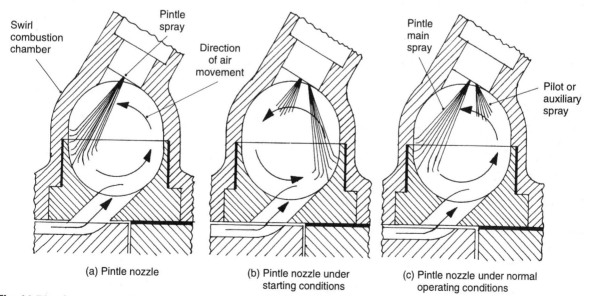

Swirl combustion chamber

Pintle spray

Direction of air movement

Pintle main spray

Pilot or auxiliary spray

(a) Pintle nozzle

(b) Pintle nozzle under starting conditions

(c) Pintle nozzle under normal operating conditions

Fig. 11.59 Comparison of pintle and pintaux injectors under starting and driving conditions

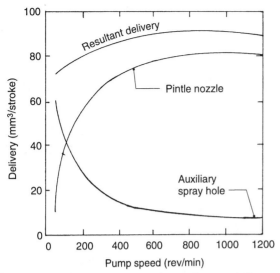

Fig. 11.60 Pintaux injection delivery characteristics over operating speed

discharge is at its maximum and the pintle orifice discharge is at a minimum, but this quickly changes with the auxiliary hole delivery rapidly decreasing while, at the same time, the pintle delivery equally rises rapidly. Thus, the operating conditions are such that the bulk of the spray quickly changes from an upstream delivery to a downstream discharge, from cranking to idling speed and higher, respectively. Consequently, the resulting fuel delivery from the combined discharge very slightly increases from cranking to idling speed but then remains approximately constant with any further increase in speed.

11.7.4 Needle valve seat contact

Single-angle needle valve
(Figs 11.61(a) and 11.62(a))
With a single-angle needle valve a conical seat is ground on the reduced diameter portion to an included angle of 60°, whereas the nozzle seat is nominally ground to between 58° and 59° thus providing something like half to one degree difference between the needle and nozzle seat angles. Thus, when the needle valve closes, the circumferential line contact (formed between the parallel and conical portion of the needle) contacts the nozzle body seat (Fig. 11.61(a)). The line contact provides a sharp snap opening and closing of the seat as the fuel and any contaminating solid particles (which may have entered the

nozzle sac via the spray holes) are pushed away from the knife edge contact. Conversely, a wide contact seat would tend to wedge and jam the seats when the valve closes and, if solid particles are present, will trap, crush and embed them between the seat faces, fuel could therefore seep out between the seats.

The opening pressure area (Fig. 11.62(a)) is therefore the annular projected area between the needle's sliding fit circumference and the line contact circumference.

Double-angle needle valve
(Figs 11.61(b) and 11.62(b))
With the double-angle needle valve, the nozzle body seat angles are still the same as the single-seat angle needle, but the needle's conical tip is ground with a normal seat angle (half to one degree larger than the nozzle seat angle) and with a second angle which intersects the first (Fig. 11.61(b)). When the needle valve seats, the circumference line formed where the two tapers meet now becomes the nozzle body seat contact line. The pressure area for the double-angle needle valve is now the annular projected area between the needle sliding fit circumference and the double taper contact line (Fig. 11.62(b)).

Nozzle life with the double-angle needle is considerably increased compared with the single-angle needle. This is because the erosion caused by the high-velocity turbulent flow over the blunt double-taper contact line is far less than with the sharper parallel-to-taper contact line with the single-angle needle. A further factor in favour of the double-angle needle valve is that its second taper acts as a hydraulic cushion, thereby reducing the needle impact when it closes. However, the double-taper angle is difficult to recondition and therefore servicing is not recommended, but this is more than compensated for by the prolonged operating life.

Differential opening and closing needle valve pressure
(Fig. 11.62(a and b))
The ratio of the needle valve outside diameter to line-seat contact diameter (known as the *differential diameter ratio*) (Fig. 11.62(a and b)) is critical as it demands a very high fuel pressure to commence needle valve lift. Once the valve tip leaves its seat the projected area is suddenly enlarged, and correspondingly there will be a rapid increase in needle upthrust and lift so that a sharp clean

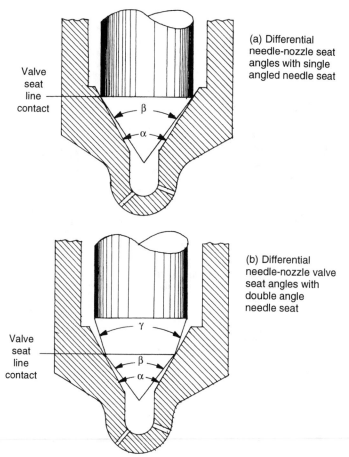

Valve
seat
line
contact

β

α

(a) Differential
needle-nozzle seat
angles with single
angled needle seat

Valve
seat
line
contact

γ

β

α

(b) Differential
needle-nozzle valve
seat angles with
double angle
needle seat

Fig. 11.61 Differential needle and nozzle valve seat angles

opening of the orifice is achieved. When the fuel delivery from the injection pump ceases, the needle valve will not close until the fuel pressure inside the nozzle passageways has dropped well below the needle valve opening pressure. Thus, any reflected pressure waves in the pipeline will be unlikely to exceed the opening pressure for each individual injection cycle. Accordingly, the needle valve snaps closed and, providing there is no spring surge or valve bounce, will eliminate fuel dribble. The needle valve differential diameter ratio may be studied in the following way, by referring to Fig. 11.62 (a and b).

The needle valve projected opening pressure area is the difference between the large sliding fit cross-sectional area and the line contact circumference area. Thus the projected opening cross-sectional area A_O will be

$$A_O = \frac{\pi D^2}{4} - \frac{\pi d^2}{4} = \frac{\pi}{4}(D^2 - d^2) \qquad (1)$$

Once the valve lifts off its seat the projected pressure area becomes the whole cross-sectional area of the large diameter portion of the needle valve. Thus, the projected cross-sectional area with the valve lifted A_L will be

$$A_L = \frac{\pi}{4}D^2 \qquad (2)$$

Consequently, the hydraulic opening upward thrust must equal the downward spring thrust if the valve is to open. Thus,

spring opening thrust = hydraulic opening thrust

$$F_{SO} = F_{HO}$$

$$= P_O A_O$$

$$= P_O \frac{\pi}{4}(D^2 - d^2) \qquad (3)$$

this means that the projected needle valve area is

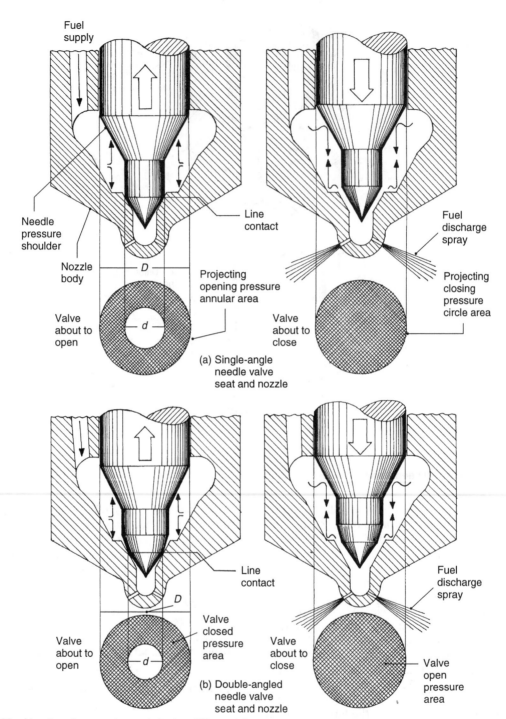

Fig. 11.62 Nozzle valve opening and closing differential pressure areas

an annulus and not a complete circle so that a high opening pressure P_O is needed to commence the needle valve lift. Conversely, when the valve is on the point of closing, the whole cross-sectional area of the needle valve is subjected to the fuel pressure, and therefore the closing spring thrust will be equal to the fuel pressure upthrust. Thus, spring closing thrust = hydraulic closing thrust

$$F_{SO} = F_{HC}$$

$$= P_C A_C = P_C \frac{\pi}{4} D^2 \quad (4)$$

and since the closing projected area A_C is larger than the opening projected area A_O then the closing pressure P_C will be lower than the opening pressure P_O. The needle valve differential diameter ratio is generally of the order of $2:1$.

11.7.5 Two-stage injector with multi-hole long-stem nozzle

1 Introduction (Figs 11.63 and 11.64)
Engine noise can be reduced by making the cylinder pressure rise slowly at the beginning of combustion and by decreasing the peak pressure which would normally be reached during each power stroke. If the combustion pressure rise is initially very gradual and the maximum cylinder pressure restricted, then diesel knock at low speed and load will be minimized. Moreover, the higher the temperature at the commencement of injection, the shorter will be the delay period before ignition takes place. It therefore leaves a longer time span for the completion of combustion. During the first part of the injection pump's phase, there is only a relatively low pressure increase in the pipeline, but it is adequate to compress the first stage injector spring to its prelift position. The initial low-pressure fuel delivery lifts the needle valve earlier than usual. Hence, it ignites the mixture and commences to increase the air charge compression pressure and temperature. The early burning therefore greatly accelerates the burning of the main fuel–air mixture so that, in fact, the rate of burning is so fast that it practically burns as it is discharged from the nozzle spray holes.

At idling speed, sufficient fuel can flow through the first prelift opening to meet the engine's fuel demands without resorting to further needle lift. Smooth and quiet combustion can therefore be achieved by retarding the point of injection. When the engine is accelerated, the prelift valve opening cannot cope with the fuel delivery flow rate needed to match the engine's power requirements. This causes the fuel pressure to build up so that the needle valve is lifted to the fully open position.

With rising speed and load the pressure and

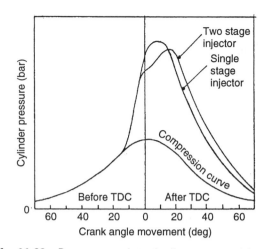

Fig. 11.63 Pressure crank-angle diagram comparing the cylinder pressure rise for both single and two-stage injectors

flow rate increase takes place so rapidly that the first and second-stage opening become as one, and subsequently perform similar to a conventional single-stage injector. Thus, the low flow rate of the first stage lift virtually disappears. However, in the high-speed and high-load range, the increased combustion chamber temperature tends to reduce the ignition delay period, speed up the combustion process and minimize knock. Because of the staged injection rates, there will be a reduction in the steepness of the pressure rise just before peak pressure is reached, compared with the conventional single-stage injector (see Fig. 11.63).

A reduction in noise from 89 to 85 dB, measured a metre away from a 2 litre direct-injection engine, has been obtained within the speed range 1000 to 1500 rev/min (Fig. 11.64). However, the rising noise level difference between the single-stage and two-stage injectors becomes progressively smaller with increasing speed until, eventually, the same noise level is reached at approximately 4000 rev/min.

These injectors use similar needles and nozzles as the single spring injectors. However, the design differs in the way in which the valve opening load is actuated to accommodate the primary and secondary spring preload stages which control the needle valve lift.

2 Construction of two-stage CAV injector (Fig. 11.65)
The needle valve primary preload is transmitted from the primary spring enclosed in the bottom

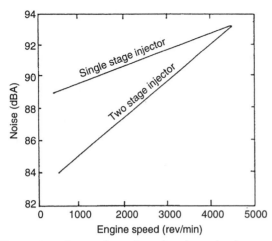

Fig. 11.64 Comparison of combustion noise for single and two-stage injectors when tested in a two litre engine

end of the injector body by the primary spring plate, which fits over the needle valve stalk. Beyond the primary spring preload needle valve lift, a secondary spring operates in parallel with the primary spring to provide a second stage of needle valve lift preload. The secondary spring is housed in the adjustment sleeve which screws into the upper end of the injector body. Additional valve lift preload is relayed from the secondary spring via the secondary spring plate and spindle to the primary spring plate that sits on top of the needle valve stalk. Shim adjustment can be made to the primary spring whereas the secondary spring preload can be set by screwing the adjustment sleeve in or out.

Operation

At low engine speed, the pump pressure build-up in the nozzle fuel gallery bears against the annular shoulder of the needle valve and lifts the needle valve until the spindle top face strikes the underside of the secondary spring plate. No further lift is possible during the pre-lift stage where the injection pump discharge pressure is relatively modest (180 to 220 bar). With rising engine speed and load, the injection delivery pressure increases until the fuel pressure upthrust acting on the annular shoulder of the needle valve overcomes the stiffness of both the primary and secondary springs. At this stage, the fuel pressure upthrust commences to compress the secondary spring in addition to the primary spring, this causes the needle valve to lift further until it is stopped by

the needle's upper shouldered face hitting the underside of the adaptor plate. The extra needle valve lift for the second stage occurs when the fuel pressure has reached something of the order of 350 to 450 bars.

3 Construction of two-stage Bosch injector (Fig. 11.66)
With this arrangement, the primary spring is in the upper portion of the injector body whereas the secondary spring is in the lower region next to the needle valve. Spring thrust from the primary spring is transmitted through the primary spring plate and spindle directly against the upper end of the needle valve. The actual preload setting is obtained by altering the thickness of the shims located between the primary spring and the blind end of the chamber which houses it. Secondary needle valve preload is obtained with a secondary spring which acts in parallel with the primary spring once the needle valve has lifted sufficiently. The secondary spring thrust is relayed (in addition to the primary spring load) to the upper end of the needle valve via the secondary spring plate and the stop-sleeve when the pressure in the nozzle fuel gallery is sufficient to lift the needle beyond the prelift stage. Adjustment of the secondary preload stage is achieved by placing shim washers between the guide collar and the upper end of the secondary spring.

Operation

At low engine speed, pump pressure in the nozzle fuel gallery pushes against the annular shoulder of the needle valve until it compresses the primary springs. The needle lifts until its upper face strikes the stop-sleeve, and further lift is prevented by the downward thrust of the secondary spring holding the stop-sleeve in its lowest position against the nozzle body pressure face. With rising engine speed and load, the injection pump delivery pressure builds up until the upthrust against the annular shoulder of the needle valve is adequate to compress both the primary and secondary spring. The needle valve lift continues to rise to the point where the stop-sleeve strikes the lower shoulder of the adaptor plate recess formed on its underside.

First stage opening pressure is typically between 180 and 220 bars whereas the second stage needle lift takes place when the fuel pressure has risen to something between 350 to 440 bars.

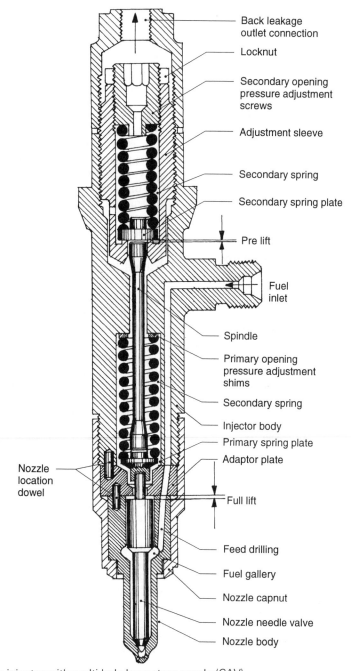

Fig. 11.65 Two-stage injector with multi-hole long-stem nozzle (CAV)

4 Comparison of single and two-stage multi-hole injectors

Single-stage injector characteristics
(Fig. 11.67(a))
Needle valve opening commences when the noz-

zle fuel gallery pressure has risen to the valve spring preload pressure setting. Initially, the cam contacts the follower on the slow lift stage of the cam profile but, near the beginning of needle lift, the cam lift rate speeds up which can cause the pressure in the nozzle passageways to rise rapidly to its peak value. The start of the needle lift is

569

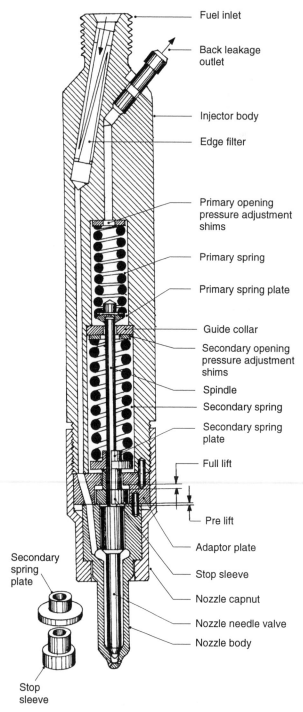

Fuel inlet

Back leakage outlet

Injector body

Edge filter

Primary opening pressure adjustment shims

Primary spring

Primary spring plate

Guide collar

Secondary opening pressure adjustment shims

Spindle

Secondary spring

Secondary spring plate

Full lift

Pre lift

Adaptor plate

Stop sleeve

Nozzle capnut

Nozzle needle valve

Nozzle body

Secondary spring plate

Stop sleeve

Fig. 11.66 Two-stage injector with multi-hole long-stem nozzle (Bosch)

imum opening. The needle then oscillates about its fully open position for the next 8° or so of crankshaft movement, during which time the fuel being pumped out from the plunger cannot keep up with the fuel discharging from the spray holes. Hence, the fuel pressure in the nozzle passages rapidly decreases until eventually it cannot support the open needle spring load, at which point the needle valve begins to close. Instantly, the restricted valve opening produces a slight surge in fuel pressure, but this quickly peters out as fuel delivery from the pump ceases. Note that when the pressure has fallen to the enclosed pipeline and nozzle passageway residual pressure, a small amount of secondary closure may occur due to needle bounce.

Two-stage injector characteristics
(Fig. 11.67(b))
With the two-stage valve opening injector the pump pressure builds up to a point where the primary spring contracts, thus causing the needle valve to open to its prelift position over a crankshaft movement of approximately 4°. At the same time, the nozzle passageway fuel pressure continues to rise steadily until it overcomes the stiffness of both primary and secondary valve springs. Thus, the needle commences its second stage rise until its full lift is reached, at about 8° after the beginning of the first instant of needle lift. However, the fuel delivery from the pump cannot maintain the discharge spray exit from the nozzle holes. Consequently, the fuel pressure rapidly decreases until there is insufficient pressure to hold open the needle. The needle at this instant will therefore snap down on its seat. Primary closing of the needle valve under certain operating conditions will be followed by the needle bouncing up again, this being known as secondary opening.

11.7.6 Cleaning and inspecting injector nozzle and needle

After dismantling the injector, examine all components for signs of scoring wear, corrosion, pitting, colouring or other forms of damage. Check the conditions of all internal and external threads. Renew any part or parts as necessary.

Cleaning procedure (Figs 11.68 and 11.69)
All parts should be cleaned in benzine and then dried off with compressed air. A rag or fluffy

gradual, but as the total projected pressure area of the needle is then exposed to fuel pressure upthrust, the needle sharply moves to its max-

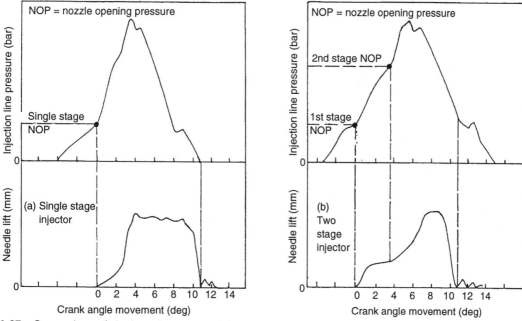

Fig. 11.67 Comparison of single and two-stage injector pressure rise and needle lift during the injector period

cloth must not be used when cleaning and examining injectors as fibres might find their way into the nozzle fuel gallery, sac or spray holes.

1 Remove carbon deposits from the outside of

the nozzle and from the needle with a brass wire brush (A). Polish the needle with a piece of soft wood. Do not use an abrasive cleaning compound.

2 Clean the feed hole or holes with a drill or

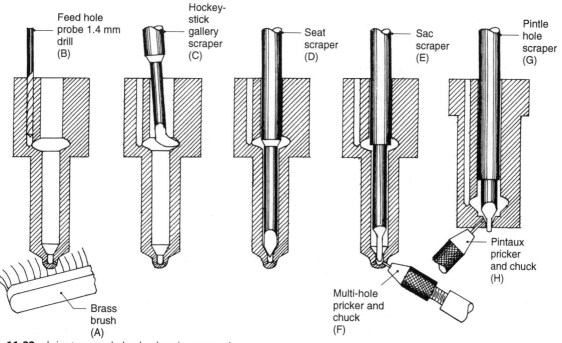

Fig. 11.68 Injector nozzle body cleaning procedure

571

wire of the same diameter as the hole, by moving in and out and rotating it to remove carbon (B).

3 Clean the fuel gallery with a hockey-stick gallery scraper, by pressing hard against the side of the cavity and rotating to clear carbon deposits from the walls (C).

4 Clean the nozzle seat by inserting a seat scraper in the bore, pressing down and rotating to remove carbon (D).

5 Clean the nozzle sac by inserting a correct-size sac scraper, according to nozzle type, pressing down and rotating to remove carbon (E).

6 With multi-hole nozzles fit an appropriate sized pricker wire in the holder so that it protrudes almost 1.5 mm (this gives the greatest resistance to bending). Grind a chamfer about 45° on its end. Enter the pricker into each nozzle spray hole, and push and rotate gently until the hole is cleared (F). Do not break the wire in the hole as it is almost impossible to remove.

7 With the pintle nozzles, insert an appropriate-size pintle hole scraper until it protrudes through the orifice, then turn to and fro to remove carbon (G).

8 With the pintaux nozzle insert the pricker wire into the auxiliary hole, and push and rotate slowly until the hole is cleared (H).

9 Observe if the needle tip is blued, burnt or the stem is scratched.

10 Reverse flush the nozzle spray hole or holes; that is, force test fuel into the nozzle spray holes from the outside so that it clears any carbon particles which have become dislodged during cleaning and have then jammed themselves in the spray holes on the sac side (Fig. 11.69).

11 Examine the pressure faces of the injector holder and the nozzle body joint, and ensure the contact faces are clean and accurately mated.

12 Examine the dowels for distortion as they may have become partially sheared by excess tightening of the nozzle capnut.

13 Examine the condition of both needle and nozzle body seats using a nozzlescope which incorporates a light beam. If the seats are pitted or worn then they must be reconditioned.

14 Thoroughly clean the needle and nozzle body interior: wet the needle, hold the nozzle vertically and then allow the needle to fall. If there is no carbon, grit, lapping paste or fluff be-

tween the nozzle and needle-lapped sliding pair, then the needle should drop freely under its own weight onto the nozzle seat. Note that the slightest trace of foreign matter can score and damage the rubbing surface of the needle stem and nozzle bore.

15 Assemble the injector holder, valve spring, needle valve and nozzle body together, taking care that the joint pressure faces and dowels are correctly positioned, and secure the assembly by tightening the capnut to the recommended torque.

11.7.7 Testing the injector for serviceability

If the injector has been removed from the engine due to suspected malfunction or if it has been dismantled, cleaned and reassembled, it can then be tested for effectiveness. The testing procedure should be carried out in the following sequence:

1 pressure setting
2 seat dryness
3 back leakage
4 spray formation and needle chatter

Pressure setting test (Fig. 11.69)

1 Close the pressure gauge valve so that the pressure gauge is isolated from the pump and injector (only finger-tip pressure is necessary).

2 Open the control valve one full turn and switch on the fume extractor.

3 Rapidly operate the hand lever several times to flush and expel the air.

4 Close the control valve to a light finger-tightness then set the pointer to zero.

5 Open the control valve to the minimum amount, something like a quarter of a turn, which will allow the injector to operate when pressing down hard on the lever.

6 If, necessary, the control valve can be re-adjusted to obtain the best minimum rate of flow. Record the amount the control valve is open for future tests on similar types of nozzles.

7 Open the pressure gauge valve a quarter of a turn, depress the lever and note the highest pressure reading on the gauge before the needle flicks, indicating that the injector nozzle needle valve pops open. If necessary, adjust the injector pressure setting to the recommended value.

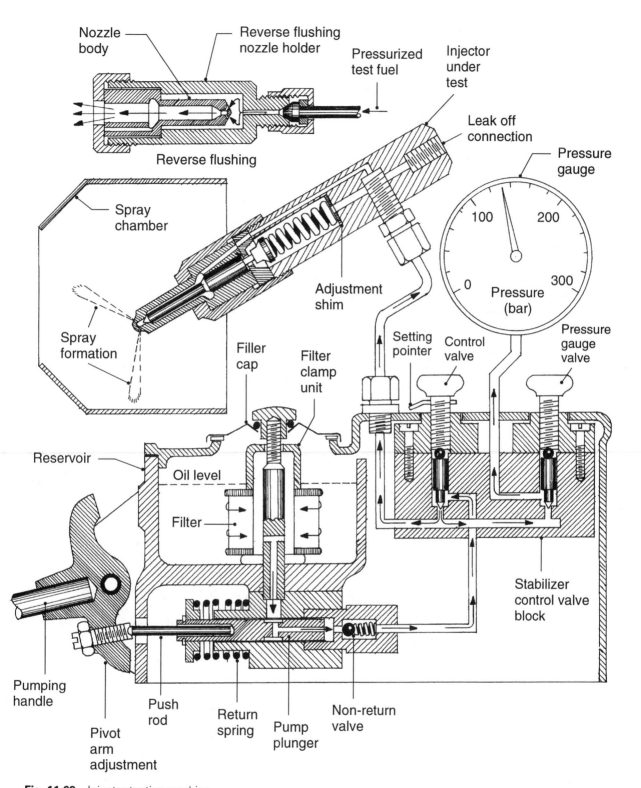

Nozzle body

Reverse flushing nozzle holder

Pressurized test fuel

Injector under test

Leak off connection

Pressure gauge

Reverse flushing

Spray chamber

Spray formation

Adjustment shim

100   200

0   300

Pressure (bar)

Setting pointer

Control valve

Pressure gauge valve

Filler cap

Filter clamp unit

Reservoir

Oil level

Filter

Stabilizer control valve block

Pumping handle

Push rod

Return spring

Pump plunger

Non-return valve

Pivot arm adjustment

Fig. 11.69   Injector testing machine

573

Seat dryness test (Fig. 11.69)

1 Switch the fume extractor off and set the control valve to the required flow rate, as under the pressure setting test procedures 4 to 7.
2 Open the pressure gauge valve a quarter of a turn and wipe the nozzle nose dry.
3 Depress the lever until pressure builds up to something like 10 bar below opening pressure. Thus, if the set opening pressure is 240 bar then the seat dryness test pressure would be 230 bar.
4 Hold this pressure by maintaining a steady effort on the lever for about 10 s.
5 Observe if the nozzle nose is reasonably dry. There must be no tendency to dribble but a slight dampness can be ignored.

An alternative method of observing seat dryness is to place a piece of blotting paper against the multi-hole or pintle nozzle and note how much the damp patch spreads in a given time. Normally, the diameter of the patch should not expand to more than 13 mm when the blotting paper is maintained in contact with the nozzle tip for roughly one minute.

Back leakage test (Fig. 11.69)
The timed back leakage pressure drop is a measure of the clearance of the needle stem in the bore of the nozzle body.

1 Open the control valve one turn and switch on the fume extractor.
2 Depress the lever so that the nozzle operates three times.
3 Open the pressure gauge valve approximately a quarter of a turn.
4 Depress the lever until pressure builds up to something like 10 bar below opening pressure.
5 Close the control valve and start the (seconds) clock.
6 Observe the time for the pressure to fall between two predetermined readings.

For multi-hole nozzles the pressure fall from 150 to 100 bar should not take less than 6 s; while for pintle nozzles the pressure fall from 100 to 65 bar should not take less than 6 s.

These figures are only a guide and the manufacturer's recommendation should be followed. The nature of the test oil and its temperature greatly influences the back leakage fall times.

An over-tight needle will give a high leakage time, a wet seat and poor atomization. Con-versely, a needle which is loose in the nozzle will give a low back leakage time, a wet seat and poor atomization.

Spray formation and needle chatter test (Fig. 11.69)

1 Close the pressure gauge valve and switch on the fume extractor.
2 Close the control valve.
3 Apply load to the lever and slowly open the control valve so that the minimum amount of fuel is delivered to open the needle valve.
4 Fully depress the lever at a uniform rate and observe if the nozzle continuously chatters and atomizes with an even buzzing sound.

Note: needle chatter can be affected by a sticky or tight needle.

The discharge from the nozzle should be completely atomized, free from irregular streaks, hosing and dribble.

A sticky needle can cause hosing. It is usually due to a tight needle in its nozzle bore, caused by distortion of the nozzle body by the capnut being overtight, a dirty nozzle bore or damaged surfaces of the bore or needle.

For multi-hole nozzles, the spray formation from each hole should be identical, they should be of equal intensity, length, spread and texture.

For pintaux nozzles, the auxiliary hole discharge spray can be checked by opening the control valve the minimum amount and slowly operating the lever. The main spray is checked by opening the control valve a full turn and quickly depressing the lever. Sprays should be well formed and free from splits or distortion. A slight centre core may be disregarded.

Warning when testing injectors
Always direct the spray away from the operator and never bring hands into contact with the spray as it can easily penetrate the skin and can result in infection.

11.7.8 Needle lift

The continual opening and closing of the needle valve results in the repeated impact of the thoroughly hardened needle onto its softer case—hardened nozzle seat. It therefore eventually wears away the nozzle seat so that the original line contact changes into a wide seat contact area.

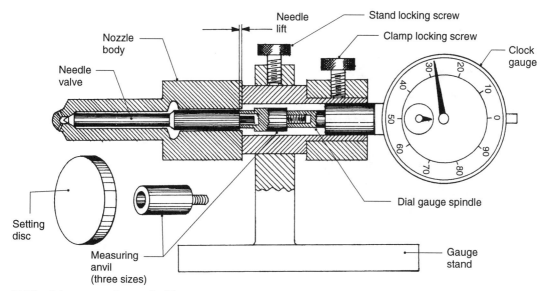

Fig. 11.70 Injector nozzle-needle lift gauge

holder until both the small and large gauge
pointers are set to zero.
6 Tighten the clamping ring lock screw.

The lapping of a worn needle and nozzle seat will
increase the needle lift so that, to bring the needle
lift back to specifications, the nozzle pressure face
must be lapped.

A nozzle valve dial gauge is used to measure
the needle lift and it should be set up and used in
the following way.

Gauge setting (Fig. 11.70)

1 Screw the appropriate measuring anvil for the
   needle size into the clock-(dial-)gauge spindle.
2 Loosen the dial-gauge lockscrew and rotate the
   dial scale until zero is approximately in line
   with the protruding stem at the top.
3 Slacken the clamping ring's knurled locking
   screw.
4 Place the circular setting disc against the front
   ground face of the gauge-holder sleeve making
   sure that both faces are clean.
5 Shift the position of the dial gauge in the gauge

Measuring valve lift (Fig. 11.70)

1 Fit the nozzle needle into the nozzle body,
   ensuring that it is fully seated.
2 Insert the stalk of the needle valve into the
   hole of the measuring anvil and press the
   nozzle forward until the pressure face of the

nozzle body firmly contacts the gauge holder
sleeve end face.
3 Observe the maximum needle lift by reading
   the clock gauge, one division on the larger dial
   gauge is equal to 0.01 mm and one division on
   the small dial gauge is equal to 1.00 mm.

Note that every time the measuring anvil is
changed to accommodate a different sized needle
the gauge must be reset.

11.7.9 Reconditioning the needle valve and nozzle body seats

If the nozzle body seat is pitted or badly scored
then the seat must be reclaimed by lapping and
the needle valve seat reground. The servicing of
the needle and nozzle seats can easily be carried
out with the aid of a Hartridge nozzle reconditioning machine. The following procedure
should be adopted.

Grinding the nozzle needle valve (Fig. 11.71)
Check the manufacturer's specifications for the
nozzle body and needle seat angles. Assume the
nozzle body seat angle to be 60°.

1 Loosen the quadrant knurled lock nut sufficiently to allow the quadrant and dial-gauge
   to move freely.
2 Adjust the quadrant's fine-setting knurled-

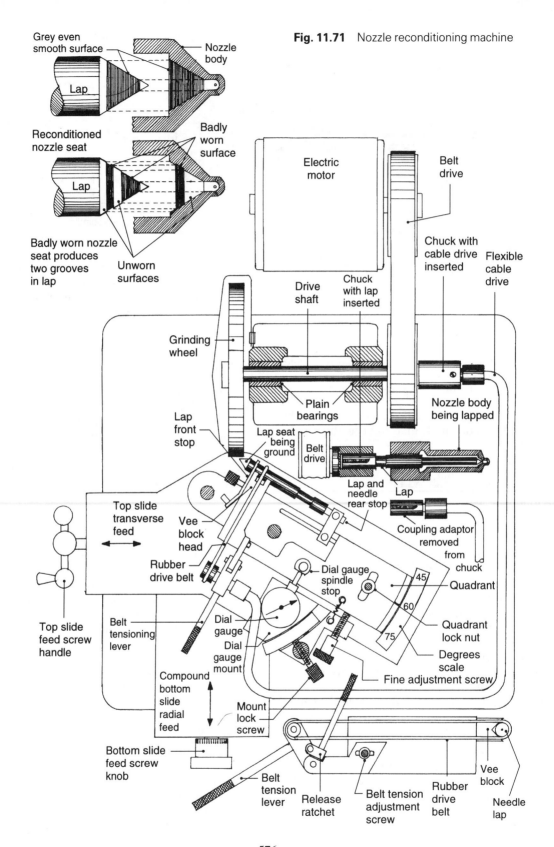

Grey even smooth surface

Nozzle body

Lap

Reconditioned nozzle seat

Lap

Badly worn surface

Badly worn nozzle seat produces two grooves in lap

Unworn surfaces

Fig. 11.71 Nozzle reconditioning machine

Electric motor

Belt drive

Chuck with cable drive inserted

Flexible cable drive

Drive shaft

Chuck with lap inserted

Grinding wheel

Plain bearings

Nozzle body being lapped

Lap front stop

Lap seat being ground

Belt drive

Lap and needle rear stop

Lap

Coupling adaptor removed from chuck

Top slide transverse feed

Vee block head

Rubber drive belt

Dial gauge spindle stop

Quadrant

45

60

75

Quadrant lock nut

Degrees scale

Top slide feed screw handle

Belt tensioning lever

Dial gauge

Dial gauge mount

Fine adjustment screw

Compound bottom slide radial feed

Mount lock screw

Bottom slide feed screw knob

Belt tension lever

Release ratchet

Belt tension adjustment screw

Rubber drive belt

Vee block

Needle lap

headed screw until it is approximately halfway along its threaded length and the clock-gauge pointer is at the top; that is, in front.

3 Move the quadrant and clock-gauge mount so that the zero line on the quadrant coincides with 59° 25' on the scale and then lock the clock-gauge mount to its base by tightening the mount's knurled-headed lock screw.

4 Check the setting of the quadrant zero line and scale and, if necessary, readjust the quadrant's fine setting knurled-headed screw.

5 Rotate the clock-gauge scale until its zero coincides with its pointer.

6 Adjust the quadrant's fine setting screw clockwise until the clock-gauge pointer has moved 25 minutes and then retighten the quadrant knurled lock nut.

7 Insert the nozzle in the vee block of the grinding attachment and adjust the end-stop to project the needle stem out sufficiently.

8 Tension the belt by depressing the engagement lever until the ratchet has engaged.

9 Switch on the electric motor and slowly turn the knurled graduated knob (of the compound slide rest) in a clockwise direction until the needle valve stem just makes contact with the grinding wheel.

10 Rotate the handle of the top slide so that the needle is passed slowly backwards and forwards across the surface of the grinding wheel, very gradually feeding in the needle until the tapered seat has been entirely skimmed.

Grinding the nozzle body lap (Fig. 11.71)
Select the nozzle body lap so that it is a good sliding fit in the bore of the body. There are three sizes of laps for each diameter of needle; for instance, for a nominal nozzle bore diameter of 6 mm there are 6.00 mm, 6.01 mm and 6.02 mm diameter nozzle laps available.

Refer to the manufacturer's needle and nozzle seat angles. The angular difference between the needle and body must not be less than that recommended. It is therefore advisable to grind the lap 15' to 30' less than the desired finish angle. Thus, if the nozzle finish angle is to be 59° the lap should be ground between 58° 30' and 58° 45'.

The procedure for grinding the nozzle body lap is identical to grinding the needle valve, which has already been explained.

Lapping nozzle body (Fig. 11.71)

1 Inspect the condition of the nozzle body with the nozzle viewer.

2 Place the prepared nozzle lap in the lapping chuck.

3 Select and deposit the correct grade of paste on the tip of the lap (either from the tube or with a pointed matchstick) taking care that the paste does not extend to the edge of the cone—that is, the parallel portion of the lap stem.

4 Place the tip of the left-hand middle finger under the end of the lap to act as a guide, and place the edge of the nozzle body bore against the underside clean parallel portion of the lap, and the finger-tip. Thus, the nozzle body can be tilted up and passed over the lap tip without it touching the nozzle bore. The procedure is necessary to avoid impregnating the bore of the body, as this would cause rapid wear and back leakage. If any paste is apparent on the lap stem, wipe it clean and reverse flush the nozzle body.

5 Switch on the electric motor and press the seat against the lap tip for about three seconds, withdraw slightly and again press home and release. After about 30 s lapping withdraw the body, wipe the sludge off the lap, recharge with fresh lapping paste and repeat the procedure. If the lapping of the seat has been sufficient, and even grey circumferential ring—having a width equal to the depth of the seat—will be apparent in the lap.

If the nozzle body seat is badly worn, the unworn portions on either side will produce two grooves in the lap. When the untouched portion of the lap makes contact with the bottom of the groove in the nozzle body seat, the lap must be reground and the process continued until the untouched portion again makes contact. When relapping a badly worn seat it may be necessary to regrind the lap several times.

Lapping needle nozzle seat (Fig. 11.71)
If the nozzle is leaking slightly when under test, fix the needle on the motor driven chuck and, with a small quantity of fine lapping paste, lap in the needle with the nozzle body for about 5–10 s, using a very light pressure, until a very narrow marking occurs in the needle seat at the cut-off edge. This should only be done when the nozzle body has been lapped and the whole assembly tested.

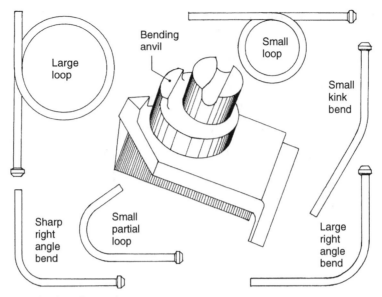

Fig. 11.72 High pressure pipe bending tool

11.7.10 Replacement of high-pressure injector pipes
(Fig. 11.72)

Unserviceability of high pressure injector pipes can be due to the following reasons.

1 When high pressure pipes are manufactured they are drawn out and then bent to various shapes. This cold working of the steel produces strain hardening which, if severe in places, may eventually cause hair-line splitting of the pipes.
2 In service, these pipes are subjected to a large and rapid variation of fuel pressure which could exceed 1000 bar. This tends to introduce periodic tensile and circumferential (hoop) impact stresses within the pipe, which eventually may lead to metal fatigue.
3 Any external vibration and resonance of the pipeline will work-harden the steel pipe walls, which again may cause metal fatigue failure.
4 With the continued removal and refitting of the high pressure piping during servicing, the pipe union nipple tends to distort and lose its sealing ability so that it will leak under pressure.
5 Under certain conditions, the nipple, washer and nut will bind together so that when the coupling nut is unscrewed, instead of the nut slipping around the pipe, the pipe twists with the rotating nut.
6 During removal of the pressure pipes it is sometimes difficult to refit the pipes into their original position so that the pipe may be bent and, in the extreme case, may kink.

High-pressure steel pipes are bent to shape cold using a bending tool as shown in Fig. 11.72.

11.7.11 High-pressure pipe nipple forming
(Figs 11.73 and 11.74)

If the high-pressure steel pipe is kinked, twisted

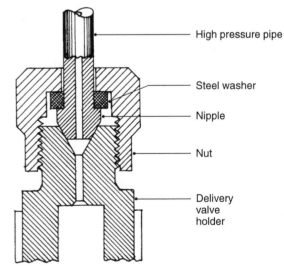

Fig. 11.73 High pressure pipe union assembly

578

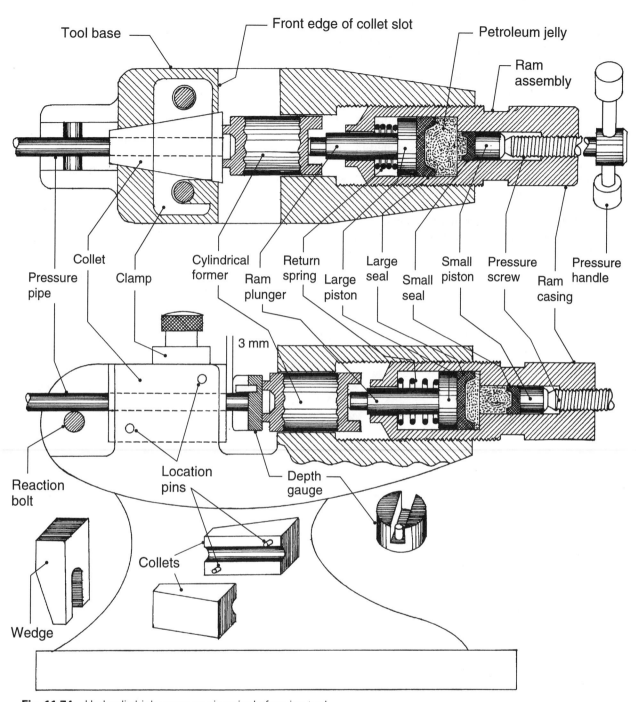

Fig. 11.74 Hydraulic high pressure pipe nipple forming tool

or damaged in any way, or if the nipple is badly distorted, then new pipe stock can be bent to shape, cut to length and finished off by forming nipples at both exposed ends of the pipe. Thus, when the coupling nut is tightened onto either the injection pump delivery holder or the injector screw adaptor, the nipple is squeezed squarely into the tapered mouth of the receiving connection (Fig. 11.73). The nipples are produced with a hydraulic nipple forming tool, as shown in Fig. 11.74, using the following procedure.

1. Select the collets, former and depth gauge, marked with the outside diameter of the pipe on which the nipple is to be formed.
2. Insert the former in the counter-bore in the body of the tool with the forming end (cavity) facing the collets, which are positioned in the tapered slot.
3. Screw in the ram until the former protrudes slightly out of its bore and place the collets in the taper slot.
4. Pass the new pipe through the collets, it should extend beyond the face of the collets by 3 mm, in excess of the depth of the slot in the depth gauge.
5. Swing the clamp over the collets and tighten the clamp screws, place the depth gauge over the pipe with the pin nesting on the pipe, and screw in the ram until the depth gauge contacts the face of the collets.
6. Apply hydraulic pressure by turning the handle in a clockwise direction until the faces of the collets are level with the front edge of the collet slot.
7. Release the hydraulic pressure and remove the depth gauge. Apply petroleum jelly to the former profile and the end of the pipe. Screw in the ram until the former contacts the end of the pipe.
8. Apply hydraulic pressure again by rotating the pressure screw. On 6 mm outside diameter pipes, the pressure can be applied until the former contacts the face of the collets and the nipple is formed in one operation. For 8 and 10 mm outside diameter pipes the operation is performed in two stages.

12

Diesel Rotary and Unit Injector Fuel Injection Pump Systems

12.1 Distributor fuel injection rotary pump (Bosch Type-VE)

12.1.1 Construction
(Figs 12.1 and 12.2)

An input shaft rotates the camplate via the rigid-slip Oldham-type coupling, consisting of a pair of input shaft drive lugs and a pair of camplate drive lugs. Both pairs of drive lugs mesh with a slotted coupling yoke located between these half couplings. The coupling is housed in the roller ring carrier which supports four rollers and their pins, the cylindrical carrier is permitted partially to rotate to and fro to facilitate the automatic advance device.

Simultaneously, the camplate applies both axial and rotary movement to the pumping plunger—the axial movement being provided by the cam action and the rotary movement through the index-pin and slot coupling (Fig. 12.2). A spring carrier plate and a pair of diametrically opposed return springs held in position by a pair of guide-pins returns the plunger to its outer position after each pumping stroke. The plunger reciprocates in its barrel, which is itself secured by a force fit in the cast-iron distributor head. Positioned between the camplate and the barrel is the control-sleeve which is a sliding fit on the plunger, its function is to determine the end of each injection stroke by uncovering the spill port, thereby releasing the trapped fuel (Fig. 12.2).

Mounted on the input shaft next to the lug drive coupling is the governor step-up gear, its drive being transmitted through two rubber pads which absorb and damp the acceleration and deceleration shock loads imposed by the flyweights and cage assembly to which it is geared.

Four flyweights, a thrust sleeve, leverage assembly and governor spring all interact to shift the control sleeve spill port to suit the speed and load demands of the engine (Fig. 12.1).

Positioned behind the governor step-up gear is the eccentric vane supply pump which keeps the camplate and governor apartment filled with fuel and provides the feed pressure for filling the plunger and barrel pressure chamber. The fuel supply pressure is regulated by a pressure control valve which can be externally removed.

A mechanical or electromagnetic fuel shut-off device is incorporated. These devices can either override the governor leverage system and provide permanent control sleeve spill, or the fuel supply to the plunger/barrel pressure chamber can be cut off.

12.1.2 Operating principle
(Fig. 12.3(a–c))

The plunger has one longitudinal distributor slit approximately mid-way along its working length

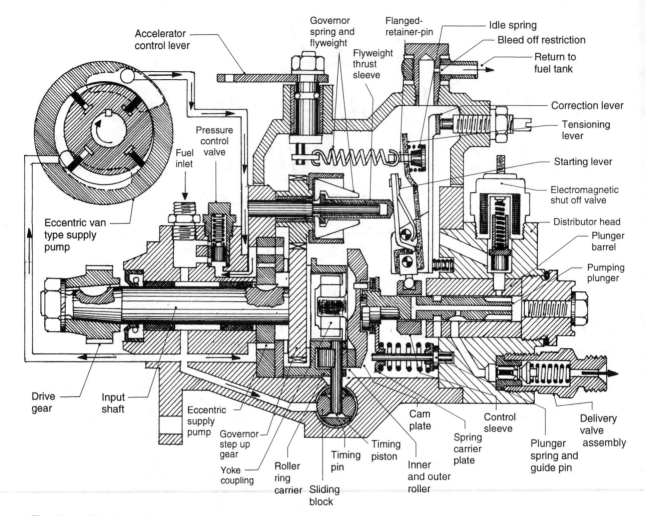

Fig. 12.1   Distributor fuel injection pump (Bosch Type VE)

and as many metering slits as there are cylinders equally spaced around the rim of the plunger at its pressure end. An axial passage enters the plunger at its pressure end and goes back to where it is intercepted by a radial spill port. The plunger barrel has a single supply port on its top side at the pressure chamber end and as many discharge ports as there are cylinders equally spaced around the barrel at the spill port end.

1 Fuel filling of plunger and barrel chamber (Fig. 12.3(a))

With the roller carrier held stationary and the rollers in their lowest position on the cam profile, the plunger will be at outer dead-centre position with the control sleeve cutting-off the spill port. One of the metering slits will now align with the supply port whereas the distribu-

tor slit will be between adjacent discharge ports. Fuel from the governor chamber therefore passes through the supply port and enters and fills both the barrel pressure chamber and the central plunger passage.

2 Fuel cut-off—beginning of fuel delivery (Fig. 12.3(b))

Further rotation of the input shaft makes the camplate move around and outwards as the cam flanks climb over the rollers. Consequently, the plunger revolves and, simultaneously, moves towards inner dead-centre. Firstly, the metering slit leading edge cuts-off the supply port and, at the same time, one of the distributor slit's leading edges opens the discharge port. Secondly, the plunger moves along its stroke, the trapped fuel is initially

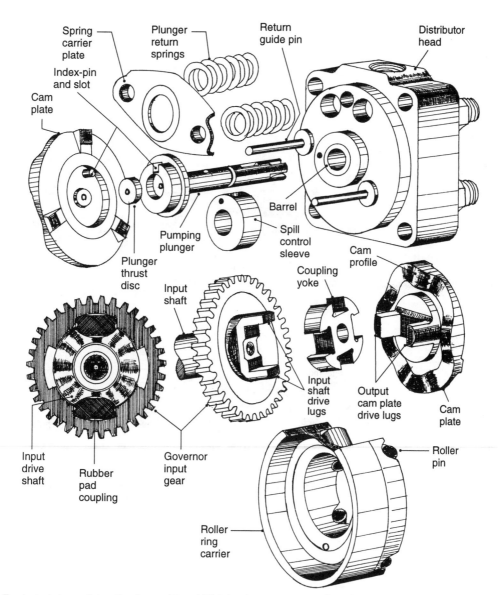

Fig. 12.2 Exploded view of the distributor (Type VE) injection pump pumping elements

pressurized until the delivery valve opens, and fuel will then be displaced from the pressure chamber out to the high-pressure pipeline via the discharge port and discharge passage.

3 *Fuel spill—end of fuel delivery (Fig. 12.3(c))*
As the plunger approaches the end of its inward injection stroke the spill-port moves beyond the control-sleeve. Instantly, fuel will spill from the open port causing the pressure in the barrel and plunger passage to collapse. The delivery valve instantly snaps closed and fuel delivery to the high pressure pipeline ceases.

12.1.3 Combustion delay period

The period between the instant the injection pump plunger traps the fuel in the barrel pressure chamber by cutting off the supply port and the moment combustion initially occurs, is known as the *combustion process delay period*. This time interval may be divided into two components: (1) the injection lag; and (2) the ignition lag.

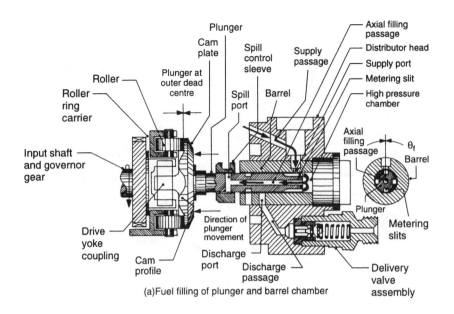

(a) Fuel filling of plunger and barrel chamber

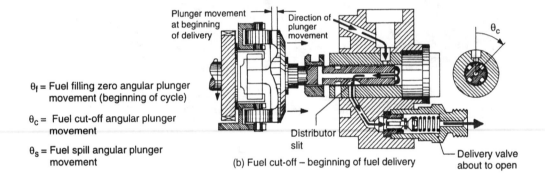

θ_f = Fuel filling zero angular plunger movement (beginning of cycle)

θ_c = Fuel cut-off angular plunger movement

θ_s = Fuel spill angular plunger movement

(b) Fuel cut-off – beginning of fuel delivery

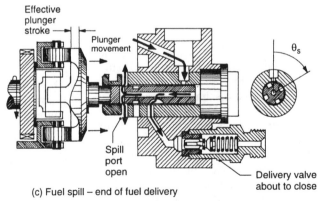

(c) Fuel spill – end of fuel delivery

Fig. 12.3 Filling and delivery cycle of operation (for Bosch Type VE injection pump)

Injection lag

The time interval between the instant the supply port is cut off by the plunger on its pumping stroke to the point where the injector needle valve opens and fuel commences to inject into the combustion chamber in the form of a fine spray is known as *injection lag*. This period is determined by the injector needle valve opening pressure, the rate at which the cam increases the plunger velocity, the pipeline residual pressure, and the size and number of spray holes.

Ignition lag

The time interval between the instant fuel is sprayed into the combustion chamber and the point when combustion commences, indicated by a rise in cylinder pressure above that of the compression curve, is known as the *ignition lag*. This period is determined by the ability of the fuel to atomize, the quality of the fuel, the amount of combustion chamber pre-heating, the cylinder's compression ratio, the thoroughness of the injected fuel and its surrounding air-charge mixing.

The delay period, composed of both the injection and ignition lag, is relatively constant in time and independent of engine speed provided there is no change in the combustion chamber pre-heating. Therefore, if the delay period is equal to 15° crank angle movement at 1000 rev/min, then the crankshaft will cover 30° when the engine speed has increased to 2000 rev/min. Consequently, if the point of injection is fixed, the peak cylinder pressure would become more and more retarded from the optimum position, which is something of the order of 10° after TDC.

As a result, the rise in power with increasing engine speed will be relatively smaller than if peak cylinder pressure is maintained at the optimum crank-angle movement after TDC.

To compensate for the increased delay-period crank-angle movement with rising engine speed, the point of injection must therefore be advanced. An automatic advance device is used to bring forward the point of injection with rising engine speed.

12.1.4 Automatic timing device
(Fig. 12.4(a and b))

Construction

The hydraulic timing device has a transverse timing piston subjected to the eccentric-vane supply-pump pressure on one side, and the thrust of a return spring on the other. The camplate rollers are mounted on a roller-ring carrier which is able to rotate within the limit of 24° crankshaft angular movement. The timing piston is interlinked to the roller ring carrier via a timing-pin which is articulated in the piston's block.

Operation

With the engine switched off, the return spring preload moves the roller-ring carrier anticlockwise and the timing piston to the right so that the advance mechanism is in the fully retarded position (Fig. 12.4(a)).

As the engine speed is increased, the hydraulic fuel pressure rises until, at about 3000 rev/min, the pressure equals the spring pre-load. A further speed increase then pushes the timing piston against the spring and, accordingly, the roller-ring carrier rotates in a clockwise direction in proportion to the fuel pressure, which is also a measure of the engine speed (Fig. 12.4(b)). The effect of moving the roller-ring carrier in the opposite direction to the camplate causes the four rollers to contact the cam flanks earlier. Accordingly, the roller pushes the camplate outwards and the plunger towards inner dead-centre on its pumping stroke. Injection is therefore advanced relative to the crankshaft angular movement. At all times there is a small circulation of fuel through the piston restriction between the timing-piston fuel-supply pressure side, and the camplate assembly chamber (governor chamber); thus, a drop in engine speed produces a fairly quick pressure reduction response on the timing piston pressure face.

12.1.5 Manual cold-start injection advance device

Reason for cold-start advances

For some indirect combustion chambers the engine starts more easily and produces less blue smoke in the exhaust gas if the fuel is injected into the hot compressed air slightly earlier than the normal warm start optimum injection setting.

Construction and operation of eccentric ball-pin cold-start device
(Fig. 12.5)

The cold start injection advance device is mounted on the side of the injection pump hous-

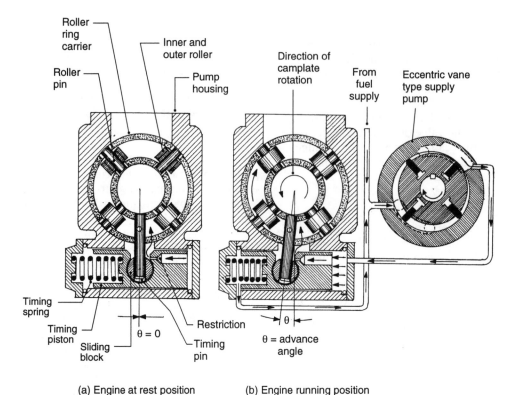

Timing spring
Timing piston
Sliding block
Roller ring carrier
Roller pin
Inner and outer roller
Pump housing
θ = 0
Restriction
Timing pin

Direction of camplate rotation
From fuel supply
Eccentric vane type supply pump
θ = advance angle

(a) Engine at rest position   (b) Engine running position

Fig. 12.4 Timing advance device

ing in line with the roller-ring carrier. It consists of an external lever, actuating shaft and an eccentric ball-pin attached to it. The ball-pin projects inside an elongated hole made in the side of the roller-ring carrier.

When a cold-start is required, a knob inside the cab is pulled, this movement is relayed via a cable to the cold-start lever which twists the shaft with the eccentric ball-pin at its end. The offset ball-pin, which is initially in its lowest position, will

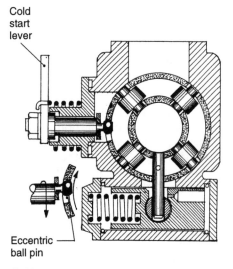
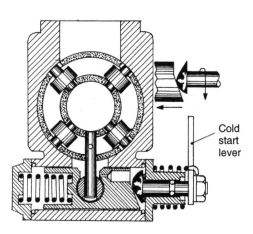

Cold start lever
Eccentric ball pin
Cold start lever

Fig. 12.5 Cold start injection advance devices

586

now rotate with the shaft. It therefore moves to a higher position and, in so doing, will rotate the roller-ring carrier in the direction opposite to that in which the camplate revolves. Accordingly, the flanks of the cams contact the rollers earlier, thereby advancing the commencement of the plunger pumping stroke. The elongated hole in the roller-ring carrier permits the hydraulically controlled piston to continue rotating the roller-ring carrier, thus advancing the beginning of injection at higher engine speeds.

12.1.6 Load-dependent injection timing device
(Fig. 12.6(a and b))

Purpose
Diesel engine performance is improved, with an injection timing advance, as speed increases. The engine also operates smoother, with a reduction in injection advance, when the engine load decreases.

The load-dependent injection timing device utilizes the flyweight support shaft, sliding sleeve and flyweights to sense the engine load and to vary the injection timing to match the engine load.

Construction
(Fig. 12.6(a))
This device consists of an axial passage in the flyweight support shaft, with radial cross-holes intersecting this passage and an annular groove machined around the shaft. In addition to the side hole in the sleeve near its blanked-off end (which releases fuel that might become trapped) there are two radial spill-ports further back towards the open end of the sleeve. The spill-port's job is to reduce the eccentric vane supply pump pressure under certain engine speeds and load conditions. This, therefore, allows the return spring thrust to exceed the hydraulic thrust and thus moves the timing piston in a retarding direction.

Operation
With the accelerator-control lever moved to any setting, the engine speed will rise until its load matches the fuel delivery. If the engine load is reduced, and its speed increases, the flyweights will swing outwards thereby pushing the sliding-sleeve away from the support shaft (Fig. 12.6(b)). When the sliding-sleeve spill ports align with the

support shaft annular groove, fuel escapes so that the eccentric vane supply pump mean-pressure decreases. Consequently, the reduced fuel supply pressure acting on the timing piston permits the return spring to push the piston towards the injection timing retard position. Conversely, if the engine load rises, the flyweights will swing inwards so that the sliding-sleeve moves towards the support shaft until the sleeve spill-ports cut off the support shaft's annular groove and radial cross holes (Fig. 12.6(a)). The supply pump pressure will immediately commence to rise, thereby causing the timing piston to move in the injection advance direction and thus move the roller-ring carrier in the opposite direction to that of the rotating camplate.

12.1.7 Fuel cut-off

The engine can be shut down by cutting off the fuel supply to the engine. There are two methods of stopping the engine:

1 mechanical shut-off lever controlled by a stop cable knob in the cab
2 electromagnetic shut-off valve (Fig. 12.1)

When the key switch is turned on, the electromagnetic solenoid is energized. This draws up the plunger-type valve against the coil return spring. Fuel can now pass from the governor chamber to the supply passage, supply port, metering slits and finally into the high-pressure barrel chamber. If the key switch is turned off, the electromagnetic solenoid magnetic field collapses. Immediately, the return spring pushes down the plunger valve onto its seat. The fuel supply is now separated from the barrel high-pressure chamber, and therefore fuel delivery ceases.

This type of shut-off device is particularly suitable for car and van applications since no mechanical linkage is required.

12.2 Governing

12.2.1 Variable speed mechanical governor
(Fig. 12.7(a–d))

Construction
(Figs 12.1 and 12.7(a–d))
The flyweight cage and sliding-sleeve are

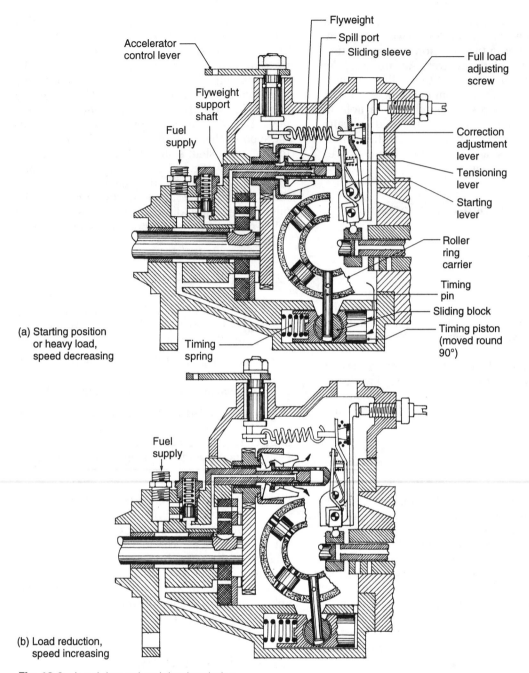

Fig. 12.6 Load dependent injection timing

mounted on the support shaft, and the input shaft drives the weight cage and weights via the step-up gears (23/37 teeth).

The tensioning and starting levers, which are pivoted together at the lower end onto a correction-lever, relay the opposing governor spring tension to the flyweights via the sliding-sleeve. The underside of the starting-lever is connected to the spill-control sleeve via a ball-pin which fits into a socket so that any tilting movement of the levers will shift the spill-sleeve in the opposite direction on the injection plunger. The initial positioning of the spill-sleeve is set by the stop screw, which tilts the correction-lever about its pivot and, at the same time, moves the tensioning and starting lever's pivot point slightly to the left or right; that is, in order to produce an earlier or later pumping stroke spill point.

Starting position
(Fig. 12.7(a))

With the accelerator lever moved to the full-load fuel delivery position, the governor spring will stretch and pull the tensioning and starting levers to the left, against the sliding-sleeve, until the flyweights are swivelled fully inwards. Accordingly, the anticlockwise tilt of the tensioning and starting levers moves the spill-control sleeve to

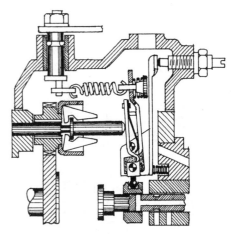

(b) Low idle speed position

Fig. 12.7(b)

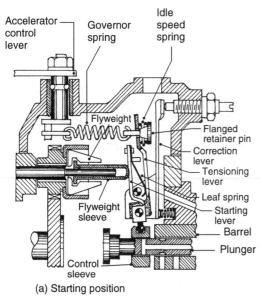

Fig. 12.7 Variable speed mechanical governor

the right so that the plunger has to move further along its stroke before spill occurs. Once the engine fires and the speed increases, the flyweights will push the sliding-sleeve outwards against the leaf-spring until the starting-lever peg contacts the tensioning-lever. In this position the spill-sleeve will move to the right, thus reducing the effective pumping stroke of the plunger and therefore its fuel delivery.

Idle speed position
(Fig. 12.7(b))

With the accelerator lever released so that it moves back against the idle speed stop, the

engine speed will be such as to swing the flyweights out against the tension of the idle-spring (which is trapped between the tension-lever and the flanged idle return pin). This light spring is sensitive to the slightest change in engine speed and can therefore rapidly respond to load changes during idle speed conditions. Thus, any small change in engine speed will tilt the tensioning and starting levers one way or the other and, likewise, shift the spill-sleeve. Hence, this alters the effective pumping stroke so that fuel delivery automatically adjusts itself to the varying engine load, thereby maintaining a relatively stable and constant idle speed.

Intermediate speed position
(Fig. 12.7(c))

As the accelerator-lever is moved to increase fuel delivery, the governor spring draws the idle retainer-pin against the idle-spring until its shoulder contacts the tensioning-lever. Beyond this point the main governor spring stretches so that the tensioning and starting lever is pulled to the left. Eventually, the tensioning and starting levers tilt in an anticlockwise direction to a point where the flyweight sliding-sleeve thrust equals the governor spring tension; at the same time, the ball-pin tilt will have moved the spill-sleeve so that spill occurs later in the plunger pumping stroke. The fuel delivery will now raise the engine speed until the generated torque matches the external load. Any change in engine load for a given accelerator-lever setting will swing the flyweights either inwards or outwards. Thus, for a given governor

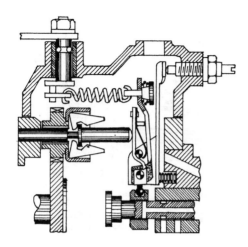

Fig. 12.7(c) Any intermediate speed position

spring tension the spill-sleeve will move to a new equilibrium position, and this will adjust the fuel delivery to bring the engine's speed back to its original setting.

Maximum speed governor cut-in
(Fig. 12.7(d))
With the accelerator lever set to full-load fuel delivery, the engine speed will increase until the engine torque equals the external torque load. However, if the external load opposing the engine's torque decreases sufficiently the engine

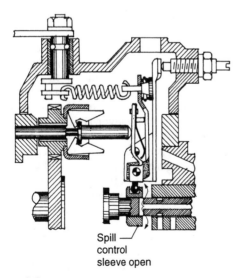

Spill
control
sleeve open

Fig. 12.7(d) Maximum speed cut-in position

speed will continue to rise until the maximum rated speed is reached. Just before this speed is reached, the outward swing of the flyweights will increase the sliding-sleeve thrust until it exceeds the governor spring tension. As a result, the tensioning and starting levers swivel clockwise, thus moving the spill-sleeve towards the no-load position. Consequently, the engine speed will rapidly decrease.

12.2.2 Minimum and maximum governor
(Fig. 12.8(a–d))

Construction
(Figs 12.1 and 12.8(a–d))
This flyweight two-speed governor is similar to the variable-speed governor but differs in the way the governor springs are arranged. With the variable-speed governor, the main spring operates in tension, whereas with the maximum–minimum speed governor, its maximum speed spring works in compression. In the governor described, a second torque control spring has been incorporated in series; however, for some applications torque control is excluded so that only the maximum speed spring is housed in the spring carrier frame.

Starting position
(Fig. 12.8(a))
With the engine stationary the accelerator-lever is moved to its maximum fuel delivery position. This tilts the tensioning-lever and starting-lever towards the sliding-sleeve until the flyweights have swung to their innermost position. Under these conditions the ball-pin swings the spill-control sleeve towards the right so that the effective pumping stroke is maximized before fuel spill takes place. Once the engine reaches cranking speed the flyweights' centrifugal force is sufficient to overcome the leaf-spring tension, this causes the starting-lever to tilt in the opposite direction to the tensioning-lever until the floating primary idle spring contacts the tensioning lever. The spill sleeve accordingly moves to a reduced fuel delivery position.

Idle speed position
(Fig. 12.8(b))
After the engine has started and the accelerator-lever has moved to the idle speed stop, the engine

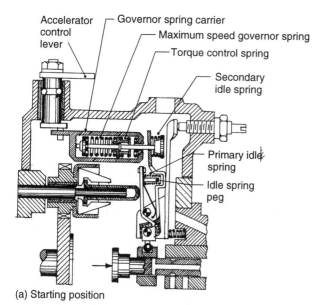

Accelerator control lever — Governor spring carrier — Maximum speed governor spring — Torque control spring — Secondary idle spring — Primary idle spring — Idle spring peg

(a) Starting position

Fig. 12.8 Maximum and minimum speed mechanical governor

speed will adjust itself to the reduced fuel delivery. The tension now opposing the flyweights pushes the starting-lever towards the tensioning-lever until the primary idle-spring slack is taken up and it is in partial compression; a state of equilibrium is therefore initially established. Any change in external load, causing a sudden speed change, will automatically be corrected by the starting-lever tilting one way or the other relative to the tensioning-lever. Thus, the spill-sleeve

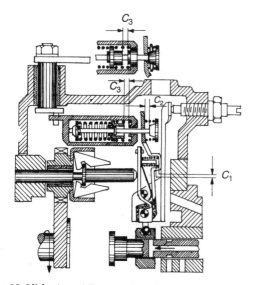

C_3

C_3

C_2

C_1

Fig. 12.8(b) Low idle speed position

corresponding movement corrects the plunger's effective pumping stroke to match the demands of the external load variations.

At slightly higher speeds, the primary idle-spring peg contacts the tensioning-lever, hence the effective spring tension opposing the flyweight thrust is now provided by the slightly stiffer secondary idle-spring. The governors' action is now between the flyweight sliding-sleeve and the tilting tensioning-lever which, under these conditions, is held rigid to the starting-lever. Consequently, the spill-control sleeve adjusts the spill point and hence the fuel delivery at any instant.

Torque control operating stage
(Fig. 12.8(c))

The torque control mechanism is built into the governor spring carrier, and the torque control spring is connected in series with the maximum speed governor spring. The space between the torque spring collar and spring carrier frame is the torque control travel.

Without the torque control spring the maximum–minimum governor plays no part in regulating fuel delivery between idle speed and the engine's maximum rated speed.

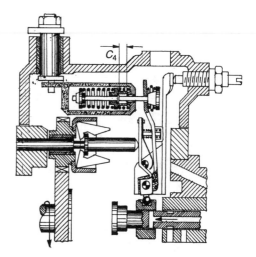

C_4

Fig. 12.8(c) Full load at maximum speed–end of torque control

Above idle speed, the idle-spring will be fully compressed as the pin-retainer shoulder contacts the tensioning-lever and the maximum speed spring.

With the accelerator-lever in the full-load fuel delivery position, at relatively high-load and low engine speed, the torque control spring is stiff enough to hold the flyweights together. However, as the engine load is reduced, its speed increases until the flyweights are able to oppose and progressively compress the torque control spring, thereby tilting the tensioning and starting lever in a clockwise direction. Accordingly, the spill-sleeve will move to an earlier plunger stroke spill position.

Thus, the torque control spring, as it compresses, will reduce the effective plunger stroke and hence decrease fuel delivery—this being necessary to compensate for the reduction in volumetric efficiency as the engine speed rises. Conversely, the torque control spring expands and moves the spill-sleeve to delay the plunger spill on its pumping stroke when the engine load is increased and its speed decreases.

Maximum speed governor cut-in
(Fig. 12.8(d))
As the engine approaches its maximum rated speed, the centrifugal force created by the flyweights will be sufficient to resist and exceed the maximum speed governor spring. The weights will therefore swing outwards forcing the sliding-

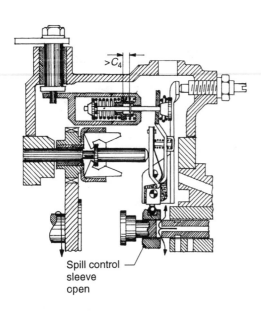

Spill control
sleeve
open

Fig. 12.8(d) Maximum speed cut-in position

sleeve to swivel the tensioning and starting lever clockwise, up to the point at which the spill-sleeve moves to the no-delivery position. Immediately,

fuel ceases to be delivered and the engine's speed decreases as shown in Fig. 11.40.

12.2.3 Boost control device for manifold pressure compensation
(Fig. 12.9(a and b))

Purpose
A boost control unit is required to match the full-load fuel delivery to the quantity of air entering the cylinders over the engine's speed range. Thus, at low engine speed, when the turbocharger discharge amounts to very little extra air supply for the cylinders, the boost control prevents the spill-sleeve moving to maximum fuel delivery. However, as the turbocharger begins to increase the quantity of air delivery to the cylinders, the boost control automatically allows the spill-sleeve to move progressively along the plunger to provide a later plunger spill and, hence, increase fuel delivery.

Construction
(Fig. 12.9(a))
The boost pressure control unit uses a diaphragm to sense changes in manifold pressure and to provide a full-load variable fuel delivery stop. This is achieved through a diaphragm which is subjected to manifold pressure on its top side and both atmospheric pressure and coil spring thrust on the other. A spindle with a conical wedge at its lower end is attached to the underside of the diaphragm. The maximum fuel delivery tilt of the tensioning-lever is set by a double-armed lever stop, which is linked to the diaphragm-spindle conical-wedge via a horizontally positioned relay-pin. The return spring pre-load can be adjusted by rotating the lower spring plate while the initial position of the conical-wedge is set by the end stop screw.

Operation
In the lower speed range, when there is no boost pressure, the preloaded return spring forces the diaphragm assembly to its highest position (Fig. 12.9(a)). As the diaphragm assembly moves to its upper position the rising conical-wedge pushes the relay-pin outwards—this motion swivels the double-armed lever stop anticlockwise, thereby reducing the maximum tensioning-lever tilt. Accordingly, the spill-sleeve is shifted to the

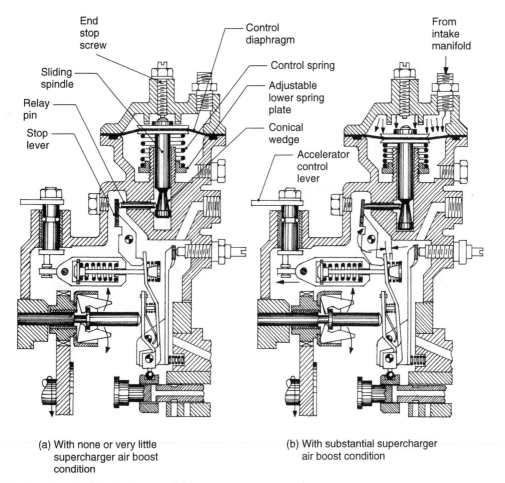

| (a) With none or very little supercharger air boost condition | (b) With substantial supercharger air boost condition |

Fig. 12.9 Boost control device for manifold pressure compensation

Labels in figure:
End stop screw
Sliding spindle
Relay pin
Stop lever
Control diaphragm
Control spring
Adjustable lower spring plate
Conical wedge
Accelerator control lever
From intake manifold

extreme right so that the effective plunger pumping stroke is maximized for the no-boost manifold pressure conditions.

With rising engine speed, the manifold boost pressure increases, and the corresponding downward pressure thrust on the diaphragm will now cause the diaphragm assembly to be progressively pushed down (Fig. 12.9(b)). As a result, the downward movement of the spindle aligns the relay-pin with the narrow part of the conical-wedge incline. Thus, as the tensioning-lever approaches full-load fuel delivery, the double-armed lever stop will be able to swivel slightly further clockwise as the relay-pin moves against the contracting portion of the conical-wedge. The outcome is that the tensioning-lever tilt at full-load delivery is able to increase as the diaphragm is moved further down with rising manifold boost pressure.

12.3 Distributor fuel injection rotary pump (CAV Type-DPC)

12.3.1 Description of pumping unit
(Figs 12.10, 12.11, 12.12(a and b) and 12.13)

Rotor and sleeve
(Fig. 12.10)
The essential component of the pumping unit is a single Nitralloy rotor shaft which functions both as the pumping element and distributes the fuel. It rotates in a Nitralloy cylindrical sleeve which is itself supported in an outer low-carbon steel casing known as *the hydraulic head*.

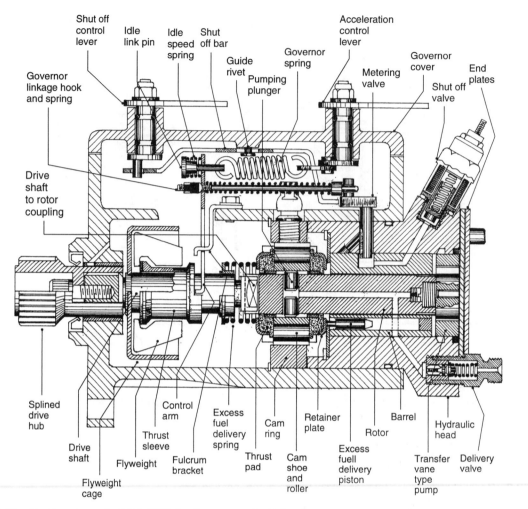

Fig. 12.10 Section view of a CAV–DPC distributor type fuel injection pump

Pumping section of rotor
(Figs 12.10 and 12.11)
The rotor consists of a cylindrical distributor section and an enlarged double-slotted pumping section. The pumping section has a transverse cylinder bore which locates and guides two opposing pumping plungers.

A stationary interval cam profile ring is positioned in the pump housing and actuates the plungers through rollers and shoes, which are carried in slots machined in the enlarged section of the rotor. The cam-ring has as many lobes as there are engine cylinders, these being equally spaced around the ring. Metering fuel pressure controls the outward travel of the plungers and hence the quantity of fuel discharged on their return stroke.

Distributor section of rotor
(Figs 12.10, 12.12(a and b) and 12.13)
A central passage intersecting the cylinder space between the pumping plungers extends to the hydraulic-head portion of the rotor where it merges with a single radial distributor port. As the rotor revolves, the distributor port aligns, in succession, with each of the distributor ports located radially at equal intervals around the hydraulic head. There is one discharge port for each engine cylinder and each one of these ports is connected to its respective injector via a high-pressure pipe.

Along the rotor's cylindrical section, between the pumping plungers and distributor-port, are radial charge-ports all in the same plane, equally spaced and equal in number to the discharge-

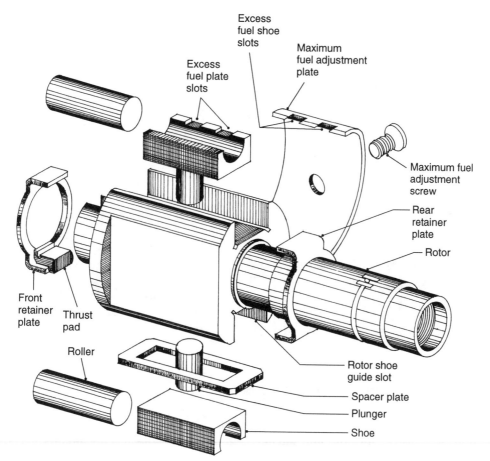

Fig. 12.11 Exploded view of the rotor and plunger pumping elements (CAV–DPC)

ports. As the rotor revolves, different pairs of charge ports align, in turn, with a pair of metering ports, radially located in the rotor-sleeve, and these metering ports have the same angular spacing between them as the charge ports. Only one metering-port is shown in Fig. 12.13 as it is easier to explain the operating principle for the filling and pumping cycle this way. The metering ports supply the metered fuel to the cylindrical space between the plungers.

12.3.2   Rotor and plunger charging and pumping cycle
(Fig. 12.12(a and b))

The cycle of operation can be described by considering a single-cylinder injection pumping element in two stages: (1) filling or charging phase; (2) injection phase.

Filling or charging phase
(Fig. 12.12(a))
Assume the rotor is revolving: each charge-port and the single distributor-port sweep around aligning and misaligning with the metering-port and discharge-ports respectively.

When a charge-port aligns with the metering-port, fuel at metering pressure enters the rotor's central passage and from there it fills the cylinder space and forces the plungers apart. At this point, both roller and shoe assemblies will be between cam-lobes. The amount of plunger outward displacement is determined by the quantity of fuel which can flow into the rotor passages for the short time the metering and charging ports coincide with each other.

Injection phase
(Fig. 12.12(b))
Further rotation of the rotor will misalign all

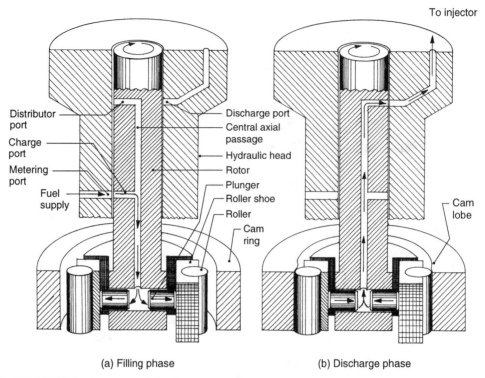

Distributor port
Charge port
Metering port
Fuel supply

Discharge port
Central axial passage
Hydraulic head
Rotor
Plunger
Roller shoe
Roller
Cam ring

Cam lobe

(a) Filling phase

(b) Discharge phase

Fig. 12.12 CAV–DPC distributor type injection pump cycle of operation

charge-ports with the metering-port and, simul-taneously, bring the distributor-port into align-ment with one of the discharge-ports in the hyd-raulic head. At the same time, both rollers con-tact the flanks of the diametrically opposed cam-lobes so that, in the next instant, the plungers are forced together. The reduced passageway volume immediately causes the fuel passage to increase, and fuel will therefore be displaced through the discharge-port and delivery-valve to its respective injector.

With a four or six-cylinder distributor (DPC) injection pump, the cycle of events is similar. There are still one pair of opposing plungers, one metering-port and one distributor-port: however, there are four or six cam-lobes, charge-ports and discharge-ports, respectively. A charge or dis-charge event occurs at rotor intervals of 90° or 60° for four and six-cylinder injection pumps, respec-tively.

12.3.3   Fuel circulation system
(Figs 12.13 and 12.14)

Fuel is drawn from the fuel tank via the filter to

the banjo-bolt mesh filter on the hydraulic head, it then enters the injection pump and passes to the vane-type transfer-pump which is screwed on the end of the rotor. This pump increases the fuel pressure in proportion to the speed of rotation (Fig. 12.13). The fuel then flows through the transfer pressure regulator valve, which maintains the output from a pump at a predetermined level in relation to engine speed—this being known as transfer pressure. Fuel then passes to the transfer pressure annular groove, and from there to the shut-off valve. With this valve open, it now passes to the metering-valve. The metering-valve vari-able orifice size controls the quantity of fuel delivery to the pumping plungers. Fuel now at metering pressure flows to the metering pressure annular groove, it passes to the two metering-ports (Fig. 12.14) and then transfers to the rotor's central passage each time pairs of charge-ports align with these equally spaced ports formed in the stationary rotor-sleeve. Fuel now trapped in the central passage pushes the pumping plungers apart, thereby filling the cylinder space between the plungers with fuel in readiness for the next pumping stroke.

Fuel from the transfer pressure annular groove

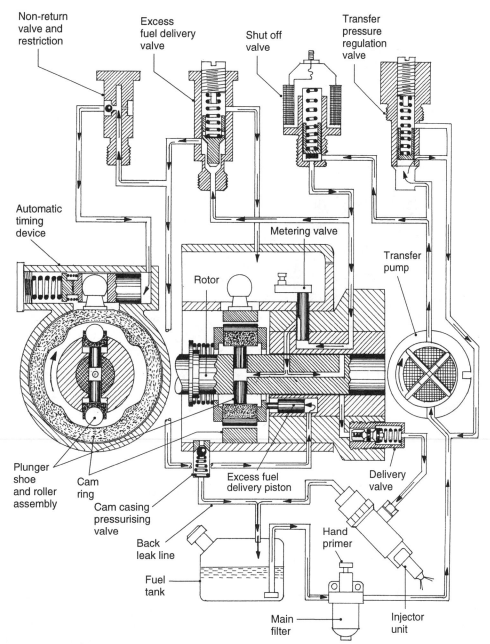

Fig. 12.13   CAV–DPC distributor type fuel injection pump system layout

(Fig. 12.14) also flows via the excess fuel delivery annular groove to the excess-fuel valve screwed into the hydraulic head. This valve controls the flow of fuel to the rear of the excess-fuel delivery plungers. When the transfer-pressure is low, the valve is closed thereby cutting-off the flow to the excess fuel delivery position. This causes the stroke of the pumping plungers to be extended to

the excess fuel delivery position. With rising transfer pressure the valve opens and the stroke of the pumping plungers will be limited to the normal maximum fuel delivery position.

With the excess fuel-delivery valve open (Figs 12.13 and 12.14) fuel also passes to the automatic timing-piston via the excess fuel delivery annular groove and the non-return ball-valve and restric-

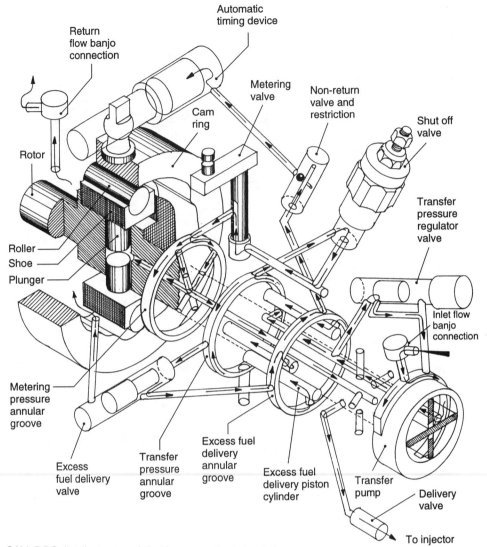

Fig. 12.14 CAV–DPC distributor type injection pump fuel circulation system

tion. Rising pump speed and therefore increasing transfer-pressure will progressively advance the point of fuel injection.

12.3.4 Vane-type transfer pump
(Fig. 12.15(a and b))

The positive displacement vane-type pump consists of a rotor which is screwed on to the end of the distributor-rotor. To prevent it from loosening in operation, it has a screw thread which tightens up in the opposite direction to the rotation of the injection pump. A pair of rigid blades at right angles to each other are free to slide in the cross-over rotor grooves, as they are pushed to and fro by the eccentrically positioned internal liner profile, as the rotor revolves. The liner eccentricity position is fixed relative to the inlet and outlet ports by the notch in the liner's side, which is located by a peg in the hydraulic head.

Operating principle
(Fig. 12.15(b))
As the rotor revolves, the enlarging semi-crescent shaped cell formed between the adjacent blades induces fuel to enter this space via the inlet port. Further, rotor rotation clears the inlet port and traps the fuel between the adjacent blades, the

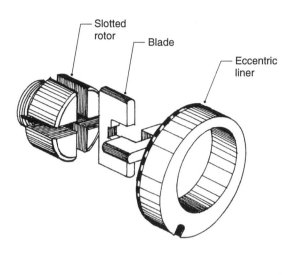

(a) Transfer pump components

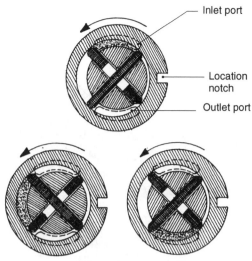

(b) Transfer pump cycle of operation

Fig. 12.15   Vane type transfer pump

eccentric liner and rotor walls. With even further rotor rotation, fuel is swept around in the decreasing semi-crescent shaped cell until the sides of the cell coincide with the outlet port. Pressurized fuel will now be discharged through the outlet port.

Output pressure from the pump increases in direct proportion to the speed of rotor rotation.

12.3.5   Transfer pressure regulator valve operation conditions
(Figs 12.16(a–c))

The regulator-valve senses the transfer pump pressure and controls this pressure so that its rise increases directly with rotor speed but at a reduced rate. The regulator-valve also provides a way for fuel to bypass the transfer pump and to prime the passageways in the hydraulic head when the engine is stationary or being cranked.

Engine at rest position
(Fig. 12.16(a))
With the engine stationary, the regulator-spring and priming-spring will hold the regulator valve plunger flush with the open end of the valve-sleeve. Under these conditions, fuel cannot be transmitted through the passages in the hydraulic head.

Priming condition
(Fig. 12.16(b))
To obtain an effective fuel feed to the hydraulic-head when the engine is stationary, a hand primer supplies fuel to the transfer-pump inlet port and to the regulator valve, the relatively low fuel pressure compresses the priming-spring and pushes out the regulator-plunger until its radial primary ports are uncovered. Fuel will now pass through the primary spring chamber to both the outlet port of the transfer-pump and the passages feeding into the hydraulic head.

Regulating condition
(Fig. 12.16(c))
As soon as the engine starts, pressure builds up from the transfer-pump, and exceeds the inlet feed pressure and regulator spring thrust. The regulator-plunger now moves inward until it exposes the regulator-ports. As a result, the pressurized fuel will be discharged and returned to the inlet side of the pump. A modified and controlled transfer-pressure is therefore diverted to the metering-valve via the transfer pressure annular groove and shut-off valve.

12.3.6   Metering
(Figs 12.10, 12.13 and 12.14)

The volume of fuel entering the injection pump is

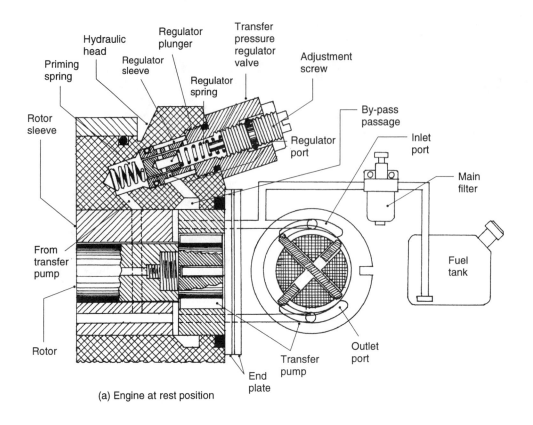

(a) Engine at rest position

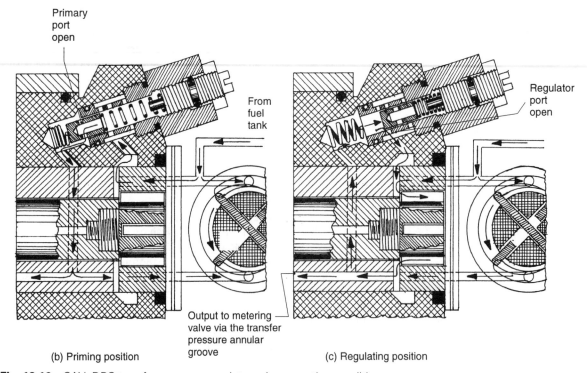

(b) Priming position

Output to metering valve via the transfer pressure annular groove

(c) Regulating position

Fig. 12.16 CAV–DPC transfer pressure regulator valve operating conditions

equal to the fuel discharged from the injectors. The quantity of fuel injected can therefore be varied by controlling the volume of fuel filling the cylinder space between the two pumping plungers. The volume of fuel entering the distributor rotor is dependent on the following:

1 the level of a transfer-pressure reaching the metering valve. This pressure is determined by the combination of rotor speed and the controlling influence of the regulator-valve;

2 the time of hydraulic head metering-port and the rotor charge-port coincide with one another. This will be determined by the port passageway cross-sectional area and the rotor's speed of rotation;

3 the metering pressure which is produced on the output side of the metering-valve. This pressure is produced by the throttling which occurs across the variable metering orifice when the valve is partially rotated by a combination of the accelerator-control lever opening and the governors' interaction. The cylindrical valve has a longitudinal slot formed half-way along its length, and it fits radially into a bore on the top side of the hydraulic head. A passageway underneath the valve supplies fuel at transfer-pressure, the fuel passes up the slot and then flows down the inclined passageway, which intersects the metering-valve bore. The fuel then flows to the rotor's central passage via the metering-port and charge-port, every time they coincide. Metering is achieved by turning the metering-valve so that the metering valve's slot alignment with the inclined passage changes its effective flow area. This produces a pressure drop across the restriction formed by the controlling edge of the metering-valve partially cutting off the inclined passage entrance. Consequently, the output flow from the metering orifice causes a lowering of fuel pressure. This is known as *metering-pressure*.

12.3.7 Maximum fuel delivery adjustment device
(Figs 12.11 and 12.17(a and b))

The rotor at the enlarged end contains a pair of pumping plungers located in a transverse cylinder. Radial movement from the internal-cam ring-lobes to the plungers is transferred through a pair of rollers supported in rectangular shoe-

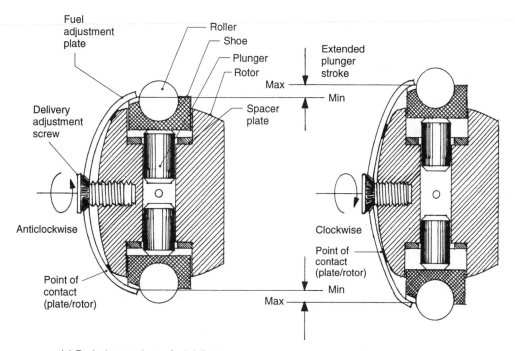

(a) Reducing maximum fuel delivery

(b) Increasing maximum fuel delivery

Fig. 12.17 Maximum fuel delivery adjustment device

blocks. These are guided in slotted longitudinal grooves machined diametrically on opposite sides of the enlarged portion of the rotor. The axial location of the shoes and rollers (in their respective slots formed in the rotor-head) is determined by the spacer-plates positioned between the underside of each shoe and the flat floor of each rotor-head slot. The four thrust-pads and the two retainer-plates clamp together the shoes and spacer-plates.

A curved spring strip forms the maximum fuel adjustment plate, this is secured to one side of the rotor-head by a self-locking screw (Fig. 12.11). The only contact between the rotor head and the maximum fuel-adjustment plate is on the corner of the rotor head (Fig. 12.17(a and b)). The ends of the plate are slightly curled so that the outward shoe movement is restricted.

Turning the screw clockwise (Fig. 12.17(b)) draws the central curved portion of the spring plate towards the rotor-head, which deflects the outer ends of the plate so that the curled ends move further apart. The plunger, shoe and rollers are therefore permitted to move further outwards when the metered fuel pushes the plungers apart. Conversely, rotating the adjustment screw anti-clockwise (Fig. 12.17(a)) bows the middle region of the spring plate outwards, this causes the curled ends to move slightly together. This, therefore, reduces the plunger's outward movement.

12.3.8   Engine stop devices

To shut down the engine, a fuel shut-off device is provided which can either be in the form of a solenoid-valve, which cuts-off the fuel flow between the transfer pressure regulator-valve and metering-valve, or by a mechanical link between the stop-lever and the metering-valve, which cuts-off the fuel passing out from the metering-valve.

Solenoid shut-off valve
(Fig. 12.10)
This assembly takes the form of an annular-shaped solenoid surrounding an armature–plunger valve, which has a rubber seat at its end. The unit is designed to be screwed into the hydraulic head.

When the engine is switched on, the solenoid is energized causing its magnetic pull to draw back the plunger. Fuel can now pass from the transfer pressure-regulator valve, at transfer-pressure, to

the metering-valve, and the hydraulic circuit is now completed and ready to be operated. The engine will now start when the engine is cranked so that the injection pump is able to provide timed injection to the engine's cylinders.

When the start switch is turned off, the collapse of the magnetic field permits the return-spring to push the plunger-valve back against its seat, and immediately the fuel flow to the metering-valve will be interrupted. The pump ceases to inject and the engine will be brought to a rapid standstill.

Mechanical shut-off device
(Fig. 12.10)
This method of cutting-off the fuel supply is only used with an all-speed governor. It consists simply of a shut-off control-lever and a shaft which has an eccentrically mounted pin fixed to its lower end. A shut-off bar in the form of a flat steel strip has a transverse elongated slot at one end which receives the eccentric-pin. A longitudinal elongated slot near the middle of the bar fits over a guide-rivet attached to the underside of the governor cover; this guide-rivet supports the shut-off bar and permits it to slide to and fro.

When the shut off lever is moved to the stop position the eccentric-pin pushes the shut-off bar against the metering-valve arm, this causes the valve to rotate until its vertical slot is completely cut-off by the metering-port. Instantly, fuel at metering-pressure is prevented from flowing to the metering-port, and therefore the injection pump ceases to inject and the engine stops.

Moving the shut-off control-lever in the opposite direction to the shut-off position, draws the shut-off bar away from the metering-valve so that it can function as normal. The engine can now be cranked and started as in the usual manner.

12.3.9   Excess fuel delivery device
(Fig. 12.18(a and b))

This device ensures that extra fuel is delivered to the engine cylinders when the engine is cranked and first starts. Above a predetermined engine speed the maximum fuel delivery will automatically be reduced to full-load fuel delivery.

The excess fuel delivery device has slots formed in the curled ends of the maximum-fuel adjustment-plate and on one side of the top face of each shoe. The maximum-fuel adjustment-plate slots have the same pitch and width as the slots in the roller shoes.

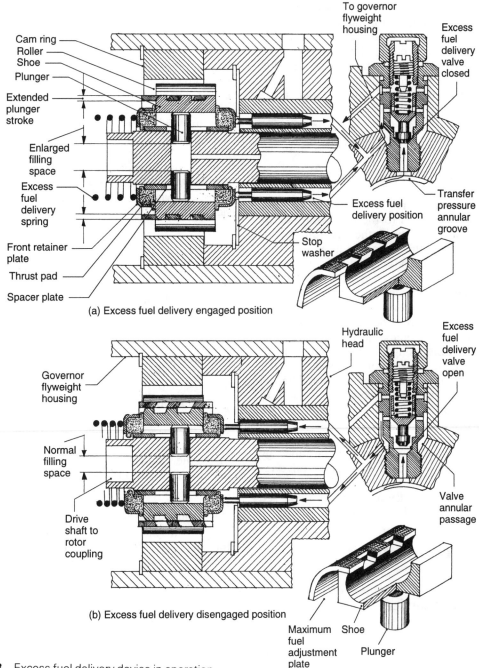

Fig. 12.18 Excess fuel delivery device in operation

Excess fuel delivery position
(Fig. 12.18(a))

When the engine is stationary there will be no transfer pressure. This allows the excess-fuel delivery-spring to exert an end thrust which slides the roller shoe assembly and the excess-fuel deliv-ery pistons in the direction of the hydraulic head (to the right). The excess-fuel delivery-piston therefore displaces fuel to the excess-fuel deliv-ery-valve surrounding the annular passage, through the plunger side ports and out to the governor housing.

603

The shoe assemblies are forced by the spring to the right until the slotted top faces of the shoes align with slots on the outer edges of the fuel adjustment-plate. The outward movement of the plungers then pushes the shoe slots into full engagement with the corresponding slots in the maximum fuel adjustment-plate. Under these conditions the plunger's outward movement is extended to lengthen the effective pumping strokes to obtain excess fuel delivery.

Maximum fuel delivery position
(Fig. 12.18(b))
Once the engine starts, transfer-pressure increases, and this passes to the excess-fuel delivery-valve's underside, until the pressure is sufficient to compress the valve spring and to push open the plunger-type valve. Fuel will now flow to the excess-fuel delivery-piston horizontal cylinders; and the piston and shoe assemblies are therefore pushed towards the spring until the piston reaches the stop washer. At this point, the shoe and maximum-fuel adjustment-plate slots misalign so that the plunger's outward movement is restricted by the top face of each shoe contacting the maximum-fuel adjustment-plate. The full stroke of the plungers now corresponds only to full-load fuel delivery.

The transfer pressure at which the excess-fuel delivery phase cuts out and the maximum full-load delivery phase takes over is determined by the pre-load of the excess-fuel delivery-valve return spring. Thus, the adjustment of the valve screw sets the spring tension and hence the speed at which the excess-fuel phase cuts out.

12.4 Injection timing

12.4.1 Automatic start—retard and speed advance device
(Fig. 12.19(a–c))

Injection advance or retard is obtained by arranging the cam-ring to be partially free to rotate in the injection pump housing. Rotation of the cam-ring in the opposite direction to that of the distributor rotor brings the cam-lobe into contact with the plunger roller and shoe assemblies earlier, thereby advancing the point of injection. Conversely, rotating the cam-ring in the same direction of rotation as the rotor retards the beginning of injection.

Construction (Fig. 12.19(a))

Speed advance
Angular movement of the cam-ring is provided by a speed-advance piston, located in the pump housing in the same plane as the cam-ring, with one side of the piston being subjected to transfer-pressure (which is a direct measure of engine speed) while on the opposite side, opposing the pressure build-up, is an advance coil-spring. Movement of the piston is relayed to the cam-ring through a cam ball-pin, which is screwed into the cam-ring, whereas the ball-pin itself is located in a slot formed in the piston.

Start retard
Suspended between the speed advance spring and the piston is a double-flanged moving stop and a start–retard spring which functions as the start–retard device.

Full retard position (Fig. 12.19(a))
With the engine stationary the excess-fuel delivery-valve closes and, therefore, only a relatively low residual pressure will be imposed on the speed-advance piston. Under these conditions, both the speed and the start retard spring are permitted to extend, thereby pushing the piston against its end plug (to the right). Correspondingly, the cam-ring will be tilted clockwise to its maximum retard position.

Zero advance position (Fig. 12.19(b))
When the engine is cranked and starts, the transfer-pressure builds up and forces open the excess-fuel delivery-valve. Fuel at transfer-pressure will now act on the speed-advance piston and will initially compress the start–retard spring until the shoulder of the moving-stop and piston contact. This brings the point of injection back to its normal timing position relative to idle speed.

Speed advance position (Fig. 12.19(c))
With rising engine speed there is a relative increase in transfer-pressure until the fuel pressure thrust equals the speed advance spring pre-load. A further rise in speed produces a corresponding transfer-pressure increase which progressively compresses the speed advance spring, thus pushing the piston towards the maximum-advance stop-sleeve. This movement rotates the cam-ring in the opposite direction of rotation to

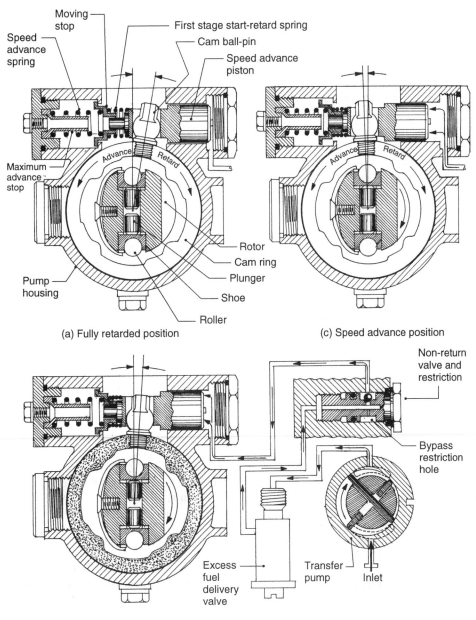

(a) Fully retarded position

(c) Speed advance position

(b) Zero advance position

Fig. 12.19 Automatic start–retard and speed advance device

the distributor rotor so, accordingly, the point of injection is proportionally advanced.

Likewise, speed reduction decreases the transfer-pressure, which causes the speed-advance spring-thrust progressively to return the piston to the no-advance position.

12.4.2 Cam profile characteristics
(Fig. 12.20)

The internal cam ring, which is made from high-speed steel, has its cam-lobes facing radially inwards. Each cam-lobe profile consists of a straight rise flank, a nose or peak, a retraction plateau, and a straight fall flank.

605

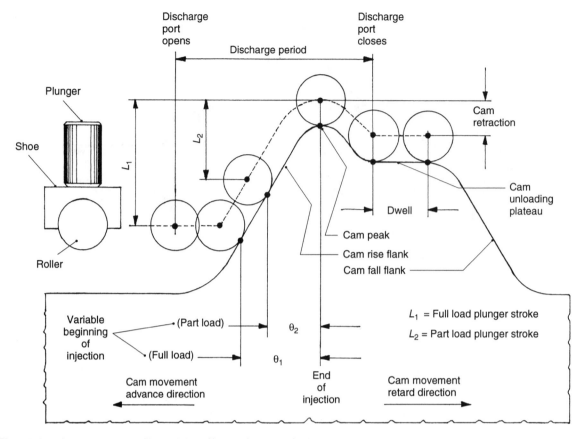

Fig. 12.20 Internal cam profile and the effects of varying fuel delivery output

When the rollers approach and contact the straight rise flank they ride up the incline, and instantly the plunger commences to accelerate on the inward pumping stroke, and injection begins.

Towards the end of the inward stroke of the plungers, the rollers begin to ride over the profile nose. The plungers then decelerate up to the point of inner dead-centre. Once the rollers reach the peak of the nose profile they quickly commence to ride down the other side, and simultaneously the plungers will accelerate outwards until the rollers reach the plateau. Correspondingly, the sudden enlargement of the space between the plungers causes a rapid reduction in pressure in the injection pipeline. This, therefore, prevents dribble and carbon formation at the injector nozzle at the end of the injection. The plateau dwell-period prevents the pipeline pressure collapsing completely for long enough for the distributor port to be cut-off from the discharge port. Residual pressure will then be maintained in the pipeline between the discharge-port and the injector. Finally, the rollers ride down the return flank, thus enabling the plungers to move to their outer dead-centre position ready for the next fuel charging phase.

12.4.3 Fuel delivery characteristics to match changes in engine load
(Figs 12.21 and 12.22)

The effects of varying the fuel delivery to match engine load for rotary injection pumps utilizing a cam-ring with internal cam-lobes and radial plungers are illustrated in Fig. 12.21. It can be seen that, as the quantity of fuel delivered decreases, the point of injection is progressively moved further back. However, the point of fuel cut-off, that is, when the metering port and charge port move out of alignment, is constant for all load conditions.

In contrast, the in-line injection pumps (which have a helix control-edge method of terminating

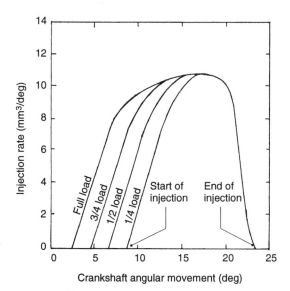

Fig. 12.21 Rotary pump with variable beginning and constant end of injection

the plunger's effective pumping stroke) have the opposite characteristics to that of a rotary pump using an internal cam-ring. Thus, with the in-line injection pump plunger action, the commencement of injection is normally constant, but the end of injection where the helix edge reaches spill point is variable and is prolonged as the fuel output increases from part-load to full-load fuel delivery (Fig. 12.22).

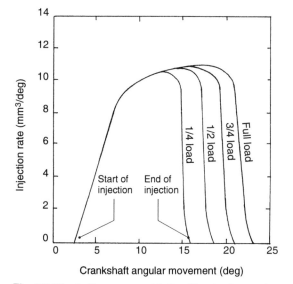

Fig. 12.22 In-line pump with fixed beginning and variable end of injection

12.4.4 Low-load advance with metering valve control
(Fig. 12.23(a–d))

Construction
(Fig. 12.23(a))
The low-load advance-piston assembly consists of a hollow low-load advance-piston, transversely positioned above the cam-ring on the opposite side to the speed advance-piston. The speed advance-spring contained inside the piston is permitted to expand between the blind end of the piston and the double shouldered moving-stop, which is itself retained in the piston by an internal circlip.

Fuel circuit
(Fig. 12.23(c))
Fuel transfer-pressure passes from the excess-fuel delivery-valve, when it is open, to the low-load advance-valve. It then flows through the calibrated-orifice and around the annular space formed between the cam-ring and low-load advance-valve. From here, it passes to the rear of the load-load advance-piston.

At the same time, some of the fuel flows into the low-load advance-valve around the waisted portion of its piston and out through a side hole. It then flows along a passage to the metering-valve's inclined-groove, which forms a variable-orifice. Rotation of the metering-valve may open or close the orifice to the passage leading to the cam-ring and governor housing chamber and, at the same time, may control the amount of fuel passing to the rotor and charging space between the pumping plungers. With the variable-orifice partially open, some of the fuel at transfer-pressure bleeds off; that is, vents to the cam-ring chamber. The effect of this escaping fuel is to reduce the pressure reaching the low-load advance-piston, and therefore reduce the hydraulic thrust opposing the speed advance-spring. Thus, the controlled venting of the fuel passing to the low-load advance-piston can be synchronized to the angular position of the metering-valve to meet the engine load and speed demand.

The area of the variable-orifice, for a given fuel delivery and hence engine load, can be varied by adjusting the height of the metering valve through the metering valve adjustment screw, which either increases or decreases the metering-valve stop-spring tension.

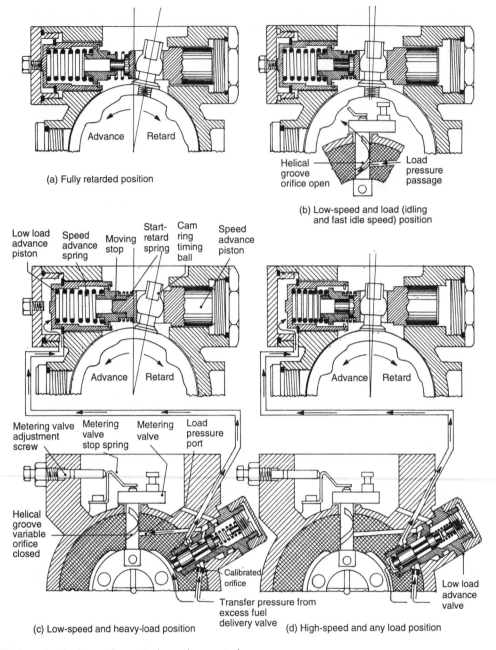

(a) Fully retarded position

Helical groove orifice open — Load pressure passage

(b) Low-speed and load (idling and fast idle speed) position

Low load advance piston | Speed advance spring | Moving stop | Start-retard spring | Cam ring timing ball | Speed advance piston

Advance | Retard

Advance | Retard

Metering valve adjustment screw | Metering valve stop spring | Metering valve | Load pressure port

Helical groove variable orifice closed

Calibrated orifice

Low load advance valve

(c) Low-speed and heavy-load position

Transfer pressure from excess fuel delivery valve

(d) High-speed and any load position

Fig. 12.23 Low load advance by metering valve control

Fully retarded engine stopped position
(Fig. 12.23(a))
With the engine stopped the excess-fuel delivery-valve will be closed so that there will be no transfer-pressure transferred to the speed-advance or to the low-load advance-pistons. Under these conditions, both the speed-advance and start-retard springs are forced to expand and, in so doing, the ball-pin rotates the cam-ring to the fully retarded injection timing position.

Low-speed and low-load (idle and fast idle speed) operating conditions
(Fig. 12.23(b))

At low engine speed and load the low-load advance-valve plunger will be forced against the cam-ring, which is in its open position.

Under these conditions the metering-valve is in the low-load fuel delivery position, with the inclined groove (variable orifice) in the open position.

With the variable-orifice open, the pressure acting on the low-load advance-piston (known as the load-pressure) will be at cam-ring chamber-pressure. The low-load advance-piston will therefore be pushed back against the end-plug stop by the speed-advance spring.

At the other end, the speed advance-piston will be pushed by transfer pressure in the injection advance direction.

Low-speed and heavy-load operating conditions
(Fig. 12.23(c))

At these operating conditions the transfer-pressure is still insufficient to push back the low-load advance-valve plunger to the valve closed position. Also, under these conditions, the metering valve is in the heavy-load fuel delivery position which corresponds to the variable-orifice being closed. The load pressure acting on the low-load advance-piston will therefore be at transfer-pressure.

The speed-advance piston will also be subjected to transfer-pressure; however, the cross-sectional area of the low-load advance-piston is larger than the speed-advance piston so that the resultant pressure will force both pistons towards the full-load injection timing position, to the point where the low-load advance piston circlip contacts the end of the cylinder. At low engine speeds, the transfer-pressure is not yet able to compress the speed-advance spring.

High-speed and any load operating conditions
(Fig. 12.23(d))

With rising engine speed, the transfer-pressure is eventually sufficient to push back the low-load advance-valve plunger until it closes the valve. The transfer-pressure will now pass directly to the low-load advance-piston without any of it being dumped to the cam-ring chamber via the metering-valve.

The low-load advance-piston is therefore thrust against the speed-advance spring until the circlip attached to the rear of the piston butts against the left-hand end of the piston's cylinder. Simultaneously, transfer-pressure acts on the speed-advance piston and, consequently, the speed-advance spring will be compressed inside the low-load advance-piston in proportion to the magnitude of the transfer-pressure; that is, it is related directly to engine speed.

12.5 Governing

12.5.1 Mechanical flyweight governor
(Figs 12.10 and 12.24(a–c))

The injection pump incorporates a centrifugal flyweight mechanical class of governor which can be of two types: (1) 'all-speed'; or (2) 'two-speed'. Both types of governor operate on similar principles but differ only in detail. The 'all-speed' governor accurately controls the metering-valve, and hence the amount of fuel injected at any load and throughout the engine speed range. The 'two-speed' governor, sometimes known as a 'minimum–maximum' speed governor, only controls the metering-valve at idle and fast idle speed and at the engine's maximum-rated speed. Between these extremes the metering-valve is controlled directly by the accelerator-control lever.

Construction
(Figs 12.10 and 12.24(a))

Both types of governor have four or six flyweights for high and low-speed engines respectively. These flyweights are constrained in a cage mounted on the driven shaft. The flyweights are held in position by a sliding-sleeve which is itself held in position by the governor spring thrust. The governor-spring is linked between the accelerator-lever and the flyweight sliding-sleeve through a control-arm that is permitted to tilt on a fulcrum-bracket. The governor-spring tension is determined by the accelerator-lever's partial rotation, which stretches the spring. Control-arm tilt, which is a measure of flyweight and governor spring interaction, is conveyed to the metering-valve through a hooked-rod and spring. For more sensitive idle speed governing, a coil idle-spring is used with the 'all-speed' governor, or a leaf-spring in the case of the 'two-speed' governor.

Rotation of the metering valve varies the flow area between the valve's vertical slot and metering port, and therefore controls the fuel delivery output of the pump.

12.5.2   Two-speed maximum–minimum mechanical-type governor
(Fig. 12.24(a–c))

Idle and fast-idle speed operating conditions
(Fig. 12.24(a))
At idling speed, governing is achieved by balancing the forces created by the deflection of the idle leaf-spring and also by the rotating flyweights' centrifugal force producing an end thrust as the flyweights swing outwards. The leaf-spring preload tension is set by the idle stop-lever adjustment-screw (not shown).

This idle-spring preload counteracts the changing flyweight centrifugal force end-load on the sliding-sleeve through the control-arm, which pivots about the fulcrum-bracket.

With the engine running at idle speed, any change in engine load will cause a corresponding change in engine speed. Thus, a sudden decrease in engine load produces an increase in engine speed. Accordingly, the flyweights swing further apart, causing the sliding-valve to move to the right—this is in opposition to the idling-spring which becomes partially flattened until its stiffness equals the thrust of the flyweights.

At the same time, the anticlockwise tilt of the control-arm through the hook-rod and spring-linkage rotates the metering-valve in the reducing opening direction. Immediately, the reduced fuelling brings the engine speed back to its original speed setting.

Conversely, if the engine speed should drop due to an increase in engine load there will be a corresponding reduction in centrifugal force and, as a result, the leaf-spring load will now exceed the flyweight thrust. The control-arm will now tilt clockwise so that the metering-valve will be moved to an increased fuelling position. The response is a rapid rise in engine speed to its normal setting.

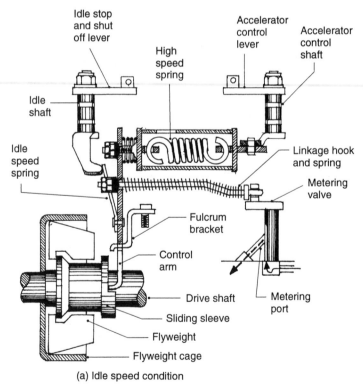

(a) Idle speed condition

Fig. 12.24   CAV–DPC maximum–minimum mechanical type governor

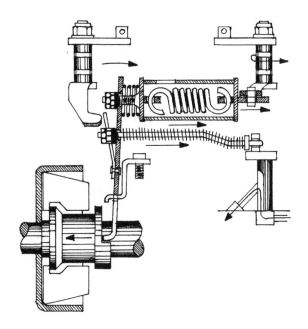

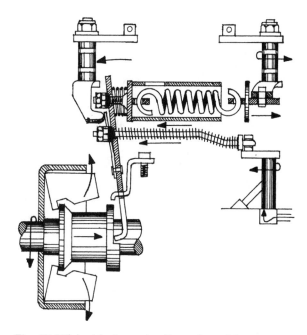

Fig. 12.24(b)   Intermediate speed condition

Fig. 12.24(c)   Maximum load/speed condition

Intermediate speed operating conditions
(Fig. 12.24(b))
With engine speed between idle and maximum rated speed, the governor high-speed spring preload is greater than the sliding-sleeve thrust produced by the flyweight centrifugal force. The high-speed spring assembly therefore acts now as a rigid-link between the accelerator-lever and shaft and the control-arm. Thus, any accelerator-lever movement is positively relayed first through the rigid governor high-speed spring and secondly through the hook-rod and spring-linkage. Under these conditions, the flyweight thrust is ineffective since it is overridden by the inflexible rigidity of the high-speed spring; the metering-valve rotation is therefore moved directly proportional to the accelerator lever setting.

Maximum rated engine speed operating conditions
(Fig. 12.24(c))
As the engine approaches its rated speed, the centrifugal force producing the flyweight outward swing increases to a point where the sliding-sleeve end thrust equals the high-speed spring preload tension. Thus, any further speed increase causes the centrifugal force thrust to exceed the oppos-

ing spring preload. It therefore commences to stretch the relatively stiff spring sufficiently to tilt the control-arm, which causes the metering-valve to turn to a much reduced fuel delivery output position. Subsequently, there will be a fall in engine power with no further rise in engine speed.

12.5.3   All-speed or variable speed mechanical-type governor
(Fig. 12.10)

Idle speed operating conditions
(Fig. 12.10)
At idle speed the flyweight centrifugal force is balanced by the light idle-speed compression-spring which is connected in series with the main governor-spring via the shouldered link-pin.

Under idle speed conditions, a reduction in engine load increases its speed and correspondingly swings the flyweights apart. Immediately this movement is relayed to the metering-valve through the sliding-sleeve, which then swivels the control-arm about its fulcrum. It therefore reduces the length of the idle-spring until its compressive load equals the centrifugal force load. The rotation of the metering-valve reduces the metering-port flow area in proportion to the speed increase. Accordingly, the reduced amount of flow injected brings the engine's speed back to its original setting.

In contrast, if the engine's load increases, its speed decreases. At this instant, the opposing idle-spring load is just greater than that generated by the flyweight centrifugal force and will (through the control-arm and sliding-sleeve) force the flyweights to swing closer together. The tilt of the control-arm will therefore rotate the metering-valve to enlarge the metering-port orifice, and the increased fuelling then raises the engine speed again to its normal value.

Intermediate and rated engine operating conditions
(Fig. 12.10)
With rising engine speed the idle-spring will be compressed until the shouldered link-pin butts against the control-lever. Any further speed increase, however, produces enough centrifugal force at the flyweights to counteract and stretch the main governor-spring. Thus, the accelerator-lever can be moved to the 'any load' fuel delivery position, and this stretches the main-spring until it is balanced by an equivalent flyweight centrifugal thrust with the engine running at the desired speed.

An increase in engine load reduces its speed and correspondingly decreases the flyweight centrifugal force. Immediately, the control-arm will tilt to a new position which, in turn, moves the metering-valve to an increased metering-pressure position until the engine speed recovers.

On the other hand, a reduction in engine load speeds up the engine until the flyweight centrifugal force builds up and forces the control-arm in the reducing fuel direction against the tension caused by the stretched spring. The metering-valve therefore adjusts its metering-pressure to match the engine load and speed condition.

Governing of this nature can be carried out at any accelerator-lever setting from just above idling speed to the maximum-rated speed of the engine.

When the engine has almost reached its maximum-rated speed, the flyweights' centrifugal force exceeds the spring stiffness and they will then swing sufficiently apart to rotate the metering-valve through the sliding-sleeve and control-arm to the low load position.

12.5.4 Boost pressure controller unit
(Fig. 12.25(a–c))

The method used to match fuel delivery to turbocharger boost-pressure at various speeds incorporates an air-boost pressure-sensitive double-diaphragm unit screwed into the hydraulic-head and a modified maximum-fuel adjustment-plate and shoes.

Slots which are normally machined in the outer ends of the maximum-fuel adjustment-plate and on the top side of the roller shoes are now replaced by inclined-planes or ramps. The rest of the excess-fuel delivery-device is similar to that used when the injection pump is to be fitted to a naturally aspirated engine.

Construction
(Fig. 12.25(a))
The controller unit consists of a circular casing that houses (in series) a large and a small diaphragm which are spring loaded on one side and subjected to transfer-pressure thrust on the other through a plunger attached to the smaller diaphragm. The small diaphragm, on its outer face, is exposed to fuel leakage from the plunger and atmospheric air, whereas the large diaphragm's outer face is subjected to both spring thrust and atmospheric pressure. The chamber formed between the inner adjacent faces of the diaphragm is set apart by a central spacer, this space is subjected to boost air pressure from the intake manifold.

Starting condition
(Fig. 12.25(a))
With the engine stationary there will be no transfer-pressure so that the excess-fuel delivery-valve will be closed. Under these conditions the excess-fuel delivery-spring pushes the shoe assemblies (carriage) to the right so that the shoe-ramp, and the fuel adjustment-plate ramp, contact each other at an earlier part of their inclines. This permits the plunger outward movement to extend to the excess-fuel delivery outer dead-centre position. Therefore, with the excess-fuel valve closed, no actuator-pressure (pressure between the excess-fuel delivery valve and pistons) reaches the excess-fuel delivery-pistons. Hence, the boost controller-unit does not operate during engine cranking speed.

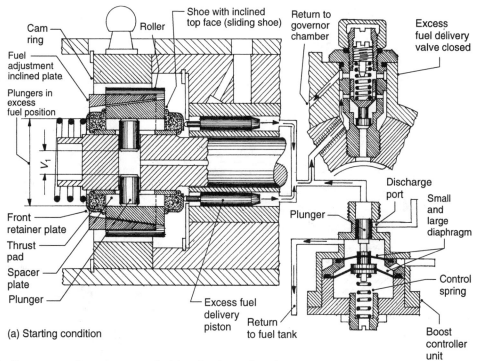

Fig. 12.25 Boost controller system used with turbocharged engines

Labels in figure:
- Cam ring
- Roller
- Shoe with inclined top face (sliding shoe)
- Return to governor chamber
- Excess fuel delivery valve closed
- Fuel adjustment inclined plate
- Plungers in excess fuel position
- V_1
- Front retainer plate
- Thrust pad
- Spacer plate
- Plunger
- Excess fuel delivery piston
- Return to fuel tank
- Plunger
- Discharge port
- Small and large diaphragm
- Control spring
- Boost controller unit
- (a) Starting condition

Engine started and operating in the low-speed region
(Fig. 12.25(b))

Once the engine is running at low-speed, the transfer-pressure rise will open the excess-fuel delivery-valve. Fuel now passes to the excess-fuel delivery-pistons and to the controller-unit plunger. The pressure of the fuel in these passages is known as *actuator-pressure* and, under these conditions (without almost any air boost-pressure) there is sufficient actuator-pressure to force the excess-fuel delivery-pistons and shoe assemblies (carriage) to the left. This will cause the fuel adjustment-plate and shoe inclined-planes to overlap one and other considerably, so that the plunger's outward stroke is restricted to its minimum full-load fuel delivery position.

Engine operating in the upper low-speed to high-speed region
(Fig. 12.25(c))

With rising engine speed there will be an increase in both transfer-pressure and boost air-pressure. However, as the rising boost air-pressure reacts between the small and large diaphragms, it pro-

duces a resultant thrust which acts in the same direction as the actuator-pressure on the controller plunger. Thus, it will progressively compress the control-spring until the discharge-port begins to open. Uncovering the discharge-port therefore reduces the actuator-pressure imposed on the excess-fuel delivery-pistons and, accordingly, the excess-fuel delivery-spring thrust will exceed the excess-fuel delivery-piston thrust and a new equilibrium position is established. Consequently, the shoe assemblies (carriage) will slide away from the minimum full-load delivery position to some new point of balance between the minimum and maximum full-load extreme position. It therefore limits the outward movement of the plungers so that fuel delivery matches the amount of air-boost entering the engine cylinders. The actual position of the sliding-shoes being determined by a combination of fuel transfer-pressure and air-boost pressure, which produces the resultant actuating-pressure acting against the excess-fuel delivery pistons.

613

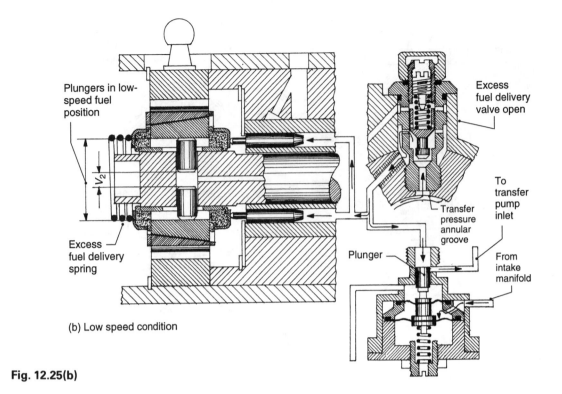

Plungers in low-speed fuel position

V_2

Excess fuel delivery spring

(b) Low speed condition

Excess fuel delivery valve open

To transfer pump inlet

Transfer pressure annular groove

Plunger

From intake manifold

Fig. 12.25(b)

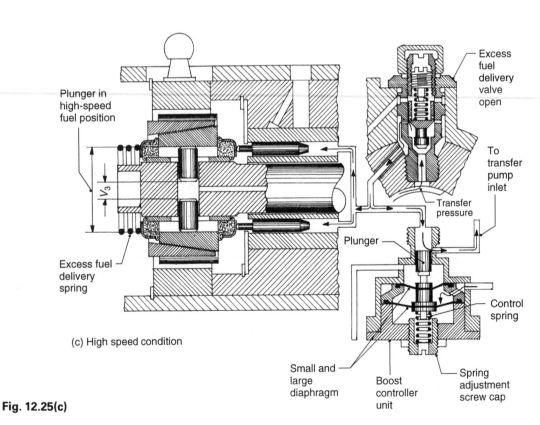

Plunger in high-speed fuel position

V_3

Excess fuel delivery spring

(c) High speed condition

Excess fuel delivery valve open

To transfer pump inlet

Transfer pressure

Plunger

Control spring

Small and large diaphragm

Boost controller unit

Spring adjustment screw cap

Fig. 12.25(c)

12.6 Cummins diesel engine fuel injection system

12.6.1 Fuel circulation system
(Fig. 12.26)

A spur-gear type pump, mounted on the PT fuel pump housing, draws fuel from the tank through the governor and pressure regulating assembly, and through the throttle and shut-down valve. Fuel then passes along a steel pipe to the common fuel supply drilling in the cylinder-head (for the in-line engine) or to each of the cylinder banks in the case of the vee-eight engine. The fuel then continues to flow along a single manifold drilling in the cylinder-head. This ensures that pressure equalization exists at the entrance of each injector inlet so that equal metering and, subsequently, equal power output from each cylinder is achieved. Some of the fuel reaching each injector is circulated through the injector, its function being to cool and lubricate the system and to purge air. Surplus heated fuel is returned to the fuel tank, where it provides a moderate degree of warming to the fuel being drawn to the spur-gear pump. This, therefore, helps to prevent crystallization in cold climate operating conditions.

12.6.2 Fuel delivery control
(Figs 12.27, 12.28, 12.29 and 12.30)

Fuel discharged from the injectors into the cylinders is determined by the rate of flow through a fixed-inlet orifice into the injector. At the same time, the rate of flow depends on the fuel pressure at the orifice and varies as the square root of the pressure; that is, $Q \propto \sqrt{P}$ where Q = rate of flow (ml/min) and P = fuel pressure (bar) (Fig. 12.27). However, the time for metering of the fuel is reduced inversely to the engine's speed; that is $t \propto (1/N)$ where t = metering time (s) and N = engine speed (rev/min) (Fig. 12.28). Therefore, the quantity of fuel entering each injector cup just before the plunger pumps the fuel through the injector nozzle holes becomes less as the engine speed rises.

To compensate for the reduced filling time at high engine speed it is therefore necessary to increase the fuel pressure.

It has been established that to maintain a constant engine torque the fuel pressure must vary as the square of the engine speed; that is,

$P \propto N^2$ for a constant engine torque (Fig. 12.29). Likewise, for a constant engine speed the engine torque varies approximately as the square root of the fuel pressure at the metering orifice that is $T \propto \sqrt{P}$, for a constant engine speed, where T = engine torque (Nm) (Fig. 12.30).

These requirements are accomplished by the combination of a spur-gear pump driven at engine speed and a centrifugal flyweight governor which produces its servo thrust with the square of the engine's speed. Consequently, the fuel injection pump and injector system are operated under the key factors 'fuel pressure and metering time'. Hence the Cummins diesel fuel injection system is known as the *Pressure–Time* (PT) injection system and the fuel pump and governor combination is also known as the *Pressure–Time–Governor* (PTG) unit.

The PT fuel pump provides the fuel pressure at the injector, as demanded by the load and speed requirements of the engine.

12.6.3 Cummins PTG idle and maximum speed governor

Construction
(Figs 12.26 and 12.31(a–c))
The governor consists of a pair of flyweights supported on a pressed steel carrier, which is itself mounted on a governor shaft. This is driven by the PT fuel pump input shaft through a pair of step-up gears, which approximately double the governor shaft speed. The flyweights' outward swing movement is transferred to the governor plunger via the bell-crank lever fingers to produce an axial movement to the plunger which is free to slide in the plunger barrel. A plunger and spring in the hollow governor shaft provide an initial bias to the flyweights' outward swing in the low-speed range. The plunger not only shifts to and fro but it also rotates with the governor weights. The middle region of the plunger has an annular recess and there is a central hole which is intersected by a radial drilling in the recessed portion, and also one excess-fuel cross-drilling in the full diameter region nearer to the open end of the central hole.

Aligning with the plunger ports and control edges at different phases of operation are three cross-holes formed in the plunger barrel. These are the idle-speed port, high-speed port and excess-fuel port.

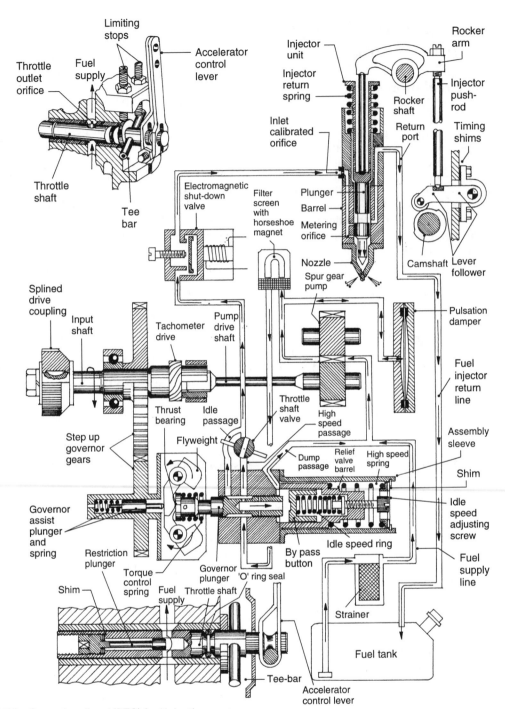

Fig. 12.26 Cummins diesel (PTG) fuel injection system

At the end of the plunger is a bypass button preloaded with an idle-spring; this pressure relief valve assembly is housed in a relief-valve barrel, which is itself preloaded by a high-speed spring and is supported in a relief-valve assembly-sleeve.

Idle speed operating conditions (Fig. 12.31(a))
With the throttle valve closed (accelerator foot-pedal released) the governor plunger moves to the extreme left-hand position; fuel now flows from the governor recess groove to the governor

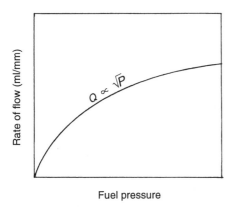

Fig. 12.27 Relationship between the rate of fuel flow and fuel supply pressure

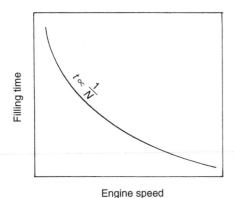

Fig. 12.28 Relationship between fuel filling time and engine speed

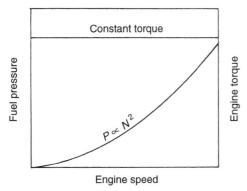

Fig. 12.29 Relationship between fuel pressure and engine speed for a constant engine torque

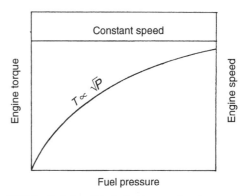

Fig. 12.30 Relationship between engine torque and fuel pressure for a constant engine speed

housing through the idle-speed port, around the plunger annular recess groove, where it then flows into the idle semi-circular passage coming out at the throttle-shaft valve outlet passage. Fuel also flows into the high-speed passage leading to the throttle-valve but, under idle speed operating conditions, the closed throttle blocks-off the flow path going to the shut-down valve. However, fuel from the idle-passage does flow to, and through, the shut-down valve. It then passes through a steel pipe to the cylinder-head manifold drilling, and from there it passes to the individual injector units.

Idle speed bypass control
(Fig. 12.31(a))

With the reduced fuel demand, the excess-fuel supplied by the spur-gear pump flows through the centre of the governor-plunger, pushing back the bypass button against the idle-spring so that fuel will escape via the dump-passage back to the inlet side of the spur-gear pump. The quantity of fuel bypassed, controls the fuel supply pressure reaching the injectors and hence the quantity of fuel entering the cylinders. The amount of bypass flow is determined by the diameter of the counter recess in the bypass button (a large cross-sectional area recess in the button reduces the blow-off pressure and vice versa) and the stiffness of the idle-speed spring, which can be varied by the idle-speed adjustment-screw. Increasing the idle-spring preload raises the fuel pressure passing through to the injectors so that more fuel in a given time is injected into the cylinders. Consequently, it increases the engine's speed but, conversely, reducing the spring preload increases the quantity of fuel bypassed to the inlet side of the spur-gear pump for a given engine speed. The

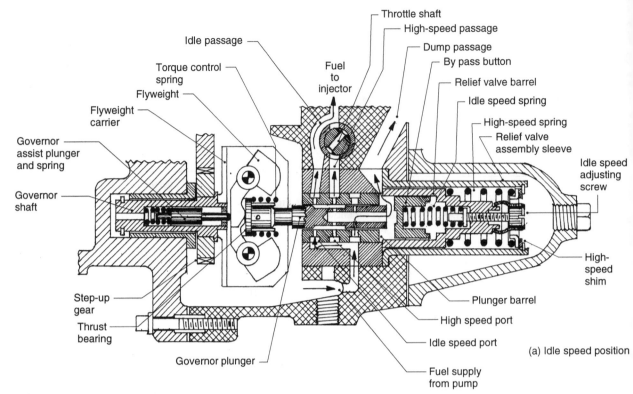

Fig. 12.31 Cummins diesel PTG fuel pump and maximum–minimum mechanical governor unit

idle-speed adjustment is achieved by removing the screw-plug, inserting a screwdriver up against the idle-screw and turning the screw either way until the desired idle speed is obtained.

Idle speed governing
(Fig. 12.31(a))
Idle-speed governing is achieved by the control-shoulder of the governor plunger partially cutting-off the idle-speed port so that there is just sufficient fuel reaching the injectors to produce a stable idle speed. Adjustment to idle-speed is obtained by altering the stiffness of the idle-speed spring, which determines the amount of fuel dumped back to the inlet side of the spur-gear pump.

Should the engine load suddenly increase, then there will be a corresponding reduction in engine speed. This causes the idle-speed spring and the fuel pressure in the button-recess to push back the plunger until the plunger control-shoulder uncovers more of the idle-speed port. Instantly, the increased quantity of fuel permitted to flow to the injectors brings the engine speed back to the normal speed setting. Conversely, should the engine speed suddenly rise due to a reduction in engine load, then the increased centrifugal force acting on the flyweights forces the governor plunger to move slightly to the right-hand side, thereby restricting the idle-speed port opening: accordingly, the engine speed is brought back to its original setting.

Intermediate speed operating conditions (Fig. 12.31(b))
When the accelerator-pedal is pressed down, the throttle-valve shaft is rotated causing the valve-orifice to open proportionally, fuel will now flow through the valve-orifice on its way to the injectors. The amount of fuel passing to the injectors depends upon how much the accelerator-pedal is pressed down and this will, in turn, determine the speed and torque output produced by the engine.

As the engine speed rises the flyweights swing outwards and, at the same time, this pushes the governor-plunger in the direction of the bypass button. Very early on, the idle-speed port is cut off by the control-shoulder of the plunger so that fuel now has only one flow path—that is, via the

618

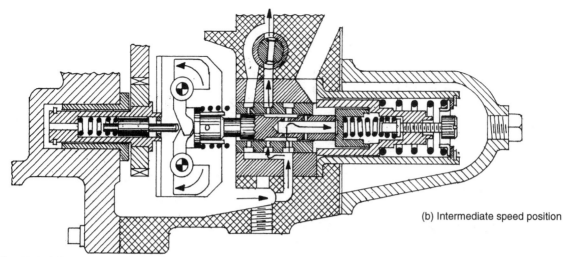

(b) Intermediate speed position

Fig. 12.31(b)

high-speed port, high speed-passage and throttle-valve—on its way to the injectors.

Throughout the intermediate speed range, the fuel pressure is regulated by the bypass button and the idle-speed spring, which becomes stiffer as it is compressed with the plunger's movement to the right (due to the flyweight thrust) and which, in so doing, increases the fuel pressure reaching the injectors.

Once the bypass button is fully pushed back against the relief-valve sleeve, then further fuel pressure regulation is obtained by the existing fuel pressure on the bypass button counter-bore producing an opposing thrust to the flyweights' swing-thrust acting on the opposite end of the governor-plunger. When fuel-pressure thrust created in the button counter-bore exceeds the flyweight-opposing thrust, the governor-plunger and the bypass button are forced apart, allowing the excess-fuel to escape. This surplus fuel is then dumped back to the inlet side of the spur-gear pump.

Consequently, there will be an increased quantity of fuel entering the injector nozzle-cups to match the engine's torque characteristics over the normal speed range of the engine.

Note that there is no governor control in the intermediate speed range once the governor plunger covers up the idle-speed port. This operating condition of course occurs very early in the engine's speed range.

Maximum governor cut-in speed operating conditions (Fig. 12.31(c))
As the engine approaches its maximum-rated speed, the outward swing of the flyweights initially pushes the governor-plunger hard against the bypass button and then commences to compress the high-speed spring so that the relief-valve sleeve moves to the right. Under these conditions, the control-shoulder of the plunger cuts off the high-speed port, and the flow of fuel to the injector therefore ceases. At the same time, the exposed end of the plunger opens the high-speed dump-port, thus permitting the excess-fuel to be returned to the inlet side of the spur-gear pump. As a result, the engine speed decreases, and this permits the governor-plunger to move back to the left-hand side until the control-shoulder of the plunger uncovers the high-speed port. Fuel again commences to flow to the injector via the high-speed passage.

The actual cut-in speed of the governor is determined by the strength of the high-speed spring. Springs are available with more or fewer coils or different wire gauge. However, for fine adjustment the cut-in speed of the governor is predetermined by the high-speed shims (which are sandwiched between the high-speed spring and the circlip situated at the open end of the relief-valve assembly sleeve). The maximum speed setting should be carried out with the pump installed on the test machine.

Spur-gear pump-pulsation damper and magnetic filter screen
(Figs 12.26 and 12.32)
This positive displacement pump is positioned at the rear of the fuel pump housing and is driven

619

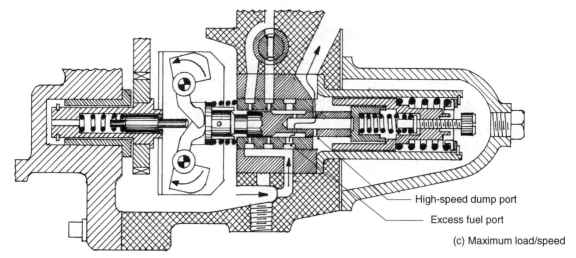

High-speed dump port

Excess fuel port

(c) Maximum load/speed

Fig. 12.31(c)

from the engine gear train, at crankshaft speed, through the main drive-shaft of the fuel pump unit.

The gear-pump is made up of a pair of intermeshing spur-gears, which draw fuel from the fuel-tank and deliver it throughout the fuel system. The gears transfer the fuel by carrying it between their teeth around the outside wall of the housing to a discharge-port. From the gear-pump, fuel flows to the magnetic-filter screen so that any metallic particles picked up in the fuel stream are prevented from moving further down the line. Pulsations in the pressure from the gear teeth cyclic action are absorbed and smoothed out by a steel diaphragm damper unit located on the pressure side of the gear pump. The faster the gear pump is driven the greater the volume of fuel delivered: the size of the pump is normally chosen so that it is capable of supplying two to three times more fuel than is actually required. The PT fuel-line pressures are normally less than 16 bar and the injectors receive fuel from the fuel pump at an average pressure of around 10 bar but this, of course, varies with engine speed and throttle opening.

Shut-down valve
(Figs 12.26 and 12.32)
This valve separates the fuel discharge from the throttle, and the idle-passageway from the injector, when the engine is to be shut-down (stopped). The valve unit consists of a steel-disc valve, a solenoid and a manual-override stop-screw.

When the engine is to be started the solenoid winding is energized, and the magnetic-field now draws the disc-valve towards the winding core, thus opening the valve. Fuel now flows unrestricted from the fuel-pump and governor-unit to the injector fuel manifold. Hence, when the starter motor cranks, the engine fuel will be injected into the cylinders, thus permitting the engine to start and operate as normal. Shutting down the engine de-energizes the solenoid, and the disc-valve then snaps closed, thereby preventing any more fuel reaching the injectors. It also holds fuel in the fuel manifold to prevent syphoning, therefore ensuring immediate availability of fuel for starting the engine. In the event of electrical failure the override stop-screw can be manually screwed in to open the shut-down valve.

12.6.4 Special variable-speed governor
(Fig. 12.32)

This special variable-speed governor provides speed control under conditions of varying load, such as for trucks with cargo discharge pumps, and other machines which have power take-off drives requiring a controlled maximum rotational speed. There are more sophisticated mechanical-type governors, which give closer speed variation control, but they are more elaborate and expensive.

The special variable-speed governor is an auxiliary control assembly which is attached

620

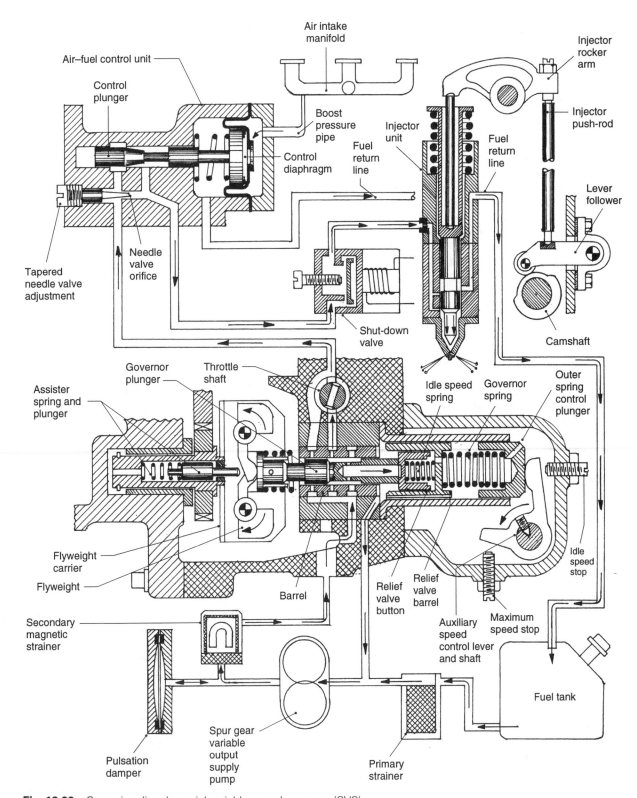

Fig. 12.32 Cummins diesel special variable speed governor (SVS)

to the back of the PTG fuel pump and governor unit. This auxiliary assembly dispenses with the idle speed adjustment screw on the idle and high-speed type mechanical governor (PTG) and replaces the fixed high-speed spring with an outer-spring control-plunger. The stiffness of the high-speed spring can be varied by rotating the auxiliary-speed control-shaft and lever, which pushes the high-speed spring closer together, or permits it to expand further apart.

Idle speed is set by allowing the outer-spring control-plunger to be pushed out hard against the auxiliary control-lever and its adjustable idle-speed stop. The idle-speed screw, located in the auxiliary assembly housing, can thus be screwed in or out in the usual way until the quantity of fuel bypassed back to the spur-gear pump inlet adjusts the fuel pressure at the injector's inlet calibrated-orifice to match the idle-speed requirements, enabling the engine's idling speed to be predetermined.

The engine for normal road driving applications is controlled by the idle and high-speed throttle-lever; however, for power take-off purposes the throttle-lever is locked fully open and the speed setting control is obtained via the auxiliary-speed control-lever. This auxiliary assembly device varies the spring force opposing the governor-plunger and flyweight swing thrust. Hence, the engine speed at which the high-speed port is cut off by the plunger's control-shoulder can be varied to suit the operating requirements.

12.6.5 Cummins diesel PTD injectors

The Cummins injectors serve a dual function, they perform both as a metering device and as a pump which injects fuel into the cylinder.

Description of unit
(Fig. 12.33(a–d))
The injector consists of a two-piece body known as the adaptor and barrel—an injector plunger and spring-retainer sleeve, and a nozzle-cup with a cup-retainer nut. The plunger downward motion is produced by a push-rod and rocker-arm which are actuated by the engine's camshaft and return-spring. The only injector adjustment necessary is setting the rocker-arm tappet clearance at regular maintenance intervals. The quantity of fuel metered into the nozzle cup is determined by the pressure at the inlet calibrated-orifice screwed in the side of the adaptor, the size of the orifice and the time period that the metering-orifice in the barrel is open. Individual injectors can be calibrated by broaching each inlet orifice bore (enlarging) until a balance is achieved between the injectors.

Barrel construction
(Fig. 12.33(a–d))
The thick-walled barrel has a vertical supply passage which passes straight through from top to bottom. The top end of the passagway links up with an offset recess formed to house a non-return ball-valve. The bottom end of the barrel has a second short vertical passage which is intersected by a pair of horizontal holes—the upper one goes directly into the inner barrel wall, whereas the lower one does not quite go into the inner wall. However, a very small metering-orifice does connect this passageway with the inside of the barrel. Both horizontal holes are plugged from the outside. A lapped pressure joint is formed between the upper face of the barrel and the lower face of the adaptor and there are two roll-pin dowels inserted in short blind holes to align the inlet passage in the adaptor body with the supply passage in the barrel. The lower face of the barrel also forms a lapped pressure joint with the nozzle-cup; however, the positioning of the barrel and cup is not important as there is an annular groove machined in the pressure face of the nozzle-cup which permits fuel to flow from the supply passage to the vertical metering-orifice and into the barrel and cup chamber.

Adaptor body construction
(Fig. 12.33(a–d))
The adaptor body has two large annular grooves: the lower one is for the fuel inlet whereas the upper one is for the fuel outlet. In addition, there are three small grooves to locate three 'O' ring seals: one groove being situated between the wide inlet and outlet fuel passageway grooves, while the other two are machined on either side of these wide fuel grooves. The wide lower inlet groove feeds into a vertical drilling via an inlet calibration-orifice, which supplies the barrel with fuel, whereas a fuel return vertical passage is connected to the wide upper output groove.

The adaptor body which supports the barrel and nozzle cup assembly is housed in the cylinder-head in counter-bores positioned above and in the centre of each cylinder. The adaptor body forms a push fit into the cylinder-head counter-bore so that the 'O' rings produce a fuel-tight seal on both sides of the inlet and outlet grooves. Each injector inlet groove is connected to the fuel inlet manifold drilled in the cylinder-head; likewise, the injector's outlet groove is connected to an outlet manifold drilling which returns the fuel back to the fuel tank.

Circulating, filling and injecting cycle of operation (Figs 12.33(a–d) and 12.34(a and b))

1 Upstroke commences—fuel circulation phase (Figs 12.33(a) and 12.34(a and b))

Fuel at regulated supply pressure enters the inlet calibrated orifice and then flows to the underside of the non-return ball-valve. It dislodges the ball and thus permits fuel to continue its flow down the barrel fuel passages to the annular groove in the upper face of the nozzle cup. Fuel now flows upwards, past the metering-orifice to the upper horizontal short passage. At this instant, the inlet circular port is aligned with the plunger's annular recess, and the fuel is therefore permitted to pass around this annular groove in the plunger before continuing upwards to the wide fuel outlet groove formed in the adaptor body. Fuel therefore flows back to the fuel tank via the fuel return manifold drilled in the cylinder-head.

The quantity of fuel circulating through the injector is dependent upon the fuel pressure at the inlet calibration-orifice. This, in turn, is dependent upon the engine speed, the plunger groove position and the throttle opening.

2 Upstroke completed—nozzle cup filling phase (Figs 12.33(b) and 12.34(a and b))

Further upward movement of the plunger first causes the lower shoulder of the plunger groove to cut off the inlet circulation port and, secondly, to uncover the metering-orifice in the side of the barrel inner wall. Fuel now enters the space between the nozzle cup and the conical end of the plunger. The quantity of fuel actually filling the cup-chamber is determined by the time the metering-orifice in the inner barrel wall remains open and also by the fuel pressure at the inlet calibration-orifice.

3 Downstroke—fuel injection phase begins (Figs 12.33(c) and 12.34(a and b))

Just before the downward moving plunger closes the metering-orifice, an impulse pressure wave

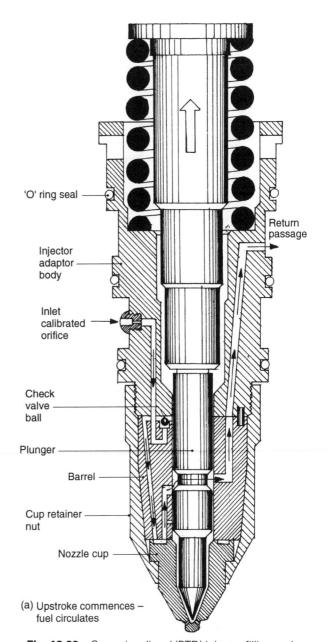

'O' ring seal

Return passage

Injector adaptor body

Inlet calibrated orifice

Check valve ball

Plunger

Barrel

Cup retainer nut

Nozzle cup

(a) Upstroke commences – fuel circulates

Fig. 12.33 Cummins diesel (PTD) injector filling and delivery operating cycle

snaps the non-return ball-valve against its seat, the control-shoulder of the plunger then cuts off the metering-orifice from the cup-chamber, such that the plunger will continue to travel down until the conical end of the plunger moves hard against the head of fuel trapped in the cup. Fuel will now be compressed and squeezed through the nozzle-cup nozzle-holes. The injected fuel now enters the engine's cylinders in the form of a high-pressure

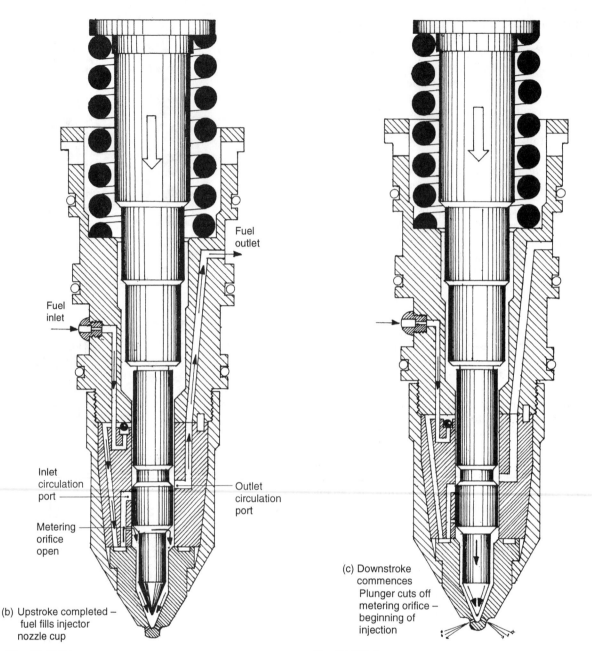

Fig. 12.33(b)

(b) Upstroke completed – fuel fills injector nozzle cup

Fuel outlet

Fuel inlet

Inlet circulation port

Metering orifice open

Outlet circulation port

Fig. 12.33(c)

(c) Downstroke commences Plunger cuts off metering orifice – beginning of injection

high-velocity fine spray which penetrates the dense compressed air charge. Peak injection pressures exceeding 1200 bar are not uncommon. The purpose of the ball-valve is to prevent fuel back-pressure returning to the fuel inlet manifold and pressurizing other injectors.

Towards the end of the downward plunger stroke, the lower shoulder of the plunger annular

groove uncovers the inlet circulating port and fuel will again commence to circulate through the plunger and barrel, after which it is returned to the fuel tank.

4 Downstroke completed—fuel circulation phase (Figs 12.33(d) and 12.34(a and b))
The instant the plunger reaches its lowest posi-

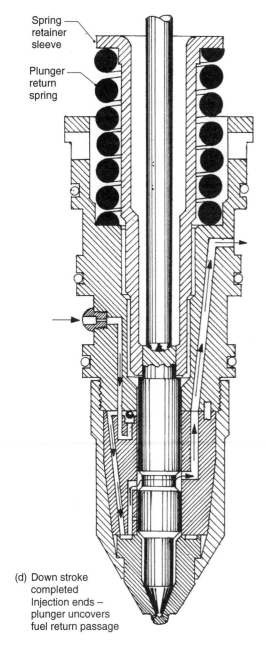

Spring
retainer
sleeve

Plunger
return
spring

(d) Down stroke
completed
Injection ends –
plunger uncovers
fuel return passage

Fig. 12.33(d)

tion, injection ceases. The plunger will then re-main seated until the next metering and injection cycle. Fuel cannot now enter the cup-chamber until the plunger rises sufficiently for the plunger-control shoulder to open the metering-orifice. However, the fuel will flow unrestricted through the injector and is returned to the fuel tank via the fuel outlet annular groove in the adaptor-body.

Fuel circulation provides cooling for the injectors and warms the fuel in the tank, which is beneficial during cold-weather operating conditions. The plunger is designed to remain seated to prevent combustion products being forced into the nozzle-cup via the injection-holes, where they would eventually clog the nozzle-sac and injection-holes during the latter part of the power stroke, the exhaust stroke and the compression stroke of the engine (Fig. 12.34(a and b)).

Full and part-load injection timing advance and retard
(Fig. 12.33(c))
It is desirable to retard the point of injection under part-load operating conditions so that fuel is only injected when the combustion chamber air-charge temperature is near its maximum, as the piston approaches TDC. Conversely, the be-ginning of injection needs to be advanced as the amount of fuel being injected per cycle increases with rising engine load. This is because the com-pressed air-charge temperature peaks earlier in the compression stroke since there is more heat energy released at any one time.

The inherent reduction in fuel entering the nozzle-cup as the engine load is reduced means that the injector plunger moves further down its pumping stroke (late injection) before the fuel is actually pressurized and expelled by the plunger through the nozzle-cup spray-holes. However, with increased throttle opening, the amount of fuel per cycle entering the nozzle-cup increases. It therefore causes the fuel level in the cup-chamber to rise so that the injector plunger begins to squeeze and inject fuel from the nozzle-cups earlier on the compression stroke.

This inherent feature of this type of injector eliminates the need for a separate automatic advance and retard device, which is normally necessary with other types of high-pressure fuel injection systems.

12.6.6 Air–fuel control (AFC) for use with turbocharged engines
(Figs 12.32 and 12.35(a and b))

This device is necessary to limit the amount of fuel applied to the injectors in the low engine speed range, when the turbocharger boost-air supply is practically non-existent, but also to permit proportionally more fuel to flow to the

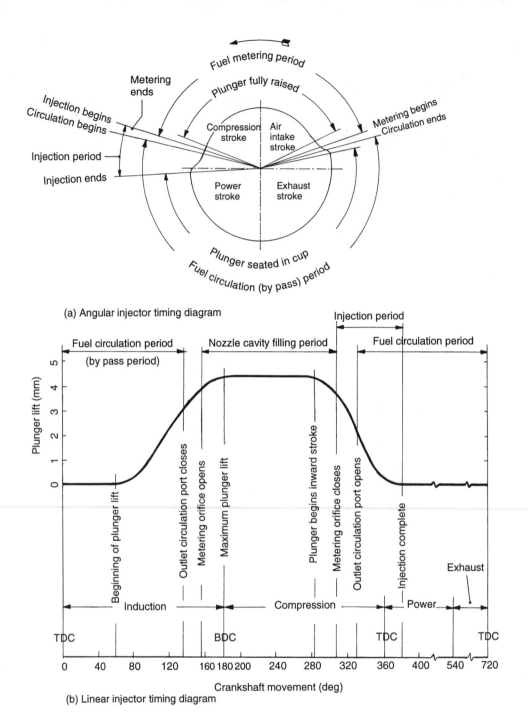

(a) Angular injector timing diagram

(b) Linear injector timing diagram

Fig. 12.34 Cummins injector cycle of operation

injectors as the turbocharger air supply builds up with rising engine speed. The 'air–fuel control' (AFC) unit therefore controls fuel delivery to match the relative amount of air entering the cylinders over the engine speed range.

The AFC unit consists of a boost-pressure controlled diaphragm-actuator, which is attached to a conical control-plunger and an adjustable tapered needle-valve and orifice passage, which bypasses the conical control-plunger fuel supply

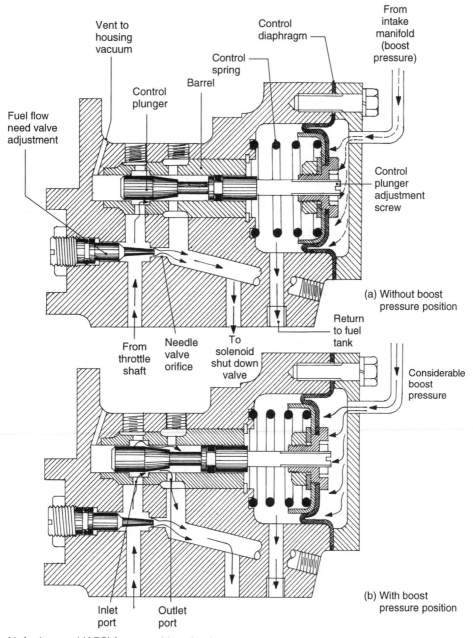

Fig. 12.35 Air fuel control (AFC) for use with turbocharged engine

passages leading to the solenoid shut-down valve.

Low boost pressure operating conditions
(Figs 12.32 and 12.35(a))
Under low operating speed conditions the diaphragm return-spring pushes the diaphragm and plunger to the right where it cuts off the inlet port. Fuel therefore flows from the throttle-valve

outlet passage to the tapered needle-valve. It then passes between the tapered-needle and its orifice on its way to the solenoid shut-down valve and from there it continues on its way to the injector; that is, the injector fuel inlet manifold. The actual quantity of fuel reaching the injector from the air–fuel control (AFC) unit is determined by the screw adjustment setting of the tapered needle-valve.

627

Increasing boost pressure operating conditions

(Figs 12.32 and 12.35(b))

With rising engine and turbocharger speed there will be an increase in the air inlet manifold (boost pressure) which will build up and oppose the diaphragm return-spring thrust. Eventually, the diaphragm and plunger will commence to move against the return-spring and, in doing so, uncover the AFC fuel inlet port. Fuel now not only passes through the tapered needle-orifice but it also flows between the conical-plunger and its barrel, it then escapes from the barrel outlet port where it rejoins the fuel flow from the tapered needle-orifice as it moves on its way to the injector via the solenoid shut-down valve.

Consequently, more fuel will be delivered to the injector cylinder-head fuel manifold and, therefore, an increased amount of fuel will be discharged from each injector per operating cycle.

12.6.7 Cummins engine injection timing procedure

(Fig. 12.36)

Stage 1—determining TDC (Fig. 12.36)

a) Install the timing adaptor.
b) Rotate the crankshaft (in the direction of engine rotation) to TDC (the piston travel plunger will be near the full upward position).

 Adjust both indicator clock gauges on the adaptor to their fully compressed position. To prevent damage, raise both indicators approximately 0.5 mm. Lock in place with the set screw.
c) Rotate the crankshaft back and forth to ensure the piston is precisely at TDC on the compression stroke. (Both indicators move in the same direction when the piston is on the compression stroke.) TDC is found by the maximum clockwise position of the piston travel indicator pointer. Zero the piston travel indicator clock gauge with the pointer and then lock it.

Stage 2—90° ATDC (Fig. 12.36)

a) Rotate the crankshaft in the direction of engine rotation to 90° ATDC (the piston travel plunger will be near the bottom of its travel).
b) Zero the injector push-rod indicator clock gauge with the pointer and lock it.

Stage 3—45° BTDC (Fig. 12.36)

Rotate the crankshaft in the opposite direction of engine rotation through TDC to 45° BTDC (the piston travel plunger indicator should be aligned with the adaptor scale mark). This stage is necessary to remove the gear-train backlash and provide more accurate indicator readings in the following stage.

Stage 4—19° BTDC (Fig. 12.36)

a) Rotate the crankshaft in the direction of engine rotation until the recommended reading BTDC is reached on the clock gauge of the piston travel plunger. (A typical reading being −0.2032 inches BTDC.)
b) At this position compare the reading of the push-rod travel indicator with the specifications listed for that timing code. Readings should fall within these specifications (a typical reading being −0.036 inches: note that the Cummins engines still use imperial measurements).

Flanged follower support gasket adjustment (in-line six-cylinder engine)

(Fig. 12.36)

If the reading is out, correct it by increasing shim thickness between the finger follow pivot support-flange and the cylinder-block in order to advance injection timing (decrease the push-rod travel indicator clock gauge reading); or decreasing shim thickness gaskets in order to retard injection timing (increase push-rod travel indicator clock gauge reading).

Stepped Woodruff key adjustment (vee eight-cylinder engine)

(Fig. 12.36)

An alternative method used to alter the injection timing on vee eight-cylinder engines is having a selection of different-sized stepped keys which change the relative angular position of the camshaft to that of the camshaft drive gear.

12.7 Detroit diesel combined injection pump and injector unit

12.7.1 Injector unit function

The fuel injector performs four functions:

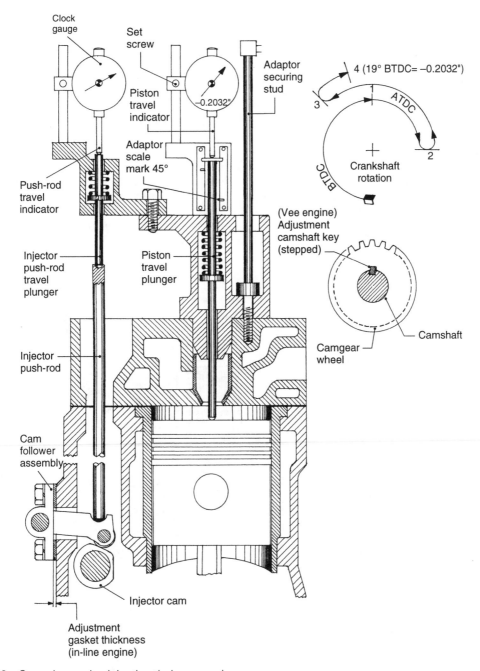

Fig. 12.36 Cummins engine injection timing procedure

1 it generates a sufficiently high fuel pressure to penetrate the dense air charge in the combustion chamber;
2 it meters and discharges the precise quantity of fuel needed to match the engine's demands;
3 it discharges the fuel in a finely atomized form so that it is able to be thoroughly mixed with the air charge in the combustion chamber;
4 it provides a continuous circulation of fuel through the injector unit for lubricating and cooling purposes.

12.7.2 Description and construction
(Figs 12.37, 12.38, 12.39(a–c) and 12.43)

This unit, known as the fuel injector, consists of a pumping plunger which reciprocates and rotates in its barrel, a plunger-type follower with its return-spring, which imparts its reciprocating motion to the pumping plunger via a rocker-arm, a push-rod, a roller-follower and a cam (Fig. 12.43). The cam profiles are formed on the same shaft as the exhaust-valve cams. Note that as the engine works on the two-stroke principle, air intake ports formed in the cylinder liner wall provide the means of delivering the air charge into the cylinder so that no inlet-valves and cams are required. The lower end of the plunger has an annular groove with a helix control edge machined partially around it. A cross-hole drilled in the grooved portion of the plunger intersects a central hole drilled in the base of the plunger (Fig. 12.38).

The barrel has an upper port, and a lower funnel-shaped port which feeds into the barrel-chamber. The ports' angular position relative to the plunger-helix is critical so that a peg protruding from the reduced diameter portion of the barrel indexes the barrel to the injector-body.

In addition to the reciprocating motion of the plunger, the plunger can be rotated about its axis by a control pinion-gear which meshes with the control-rack (Fig. 12.39).

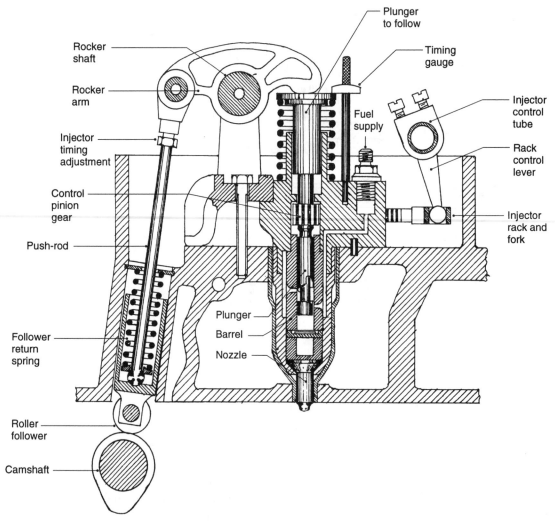

Fig. 12.37 Detroit diesel injector unit timing

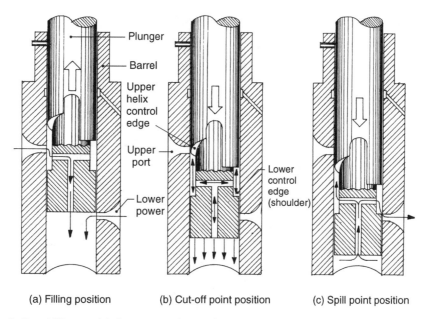

- Plunger
- Barrel
- Upper helix control edge
- Upper port
- Lower power
- Lower control edge (shoulder)

(a) Filling position   (b) Cut-off point position   (c) Spill point position

Fig. 12.38 Detroit diesel filling and delivery operating cycle

To permit the plunger's vertical motion to take place and, at the same time, to enable the plunger to be twisted either clockwise or anticlockwise, the internal cylindrical wall of the pinion-gear has a flat portion which aligns with a flat machined on the upper end of the pumping plunger (Fig. 12.39). The correct position of the rack relative to the pinion is where the punch dots on the rack are on either side of the pinion's single-punch dot. A nozzle cup-nut supports and clamps together the barrel and the lower assembly which consists of the disc-valve and body, the injector needle-spring and its retainer, and the injector needle-valve and its discharge-nozzle.

12.7.3 Fuel circulation system
(Figs 12.39 and 12.40)

Fuel is drawn from the fuel tank by the engine driven spur-gear pump via the fuel strainer. It then flows, under pump pressure, through a fine filter to a fuel manifold drilling formed in the engine cylinder-head, where it passes to the individual injector units via short steel pipes (Fig. 12.38). Fuel now enters the injector-body via the fuel supply gauze-filter cup-nut, which is screwed into the top of the body. After passing through the gauze-filter in the inlet passage, the fuel fills the annular supply chamber formed

between the barrel and the spill deflector tube (Fig. 12.39). The pumping plunger reciprocates in its barrel—the bore of which is connected to the fuel supply in the annular supply chamber by the two funnel-shaped ports. Every time the plunger reaches its outermost position, fuel flows through the upper and lower ports thus filling the barrel pressure chamber. As the plunger moves downwards, fuel in the barrel chamber is displaced back into the annular supply chamber via the upper and lower ports until the plunger lower land closes the lower port. The remaining fuel in the barrel-chamber is now forced upwards through the central drilling in the plunger, and into the recess formed between the upper helix-edge and the lower control-edge (shoulder), from which it can still flow back to the annular supply chamber until the upper helix closes the upper-port (Fig. 12.39).

12.7.4 Check valve and needle valve action
(Fig. 12.39)

Further downward movement of the plunger forces the disc-valve down in the top face of the spring-retainer, down the vertical feed-hole to a second annular-groove in the bottom face of the spring-retainer. It then passes down to the

631

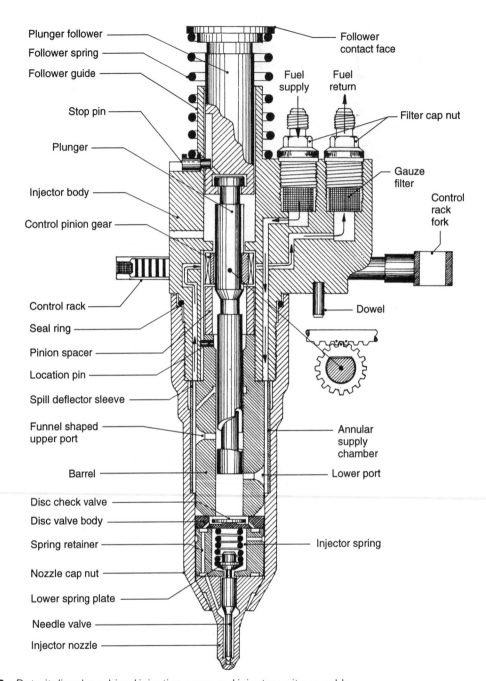

Plunger follower

Follower spring

Follower guide

Stop pin

Plunger

Injector body

Control pinion gear

Control rack

Seal ring

Pinion spacer

Location pin

Spill deflector sleeve

Funnel shaped
upper port

Barrel

Disc check valve

Disc valve body

Spring retainer

Nozzle cap nut

Lower spring plate

Needle valve

Injector nozzle

Follower
contact face

Fuel
supply

Fuel
return

Filter cap nut

Gauze
filter

Control
rack
fork

Dowel

Annular
supply
chamber

Lower port

Injector spring

Fig. 12.39   Detroit diesel combined injection pump and injector-unit assembly

needle-valve fuel-galley via the slightly inclined feed passages—there are, in fact, three feed-holes although only one is shown in Fig. 12.39.

When sufficient fuel pressure is built up on the underside of the annular tapered-area of the needle-valve, the needle-valve is pushed up. This permits fuel to be squeezed through the now open needle-valve and out of the very small discharge-holes formed in the nozzle-nose. Its high pressure and velocity means its exit is in the form of a fine atomized spray which directly enters the combustion-chamber. Towards the end of the plunger's downward stroke, the lower control-edge uncovers the lower-port, and injection immediately

632

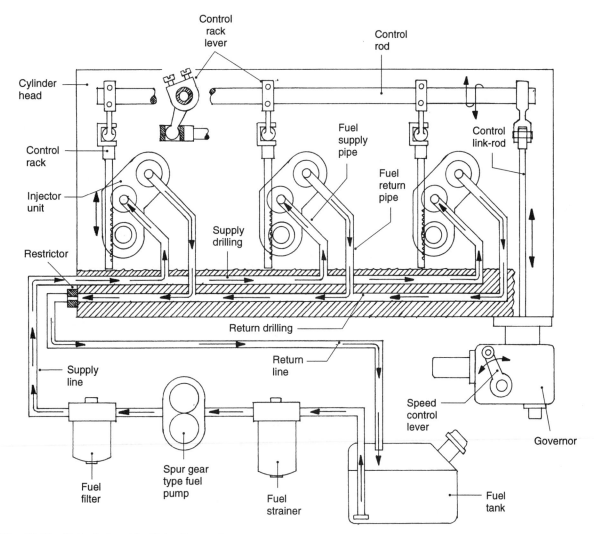

Fig. 12.40 Detroit diesel fuel injection system

ceases, as fuel will now flow back from the barrel-chamber to the annular supply-chamber. Instantly the injector needle-valve snaps down on its seat.

The excess-fuel now flows up from the annular supply-chamber, to the return cup-nut by way of the return-passage in the injector-body. After this, it is returned to the fuel tank via the return fuel manifold drilling in the cylinder-head and the restriction-orifice. The constant circulation of fresh, cool fuel through the injector passages renews the fuel supply in the annular-chamber, helps to cool the injector, and also effectively removes traces of air which might otherwise accumulate in the system and which might inter-

fere with accurate metering of the fuel. At the same time as fuel spill occurs, the collapsed fuel pressure in the barrel pressure-chamber permits the injector spring load to close the needle-valve rapidly. Instantly, the existing fuel pressure in the injector-nozzle passages jerks up the disc check-valve against its barrel-seat (Fig. 12.39). As a result, a residual-pressure will exist in the trapped fuel remaining in the injector-nozzle passages. Thus, each successive pumping stroke can, without delay, transfer pressure waves to the needle-valve so that the valve lifts and opens without hesitation. The disc check-valve also prevents leakage from the combustion-chamber into the fuel injector in case the valve is accidentally held open by (possibly small) particles of solid combustion products.

A pressure-relief passage has been provided in the spring-retainer to permit bleed-off of fuel leakage, which takes place between the needle-stem and nozzle-wall.

12.7.5 Filling and pumping operating cycle
(Fig. 12.38(a–c))

Filling phase
(Fig. 12.38(a))
The plunger moves to its outermost position uncovering both the upper and lower funnelled ports. Fuel from the annular supply-chamber surrounding the barrel now enters and fills the barrel pressure-chamber. The fuel flows directly through the lower-port and indirectly through the upper-port via the recessed upper helix-groove, the cross drilling and the central passage leading to the barrel pressure-chamber.

Cut-off phase—injection begins
(Fig. 12.38(b))
The plunger moves on its downward stroke, and the lower land of the plunger first closes the lower-port; then the plunger continues to move down until the helix control-edge just cuts-off the upper-port. This cut-off point, where the inclined control-edge of the helix just closes the upper-port, also becomes the point where fuel injection commences. The fuel in the barrel pressure-chamber is now separated from the annular supply-chamber so that any further movement of the plunger pressurizes the trapped fuel underneath the plunger-crown. Fuel under pressure now dislodges the disc-valve, thus permitting fuel to pass down to the needle-valve fuel-gallery surrounding the tapered-shoulder of the needle-valve. Eventually, the build-up in pressure compresses the injector-spring, thus causing the needle-valve to lift off its seat. Instantly, fuel will be squeezed through the nozzle-holes, it then emerges as a fine atomized spray as it enters the combustion-chamber.

Spill-point—injection ends
(Fig. 12.38(c))
As the plunger continues to move towards the end of its pumping stroke, the lower control-edge

of the plunger's annular recess uncovers the lower-port. Fuel will immediately commence to spill back into the annular supply-chamber. Hence, the collapse of fuel pressure prevents any further delivery of fuel to the injector needle-valve fuel-gallery. Correspondingly, the injector-spring rapidly returns the needle-valve to its seat, thereby effectively ending the injection of fuel.

12.7.6 Fuel delivery output control
(Fig. 12.41(a–c))

The plunger stroke is always constant (being determined by the cam-profile lift), but that part of it which is actually pumping is variable. The quantity of fuel delivered is controlled by the edge of the helix-groove (inclined recess) which is cut about half-way around the plunger. Varying the angular position of the plunger helix edge relative to the upper feed-port extends or contracts the effective pumping stroke of the plunger.

No-load position
(Fig. 12.41(a))
For the no-load position, the plunger is rotated so that the upper-port is not closed by the helix until after the lower-port is uncovered. Therefore, it does not matter where the plunger is along its stroke because fuel will always be forced back from the barrel pressure-chamber to the annular supply-chamber. Hence, no injection of fuel can take place.

Part-load position
(Fig. 12.41(b))
Rotating the plunger so that the portion of the helix-edge which aligns vertically with the upper-port is in its highest position (the beginning of the helix) means that the plunger has to move further down before the inclined control-edge cuts-off the upper-port. Thus, the remainder of the plunger's effective stroke will be reduced, which produces a corresponding decrease in fuel discharge from the injector.

Full-load position
(Fig. 12.41(c))
Conversely, rotating the plunger in the opposite direction lowers the portion of the helix-edge,

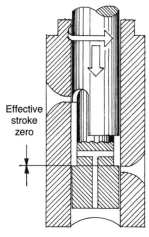

Effective stroke zero

(a) No load position

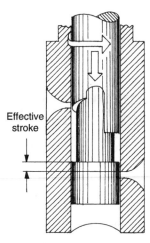

Effective stroke

(b) Half load position

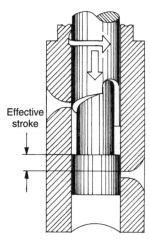

Effective stroke

(c) Full load position

Fig. 12.41 Detroit diesel delivery output control

which vertically aligns with the upper-port so that the beginning of injection occurs higher in the plunger's stroke. Consequently, the plunger stroke is extended before the spill point ends injection. Accordingly, the fuel output delivery will be increased.

12.7.7 Injection timing variation
(Figs 12.41(a–c) and 12.42(a))

Engine performance can usually be improved by retarding the injection timing under part-load running, and advancing the point of injection when full-load conditions apply. These changes in injection time are inherently obtained within the Detroit diesel injector unit.

Changing the position of the helix by rotating the plunger either retards or advances the closing of the upper feed-port (Fig. 12.41(a–c)). Rotating the plunger in the reduced load position extends the downward movement of the plunger before the upper-port is cut-off—therefore the point of injection is retarded (Fig. 12.42(a)). Alternatively, rotating the plunger towards the full-load position moves the helix so that the plunger cuts-off the upper-port earlier in its pumping stroke and, therefore, the point of injection is advanced (Fig. 12.42(a)). However, the end of injection remains constant from no-load to full-load fuel delivery since the lower control-edge or shoulder of the groove is symmetrical (Fig. 12.42(a)).

12.7.8 Upper and lower helix plungers
(Figs 12.41(a–c), 12.42(a–b) and 12.43(a–b))

Note that some plunger designs not only have an upper fuel cut-off helix (Fig. 12.41(a–c)) but also utilize a lower fuel spill helix (Fig. 12.43(a and b)). This type of plunger shares the fuel delivery output control between the upper and lower helices so that the upper helix pitch (steepness of incline) can be reduced. Normally, engines operate better with a moderate amount of injection retardation and advance with part-load and full-load operating conditions respectively; however, the single cut-off upper-helix (Fig. 12.42(a)) generally provides too much injection timing variation with changes in fuel delivery output. The double-helix plunger (Fig. 12.42(b)) therefore permits a much smaller upper helix (less inclined) to be used, such that the injection

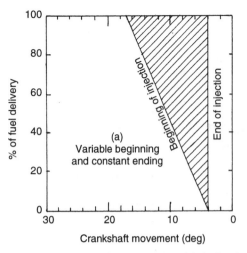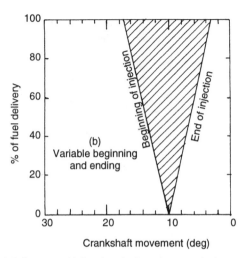

Fig. 12.42 Comparison of single and double helix plunger fuel delivery and injection timing characteristics

retardation (with the plunger rotated to a reduced fuel delivery output) can be better matched to suit the engine's optimum performance requirements.

12.7.9 Fuel injector timing procedure
(Figs 12.37 and 12.44)

For correct injector timing the injector follower must be adjusted to a predetermined height in relation to the injector-body. This measurement being set to the manufacturer's specifications.

All the injectors can be timed, in firing order sequence, during one full revolution of the crankshaft for a two-stroke cycle engine.

1 Remove the valve rocker-cover and position the governor stop-lever in the no-fuel position.
2 Rotate the crankshaft until the exhaust-valves are fully depressed on the particular cylinder to be timed. The roller-follower will now be on the base circle of the cam.
3 Place the small end of the injector timing-gauge in the hole provided in the top of the injector body, with the flat on the gauge above the injector follower (Fig. 12.37).
4 Loosen the push-rod lock-nut, turn the push-rod and adjust the injector rocker-arm until the extended part of the gauge will just pass over the flat top of the injector follower.
5 Hold the push-rod and tighten the lock-nut. Check adjustment and, if necessary, readjust the push-rod length.
6 Time the remaining injectors in the same manner as outlined in procedures 1 to 5.

A typical port, valve and injector timing diagram is shown in Fig. 12.44.

12.8 Unit injector electronically controlled fuel injection system (Lucas CAV)

12.8.1 The benefit of a unit injector

The unit injector combines the traditional injection pump and the separate injector into a compact rigid single unit. Through these means there is far less dead fuel volume space since there are no pipe tubes joining the injection pump to the injector. Consequently, the reduction in compressible fuel volume and increased rigidity of the unit injector enable much higher injection pressure and higher rates of injection with more spray-holes to be utilized. As a result, proportionally more of the energy for the fuel–air mixing comes from the injected fuel spray; therefore, lower rates of air swirl can be tolerated for medium-speed engines so that shallow quiescent chambers can be used effectively, as these tend to reduce exhaust emissions and consume less fuel. Conversely, higher engine speeds can be tolerated with direct-injection small diesel engines, since the thoroughness of mixing the fuel and air with the higher injection rates and pressures can be sustained at high engine speeds, even though injection and mixing time are considerably reduced.

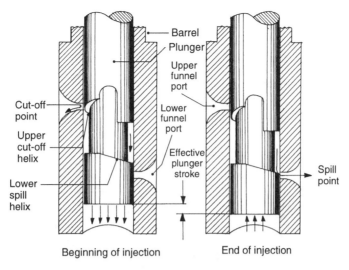

(a) Part load fuel delivery position
retarded injection timing

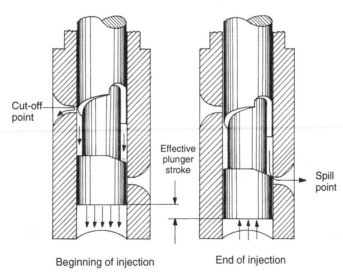

(b) Full-load fuel delivery position
advanced injection timing

Fig. 12.43 Upper and lower helix plunger operating between part and full load

12.8.2 Unit injector electronically controlled system

(Fig. 12.45)

A typical arrangement for the complete diesel injection electronic system is illustrated in Fig. 12.45.

A number of transducers are used to detect a range of engine variables, which are then trans-

mitted to the electronic control unit. These primary sensors are described as follows.

1 Accelerator pedal position which monitors the driver's demand. The pedal position is monitored by a potentiometer which provides a dc voltage signal proportional to the pedal pivot angle.

2 A crankshaft driven, 60-tooth wheel with an inductive transducer which calculates the en-

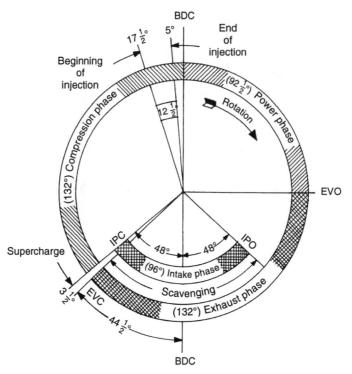

Fig. 12.44 Port valve and injector timing diagram two stroke cycle

gine's speed; while a unique tooth indicates cylinder No. 1 at TDC on its compression stroke, for injection timing purposes. Engine crankshaft angular data are used to derive the start of the solenoid actuation electric drive pulse and its duration, in order to achieve the required fuel delivery and timing.

3 A camshaft driven $n+1$ wheel where n = number of cylinders, which signals (via an inductive transducer) the sequence of the engine's firing order to the electronic control unit.

4 Temperature transducers, which monitor such things as coolant temperature, air temperature and oil temperature, and relay their information to the electronic control unit. The coolant temperature signals the cold start and normal operating condition fuel delivery and timing requirements. The air temperature is a measure of the air charge density entering the cylinder and, accordingly, signals the matching fuelling needs, and finally, the oil temperature sensor, if fitted, relates to engine load and provides information to the engine's protection system.

5 A boost pressure transducer, which monitors the intake manifold air charge pressure and relays this in the form of an electric signal to the electronic control unit. Accordingly, corrections will continuously be made to the fuelling to match the turbocharger's characteristic boost pressure contribution over the engine's speed range.

12.8.3 Characteristics of the electronic control unit and the solenoid actuator

The microprocessor in the electronic control unit receives information from the transducers and compares it with the pre-programmed data stored in its memory. It then computes appropriate fuelling and timing instructions and transmits these to the unit injector solenoid (Colenoid) actuators. The accuracy necessary for metering and timing the fuel deliveries means that the actuators have to respond to commands well within 5 μs.

A high degree of timing and torque shaping flexibility is provided with electronic control. Governing strategies for both two-speed or all-speed operations can be implemented and, for power take off or generator drives, steep slope governing can be programmed. Since the quantity

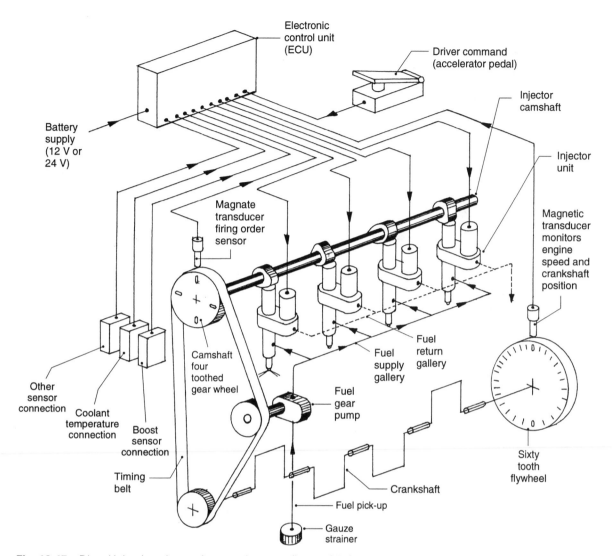

Fig. 12.45 Diesel injection electronic control system (Lucas CAV)

of fuel to be delivered for each individual injection is calculated only a fraction of a second before the event, the spill-valve solenoid-actuator is able to respond exceptionally quickly to changing demands between consecutive injections. Subsequently, the rapid and precise fuelling adjustments which are continuously taking place, provide a smooth and progressive engine response with both changing road conditions and driver's demands, and these adjustments are generally not possible with the conventional mechanically controlled fuel injection systems.

12.8.4 Operating phases of the unit injector
(Fig. 12.46(a–d))

Filling phase
(Fig. 12.46(a))

The spill-valve is opened and the plunger has moved to its outer dead-centre position. Fuel from the supply pump passes into the barrel via the feed-port, it then flows along a passage to the open spill-valve. Here, it is expelled via the spill-port to the return fuel gallery. Fuel is therefore continually circulating so that it is purged of

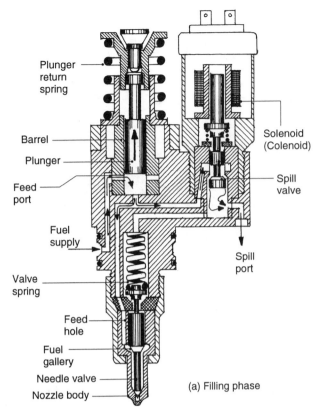

Plunger return spring

Barrel

Plunger

Feed port

Fuel supply

Valve spring

Feed hole

Fuel gallery

Needle valve

Nozzle body

Solenoid (Colenoid)

Spill valve

Spill port

(a) Filling phase

Fig. 12.46 Electronically controlled injection (Lucas CAV)

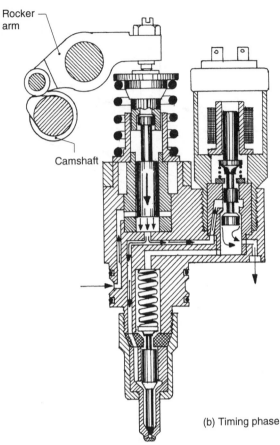

Rocker arm

Camshaft

(b) Timing phase

Fig. 12.46(b)

air and its temperature is kept within a reasonable upper limit.

Timing phase
(Fig. 12.46(b))

With the spill-valve still open, fuel will continue to circulate until the downward-moving plunger cuts-off the feed-port so that no more fuel enters the barrel pressure-chamber. The plunger's continuing inward movement is now responsible for displacing fuel from the barrel-chamber to the open spill-valve; consequently, no injection can occur at this stage. The timing phase is therefore the variable duration from after the feed-port closure to the instant where the spill valve closes.

Injection phase begins
(Fig. 12.46(c))

The plunger continues to move towards its inner dead-centre position after cutting-off (closing) the feed-port until the electronic control unit signals the solenoid to close the spill-valve. Instantly, the de-energized solenoid return-spring closes off the spill-valve. Note that there is no hydraulic opposition to closing off the spill-valve since the annular projected area around the waist of the spill-valve and the annular valve seat orifice are the same. The trapped fuel in the passageway between the plunger and needle-valve is now pressurized to the point where the valve spring collapses; it therefore causes the needle-valve to lift. Accordingly, the open valve now permits fuel to be discharged from the spray-holes via the nozzle-sac. The actual commencement of injection during the plunger's down stroke will take place somewhere beyond the feed-port closure position, this being determined by the engine's requirements, which are monitored by the electronic control unit (ECU) through the signals supplied by the control system sensors. Thus, the timed point of injection will be continuously altering to match the changing engine demands.

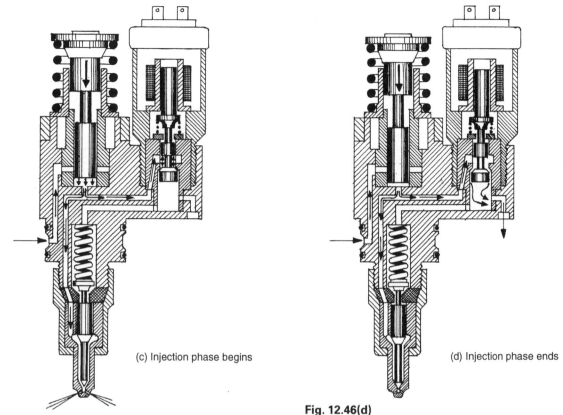

(c) Injection phase begins

Fig. 12.46(c)

(d) Injection phase ends

Fig. 12.46(d)

Injection phase ends
(Fig. 12.46(d))
Fuel will continue to be injected until the computed quantity of fuel to be discharged per cycle is reached, at which point the solenoid is signalled to open the spill-valve. Immediately, the spill-valve opens and the fuel pressure collapses and injection ceases. Thus, the information input received by the electronic control unit from the various sensors automatically calculates the time the plunger effectively injects at a given engine speed. This, therefore, determines the distance the plunger moves down its stroke from the commencement of injection to the termination of injection. In other words, the electronic control unit constantly monitors the quantity of fuel needed to meet the demands made on the engine so that the vehicle responds to the driver's commands.

12.9 Unit injector electronically controlled fuel injection system (Cummins Celect)

12.9.1 Electronic control module (ECM)
(Fig. 12.47)

At the centre of the electronically controlled fuel injection system is the electronic control module. This contains a memory element, engine control microprocessor, a fault code microprocessor and a solenoid driver circuit. The engine control microprocessor receives and processes information signalled from a number of transducers (sensors), which monitor the engine's variables, as well as from switches distributed throughout the vehicle and from information pre-stored in the memory element.

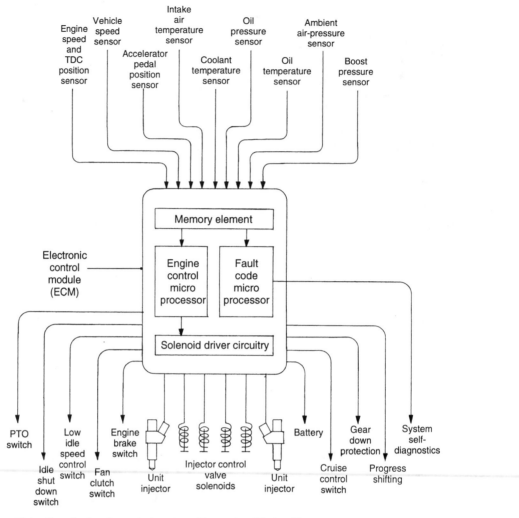

Fig. 12.47 Diesel injection electronic control system (Cummins "Celect")

Input signals from the transducers to the engine control microprocessor are computed so that calculated fuel delivery and timing demands are sent to the solenoid driver circuitry. The solenoid driver, which uses an internally generated step-up voltage source of something like 90 V, provides pulsed power to the injector control solenoids in response to the timing and fuelling demands, in order to distribute the pulses in the correct cylinder firing order. The second microprocessor performs self-diagnostic functions such as broadcasting fault code information. This system continuously monitors the electronic sub-systems for problems. If a problem is detected, a fault is logged in the electronic control module memory. The resulting fault code can then be read out on the testing equipment screen or as a flashing light code on the instrument-panel mounted fault lights. The various sensors and their functions are listed as follows:

1 engine speed and TDC position sensor—provides a reference datum point for injection timing purposes;
2 accelerator pedal position sensor—provides input information to monitor the driver's demands;
3 intake air temperature sensor—provides input information to the engine protection system as well as monitoring the air density;
4 coolant temperature sensor—provides input

information to control the engine timing and cold start fuelling;

5 oil pressure sensor—provides input information to the engine protection system;

6 oil temperature sensor—provides input information to the engine protection system;

7 ambient air pressure sensor—provides altitude input information for derating purposes;

8 boost pressure sensor—provides manifold pressure input information for matching the fuelling charge pressure entering the cylinders.

12.9.2 Output control by the electronic control module
(Fig. 12.47)

The electronic control module provides for the electronic control of the vehicle performance and vehicle speed, and provides engine protection through the following components.

Power take-off (PTO) control

This feature enables the PTO to be operated when the vehicle speed is less than 6 miles/h (4.7 km/h). When in the PTO mode, the engine speed's lower limit is set at 850 rev/min. If the accelerator pedal is depressed the PTO mode is overriden so that it is cut-off. Consequently, the select switch must be toggled to 'set', or 'resume', to re-engage the PTO mode after the accelerator pedal is released.

Idle shut down

This feature automatically shuts the engine down after a predetermined period of engine idling providing the clutch, service brake or accelerator pedal are not actuated, if these controls are operated the shut down timer will reset itself and begin a new timing interval.

Low idle speed control

This feature permits the driver to raise or lower the idle speed within a predetermined range in 25 rev/min increments by moving a toggle switch on the dash.

Fan clutch and engine brake control

This feature provides output to the fan clutch and engine brake circuitry under suitable operating conditions.

Cruise control

This feature improves the fuel economy by setting the maximum cruise control speed to something like 2 to 4 miles/h (3.2–6.4 km/h) below maximum vehicle speed in top gear.

Gear down protection control

This feature prevents the driver from operating at maximum vehicle speed when operating one gear down from top gear, this is accomplished by restricting the road speed in all lower gears.

Progressive shifting control

This feature operates, when the vehicle is operating in the lower gears, to restrict the rate of engine acceleration if the driver attempts to operate beyond the limits imposed by the progressive shift curves. It is particularly suitable in promoting fuel economy when a large proportion of the time is spent in low gears, such as inter-city driving.

12.9.3 Operating phases of the unit injector
(Fig. 12.48(a–d))

1 Filling and metering phase (Fig. 12.48(a))

At the commencement of filling, the injector control-valve is closed by the energized solenoid and the cam-lobe lift has moved both the metering plunger and the timing plunger to their lowest positions. As the camshaft rotates the cam lift decreases, which causes the return-spring to raise the timing plunger in proportion to the fall in cam lift. Fuel from the engine-driven spur-gear pump will now flow past the open metering ball-valve and into the metering-chamber. The fuel supply pressure acting underneath the metering-plunger forces it against the timing-plunger so that the flow of fuel into the metering-chamber will continue as long as the timing-plunger is rising, the metering-chamber is expanding and the injector control-valve remains closed. The quantity of fuel actually permitted to enter the metering-chamber during each cycle (that is, 'the metered amount of fuel') will be determined by the electronic control module (ECM), which contains the memory elements, two microprocessors and the solenoid driver circuitry. When the metering-chamber has been filled with the correct amount of fuel to match the operating conditions (metered quantity

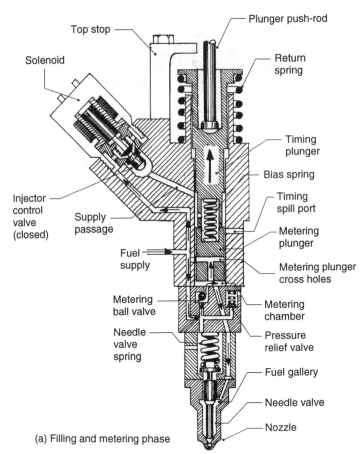

Top stop

Solenoid

Plunger push-rod

Return
spring

Timing
plunger

Bias spring

Timing
spill port

Metering
plunger

Metering plunger
cross holes

Metering
chamber

Pressure
relief valve

Fuel gallery

Needle valve

Nozzle

Injector
control
valve
(closed)

Supply
passage

Fuel
supply

Metering
ball valve

Needle
valve
spring

(a) Filling and metering phase

Fig. 12.48 Electronically controlled injection (Cummins "Celect" System)

of fuel) the injector control-valve is signalled to open (solenoid is de-energized). Immediately, fuel at supply pressure flows into the timing-chamber above the metering-plunger, thereby preventing further metering-plunger upward movement. The bias-spring also contributes by holding the metering-plunger still while the timing-plunger continues to rise.

2 Timing phase (Fig. 12.48(b))
The cam lift continues to move the timing-plunger upwards, causing the timing-chamber to be filled with fuel. When the cam lift has reached its peak the timing-plunger begins its downward stroke. The slight rise in the metering-chamber pressure causes the metering ball-valve to snap closed.

During the first stage of the pumping stroke the injector control valve remains open and, since the metering-plunger will not move due to the trapped fuel in the metering-chamber, fuel will be forced from the timing-chamber back to the fuel

supply passages via the open injector control-valve.

When the incoming feeds from the various sensors have been computed by the electronic control module (ECM) the exact point when injection commences, relative to the crank angle movement, will be determined by a signal to the injector control-valve solenoid, thus causing the injector control-valve to close.

The trapped incompressible fuel in the timing-chamber now acts as a solid strut between the timing-plunger and metering-plunger. Thus, further downward movement of the plunger push-rod transfers an equal thrust via the timing-plunger, timing-chamber, hydraulic-lock and metering-plunger to the metering-chamber.

3 Injection beginning phase (Fig. 12.48(c))
Once the fuel pressure in the metering-chamber approaches the thrust under the annular area of the middle-valve, it overcomes the stiffness of the

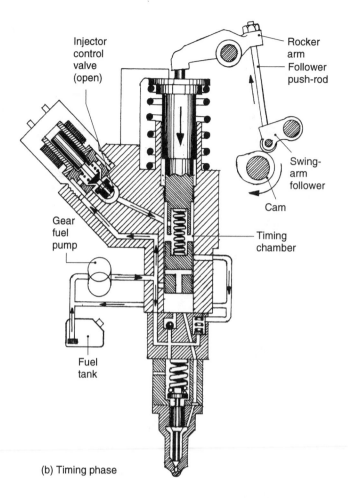

Injector control valve (open)

Rocker arm

Follower push-rod

Swing-arm follower

Cam

Timing chamber

Gear fuel pump

Fuel tank

(b) Timing phase

Fig. 12.48(b)

needle-valve spring, and the needle therefore lifts, thus opening the nozzle-valve. With the continued displacement of fuel from the metering-chamber to the injector-nozzle fuel-gallery and the annular space surrounding the small diameter portion of the needle-valve, fuel pressure in the nozzle-sac will continue to rise even though fuel is being blasted into the combustion-chamber via the spray-holes in the nose of the nozzle.

4 Injection ending phase (Fig. 12.48(d))
As the push-rod and plunger continues to move down their stroke, a point will be reached when the cross-hole in the metering-plunger aligns with the metering spill port. Abruptly, the pressurized fuel in the metering-chamber collapses as it partially escapes back to the output side of the gear fuel pump; however, the peak pressure wave produced when the metering spill port opens is released by the pressure-relief valve blow-off.

With slightly further downward movements the metering-plunger uncovers the timing spill port, and it therefore permits fuel in the timing-chamber to drain back to the inlet side of the fuel pump as the timing-plunger completes its downward stroke.

12.10   Diesel engine cold starting

12.10.1   Operating environment for ignition to occur
(Fig. 12.49)

For ignition to take place there must be sufficient pressure and temperature generated in the cylinder when the piston approaches TDC so that the surface surrounding each droplet of fuel injected into the combustion chamber commences to

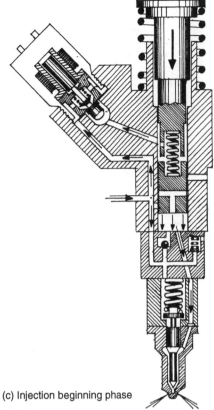

(c) Injection beginning phase

Fig. 12.48(c)

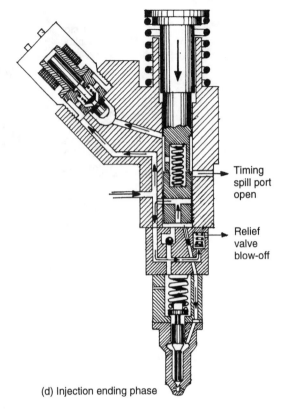

(d) Injection ending phase

Fig. 12.48(d)

vaporize. This is followed briefly by an interaction period, after which the hydrocarbon fuel and oxygen ignite. Ignition can only occur if the following conditions are satisfied.

1. The temperature of the air must be high enough to exceed the self-ignition temperature of the fuel by a sufficient threshold (Fig. 12.49). For most diesel fuels this is generally between 220° and 250°C.

2. The pressure must be high enough to ensure that the molecules of the air and fuel mixture are sufficiently close for heat to be rapidly transferred from the air to the surface of the liquid-fuel particles.

12.10.2 Optimum cold start cranking speed
(Fig. 12.50)

With very low cranking speed the heat loss to the cold cylinder walls and the leakage of air past the pistons and rings may prevent the compression temperature reaching the self-ignition threshold

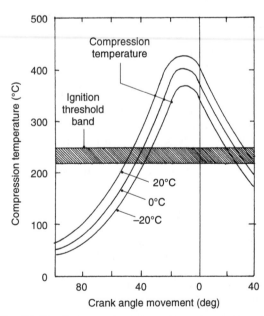

Fig. 12.49 Compression temperature rise during cold start cranking relative to the crank-angle movement for different ambient air temperatures

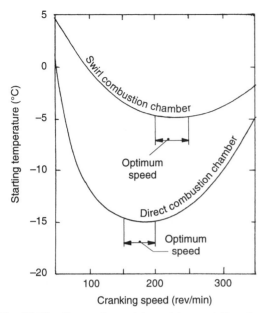

Fig. 12.50 Comparison of the cold start ability of direct and indirect combustion chamber engines for different cranking speeds

temperature. Conversely, if the cranking speed is very high, the interval in the TDC region for peak pressure and temperature to occur will be far too short for ignition to be established. Consequently, there must be an optimum speed which is fast enough to prevent the compressed and hot air charge transferring too much heat through the cold cylinder walls yet slow enough for the intensity of temperature and pressure to initiate and consolidate ignition (Fig. 12.50). The optimum cranking speed will depend upon the following.

1 The surface/volume ratio: thus, the larger the cylinder, the proportionally smaller will be the heat losses, whereas small cylinders have proportionally more surface-area to cylinder-volume, and therefore incur greater heat losses. Thus, small engines (particularly indirect injection chambers) tend to have much higher compression ratios than larger direct injection chamber engines to boost their compression pressure and temperature as a means of compensating for the increased heat losses during the compression stroke.

2 The intensity of air swirl in the cylinder and combustion chamber: the greater this is, the greater will be the rate of heat loss by convection during the compression stroke.

3 The engine's mechanical condition: for example, the piston and ring sealing efficiency in the cylinder bores, and the valve seat tightness when the valves close.

4 Open direct-injection chambers will give the easiest cold starting since they have both the smallest surface/volume ratio and the lowest intensity of air swirl compared with indirect injection chambers.

12.10.3 Cranking speed cyclic fluctuation
(Fig. 12.51)

For cold starting it is not the mean rotational speed that is important but the angular crankshaft speed from about 60° before TDC on the piston's compression stroke, because it is during this crankshaft movement that most of the compression work is done and the majority of the temperature and pressure rise occurs. Thus, with a small flywheel, the crankshaft speed fluctuation over each operating cycle can be considerable (Fig. 12.51). Also to be appreciated is the fact that the last 60° crank angle movement before the piston reaches TDC can be so slow during the critical piston compression period that a higher mean cranking speed has to be employed if the engine is to have an adequate startability margin.

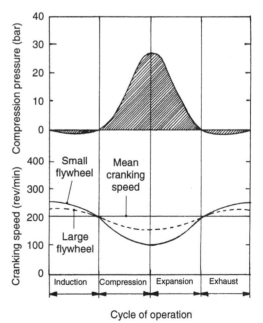

Fig. 12.51 Comparison of compression pressure rise and fall and the corresponding speed variation during a cranking cycle

12.10.4 Battery capacity

The electric starter motor cranking time is limited by the amount of energy stored in the battery; that is, its capacity to produce continuously the chemical reaction at the rate demanded, and the restriction imposed on the battery's chemical activity in cold weather. Batteries operating at $-18°C$ are only 40% as efficient as at $27°C$. However, at the lower temperature they are expected to furnish roughly 2.5 times the cranking power at normal temperature, due to the increased frictional drag which further reduces their efficiency to less than 10% of the original efficiency, which itself may be as low as 65% under heavy load operating conditions.

12.10.5 Excess fuel
(Fig. 12.52)

The purpose of introducing excess fuel into the combustion-chamber is as follows:

1 to produce a liquid film of fuel around the piston rings and the upper portion of the cylinder wall so that compressed air leakage is at a minimum;
2 to reduce clearance volume by injecting an excess amount of fuel in the combustion chamber so that the engine's compression ratio is temporarily raised;
3 to increase the number of injected droplet sites in the combustion chambers so that the air swirl has a better opportunity to bring as many fuel particles as possible in contact with the air in the hottest region in the chamber (see Fig. 12.52).

12.10.6 The need for glow-plugs
(Fig. 12.53)

To make cold starting easier with indirect combustion chambers (swirl and precombustion chambers), which experience greater heat losses than the compact direct injection combustion chambers, a heating element (known as a glow-plug) is generally installed in each of the small chambers that receive the initial spray of injected fuel (Fig. 12.53). A glow-plug when switched on dissipates heat to the cooler air surrounding its pencil sheath. It therefore supplements the

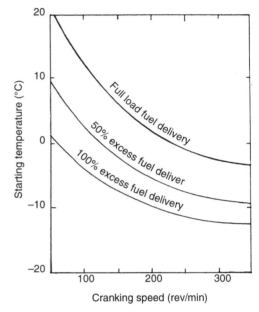

Fig. 12.52 Comparison of the cold start ability of a direct injection diesel engine with various amounts of excess fuel relative to cranking speed

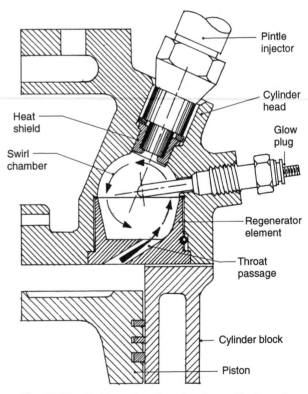

Fig. 12.53 Swirl combustion chamber with glow-plug cold start

648

ambient air temperature trapped in the small chamber.

With the trend towards direct-injection combustion chambers for some small diesel engines, which inherently suffer from high heat losses, glow-plugs are sometimes mounted so that they project at an inclined angle directly into the combustion chamber formed in the crown of the piston.

Subsequently, when the engine is cranked the peak compression temperature is able to rise to a value well above the ignition threshold temperature of the fuel so that combustion can readily be established.

12.10.7 Developments in glow-plug circuitry
(Figs 12.53, 12.54 and 12.55)

Very early cold start preheating glow-plugs, were of the double-pole (two terminal) type, where all the glow-plugs are connected in series. This arrangement was essential to increase the overall resistance of the circuit, but it limited the current passing through each exposed heating filament (made from relatively few turns of thick wire) as the filaments would otherwise consume too much electrical energy and destroy themselves—that is, burn out.

The next glow-plug design was of the exposed two or three-loop thick heating wire type, but these were now of the single-pole (one terminal) type (Fig. 12.54) with each pole being linked to a common conducting wire so that the heating elements were in parallel with each other (Fig. 12.55). Current flow with these glow-plugs was restricted by a limiting resistance connected in series with the glow-plug and the battery. A considerable improvement in glow-plug design

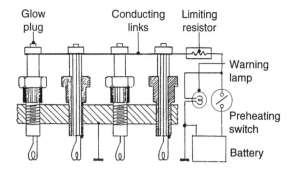

Fig. 12.55 Basic preheating parallel circuit with single pole glow plug

came about with the introduction of the sheathed heating element-type which protected the filament wire from the products of combustion (Fig. 12.53). This design uses a thin nickel-chrome wire filament wound into a coil of about 25 turns: these loops are supported in an insulating and refractory magnesia powder, which makes them relatively robust and long-lasting.

In the simplest electrical circuit the glow-plugs are supplied with current only when the ignition key is moved to its preheating position and then switched off when the engine starts and the ignition key is moved to its off-position. With this elementary system the driver has to judge the preheating time before he turns the ignition key to the start-position. Also, once the engine starts, the glow-plugs immediately switch-off as the ignition key is returned to the off position; hence, there is no further heating provision during the engine's warm-up period.

More sophisticated electrical circuitry which provides different degrees of refinement to preheating and postheating control are now in common use for various applications. The functions of these components are listed and briefly described as follows.

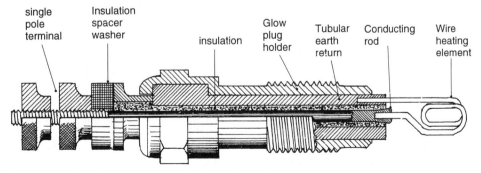

Fig. 12.54 Single pole glow plug with exposed wire heating element

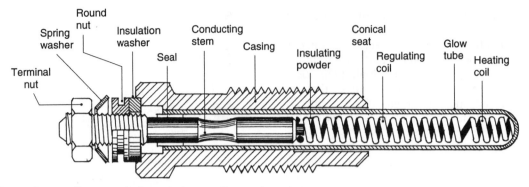

Fig. 12.56 Double filament winding tubular pencil type glow-plug

1 A preheating warning-lamp is illuminated during the preheating period, which extinguishes itself when the glow-plugs have reached their operating temperature. This light indicates to the driver that the engine is ready to be cranked.

2 A glow-plug preheat relay enables a relatively light wiring and low current supply to switch on and off the heavy glow-plug load as required.

3 A time control device controls the duration of the preheating and post-heating periods.

4 A coolant temperature thermo-switch breaks the glow-plug circuit above some predetermined coolant temperature so that the glow-plugs do not heat up during warm start-and-stop operating conditions, or when the surrounding air temperature is very high.

5 An ambient temperature sensor signals to the time control unit the surrounding air temperature so that the unit can adjust the preheating and post-heating durations before the timer switches off the glow-plugs.

6 A post-heating relay switches a reducing resistance in series with the glow-plugs once the engine starts. It therefore prevents the glow-plugs from overheating when the generator output exceeds a predetermined voltage.

7 A fused glow-plug or overvoltage/short circuit cut-out device protects the glow-plug against absorbing an excess of electrical energy, which might then burn the filament out.

12.10.8 Glow-plug design and performance characteristics
(Figs 12.53, 12.56 and 12.57)

Early glow-plugs were of the exposed wire filament type, and tended to corrode when subjected to combustion over a period of time. They also suffered from a low electrical resistance which produced a high current draw. These problems have largely been overcome by utilizing a sheathed nickel–chrome fine-wire filament embedded in an inert refractory material, such as magnesia powder (an insulator), inside the corrosion resistant thermally conductive pencil tubing. This fine-wire filament draws less current than the exposed coarse-winding type of glow-plug, for a similar preheating focal source, and it is protected from chemical reaction by the pencil tubing so that, compared with the exposed element glow-plug, its life expectancy is increased.

An improved version of the single filament-winding tubular-pencil type glow-plug uses a heating filament whose resistance is practically independent of temperature and a control winding whose resistance increases with temperature; that is, it has a positive temperature coefficient, with both windings being connected in series to each other (Fig. 12.56). With a matched pair of windings the glow-plug can reach its preheating temperature within 4 to 10 seconds, as opposed to the single-winding filament, which requires between 20 and 30 seconds to reach a similar working temperature.

The rapid preheating to ignition temperature conditions is achieved without causing the winding filament to become excessively hot since the control winding resistance increases with rising temperature, and thus restricts the flow of current if the winding filament is left switched on.

Glow-plug performance characteristics are shown in Fig. 12.57. With both the single and the double-winding types of glow-plugs the initial cold current flow is very high but, with the double-winding glow-plug, this quickly decreases with heating time. The single-winding glow-plug,

650

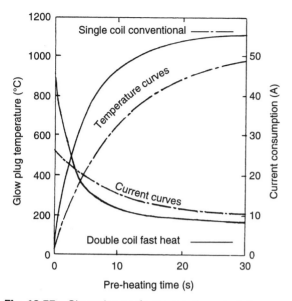

Fig. 12.57 Glow-plug performance characteristics

on the other hand, does not consume so much current when first switched on and, for both types, the current only gradually decreases with heating time. The final current flows are about 8 A and 10 A for the double and single-windings, respectively.

Conversely, the single-winding glow-plug heats up relatively slowly compared with the double-winding glow-plug, which has a high initial rate of temperature rise with respect to heating time. However, its rate of temperature increase slows down until there is no further rise in temperature beyond about 30 seconds.

When installed in the combustion-chamber the glow-plug conical seat is screwed tightly against a corresponding seat formed in the stepped plug-hole (Fig. 12.53). With the glow-plug in position its pencil-stem projects well into the centre of the combustion-chamber to a point where a portion of the injected fuel spray impinges on the hot glow-plug tip. Liquid fuel will then spread over the partial cylindrical and spherical surface of the sheath and, in the process, it vaporizes so that the dense surrounding air and the intense heat in the centre of the chamber readily ignites the air and fuel vapour. Cold starting would be prolonged and inconsistent without the fuel spray impinging on a localized temporary heat source, such as the smooth surface of the glow-plug sheath.

12.10.9 Cold start preheating circuitry
(Fig. 12.58)

With the ignition-key in its first position (1) there is a voltage supply from the battery to the ignition switch, the alternator and preheat control unit. When the coolant temperature is below a predetermined value (normally below 40°C) the coolant thermo-switch contacts close, enabling current to flow to and from the control unit.

With the ignition switch-key moved to the second position (2), battery voltage is applied to the preheat glow-plug relay winding, energizing its solenoid winding and causing the contacts to close. This allows a heavy current to pass from the battery to the four glow-plugs and, at the same time, to illuminate the glow-plug warning-lamp and the no-charge warning-lamp. After a preheating period of between 8 to 20 seconds (depending upon ambient temperature) the time control unit opens the circuit which supplies the voltage to the glow-plug warning-lamp. It therefore extinguishes the light, which informs the driver that the engine is ready to be started. The higher the surrounding air temperature the shorter will be the duration of this preheat period.

The ignition-key should now be twisted to position (3) and a voltage will be applied to the starter-motor solenoid-relay causing the starter motor to crank the engine with the glow-plugs still energized. Once the engine fires, its speed increases until it is self-sustaining. The generator then commences to generate a charge voltage which opposes the battery-voltage and increases with rising engine speed. This diminishes the no-charge warning-lamp which is in series with the battery and generator.

At a predetermined generated voltage, the glow-plug preheating relay contacts open and, at the same time, the post-heating relay contacts close, so that a reducing resistance is placed in series with the supply voltage and the glow-plugs. This safeguards the glow-plugs—which have a nominal voltage rating of only 11 V but would be subjected to a maximum regulated generator voltage of approximately 14.4 V. Note that the 11 V rating of the glow-plugs is required to maximize the heating capability of the filament during the preheating period before the engine is cranked.

The post-heating period (that is, the length of time during which current continues to flow to the glow-plugs once the engine has started and is running) is determined by the ambient tempera-

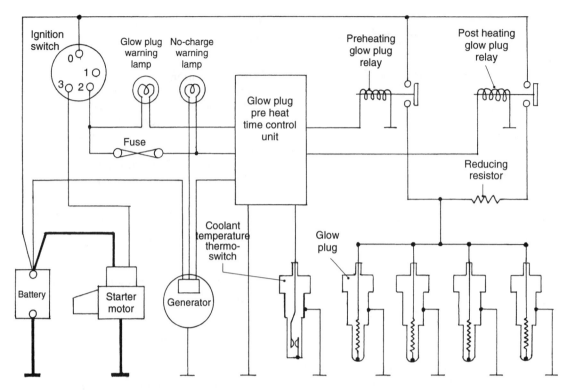

Fig. 12.58 Cold start glow plug preheating electrical system

ture air sensor which is usually contained inside the preheat time control unit. The post-heating relay contacts will remain closed if the air temperature is low for something like 90 seconds, as this ensures smooth combustion during the engine's warm-up period. Once the engine has reached its normal working temperature, the post-heating of the glow-plugs ceases within 20 seconds.

If the ignition switch is left in the preheating position (2) without the engine starting, the time-control will interrupt the preheating relay switch circuit so that current flowing to the glow-plugs will be cut-off after about 30 seconds. The relay can be actuated again by moving the ignition key to position (1) and then back to position (2).

12.10.10 Glow-plug cold start cycle of operation
(Fig. 12.59)

First, the ignition-key is moved to the preheat position. Current now passes to the glow-plug over a period of about 10 seconds depending upon glow-plug element design and the ambient air temperature, the heating element temperature consequently rises to approximately 1100°C and the battery voltage gradually decreases due to the current draw from the glow-plugs. The power rating for a typical glow-plug is 90 W so that the current consumption from a 12 V supply would be 7.5 A, and for four glow-plugs this amounts to 30 A, which is a significant drain on the battery over a period of time.

The ignition-key is now moved to the starter-motor cranking position. This causes the starter-solenoid to be energized and something like 150 to 300 A (depending upon the size of the engine) passes to the motor before it commences to rotate. This large increase in current draw causes a sharp fall in battery-voltage and a corresponding drop in glow-plug temperature; in addition, the fall in glow-plug temperature is also due to the cooling effect of the air swirl and the heat transference losses to the combustion-chamber walls.

Once the engine fires, its speed quickly rises to either cold or hot idle. Similarly, the supply voltage increases as the generator output supplements the battery voltage. However, to protect the glow-plug heating filament from the excess

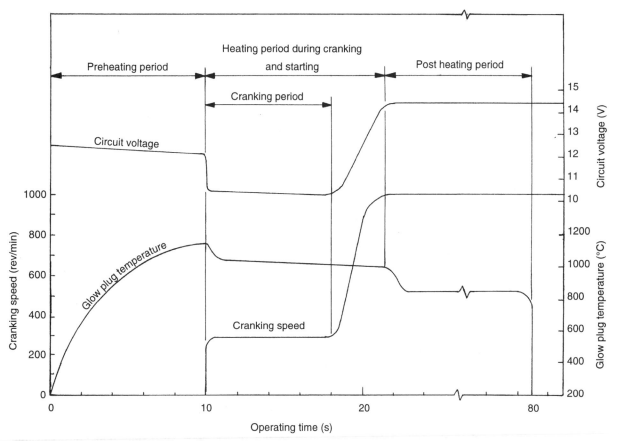

Fig. 12.59 Cold start cycle of operating characteristics

power supplied, the time control unit now switches in a reducing resistance in series with the glow-plug circuit. Thus, during the post-heating period, the glow-plugs operate at a lower power output, which is reflected by the reduced glow-plug temperature. After a predetermined time, the glow-plugs will automatically be switched off by the time control unit.

12.10.11 Direct-injection engine cold-start manifold preheating devices

Direct-injection combustion chambers have far less surface area and therefore have proportionally less heat loss than their counterpart indirect-combustion chambers. As a result, it is easier to attain the ignition threshold temperature when cranking the direct-injection open chamber engine, than it is with the indirect-swirl or pre-combustion chamber engine during a cold start.

Consequently, combustion-chamber glow-plugs are not essential with large direct-injection combustion-chamber engines. However, single-point cold-start preheating-devices situated in the air intake of such engines ensure shorter cold-start cranking periods and consistent starting.

The following manifold preheating devices will now be described:

1 heat operated thermostart
2 flame glow-plug
3 solenoid operated ether injection

12.10.12 Heat operated thermostart (CAV)
(Fig. 12.60)

This thermostart comprises a tubular valve body supported in a casing, which screws into the inlet manifold horizontally, as near to the air-intake as possible. Inside the tubular valve-body is a ball-

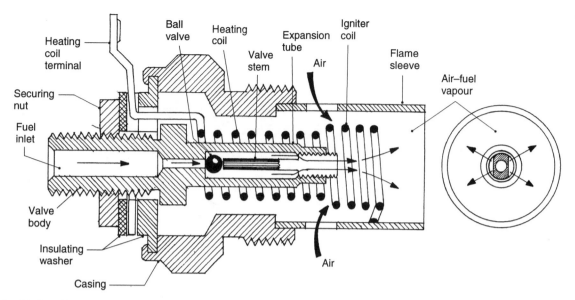

Fig. 12.60 Heat operated thermostart (CAV)

valve, which is held on its seat (when the engine is cold) by a valve-stem with a small diameter but proportionally long length, which is screwed into the discharge end of the tube. The larger diameter screw portion of the valve-stem has two parallel flats machined so that fuel is discharged through the segment-shaped passageways formed on either side of these flats. Surrounding the tubular-valve is a small diameter heating-coil which, at its outer end, enlarges to form an igniter-coil—the end of which is attached to a perforated eight-hole flame-sleeve (shield). This gives support to the winding and also functions as the earth return. The inner end of the small diameter coil extends rearward through a flanged insulation washer and comes out at the back to form the live terminal for the electrical wire connection.

For cold starting purposes the igniter-key should be moved to its second preheat position and held there for something like 15 to 20 seconds before moving the ignition-key to the third starter-motor cranking position.

During the time the heating-coil and the igniter-coil heat up, some of this heat transfers to the valve-tube, which therefore expands. However, the cooler valve-stem inside the tube, which also has a lower coefficient of linear expansion does not expand to the same extent on the tube. Consequently, the ball-valve is freed from its seat and thus opens the passage from the fuel inlet to

the annular passage formed between the valve tube and stem.

A small gravity tank (25 ml capacity) which has a gravity head of something like 100–150 mm is continuously topped up by a pipe connected to the injection pump filter. This tank supplies fuel to the thermostart, and thus fuel flows through the valve where it drips onto the igniter-coil at a flow rate not exceeding 0.15 ml/s. The fuel is vaporized by the heat of the valve-body and is ignited by the outer large loop igniter-winding.

When the engine is cranked, air is drawn into the manifold. The vapour ignited by the outer igniter-coil continues to burn as it mixes with the incoming air stream on its way to the engine cylinders. It is claimed that the minimum starting ambient temperature can be approximately 10°C lower than normal when using this device and the conventional excess fuel mechanism installed as part of the injection pump.

12.10.13 Flame glow-plug
(Fig. 12.61)

This preheating device consists of a coil heating element enclosed in a pencil-type tubular sheath, which is mounted at one end inside the heater unit body, and whose other end protrudes into the perforated flame-sleeve. The middle portion

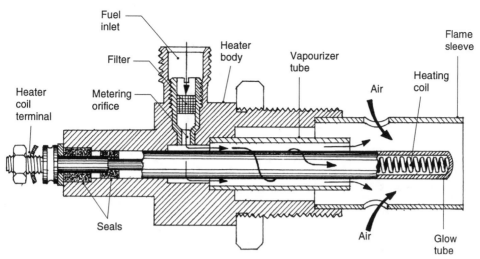

Fig. 12.61 Flame glow plug (Bosch)

of this pencil-sheath is surrounded by a vaporizer tube or sleeve.

Fuel from the injection pump is usually supplied via a solenoid-valve to the inlet connection of the heater unit body and a metering-orifice and filter is positioned inside the inlet passage bore. When starting the engine from cold the ignition-key is moved to the second preheating position and held there for approximately 20 seconds during which time current is supplied to the pencil heating-element. This allows time for the pencil-sheath to reach its operating temperature of just above 1000°C. At the same time, the solenoid-valve opens permitting fuel to flow to and through the flame glow-plug. As the fuel passes from the metering-orifice to the pencil-sheath it spreads over the hot surface and commences to vaporize as it moves towards the mouth of the vaporizer tube. By the time the fuel vapour reaches the exposed glowing element tip, it is at ignition point and burning begins.

When the ignition-key is moved to the third start position, the engine is cranked by the starter-motor, and air is induced, via the intake manifold, into the cylinder. The air movement therefore sweeps past the thermostart flame-sleeve and thus takes the burning vapour and unburnt fuel droplets with it into the cylinders: it thus raises the temperature of the air entering the cylinders.

When the engine fires and the engine develops its own power, the ignition-key is returned to the off-position, which then closes the solenoid-valve;

the flow to the flame glow-plug is thus interrupted.

12.10.14 Ether ignition promotion agent
(Fig. 12.62)

In very cold weather conditions and for high mileage worn engines, additional ignition accelerating-agent fluids can be beneficially used to ease and speed-up the cold-start process. Diethyl ether has proved to be the most satisfactory cold-start ignition promoting agent. Its composition is $[(C_2H_5)_2O]$, it has a boiling point of 35°C, a cetane number of 53, a latent heat of 383.79 kJ/kg and it is an ignition accelerator—that is, it reduces the delay-period in the combustion process. The ignition point of a typical ether compound fluid is as low as 180°C. Ether injected into the incoming air-stream considerably reduces the starter-motor's cranking time before the engine fires and commences to develop its own power. Thus, the normal start times of 30 to 50 seconds can be reduced to something like 5 to 10 seconds or less depending upon the ambient air temperature.

Engine tests have shown that 0.1 to 0.2 ml of ether fluid per litre of piston displacement per second of air intake, on a four-stroke cycle engine is approximately the optimum discharge rate into the intake air stream in cold weather. Thus, for a 10 litre four-stroke engine this would amount to

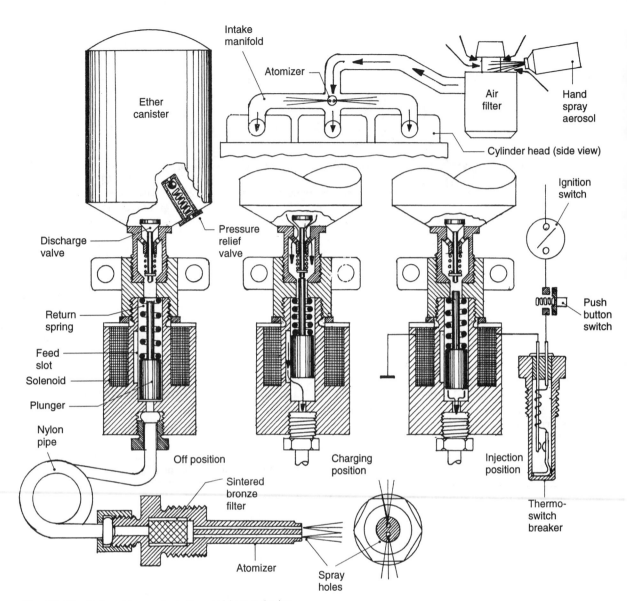

Fig. 12.62 Solenoid operated ether cold start device

between $(0.1 \times 10$ to $0.2 \times 10)$ 1.0 to 2.0 ml/s of ether fluid injected into the air intake.

If aerosol ether sprays are to be used (Fig. 12.62), they should be squirted into the air intake for between 3 to 5 seconds. A standard aerosol can discharge approximately 12 ml of start fluid per second and most operators fully depress the aerosol for at least 3 seconds, or something like 36 ml of start fluid: that is, at least six times the amount of ether required.

Manually operated ether spray devices are in common use, and these are normally operated by pulling out the ether start knob for 1 to 2 seconds and then pushing it in again; this releases the spring-loaded plunger which discharges the fluid into the intake manifold.

12.10.15 Ether cold start device

Because of the danger of producing spontaneous and excessively high pressure rises in the combustion-chambers when using an ether–fluid mixture aerosol, which can and will damage the engine components over a period of time, it is very important that a precisely controlled quantity of starting fluid is introduced into the intake air stream when the engine is cold started. Therefore, ether cold-start devices installed in the manifold entrance have become standard equipment in cold climate countries, and are offered as optional equipment in those regions of the world which do not always suffer from cold weather conditions.

12.10.16 Solenoid operated ether cold-start device
(Fig. 12.62)

These units consists of a pumping unit, a screw-on dispensable ether canister and an atomizer screwed into the intake manifold at a central point (or as near as possible to where the branch pipes merge) so that approximately equal quantities of ether mixture will be drawn into each branch pipe during the cranking period.

The dispensable canister has both a pressure relief valve which is normally set to a burst pressure of 69 bar and a discharge conical stalk valve which screws down into a union member and which, in turn, screws into the pump unit body. In the off-position the return-spring pushes the plunger down to its lowest position.

When the push-button is pressed, current passes to the solenoid via the thermo-switch and the magnetized winding then pulls up the plunger (armature) against the return-spring, until its stem contacts and pushes back the discharge valve-stop. The open canister-valve now permits the ether liquid to discharge into the chamber below the plunger via the feed slot. Releasing the push-button breaks the solenoid circuit so that the return-spring is free to force the demagnetized plunger down its stroke. Once the plunger cuts-off the feed slot, fuel will be displaced into the pipe and into the manifold, in the form of a fine spray, via the atomizer's two calibrated holes. The atomized ether spray then mixes with the incoming stream so that it enters the cylinders as part of the air charge. A thermo-switch is located in the cooling system and, at temperatures above 40°C, breaks the solenoid circuit. This prevents the ether spray device operating when the engine is warm.

Because the ignition temperature of an ether compound is so low, approximately 180°C, the compression temperature when the engine is cranked far exceeds this so that ignition, followed by combustion, is ensured with ambient temperatures down to about −20°C and even lower.

13

Emission Control

Vehicle pollutants are emitted from the engine (Fig. 13.1) by three main sources:

1 the crankcase where piston blow-by fumes and oil mist are vented to the atmosphere;
2 the fuel system where evaporative emissions from the carburettor or petrol injection air intake and the fuel tank are vented to the atmosphere;
3 the exhaust system where the products of incomplete combustion are expelled from the tail pipe into the atmosphere.

13.1 Crankcase emission

13.1.1 Piston ring blow-by
(Figs 13.2 and 13.3)

The piston and its rings are designed to form a gas-tight seal between the sliding piston skirt and the cylinder walls. However, in practice there will always be some compressed charge and burnt fumes which manage to escape past the compress-

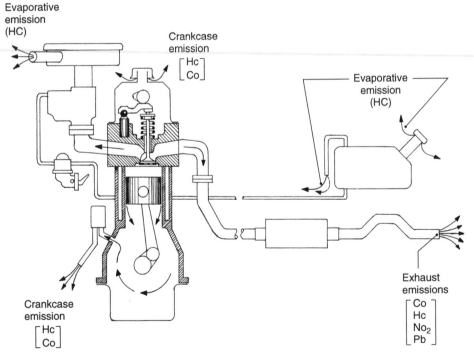

Fig. 13.1 Emission points of the motor vehicle

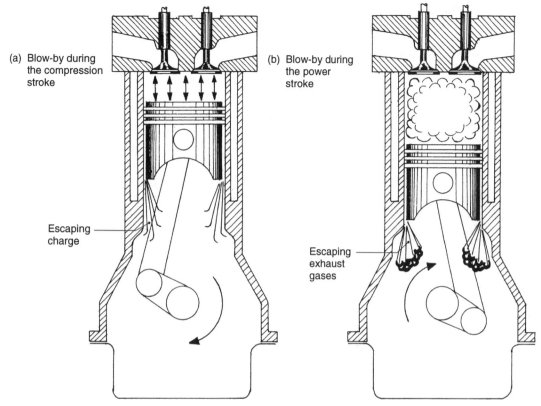

(a) Blow-by during the compression stroke

(b) Blow-by during the power stroke

Escaping charge

Escaping exhaust gases

Fig. 13.2 Piston blow-by

ion and oil control piston rings and therefore enter the crankcase (Fig. 13.2). These gases which find their way past the piston ring belt may be unburnt air–fuel mixture hydrocarbons, or burnt (or partially burnt) products of combustion, carbon dioxide, water (steam) or carbon monoxide.

Piston blow-by increases with engine speed and, in particular, as the piston rings and cylinder bore wears (Fig. 13.3), the blow-by becomes more noticeable in the upper speed range.

Blow-by takes place between the piston ring gap, piston-ring to piston-groove clearance and in the TDC region where the piston ring circumferential shape cannot accurately follow the contour of an oval or bell-mouthed cylinder wall. Piston blow-by can become a major problem above some critical engine speed if ring flutter occurs due to the ring directional inertia becoming out of phase with the directional movement of the piston. This problem has largely been reduced by decreasing the width of the piston rings (and therefore the weight of each ring) and by increasing ring radial cylinder-wall pressure. Neverthe-

less, there is one critical speed for each width of piston ring. Fortunately, however, a good design is able to restrict flutter so that it occurs outside

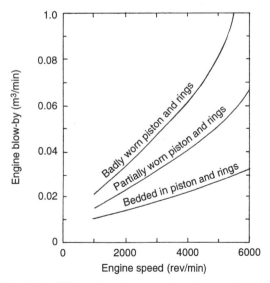

Fig. 13.3 Effect of engine speed on piston and ring blow-by

659

the normal speed range of the engine, provided that the ring-to-groove clearance does not become excessive.

Besides effectively reducing the engine compression ratio and the power developed, the effects of piston blow-by are twofold. Firstly, it can lead to a high concentration of combustible air–fuel mixture which could cause an explosion in the crankcase. Secondly, the air–fuel mixture, partially burnt and fully burnt vapour fumes, will condense and contaminate the engine's lubricating oil. This may cause the lubricants' molecular structure to break down, thereby reducing its oiliness and oxidizing resistance and, at the same time, produce sludge with the result that the bearings are subjected to much higher wear rates.

Since it is impossible to eliminate piston blow-by completely, an organized convection current is deliberately created which circulates the crankcase and rocker or camshaft cover spaces and consequently carries the unwanted fumes out with it. The removal of blow-by gases and vapour fumes from the crankcase is obtained by creating a partial depression at one outlet location so that the blow-by gases under pressure (escaping between the piston and cylinder wall) are attracted towards the lower pressure region of the crankcase, at which point they are expelled.

There are two methods of creating the extraction depression:

1 the road draught crankcase ventilation system;
2 the induction manifold vacuum positive crankcase ventilation system.

13.1.2 Road draught crankcase ventilation system
(Fig. 13.4(a and b))

This method of ventilating the crankcase incorporates a pipe whose outlet is exposed to the air stream passing between the underside of the vehicle and the road surface when the vehicle is in motion. Consequently, as the vehicle moves, the relative velocity between the air and the moving pipe creates a depression (vacuum) which increases with the speed of the vehicle. At the same time, if there is a head-on or side-wind blowing, this will further increase the magnitude of the depression at the draught pipe exit.

Internal gas circulation is achieved by having a vented filler and breather cap in the top of the rocker cover and an outlet pipe low down in the side of the crankcase (Fig. 13.4(a and b)). Thus, when the vehicle movement is sufficient to create a draught depression at the exit pipe, air will enter the rocker cover through the filler cap, pass down the push-rod and camshaft passages, where it then enters the crankcase. The rotating crankshaft webs and balance weights then establish a circular partial vacuum path for the fresh air and blow-by gases to follow, and these gases are then ejected into the atmosphere via the oil and mist separator and draught pipe (Fig. 13.4(a)). Oil mist, which is caught up with the air and fumes on their way through to the draught pipe, passes through baffles installed in the separator container which cause most of the oil mist to condense. It is then drained back to the engine's sump, whereas the majority of the air and gases is permitted to escape.

At high engine speeds, and particularly with a worn engine, the pressure build-up in the crankcase may exceed the extraction capability of the draught pipe, this causes the air and gas in the crankcase to reverse their flow direction. Some of the air, gas and oil mist are therefore forced to move up to the rocker cover, where they are then expelled to the atmosphere by way of the filler cap wire-mesh (Fig. 13.4(b)). An indication that there is a significant amount of piston blow-by, possibly due to a worn engine is the excessive amount of oil splashed outside the filler cap and rocker cover.

At speeds below about 40 km/h there is insufficient relative air movement around the draught pipe to create the necessary air circulation within the engine's crankcase. Therefore, the gas and vapour build-up and its condensation in the crankcase and rocker cover will contaminate the lubricant, thus reducing the oil's effectiveness in minimizing wear between bearing rubbing pairs.

The unacceptable limitation of this system of crankcase ventilation, due to the expulsion of gas and fume vapour (HC and CO) into the atmosphere thereby contributing to pollution, has made this method of internal purging of the engine obsolete.

13.1.3 Positive crankcase ventilation (PCV) system
(Figs 13.5(a and b) and 13.6(a and b))

This second method of ventilating the crankcase and removing piston blow-by gases and fumes utilizes the manifold vacuum to establish air cir-

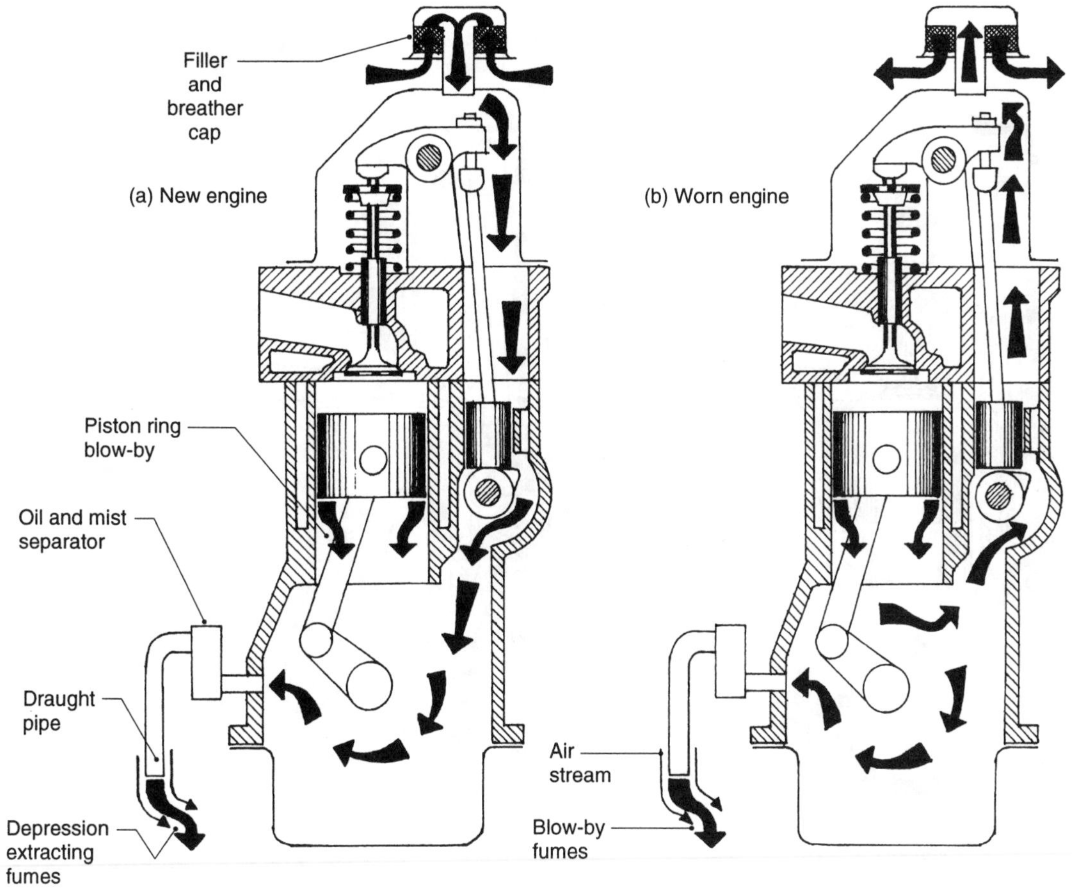

Filler
and
breather
cap
(a) New engine
Piston ring
blow-by
Oil and mist
separator
Draught
pipe
Depression
extracting
fumes
(b) Worn engine
Air
stream
Blow-by
fumes

Fig. 13.4 Road draught crankcase ventilation system

culation and, with it, the removal of the unwanted vapour mixtures and burnt and partially burnt products of combustion.

In-line engine PCV systems

For the in-line engines there are, broadly, two circulation layouts.

Firstly, there is the crankcase to induction manifold system (Fig. 13.5) where air is drawn into the rocker cover via a rubber pipe connected to the air filter. The air then flows from the rocker cover to the crankcase by way of the push-rod and camshaft passageways, it is then drawn from the side of the crankcase to a central point on the induction manifold via the oil and mist separator. The gas and vapour fumes are then drawn into the cylinders on each of their induction strokes so that, in effect, the engine is consuming its own blow-by gases and fumes. Some of these fumes will contribute to the combustion process and will then pass through the normal exit of the exhaust port to the exhaust system, where they may be given further after-burning treatment before being ejected into the atmosphere.

At very high engine speesds, or slightly lower speeds if the engine is worn, the crankcase to induction-manifold flow path is insufficient to accommodate the amount of blow-by. Hence, the direction of flow in the rocker cover switches from rocker-cover to crankcase, to rocker-cover to air filter, where it is again drawn (via the induction venturi) into the induction manifold. Thus, the blow-by gases are now drawn into the combustion chamber by way of the crankcase and rocker-cover.

Secondly, there is the camshaft-cover to induction-manifold circulation system (Fig. 13.6(a and b)) where air is drawn from the air filter directly to the crankcase via a rubber pipe, the air then circulates around the crankcase, picks up the blow-by gases and then passes to the camshaft

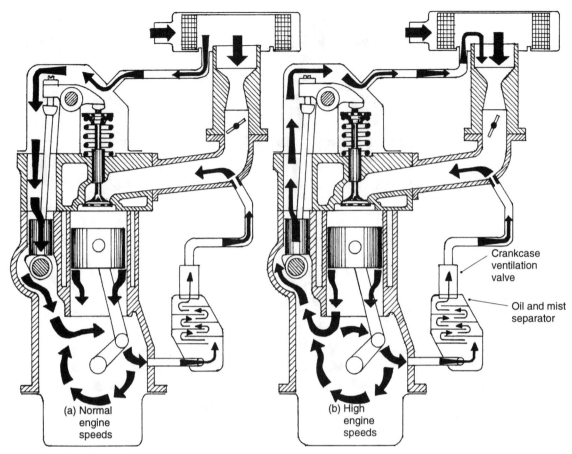

Fig. 13.5 Positive crankcase ventilation (PCV) system (crankcase to induction manifold circulation)

cover space. The air, gases and fumes then move from the camshaft cover to a central point on the induction manifold, due to the existing vacuum in the manifold, where they are immediately drawn into the cylinders as part of the normal induction charge.

Again, at very high engine speeds or, if the engine is worn, at much lower speeds, the camshaft-cover to induction-manifold passageway is inadequate to circulate all the air and blow-by products. Thus, a part of the air and fumes in the crankcase will be forced up the crankcase to the air filter pipe, it is then drawn into the induction manifold via the carburettor or petrol injection venturi. Note that at very high engine speeds there will only be a small to moderate amount of vacuum created in the intake of the venturi.

With the camshaft cover to induction manifold circulation system, oil and mist separation is not required, as would be the case with the crankcase exit system, where a considerable amount of oil splash and mist blow-by accumulates.

Open and closed positive crankcase ventilation systems

Earlier positive crankcase ventilation systems used the open method of circulating air and did not have a pipe between the air filter and rocker cover. However, when the engine becomes worn, piston blow-by pushed the gas and vapour out from the rocker filler cap directly into the atmosphere.

The present positive crankcase ventilation (PCV) systems have adopted the closed system where excess blow-by gas in the upper speed range is drawn back into the cylinder by way of the air filter and venturi route.

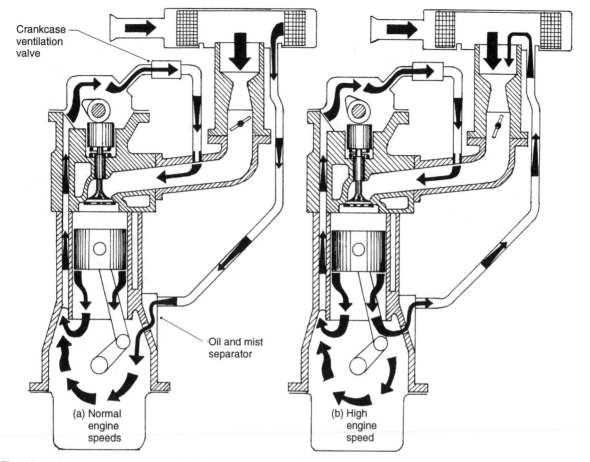

Oil and mist
separator

(a) Normal
engine
speeds

(b) High
engine
speed

Fig. 13.6 Positive crankcase ventilation (PCV) system rocker cover to induction manifold circulation

13.1.4 Vee banked-engine PCV systems
(Figs 13.7(a and b) and 13.8)

A common PCV system arrangement (Fig. 13.7(a and b)) connects the air filter to the right-hand rocker cover causing air to enter this cylinder bank first. It is then drawn downwards to the crankcase where the rotating crankshaft forces the air and blow-by gases to flow up to the left-hand rocker cover via the vertical push-rod passage. The trapped air and gases are then forced into the induction manifold via the return pipe and PCV valve. Finally, they flow with the fresh charge into the cylinders during the engine's induction stroke periods.

If the engine is worn and it is rotating in the upper speed range, then the extra piston blow-by causes the right-hand cylinder-bank air circulation to change its direction so that some of the crankcase circulating air and blow-by gases will be drawn up the push-rod passage directly to the intake air filter and venturi. This is at the same time as the left-hand cylinder bank rocker-cover air and gases are being forced into the left-hand induction manifold. Thus, the crankcase circulation system is able to cope adequately with any additional blow-by pressure build-up due to excess piston blow-by.

An alternative method of circulating air, adopted extensively on diesel engines, uses the rocker-cover to induction-manifold connections for each cylinder bank (Fig. 13.8). Circulation is provided by a pipe joining the air filter to the centre of the crankcase. Air and its collected blow-by products are therefore drawn upwards through the ventilating passages provided in each cylinder bank. The air and gases are then expelled from the rocker cover spaces through pipes which merge together at a twin port positive crankcase

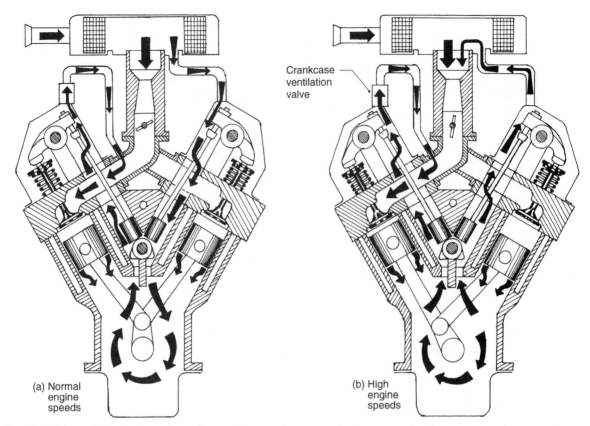

Crankcase ventilation valve

(a) Normal engine speeds

(b) High engine speeds

Fig. 13.7 Vee cylinder bank configuration positive crankcase ventilation system (rocker cover-crankcase-rocker cover-induction circulation)

ventilation valve situated at the central riser of the air intake manifold.

The colour of the rocker shaft and rockers will give an indication if there is adequate air circulation. If the oil splash and mist are clean, then the circulation is operating normally but if the rockers and shaft take on a chocolate-brown appearance then there is some sort of circulation restriction which could be due to a blockage in either the PCV valve or vacuum return pipe.

13.1.5 Positive crankcase ventilation valves

(Figs 13.9(a–c) and 13.10(a–c))

The air circulation flow-rate through the crankcase and rocker-cover or camshaft-cover is controlled by a positive crankcase ventilation valve (PCV valve). The PCV valve opening is vacuum sensitive, that is, the vacuum increase or decrease in the induction manifold respectively decreases

or increases the amount the valve opens. Therefore, at idle speeds when the throttle is closed and the vacuum in the induction manifold is high, the valve will be almost closed, thus providing only a small flow path whereas at high engine speeds with a wide open throttle, the vacuum is low, causing the valve to move to a fully open position thereby enabling a maximum air flow-rate to be achieved through the ventilating system. Should the manifold vacuum become very high, the valve closes against its spring, cutting off the air circulation. Conversely, if a positive backfire pressure should occur in the induction manifold, the valve will automatically move in the opposite direction until it hits the end stop, thereby completely cutting off the air circulation. Without the positive crankcase ventilation valve, too much air would bypass the throttle and enter the cylinders during engine idling conditions causing the engine to race and run erratically.

Two types of PCV valves will now be described.

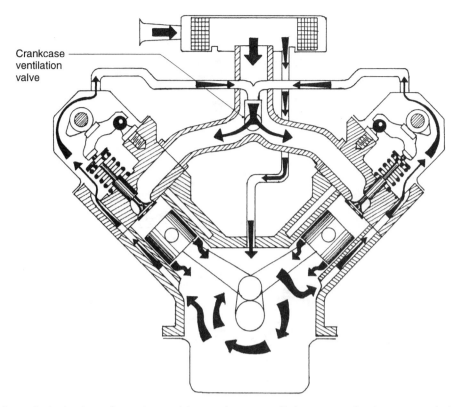

Fig. 13.8 Vee cylinder bank configuration positive crankcase ventilation system (rocker cover to induction manifold circulation)

Slit-type positive crankcase ventilation (PCV) valve
(Fig. 13.9(a–c))

The valve consists of a plunger with a helical slit which slides in the bore of the casing. As the manifold vacuum increases, the valve plunger is drawn further into its bore, thus decreasing the effective size of the valve opening orifice and, consequently, reducing the amount of air circulation (Fig. 13.9(a)). If there is a high surge of vacuum, such as during engine overrun (Fig. 13.9(b)), the slit end of the plunger contacts the loose spring, and this raises the plunger spring force opposing the vacuum, which therefore prevents the valve fully closing. However, under normal driving conditions (Fig. 13.9(a)) the manifold vacuum will be far less, so that the valve's return-spring is able to push the valve backwards, thereby increasing the helical slit orifice passageway opening. When the engine is switched off the return-spring pushes the valve fully out until the end-stop flange closes the passageway (Fig. 13.9(c)); similarly, if there is an induction man-

ifold backfire the valve is jerked outwards against its stop and closes.

Taper-type positive crankcase ventilation (PCV) valve
(Fig. 13.10(a–c))

This valve has a parallel body flanged at its rear end and tapered at the variable orifice end (Fig. 13.10(a)). The flanged rear end acts as the spring retainer and blocks off the flow passageway if the engine should backfire, whereas the front tapered position of the plunger acts as a variable orifice. Sometimes a jiggle-pin is located inside the hollow plunger, so that during idling conditions, when the vacuum is high and the tapered valve contacts its seat, a small supply of air will still be able to circulate through the loose jiggle-pin hole (Fig. 13.10(b)). The assumption is that the fluctuating vacuum makes the jiggle-pin vibrate and therefore helps to prevent the central idle orifice from becoming blocked. In fact, water contamination and heat oxidation may still cause this relatively small orifice to become choked. If

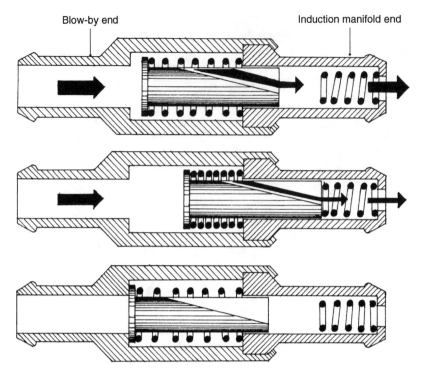

Fig. 13.9   Slit type positive crankcase ventilation valve

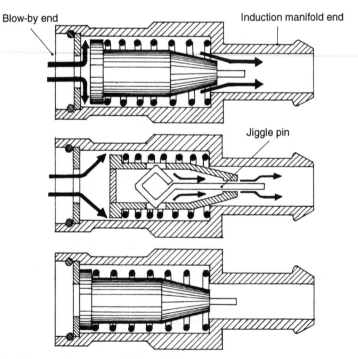

Fig. 13.10   Taper type positive crankcase ventilation valve

the engine should backfire or the engine is stopped (Fig. 13.10(c)) the return-spring rapidly moves the valve against the stop washer and this, therefore, completely seals-off the flow passage between the crankcase or rocker cover and the induction manifold.

13.2 Fuel evaporation control

In countries which are hot most of the year and in cooler climates where, in the summer, it can be reasonably hot, a serious source of atmospheric pollution is the evaporation of hydrocarbon petrol from the fuel tank and carburettor. For example, in California it has been estimated that a car left out all day without special provision to prevent the loss of petrol vapour fumes from the fuel tank, could lose as much as 4 litres of liquid fuel by evaporation.

Fuel evaporation control is generally achieved by using activated charcoal or active carbon granules housed in a ventilated canister. Most evaporation control systems reduce the emission of fuel vapour during the time the vehicle is idling in traffic, is stationary or, in particular, if it is parked in strong sunshine, by absorbing the vapour fumes in the carbon-rich container. Once the engine is operating under normal driving conditions, these collected and stored hydrocarbons are released to the induction manifold where they are drawn into the combustion chambers and consumed as part of the combustible air–fuel mixture charge.

13.2.1 Fuel evaporation control for a petrol injection air–fuel mixture system
(Fig. 13.11(a and b))

Engine idling or stationary
(Fig. 13.11(a))
With the engine either idling or stationary there will be no vacuum signal in the cut-off diaphragm chamber so that the vapour fumes in the canister are prevented from entering the induction manifold and cylinders. Fuel vapour formation in the fuel-tank rises to the highest point in the specially enlarged filler tube, which also acts as a vapour–liquid separator. The relatively large and cool surface area of the tube therefore condenses some of the virtually stagnant and heavier vapour fractions trapped in the upper region, which then

drain back to the fuel-tank. The remainder of the vapour will be forced along the vapour pipeline, due to the vapour pressure build-up, until it enters the active carbon canister. Here, the active carbon soaks-up and absorbs the fumes in the same way as a sponge absorbs water.

Engine operating during normal driving conditions
(Fig. 13.11(b))
With the engine running above idling speed there will be a relatively high vacuum in the vacuum signal pipeline. This causes the vacuum cut-off valve to lift off its seat. Fresh clean air will now be drawn to the bottom of the canister, through the central air intake tube, where it then spreads over the base of the canister, moves through the detachable filter and purges (clears out) the vapour fumes through the open cut-off valve and vapour purge pipeline. It is then drawn into the induction manifold where it is consumed with the normal air–fuel mixture charge in the various combustion chambers. Carbon canister air filters in the base of the container should be replaced every 48 000 km (30 000 miles) or every two years.

13.2.2 Fuel tank ventilation valve with spill prevention
(Fig. 13.12(a–d))

Fuel tanks for cars are nowadays made from plastic materials, and the filler tube may be integral or attached separately to the fuel tank. Likewise, the ventilation valve assembly may be manufactured as an integral part of the filler tube or it may form a separate unit. Sometimes it is incorporated in the filler cap itself.

The ventilation valve assembly incorporates a two-way breather valve in the upper half, and a gravity valve is situated in the lower half casing (Fig. 13.12(a)). With this arrangement there are three external connections, a ventilation pipe exit, a filler tube pipe connection and an overflow pipe connection.

Fuel tank spill prevention
(Fig. 13.12(b))
When the tank is being filled rapidly, and particularly when it is almost full, air and vapour locks tend to make the top part of the filler tube flood and spill from the neck of the tube, thus releasing unwanted hydrocarbons, liquid and vapour into the atmosphere. The aim of the gravity valve is to

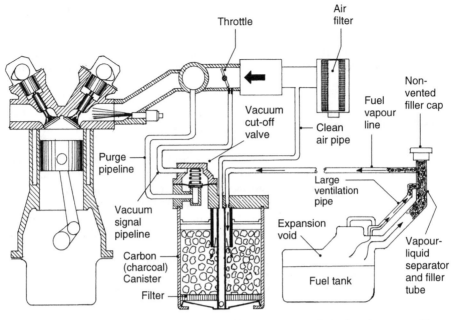

(a) Engine idle and stationary conditions

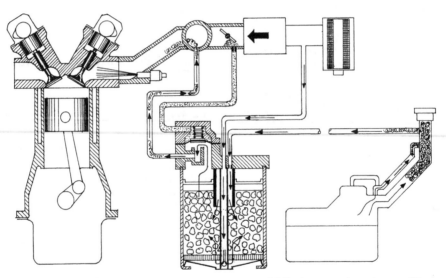

(b) Engine normal running conditions

Fig. 13.11 Fuel evaporation control petrol injection air–fuel mixture system

permit excess fuel, which is accumulated in the upper filler tube region when the tank is being topped up, to lift and hold the gravity valve float and tapered valve against its seat. This, therefore, prevents fuel escaping by the breather valve to the atmosphere, and it also provides an overflow path for the surplus fuel to be rapidly returned to the fuel tank.

Warm-up tank operating conditions
(Fig. 13.12(c))

With most fuel supply systems, the fuel is pumped to the carburettor or petrol injection system and, to prevent vapour locks due to stagnant fuel becoming very hot, a portion of this fuel is returned to the fuel-tank by a separate pipe. Thus, there is a continuous circulation of fuel

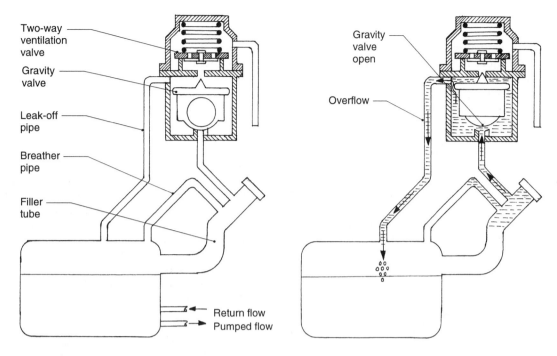

(a) Normal operating tank condition

(b) Rapidly overfilled filler tube condition

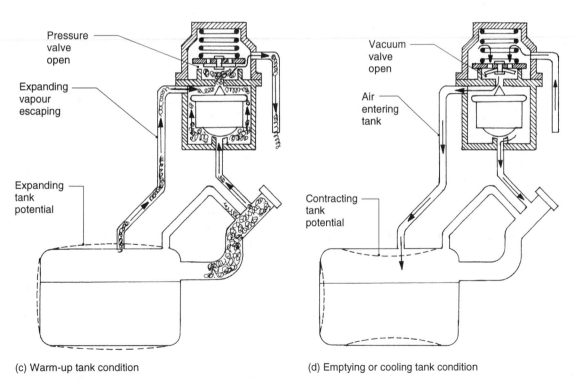

(c) Warm-up tank condition

(d) Emptying or cooling tank condition

Fig. 13.12 Ventilation valve with spill protection

between the engine and fuel-tank that tends to stabilize the fuel temperature. However, the effect of this fuel circulation is that the surplus, slightly aerated, fuel flows in a warmed-up state back into the fuel tank, which causes the fuel volume to increase over a period of time and, consequently, the pressure within the fuel-tank to rise. At the same time, excessive splashing of the fuel within the tank and the ambient rise in temperature may cause a portion of the fuel to vaporize, again raising the pressure in the fuel-tank. When the fuel tank pressure exceeds a predetermined value, something like 0.3 bar, the ventilation valve overcomes the tension of the return-spring and pushes the valve open to relieve the excess pressure into the atmosphere.

Emptying and cooling tank operating conditions
(Fig. 13.12(d))
During a vehicle's journey, fuel will be consumed by the engine, particularly when the engine is first started from cold, so that the fuel tank progressively empties, which thus causes its volume to decrease. Similarly, after a journey, particularly if the weather turns cold in, say, the evening or night, the fuel temperature will drop, thereby causing the volume of the liquid fuel to decrease again.

The net result is that the pressure in the tank will decrease, thus establishing a vacuum in the fuel tank. If this vacuum is not removed the whole tank will collapse and cave inwards, severely and permanently damaging the tank. To prevent this happening, the ventilation valve has an inward opening rubber disc valve which permits air to enter the tank when the internal pressure drops below about 0.1 bar, this relieves the pressure differential between the outside and inside of the fuel tank.

13.3   Air intake temperature control

The air intake temperature greatly influences the uniformity of the air–fuel mixture, its evenness and the degree of atomization; that is, the size and number of fuel particles or droplets suspended in the intake air stream.

If the ambient (surrounding) engine temperature is very low, the liquid fuel finds it difficult to mix with the incoming air stream, especially if there is a very low intake air speed, as there is then insufficient air movement to suspend and support the larger and heavier fuel droplets. Consequently, the larger particles of liquid fuel will tend to gravitate to the floor of the intake passages or cling to the walls of the passages so that the overall mixture distribution becomes very uneven.

The result of poor quality intermingling of the air and fuel under these conditions will almost certainly cause engine hesitation and misfiring during acceleration. This will continue until the engine coolant circulation temperature has risen sufficiently to pre-heat the surrounding induction walls and incoming air charge. This means that every time a cylinder misfires or a cylinder charge only partially burns, hydrocarbons and carbon monoxides are released to the atmosphere.

Conversely, if the intake air temperature increases, the charge expands causing its density to decrease. Accordingly, the mass of the air–fuel mixture entering the cylinder with every stroke decreases with rising air temperature.

Now, the power developed per cylinder is proportional to the mass of the homogeneous mixture drawn into the cylinder; therefore, the maximum attainable power output becomes less as the intake air and fuel's temperatures rise.

Consequently, between the extremes of ambient air temperature, the optimum air stream temperature can be chosen to produce regular firing when the engine is warming up and, at the same time, the air intake temperature is prevented from rising above some predetermined maximum so that power output does not suffer.

Thus, the purpose of an air intake temperature control system is to regulate and maintain the temperature of the air entering the induction system within some attainable temperature band.

13.3.1   Bimetallic leaf-spring operated flap-valve air intake temperature control
(Fig. 13.13(a))

The object of this air intake temperature control arrangement is to speed up the engine's warm-up period and to maintain an almost constant temperature inside the air cleaner assembly during normal operating conditions, even when the outside temperatures may reach extreme values. Consequently, the air–fuel mixture distribution and quality of atomization will be greatly im-

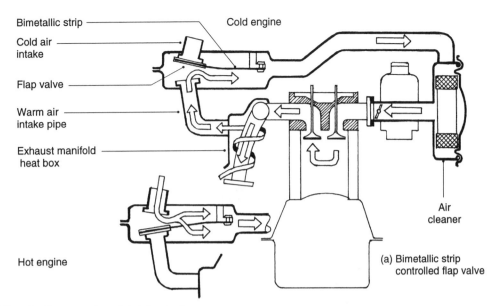

Bimetallic strip ——————— Cold engine

Cold air ———
intake

Flap valve ———

Warm air ———
intake pipe

Exhaust manifold ———
heat box

Hot engine

Air
cleaner

(a) Bimetallic strip
controlled flap valve

Fig. 13.13(a–c)   Pre-warming air intake devices

proved under a wide range of driving conditions.

This concept incorporates a large bore flexible pipe joining the exhaust manifold and the air intake to the air cleaner assembly. A sheet steel casing partially encloses the outside of the exhaust manifold so that heat radiated from the exhaust manifold preheats the surrounding air. At the same time it creates a convection current of air which spirals upwards and around the central down-pipe, merging the individual exhaust manifold branch passages. Dividing the pipe between the exhaust heat box and the air intake cleaner assembly is a hot/cold temperature-sensitive flap-valve mounted in a plastic container. The hot/cold valve assembly consists of a flap-value supported on a bimetallic leaf-spring, which is attached to the container at one end, whereas the flap-valve end floats between the cold and hot inlet passage opening.

When the engine is cold, the bimetallic spring distorts and curls in a way which closes the flap valve against the cold air intake opening. Consequently, this opens the hot air intake passage enabling preheated air from the exhaust heat box to be drawn into the air cleaner assembly.

With rising intake air temperature, the bimetallic heat sensor spring distorts in the opposite direction, thus opening the cold inlet passage and closing the hot intake passage. Cooler air from the engine's surroundings will now enter the air cleaner assembly and subsequently the cylinders.

In practice, the flap-valve, supported by the bimetallic leaf-spring, will oscillate to and fro between the cold and hot intake openings thereby constantly adjusting the mix or blend of cold and hot air to within a narrow temperature band of something like 28°C to 32°C.

13.3.2   Wax thermostat operated flap-valve air intake temperature control
(Fig. 13.13(b))

This arrangement permits close control of the air intake temperature to be achieved over the various driving conditions such as cold starting, warm-up driving in traffic and at high speed in cold weather. The benefits are that the mixture distribution and thoroughness of atomization can be maintained under the majority of driving conditions and engine warm-up time will be at a minimum.

The wax heat sensor method of controlling the air intake temperature consists essentially of a flexible large-bore pipe connecting the exhaust manifold preheat box to the intake air cleaner via a thermostat-controlled flap-valve.

When the engine is cold, the wax sensor plunger will be fully retracted by the wax capsule due to the return-spring tension, this causes the flap-valve to close off the cold air-intake and open the hot air-passage. Preheated air circulating the ex-

671

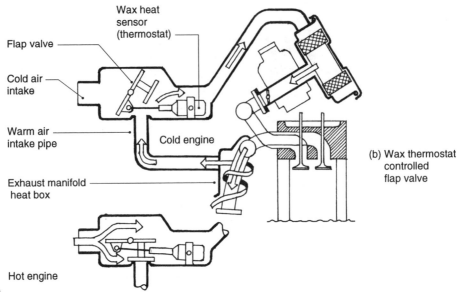

Wax heat sensor (thermostat)

Flap valve

Cold air intake

Warm air intake pipe

Cold engine

Exhaust manifold heat box

(b) Wax thermostat controlled flap valve

Hot engine

Fig. 13.13(b)

haust manifold under the cover of the heat box will now be drawn into the intake air cleaner assembly, which is attached to the carburettor or air intake venturi.

In contrast to cold running conditions, when the engine has fully warmed-up and the surrounding air temperature is fairly high, the wax in the heat sensor expands and pushes out fully the plunger against the return-spring, thus causing the flap-valve to swing over to open the cold intake passage and to close the hot air intake.

However, under normal operating conditions, the wax-sensor will position the flap-valve so that hot and cold passages will be partially open so that the hot and cold blend of incoming air will maintain the intake air temperature within the optimum working range (28°C to 32°C).

13.3.3 Bimetallic vacuum-operated flap-valve air intake temperature control
(Fig. 13.13(c))

This system of controlling the air intake temperature, operates by diverting air (which previously circulated the exhaust manifold) to the entrance of the air cleaner when the engine is cold, and by automatically adjusting the blend of hot and cold air drawn into the air cleaner assembly when the engine is hot so that a uniform and optimum air intake temperature (28–32°C) is obtained during normal operating conditions.

The air intake temperature control system comprises a bimetallic heat sensor located inside the air cleaner (filter) and an air intake hot and cold flap-valve container which is operated by a vacuum diaphragm unit. Basically, the air intake to the air cleaner assembly is serviced by both a cold intake passage exposed directly to the surrounding atmosphere and a hot air intake which is supplied by preheated air coming from the exhaust manifold heat box.

When the engine is cold, the heat sensor bimetallic strip and ball closes the air bleed hole, thus subjecting the upper diaphragm chamber to an induction manifold vacuum. Consequently, the vacuum lifts up the diaphragm against the tension of the return-spring, causing the flap-valve to close the cold air intake passage and to open the hot intake. This results in the hot air collected from around the exhaust manifold being drawn up the hot air pipe and through the air cleaner assembly to the carburettor or directly into the induction system if petrol injection is employed. The hot air intake will be fully open when the depression acting on the diaphragm exceeds 100 mm Hg.

As the engine warms up and the ambient air temperature rises, the heat sensor bimetallic strip curls downwards, causing the ball to become dislodged from its seat. Subsequently, air enters the heat sensor chamber neutralizing the vacuum transmitted from the induction manifold to the diaphragm unit. As a result, the diaphragm return-spring pushes down the flap-valve so that the

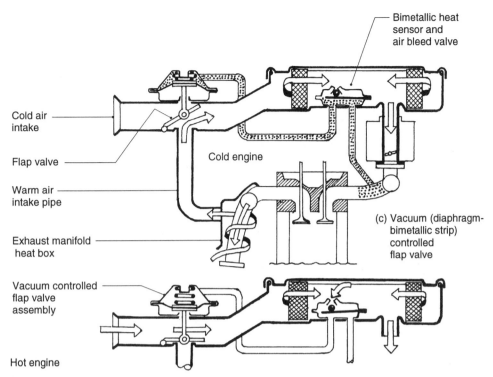

Cold air intake

Flap valve

Warm air intake pipe

Exhaust manifold heat box

Vacuum controlled flap valve assembly

Cold engine

Bimetallic heat sensor and air bleed valve

(c) Vacuum (diaphragm-bimetallic strip) controlled flap valve

Hot engine

Fig. 13.13(c)

cold air intake opens, and the hot air passage closes. It thus permits cooler air to be drawn into the induction system. The change from hot to cold air supply is progressive as the depression in the diaphragm chamber controlled by the bimetallic strip falls from 100 mm Hg to 50 mm Hg, at which point the hot air intake will be fully closed.

Between the extremes of cold and hot operating conditions, particularly during cold weather, the bimetallic heat sensor will sense the changing temperature conditions and automatically bleed more or less air through the bleed hole. This causes the diaphragm and flap-valve to open and close frequently, thus maintaining the air intake temperature within the desired operating range.

An additional feature of this system is that if the engine is subjected to full-load wide-open throttle conditions (even when the weather is cold) the induction manifold depression will automatically decrease, hence it allows the diaphragm return-spring to move the flap-valve to the cold intake opening position, thus permitting the maximum mass of cool air to enter the cylinder.

13.3.4 Butterfly valve hot-spot control
(Fig. 13.14(a and b))

One of the major problems when the engine is started from cold is that the cold induction manifold internal passage walls quench and condense the atomized air–fuel mixture passing through the carburettor venturi, resulting in globules of fuel droplets washing the floor of the central riser and branch pipes. The poor mixture distribution and coarse atomization therefore produce hesitation, irregular running and excessive exhaust emission during the time the engine is warming up.

One method of reducing the warm-up period employs an offset butterfly-flap valve mounted on a spindle and situated below the induction manifold central riser and inside the exhaust manifold exit down-pipe (Fig. 13.14(a and b)). The butterfly-valve spindle is positioned to one side of the downpipe and, in the cold start position, the bimetallic spiral spring rotates the butterfly-valve, positioning it in an approximately vertical position (Fig. 13.14(a)). The normal exhaust gas passageway flow is therefore blocked by the larger of the two butterfly flaps; correspondingly, the short flap now permits the exhaust gases to

673

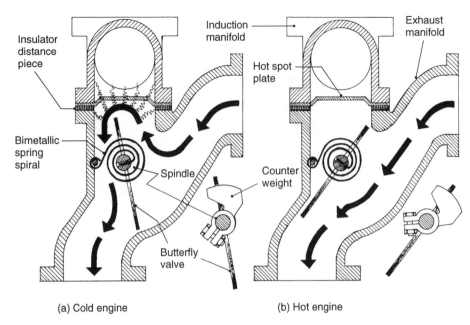

Fig. 13.14 Butterfly controlled induction–exhaust manifold hotspot

flow underneath the hot-spot plate which divides the induction and exhaust manifold vertical risers. Immediately, within only a few cycles of operation, exhaust gas heat soaks the hot spot so that the upper floor of the hot-spot plate is preheated and vaporizes any liquid droplets landing on top of it on their way to the induction manifold's branch pipes. The amount of preheat provided by the exhaust gases moving underneath the hot-spots is possibly more than is necessary. It therefore vaporizes a large proportion of the charge mixture but nevertheless enables most of the mixture to be effectively burnt when the engine is cold or only partially warmed-up.

As the exhaust manifold absorbs the heat from the expelled gases the bimetallic-spiral spring progressively increases its tension so that it curls, causing the butterfly-valve spindle to rotate until the butterfly-valve completely separates the exhaust manifold central riser from the out-flowing exhaust gases (Fig. 13.14(b)). This redirected exhaust gas flow therefore shields the hot-spot from the majority of the outgoing exhaust gas heat so that the fresh incoming air–fuel mixture is not now vaporized to any extent but remains highly atomized; this butterfly-controlled hot-spot therefore provides the conditions for maximizing the mass of air–fuel charge entering the cylinders per stroke.

During the warm-up period, when the engine

accelerates, the large quantity of exhaust gas expelled from the cylinders applies a resultant clockwise twisting effect (torque) to the butterfly-valve, causing the valve to rotate against the tension of the bimetallic spiral spring. Accordingly, the butterfly-valve opens the main exhaust gas exit passage to cope with the increased gas flow. A counterweight attached to the valve-spindle helps to steady the valve opening action during engine acceleration, when the engine has warmed-up, and when the bimetallic spring has fully opened the valve.

13.3.5 Electrically heated hot spot control
(Fig. 13.15)

This method of pre-heating the incoming air–fuel mixture reduces the quench and condensation effect of the cold induction manifold walls as the atomized mixture flows down the induction manifold riser. The mixture then divides as it is drawn through each individual manifold branch before entering the cylinders.

The hot spot heater unit is fitted in the base of the induction manifold riser (and is utilized with cross-flow cylinder-head engines where the induction and exhaust manifolds are mounted on opposite sides of the cylinder-head) making it difficult

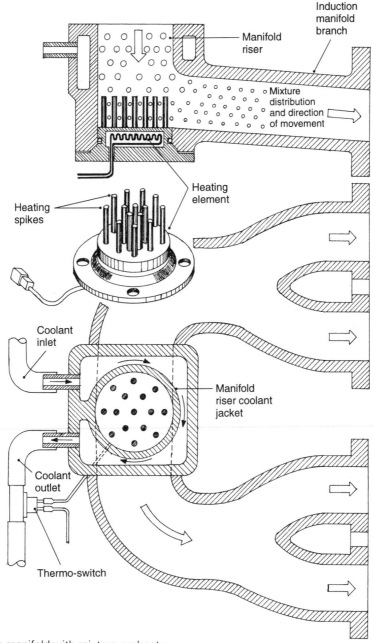

Fig. 13.15 Induction manifold with mixture preheater

to supply exhaust gas to pre-heat the induction manifold (Fig. 13.15).

The electrical hot-spot heater unit takes the form of many vertical rods or spikes mounted in a circular flanged base. Housed inside the flange base is an insulated electrical heating element, which is earthed at one end and whose other end

is connected to a thermo-switch mounted in the manifold coolant jacket exit pipe.

When the ignition is switched on, current flows through the heater element via the thermo-switch and rapidly heats the base and spikes of the heater unit. Thus, when the engine is started from cold, the cool air–fuel mixture stream entering the induction riser very quickly warms-up as it flows between the heater spikes on its way to the various branch pipes and cylinders. Consequently, within a very short time there is a homogeneity in the atomization and mixture distribution of the incoming charge which greatly improves cold driveability and minimizes exhaust gas carbon monoxide and hydrocarbon emission.

Once the induction manifold circulating coolant reaches a temperature of 65°C, the thermo-switch automatically cuts out the hot spot heater unit.

13.4 Exhaust gas recycling

(Figs 13.16 and 13.17)

The three offending pollutants in the atmosphere released by the engine are carbon monoxide (CO), hydrocarbon (HC) and oxides of nitrogen (NO_x), and generally the method of altering the air–fuel mixture strength to reduce the exhaust gas emission has conflicting results. Thus, weakening the air–fuel mixture ratio from rich to stoichiometric (chemically correct) considerably reduces both the carbon monoxide and hydrocarbon products of combustion, but this is at the expense of oxides of nitrogen which rapidly increase to a maximum as the mixture approaches the stoichiometric (15:1) ratio and just beyond. Only by leaning the mixture still further, to about 18:1 to 20:1, can there be a marked reduction in the oxide of nitrogen content of the exhaust gas, but then this may produce instability, hesitation and a reduction in performance under certain driving conditions which cannot be tolerated.

The air–fuel mixture entering the cylinders contains a large amount of air which mainly consists of 75.5% nitrogen and 23% oxygen by mass. When these two elements, nitrogen and oxygen, are subjected to high temperatures and pressures they react to form oxides of nitrogen (NO_x) and, consequently, they are expelled with the other products of combustion to the atmosphere.

Oxygen and nitrogen tend to combine and form oxides of nitrogen when the immediate surrounding temperatures exceed 1370°C and, during the normal combustion process, the temperature in the cylinders will go well above this value.

One method of reducing the amount of nitrogen oxides in the exhaust emission is to lower the combustible charge temperature in the cylinder, so that the reaction between the nitrogen and oxygen does not occur or is only partially completed.

This can be achieved by returning some of the exhaust gas back into the cylinders or introducing inert gas, such as carbon dioxide, water vapour or nitrogen, to the combustible mixture since this will dilute the overall mixture and therefore reduce the peak temperature reached during the combustion process. The amount of nitric oxides created in the cylinders from the nitrogen and oxygen contained in the air is an exponential function of combustion temperature, so that even a small decrease in the combustion temperature will produce a significant reduction in nitric oxide production. It has been estimated that a 16% lowering of peak temperature would produce roughly a 85% reduction of nitric oxide concentration in the ejected exhaust gases.

The consequence of recirculating a portion of the exhaust gas expelled from the cylinder has two effects:

1 to reduce the overall temperature of the burning air–fuel mixture because the recirculated exhaust gas in the cylinder does not take part in the burning process but does absorb a portion of the heat generated by the combustion;
2 to reduce the overall quantity of fresh air–fuel mixture drawn into the cylinder due to the proportion of recirculated exhaust gas also entering the cylinder for a given load and speed condition.

The direct effects of exhaust gas recirculation on power, specific fuel consumption and nitrogen oxides emissions, relative to the percentage of recirculated exhaust gas, have been studied experimentally by testing an engine operating with a fixed throttle setting, constant ignition timing and constant air–fuel ratio. These results (Fig. 13.16), plotted against percentage of exhaust gas recirculation, show that as the percentage of recycled exhaust gas in the cylinder increases, the engine's power progressively decreases, whereas the specific fuel consumption increases. However, the nitrogen oxides emission decreases, up to approximately 15% recirculation, at which point there is a tendency for the nitrogen oxides emissions to level out. This indicates that there will be very little advantage in increasing the recirculation of burnt gas beyond about 15%. The actual reduction in nitrogen oxides with 15% recirculation amounted to roughly 88% with a power loss of 16% and a rise in fuel consumption of around

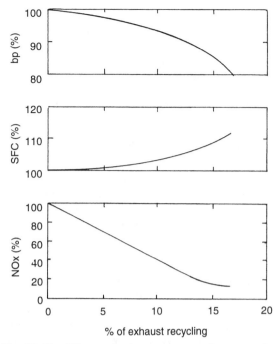

Fig. 13.16 Effects of exhaust gas recycling on engine power, brake specific fuel consumption and nitrogen oxides

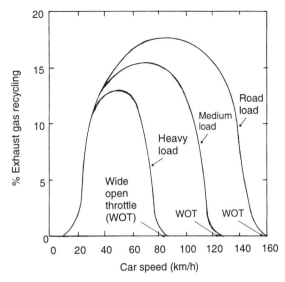

Fig. 13.17 Exhaust gas recycling characteristics relative to car cruising speed

14%. However, the deterioration in power and fuel consumption can almost be recuperated by optimizing the ignition advance and air–fuel ratio for the different quantities of exhaust gas being fed back into the cylinder. Note that the benefits of reducing the nitrogen oxides by 88%, with an exhaust gas recirculation of 15%, are now reduced to roughly a 60% decrease in nitrogen oxides. Thus, a workable compromise can be obtained which can justify the extra equipment and modified performance.

The characteristics of exhaust gas recirculation over a speed range from standstill to 160 km/h for heavy, medium and road (light) loads can be seen in Fig. 13.17. Generally, there is no exhaust gas recirculation at idling engine speed and low vehicle speed as this could cause driving instability. At the other extreme the amount of exhaust gas recirculation is made to fall rapidly towards maximum speed for light road conditions, but this occurs much earlier as the engine load rises so that power output does not suffer when greater demands are imposed on the engine.

13.4.1 Comparison of exhaust emissions using conventional and exhaust gas recycled carburation systems
(Fig. 13.18)

A comparison between an exhaust emission with a conventional carburation system and with an enriched carburation system having exhaust gas recycling, for a car travelling at different speeds, is shown in Fig. 13.18. It can be seen that the nitrogen oxides were considerably reduced with increasing car speed when exhaust gas recycling was adopted, as opposed to the conventional carburation layout.

Likewise, the recycled exhaust gas system showed a fairly consistent level of hydrocarbon emission relative to the conventional carburation system, which produced a classic hydrocarbon hump up to a speed of roughly 60 km/h. However, beyond this speed, the conventional carburation system produced a much reduced flat emission response, which undermines the exhaust gas recycling system in the upper speed range.

In contrast to both oxides of nitrogen and hydrocarbon emissions the carbon monoxide emissions showed a marked increase in exhaust emission when employing exhaust gas recirculation, compared with the conventional carburation system, when the car speed exceeded approximately 20 km/h. However, this would be as ex-

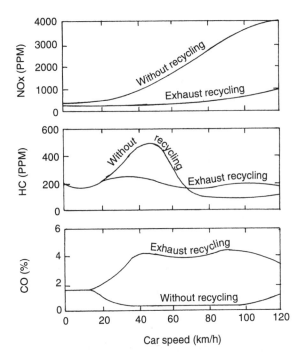

Fig. 13.18 Comparison of emission resulting from engine without and with exhaust gas recycling

pected since the recycled exhaust gas system has its mixture strength set on the rich side of the stoichiometric ratio.

13.4.2 Exhaust gas recirculation system vacuum regulated
(Figs 13.19(a–d) and 13.20(a–b))

This exhaust gas recirculation system transfers exhaust gas from the exhaust manifold to the induction manifold via a steel pipe and a vacuum operated cut-out valve, which is activated by a vacuum signal provided by a vacuum pick-up at the throttle valve. A thermo-vacuum lock-out valve (Fig. 13.20(a–b)) prevents the vacuum signal reaching the EGR valve when the coolant temperature is below approximately 40°C, as this would introduce a measure of engine instability. Thus, once the coolant temperature rises to 40°C the bimetallic disc distorts and snaps downwards from the 'O' ring mouth, thus opening the vacuum passage.

Idle running conditions
(Fig. 13.19(a))
With the throttle valve in the closed position the

throttle valve vacuum pick-up will be on the atmospheric side of the air intake. Under these conditions there will be no, or very little, vacuum conveyed to the exhaust gas recirculation cut-out valve-unit diaphragm chamber. Consequently, the return spring maintains the valve in the closed position, thus preventing exhaust gas recycling.

Light to medium load driving conditions
(Fig. 13.19(b and c))
As the engine speed increases, the throttle valve moves to a part open position causing the throttle valve plate to cover the vacuum pick-up partially. Under these driving conditions the relatively high vacuum in the throttle valve region is relayed to the exhaust gas recirculation valve diaphragm chamber. Accordingly, there will be an upthrust causing the exhaust cut-out valve to open partially. With a further speed and load increase, the combination of air intake speed and throttle opening causes the throttle valve pick-up vacuum signal to reach a maximum value. This results in the exhaust gas recirculation diaphragm lifting the exhaust cut-out valve to its maximum open position.

Full load driving conditions
(Fig. 13.19(d))
When the engine is subjected to full load driving conditions the throttle valve moves to the wide open throttle. Under these operating conditions the vacuum in the throttle valve vacuum pick-up region considerably decreases, it therefore permits the diaphragm return spring to overcome the reduced vacuum upthrust, and instead now moves the diaphragm and exhaust gas cut-out valve downwards to the closed position. This desirable feature prevents loss of engine performance under severe operating conditions.

Exhaust gas recirculation response time damping
(Fig. 13.19(a))
To obtain the best overall exhaust gas recirculation response without compromising too much engine performance, it is necessary to slow the rate at which the throttle valve vacuum pick-up signal reaches the exhaust gas recirculation (EGR) valve, and it is also desirable for the EGR valve to prolong its operating time deliberately when responding to changing conditions.

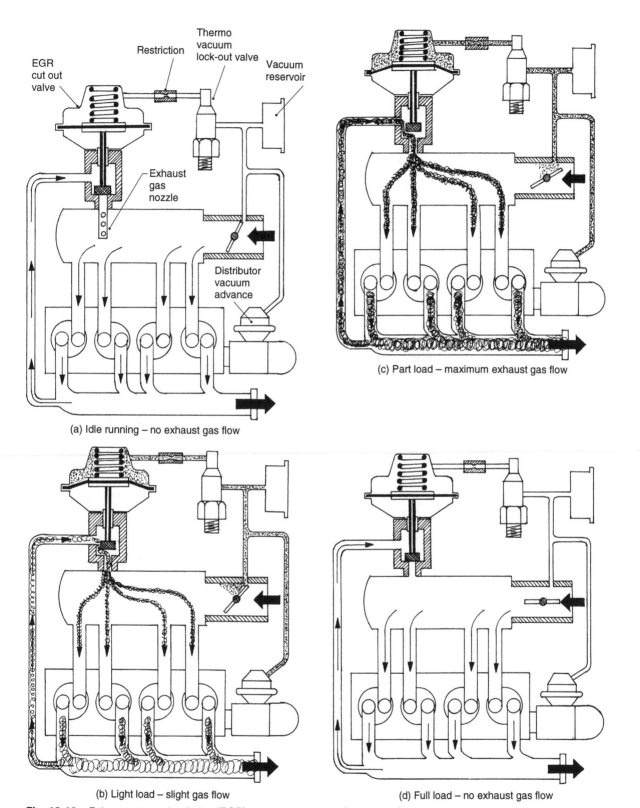

Label callouts in the figure:

EGR cut out valve

Restriction

Thermo vacuum lock-out valve

Vacuum reservoir

Exhaust gas nozzle

Distributor vacuum advance

(a) Idle running – no exhaust gas flow

(b) Light load – slight gas flow

(c) Part load – maximum exhaust gas flow

(d) Full load – no exhaust gas flow

Fig. 13.19   Exhaust gas recirculation (EGR) system vacuum only operated

679

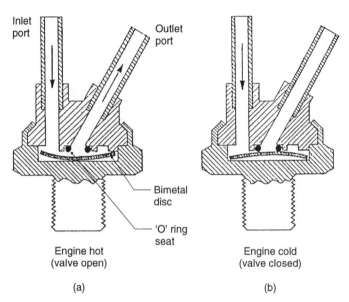

Inlet port

Outlet port

Bimetal disc

'O' ring seat

Engine hot
(valve open)

(a)

Engine cold
(valve closed)

(b)

Fig. 13.20 Thermal vacuum (bimetal disc) valve

A delay in vacuum signal transfer from the throttle valve vacuum pick-up to the EGR valve diaphragm chamber is achieved by inserting a restrictor (calibrated orifice) in the connecting line between the throttle valve vacuum pick-up and the diaphragm chamber (Fig. 13.19(a)). This restrictor slows down the rate at which vacuum is conveyed from the throttle pick-up to the diaphragm chamber and also retards the rate at which the vacuum in the diaphragm chamber escapes when the pick-up vacuum signal is reduced.

An extended response time for the exhaust gas recirculation cut-out valve to open or close is achieved by incorporating a plastic reservoir between the restrictor and exhaust gas recirculation valve so that the effective volume of the internal pipeline passage and diaphragm chamber space is enlarged (Fig. 13.19(a)). As a result, more air will have to pass through the restrictor before the exhaust gas recirculation valve commences either to open or close. Thus, the response time for the valve to operate is prolonged. A delayed action period is sometimes referred to as 'damping the valve's response'.

13.5 Air injected exhaust systems

13.5.1 The need for air injection

To reduce the exhaust emission of carbon monoxide and hydrocarbons, compressed air supplied from an engine-driven vane-type air-pump is sometimes injected into each exhaust port slightly downstream of the exhaust valve or valves. The injected air mixes with the very hot, burnt and partially burnt exhaust gases passing out of the exhaust ports, and therefore promotes further oxidation (combustion) of the partially burnt carbon monoxide and hydrocarbon; thereby considerably reducing these unwanted pollutants.

The effectiveness of an air injection system depends upon the following conditions:

1 the strength of the air–fuel mixture. Generally a marginally rich mixture improves the exhaust gas–air injected mixture oxidation reaction process;
2 the temperature and pressure of the exhaust gas–air mixture;
3 the time for the exhaust gas–air mixture reaction;
4 the heat losses from the exhaust reaction zone, these should be minimized for efficient conversion;
5 an adequate reaction volume to enable the

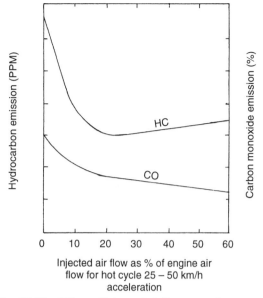

Fig. 13.21 Effect of injected air flow on carbon monoxide and hydrocarbon emission

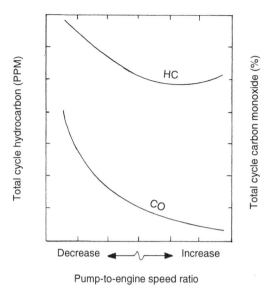

Fig. 13.22 Effect of air pump drive ratio on engine exhaust gas emission

exhaust gas–air mixture to complete the continuous oxidation process.

13.5.2 The effects of air-injection on carbon monoxide and hydrocarbon emission
(Figs 13.21, 13.22 and 13.23)

The air injected flow-rate, relative to the engine air consumption, must be optimized to produce the minimum combination of carbon monoxide and hydrocarbon emissions. As can be seen in Fig. 13.21, this occurs with roughly 20% injected air flow, as a percentage of the engine's air consumption for hydrocarbon. However, increasing the injected air flow-rate further would still reduce the carbon monoxide emission but at the expense of a slight rise in the hydrocarbon emission. Consequently, a compromise is normally arrived at to minimize the overall emission effects. The quantity of air supplied for a given size of air-pump is dependent upon the air-pump speed which, accordingly, is related to the air-pump to engine-pulley speed ratio. Figure 13.22 shows that with an increased speed ratio, both the carbon monoxide and hydrocarbon emissions reduce, but the hydrocarbons start to rise. The reason for this is most likely due to the increased cooling effect of the greater quantity of air sup-

plied, causing a reduction in the hydrocarbon oxidation reaction rate. However, the carbon monoxide emission continues to decrease with a higher step-up engine-crankshaft to air-pump speed ratio. Thus, there cannot be an optimum speed ratio and a compromise ratio is usually chosen which puts neither the carbon monoxide

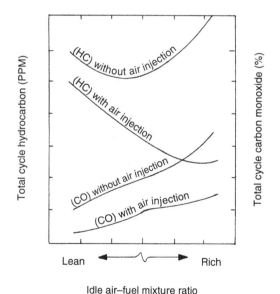

Fig. 13.23 Effect of idle air–fuel mixture ratio on carbon monoxide and hydrocarbon exhaust emission without and with air injection

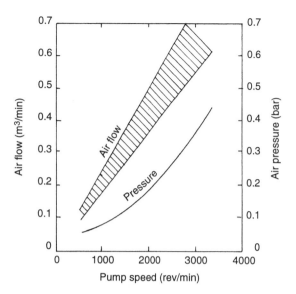

Fig. 13.24 Relationship between air delivery and air pressure relative to pump speed

reduction in both carbon monoxide and hydrocarbon emission when air injection is introduced into the exhaust system (Fig. 13.23). It can be seen that the reduction in hydrocarbon is considerable as the air–fuel mixture is enriched. However, although there is a great improvement in carbon monoxide emission with air injection, the trend for engines with and without air injection is of increasing emissions as the air–fuel mixture is enriched (Fig. 13.23). Thus, there is a conflict of interest with carbon monoxide and hydrocarbon emission, and it can be seen that lean mixtures benefit carbon monoxide emission, whereas rich mixtures benefit the hydrocarbon emissions. Again, a compromise is reached with the mixture strength usually set slightly to the rich side of the stoichiometric air–fuel ratio.

nor the hydrocarbon emissions at a great disadvantage.

A comparison of carbon monoxide and hydrocarbon emission with and without air injection, when varying the air–fuel mixture strength during idling and low engine speeds shows a marked

13.5.3 Air-pump supply
(*Figs 13.24 and 13.25*)

The delivery and pressure increase characteristics for a typical air-pump are shown in Fig. 13.24. Here, it can be seen that with rising pump speed the delivery discharge increases directly with a widening tolerance band. In contrast, the press-

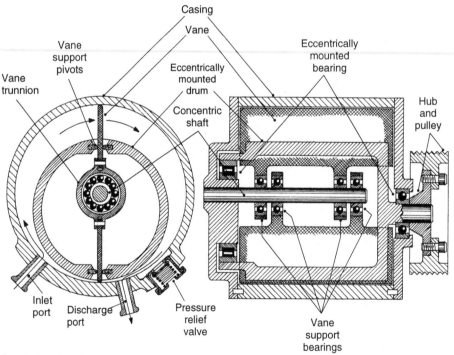

Fig. 13.25 Semi-articulated vane type air pump

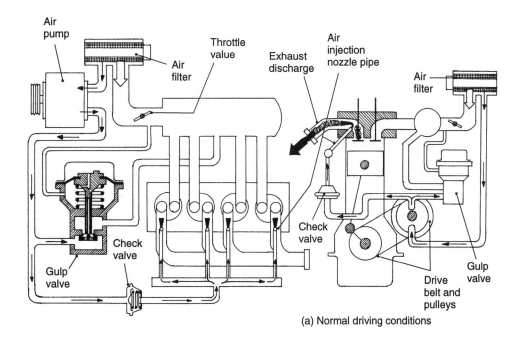

(a) Normal driving conditions

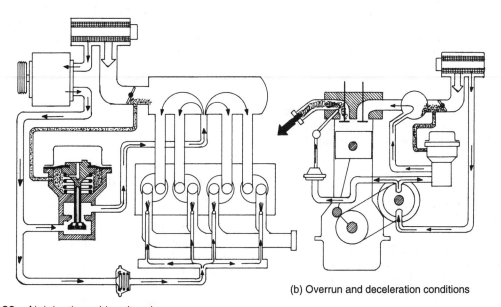

(b) Overrun and deceleration conditions

Fig. 13.26 Air injection with gulp valve

ure build up increases roughly with the square of the rotational speed.

The most suitable type of air-pump used for injection into the exhaust system is the semi-articulated vane rotary pump (Fig. 13.25). This pump has an outer aluminium cylindrical casing, a central mounted pair of vanes supported on a concentric shaft and an eccentrically mounted aluminium drum. When the drum is rotated, the centrally mounted vanes will appear to slide in and out from the vane support pivots located in the drum, without the vane tips actually touching the internal walls of the casing. This construction results in a semi-crescent shaped space being formed on either side of the vanes between the casing and eccentrically mounted drum. Rotating the drum causes the crescent space on the left-hand side of the drum initially to increase, thus drawing in atmospheric air. As the lower vane passes the inlet port the induced air will be trapped so that, with further drum rotation, the crescent space decreases and thereby pressurizes the air, which is then expelled out of the discharge port on the right-hand side of the drum. This continuous process of induction compression and discharge, eventually builds up to a relatively steady delivery of air to the exhaust system. The amount of air being discharged per unit time is proportional to the engine crankshaft speed and, correspondingly, the air pump drum rotation.

Relief valve
(Fig. 13.25)
Air pump relief valves are necessary to regulate the maximum quantity of air supply and the corresponding pressure rise. The controlling effects of the relief valve are as follows:

1 to limit the exhaust manifold system's maximum internal wall temperature particularly at high engine speed;
2 to minimize power consumption in driving the air pump;
3 to limit the internal work done by the pump;
4 to limit the exhaust system's back pressure caused by the gas volume and temperature increase.

13.5.4 Anti-backfire prevention
(Figs 13.26 and 13.27)

Objectionable and damaging exhaust backfiring may occur during the transition period when the driving mode changes from steady normal to overrun and deceleration conditions. Backfiring in the exhaust system is generally caused by the following.

1 A momentary enrichment of the air–fuel mixture reaching the cylinders, due to the sudden reduction in throttle opening, causes the increased manifold vacuum to evaporate the fuel from the internal walls of the manifold.
2 A momentary enrichment of the air–fuel mixture reaching the cylinders due to a delay in the

fuel flow reduction from the carburettor or petrol injection system as the throttle is shut down.

This brief enrichment period during the transition from steady driving to deceleration conditions produces incomplete combustion in the combustion chamber which is then expelled into the exhaust port and manifold system. This very hot, partially burnt (and burnt) charge then enters the exhaust system where it mixes with the injected air causing further after-burning, which results in backfiring in the exhaust system.

The elimination of this exhaust backfiring when the throttle is suddenly closed can be achieved by two methods: (1) supplying additional air to the induction manifold to maintain a nearly stoichiometric air–fuel ratio; (2) cutting off the air-supply momentarily to the exhaust system.

An anti-backfire valve senses the sudden change from some intermediate value of intake depression to the very high depression that occurs during deceleration, it then diverts the air-pump flow momentarily to either:

1 the induction manifold where the air is then gulped (sucked) into the cylinders (this anti-backfire valve is therefore known as a gulp-valve (Fig. 13.26))
2 the atmosphere via the air cleaner so that it bypasses the exhaust system (this anti-backfire valve is therefore known as a bypass valve (Fig. 13.27)).

13.5.5 Air injected exhaust system with a gulp-valve device
(Fig. 13.26(a and b))

This layout utilizes a gulp-valve device to introduce, momentarily, additional air into the induction manifold whenever the throttle is suddenly closed during engine deceleration, in order to compensate for an over enrichment tendency of the air–fuel mixture during this transitional period.

Under normal steady driving conditions (Fig. 13.26(a)), the air-pump generates a supply of pressurized air, which is delivered to each exhaust port slightly downstream of the exhaust valve or valves, via a non-return check valve and a common air delivery.

When the engine decelerates fairly fast, the throttle valve tends (partially or almost) to close,

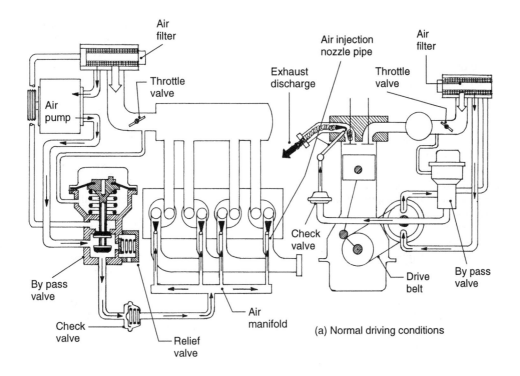

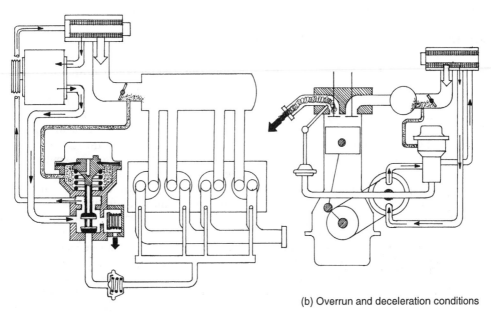

Fig. 13.27 Air injection with by-pass valve

this causes a high depression to occur within the induction manifold (Fig. 13.26(b)). A high vacuum signal is immediately produced underneath the gulp-valve diaphragm, causing the diaphragm assembly to move downwards against the

resistance of the return-spring: it therefore opens the conical valve. The very high induction manifold vacuum will momentarily divert a large proportion of the compressed air supply through the open gulp-valve to some central point in the induction manifold, where it is immediately gulped up by the individual cylinders during their respective induction strokes. As a result, the excess air counteracts and corrects the momentary over-enrichment of the air-fuel mixture when the throttle has been rapidly moved to the closed or nearly closed position. Subsequently, it eliminates any backfiring tendency which might otherwise occur while the engine is on overrun.

After a short time interval (2 to 4 seconds) the vacuum is transferred from below the diaphragm to the upper diaphragm face via the restriction passage in the central diaphragm stem. Accordingly, since both sides of the diaphragm are subjected to similar intensities of vacuum, the return-spring has no difficulty in moving the diaphragm assembly to its upper position, thereby closing the conical gulp-valve. At this instant, air will again be delivered and injected into the exhaust port and manifold system. If backfire should occur in the exhaust system for any reason, the non-return check valve will automatically close. This prevents instantaneous pressure waves interfering and damaging the air-pump system.

13.5.6 Air injected exhaust system with a bypass-valve device
(Fig. 13.27(a and b))

Air injection exhaust systems provide a variable-output engine-driven air-pump, to discharge air continuously into each exhaust port through a nozzle pipe situated downstream of the exhaust valve (or valves) whenever the engine is operating under steady driving conditions (Fig. 13.27(a)).

However, during transition conditions, such as on engine overrun and deceleration, there is a tendency for the air–fuel mixture to become momentarily over-rich. As a result, unburnt exhaust products passing out of the exhaust ports mix with the injected air and oxidize, and some of these products are active in producing unpredictable exhaust backfire. To overcome deceleration backfire, this system incorporates a bypass valve which momentarily diverts the compressed air from the exhaust ports directly back to the intake air cleaner.

When the air intake throttle is rapidly closed to reduce engine speed and, correspondingly, vehicle speed; a high vacuum on the engine side of the throttle will be created. This momentary large vacuum will be felt below the bypass diaphragm, causing it to move downwards against the tension of the return-spring until the lower valve shuts off the air supply to the exhaust nozzle pipes (Fig. 13.27(b)). At the same time, the upper valve opens permitting the compressed air to bypass back to the intake air cleaner where it is again drawn into the air pump. The purpose of the return system is to produce a closed loop system which prevents noisy air exhausting directly into the atmosphere.

Shortly after the diaphragm is pulled downwards, a vacuum will also enter the upper diaphragm chamber via the equalizing passage in the diaphragm stem so that both upper and lower diaphragm chambers will be subjected to the same amount of vacuum. Accordingly, the diaphragm assembly will again be able to move to its highest position due to the upthrust from the return spring. Therefore, the air delivery from the pump will again pass directly to the exhaust port nozzle pipes to assist the exhaust system's oxidation process of the unburnt exhaust gases.

13.5.7 Air induction exhaust system with a bypass valve
(Fig. 13.28(a and b))

Carbon monoxide and hydrocarbon exhaust emission can be reduced substantially by utilizing exhaust gas pulses to induce fresh air to enter each exhaust port just downstream of the exhaust valves (Fig. 13.28(a and b)). The exhaust ports act as venturis to the periodically ejected exhaust gases so that they rush through the ports at very high speeds. This periodic expulsion of the exhaust gases from the cylinders due to the valves' opening and closing causes a pulsating exhaust pressure which alternates above and below atmospheric pressure. Consequently, the air supply nozzles protruding into the exhaust ports are subjected to a pulsating vacuum, which thereby induces air from the air filter via the bypass valve and the twin check valves to enter the exhaust ports. The induced air entering the ports mixes with the hot gases, thus causing further oxidation of unburnt products of combustion being expelled from the cylinders.

By having twin check-valves, one feeding the two inner exhaust ports and the other feeding the outer exhaust ports, air will be pulsed into each exhaust port in the normal firing order sequence from each check-valve alternately.

During normal driving conditions, air will be

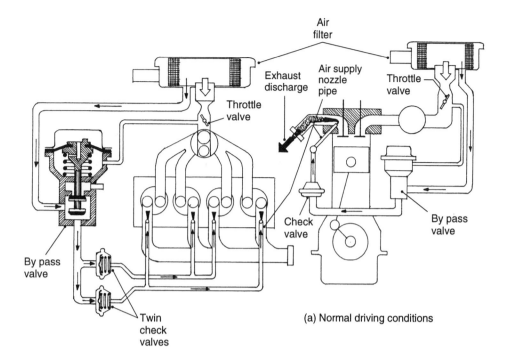

(a) Normal driving conditions

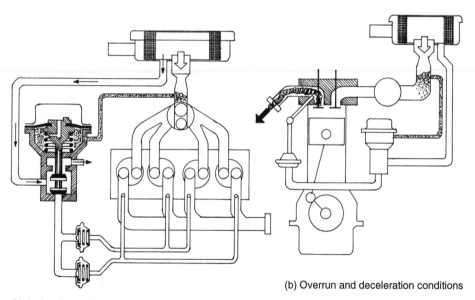

(b) Overrun and deceleration conditions

Fig. 13.28 Air induction exhaust system with by-pass valve

continuously pulsed into the exhaust ports (Fig. 13.28(a)). However, during transient driving when the engine is rapidly decelerated, a high vacuum on the engine side of the throttle signals the bypass valve diaphragm to close the valve momentarily (Fig. 13.28(b)).

This is done to prevent the combustion products from the momentarily enriched mixture oxidizing and causing uncontrollable burning, which would result in exhaust backfiring. Within about two to three seconds after closing, the bypass valve will reopen, thus permitting air to be

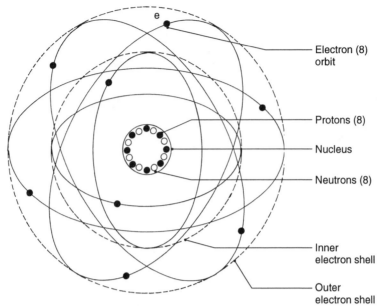

e

Electron (8) orbit

Protons (8)

Nucleus

Neutrons (8)

Inner electron shell

Outer electron shell

Fig. 13.29   Three-dimensional model of an oxygen atom

drawn again into the exhaust ports to facilitate exhaust afterburn treatment.

The air supply to the exhaust ports by the induction exhaust gas pulsation method is effective provided that the amounts of air required are relatively low. However, if the air demands are greater, then the more expensive air-pump injection system would be preferred.

13.6   Combustion of air–fuel mixtures

Before studying the combustion equations it will be useful to have a basic insight into the structure of the atom and how the mass of a substance is determined.

13.6.1   Atomic structure
(Figs 13.29 and 13.30)

Atoms are formed from three small particles known as protons, electrons and neutrons. The proton is a positively charged particle of mass approximately equal to that of a hydrogen atom. The electron is negatively charged, its charge being equal but opposite to the charge of a proton. It has a very small mass about 1/1836 of the mass of the proton. The neutron has no

charge and its mass is about equal to the mass of a proton.

In the centre of the atom is a small heavy nucleus consisting of protons and neutrons and, generally, there are as many neutrons in the nucleus as there are protons. This nucleus of protons and neutrons is held tightly together.

Orbiting around the nucleus are one or more electrons (Fig. 13.29), but nearly all the mass of the atom is concentrated in the small nucleus. Each atom is electrically neutral and the positive charges of the protons are exactly counterbalanced by the negative charges of the electrons—that is, there are as many rotating electrons as there are protons in the nucleus.

The orbiting electrons can be considered to revolve in concentric rings or shells about the nucleus (Fig. 13.30). Electrons in each shell are associated with different energies, and therefore each shell has a different energy level. The shell nearest to the nucleus has the lowest energy level and electrons fill this shell first. The closer the electron shell is to the nucleus, the more tightly its electrons are held in their orbit, and therefore they are more difficult to remove. The outer shell has valency electrons which are weakly held in their orbit and may easily be dislodged.

Atoms are held together by forces which are called bonds, these bonds are produced by the outer-shell electrons of individual atoms being

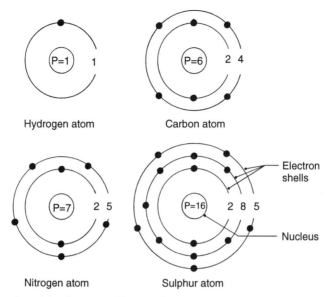

Fig. 13.30 Two-dimensional models of atoms taking part in the combustion process

attracted to each other so that they form a more stable configuration of electrons. The ease of making and breaking bonds is known as *chemical reactivity*. The most stable electronic configuration of all is the noble-gas structure with its outer shell of eight electrons. Atoms of this character are helium, neon, argon, krypton, xenon and radon. They do not combine and are therefore able to form inert gases. Other atoms attempt to achieve a stable electronic structure (outer shell with eight electrons) by either electron transfer or electron sharing.

13.6.2 A covalently bonded carbon dioxide molecule
(Fig. 13.31)

The attachment together of two or more atoms by shared pairs of electrons is known as *covalent bonding*. Here, electrons are not actually gained or lost by the atoms concerned. In a carbon dioxide (CO_2) molecule the carbon atom is bonded to each oxygen atom by two pairs of electrons in covalency. The molecule electron configuration encourages mutual repulsion, and this therefore produces a linear molecule.

Oxygen atoms have six electrons in their outer

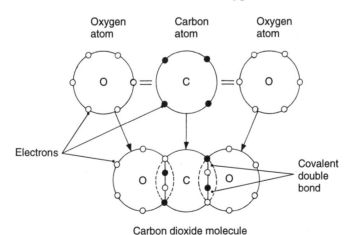

Carbon dioxide molecule

Fig. 13.31 Formation of a carbon dioxide molecule

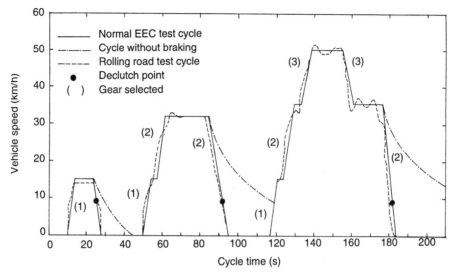

Fig. 13.32 EEC 15-cycle driving pattern

shell whereas a carbon atom has an outer shell of only four electrons. In order to achieve a stable octet (eight) electron outer shell, two pairs of electrons are shared by two atoms, it is called double bonding (Fig. 13.31). The bonding of the carbon and oxygen atoms can be represented in the following way:

Double bonds can be represented by two straight lines; thus, a carbon dioxide molecule is often shown as $O{=}C{=}O$, where each line represents a pair of shared electrons.

13.6.3 Atomic theory definitions

Matter
This is anything which occupies space and possesses weight. The amount of matter contained in any object is known as its mass.

Atoms
This is the smallest possible particle of an element that can exist.

Element
This is a substance which cannot be split up into anything simpler by a chemical change. It consists entirely of atoms of the same atomic number.

Isotopes
These are elements which have the same number of protons but differ in the number of neutrons.

Compounds
These are substances which contain two or more elements combined in such a way that the properties of the individual elements are completely lost.

Molecules
This is the smallest part of a substance, whether element or compound, which is capable of existing in the free state.

Mixture
A mixture contains two or more different substances, either an element or compound, which are not chemically joined together.

Standard mass
The standard mass is the mass of a proton (1 unit), and the mass of the neutron and the electron can be compared with this standard.

Atomic number (Z)
The number of protons in the atomic nucleus is called the atomic number and is denoted by the symbol (Z).

Mass number (A)
The sum of the protons and neutrons in the

690

nucleus is called the mass number and is denoted by the symbol (A).

Relative atomic mass

Atomic masses are compared with the mass of a standard atom. The carbon 12 atom which has a mass of 12 units is used as the standard for comparison. The atomic masses of all elements are determined relative to one-twelfth the mass of the carbon 12 isotope. This is termed the *relative atomic mass*, and was originally known as *atomic weight*.

Table 1 Atomic numbers and relative atomic masses for the elements most commonly used to solve the combustion equations

Element	Symbol	Atomic number	Relative atomic mass
Hydrogen	H	1	1
Carbon	C	6	12
Nitrogen	N	7	14
Oxygen	O	8	16
Sulphur	S	16	32

When the molecule is made up of atoms of different elements, the relative molecular mass can be calculated by adding the relative atomic mass of the elements concerned.

For example one molecule of carbon dioxide (CO_2) consists of one atom of carbon plus two atoms of oxygen.

Therefore, the relative molecular mass of carbon dioxide is

$$CO_2 = 12 + 2 \times 16 = 44$$

The relative molecular mass of carbon monoxide is

$$CO = 12 + 16 = 28$$

The relative molecular mass of steam is

$$H_2O = 2 \times 1 + 16 = 18$$

The relative molecular mass of sulphur dioxide is

$$SO_2 = 32 + 2 \times 16 = 64$$

The relative molecular mass of methane is

$$CH_4 = 12 + 4 \times 1 = 16$$

The relative molecular mass of heptane is

$$C_7H_{16} = 7 \times 12 + 16 \times 1 = 100$$

The relative molecular mass of octane is

$$C_8H_{18} = 8 \times 12 + 18 \times 1 = 114$$

13.6.4 Fuels and combustion

Combustion is a chemical reaction in which a substance reacts rapidly with oxygen to produce heat and light. Such reactions are free radial chain reactions in which the oxidation of carbon (to form its oxides) and the oxidation of hydrogen (to form water) occur.

For combustion to be complete it is assumed that the reaction between the carbon and hydrogen of the fuel, and oxygen of the air, produce, respectively, carbon dioxide (CO_2) and water vapour (H_2O).

This oxidation produces two simple equations

$$C + O_2 \rightleftharpoons CO_2 \tag{1}$$
$$2H_2 + O_2 \rightleftharpoons 2H_2O \tag{2}$$

If there is sulphur in the fuel it will oxidize to sulphur dioxide

$$S + O_2 \rightleftharpoons SO_2 \tag{3}$$

In practice, some of the exhaust gases contain products of incomplete combustion such as carbon monoxide (CO), unburnt hydrocarbons (HC) and the oxidation products of nitrogen such as nitrogen dioxide (NO_2).

This further oxidation produces two more equations

$$CO + O_2 \rightleftharpoons 2CO_2 \tag{4}$$
$$N + O_2 \rightleftharpoons NO_2 \tag{5}$$

Burning carbon to carbon dioxide
When carbon is burnt with an ample supply of oxygen

$$C + O_2 \rightleftharpoons CO_2$$

This equation implies that 1 molecule of carbon combines with 1 molecule of oxygen and produces 1 molecule of carbon dioxide.

Putting in the relative atomic masses

$$\begin{matrix} 12 \text{ units by mass} \\ \text{of carbon} \end{matrix} + \begin{matrix} 32 \text{ units by mass} \\ \text{of oxygen} \end{matrix}$$

$$\rightleftharpoons \begin{matrix} 12 + (2 \times 16) \text{ units by mass} \\ \text{of carbon dioxide} \end{matrix}$$

dividing through by 12

$$\frac{12}{12} \text{ kg of C} + \frac{32}{12} \text{ kg of O}_2 \rightleftharpoons \frac{44}{12} \text{ kg of CO}_2$$

therefore

$$1 \text{ kg of C} + \frac{8}{3} \text{ kg of O}_2 \rightleftharpoons \frac{11}{3} \text{ kg of CO}_2$$

that is,

1 kg of C needs $\frac{8}{3}$ kg of O_2 to produce $\frac{11}{3}$ kg of CO_2.

Incomplete burning of carbon to carbon monoxide
When carbon is burnt with a deficiency of oxygen

$$C + O_2 \rightleftharpoons CO$$

balance the equation by multiplying carbon by 2 and carbon monoxide by 2.
  Putting in the relative atomic mass

$$2 \times 12 + 2 \times 16 \rightleftharpoons 2(12 + 16)$$

dividing through by 24

$$\frac{24}{24} \text{ kg of C} + \frac{32}{24} \text{ kg of O}_2 \rightleftharpoons \frac{56}{24} \text{ kg of CO}$$

therefore

$$1 \text{ kg of C} + \frac{4}{3} \text{ kg of O}_2 \rightleftharpoons \frac{7}{3} \text{ kg of CO}$$

that is,

1 kg of C needs $\frac{4}{3}$ kg of O_2 to produce $\frac{7}{3}$ kg of CO.

Burning carbon monoxide to carbon dioxide
When CO is exposed to additional oxygen

$$CO + O \rightleftharpoons CO_2$$

balance the equation by multiplying carbon monoxide by 2 and carbon dioxide by 2.
  Putting in the relative atomic mass

$$2(12 + 16) + 2 \times 16 \rightleftharpoons 2(12 + 2 \times 16)$$

dividing through by 56

$$\frac{56}{56} \text{ kg of CO} + \frac{32}{52} \text{ kg of O}_2 \rightleftharpoons \frac{88}{56} \text{ kg of CO}_2$$

therefore

$$1 \text{ kg of CO} + \frac{4}{7} \text{ kg of O}_2 \rightleftharpoons \frac{11}{7} \text{ kg of CO}_2$$

that is,

1 kg of CO needs $\frac{4}{7}$ kg of O_2 to produce $\frac{11}{7}$ kg of CO_2.

Burning hydrogen to steam
When hydrogen is burnt with a sufficient supply of oxygen

$$H_2 + O \rightleftharpoons H_2O$$

balance the equation by multiplying hydrogen by 2 and water (H_2O) by 2.
  Putting in the relative atomic mass

$$2(2 \times 1) + 2 \times 16 \rightleftharpoons 2(2 \times 1 + 16)$$

dividing through by 4

$$\frac{4}{4} \text{ kg of H}_2 + \frac{32}{4} \text{ kg of O}_2 \rightleftharpoons \frac{36}{4} \text{ kg of H}_2O$$

therefore

$$1 \text{ kg of H}_2 + 8 \text{ kg of O}_2 \rightleftharpoons 9 \text{ kg of H}_2O$$

that is,

1 kg of H_2 needs 8 kg of O_2 to produce 9 kg of H_2O.

Burning sulphur to sulphur dioxide
When sulphur is burnt with a sufficient supply of oxygen

$$S + O_2 \rightleftharpoons SO_2$$

Putting in the relative atomic mass

$$32 + 2 \times 16 \rightleftharpoons 32 + (2 \times 16)$$

dividing through by 32

$$\frac{32}{32} \text{ kg of S} + \frac{32}{32} \text{ kg of O}_2 \rightleftharpoons \frac{64}{32} \text{ kg of SO}_2$$

therefore,

1 kg of S needs 1 kg of O_2 to produce 2 kg of SO_2.

Burning nitrogen to nitrogen dioxide
When nitrogen is burnt with a sufficient supply of oxygen

$$N + O_2 \rightleftharpoons NO_2$$

Putting in the relative atomic mass

$$14 + 2 \times 16 \rightleftharpoons 14 + (2 \times 16)$$

dividing through by 14

$$\frac{14}{14} \text{ kg of N} + \frac{32}{14} \text{ kg of O}_2 \rightleftharpoons \frac{46}{14} \text{ kg of NO}_2$$

therefore

$$1 \text{ kg of N} + \frac{16}{7} \text{ kg of O}_2 \rightleftharpoons \frac{23}{7} \text{ kg of NO}_2$$

that is,

1 kg of N needs $2\frac{2}{7}$ kg of O_2 to produce $3\frac{2}{7}$ kg of NO_2.

13.6.5 Theoretical air supply for complete combustion of fuel

The burning of the individual elements of a fuel can be summarized as follows:

$$1 \text{ kg of C} + \frac{8}{3} \text{ kg of O}_2 \rightleftharpoons \frac{11}{3} \text{ kg of CO}_2$$

therefore, carbon requires $\frac{8}{3}$ times its own mass of oxygen to burn to CO_2.

$$1 \text{ kg of H}_2 + 8 \text{ kg of O}_2 \rightleftharpoons 9 \text{ kg of H}_2\text{O}$$

therefore, hydrogen requires 8 times its own mass of oxygen to burn to H_2O.

$$1 \text{ kg of S} + 1 \text{ kg of O}_2 \rightleftharpoons 2 \text{ kg of SO}_2$$

therefore, sulphur requires its own mass of oxygen to burn to SO_2.

Thus, the complete burning of 1 kg of fuel is composed of C kg of carbon, H kg of hydrogen, S kg of sulphur and O kg of oxygen.

Hence, oxygen required to burn C kg of carbon = $\frac{8}{3}$C

Hence, oxygen required to burn H kg of carbon = 8H

Hence, oxygen required to burn S kg of carbon = 1S

Therefore, total oxygen required = $\frac{8}{3}$C + 8H + S

If the fuel already contains O kg of oxygen, then the theoretical oxygen required for complete combustion of 1 kg of fuel is expressed as

$$\text{oxygen supply } (O_S) = \frac{8}{3}\text{C} + 8\text{H} + \text{S} - \text{O}$$

Air contains 23% by mass of oxygen and approximately 77% of nitrogen ignoring the very small quantities of other elements. Therefore, the theoretical air required for the complete combus-

tion of 1 kg of fuel may be expressed in terms of the air–fuel ratio

$$\text{A/F} = \frac{100}{23}\left(\frac{8}{3}\text{C} + 8\text{H} + \text{S} - \text{O}\right)$$

Example
Calculate the theoretical air–fuel ratio of a petrol fuel containing 86% carbon and 14% hydrogen. Any sulphur and free oxygen may be neglected.

$$\text{A/F} = \frac{100}{23}\left(\frac{8}{3}\text{C} + 8\text{H} + \text{S} - \text{O}\right)$$

$$= \frac{100}{23}\left(\frac{8 \times 0.86}{3} + 8 \times 0.14\right) = \frac{100}{23}(2.2933 + 1.12$$

$$= \frac{100 \times 3.4133}{23}$$

$$= 14.84 : 1$$

Example
Determine the theoretical air–fuel ratio of an octane C_8H_{18} fuel.

Molecular mass of $C_8 = 8 \times 12 = 96$
Molecular mass of $H_{18} = 18 \times 1 = 18$
Molecular mass of $C_8H_{18} = 96 + 18 = 114$

Fraction of carbon by mass $= \dfrac{96}{114} = 0.8421$

Fraction of hydrogen by mass $= \dfrac{18}{114} = 0.1579$

$$\text{A/F} = \frac{100}{23}\left(\frac{8}{3}\text{C} + 8\text{H}\right)$$

$$= \frac{100}{23}\left(\frac{8 \times 0.8421}{3} + 8 \times 0.1579\right)$$

$$= \frac{100}{23}(2.2456 + 1.2632)$$

$$= 15.255 : 1$$

13.6.6 Exhaust gas emissions

Carbon dioxide (CO₂)
This is a colourless gas with a faint pungent odour and taste but it is not poisonous.

The combustion of air and fuel produces carbon dioxide (CO_2), which then rises to the upper atmosphere and thereby forms a layer of carbon dioxide around the earth's atmosphere.

Plants and animals consume carbohydrates, which contain carbon and water, and other foodstuff and oxygen from the air. Some of this is used to replace wastage and maintain growth of the plant or animal whilst the remainder combines to form carbon dioxide (CO_2) and is breathed into the air where it then rises to the upper atmosphere in a similar manner to the carbon dioxide (CO_2) released by combustion.

Conversely, green plants and trees absorb carbon dioxide through the underside of their leaves, when exposed to solar energy; that is, sunlight. The absorbed carbon dioxide (CO_2) combines with water in the leaf to produce carbohydrates, which are compounds containing carbon, water and oxygen, and which are then released into the atmosphere. This process, where carbon dioxide is removed from the air and reacts with water to produce (1) carbohydrates in the leaf and (2) oxygen, which is released to the atmosphere, is known as *photosynthesis*.

Thus, the plant life of the world is active in restoring the balance between the amount of carbon dioxide produced from plants, animals and the burning of fossil fuels in power stations and internal combustion engines, and photosynthesis. Photosynthesis thus absorbs carbon dioxide and converts it into carbohydrates (starch and sugar), which are contained in the plant, and oxygen, which is released into the atmosphere.

However, the extensive burning of fossil fuels in power stations, factories and the internal combustion engine has increased the carbon dioxide concentration in the atmosphere beyond that which can be absorbed by plant life, so that the upper carbon dioxide atmospheric layer is becoming thicker.

As a result, heat radiated from the ground cannot penetrate this carbon dioxide (CO_2) gas rim around the earth so that it raises the average temperature of the earth. Consequently, this global warming—the greenhouse effect—is causing the polar caps to melt and therefore, it is claimed, the levels of the oceans are rising.

Carbon monoxide (CO)

This is a colourless, odourless and tasteless gas which is poisonous when inhaled.

If inhaled into the lungs it combines with the blood and prevents the blood absorbing oxygen.

Low concentrations of carbon monoxide cause headaches and slows down mental and physical activity, whereas high concentrations cause unconsciousness and death.

When in fresh air, the human body is able to purge itself of carbon monoxide provided that the exposure to CO has not been excessive.

Hydrocarbons (HC)

Exhaust gases contain many kinds of hydrocarbon compounds, which are generally harmless but combustible. However, some hydrocarbons are known carcinogens; that is, they are cancer-producing. In addition, some hydrocarbons tend to irritate the eye and throat mucous membranes. Hydrocarbons contribute to the formation of acid rain and some hydrocarbon compounds react with ultra-violet light, which encourages the formation of photo-chemical smog.

Oxides of nitrogen (NO_x)

At extremely high combustion temperatures, some of the burning nitrogen and oxygen combines to form oxides of nitrogen. Oxides of nitrogen can exist as nitrogen monoxide (NO) which is a colourless, odourless and tasteless gas. With further oxygen exposure it will readily change to nitrogen dioxide (NO_2), which has a reddish-brown colour. This gas is poisonous with a penetrating odour which can destroy lung tissue. Both of these gases exist together in a combined form so that these oxides of nitrogen are given a chemical formula of NO_x, where the suffix 'x' ranges between one and two atoms. Oxides of nitrogen combine with water to form acid rain, which is a very dilute nitrous acid and which is particularly harmful to the environment. Under certain conditions, such as altitude, strong sunlight and high humidity, oxides of nitrogen and hydrocarbon combine to form smog. Lean burn engines tend to operate at higher temperatures and so combine the nitrogen present in the air with any available oxygen to form relatively large amounts of nitrogen oxides, and thus more acid rain in the atmosphere.

Lead (Pb)

Lead is a poisonous heavy metal which is neuro toxic, causing it to affect the brain and nervous system. Lead is associated with nervous and mental disorders and is harmful to plants and animals.

The body absorbs lead in two ways:

1 through the lungs during breathing
2 through the stomach in food contaminated by lead and also via the atmosphere.

Lead is an anti-knock additive which is introduced into the fuel in the form of tetraethyl lead (TEL) or tetramethyl (TML) with the addition of a scavenger in the form of bromide. About 10% of the lead is emitted as particles of less than 25×10^{-6} mm diameter, which are likely to remain airborn. Another 25% of the lead becomes attached to the lubrication oil or remains in the exhaust system. The remainder of the lead tends to separate rapidly from the air and settles.

13.6.7 Engine exhaust emission regulation tests

European regulations governing exhaust emission for motor vehicles come under two equivalent directives:

Non-EEC countries except Sweden	EEC countries
Original regulation ECE R15	70/220/EEC
Original regulation ECE R15.01	74/290/EEC
Original regulation ECE R15.02	77/102/EEC
Original regulation ECE R15.03	78/665/EEC
Original regulation ECE R15.04	83/351/EEC
Regulation F23 Regulation F40	Sweden only

ECE R15, R15.01, etc are regulations approved by the Economic Commission of Europe.

70/220/EEC, 74/290/EEC, etc are the equivalent regulations adopted by the European Economic Community. The EEC regulations are based on ECE R15 while the Swedish regulation F40 is based on the 1973 USA requirements.

With the European regulation ECE R15.04 the European driving cycle emissions are quoted in grams per test and are measured on the 15-cycle mode (see Table 2).

Table 2 ECE R15.04 European driving cycle emissions in grams per test

Constituents	g/test
Carbon monoxide	58
Hydrocarbons	19
Nitrogen oxides	(HC + NO$_x$)

The limits established from 1 April 1991 (88/77/EEC) will be quoted in grams per kilowatt-hour and are measured on the 15-cycle mode (see Table 3).

Table 3 88/77/EEC European driving cycle emissions in grams per kilowatt per hour (Fig. 13.32)

Constituents	g/kWh
Carbon monoxide	11.2
Hydrocarbons	2.4
Nitrogen oxides	14.4

The three basic tests necessary to meet the engine emission regulations in Europe are as follows.

1 Driving cycle exhaust emission test (Fig. 13.32)
A predetermined drive cycle known as the 15-mode cycle has been constructed to simulate a typical driving pattern when pulling away from a standstill and driving through the first three gear ratios (Fig. 13.32). However, the actual driving cycle testing is carried out on a rolling road machine, and the exhaust gases emitted from the vehicle during the test are collected in three separate bags—one for each main stage of the test.

The first part of the test is a cold transient stage which represents the beginning of a journey starting from cold, it consists of cycles 1 to 5 of the test at which point the collected gas is analysed.

The second part of the test consists of completing the journey with a further 13 cycles of the driving sequence with the engine now warmed so that the cylinders reach peak working temperatures at some time in the individual cycles. Again the collected exhaust gas composition is analysed.

The third part of the test, known as the hot transient stage, takes place after the engine has stopped for 10 minutes to produce a hot soak, it is then restarted and the first five cycles are repeated again. This final stage symbolizes a hot transient stage with a thoroughly warmed engine so that the collected exhaust gas composition can be compared with the initial test stage.

2 Idle exhaust emission check
The level of exhaust emission at idle speed with the engine at normal operating temperature is measured with an exhaust gas analyser.

3 Crankcase emission check
With this test, crankcase emissions are collected from the crankcase ventilation system under pre-

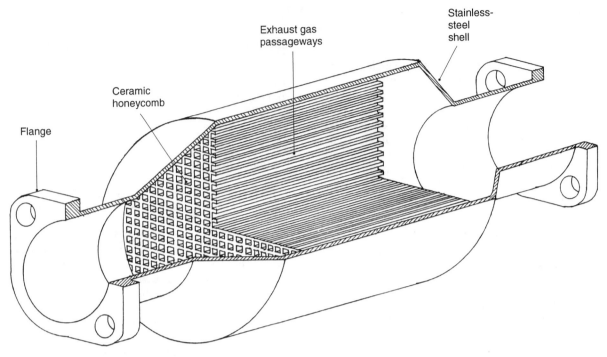

Fig. 13.33   Sectioned view of a honeycomb type catalytic converter

determined controlled conditions of engine speed and load.

Recent legislation proposals for exhaust emission limitations have included particulate emissions. The limits for the individual pollutants are to be phased in and are shown in Tables 4 and 5.

Stage I
Proposed emission legislation was implemented on 1 July 1992 for new models (Type Approved) and 10 October 1993 for new registrations.

Table 4   July 1992 maximum exhaust emissions

Constituents	g/kWh
Carbon monoxide	4.5
Hydrocarbon	1.1
Nitrogen oxides	8.0
Particulates	0.36

Stage II
Proposed emission legislation will be implemented on 1 October 1995 for new models

(Type Approved) and 1 October 1996 for new registrations.

Table 5   October 1995 proposed maximum exhaust emissions

Constituents	g/kWh
Carbon monoxide	4.0
Hydrocarbon	1.1
Nitrogen oxides	7.0
Particulates	0.15

13.6.8   Comparison of exhaust gas composition during different modes of driving

An overall picture of the various pollutants coming from petrol engine exhausts can be obtained by studying Table 6. The range of variation for each constituent under similar operating conditions is wide for different engines tested, but nevertheless the trends are plain.

Table 6 Typical petrol engine exhaust gas compositions

Exhaust constituents	Driving mode			
	Idle	Acceleration	Cruise	Deceleration
Hydrocarbons (PPM)	300–1000	300–800	250–550	3000–12 000
Carbon monoxide (%)	4–9	1–8	1–7	3–4
Carbon dioxide (%)	10	12	12.5	6
Nitrogen oxides (PPM)	10–50	1000–4000	1000–3000	5–50
Oxygen (%)	2	1.5	1.5	8
Exhaust flow (m³/min)	0.185–0.95	1.5–7.5	0.95–2.25	0.185–0.95
Exhaust gas temperature at entrance to silencer (°C)	150–300	450–700	400–600	200–400

The level of carbon monoxide at idle and deceleration speeds tends to be high because it is difficult to burn weak mixtures, which are diluted with exhaust gas, under both of these conditions. It will also be observed that the hydrocarbon emission on deceleration is at a maximum. This is because the lean burning mixtures on overrun produce intermittent firing so that partially burnt products of combustion and hydrocarbons will pass through to the exhaust system. With the third major pollutant, it can be seen that high concentrations of nitrogen oxides occur during acceleration and cruising. This is due to the high combustion temperatures produced during these modes of driving, which encourage the formation of nitrogen oxides in the combustion chamber.

13.7 Three-way catalytic converter exhaust treatment

(Figs 13.33, 13.34 and 13.35)

The purpose of the three-way catalyst converter is to remove the unwanted pollutant gases—carbon monoxide (CO), hydrocarbons (HC) and the oxides of nitrogen (NO_x)—from the exhaust gas stream by converting them, by chemical reactions, to carbon dioxide (CO_2), steam (H_2O) and nitrogen (N_2). This is achieved by oxidizing the carbon monoxide (CO) and hydrocarbons (HC) so that carbon monoxide and hydrocarbon molecules are rearranged to form carbon dioxide (CO_2) and steam (H_2O), whereas oxides of nitrogen (NO_x) are reduced (oxygen removed) to form carbon dioxide (CO_2) and nitrogen (N_2). These after-burn chemical reactions in the exhaust system would take a very long time and would, therefore, pass out from the exhaust tail pipe incomplete. However, the rate of chemical reaction can be considerably speeded up if it takes place in the presence of a catalyst; this is a substance that increases the rate of chemical reaction without itself undergoing any permanent chemical change. The catalyst materials chosen for treating the exhaust gases are usually noble metals: those in question being platinum (Pt) and rhodium (Rh). Platinum has a high activity during the oxidation process of carbon monoxide (CO) and hydrocarbons (HC) under stoichiometric and slightly rich conditions, whereas rhodium is very active during the reduction of oxides of nitrogen (NO_x); that is, oxygen atoms are readily disassociated from the nitrogen atoms under similar air–fuel ratio mixture strengths. Since all three pollutants are removed simultaneously, this type of catalyst is known as a three-way catalyst (Fig. 13.33).

Catalysis involves the continuous absorption (formation of a layer of gas) of the reactants onto surface sites of high activity, followed by chemical reaction, then desorption (removal of absorbed gases from the surface) of the products (Fig. 13.34). For oxidization and reduction of the carbon monoxide, hydrocarbons and nitrogen oxides to occur, the exhaust gas flow temperature must be above 250–300°C. For the catalyst to work effectively, the active noble metals must be minutely spread over a very large surface area, which is exposed to the exhaust gas path. To maximize the flow path surface area, a ceramic or metallic multi-passageway monolith substrate is positioned in the flow path (Fig. 13.35). These passageway or channel walls are covered with a washcoat of inert alumina (Al_2O_3), this is porous and, when applied, has an irregular surface finish which considerably increases the active surface area. It also prevents particle-to-particle metal contact of the noble metals which are vapour deposited onto the washcoat.

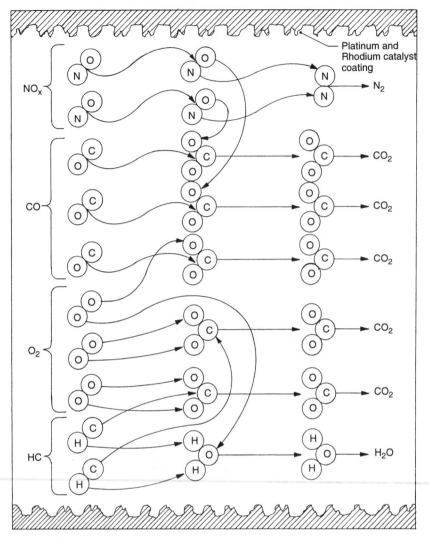

Fig. 13.34 Exhaust gas composition catalytic reaction process

13.7.1 Catalyst converter operating conditions
(Figs 13.36, 13.37 and 13.38)

The outgoing exhaust gases in a petrol engine can range from 300°C to 400°C while idling, and may reach 900°C under full-load operating conditions. A typical operating temperature range would be between 500°C and 600°C. To maintain high conversion efficiency over a long period of time, converters should be made to operate within a temperature band of between 400°C and 800°C. If the exhaust temperature in the converter should move into the 800°C–1000°C range for any length of time, the noble metals and the substrate wash-

coat will tend to sinter—thus considerably increasing the rate of thermal aging.

A converter operating under ideal conditions can expect to have a moderately high conversion efficiency life of roughly 100 000 km (60 000 miles). However, if the engine should backfire or misfire, maybe due to operating with a very lean mixture at some particular speed and load condition, this may cause the exhaust gas temperature temporarily to rise and, if it should exceed 1400°C, the substrate material will melt, thus completely destroying the catalyst activity in the honeycomb passages.

With temperatures in excess of 300°C the steady-state conversion efficiencies of a new converter can be between 98% to 99% for carbon

698

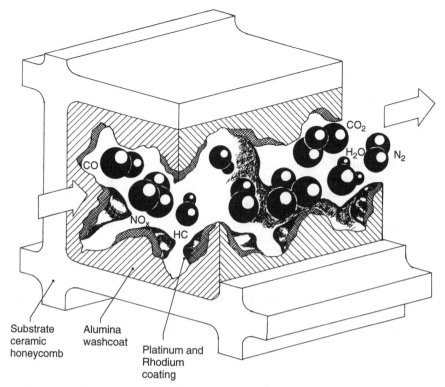

Substrate ceramic honeycomb

Alumina washcoat

Platinum and Rhodium coating

Fig. 13.35 An enlarged section view of a catalytic converters honeycomb active passageways

monoxide and above 95% for hydrocarbon. However, for temperatures much below 300°C the catalyst is practically ineffective (Fig. 13.36).

The temperature at which the catalyst is 50%

effective is known as the *light-off temperature*, this temperature is sometimes used by the manufacturers as a specification.

Catalysts lose their effectiveness due to the

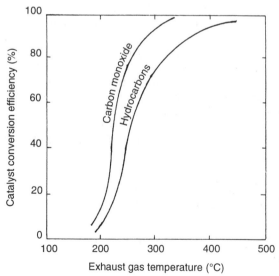

Fig. 13.36 Conversion efficiency for CO and HC relative to exhaust gas temperature

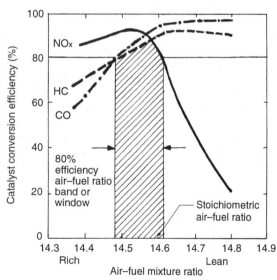

Fig. 13.37 Conversion efficiency for CO, HC and NO_2 relative to exhaust gas air/fuel ratio using a three catalyst

deterioration of the active materials' exposure to the hot exhaust gases. This degradation of the converter is basically caused by the active sites becoming contaminated and poisoned, and also due to the noble metals sintering, caused by operating at very high temperatures for prolonged periods. This, in effect, reduces the effective surface area of the active sites so that there is insufficient time to convert all the gases as they flow through the passageways.

Prolonged contact with interfering elements, such as the antiknock agent (lead) and the phosphorus-in-oil additives which block the active sites, prevents the exhaust gases from chemically interacting with the active material. This condition is known as *poisoning* of the catalyst active material. Small amounts of contaminated leadless petrol can also poison the catalyst converter over a long period of time.

The catalyst should have low thermal inertia for rapid warm-up and low light-off temperature for carbon monoxide and hydrocarbons so that the active materials become effective quickly—this should be normally under one minute and preferably should be as short as 30 seconds. This can partially be achieved by placing the converter very close to the engine's manifold as this will speed up the time for the light-off temperature to be reached. However, having the converter very near to the exhaust manifold can cause the exhaust gases (under certain operating conditions) to exceed the safe working temperature of the substrate and noble metals, thus drastically shortening the life of the converter. For the converter conversion efficiency to remain reasonably high, the engine must operate very close to the stoichiometric air–fuel ratio (Fig. 13.37).

Oxides of nitrogen increase dramatically with rising engine load (Fig. 13.38), particularly if the air–fuel mixture is just on the lean side of the stoichiometric ratio. This, therefore, highlights the importance of maintaining the mixture strength to be slightly on the rich side of the stoichiometric point, in order to minimize these emissions.

13.7.2 Three-way catalytic converter construction
(Fig. 13.39)

The three-way catalytic converter is situated ahead of the noise suppression silencer, as near to the exhaust manifold as possible (Fig. 13.39). The

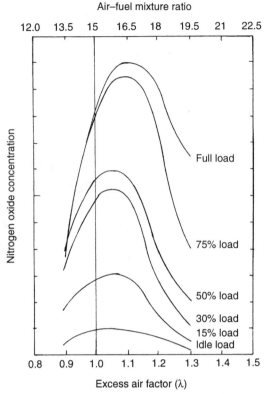

Fig. 13.38 Effect of engine load on oxides of nitrogen concentration in the stoichiometric mixture region

three-way catalytic converter normally takes the form of a stainless-steel cylindrical or oval-shaped casing with conical or semi-conical front and rear ends which merge with short flanged tubes to provide the inlet and exit for the polluted and treated exhaust gases, respectively. Mounted within and between the cylindrical or oval-shaped casing is a catalytic bed comprising a substrate (that is, surfaces which are treated with an active material) in the form of a monolith or matrix core which has a large number of very small passageways passing through its bulk structure.

The catalytic bed may take three different forms:

1 ceramic pellets
2 ceramic honeycomb (monolith)
3 metallic honeycomb (monolith)

13.7.3 Ceramic pellets
(Fig. 13.40)

This type of catalyst has layers of spherical pellets resting on top of each other, with the pellets being

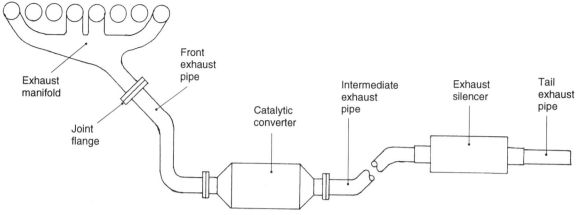

Fig. 13.39 Exhaust system with catalytic converter

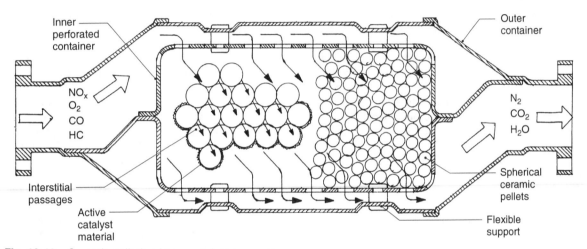

Fig. 13.40 Ceramic pelletised type catalyst converter

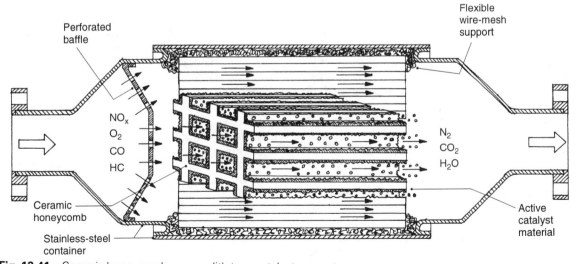

Fig. 13.41 Ceramic honeycomb or monolith type catalyst converter

made of a ceramic such as high-temperature resistant magnesium-aluminium silicate. The passageways and large surface area exposed to the exhaust gases are created by the interstitials, or spaces surrounding the spherical contact points of the pellets. The pellet-type catalyst dispenses with the washcoat (which is necessary with the honeycomb monolith) and therefore has the active noble metals platinum (Pt) and rhodium (Rh) impregnated directly onto the highly porous surface of the spherical alumina pellets (which are about 3 mm in diameter) to a depth of around 250×10^{-6} m. These ceramic (alumina) pellets have good crush and abrasion resistance after they have been subjected to temperatures of at least 1000°C. The pellet pack is stacked within an inner perforated container which is suspended inside the outer converter casing. One half side of the perforated container is exposed to the untreated flow exhaust gas while the other half side acts as the exit for the treated exhaust gas. This construction partially protects the pellets from damage should the outer container be subjected to external road shock and distortion, which may leave permanent dents in the outer casing.

13.7.4 Ceramic honeycomb (monolith)
(Fig. 13.41)

The matrix of the converter resembles a honeycomb structure consisting of thousands of parallel channels through which the exhaust gas flows. This substrate honeycomb is made from a magnesium–aluminium silicate ceramic material, which remains stable at high working temperatures. These passageways are covered with a highly porous alumina (Al_2O_3) washcoat roughly 20×10^{-6} m thick which increases the effective surface area of the catalyst channels by a factor of roughly 700. This washcoat is impregnated by vapour deposition with the noble metals platinum (Pt) and rhodium (Rh). The honeycomb structure has approximately 1 mm square passageways with porous walls roughly 0.15 to 0.3 mm thick. There are about 30 to 60 passageways per square centimetre of frontal area. The washcoat covers a surface area of about 100 to 200 m^2/g and has a mass of about 5 to 15% of the honeycomb monolith.

Ceramic honeycomb structures are fragile and are therefore suspended inside the casing between a high alloy steel wire-mesh. This protects the honeycomb from thermal expansion stresses and external impacts which may deform the outer casing.

13.7.5 Metallic honeycomb (monolith)
(Fig. 13.42)

The honeycomb substrate comprises thin steel foils of alternating layers of flat and corrugated foils of thickness ranging from 0.04 to 0.05 mm which are rolled over a width (corresponding to the specified length of the substrate) into a spiral or 'S'-shape configuration. There are usually two separate honeycomb substrates arranged end on with a small gap between them. This enables thorough mixing of the exhaust gas flow and a fresh formation of the laminar flow in the second substrate, thus improving the pollutant conversion to harmless gases.

The metallic substrate is wound in a spiral or 'S' shape, as this provides a better distribution of the tension generated by thermal expansion so that both mechanical stability and life are increased. The contact zone between the flat and corrugated foil are subjected to a special high-temperature soldering technique which provides the support and rigidity necessary for the many hundreds of channel-ways. The metallic substrate is coated with a highly porous alumina (Al_2O_3) washcoat which is itself very sparsely impregnated with the active noble metals platinum (Pt) and rhodium (Rh). The metallic honeycomb structure is mounted inside the casing, directly against the internal stainless-steel walls. This enables the effective frontal area of the substrate channels to be increased by about 15% compared with the ceramic honeycomb which has to be supported in a wire mesh or fibre mat to avoid brittle damage. The significantly thinner metal walls and larger frontal area for the passageways for a similar-sized converter casing offers a much lower resistance to the exhaust gases.

This steel foil construction has a very high mechanical durability and fatigue strength, with a relatively low exhaust-gas back pressure, and it can also absorb considerable outside deformation without causing severe damage to the substrate flow path. The disadvantages of the metallic honeycomb structure are that it is heavier than the ceramic honeycomb and is roughly 15% more expensive.

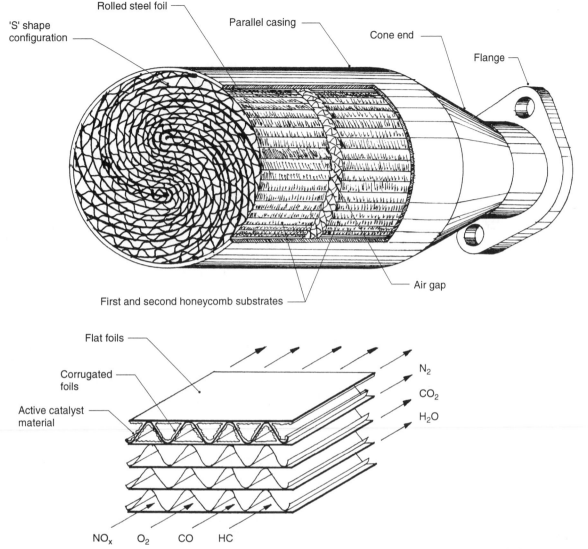

Fig. 13.42 Metallic honeycomb or monolith type catalyst converter

13.8 Air–fuel ratio control

13.8.1 Stoichiometric air–fuel ratio and excess air factor
(Fig.13.43)

With the catalytic converter it is vital for the effective conversion of carbon monoxide and hydrocarbon to operate the engine with a stoichiometric air–fuel mixture (Fig. 13.43). This stoichiometric air–fuel ratio may be defined as the number of parts of air theoretically necessary to burn completely one part of fuel by mass. For petrol, the air–fuel ratio required for complete combustion is normally taken as 14.7 kg of air to 1 kg of petrol.

If the actual air–fuel ratio entering the cylinders differs from the stoichiometric ratio, then there will be a deficiency (or an excess) of air in the cylinders. It is therefore usual to express a deficiency (or excess) of air consumed by the engine in terms of an excess air factor which is given the symbol lambda (λ). This air ratio is defined as:

$$\lambda = \frac{\text{actual air consumed}}{\text{theoretical air required}} = \frac{A_a}{A_t}$$

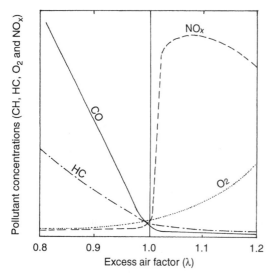

Fig. 13.43 Lambda sensor pollutant control characteristics

Excess air factor examples.

For a stoichiometric mixture having no deficiency or air excess

$\lambda = 1.0$

For a rich mixture having a 5% deficiency of air

$\lambda = 0.95$

For a lean mixture having a 5% excess of air

$\lambda = 1.05$

Note: the excess air factor is sometimes known as the *excess air coefficient*.

13.8.2 Unheated lambda sensor description
(Fig. 13.44(a))

Oxygen components exist in the exhaust gas in various amounts depending upon the air–fuel mixture strength. In fact, the oxygen content in the exhaust system is a direct measure of the mixture composition.

The lambda sensor (Fig. 13.44(a)) when installed in the exhaust system under operating conditions, generates a voltage proportional to the oxygen concentration in the outgoing exhaust gas. The magnitude of this voltage is then signalled to the engine's electronic control unit.

The lambda (λ) sensor is centred around a solid-state electrolyte formed in the shape of an extended thimble and made from ceramic zirco-nium oxide (ZrO_2) and stabilized with yttrium oxide (Y_2O_3). The internal and external surfaces of this thimble-shaped ceramic are coated with gas-permeable microporous platinum, and these layers represent the electrodes between which electrical activity takes place.

The inner electrode surface of the solid-state electrolyte thimble is exposed to the atmosphere, whereas the outer electrode surface is surrounded by the stream of hot outgoing exhaust gas and is therefore protected by a coating of porous cera-mic—this prevents erosion of the platinum layer by solid particles in the exhaust.

This outer platinum electrode simultaneously becomes a small catalytic converter once operating temperatures have been reached, and therefore brings the immediate surrounding exhaust gas into stoichiometric equilibrium. Physical support for the ceramic moulding is provided by a steel housing which is partially threaded on the outside for locating and securing the unit in the exhaust system. The lower half of the thimble is further protected by a steel slotted shield, whereas the upper part of the thimble is extended by a ceramic sleeve and an outer protection steel sleeve. A hollow, central-flanged conductor is sandwiched between the ceramic thimble and the upper ceramic sleeve, and makes electrical contact with the inner platinum electrode coating. The outer platinum electrode coating, on the other hand, is earthed to the steel housing. An insulator holds the central protection steel sleeve and the central cable connector into position.

13.8.3 Heated lambda sensor description
(Fig. 13.44(b))

With the heated lambda sensor, heat is provided by the heating element under low engine load and speed conditions, when exhaust gas temperatures are relatively low, and by the increased exhaust gas heat under higher load and speed conditions (Fig. 13.44(b)). The heater element will therefore permit the sensor to be installed further away from the engine so that it is not subjected to excessive temperature (above 850°C) for any length of time, otherwise severe aging of the sensor's outer electrode will occur. The heater element ensures that the sensor reaches its minimum operating temperature of about 280°C within 20 to 30 seconds after the engine has started, at which point the lambda sensor can be effectively

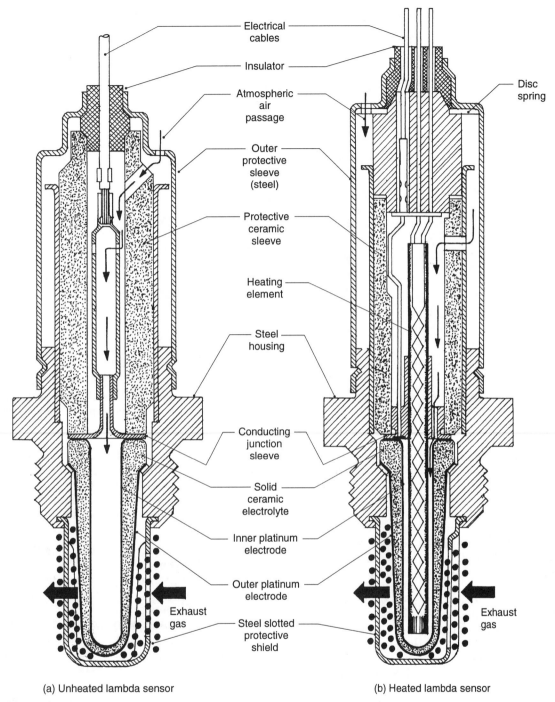

Electrical cables	
Insulator	
Atmospheric air passage	Disc spring
Outer protective sleeve (steel)	
Protective ceramic sleeve	
Heating element	
Steel housing	
Conducting junction sleeve	
Solid ceramic electrolyte	
Inner platinum electrode	
Outer platinum electrode	
Exhaust gas	Exhaust gas
Steel slotted protective shield	

(a) Unheated lambda sensor (b) Heated lambda sensor

Fig. 13.44 Lambda sensors

switched into the closed-loop mixture correction system.

The heated zirconia exhaust gas oxygen sensor is basically identical to the unheated sensor. Both sensors use the same active oxygen sensor element, made of an yttrium oxide (Y_2O_3) partially stabilized zirconium (ZrO_2) ceramic, which is activated by the difference in oxygen concentra-

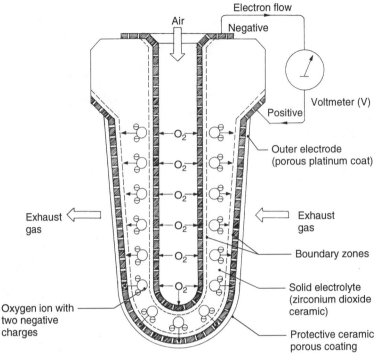

Air · Electron flow · Negative · Voltmeter (V) · Positive · Outer electrode (porous platinum coat) · O$_2$ · Exhaust gas · Exhaust gas · Boundary zones · Oxygen ion with two negative charges · Solid electrolyte (zirconium dioxide ceramic) · Protective ceramic porous coating

Fig. 13.45 Principle of the lambda sensor

tion between the inner and outer boundary layers. The greater the difference in oxygen concentration the greater will be the generated sensor voltage. The inner and outer electrodes consist of high temperature resistant, firmly adhering, platinum cermet conductive strips which are attached with high pressure to a metallic seal ring, and a clamping conducting junction sleeve. The contact pressure is produced by a disc spring at the relatively cold terminal end of the sensor.

In the heated sensor, a ceramic heating rod-element projects into the inside of the zirconic thimble. Embedded into one end of the heating element is a hermetically-sealed thick film heating conductor. The heating element is flexibly held by a clamping junction sleeve, which also conducts the sensor potential from the inner electrode.

The heating element is normally designed for 13 V and exhibits a positive temperature coefficient (PTC). In the first few seconds after switching on the current, power consumption is about four times the equilibrium value for a 350°C exhaust temperature. This results in rapid heating of the sensor ceramic after a cold start. Power consumption drops to about 9 W as the exhaust temperature increases to 850°C.

In the heated sensor the open cross-section of the protective shield has been reduced to only three slots, as compared with nine slots in the unheated sensor to reduce the cooling effect of cold exhaust gas. This reduced number of slots also has a positive effect on the stability and symmetry of the response times. The protective action against fouling and deposits from oil and ash in the exhaust is also enhanced.

As compared with the conventional unheated sensor, the heated sensor offers the following advantages:

1 shorter warm-up time after starting the engine until control cuts in;
2 sensor characteristics for exhaust temperatures below 500°C are considerably less dependent on exhaust temperature;
3 reduced dependence of sensor characteristics on conditions and duration of aging;
4 lower emissions when new, and smaller changes of emission over service life;
5 greater freedom in designing the engine and emission control systems without having to give too much consideration to the exhaust gas temperature;
6 more freedom in sensor installation position;
7 more stable idling during closed loop control;
8 only one sensor needed for both banks in V-engines with merging exhaust pipes.

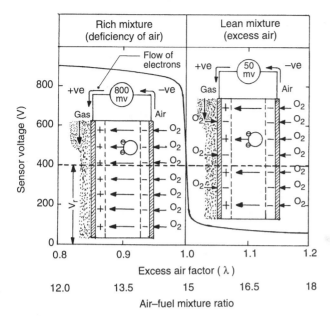

Fig. 13.46 Lambda sensor voltage characteristics

13.8.4 Lambda oxygen sensor operating principle
(Figs 13.43, 13.45 and 13.46)

When the air–fuel mixture moves to the rich side of the stoichiometric ratio there is very little free oxygen in the exhaust gas to act as an oxygen supply to feed the outer platinum electrode catalyst (Fig. 13.45). However, oxygen is made available by depleting the outer boundary region of the solid electrolyte of oxygen. This causes large numbers of negatively charged ions to migrate from the oxygen rich inner electrode, which is exposed to the atmosphere through the electrolyte, to the outer catalyst electrode once a conducting temperature (above 280°C) of the solid-state electrolyte has been reached. Accordingly, a voltage is generated between the inner and outer platinum electrodes in proportion to the rate of oxygen ions diffusing through the ceramic electrolyte.

Should the air–fuel mixture move to the lean side of the stoichiometric ratio, the excess oxygen in the exhaust gas will oxidize the carbon monoxide and hydrocarbons in the vicinity of the outer platinum electrode, in preference to the oxygen ions diffusing through the solid ceramic electrolyte. This results in very little voltage being generated across the solid electrolyte and electrodes.

The magnitude of the generated voltage is dependent upon the difference in the oxygen concentration between the inner and outer boundary regions, this being a measure of the oxygen content in the exhaust gas at any one instant, and therefore relates directly to the air–fuel mixture ratio strength entering the cylinders.

The lambda sensor is very sensitive to the slightest variation of air supply and oxygen concentration in the exhaust gas at the stoichiometric ratio; that is, when $\lambda = 1$ (Fig. 13.43).

If the composition of the exhaust gas should shift slightly into the rich air–fuel ratio band, the deficiency of oxygen in the gas causes a large flow of oxygen ions to pass from the inner to the outer electrode (Fig. 13.46). Conversely, a very small shift of air–fuel ratio into the lean band produces an excess of oxygen in the exhaust gas, which abruptly interrupts the oxygen ions migrating through the solid-state electrolyte (Fig. 13.46). The transition from conducting to non-conducting occurs at the stoichiometric mixture ($\lambda = 1$), and this switched on–off air–fuel composition produces a relatively large stepped voltage change, which instantly rises to around 800 mV if the mixture tends towards a rich mixture. Conversely, it instantly drops to roughly 50 mV when the mixture favours the beginning of the lean band.

13.8.5 Lambda closed-loop control system
(Fig. 13.47)

The effective conversion of the three principal pollutants, carbon monoxide, hydrocarbons and oxides of nitrogen, is achieved by utilizing a closed loop control which, basically, can be divided into four interconnecting parts:

1 the engine
2 the lambda sensor
3 the lambda regulator
4 the fuel metering system

1) The engine consumes the air–fuel mixture and expels the pollutants in the form of burnt and partially burnt gases. These hot exhaust gases are then subjected to after treatment to neutralize the newly created toxic gases.
2) The lambda oxygen sensor is situated along the exhaust pipe as near to the engine as is practical.
3) The regulator receives a small signal voltage from the lambda sensor. This is then amplified

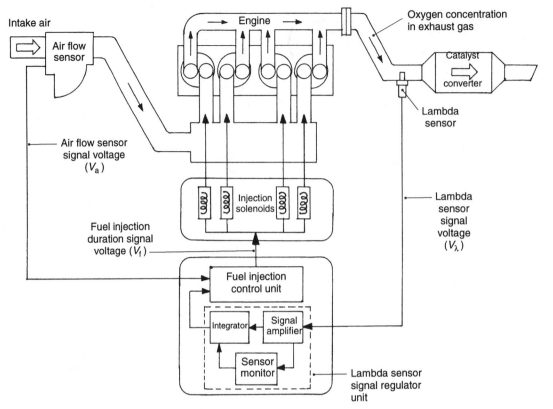

Fig. 13.47 Lambda oxygen sensor closed-loop air–fuel mixture control circuit

and modified while, at the same time, the sensor and signal amplifier is being monitored for any abnormal signals, and finally, this information is integrated before being passed on to the fuel metering control.

4) The fuel metering device is operated by an electronic control unit which takes in information from a number of sensors in addition to the lambda regulator voltage control output. It then supplies the corrected quantities of fuel to the cylinder under the various temperature, speed and load operating conditions.

Oxygen sensor
(Figs 13.47 and 13.48)

The regulator receives the lambda oxygen sensor voltage signal (V_λ) and compares it with the reference voltage (V_r), normally set at 400 mV near the middle of the steep voltage change just on the lean side of $\lambda = 1$; that is, at approximately $\lambda = 0.995$ (Fig. 13.48). Any variation of the lambda sensor voltage signals a change in the residual oxygen concentration in the exhaust gas,

which is indirectly a measure of the instantaneous air–fuel mixture strength ratio entering the engine. The characteristics of the lambda oxygen sensor are such that at $\lambda = 1.0$, there is an almost instantaneous voltage rise or fall as the air–fuel mixture (and correspondingly the oxygen content in the exhaust) changes; therefore, over time, the sensor is capable of providing a trapezium-shaped voltage pulse (Fig. 13.48). Thus, as the exhaust composition is constantly varying so will the lambda-sensor generated voltage respond by abruptly increasing to 800 mV or decreasing to 200 mV for the slightest shift towards a rich or lean mixture, from the stoichiometric position where $\lambda = 1.0$.

Signal amplifier
(Figs 13.47 and 13.48)

The signal amplifier takes in the lambda sensor's small signal voltage and converts these trapezium waveforms into a square-shaped output voltage (Fig. 13.48). When the mixture moves into the rich band, the oxygen deficiency will cause a

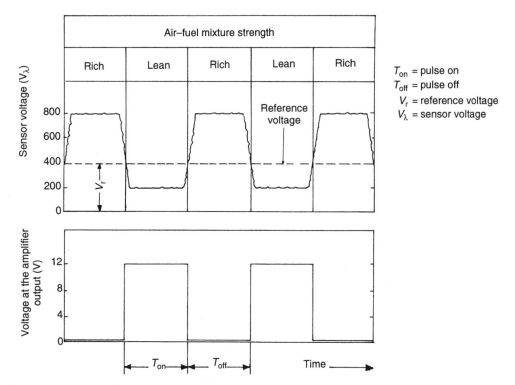

Fig. 13.48 Lambda closed-loop control cyclic voltage characteristics

sharp rise in the oxygen sensor's signal voltage (V_λ) feeding into the signal amplifier. Thus, where the sensor voltage (V_λ) exceeds the 400 mV reference voltage (V_r) the amplifier output voltage controlling the other components switches-off for a pulse-off period T_{off}. This information is relayed directly to the fuel injection computer control unit or management control system, which immediately reduces the quantity of fuel being supplied to the engine. The consequence of this reduction in the amount of fuel entering the cylinders will be a tendency to shift the air–fuel mixture towards the lean band. As a result, the excess oxygen now in the exhaust system will rapidly reduce the generated sensor voltage, so that when it drops below the 400 mV reference voltage (V_r), the signal amplifier switches on. Accordingly, the signal amplifier output voltage now has a pulse-on period, of duration T_{on}. These voltage pulses from the signal amplifier output are therefore fed to the fuel metering components via the integrator, indicating an increase in the fuel supply to the cylinders.

Integrator
(Figs 13.47 and 13.48)
This circuitry makes allowances for the time delay between the instant of fuel metering, to the time the exhaust gas created by this air–fuel mixture flows past the sensor, as well as for the reaction time of the lamba sensor itself.

Without this device, a prolonged response time would cause the output from the regulator to have shifted out of phase with the initial oxygen sensor signal. This would cause the air–fuel mixture strength to shift continuously to and fro, between rich and lean, over a very wide mixture band so that large amounts of carbon monoxide, hydrocarbons and oxides of nitrogen would not be adequately treated by the catalyst. The integration of the control pulse over a time period enables the control system to react in a direction which makes the fuel being metered match the exhaust composition requirements for minimizing untreated emission.

Sensor monitor
(Figs 13.47 and 13.48)
This component checks whether the lambda sensor is correctly measuring the residual oxygen

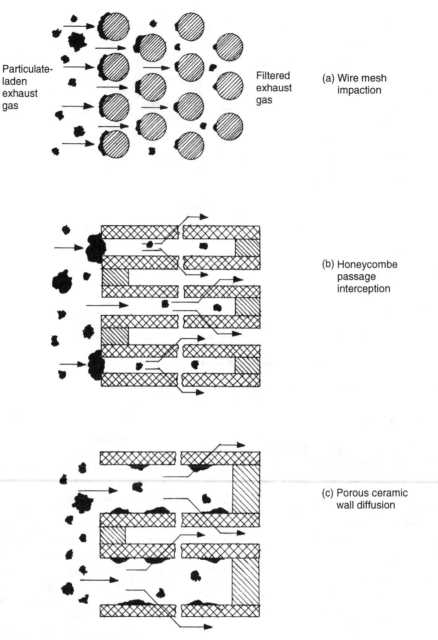

Particulate-
laden
exhaust
gas

Filtered
exhaust
gas

(a) Wire mesh
impaction

(b) Honeycombe
passage
interception

(c) Porous ceramic
wall diffusion

Fig. 13.49 Particulate trap filteration mechanisms

content and that the signals passed to the control unit are being accurately interpreted.

Thus, if the oxygen sensor is operating below its working temperature, the sensor monitor blocks the integrator and thereby opens the closed-loop system. Immediately, a constant voltage signal simulating average driving conditions will be conveyed to the integrator and the fuel metering electronic control unit. Without this constant voltage signal provided by the sensor monitor, a fault in the sensor system would shift the air–fuel mixture right over to the far side of the lean band until the engine stalls.

13.9 Diesel engine particulate traps

13.9.1 Combustion particulates

Combustion particulates are substances other than water that can be collected by filtering the exhaust gas. The amount which can be collected with a homogeneous-charge petrol engine is very small, and is therefore not a problem; however, it does become a problem with heterogeneous charge diesel engines where their emission is relatively high.

Diesel particulates basically consist of combustion-generated solid carbon particles, commonly known as soot, on which some organic compounds have become absorbed. The carbon particles become coated with absorbed and condensed organic compounds such as unburnt hydrocarbon and oxygenated hydrocarbon. However, some of the condensed matter can be inorganic substances such as sulphur dioxide, nitrogen dioxide and sulphuric acid. A portion of the organic matter originates from the lubrication oil, and may amount to anything from 25% to 75% of the organic fraction.

The carbon soot particles are created in the combustion chamber in the region of the rich fuel spray that has not yet spread out and weakened until the gases have been considerably cooled by the expansion process.

The organic particulate fraction comes from the hydrocarbons and their partial oxidation products during the dilution process, when some of these cool enough to condense or absorb the soot.

During the combustion process, the carbon soot particles (which resemble spheres of approximately 2×10^{-6} cm in diameter) collide and are joined together in the form of a cluster of spheres, or they can be branched and chainlike in appearance.

These particulates can become a health hazard if they are small enough to pass through the human nose as they are sufficiently large to be retained in the lungs.

13.9.2 The background to particulate traps

With the trend towards more stringent exhaust emission levels, the restrictions of diesel engine particulates entering the atmosphere also have to be improved.

This improvement can be achieved by passing the exhaust gas through a filter trap, where the clean gas continues on its way into the atmosphere, but the blocked particulates (which accumulate on the filter media) are removed by the hot exhaust gases oxidizing the solid particulates. The trapping and burn-up of the collected particulates produces a self-cleansing cycle which prevents the filter passageways from becoming clogged. The concept of a self-cleansing particulate trap has many limitations which make it difficult to function effectively. These shortcomings are listed as follows:

1 any filtering media in the exhaust system must increase the exhaust gas back-pressure, which correspondingly reduces the engine's power;
2 the exhaust back-pressure increases as more particulates become trapped in the filter passages;
3 ignition and oxidation of the particulates is almost impossible when operating over the normal speed and load range of the engine;
4 when oxidizing does occur it must be carefully controlled to prevent the temperature of the exhaust gases rising to dangerous levels, which would damage the filter media.

Trap regeneration of the accumulated particulates by burning them away can be improved by introducing an outside air supply to the exhaust gas so that it can continue to support combustion and, at the same time, raise the surrounding temperature to the ignition point of the particulates. Unfortunately, a major snag is that diesel particulates ignite at 500°C to 600°C, and this is well above the normal operating temperature of the exhaust gas, which generally ranges between 200°C and 500°C from idling to full load, respectively. Thus, the two options are either to raise the exhaust gas temperature or to lower the ignition point of the particulates. The first method can, to some extent, be achieved by incorporating a thermal reactor or an electric heating element enclosed in the filter passageways. The second approach, which is the most promising can be achieved by coating the trap passageway walls with a catalytic material. The result is to reduce the ignition point of the particulates by up to 200°C and, at the same time, to oxidize the carbon monoxide and hydrocarbons passing through with the spent exhaust gases. Trap oxidizers can be effective for small diesel engines, which tend to run with high exhaust temperatures. However, large diesel engines produce relatively more particulates and operate with lower exhaust temperatures, and these make regeneration of the particulates much more difficult.

13.9.3 Particulate trap filtration and its regeneration
(Fig. 13.49(a–c))

There are three basic methods of mechanically trapping combustion particulate matter without obstructing the exhaust gas flow, these may be classified as:

1 impaction
2 interception
3 diffusion

Impaction (wire mesh restriction)
(Fig. 13.49(a))
This is where large particulates suspended in the exhaust gas stream strike, spread and adhere to the surface of the wire-mesh strands, or accumulate on previously impacted particulates. Excessive blockage is prevented by the periodic oxidation (regeneration) of the collected particulated matter.

Interception (honeycomb passageway entrance restriction)
(Fig. 13.49(b))
This is where medium-to-large particulates, moving with the exhaust gas flow, collide with the mouth of individual passageways. The interception continues as more particulates bombard and partially block the entrances to the filter flow paths until enough heat is generated to burn-up the accumulated particulate.

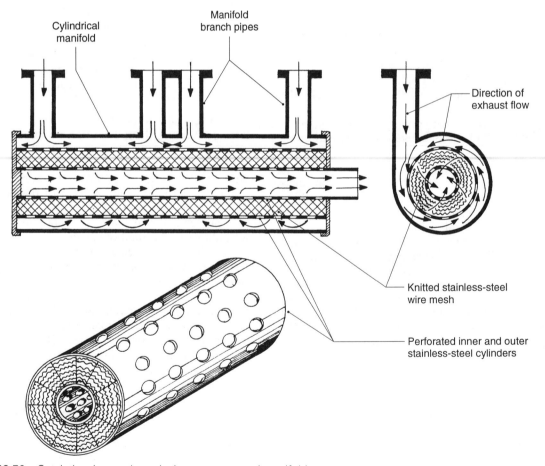

Fig. 13.50 Catalytic wire mesh particulate type trap and manifold

Diffusion (porous passageway wall restriction)
(Fig. 13.49(c))

This is where small particulates entering the filter passageways attach and spread themselves against the porous walls or diffuse onto existing particulate matter which has already become attached to the porous surfaces of the trap. As the porous channel walls become clogged, the rise in exhaust back-pressure causes more fuel to be injected to compensate for the resulting power loss. Consequently, there will be a rise in the exhaust gas temperature until the ignition point of the particulates is reached, and subsequently the accumulated particulates will be burnt off.

13.9.4 Wire mesh particulate trap and oxidizer
(Fig. 13.50)

The trap substrate is made from a knitted stainless-steel wire-mesh compressed between the perforated cylindrical inner and outer casings. The wire-mesh is woven and rolled to give a low density and surface volume ratio on the outer region of the annular-shaped substrate, with an increasing density and surface-to-volume ratio towards the inner diameter region. Thus, the larger particulates are trapped in the outer circumferential mesh, whereas the smaller particulates are obstructed towards the inner diameter. This progressive change from an open outer diameter mesh to a more closed inner diameter mesh allows different-sized particulates to be trapped at various depths. It therefore prevents all the particulates collecting only in the outer region of the mesh, where they could quickly clog the filter media. A refractory washcoat is deposited onto the wire-mesh which is then speckled with the noble metals used as the catalyst. The rough irregular surface contour of the washcoat increases the catalyst surface area many times and therefore considerably accelerates the surface activity of the catalyst.

The off-centre attachment of the manifold branch pipes to the cylindrical manifold chamber provides the exhaust gases with a tangential entry for improved exhaust flow and distribution around the doughnut-shaped filter element. At the same time, the wire-mesh gives the gas a high degree of turbulent exhaust flow, which produces an effective particulate trap and oxidizer.

13.9.5 Porous ceramic honeycomb particulate trap and oxidizer
(Figs 13.51 and 13.52)

This form of trap oxidizer consists of a porous cellular ceramic honeycomb structure which has many parallel passages of approximately square section (Fig. 13.51). Alternative passageways are blocked at either end so that their inlet and exit faces resemble the black and white squares of a chess board. The inlet passageways are plugged at the far outlet end, whereas the exit passageways are plugged at the inlet ends and are open at the outlet. The exhaust gas entering the inlet passages is compelled to pass at right-angles through the thin walls of the porous ceramic into the exit passageways, and finally the filtered exhaust gas is expelled to the rear outlet chamber of the particulate trap. As with the conventional three-way catalyst converter, the passage channel walls are treated with a refractory washcoat which is impregnated with precious metal particles to speed up the hydrocarbon and carbon monoxide oxidization process.

Typical pressure drops across a clean unclogged ceramic honeycomb filter trap range from roughly 0.02 bar to 0.15 bar, for 1000 to 4500 rev/min, respectively.

Particulate removal from the honeycomb trap is affected by the pressure drop across the filter, and as the passageways become partially clogged, this necessitates more fuel to be injected to make up for the resulting power loss. Consequently, more heat energy passes out with the exhaust gas, and it therefore raises the gas temperature, which activates the ignition and oxidation of the accumulated particulate matter.

Large heavy-duty diesel engines, which operate with much lower exhaust temperatures, require an external heat source and an additional air supply to support combustion for regeneration of the accumulated particulates. One version suitable for buses and trucks operating in urban areas and covering a distance of about 100 to 250 km per day, incorporates an electric heater element and air supply pipe at its base (Fig. 13.52). Regeneration is carried out at the vehicle's terminal, after each day's operation, by connecting an external electricity mains supply and a compressed air supply to purge the honeycomb trap. Regeneration of the exhaust-gas accumulated particulates takes roughly three hours. This trap oxidizer reduces the particulates by 80% and it

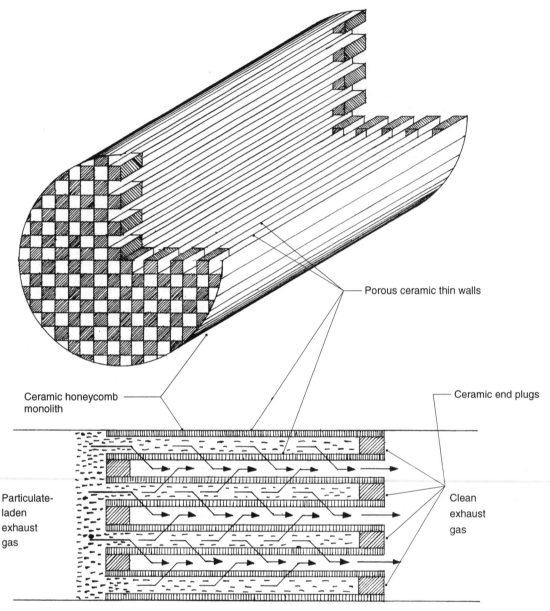

Fig. 13.51 Catalytic porous ceramic monolith particulate type trap

also reduces the hydrocarbons and carbon monoxide by at least 60% and 50%, respectively.

13.9.6 Silica fibre candle particulate trap and oxidizer
(Fig. 13.53)

This approach incorporates nine filtering elements consisting of punched sheet metal support tubes with one end closed, and around these tubes are woven layers of silicon fibre yarn in a crosswire pattern to form the filtration media (Fig. 13.53). Because these elements resemble candlesticks they are known as candle elements. The silicon dioxide yarn thread has a diameter of between 0.7 and 1.0 mm which, in turn, is made up of individual fibres of about 9×10^{-6} mm.

Catalytic oxidation can be achieved by depositing a refractory washcoat over the woven yarn

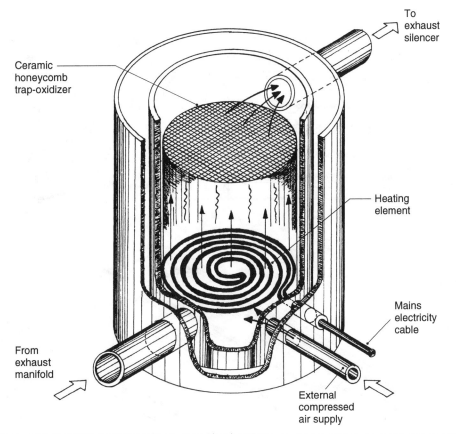

Fig. 13.52 Porous ceramic trap oxidiser and regeneration heater

and impregnating its exposed surface with precious metal.

Particulate-laden exhaust-gas flows into the conical inlet chamber and through to the cylindrical chamber which encloses the candle elements. At the same time, gas surrounding the candle elements flows radially inwards through the woven silica fibre yarn and perforated tubes. The filtered and oxidized gas then passes through the centre of the tube and out to the exit conical chamber before proceeding to the exhaust silencer.

This trap oxidizer produces very low exhaust-gas back pressure and it is subjected only to low thermal and mechanical stresses. However, it suffers from a low particulate loading capacity, and thus requires frequent regeneration.

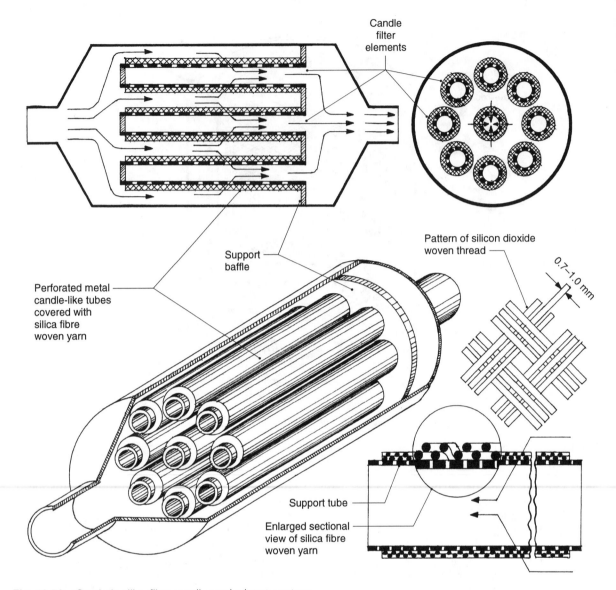

Candle filter elements

Perforated metal candle-like tubes covered with silica fibre woven yarn

Support baffle

Pattern of silicon dioxide woven thread

0.7–1.0 mm

Support tube

Enlarged sectional view of silica fibre woven yarn

Fig. 13.53 Catalytic silica fibre candle particulate type trap

14
Cooling and Lubrication Systems

14.1 Heat transference via the liquid cooling process

Heat is transferred by radiation and convection from the gases forming the products of combustion to the metal walls, and also by conduction through the cylinder walls, by convection through the coolant, and finally, heat is dissipated by air convection and radiation to the surrounding atmosphere.

Something like 20% to 30% of the heat generated in the engine's cylinders is transferred to the cylinder-bores and cylinder-head during each combustion cycle, and this heat is then conveyed through the coolant to the radiator tubes. The final phase of removing the heat from the cooling system is achieved by forcing air through the radiator's many air-cell passageways. The air now containing the heat is subsequently expanded into the atmosphere.

14.1.1 Metal-to-liquid heat transference
(Fig. 14.1)

Heat transference from a metal surface to a liquid coolant is basically dependent upon the following:
(a) thermal conductivity; (b) temperature gradient; and (c) heat flux.

(a) The thermal conductivity of the metal or alloy can be defined as the heat flow per unit area per unit time when the temperature decreases by 1° in

unit distance. Typical values are summarized as follows.

Material	Conductivity (W/mK) at 27°C
Cast iron	58
Steel	55
Stainless-steel (18/8)	14
Aluminium	202
Light aluminium alloys	194
Copper	386
Brass (63/37)	113
Water	0.614
Air	0.026
Carbon	3.45

Thus, it can be seen that aluminium alloy is approximately 3.3 times better at conducting heat than cast iron and, after allowing for a thicker section of aluminium alloy, the advantage will still be approximately 2.5 to 1. However, cast iron is still universally used for the cylinders but the cylinder-head and the cylinder-block can advantageously be made from light aluminium alloy. The other materials listed provide a comparison of thermal conductivities for other parts of the cooling system, such as pure copper or pure aluminium, brass and steel (which are used in the construction of radiator tubes, fins, end tanks and the outer mounting frame) and of course water and air as the cooling media. Carbon has also been included as carbon deposits frequently form on metal surfaces, which usually increases the heat flow resistance through the metal walls.

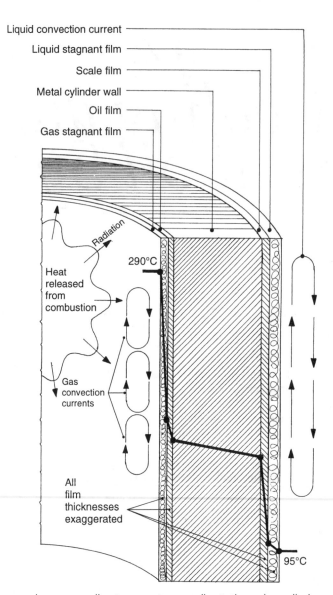

Fig. 14.1 Heat transference and corresponding temperature gradients through a cylinder wall

(b) The temperature difference across the metal walls from the internal combustion chamber side to the external coolant passage side (Fig. 14.1) is normally referred to as the *temperature gradient* and is defined as the temperature drop across a unit width of section expressed in deg C/mm. Thus, the temperature gradient will be greater with low thermal conductivity cast-iron as opposed to aluminium alloy, which has a relatively high thermal conductivity. Keeping the wall thickness to a minimum helps to reduce considerably the temperature gradient between the inter-

nal and external surfaces, which are subjected to the transference of heat.

The outer surface of a cast-iron casting tends to form a thermal barrier skin which hinders the transference of heat from the cylinder to the coolant. Cylinder liners can be machined on both sides, thus removing the thermal barrier. However, this is not possible with integral cast-iron cylinder blocks. Cast aluminium alloy does not produce thermal barriers on the outer skin.

(c) The rate of heat flow per unit of surface area,

718

this being known as the *heat flux*, can also be defined as the thermal conductivity times the temperature gradient. With a low heat flux, heat transfer from a metal surface to a liquid coolant will be by forced convection, but this mode of heat transference changes to nucleate boiling and even unstable film boiling with higher heat flux when the engine is subjected to greater loads and higher speed conditions.

14.1.2 Heat transference and corresponding temperature gradients through a cylinder wall
(Fig. 14.1)

The heat of combustion is transferred in all directions to the metal of the combustion chamber, cylinder walls and piston by direct radiation and by convection currents of the gas scrubbing against the practically stationary thin contact gas film, which always forms on metal surfaces (Fig. 14.1). A thin film of oil normally exists between the stagnant gas layer and the cylinder wall, whereas a thin layer of carbon separates the stationary gas film from the metal walls of both the combustion chamber and piston crown. Heat then flows through the metal walls with the minimum resistance shown by the very small temperature gradient or slope. Note that a perfect heat conducting material would have no temperature gradient; that is, the temperature on both sides of the wall would be similar.

A covering film of corrosion products, scale and contamination from the coolant forms on the opposite coolant passage side of the metal wall, and next to this is a more or less stationary liquid contact film which separates the bulk of the moving liquid coolant from the layer of scale surround the cylinder barrel (Fig. 14.1).

Each of these layers (be they gas, liquid or solid) forms unwanted heat barriers which severely hinder the transference of heat from the swirling and turbulent combustion products to the steady circulation of the liquid coolant.

The steepness of the temperature gradient indicates the magnitude of the heat flow resistance. Thus, it can be seen that the gas film offers by far the greater resistance to heat flow, followed in descending order by the scale film, oil film, and liquid stagnant film, whilst the least resistance to heat conduction is the metal wall itself (Fig. 14.1).

In practice, the film thickness will be greatly influenced by the load and speed of the engine at any one time, and the length of operating time between overhauls. Investigations have shown that the relative conductivity of steel and cast iron is something like 95 times that of water, while that of aluminium 330 times that of water. In addition, the conductivity of steel and cast iron is of the order of 2200 times that of air, whilst that of aluminium is 15000 times that of air.

Very rough estimates of the relative heat flow resistance of a stagnant air film on one side of a metal wall, and a corresponding stagnant liquid film on the opposite side (without taking into account oil and scale films which normally exist when an engine is operating) and taking the metal wall as having unity conductivity, are as follows.

Material	Air film	Metal wall	Water film
Cast iron	113	1	4.8
Aluminium	388	1	17

This comparison illustrates the considerable heat flow resistance experienced through the air film as opposed to the water film—both films being approximately of similar thickness—whereas the metal wall, which is at least 20 times thicker than either of the films, offers relatively very little resistance. Note the superiority of aluminium over cast iron as a heat conducting material.

14.1.3 Heat transfer by convection
(Fig. 14.2)

Heat is transferred from the cylinder bore and cylinder-head walls to the liquid coolant in a number of convection or semi-convection phases, which are dependent upon the rates of heat flow through the metal per unit area and the temperature difference between the metal surface and the liquid coolant. The different convection phases, with increasing heat flux and temperature difference between the hot metal surface and the coolant temperature, are shown in Fig. 14.2.

Convection phase
(Fig. 14.2)
At very low rates of heat flow through the metal per unit area (heat flux) no boiling occurs and movement of the liquid coolant is by simple free convection or by forced convection caused by water pump pressure. In a cooling system, the majority of heat transferred from the cylinder

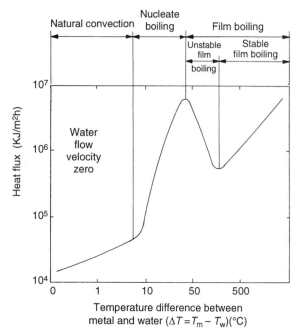

Fig. 14.2 Relationship of heat transfer with metal to liquid temperature difference

bores and cylinder-head walls to the coolant occurs by natural and forced convection currents over areas where the rate of heat flow through the metal is relatively low.

Nucleate boiling phase
(Fig. 14.2)
At higher engine load and speed the rate of heat flow through the metal (per unit area of wall surface) is increased until steam bubbles are formed in certain regions on the metal surfaces (which are subjected to much higher heat flux) such as the metal bridge between exhaust ports between valve seats, the sparking-plug boss etc. Nucleate boiling involves the nucleation and growth of vapour bubbles that originate at nucleation sites on the metal surface. In this phase, large numbers of bubbles form on the hot surfaces and travel through the bulk of the coolant to emerge at the free surface of the filler tank. This results in an increased agitation or movement of liquid so that heat transfer is considerably improved.

Unstable film boiling phase
(Fig. 14.2)
Under severe engine load and speed operating conditions the vapour bubbles become so large and numerous that liquid has difficulty in flowing

back to the hot metal surface as the bubbles rise. When this critical heat flux is reached, the hot surface suddenly becomes insulated by individual steam bubbles which are merged together to form a film or blanket. Under these conditions, the outflowing heat from the metal surface exposed to this vapour bubble blanket finds it difficult to escape. Consequently, there will be a sudden rise in temperature at the metal surfaces covered by this vapour bubble film.

These metal surfaces where the temperature is much higher than the surrounding metal are generally referred to as the *hot spot zones* and should be avoided.

Stable film boiling phase
(Fig. 14.2)
If the metal surfaces subjected to unstable film boiling can withstand even higher temperatures, the heat flux can be increased still further by increasing the temperature difference between the metal surface and the liquid boiling temperature until a stable film boiling phase is established. With stable film boiling, heat transfer is now mainly by conduction and radiation across the vapour film; however, this phase is not generally experienced in a typical cooling system and therefore is not considered a problem.

14.1.4 Engine cooling design considerations

As the power per litre of cylinder capacity increases with advancements in engine design, proportionally more heat will be rejected from the engine to the radiator so that greater demands are being made on the cooling system all the time. Cooling system efficiency improvements to meet the increased engine requirements are primarily due to the following features.

a) The incorporation of an engine-driven water pump in the system which provides a forced liquid coolant circulation and which has a flow rate roughly proportional to the engine speed.
b) The internal coolant passageways surrounding the combustion chamber and valve ports are designed with the minimum of thickness in order to reduce the temperature gradient, and they are accurately cast so that the wall width between the gas and liquid coolant interfaces is consistent. The coolant passageways should

Table 1

Ethylene glycol % by volume	10	20	30	40	50	60	70	80	90	100
Boiling point °C at atmospheric pressure	102	103	104	106	109	113	118	127	145	195

be made to extend to all critical parts of the cylinder-head such as the valve ports and seats, sparking-plug and injector bosses etc.

c) The liquid coolant delivery and discharge to and from the engine is equally distributed between the cylinder barrels. With large diesel engines and high performance engines, separate delivery and collection manifolds are attached to the cylinder block and head. However, with small-to-medium engines, the passageways are internally cast with the input delivery entrance at the front of the cylinder block. This makes it difficult to supply each cylinder barrel with equal amounts of coolant flow at similar input temperatures.

d) The coolant vertical passages between the cylinder block and head should be designed to increase in cross-section from front to rear, to compensate for the single delivery entrance being at the front of the block. Sometimes, reducing the flow resistance towards the rear is achieved by increasing the cylinder-head gasket coolant hole sizes from front to rear.

e) The cylinder-head roof passages from rear to front should slope very slightly upwards and increase in section to collect the hot liquid coolant flowing up from the cylinder barrels so that there is no bottle-neck effect which could cause temperature differences between the coolant flowing between cylinders.

f) The cylinder block and head coolant jackets' capacity should only just be adequate to remove the maximum amount of heat rejected from the cylinder and head walls under full load and speed conditions, so that the warm-up periods are minimized. This greatly influences the fuel consumption under start–stop operating conditions.

g) The cylinder bore and cylinder-head wall temperature can be suppressed within limits by increasing both the coolant system operating pressure and the coolant flow-rate passing over the hot spot regions in the cylinder-head coolant passageways.

h) The better thermal conductivity of aluminium

alloy as opposed to cast iron reduces the temperature gradient across the walls of the cylinder bores and the cylinder-head, which for similar working conditions enables the cylinder-head operating temperature to be lowered. The use of aluminium, with its smaller thermal capacity, also provides for a more rapid warm-up period.

14.1.5 Heat transference with ethylene glycol—water solutions

The use of anti-freeze solution such as ethylene glycol not only suppresses the on set of the coolant freezing but it also raises the mean radiator temperature since its boiling point is considerably higher than water. Ethylene glycol has a boiling point of 195°C and a specific heat capacity of 2.79 kJ/kg°C at 90°C increasing to 3.04 kJ/kg°C at 146°C, whereas water has a much lower boiling point of 100°C but its specific heat capacity is much higher at 4.19 kJ/kg°C. Thus ethylene glycol specific heat capacity is only two thirds that of water at 90°C. When diluted with water the boiling point falls as shown by table 14.1, this gives boiling points for corresponding ethylene glycol and water solution.

Ethylene glycol ignites at a temperature of about 125°C so if a very strong solution leaked from the cooling system near the exhaust manifold it could become a fire hazard, however the solution used for most automotive cooling system applications very rarely are stronger than a 50:50 ethylene glycol—water solution. The heat transference from liquid to air is slightly less with ethylene glycol than water due to its inferior heat transference process. The employment of ethylene glycol such as a 50:50 solution raises the metal liquid interface temperature due to its lower specific heat capacity and the inferior heat transference from metal to liquid, it therefore increases by up to 40°C the temperature of some of the metal components exposed directly to the products of combustion. However the higher working temperatures and the greater proportionate direct heat loss from the engine cooling passages, pipes and radiator permits a reduction in radiator matrix frontal area for the same engine power output.

Heat transfer from the metal interface to the liquid coolant in most parts of the engines coolant passageways is by convection but with some local regions where the heat exposure is very high the transference of heat is by nucleate boiling. Ethylene glycol solutions tend to raise the temperature at which the mode of heat transference changes from convection to nucleate boiling. In general increasing the velocity of coolant circulation reduces the temperature of the metal where heat transference is by convection, but does not reduce the metal wall temperature in the hotter regions where nucleate boiling heat transference takes place.

Because of the lower specific heat capacity of ethylene glycol solutions, the engine warm-up period with these solutions is shorter when compared with water filled cooling systems.

Increasing the coolant solution strength beyond the recommended maximum usually 50:50 can have the effect of raising the cylinder head temperature to such an extent as to promote detonation under severe load operating conditions.

14.2 Liquid cooling systems

14.2.1 Thermo-syphon cooling system

The original thermo-syphon cooling systems relied on the natural circulation of heated low-density water (surrounding the cylinder barrels and combustion chamber) rising to the roof of the internal cylinder-head cooling passages, and from there, flowing to the top tank of the radiator. Simultaneously, cool air (drawn through and around the radiator tubes and fins by the fan) moves from the front leading side to the rear output side of the radiatior matrix and, in doing so, extracts heat from the vertical columns of water in the tubes.

The cooled and therefore denser water sinks to the bottom of the radiator so that the existing water occupying the bottom tank is compelled to flow into the water jackets surrounding the cylinder barrels to replace the heated water which has moved up to the cylinder-head. Unfortunately, this form of coolant circulation is slow and cannot adequately cope with the heat released to and

from the cylinders and cylinder-head walls. Thus, under high-load and high-speed operating conditions, unless very large water jackets in the cylinder-block and cylinder-head (and a high mounted and relatively large vertical down flow radiator) are utilized, this type of cooling system would overheat when climbing hills or operating under load at low speed.

14.2.2 Vertical down-flow radiator
(Figs 14.3 and 14.4)

Traditionally, cooling systems used the vertical down-flow radiator as it was originally intended; that is, for coolant circulation based on thermo-syphon principles. The incorporation of the water pump in the system enabled a smaller heat-exchanger matrix, header-tank and bottom-tank to be used.

The benefit of the vertical down-flow radiator is that although the coolant movement takes place mainly by forced convection there is still a small amount of liquid-cooled movement caused by thermo-syphoning.

This mode of liquid coolant movement therefore provides the minimum flow resistance with the maximum heat transference effect. However, although the overall height and slimmed-down dimensions of a modern radiator make it suitable for commercial vehicle applications, it is difficult to accommodate the vertical down-flow radiator underneath the low streamlined bonnets of present-day cars.

Pressurized cooling systems
It is now standard practice to incorporate pressure relief valves in the filler cap to pressurize the cooling system, as this permits smaller sized radiators and fans to be used for a given cooling performance. Constraints on the degree of pressurization of a cooling system are the durability of the radiator and hoses and the maximum working temperatures which can be tolerated by the engine components.

Pressure cap blow-off pressures depend to some extent on the size of engine and its application, a typical list of pressure settings are as follows:

cars 0.88–1.00 bar (13–15 lbf/in^2)
light trucks 0.48–1.00 bar (7–15 lbf/in^2)
medium trucks 0.48–0.88 bar (7–13 lbf/in^2)
heavy trucks 0.27–0.61 bar (4–9 lbf/in^2)

14.2.4 Cooling system with cylinder block input flow control
(Fig. 14.3)

A different approach, which provides relatively close control of the quantity of coolant flowing to the water pump and entering the cylinder-block, is so arranged that the thermostat controls the outlet of coolant from the radiator as it passes to the inlet side of the water pump. With this layout, the radiator is subjected to the full pressure of the water pump whereas the return to the cylinder-block via the water pump is regulated by the throttling effect of the thermostat valves. This method of controlling coolant circulation tends to avoid air becoming trapped in the cylinder-block and head coolant passageways.

Attached to the wax capsule at one end is a large main flow-valve, while at the opposite end is a smaller bypass valve. When the engine is cold or warming up the main flow-valve is closed while the bypassing valve is fully open. Cold coolant flowing through the radiator under these conditions cannot return to the water pump inlet; however, a small proportion of heated coolant coming directly from the cylinder-head will flow through the bypass valve and pipe to feed the inlet to the water pump so that a restricted circulation of liquid will exist. This limited liquid movement enables the cylinder-block and cylinder-head quickly to warm up to operating temperatures without any tendency for the coolant to boil or for the engine to overheat and cause damage to critical components.

As the engine's mean temperature rises to approximately 75°C the wax in the thermostat capsule expands, which pushes it away from its thrust-pin. The main flow-valve progressively begins to open while, at the same time, the bypass valve commences to close. Thus, during this inbetween stage of reaching normal operating load coolant temperatures, both valves will be partially open. However, under heavy loads and at high ambient temperatures the main flow-valve opens fully at 95°C and, about the same time, the bypass valve completely closes so that coolant circulation is now entirely via the radiator.

Hence, controlling the throttling of coolant delivery to the water pump, and the amount of coolant bypassing the radiator, provides a precise method of adjusting the coolant flow and the heat rejection to the ambient surrounding air so that it matches the changing operating conditions imposed on the engine.

If a liquid-to-liquid oil cooler is to be incorporated into the system, this may be connected between the radiator bottom outlet tank and the input to the thermostat valve (Fig. 14.3).

14.2.5 Heavy duty cooling systems
(Fig. 14.4)

With large diesel engines, which may have more than one cylinder-head for each cylinder-block, better distribution of coolant delivery and collection can be obtained by having external coolant manifolds bolted onto the cylinder-block and to the individual cylinder-heads. Thus, for Fig. 14.4, it can be seen that a coolant delivery manifold attached directly to the water-pump lies parallel to the cylinder-block and has one branch pipe bolted centrally between each pair of cylinders. In turn, the hot coolant flow to the radiator is provided by the tapered collector manifold which has one branch pipe for each cylinder. In some arrangements there might only be one branch pipe between pairs of cylinders joining the cylinder-head passageways to the manifold. The output from the collector manifold can be seen to pass to a two-stage thermostat where there are two (sometimes three) thermostats which are phased to open at different temperatures. Attached to one of the thermostat capsules is a bypass valve which closes the bypass passageway when its main flow valve opens. This design provides a much closer control of the coolant circulation during differing operating conditions with the least flow resistance.

A further refinement to de-aerate the different parts of the cooling system employs a header expansion tank which has pipes leading to three different positions in the system—the top of the radiator, the output side of the thermostat housing, and the collector manifold. Usually, the pipe joining the thermostat housing to the expansion tank has a slightly larger diameter than the other two, as it functions as the filler pipe, whereas the other two pipes compensate for the expansion and contraction of the coolant or the expulsion of air or gas from the system.

With heavy duty engines it is usual to incorporate an oil cooler which is liquid cooled. In this example, the end of the delivery manifold supplies cool coolant to the inlet side of the oil heat exchanger. The heated coolant then passes out to the coolant collector manifold. Conversely, the hot coolant from the collector manifold is sup-

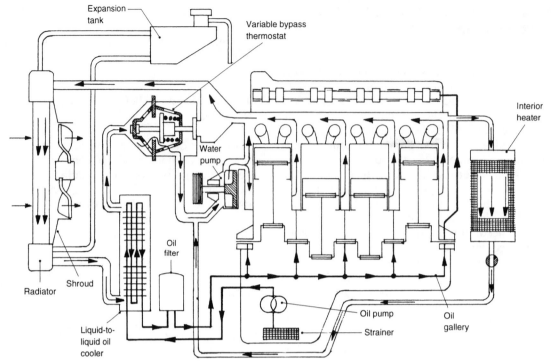

Fig. 14.3 Medium size engine vertical flow cooling system with interior heater and liquid-to-liquid oil cooler

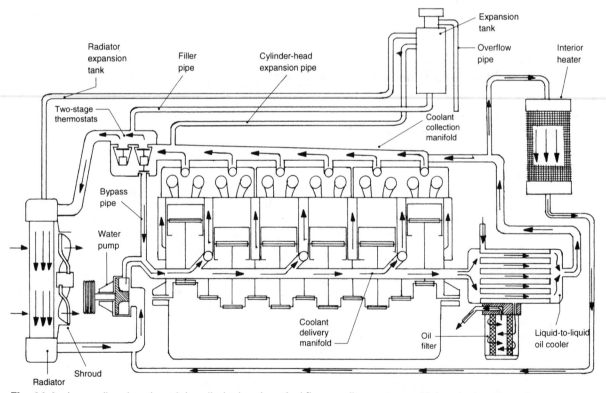

Fig. 14.4 Large diesel engine triple cylinder head vertical flow cooling system with interior heater and liquid-to-liquid oil cooler

plied to the cab's interior heater where it releases its heat and returns to the main cooling system at the input side of the water-pump.

14.2.6 Cross-flow cooling system
(Figs 14.5 and 14.6)

To reduce the overall height of radiators used in cars, the cross-flow matrix type has generally superseded the down-flow type. With these radiators there is no thermo-syphon action and coolant movement through the cooling tubes is controlled entirely by the centrifugal-force created by the water-pump. The cross-flow radiator can generally be of the single-pass (Fig. 14.5) or double-pass (Fig. 14.6) kind where the liquid flow is either one way only, or where it flows from one side to the other and back again respectively.

The single-pass radiator matrix normally has two or three rows of flattened tubes or channels in which the liquid coolant flow is unidirectional (Fig. 14.5). However, with the double-pass radiator, the liquid may flow in two ways as follows:

a) one way through the top half of the radiator and then back again through the lower half (Fig. 14.6);
b) one way through the front half-layer of the radiator and then back again through the rear half-layer (Fig. 14.21).

The double-pass cross-flow radiator has mainly been used for small cars, whereas the single-pass cross-flow radiator has been, and is, employed for all sizes of engines and is currently the most common type of radiator used.

14.2.7 Cooling system with cylinder-head output flow control
(Figs 14.5 and 14.6)

Hot operating conditions
With the cross-flow cooling system, coolant is forced out from the pump and enters the cylinder-

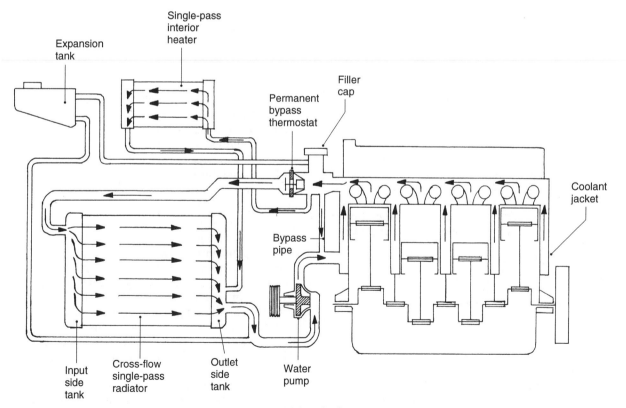

Fig. 14.5 Single pass cross-flow cooling system with interior heater

725

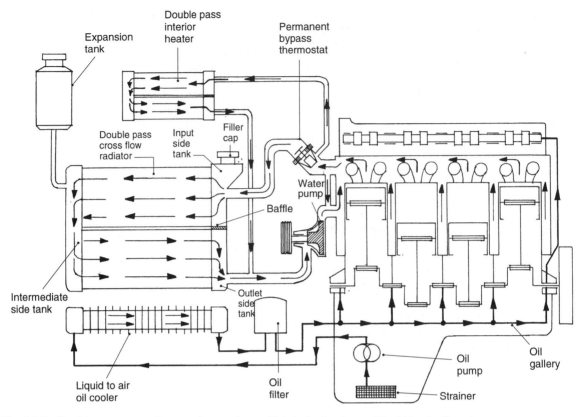

Fig. 14.6 Double pass cross-flow cooling system with interior heater and liquid to air oil cooler

block by the front of the engine where it is directed and flows towards the rear of the block and, at the same time, a proportion of the coolant flows around each cylinder barrel before it rises to the coolant passages in the cylinder-head. The hot coolant is now forced towards the front of the cylinder-head where it passes via the thermostat housing to the input side-tank of the radiator. Coolant then flows down the input side-tank and across the heat-exchanger matrix via the cooling tubes to the output side-tank and, in the process, discharges its heat to the air stream, which passes around and through the tube matrix. The coolant now passes from the radiator's outlet to the inlet side of the water-pump where the forced circulation process is again repeated.

Cold operating conditions

When the engine is cold the thermostat will be closed, coolant circulation is therefore limited entirely to the return flow from the cylinder-head exit and back to the delivery entrance in the cylinder-block by way of the bypass pipe. Coolant

movement, which may amount to only 5%–10% of the normal liquid coolant circulation under these conditions is achieved solely by thermosyphon action. However, this restricted movement of liquid is essential to prevent overheating and boiling taking place at critical areas in the cylinder-head during the warm-up period.

De-aeration
(Figs 14.5 and 14.6)

To remove air, gas or steam bubbles from the liquid coolant in the coolant passages it is desirable to have a low velocity flow region just below the radiator cap in which air, gas and (in extreme cases) steam can separate from the coolant. The formation of air and gas in a cooling system under operating conditions is continuous so that a permanent bleed system is necessary to de-aerate the passageways. This may take the form of a pipe connecting the space below the filler cap to the expansion-tank with a second pipe joining the expansion tank to either the radiator outlet side-tank or to the water-pump inlet (Fig. 14.5). Thus, with this system, air, gas and possibly steam will rise to the highest point in the cooling passages; that is, just behind the thermostat or slightly below the radiator-cap (Fig. 14.5). In turn, the condensed gas and/or steam in the expansion-tank can feed back into the inlet side of the water-pump so that it again becomes part of the main liquid coolant circulating throughout the system. An alternative coolant filling and air-bleed arrangement has the filler cap mounted on the input side-tank of the radiator, whereas a bleed-pipe connects the opposite intermediate side-tank directly to the expansion-tank (Fig. 14.6).

14.3 Fundamentals of radiator design

Factors which influence the size and effectiveness of a radiator to be installed in an engine cooling system are as follows.

1 Increasing the coolant flow-rate reduces the temperature drop between the entry and exit passageways.
2 Increasing the mean temperature difference between the coolant and the air entering the matrix core raises the heat dissipation capacity of the radiator.
3 Increasing the coolant flow-rate, proportionally raises the power required to drive the water-pump and therefore offsets the improvement in radiator efficiency at high flow-rates.
4 A suitable liquid coolant temperature drop through the radiator matrix ranges from 5°C to 10°C at full throttle over the normal working speed range of a typical engine.
5 Coolant containing up to 50% ethylene glycol reduces the heat dissipation capacity by something like 15%, owing to the lower specific heat capacity compared with that of water.
6 Increasing the cooling system's operating pressure raises the boiling point of the coolant. A 0.1 bar increase in coolant pressure raises the coolant temperature by approximately 2.5°C. The raised coolant temperature entering the radiator increases the temperature difference between the liquid coolant input to the core tubes and the air-flow temperature entering the frontal area of the radiator matrix. The thermal capacity of the radiator is increased by something like 1.8% for each 1°C increase of temperature difference between the coolant input temperature and the input air temperature.
7 The most economical radiator is the one with the narrowest matrix depth and the widest air cell spacing that will provide the required heat dissipation within the frontal area available.
8 Increasing the matrix core depth increases the heat dissipation capacity but this capacity increase becomes much smaller beyond some critical depth. For example, a 50% increase in core depth improves the heat dissipation capacity by 15%, whereas a 100% increase in depth raises its capacity by only 20%.

Radiator matrix design and construction

There are three basic radiator matrix designs which are in use today, and these are listed as follows:

a) cellular (film) and corrugated fin type
b) elongated tube and corrugated fin type
c) elongated tube and flat fin type

14.3.1 Cellular (film) and corrugated-fin type of matrix
(Fig. 14.7(a))

Vertical liquid coolant channels are formed by corrugated (wavy) strips bent into narrow rectangular box shapes. The end strip portions are folded flat whereas the long strips are corrugated and have flat depressions superimposed at right angles to the corrugated profile. The rectangular strips with corrugated sides are placed back to back so that the half depressions in each corrugated side member form a closed channel which passes through the long strips from end to end. Rows of these corrugated rectangular strips are placed side by side to make up the required width of radiator core. The surface areas of the strips are then thermally and mechanically bonded with tin–lead solder. Ribbons of electrolytic copper are folded, concertina fashion, between the corrugated strip spacings: their function being to form triangular air cells which readily transfer heat from the liquid coolant channels to the air stream passing from one side of the matrix to the other. Louvres pierced in the ribbon fins are designed to promote turbulence in the air cells as the air passes between the triangular walls. Through these means, more heat is extracted from the fins than would otherwise be possible.

The cellular matrix radiator has the highest

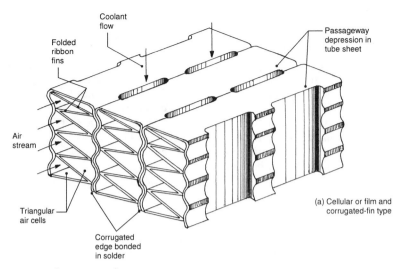

Fig. 14.7 Radiator core-matrix construction

thermal efficiency relative to its weight compared with the other two types, and the inherent flexibility of the cellular construction makes this design less likely to be damaged from road shock and vibration. These radiators are simpler to repair than the tube core types. Conversely, the flexibility of the cellular construction may make it susceptible to failure if subjected to rapid and large changes of internal pressure.

14.3.2 Elongated tube and corrugated-fin type of matrix
(Fig. 14.7(b))

The liquid coolant tubes utilized for this design of radiator matrix have a flattened elongated sec-tion, are lock-seamed at one edge and are made from high-grade brass strip 2.0 mm in thickness. The fins, on the other hand, are made from copper strip 1.2 mm thick. The elongated tube sections are placed three abreast and are spaced by a copper ribbon folded in a triangular fashion. The tubes and air-cell ribbon folds are stacked alternately to produce the radiator width, and the whole assembly is then baked to form a solder between the flat surface of the tubes and the triangular edges of the fins. Again, to increase the heat dissipation, louvres are pierced in the copper ribbons before the radiator is folded as this is the most effective method of producing turbulent movement as the air passes through the triangular-sectioned passageways.

The tube and corrugated-fin matrix radiator

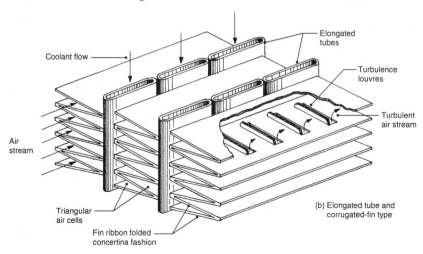

Fig. 14.7(b)

728

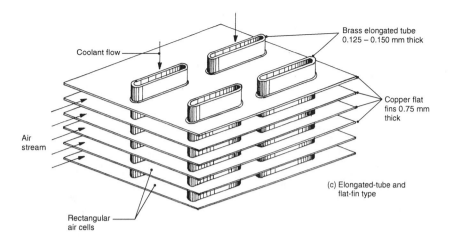

Coolant flow

Brass elongated tube
0.125 – 0.150 mm thick

Copper flat
fins 0.75 mm
thick

Air
stream

(c) Elongated-tube and
flat-fin type

Rectangular
air cells

Fig. 14.7(c)

combines the structural strength of the tube and flat-fin type (and its resistance to internal pressures) with the higher thermal efficiency of the cellular and corrugated-fin construction.

Spacing the tubes by the corrugated fins is less expensive for manufacturers than the flat-fin type of construction, and is now the most popular radiator construction in current use.

14.3.3 Elongated tube and flat-fin type of matrix
(Fig. 14.7(c))

Lock-seamed flattened tubes fit at right-angles into flanged slots pierced in each layer of copper-sheet (which makes up the fins). The flanged slots in the copper fins are deliberately made to provide more support for the tubes and to increase the surface area of the soldered joints. When the lock-seamed tubes are sealed, a heavy layer of solder is deposited so that during the baking operation following assembly, the solder melts and fuses to the flange formed around each slot. The parallel layers of copper-sheet have louvres pierced between the tube slots so that a high degree of turbulence is generated inside the rectangular air-cells, as the air stream passes through these passageways.

The tube and flat-fin type radiator has superior structural strength and can withstand high internal pressures. These radiators are therefore used for heavy-duty applications. However, the thermal efficiency for the tube and flat-fin construction is lower than that for the cellular and corrugated-fin type and it is more expensive to manufacture than the other two types described.

14.4 Thermostats
(Fig. 14.8)

Thermostats are used to block the coolant circulation to or from the radiator when the engine is cold or warming up so that the trapped coolant in the cylinder-block and cylinder-head passageways absorbs and accumulates the rejected heat of combustion, causing the engine to reach its normal working temperature rapidly. When the circulating coolant in the thermostat housing reaches the crack-opening temperature of the thermostat, the valve commences to open. However, it does not fully open until the temperature in the thermostat housing has risen to its designed operating value. The opening of the thermostat valve then permits the coolant to flow from the cylinder-head to the radiator. However, a small proportion of the uncooled coolant will continue to circulate through the permanent bypass pipe.

The basic operating principle of all thermostats is similar but the construction details may vary to suit different applications. The thermostat consists of an upper bridge pressing centrally supporting a stainless-steel thrust-pin and a lower body pressing, which has its flanged outer rim spun over to secure it to the upper bridge pressing (Fig. 14.8). A centrally mounted expansion capsule supporting the main flow annular-shaped valve fits over the thrust-pin and is itself held in position by a stainless-steel return-spring. The expansion capsule and the annular disc valve are the moving members of the unit. The brass cylindrically shaped capsule contains the wax content

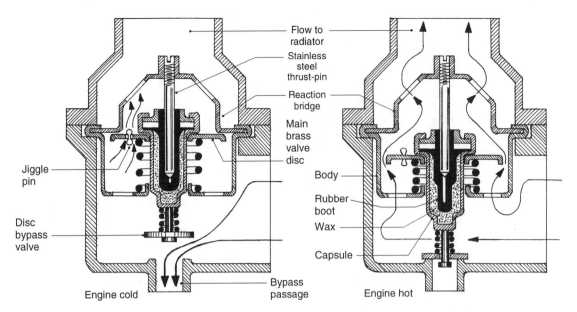

Fig. 14.8 Thermostat with disc-valve radiator and by-pass flow control

and a synthetic rubber boot seals the wax and simultaneously distorts its shape to accommodate the volumetric expansion or contraction of the wax according to the change in temperature. The wax enclosed between the capsule case and rubber-boot is a microcrystalline paraffinic type which may be mixed with copper powder to increase its thermal conductivity. Its coefficient of volumetric expansion is typically 0.9%/deg C, which provides it with a maximum lift of 9.5 mm with a thrust of no less than 150 N for a 0.9 g content.

Generally, the thermostat only controls the coolant flow to the radiator so that there is always a permanent open bypass passage. For more precise regulation of the coolant under different operating conditions, a positive controlled bypass actuated by the thermostat can be used.

Four different thermostat arrangements, which simultaneously control both the flow to the radiator and to the bypass circuit, are described next.

14.4.1 Thermostat with disc valve radiator and bypass flow control
(Fig. 14.8)

With this design of thermostat, the main return-spring takes up the space between the lower fixed thermostat body pressing and the main radiator

flow annular-shaped disc-valve, which fits over and against the flanged portion of the wax-capsule. A spring-loaded bypass disc-valve and stem is attached to the lower closed end of the capsule. When the coolant is cold, the contracted wax permits the return-spring to push the capsule up and over the thrust-pin, up to the point where the valve closes against the upper reaction bridge seat pressing. In this position the lower bypass valve is held away from the bypass entrance in the thermostat housing. Under these conditions coolant is prevented from flowing to the radiator, but a small proportion of coolant is able to circulate between the cylinder-head outlet and the cylinder-block inlet via the bypass passageway.

As the coolant temperature reaches the crack-point of the capsule and valve, the expanding wax grips the lower end of the rubber boot so that the capsule moves away from the fixed thrust-pin. Consequently, the main radiator flow-valve commences to open while the bypass valve begins to close-off the bypass entrance. Eventually, the temperature of the coolant flowing around and past the capsule will rise to some predetermined operating value, at which point the radiator flow-valve will be fully open, and the bypass valve will be fully seated.

The complete interruption of the bypass circuit enables the coolant flow from the cylinder-head and the cylinder-block to be controlled entirely by

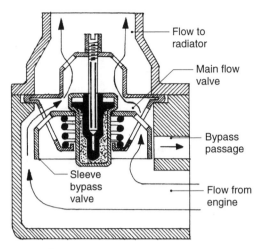

Fig. 14.9 Thermostat with disc and sleeve valve radiator and by-pass flow control

the main radiator flow-valve. Through these means the output coolant flow from the cylinder-head to the radiator can be accurately monitored and throttled to match the variation in heat released to the coolant from the cylinders under the changing operating conditions imposed on the engine.

14.4.2 Thermostat with disc and sleeve valve radiator and bypass flow control
(Fig. 14.9)

An alternative method of controlling the bypass flow when the main indicator flow passage is near closure, is achieved by forming a sleeve rim around the main flow disc-valve so that the valve becomes cup-shaped. Thus, the inward and outward movement of the capsule along the fixed thrust-pin as the main flow-valve closes and opens respectively moves the sleeve across the bypass entry port.

When the coolant is cold the return-spring pushes the wax-capsule over the thrust-pin until the main radiator flow-valve closes. At the same time the outer valve sleeve uncovers the bypass entry port. Under these operating conditions, the flow to the radiator is blocked but a small amount of coolant circulation between the cylinder-head and cylinder-block will continue by way of the bypass passage.

As the coolant temperature reaches the 'crack-point', the expanding wax squeezes the protrud-

ing end of the rubber boot inwards, this forces the capsule away from the fixed thrust-pin and, in so doing, progressively opens the main flow to the radiator. Simultaneously, the bypass port will be cut-off by the sleeve, proportional to the backward movement of the valve. The double acting cup-shaped valve therefore converts the restricted bypass flow to full radiator circulation as the coolant becomes hot. Conversely, when the coolant cools or is only warm, the valve throttles the radiator circulation and diverts the coolant flow to an uncooled and much restricted bypass circuit, it therefore speeds up the warm-up process without causing overheating of critical zones in the cylinder-head and block during this transition phase.

This design of thermostat valve can be susceptible to the sleeve face sticking if the bypass port-to-sleeve clearance is small and scale builds up around the port entrance.

14.4.3 Thermostat with ventless sleeve valve radiator and bypass flow control
(Fig. 14.10)

For heavy-duty applications a sleeve-valve thermostat can be employed in which the wax-capsule is attached to the valve seat pressing and the thrust-pin and sleeve become the moving members. The valve sleeve (in the form of an annular ring) is attached to the thrust-pin, whereas the capsule is supported by the valve seat pressing (which is sandwiched between the upper and lower halves of the thermostat housing). The sleeve valve slides between a rubber seal, which separates the bypass passage from the passage leading to the radiator. A return-spring spaced between the shoulder of the capsule and the spring cage, which is fixed to the thrust-pin and sleeve-valve, holds the valve sleeve down on its seat when the coolant is cold. Under cold coolant circulation conditions coolant flow to the radiator is blocked, but the coolant is free to pass through the centre of the sleeve-valve on its way to the bypass passage.

When the coolant reaches the capsule's 'crack-point' the expanded wax squeezes together the lower part of the rubber boot so that the thrust-pin is pushed partially out from the capsule boot until the sleeve hits the end wall of the upper thermostat housing. Coolant can now flow be-

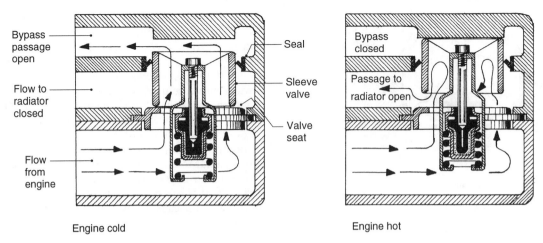

Bypass passage open

Flow to radiator closed

Flow from engine

Seal

Sleeve valve

Valve seat

Bypass closed

Passage to radiator open

Engine cold

Engine hot

Fig. 14.10   Thermostat with ventless sleeve valve radiator and by-pass flow control

tween the capsule and valve seat pressing where it then passes to the radiator passageway.

This design of thermostat has very little uncontrolled leakage between the radiator and bypass passage, as in the case of the disc and sleeve-valve type of thermostat.

14.4.4   Two-stage thermostat flow control
(Fig. 14.11)

Owing to the much greater circulation of coolant in a large diesel engine, it is usual to have two or even three thermostats of standard size mounted either separately or on a common disc-plate in the thermostat housing. The necessary flow-rate to meet the water-pump demands over the engine's speed range can therefore be easily met without causing undue flow resistance, which would be experienced with a single thermostat unit unless it was exceptionally large and costly.

Further benefits of having two or three thermostats together is that they can be phased to open at different temperatures during the warming up stage and therefore good closure control of the coolant under different operating conditions can be obtained. Generally, if there are three thermostats, one will open earlier at a lower temperature than the other two to satisfy light load and speed operating conditions, but at higher engine load and speed conditions the other two thermostats will open to cope with a greater circulation of coolant. The increased heat rejected by the cylinders can therefore be dissipated with the mini-

mum of flow resistance and power consumed from the water-pump.

With the arrangement shown in Fig. 14.11 it can be seen that normal control is achieved with the primary thermostat, which regulates the main radiator flow and provides positive bypass control. However, under heavy load and speed conditions the secondary thermostat opens at slightly higher operating temperatures to increase the flow-rate of hot coolant passing through the radiator.

14.5   Fan blade drive and shutter control

14.5.1   The need for fan blade drive control

The radiator cooling fan is wasteful in that it takes power from the engine to drive it when, in fact, the ram-air available, due to the forward movement of the vehicle or a head wind, gives adequate air-stream movement through the radiator core to cool the liquid coolant circulating through the core's cooling passages for the majority of the vehicle's operating time.

Fans are designed to cool the radiator core passages when the engine is producing power under high temperature environments, or operating at low vehicle speed under load with insufficient forward ram air effect. Generally, these conditions only occupy a small part of most car journeys; however, heavy goods vehicles do have

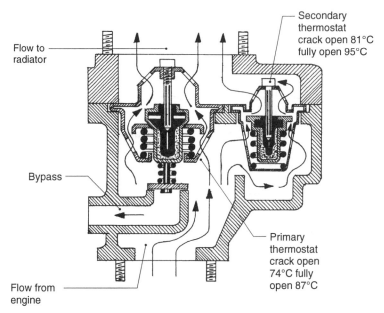

Fig. 14.11 Two-stage thermostat control

to operate for longer spells in low gear under load when ascending hills.

The engine is therefore overcooled for the greater part of its running life and as a result, does not operate at its optimum temperature to produce maximum thermal efficiency and minimum frictional resistance. In addition, the fan absorbs power (at the cube of its speed) which, at high speed is wasteful. Simultaneously, the interaction of the air with the blades generates considerable unwanted noise. The object of automatic fan-blade drive control is to idle or reduce the fan-blade speed when the engine is warming up from cold or when there is adequate external air movement owing to high-speed motoring.

14.5.2 Electric motor driven thermoswitch controlled fan
(Fig. 14.12)

Fans which do not consume too much power, for car and light van use, are frequently driven by two-pole permanent-magnet two-brush electric motors. The fan and electric motor assembly is positioned to one side and opposite the radiator matrix by a fan shroud. The shroud's functions is to concentrate the pressure drop created by the fan's rotation against the radiator core so that air on the opposite side of the radiator is drawn through the matrix air cells.

When the ignition switch is closed, a voltage will be applied across the relay winding, the energized winding then draws the contacts together so that current from the battery now passes via the relay contacts, the fuse and thermo-switch, to the electric motor. Turning off the ignition de-energizes the relay winding. It thus causes the contacts to open, thereby interrupting the current supply to the fan's electric motor.

The fan is automatically switched in and out by a bimetallic-type thermoswitch, which cuts-in and cuts-out the current supply to the electric motor when the thermoswitch mounted in the side of the radiator reaches the cut-in temperature (typically 95°C). It then cuts-out when the coolant temperature around the thermostat drops about 5°C below the cut-in temperature.

On some models, a two-speed electric motor is incorporated. Typically, the first speed cuts-in at 96°C and cuts out at 92°C, whereas the second speed cuts-in at 101°C and cuts out at 97°C.

14.5.3 Magnetically controlled thermo fan drive
(Fig. 14.13(a and b))

Construction and design
The magnetic coupling comprises a solenoid ring winding wound on an iron core which is pressed

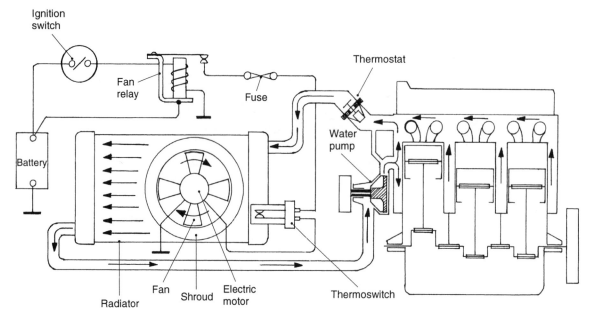

Fig. 14.12 Electric motor driven thermo-switch controlled fan

onto the water-pump pulley wheel. Mounted on the end of the water-pump shaft is a double-row ball race and hub which supports the fan-blades. Also supported by this hub is a flat annular-shaped iron armature, which is attached to the bearing hub by means of three flexible spring arm-blades (leaf springs) riveted at their ends and set tangentially to the hub and annular armature. This annular-shaped armature is so positioned that the clearance between both armature and solenoid-core contact surfaces are at a minimum without producing drag. A single insulated inclined brass slip-ring is attached to the inside of the pulley wheel on the water-pump side, it being electrically connected to one end of the solenoid winding, while the other end of the winding is earthed. To energize the solenoid winding, current is supplied to a spring-loaded carbon brush mounted on the water-pump housing. Current will therefore pass to the solenoid via the contact made between the slip ring and brush and by way of the ignition switch and thermoswitch.

Operation

Cold operating conditions
(Fig. 14.13(a))
When the engine is operating cold, the thermo-switch contacts open so that the electrical connection to the solenoid via the slip-ring and brush is interrupted. Under these conditions, the fan-belt drives the pulley, water-pump shaft and blades, but leaves the fan-blades and hub to free-wheel on the double row bearing and shaft.

Hot operating conditions
(Fig. 14.13(b))
As the engine warms up the coolant temperature reaches some predetermined value, usually between 85°C and 95°C, at which point the thermo-switch contacts close. Current will now pass to the solenoid winding via the slip-ring and brush, and the energized winding magnetic field now pulls the annular armature hard against the solenoid iron core, thereby causing the fan-blade assembly to rotate with the pulley-wheel. The revolving blades will draw cold air through the radiator matrix until the engine cools due to the forward speed of the vehicle, providing a ram effect to the air entering the radiator or when the engine is stopped. This cycle of engaging and disengaging is automatic and continues for all operating conditions.

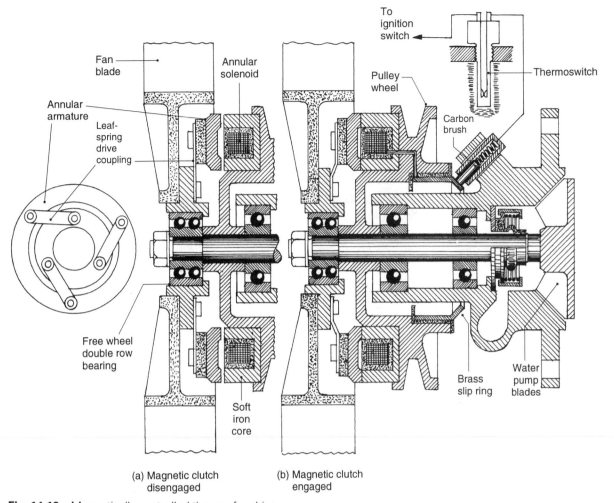

Fan blade

Annular solenoid

To ignition switch

Pulley wheel

Thermoswitch

Annular armature

Leaf-spring drive coupling

Carbon brush

Free wheel double row bearing

Brass slip ring

Water pump blades

Soft iron core

(a) Magnetic clutch disengaged

(b) Magnetic clutch engaged

Fig. 14.13 Magnetically controlled thermo-fan drive

14.5.4 Speed sensitive flexible variable pitch fan blades
(Fig. 14.14(a and b))

Construction and design
These consist of flexible one-piece plastic-moulded fan-blade assemblies bolted to the water-pump pulley hub, with their blades set to provide maximum pitch while in a free state.

Operation
When rotated, the blades (due to the material's torsional flexibility) tend to reduce their pitch. This is due firstly to the increased air drag straightening out the blade set when cutting through the atmosphere and secondly to the centrifugal stress created in the plastic tending to straighten out the blade tips radially, with increased speed. This construction thus provides the maximum blade pitch at low speed when there is very little air movement through the radiator (Fig. 14.14(a)). At higher speeds (Fig. 14.14(b)) when the air stream through the radiator core is greater, the air drag and centrifugal loading of the blades progressively reduces the pitch of the blades—this reduces the air displaced by the blades and hence their capacity to draw air through the radiator matrix.

The reduced elastic range, to match the fatigue strength of the plastic blades, restricts the pitch variation that can be expected during a blade's normal service life; however, this type of fan-

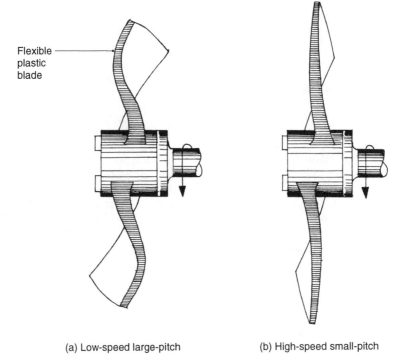

(a) Low-speed large-pitch   (b) High-speed small-pitch

Fig. 14.14   Speed sensitive flexible variable pitch fan-blades

blade does provide a certain amount of cooling control over the engine's speed range.

14.5.5   Mechanical crank-pin variable pitch temperature sensitive fan drive
(Fig. 14.15(a and b))

Construction and design
This mechanism has the blades radially mounted on a central housing bolted to the pulley wheel support sleeve. The pivot-shaped inner end of each of the four blades fits into sockets formed in the central housing, while a ring-clip located in grooves half-way along the pin and sockets prevents the fan-blades from being pulled out the central housing. The pitch of each blade is synchronized to a grooved sliding sleeve positioned inside the central housing via eccentric-pins attached to the pivot end of the blades. A return-spring is located between the central housing and one side of the sliding-sleeve, whereas a rod connects the other side of the sleeve to an expansion wax-capsule mounted at the far end of the flanged support sleeve.

Operation

Cold operating conditions
(Fig. 14.15(a))
As the coolant warms up, the wax in the capsule expands, thus causing the thrust-pin to push the rod and sliding-sleeve against the return-spring in the outward direction. Consequently, the grooved sliding-sleeve cranks the eccentric pins about their pivots, thus causing the blades to twist around to a larger pitch position. The wax-capsule therefore twists the blades an amount proportional to the rise in coolant temperature. The effect of this will be to maximize the displacement of cold air being drawn through the radiator matrix.

14.5.6   Viscous fan drives
(Fig. 14.16(a and b))

The need for viscous fan drives
Engine cooling systems normally incorporate a cooling fan driven directly from the crankshaft. This means that the fan speed will not be matched

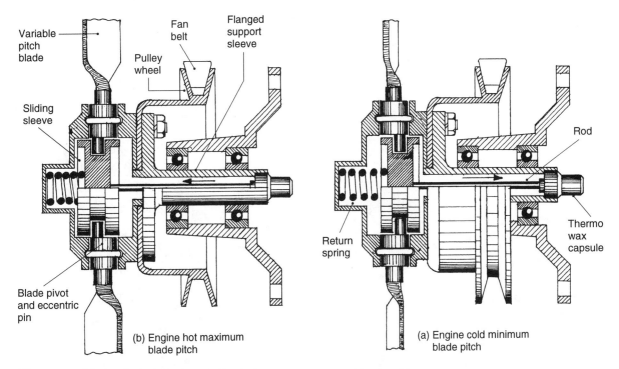

Fig. 14.15 Mechanical crank-pin variable pitch temperature sensitive fan-drive

to the cooling needs of the engine under a wide range of conditions and, at high engine speeds in particular, the fan absorbs power unnecessarily and creates undue noise. Viscous fan drives partially compensate for these limitations.

Torque limiting viscous couplings
These units operate as shear-type fluid-couplings—the drive being transmitted from the input drive disc to the output housing by a film of silicone-fluid. When the clearance between the drive disc and housing is filled with this viscous silicon fluid, the shearing action caused by the speed differential between the clutch components transmits torque from the drive pulley to the fan. As the speed of the engine increases, the torque needed to transmit the drive from the water-pump pulley to the fan-blades also increases. In operation, the fan speed increases, at first directly with engine speed, and then at a lower rate due to the rise in viscous drag until the limit of the torque capacity (which has been pre-set for the drive's application) is reached. At this point, the fan is operating at its maximum speed which will not be exceeded as the engine speed continues to rise. The torque limiting viscous coupling is very

similar to air-temperature sensing viscous coupling except that it does not have the air temperature sensing valve control and relies only on the constant amount of silicon oil in contact with the drive plate annular faces and the side walls of the casing (Fig. 14.16).

14.5.7 Air temperature sensing viscous coupling
(Figs 14.16(a and b) and 14.17)

Construction and design
The coupling comprises an aluminium housing with cast cooling fins, a steel separator-plate and a steel end-plate assembly, mounted on a double-row ball race. Enclosed between the aluminium housing and separator-plate is a disc-shaped drive plate, whereas the space between the separator-plate and end-plate acts as a reservoir for the silicon-fluid. A flanged coupling spindle bolted on the water-pump flanged shaft supports both the double-row ball race and the drive-plate. Attached to the outside of the end-plate is a bimetallic strap which is sensitive to the ambient temperature of the air, this controls the opening

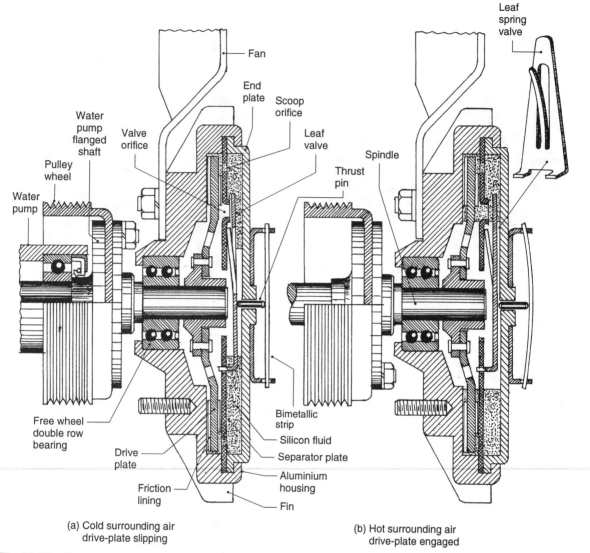

Fig. 14.16 Air temperature sensing viscous coupling

and closing of a spring leaf-valve via a central thrust-pin. The separator-plate has a large central hole to clear the drive-plate hub, one or two scoop orifices near the outer periphery, and a single valve orifice on the inside of the drive-plate active annular surface. The fan-blades are themselves bolted to the aluminium housing.

Operation

Cold operation conditions
(Fig. 14.16(a))
When the air temperature surrounding the vis-

cous drive assembly is low, the bimetallic strip (supported on a bracket) will be in a flat, relaxed, free state. Under these conditions the flat bridge of the bimetallic strip pushes the thrust-pin hard against the spring-leaf valve, thus causing the valve to close. With the pulley and viscous housing rotating, the silicon-fluid in the drive-plate chamber will be subjected to centrifugal force, hence the radial outward pressure on the fluid forces it to flow through the scoop orifice into the reservoir chamber on the opposite side of the separator-plate until the drive-plate chamber is practically empty. As a result, the clearance between the drive-plate and side members will interrupt the viscous drag so that the fan and housing assembly will freewheel while the water-pump pulley and viscous drive spindle are driven at engine speed.

Hot operating conditions
(Fig. 14.16(b))
As the air temperature surrounding the viscous coupling rises, the bimetallic strip tends to bend or bow outwards from the centre so that the leaf-spring is able to spring open the valve orifice. The stored silicone-fluid in the reservoir chamber is now able to return, via the valve orifice, to the drive-plate chamber. Thus, as the chamber fills, the submerged fluid surface area between the drive-plate and side members enlarges to increase the viscous drag between the coupling members. Consequently, the torque imposed on the outer housing and the fan assembly increases its rotational speed according to the fan air resistance and the speed of the water-pump pulley, which is driven by the crankshaft pulley via the fan-belt.

Conversely, when the vehicle is moving fast, the increased air stream moving through the radiator and over the engine cools the viscous-coupling assembly so that the bimetallic strip straightens out and causes the leaf-valve to close. Hence, the fluid circulation is blocked so that the majority of the fluid in the drive-plate chamber will be pumped, via the scoop orifice into the reservoir chamber. Again the lack of viscous drag

allows the outer housing and fan assembly to slip an amount proportional to the loss of surface area exposed to the fluid.

Merits of viscous drives

a) automatic operation
b) self-contained and sealed for life
c) increased fan-belt life
d) no maintenance required
e) saves engine power
f) reduces fan noise
g) ensures quicker warm-up
h) can be used with existing fan

Viscous fan drive characteristics for air temperature sensing types can be seen in Fig. 14.17.

14.5.9 Air-controlled cone clutch fan drive
(Fig. 14.18(a and b))

Construction and design
The fan drive coupling and its control basically consists of two units—a friction clutch fan-drive (which is pneumatically operated) and a coolant temperature-sensing air control-valve.

The fan-drive unit comprises a flanged support sleeve bolted to the water-pump pulley. This member also forms the inner cone of the clutch. A floating fan-hub is supported on the flanged

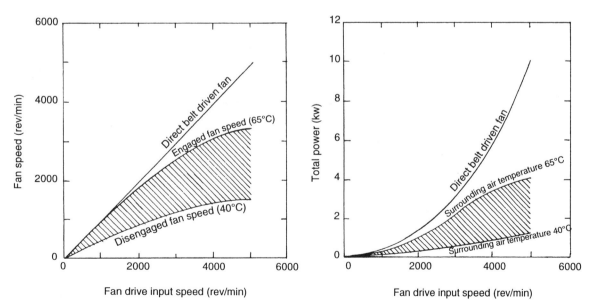

Fig. 14.17 Viscous fan-drive with air temperature sensing-speed and power absorption characteristics

support sleeve, but can revolve relative to it by means of roller bars. The rear end of the hub forms the outer cone member of the clutch, whereas the front end of the fan-hub is internally machined to form the servo-cylinder bore.

A servo-piston with an extended hollow stem fits inside this cylinder. The rear end of this hollow stem is supported at the rear on a ball-race mounted inside the flanged sleeve, and this permits it to rotate but not to move axially. An axial clutch thrust-spring is housed between the servo-piston and floating hub, and the latter forms the outer cone which, when engaging the inner cone, provides the drive from the water-pump pulley to the fan-hub and fan-blade assembly.

The air control valve mounted in the cylinder-head coolant outlet combines a spool-shuttle and poppet-valve, which control the supply of air pressure from the vehicle's air-braking system to the fan-drive. Its operation is controlled by a temperature sensing wax-capsule which activates the valve, opening or closing the air feed inlet port and exhaust port in the valve body.

Operation

Cold engine coolant
(Fig. 14.18(a))
When the engine coolant is cold, the wax-capsule thrust-pin is in its innermost position, which permits the poppet-valve and shuttle-valve to move away from the inlet port positioned at the top of the control-valve casing. Compressed air will then pass through the shuttle-valve on its way to the servo-cylinder, this then pushes the floating hub away from the fixed piston. It thus separates the inner and outer cone-clutch members and hence disengages the fan-drive. Under these operating conditions the hub and fan-blade assembly will freewheel at about 200 to 300 rev/min.

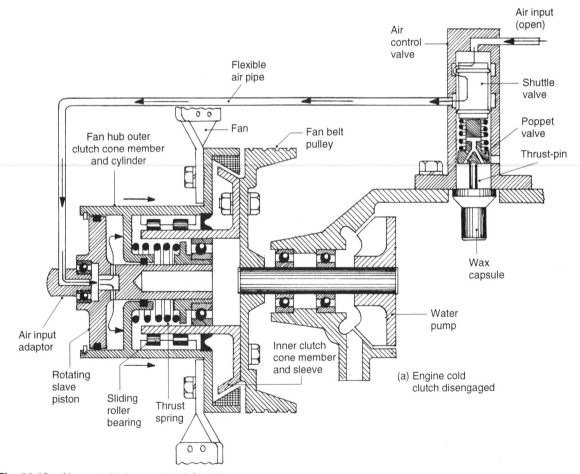

Fig. 14.18   Air controlled cone-clutch fan drive

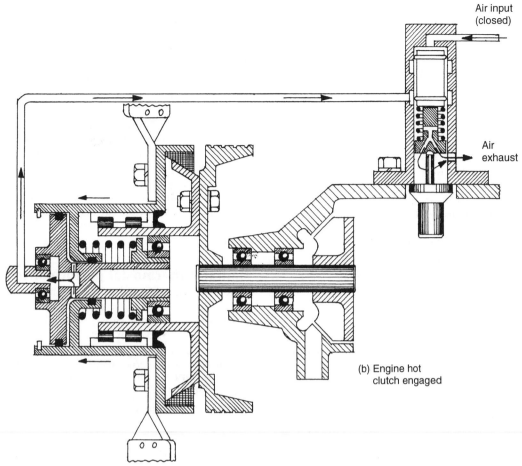

Fig. 14.18(b)

Hot engine coolant
(Fig. 14.18(b))
As the engine coolant reaches its working temperature, the wax capsule thrust-pin moves outwards, displacing the valve assembly hard against the inlet port, this closes the inlet port and prevents any more pressurized air from reaching the servo-cylinder. At the same time, the poppet-valve will uncover the exhaust port so that the residual air in the servo-cylinder can be released to the atmosphere. Under these conditions the clutch thrust-spring is free to expand and thus push the hub and outer cone assembly into engagement with the inner cone member so that the fan-drive is re-established. The fan-blades will now be positively driven at input water-pump pulley speed.

The engagement and disengagement characteristics are similar to that shown in Fig. 14.19.

14.5.10 Automatic radiator shutter control
(Fig. 14.20(a, b and c))

Construction and design
Components making up the radiator shutter system are the shutter frame and blades, an air-operated slave-cylinder actuator and a temperature sensing air control-valve (Fig. 14.20(a)). The shutter assembly is made up of a number of aluminium-alloy blades of aerofoil cross-section to provide the minimum air flow resistance. Each blade is pivoted at its end on a rectangular steel frame which is attached to the front face of the radiator matrix in the United Kingdom, but to the rear for United States' installations. Crank-arms connect each blade-pivot at one end to a link-bar, which is itself connected to the slave-piston via

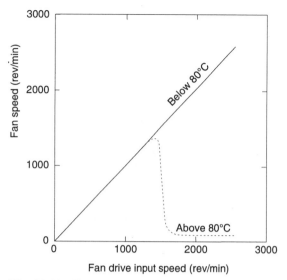

Fig. 14.19 Relationship between input drive and fan speed when coolant is below or above operating temperature

the relay-lever and ram-rod while the slave cylinder is anchored at its lower end to the shuttle frame. The shuttle-blades are held in the open position by the slave-cylinder actuator, fail-safe return-spring in case there is a failure of the air supply to the slave-cylinder.

The air control-valve is activated by a temperature sensitive wax-expansion capsule, which opens or closes an air supply port and an exhaust port via a spool-valve.

Operation

Cold engine coolant
(Fig. 14.20(b))
When the engine is cold, the wax-capsule thrust-pin will be in its innermost position, which permits the return-spring to move the spool-valve into its lowest point in its casing. In this position, pressurized air supply passes from the inlet valve port to the outlet port via the spool valve's annular groove, it then passes to the slave-cylinder. The compressed air then pushes the piston up against the fail-safe return-spring, and this causes the relay-lever to tilt and to move the link-bar so that the shutter-blades are compelled to rotate about their pivots until they are in the closed position.

Hot engine coolant
(Fig. 14.20(c))
As the engine coolant reaches its operating temperature the wax in the capsule expands causing the thrust-pin to be squeezed out, and at the same time it moves the spool-valve to its uppermost point. This movement cuts off the air-supply port and opens the exhaust port, thus preventing more air entering the servo-cylinder and releasing the existing compressed air trapped in the cylinder. The powerful fail-safe return-spring then pushes the piston to its lowest position and simultaneously cranks open the shutter-blades via the relay-lever and link-bar. It thus permits air movement through the radiator matrix to commence again.

Radiator thermostat performance
When the conventional cooling system thermostat closes, it stops nearly all coolant circulation through the radiator. Thus, coolant already in the radiator (approximately 60% of the total coolant in the system) quickly cools, and this may reduce its temperature in cold weather to something like 30°C to 40°C colder than the coolant in the engine. This stop and go thermostatic control creates an endless surge of cold quenching coolant through to the hot cylinder-block. This is capable of producing a temperature differential, from the bottom to the top of the engine, of as much as 40°C. Consequently, high thermo-cyclic stresses are created which distort the cylinder-block and head and will eventually cause damage to critical areas in the engine's cooling passageways.

The great benefit of the automatic radiator shutter system is that once the engine reaches its operating temperature, the main-flow radiator thermostat remains permanently open so that coolant circulation is continuous.

Coolant temperature control is now achieved entirely by the shutters regulating the quantity of air flowing to the radiator matrix.

Radiator shutter performance
Thermostatic fan-drives only control overheating situations, whereas the thermo-automatic radiator shutters prevent overcooling occurring. The automatic shutters transform the radiator into a variable-rate heat-exchanger by controlling and changing the air flow across the radiator core (cooling capacity) corresponding to the constantly changing load and speed of the engine.

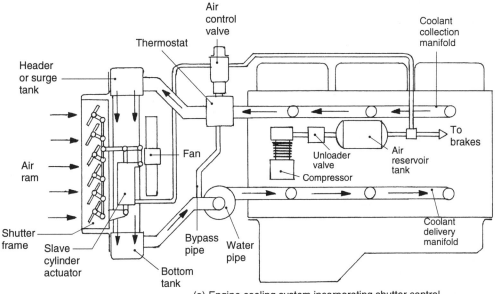

(a) Engine cooling system incorporating shutter control

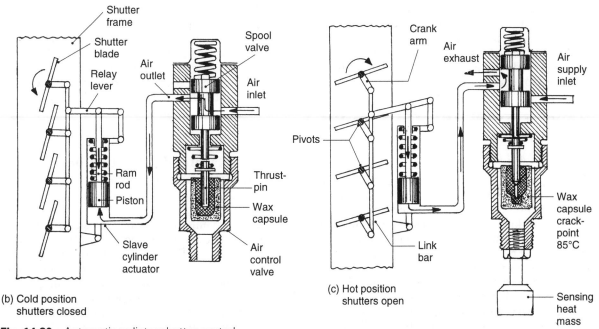

(b) Cold position shutters closed

(c) Hot position shutters open

Fig. 14.20 Automatic radiator shutter control

The sensitive action of the air control-valve provides a snap open and snap closed shutter movement, with approximately 40 second intervals under full-load, or slightly longer periods under part-load. The thermally controlled valve operates within a temperature variation of ±1.5°C.

With this system it is claimed that the tempera-ture variation between the top and bottom hoses is only 8°C, compared with a fluctuating temperature variation of up to 60°C during cold weather. In addition, the normal operating temperature is raised by about 10°C (Fig. 14.21). On average, the warm-up period is reduced from approximately 6 miles (9.7 km) down to 3 miles (4.8 km) from the start of a journey. This reduction of

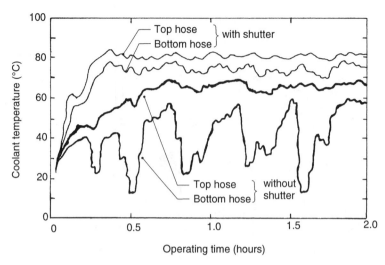

Fig. 14.21 Relationship between coolant inlet and outlet temperature for both conventional thermostat controlled and shutter controlled systems

temperature variation reduces the degree of thermal stressing and keeps the engine working at its most efficient temperature under most driving conditions. Shutters work well in conjunction with fan-blade drive controls as the latter are still required to reduce power consumption and noise.

14.6 Heater and ventilation system

(Figs 14.22 and 14.23(a))

The majority of heater-ventilation systems utilize the 'recirculation hot coolant system'. This system incorporates a tube and corrugated fin-type heat-exchanger (known as the heater unit) mounted behind the dashboard, and which is normally supplied by hot coolant from the cylinder-head outlet thermostat housing (Fig. 14.22). The engine water-pump circulates cool liquid coolant under pressure through the engine's cooling passages, where it absorbs heat from the combustion process before it passes through to the cylinder-head outlet on its way to the radiator intake tank. Thus, the ideal point to tap-off the hot coolant from the heat-exchanger is at the cylinder-head thermostat housing through a rubber hose, and once this heat has been extracted from the coolant by the air-flow passing through the heat-exchanger matrix, the cooler coolant is returned to the engine's cooling system, usually at a point somewhere near the inlet side of the water-pump (Fig. 14.23(a)). Early interior heater systems

controlled the heating capacity by restricting the amount of coolant circulating through the heat-exchanger tubes via a coolant-valve. However, this system has largely been replaced with a system which controls the proportion of cold air bypassing the heat-exchanger, so that a mix of hot and cold air then combines to give the desired warmth to the air stream entering the passenger compartment.

14.6.1 Heater distribution box heat exchanger and fan
(Fig. 14.23(a))

The heater unit consists of a plastic distribution box which contains an electric motor-driven centrifugal fan at the input end and a tube and corrugated fin-type heat-exchanger, which is supplied by hot engine coolant. The coolant gives up its heat to the interior of the passenger compartment before returning to the radiator output or water-pump input coolant flow point. The heater distribution box has a passenger compartment interior exit directed towards the feet, and a pair of demister duct exists are channelled to both sides of the windscreen via plastic ducting. There are also ventilation duct exits which are piped to each side of the dashboard but these are not shown in Fig. 14.23(a). Air input supplied either from outside the vehicle or internally recirculated is controlled by an internal/external deflector flap. The air mix then passes to the centrifugal fan

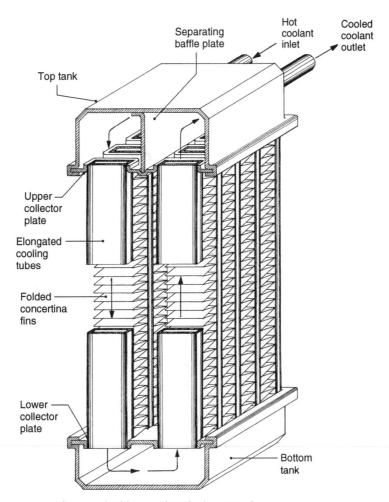

Fig. 14.22 Tube and corrugated fin type double-pass interior heater unit

input eye. Internal circulation is used if there are fumes or smells outside the vehicle.

14.6.2 Air flow control
(Fig. 14.23(a))

The speed of the vehicle, to a limited extent, determines the amount of air entering the fan and cold air chamber. However, if the vehicle is moving slowly or it is very cold or very hot outside, the air-flow rate can be considerably raised by increasing the fan speed and therefore the pumping pressure. This is usually achieved with a three or four speed (or an infinitely variable speed) electric fan motor controlled by a dashboard toggle switch or a multi-position rotary switch.

14.6.3 Temperature (interior screen) control
(Fig. 14.23(a))

Temperature control of the air stream leaving the heat distribution box and passing to the passenger compartment is controlled by the hot–cold deflector-flap. This temperature control is achieved by adjusting the deflector-flap so that a proportion of cold air-supply passes through the heat-exchanger matrix. The remainder of the cold air-supply bypasses the heat-exchanger but meets and mixes with the heated air before it is discharged, either through the interior exit directed towards the feet or through to the demister exit.

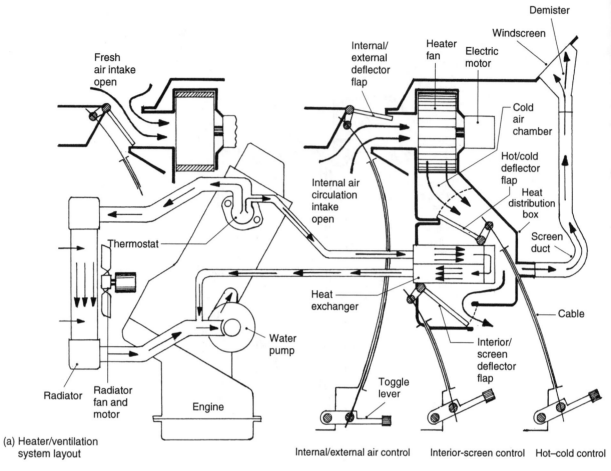

Labels in figure:
Fresh air intake open
Internal/external deflector flap
Heater fan
Windscreen
Demister
Electric motor
Cold air chamber
Hot/cold deflector flap
Heat distribution box
Internal air circulation intake open
Screen duct
Thermostat
Heat exchanger
Cable
Water pump
Interior/screen deflector flap
Radiator
Radiator fan and motor
Engine
Toggle lever

(a) Heater/ventilation system layout

Internal/external air control   Interior-screen control   Hot–cold control

Fig. 14.23 Heater-ventilation system

14.6.4 Hot–cold air mix distribution (interior–screen) control
(Fig. 14.23(a))

The distribution of the hot or cold (or hot and cold) air mix to either the interior foot exit or the demisters, or partially to both, is controlled by the interior screen deflector-flap. If the deflector-flap is positioned to block-off the interior exit, then all the air mixture passes to the demister and windscreen. Conversely, if the deflector-flap opens the interior exit, the majority of the air mix will divert to the passenger compartment at foot level while a much smaller proportion will pass through to the windscreen demisters.

14.6.5 Interior hot–cold control
(Fig. 14.23(b))

Setting the air temperature flowing to the passenger compartment is achieved by moving the interior screen deflector-flap toggle-lever to its fully open position and then adjusting the hot–cold deflector-flap via its toggle-lever to give the desired amount of cooling or warmth.

If the hot–cold deflector-flap closes the entry to the heat-exchanger matrix (Fig. 14.23(b)) (cold), then all the cold air pumped out by the fan into the cold air chamber bypasses the heat-exchanger and flows directly towards the interior exit. However, a small amount of air under these conditions is diverted to the screen duct and passes to the demisters. If, however, the hot–cold deflector-flap closes the bypass passage (Fig.

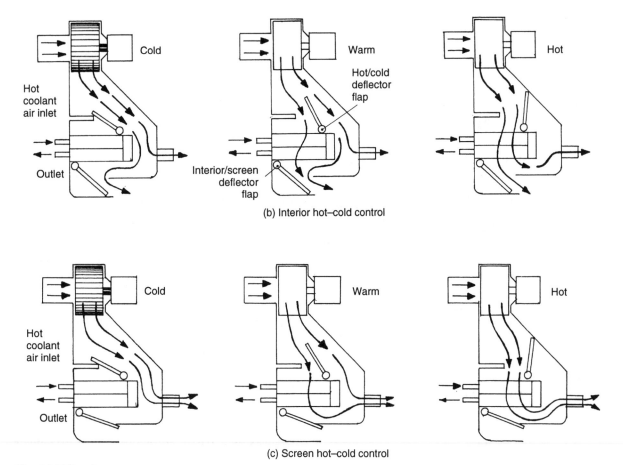

(b) Interior hot–cold control

(c) Screen hot–cold control

Fig. 14.23(b–c)

14.23(b)) (hot) leading to the demister and interior exits, then all the cold air has to pass through the heat-exchanger matrix. The heated air now diverts and a relatively small amount passes to the screen duct where it then goes to the demisters. However, the bulk of the air passes out through the interior exit. If only warm air is desired for the interior passenger compartment then the hot–cold deflector-flap is moved to its midway position (Fig. 14.23(b)) (warm). Under these settings a proportion of the cold air passes through the heat-exchanger and the rest flows via the bypass passage. Both portions then merge to form some intermediate temperature mix before being discharged to the interior and windscreen.

14.6.6 Screen hot–cold control
(Fig. 14.23(c))

Setting the air temperature flow to the demisters only is obtained by moving the interior-screen

toggle-lever to the 'screen position' and this moves the interior-screen deflector-flap to the interior exit closed position. The hot–cold deflector-flap toggle-lever is then adjusted to give the required heat to the windscreen.

If the hot–cold deflector-flap closes the entry to the heat-exchanger matrix, all the cold air displaced by the fan is now diverted to both demisters, which are positioned against each half of the windscreen (Fig. 14.23(c)) (cold). If the hot–cold deflector-flap closes the bypass passageway to the demisters and interior exits, then all the air is compelled to pass through the heat-exchanger, absorbing heat before it continues on its way to the demisters and windscreen (Fig. 14.23(c)) (hot).

If an in-between demister temperature is required, then the hot–cold deflector-flap is moved to its midway position where a hot air stream exits from the heat-exchanger matrix to meet up with the cold air flowing out from the bypass passageway (Fig. 14.23(c)) (warm). The combined hot and cold air now passes through the screen ducts to the demisters and windscreen.

14.7 Lubricating oil heat-exchanger (oil-cooler)

The amount of heat received by the lubricating oil in liquid cooled engines (due to contact with the heated engine component surfaces, caused by the combustion in the cylinders and by heat generated by friction) is 2.0% to 2.5% of all the heat released in the cylinder by the air–fuel mixture.

The lubricating oil picks up heat from the oil splashed onto the cylinder's walls and pistons, from oil flowing over the hot cylinder-head, and the heat generated between rotating camshaft and crankshaft journal bearings. This heat is then transferred to the cylinder-block, crankcase and sump-pan, where it is then dissipated between the cylinder-block cooling passages and by the air stream convection currents and radiation over the engine sump.

For high rated engines or large heavy-duty engines, the heat conduction path necessary to extract the heat from the oil is sometimes inadequate and to supplement this heat transfer, heat-exchangers (known as oil-coolers) are commonly used. An oil-cooler restricts the peak operating oil temperature and, as a result, protects the lubricating properties of the oil and extends the interval between oil changes for a given quantity of circulating oil in the system.

14.7.1 Type of oil-coolers and their flow connections
(Figs 14.3, 14.4 and 14.6)

Oil-cooler heat-exchangers can be either of the liquid-to-air type, where the hot oil is cooled by the surrounding air; or of the liquid-to-liquid type where the hot oil is cooled by the engine's liquid cooling system.

Oil-coolers take in the hot pressurized oil supply from the oil-pump outlet, and this is then passed through the cooling tubes, after which, the cooled oil is discharged to the engine's lubrication system's main oil gallery (Figs 14.3 and 14.6).

Liquid-to-air type oil-coolers pass the hot lubricating oil through finned tubes which are directly exposed to the air-stream caused by either the moving vehicle or by the engine's cooling system fan (Fig. 14.6). These liquid-to-air type oil-coolers function in the same way as the engine's cooling system radiator and therefore need no further explanation.

Liquid-to-liquid oil-coolers usually receive the cool coolant input from the bottom radiator tank if the engine has a down-flow radiator cooling system (Fig. 14.3), and from the outlet side-tank of the radiator if a cross-flow cooling system is employed. The output coolant from the oil cooler then flows to the input side of the cooling system's water-pump. An alternative coolant supply layout adopted on some large diesel engines uses the feed-in from the lower coolant delivery manifold as its input, whilst its outlet from the heat-exchanger is passed to the upper coolant collector manifold (Fig. 14.4).

14.7.2 Objective of a liquid-to-liquid heat-exchanger
(Fig. 14.24(a–g))

A heat-exchanger transfers heat from one liquid (hot oil) to another liquid (cooler coolant) when the two liquids are separated by a metal wall (usually tubes). The temperature of each liquid changes as it passes through the heat-exchanger passageway and hence the temperature of the dividing wall between the liquids also changes along its length.

A basic liquid-to-liquid heat-exchanger is com-

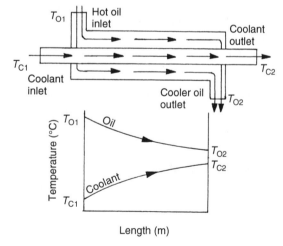

Fig. 14.24 Liquid-to-liquid heat exchangers

posed of a tube through which coolant flows, which is surrounded by a sleeve blanked off at each end with an inlet and outlet pipe joined to the sleeve at opposite ends through which the hot lubricating oil flows. In order to make the heat-exchanger more efficient and compact the single tube is replaced by many tubes—known as the tube core, bundle, pack, or stack (Fig. 14.24(c)).

Heat-exchangers are classified according to the direction of flow of one liquid relative to the other within the system. There are three basic designs, described as follows.

Parallel flow
(Fig. 14.24(a))
Where the liquid which surrounds the tubes flows parallel and in the same direction as the liquid flowing through the tube.

Counter flow
(Fig. 14.24(b))
Where the liquid which surrounds the tubes flows

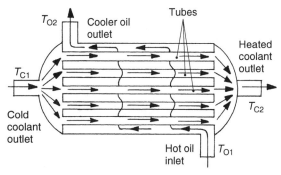

(c) Single-pass counter-flow heat exchanger

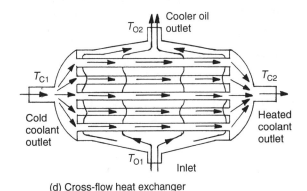

(d) Cross-flow heat exchanger

Fig. 14.24(c–d)

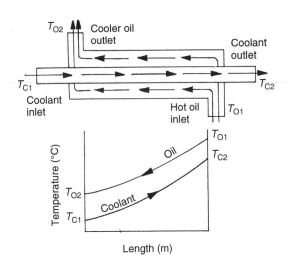

(b) Counter-flow heat exchanger

Fig. 14.24(b)

in the opposite direction to the liquid flowing through the tubes.

Cross flow
(Fig. 14.24(d))
Where one liquid flows at right angles across the tubes when the other liquid flows through the tube bores.

Cooling fins (Fig. 14.24(g))
In order to make the simple cooling tubes more efficient in transferring heat from the surrounding hot oil to the coolant flowing through the tube bores, the metal tube walls separating the two liquids can be extended on the outside in the form of fin sheets (Fig. 14.24(g)). These fins present a larger external surface area for the same internal tube bore surface area, and hence increase the cooling effect for a given mass of tube metal.

14.7.3 Heat-exchanger characteristics

Parallel flow
(Fig. 14.24(a))
With the parallel-flow heat-exchanger, the oil temperature falls as it flows from the inlet to the outlet of the cooling tubes, due to the transference of heat from the lubricating oil to the liquid coolant. Conversely, the coolant temperature rises as it absorbs heat from the oil while flowing in the same direction as the oil—from the hot oil inlet end to the cooled oil outlet end.

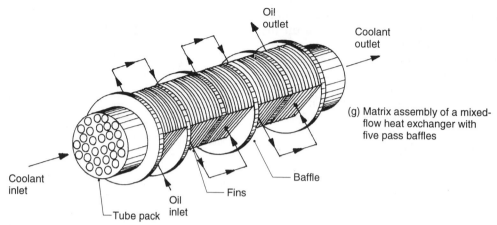

Oil outlet

Coolant outlet

(g) Matrix assembly of a mixed-flow heat exchanger with five pass baffles

Baffle

Coolant inlet

Fins

Tube pack

Oil inlet

Fig. 14.24(g)

With this arrangement, due to imperfections in transmitting heat through the walls of the tubes, the oil temperature can never be brought down to that of the coolant's highest temperature.

Counter flow
(Fig. 14.24(b))

With the counter-flow heat-exchanger the oil temperature drops similarly to that in the parallel-flow arrangement, as it flows from the inlet to the outlet ends of the tubes discharging heat. Now, with this layout, the coolant entry is at the same end as the cooled oil exit, and this causes the coolant to flow in the opposite direction to the surrounding oil. Proportionally, slightly larger amounts of heat will be extracted by the coolant compared with the parallel-flow system. It therefore compels the coolant temperature to rise relatively steeply as it approaches the hot oil inlet end of the casing. One advantage of the counter-flow system is that it permits the cooled oil temperature to be brought down below that of the highest coolant temperature reached, which is not possible with the parallel-flow system.

A second advantage of the counter-flow system over the parallel-flow heat-exchanger is that, for a given mass flow and temperature change, it requires less surface area to transfer a given quantity of heat from the oil to the coolant—hence the counter-flow heat-exchanger can be made more compact.

Cross-flow
(Fig. 14.24(d))

With the cross-flow heat-exchanger, the hot oil temperature decreases in the same way as for the parallel-flow and counter-flow systems, as the oil flows across and over the tubes from the inlet to the outlet side of the casing. At the same time, the coolant temperature increases as it flows from one end of the tubes to the other, absorbing the rejected heat from the hot oil.

As with the counter-flow system, the inlet temperature of the heated coolant in the cross-flow system can be raised to a temperature higher than the outlet temperature of the cooled oil.

The cross-flow system's ability to transfer heat from the hot oil to the coolant lies somewhere between the effectiveness of the parallel-flow and the counter-flow heat-exchangers.

The cross-flow principle of moving heat from a hot to a cold liquid is used to improve the heat-transfer for a given size of heat-exchanger. This is achieved with the use of baffle-plates spaced out along the 'tube-pack' so that the hot oil is directed to flow across the tubes several times before it is permitted to pass out from the casing (Fig. 14.24(g)). The cross-flow heat-exchanger, although slightly less economical in size than the counter-flow type is sometimes preferred when it makes the layout of ducts and piping more convenient and accessible.

Mixed flow heat-exchangers
(Fig. 14.24(e))

Heat-exchanger efficiency can be further improved by spacing baffle-plates along the length of the tube stack, which therefore divides the cooling tube casing into a number of compartments. The baffle-plates are shaped in the form of a disc with a segment section removed so that oil is permitted to pass between compartments.

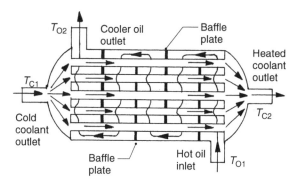

(e) Single-pass mixed-flow heat exchanger

Fig. 14.24(e)

When the hot oil enters the casing at one end, it cannot now pass straight through, but has to wind or pass across the tubes several times before it reaches the exit. Relatively more heat is transferred by these means for a given size of container. The oil movement is therefore partially cross-flow and counter-flow as it passes from one end to the other—this typical arrangement is therefore referred to as a mixed-flow heat-exchanger.

14.7.4 Double pass mixed flow heat-exchanger
(Fig. 14.24(f))

Heat-exchangers can be described by the number of passes the coolant has to make from one end of the tubes to the other—for industrial applications, double, triple and quadruple-pass exchangers are available. However, automotive heat-exchangers are generally of the single mixed-flow type, and only very large power units used for special heavy-duty purposes would justify the use of the double-pass mixed-flow heat-exchanger.

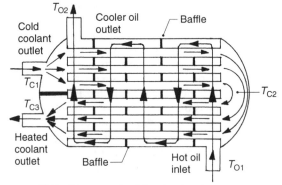

(f) Double-pass mixed-flow heat exchanger

Fig. 14.24(f)

The compactness of heavy-duty heat-exchangers can sometimes be improved by dividing one end-chamber of the cooling tubes into two compartments and making one half-chamber the entry point of the coolant and the other half-chamber its exit. The chamber at the opposite end of the cooling tubes then becomes the passageway for the coolant coming from the first forward pass and this then goes to the second return pass, through the tubes to the exit half-chamber.

At the same time, hot oil entering the tube outer casing passes to and fro across all the tubes as it winds its way between baffle-plates and moves steadily forward in the direction of the oil exit duct.

14.7.5 Thermostatically controlled oil cooler and filter unit
(Fig. 14.25(a and b))

Construction and design
(Fig. 14.25(a and b))
The heat-exchanger tube core consists of a stack of 8 mm outside diameter copper tubes supported at their ends on perforated disc plates, which are themselves enclosed to form small entry and exit chambers. Perforated copper fins are spaced over the tubes at intervals of six to seven for every centimetre of tube length, and are bonded to the tubes by dipping in high quality solder. Hence, all the copper surfaces on the outside are covered in a protective coat of solder. Also, evenly spaced along the tubing are six baffle-plates with segment pieces cut away. These baffles provide a winding flow path for the oil as it crosses to and fro over the cooling tubes on its zigzag path from one end of the tubes to the other. The tube-stack slides into a cylindrical aluminium housing where it is held in position by sealing 'O' rings and end retaining covers. Attached to the oil cooler housing at one end is a full-flow oil filter unit whereas the other end of the oil cooler has an extended retaining cover and oil-cooler support which bolts onto the engine's crankcase. Also built into the extended end retainer cover is a thermostat which diverts the oil either straight to the oil-filter unit when the engine is cold, or to the oil cooler followed by the oil-filter when the engine is hot.

Thermostatic control
(Fig. 14.25(a and b))
The purpose of the oil thermostat is to minimize

the warm-up period of the lubricating oil when the engine is cold, by closing the oil passage to the oil cooler and opening a second passage, which bypasses the oil cooler and goes directly to the oil-filter unit. Once the engine is hot, the thermostat opens the passage to the oil-cooler and closes the bypass passage leading to the oil-filter. Oil will now flow through both the oil-cooler and filter before passing to the engine's main lubrication oil gallery.

Cold lubricant operating conditions
(Fig. 14.25(a))
When the engine lubrication oil temperature is below 102°C the thermostat wax capsule thrust-pin is fully inside its rubber-boot. This positions the sleeve-valve against the capsule. The oil supply now flows through the centre of the sleeve-valve and then passes directly to the inlet and to the oil-filter unit. The filtered oil flows through the centre of the filter-element and then passes back to the end retainer cover flange where it continues on its way to the engine's main oil gallery.

Warm lubricant operating conditions
(Fig. 14.25(a and b))
As the engine lubrication oil temperature exceeds 102°C the thermostat wax-capsule expands and commences to push out the thrust-pin. This moves the sleeve-valve away from its seat and towards the back face of the thermostat housing. The oil flow is now split between the bypass passage leading directly to the oil-filter unit and to the oil-cooler inlet passageway.

Hot lubricant operating conditions
(Fig. 14.25(b))
When the lubricant temperature reaches 113°C the wax-capsule is fully expanded. This causes the sleeve-valve to be pushed hard against the back face of the thermostat housing. The bypass passage is now fully closed so that all the lubricant oil flows directly to and through the oil-cooler. The hot oil therefore circulates from one side to the other across the cooling tubes, guided by the baffle-plates, until it has worked its way to the far end of the housing, where it then passes to the oil-filter unit via the oil-cooler to oil-filter connecting passage.

14.8 Centrifugal oil filters

14.8.1 Fundamentals of filtration
(Fig. 14.26)

All the oil supply flowing from the oil pump to the main oil gallery passes through a full-flow filter, whereas something like 10% of the oil supply in the main oil galley is passed through the bypass filter.

Good filter media should be a porous material which lets through oil with very little resistance but prevents the suspended unwanted small solid particles from passing. For full-flow filters the pore and mesh size provided by the filter-element must not be too small, as this restricts the oil flow and increases the pressure-drop across the filtering media. Consequently, the full-flow depth filter media must have a transmissibility such that it can pass particle sizes of up to 8 µm to compensate for progressive partial clogging during service life if the flow resistance is not to become excessive.

Thus, the larger particles passing through the full-flow filter media could be larger than the oil film thickness under load between the bearings and crankshaft and piston rings and cylinder bores. The minimum working oil film thickness can be as little as 1 to 2 µm. Therefore, if there are enough hard particles, larger than about 3 µm reaching the bearings, the abrasive wear rate can be excessive and rapid.

A bypass plastic impregnated depth element of a finer texture will give good filtration when installed, in addition to the full-flow filter unit. However, its filtering efficiency deteriorates fairly rapidly during service (Fig. 14.26). An alternative type of filter unit which can operate either in the full-flow or bypass mode, works on the centrifugal force principle in which solid particles are separated from the liquid. This type of filter proved to be effective with no fall-off in efficiency within normal oil change intervals and the particle removal rate remains practically constant throughout the life of the oil (Fig. 14.26). Filter units of this design pass oil through a rotor bowl spinning at high speed, during which time any solid particles in the oil are flung out and deposited on the bowl wall. These filter units operate on the centrifugal principle and therefore they are known as centrifugal-type filter units.

The centrifuge is capable of separating from the oil, solid carbon and abrasive particles which are

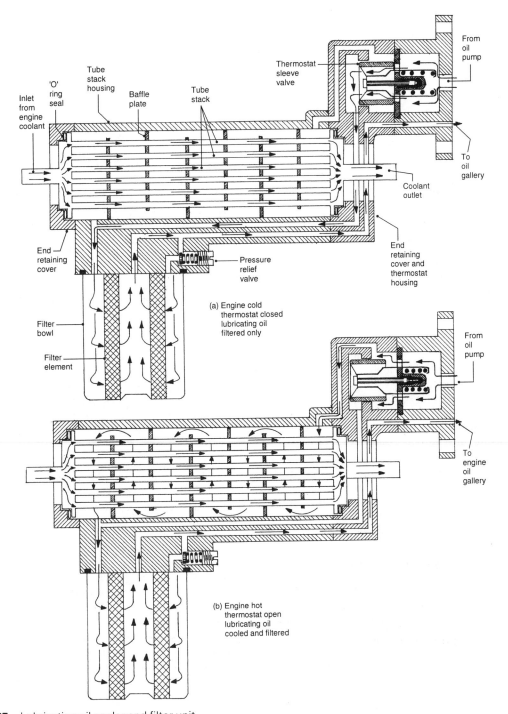

Fig. 14.25 Lubricating oil cooler and filter unit

2 μm and less in size, continuously throughout the service life of the oil.

14.8.2 Bypass centrifuge oil filter
(Fig. 14.27(a))

With this type of filter unit, an inverted rotor

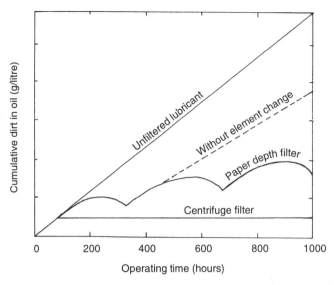

Fig. 14.26 Relationship between operating time and cumulative dirt in oil under different filter conditions

bowl mounted on a flanged sleeve rotor is permitted to spin about a sleeve spindle and plain bushing bearing, all of which is enclosed in a centrifuge body and cap.

Lubricating oil enters from the side of the centrifuge body where it then travels up through the sleeve spindle to cross-holes where it exits and is uniformly distributed around the bowl by a baffle-sleeve. The oil at supply pump pressure flows downwards and enters the drive chamber via the sleeve-screen. It then emerges from the twin-jets with a spray intensity which impacts the wall of the centrifuge body. The resulting reaction to the jet thrust spins the rotor and bowl assembly around at a speed approximately proportional to the oil supply pressure. Once circulation of the oil is established, the oil contained in the bowl at any one time whirls around with the rotor. It is subjected therefore to high acceleration resulting in the centrifugal force pushing the heavier solid particles outwards so that they impinge against the wall of the rotor bowl in the form of a dense consolidated layer. The clean oil then continues its flow-path through to the sleeve-screen, drive chamber and jet orifices. The spray discharge thus becomes the bypass oil which drains down the inside wall of the centrifuge body on its return route to the engine's sump-pan.

Periodically the rotor bowl is removed, the accumulated dirt is removed and it is then cleaned and replaced. However disposable rotor bowls on certain types of centrifuges are available.

14.8.3   Full flow centrifuge oil filter
(Figs 14.27(b) and 14.28)

This centrifuge basically consists of three components—the body, the rotor-bowl and the central spindle and passageway assembly.

The centrifuge body and cap contain the central support sleeve spindle on which the rotor bowl is permitted to revolve.

Lubricant oil discharged from the oil-pump enters the inlet port from the bottom of the mounting flange. It then passes up and between the annular passage formed between the inner and outer sleeve to where the bowl cross-holes permit the oil to enter the rotor bowl. Oil thus fills the bowl and 'stand tubes' at oil-pump supply pressure. The oil stream is then divided, a small proportion is expelled from the twin-jets formed in the base of the rotor bowl, whereas the bulk of the oil passes to the exit cross-holes joining the bowl to the central outlet passage. The oil is then permitted to pass down the outlet passage to the flanged mounting, where it continues its way to the engine's main gallery unfiltered.

Once the oil jets impinge on the wall of the centrifuge body, the reaction on the rotor bowl will be to spin it until it reaches some rotational speed which balances the oil pressure jet thrust and the resisting spin drag of the rotor bowl.

Centrifugal force, due to the oil rotating with the bowl, now separates the heavier dirt particles from the oil and slings them onto the cylindrical

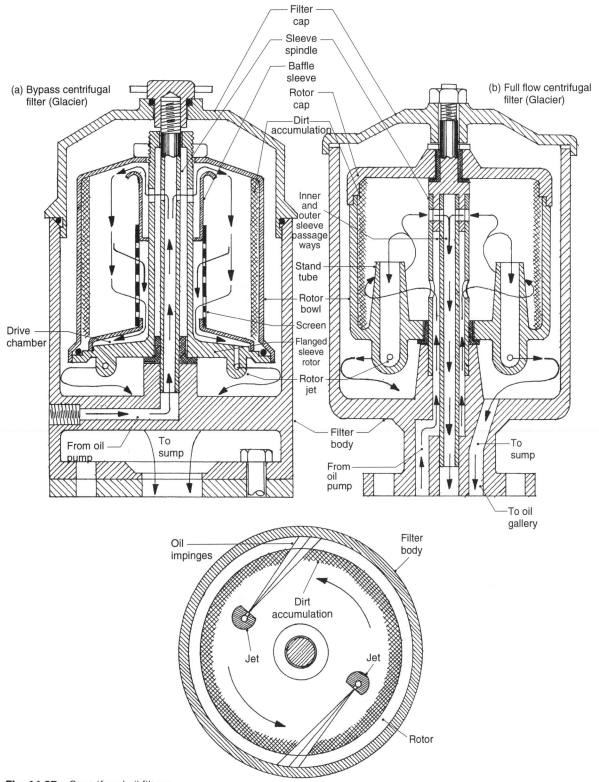

(a) Bypass centrifugal filter (Glacier)

(b) Full flow centrifugal filter (Glacier)

Filter cap

Sleeve spindle

Baffle sleeve

Rotor cap

Dirt accumulation

Inner and outer sleeve passage ways

Stand tube

Rotor bowl

Screen

Flanged sleeve rotor

Rotor jet

Filter body

From oil pump

To sump

To oil gallery

Drive chamber

From oil pump

To sump

Oil impinges

Filter body

Dirt accumulation

Jet

Jet

Rotor

Fig. 14.27 Centrifugal oil filters

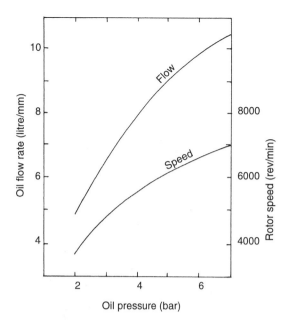

Fig. 14.28 Relationship between oil supply pressure–oil flow rate and rotor speed for a typical centrifugal oil filter

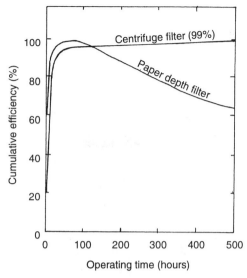

Fig. 14.29 Comparison of paper element and centrifuge filter cumulative efficiency against operating time

wall of the bowl in a compact, easy-to-remove, mass. The clean oil now leaves the bowl, via the exit cross-holes, to be discharged down the central outlet passage on its journey to the lubrication system's main oil gallery. The working speed of the rotor bowl is usually between 2000 and 7000 rev/min, largely depending upon the oil pressure circulating the oil (Fig. 14.28). Forces of up to 2000 times greater than gravity are created which are very effective in throwing the dirt and contaminates out of the oil system, where they can be stored and removed later at recommended service intervals.

14.8.4 Performance characteristics of paper element and centrifuge-type oil filters
(Fig. 14.29)

A comparison of the accumulation of contaminating dirt in the lubricating oil between an unfiltered system, paper element depth filter and centrifuge-type filter, against operating time, can be seen in Fig. 14.29. It can be seen that in an unfiltered lubrication system the accumulated dirt (consisting mostly of carbon and grit particles) increases in direct proportion to the operating time. With the conventional paper element filter, the accumulation of dirt increase with the oil, rises, at first as with the unfiltered system and then it breaks away from the unfiltered curve as the rate of dirt accumulation decreases to the first filter change point. After replacing the first filter, the rate of dirt accumulation increases slightly and then decreases again. This pattern was again repeated with the third filter element. However, it should be noticed that the accumulation of dirt in the oil is steadily increasing, as shown by the increasing height of each of the three paper element filter curves. If the first filter element is unchanged after its normal service life, the dirt accumulation in the oil continues to increase at almost the same rate as the unfiltered lubrication system (shown by the broken line). In contrast with the paper-element depth filter, the centrifuge-type filter shows an increase in accumulated dirt contained in the oil, which is initially similar to the unfiltered system but, at a very early stage, it levels off once the circulation of oil through the centrifuge reaches its minimum design rate.

15

Alternative Power Sources

The conventional reciprocating piston engine, be it spark ignition or compression ignition, still remains the most popular form of power source for the motor vehicle. Nevertheless, there are other kinds of engines which, with sufficient research and development, do have potential for driving light cars and vans, passenger vehicles or heavy goods trucks.

Justification for changing to alternative power sources would depend on a number of considerations, which may be influenced by the source and availability of fuel, its quality and price, the size of the power unit required and the operating conditions of the vehicle being propelled. Other considerations in determining which type of power source to be used will be greatly influenced by a number of factors such as: initial costs, improved fuel consumption, power output potential, smoothness of running, minimum maintenance, reliability, working life of unit, reduced pollution levels and the ability of the engine to tolerate varying qualities of fuels.

Three alternative types of engines which, under favourable economical and commercial conditions and with possibly further improvement programmes, could more than compete with (and be preferred to) the Otto and diesel engines will now be described and their merits and limitations discussed under the following headings.

1 Wankel rotary engine
2 Stirling engine
3 Gas turbine engine

15.1.1 The Wankel rotary engine

Felix Wankel of Germany invented the epitrochoidal chamber and the triangular rotor concept in 1954. Further development by Wankel and the German NSU company led by Dr Walter G. Froede introduced the eccentric rotary movement which forms the basic Wankel engine design as we know it today. Continued research and development by many major engine manufacturers, in particular Toyo Kogyo (Mazda) of Japan, has considerably improved the performance of this type of engine in terms of power, fuel consumption, life expectancy and reliability.

15.1.2 Design and construction
(Figs 15.1 and 15.2)

The Wankel rotary engine has three main parts:

1 the rotor which is a wide triangular-shaped disc with three equal slightly curved flanks (a combustion chamber recess is formed in each of these flanks);
2 the eccentric output shaft around which the rotor orbits;
3 the rotor bore wall housing which provides the working space for the cyclic phases of operation.

The four phases of operation are obtained by making the rotor wobble as it rotates so that the combustion chamber volumes periodically increase and decrease to provide the induction, compression, power and exhaust cyclic phases (Fig. 15.1). The shape of the outer housing wall surface is something like the figure-of-eight formed by two circles merging together, with the sides pinched inward slightly in the middle. The special name given to this geometric profile is an epitrochoid.

The geometry of the wall bore and the eccentric movement of the rotor is such that the apexes of

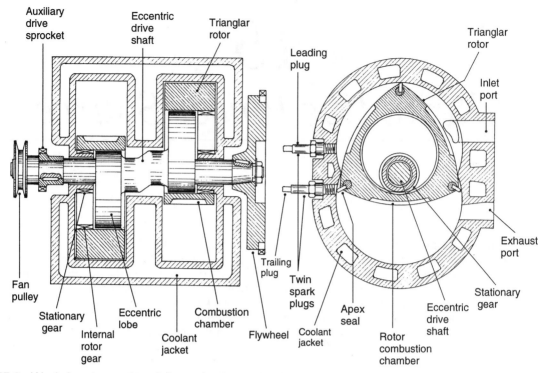

Fig. 15.1 Wankel engine sectioned views of rotary mechanism

the triangular rotor seal off three distinct spaces at all times.

To obtain the desired wobble action for the rotor, the output shaft has a perfectly round lobe (Fig. 15.2) situated in the middle of the shaft with its centre offset slightly to the shaft axis. This output shaft is therefore commonly referred to as the eccentric shaft.

In the centre of the rotor is a large circular hole to accommodate the lobe formed on the eccentric output shaft. Therefore, when the shaft turns the rotor will wobble in an orbit relative to the centre axis of the shaft, in accordance with the position of the lobe.

To provide a forward positive drive from the rotor to the eccentric shaft a small gear with teeth on the outside circumference is attached directly to the inside of one of the flat rotor housing covers—through the centre of this stationary hollow-centred gear passes the eccentric shaft (Fig. 15.2). A much larger internal ring-gear is situated inside the rotor bore hole but to one side in order to permit the rotor central bore hole to fit over the lobe on the eccentric shaft. When assembled, the eccentric shaft, rotor and rotor housing are so arranged that the rotor housing end cover

supports the eccentric shaft while the rotor is mounted on the eccentric-lobe of the shaft (Fig. 15.1).

In this position, the rotor's internal gear and the stationary external gear align in the same plane so that meshing of the two gears occurs at the point of closest approach, caused by the eccentricity of the lobe relative to the shaft axis.

Thus, when the eccentric shaft revolves, the rotor's internal ring-gear rolls or walks around the stationary gear so that the rotor is compelled to rotate and to orbit in a path set out by the eccentric shaft's lobe offset (eccentricity). In effect, this will provide a smooth pre-determined wobble at a definite speed relative to the eccentric shaft.

To synchronize the eccentric lobe movement with the rotor speed so that the three corners of the wobbling rotating rotor touches the housing at all times, the ratio of the gears must be 2:3, which means that the rotor gear must have three teeth for every two on the stationary gear. Therefore, if the stationary gear has 30 teeth, the rotor internal gear will need 45 teeth. This particular gear ratio will cause the rotor to revolve at exactly one third the speed of the eccentric shaft: in other words,

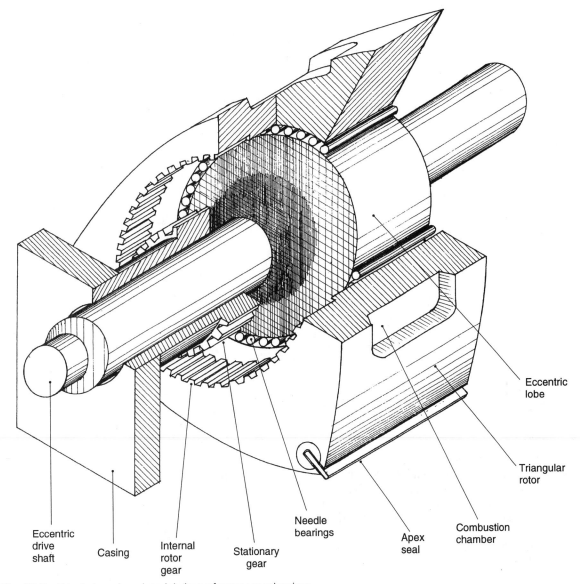

Fig. 15.2 Wankel engine pictorial view of rotary mechanism

Eccentric
lobe

Triangular
rotor

Combustion
chamber

Apex
seal

Needle
bearings

Stationary
gear

Internal
rotor
gear

Casing

Eccentric
drive
shaft

the rotor rotates only once for every three revolutions of the eccentric shaft.

The inlet port enters the combustion-chamber just above the oval waisted portion of the bore wall, through one of the side covers, while the exhaust port enters below the narrow chamber portion directly through the bore wall (Fig. 15.1). For effective combustion, the spark-plug or plugs screw into a hole or holes slightly offset from the waisted middle bore-wall on the opposite side to the inlet and exhaust ports.

15.1.3 Geometry of a two-lobe working chamber
(Fig. 15.3)

The epitrochoid shape used for the Wankel engine bore walls is the path traced by a point (P) within a circle as the circle rolls (generates) without slip around the outside periphery of another circle (stationary), known as the base circle, until it returns to its original position (Fig. 15.3). If a point is chosen within the generating circle of half its radius, and the circle is rolled

759

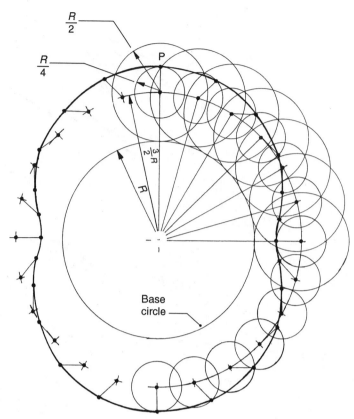

Fig. 15.3 Generation of an epitrochoid

around the base circle, the curve generated will describe a figure-of-eight shaped two-lobe epitrochoid. This two-lobe chamber demands a three-lobe rotor and a 3:1 gear ratio between the rotor annular ring gear and the corresponding stationary reaction gear.

The epitrochoid (Fig. 15.3) may be constructed by the following method.

1 Draw a base circle of radius R.
2 Divide this circle into, say, 15° increments.
3 Draw around the outside of the base circle, at 15° intervals, a generating circle of radius half that of the base circle $R/2$.
4 Draw in each generating circle a second circle of half its radius $R/4$ and in the highest position at the top inner circle mark point (P).
5 Go around each generating circle in turn and mark a 30° incremental positional change of point (P) on each inner circle until the generating circle has completed two revolutions and therefore has rolled around the base circle once. Because of the 2:1 diameter ratio of the base to generating circles, a 15° interval on the

base circle is equal to 30° angular movement of the generating circle.
6 The overall path point (P), traced as the generating circle rolls around the base circle, can be finally recognized by drawing a smooth line through the points.

15.1.4 Phases of operation

Four-phase cycle
(Fig. 15.4(a–d))
The cycle of operation may be looked at in terms of four phases (Fig. 15.4). Consider point (P) at one of the angular corners of the rotor and follow its anticlockwise movement as the rotor revolves.

Induction phase
(Figs 15.4(a–d) and 15.5)
Induction commences with the minimum volume between the rotor flank and the bore wall with the leading apex seal point (P) a short distance

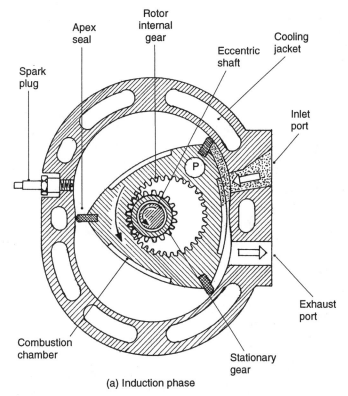

(a) Induction phase

Fig. 15.4 Wankel rotary engine four phase cycle of operation

beyond the inlet port (Fig. 15.4(a)). As the rotor revolves, the combustion chamber volume adjacent to the inlet port is enlarged, causing an average partial pressure of about 0.12 bar to be obtained during this phase (State 1 to 2 in Fig. 15.5). Immediately, a premixed charge of air and fuel will rush through the inlet port to fill up this space. The induction phase will be complete when the trailing seal is about to sweep past the inlet port.

Compression phase
(Figs 15.4(b) and 15.5)

With the trailing apex seal moved beyond the inlet port, the fresh charge becomes trapped between the rotor flank, corner seals and bore wall. Further rotation of the rotor steadily reduces the enclosed charge space—this volume becoming a minimum when the leading apex seal point (P) (Fig. 15.4(b)) has passed the spark-plug and the mid-position of the rotor flank, and is adjacent to the waisted portion of the bore wall. During this compression phase, the charge pressure will rise from State 2 to 3 (in Fig. 15.5) to

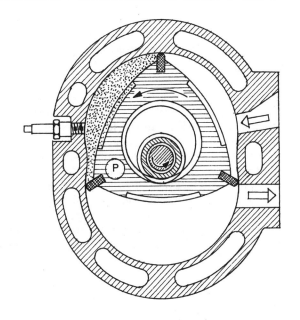

(b) Compression phase

Fig. 15.4(b)

between 8 and 14 bar with a wide-open throttle. The actual peak compression pressure will depend mainly upon the theoretical compression ratio which may be anything from 8:1 to 11:1.

Power phase
(Figs 15.4(c) and 15.5)
When the rotor flank has reduced the fresh mixture volume to a minimum point (P) (Fig. 15.4(c)) the spark-plug is timed to fire, the result-

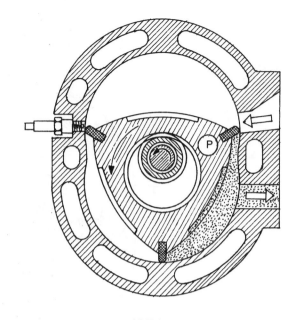

(d) Exhaust phase

Fig. 15.4(d)

remaining kinetic energy to empty the combustion chamber space. The pressure in the combustion chamber will then drop from about 3 to 4 bar, down to atmospheric pressure (State 4 to 1 in Fig. 15.5) as it escapes into the atmosphere. At the same time as the gas is being released, the rotor flank facing the exhaust port will move

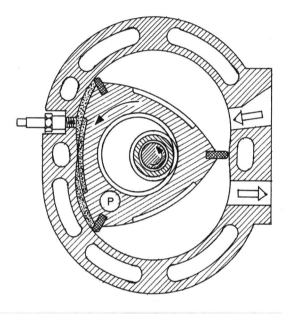

(c) Power phase

Fig. 15.4(c)

ing burning of the explosive charge causes a rapid rise in gas pressure (peak pressures of 60 bar) until the contained gas in the leading portion of the combustion chamber space expands, the pressure then decreases as combustion reaches completion (State 3 to 4 in Fig. 15.5). As a result, a forward anticlockwise driving thrust will be applied to the rotor and the internal ring gear which, through the media of the eccentric lobe and the reaction of the stationary external gear, compels the eccentric shaft to rotate at three times the speed of the rotor.

Exhaust phase
(Figs 15.4(d) and 15.5)
As the rotor continues to turn, the apex seal point (P) (Fig. 15.4(d)) eventually sweeps ahead of the exhaust port, therefore permitting the gases' own

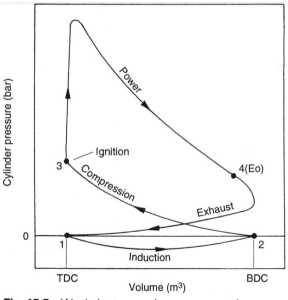

Fig. 15.5 Wankel rotary engine pressure–volume diagram

closer to this portion of the bore wall. Thus, practically all the chamber's swept volume will be expelled before the induction phase recommences.

The engine will therefore complete three four-phase cycles and three revolutions of the eccentric shaft after a single rotation of the triangular rotor.

15.1.5 Combustion and spark-plug location
(Fig. 15.1)

The combustion process can generally be improved by installing two spark-plugs: one in the lead combustion space and the other in the trailing region of the combustion-chamber (Fig. 15.1). A tilting or rocking action takes place when the rotor is near the TDC position and combustion commences. This tends to expand the gases in the leading portion of the combustion chamber but to compress them in the trailing region.

Positioning a plug in the high-pressure trailing region of the combustion chamber tends to produce slightly more power and use less fuel, whereas if the plug is located in the leading portion of the chamber, better starting and idle running is obtained.

Rapid combustion and pressure rise are necessary for high efficiency, but the backward flame propagation of the burning gases in the chamber is opposed by the forward movement of the gases as they are swept around with the rotor which, in effect, slows down the speed of combustion and therefore does not give optimum performance.

Efficient combustion is produced by having a very advanced timing, this can be partially achieved by positioning a spark-plug in the trailing region of the bore wall as far as possible from the waist or minor axis of the epitrochoid wall profile.

15.1.6 Rotor seals
(Figs 15.1, 15.6 and 15.7(a–c))

The planetary motion of the rotor within the epitrochoid bore of the rotor housing is designed to maintain a contact gap between the triangular corners of the rotor and the cylinder walls (Fig. 15.1). Peripheral radial corner blades (known as the apex seals) are necessary to prevent gas leakage between the three cylinder spaces created by the three-sided rotor (Figs 15.6 and 15.7(b)). Similarly, side seals between the flat rotor sides and the end and intermediate housing side walls (Figs 15.6 and 15.7(b)) are essential to stop engine oil reaching the cylinders, and gas from combustion escaping into the eccentric output shaft region.

The gas-tight sealing between the rotor and housing may be considered in terms of primary and secondary sealing areas (Fig. 15.7(a and b)). Primary sealing areas are those between the sealing elements (apex and side blades) and the cylinder housing bore and side walls, whereas secondary sealing areas are those between the sealing elements and the grooves, slots and trunnion counter bores. The total sealed system around the rotor is known as a sealed grid.

Both apex seals and side seals rely on gas pressure from combustion seeping behind the blades in their respective grooves (Fig. 15.7(c)) to apply an outward thrust on the blades, and lubricating oil spreading between the blade tips and the stationary walls, which are machined to have a porous texture. To ensure adequate starting conditions, light leaf-springs and washer springs are placed behind their respective sealing elements.

A three-piece apex seal (Figs 15.6 and 15.7(b)) has proven to be very durable since the centre blade is free to move radially outwards to contact the bore walls as they wear, whereas the outer triangular corner pieces are permitted to move sideways in addition to their radial movement. Through this method, automatic compensation for corner blade wear is achieved so that the gap between rotor and cylinder surfaces is sealed under all normal operating conditions. The centrifugal force acting on the apex blades contributes to the pressure the blades exert against the bore walls.

It is common practice for rotor double-side seals, supported in grooves, to be used so that combustion gas is retained in the cylinder chambers by a two-stage leakage grid (Fig. 15.7(c)). Here, the first outer blade blocks the majority of side wall leakage, with the second blade sealing off the remainder of the gas which escapes past the first blade. Gas is permitted to enter the clearance space between the groove and blade so that it pressurizes the rear of the blade. It therefore provides the blade with a self-sustaining outward pressure against the side walls enclosing the rotor.

To reduce gas leakage to a minimum at each

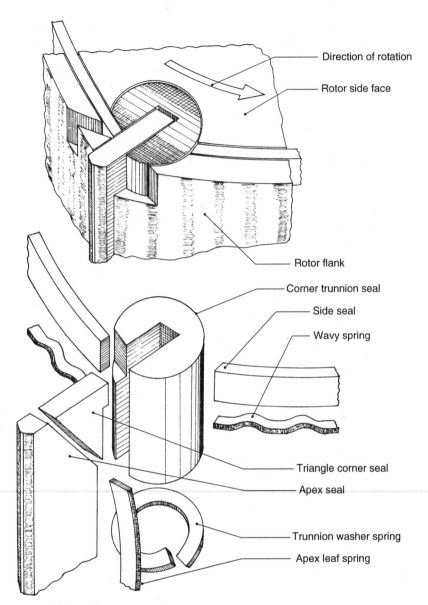

Direction of rotation

Rotor side face

Rotor flank

Corner trunnion seal

Side seal

Wavy spring

Triangle corner seal

Apex seal

Trunnion washer spring

Apex leaf spring

Fig. 15.6 Rotor apex and side seal arrangement

rotor apex, slotted cylindrical trunnion blocks (Fig. 15.6) are counter-sunk on both sides of the rotor at each rotor corner, the object being to provide a much longer leakage path at different meeting points where the apex and side blades overlap.

15.1.7 Rotor and housing materials

Rotors are generally made from high-grade malleable spheroidal graphite iron. The rotor housing is an aluminium silicon alloy, bonded to the cylinder-bore walls in thin sheet steel, the outer surfaces of which have a saw-tooth finish to improve adhesion and thermal conductivity. This lining is then given a hard chromium-molybdenum plating, which in turn is plated with more chrome but in a thin, porous and oil-retaining layer. End and intermediate rotor housings are made from unplated high silicon aluminium alloy. Apex and side seal blades can be made from cast-iron but the more popular types are made from hard carbon material. Both the

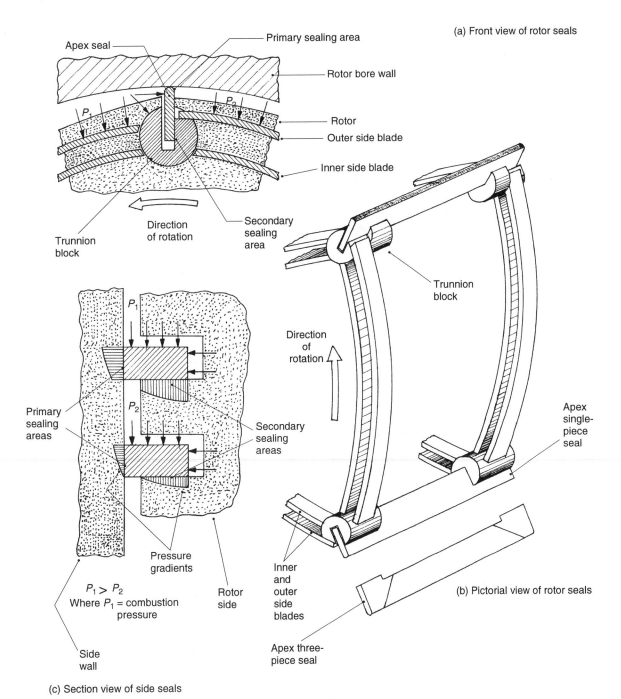

(a) Front view of rotor seals

Apex seal

Primary sealing area

Rotor bore wall

Rotor

Outer side blade

Inner side blade

P_1

P_2

Trunnion block

Direction of rotation

Secondary sealing area

Trunnion block

Direction of rotation

Primary sealing areas

P_1

P_2

Secondary sealing areas

Apex single-piece seal

Pressure gradients

$P_1 > P_2$
Where P_1 = combustion pressure

Rotor side

Inner and outer side blades

(b) Pictorial view of rotor seals

Side wall

Apex three-piece seal

(c) Section view of side seals

Fig. 15.7 Rotor apex and side sealing arrangements

leaf and washer springs can be made from beryllium copper, which has the ability to retain its elasticity when operating under working temperatures.

Excess seal-to-bore wear, due to fuel dilution of the oil film on the bores after cold starting followed by rapid acceleration or deceleration of the rotor, can be minimized by covering the chromium-molybdenum plating with Du Pont Teflon PTFE. A 20 μm thick coating is sprayed on the bore surface, heated to 80°C for 30 minutes, and cured at just over 150°C. This treatment considerably reduces the boundary friction conditions, thereby prolonging the life expectancy of both bore and seals.

15.1.8 Induction and exhaust ports
(Figs 15.8(a–c))

Peripheral inlet and exhaust ports; that is, ports which are situated in the epitrochoid bore walls will give optimum volumetric efficiency and hence good power at high speed. Unfortunately, there is a lot of overlap when both ports are open, so running at low speeds is rough and performance is poor. Port overlap can be reduced or even eliminated by using side inlet ports (Fig. 15.8(a)), but these ports have to be small to be swept by the rotor seals, and they inherently open much later than peripheral ports.

One solution to the problem of providing good performance over the entire speed range of the engine can be achieved by maintaining peripheral exhaust ports, but incorporating twin side inlet ports in conjunction with a double barrel compound carburettor. The primary barrel throttle opens to control the engine at low speed and light loads, whereas the secondary barrel throttle opens only above about one-third load. Such an arrangement ensures adequate cylinder filling over the lower and middle speed range with corresponding good acceleration response and performance.

To obtain even better performance at the upper speed range without sacrificing low and middle speed response, a triple inlet port system can be adopted (Fig. 15.8(a)). These three ports are known as the primary port, main secondary port and the auxiliary secondary port. Both the primary and main secondary port's mixture supply is controlled by primary and secondary double-barrel carburettor throttles, whereas the auxiliary-port mixture delivery is opened by an exhaust pressure slave actuator. At low speed and load, only the primary side port is in action, it overlaps little with the peripheral exhaust port so that good torque and fuel consumption at low speed and light load can be obtained (Fig. 15.8(c)). With increased load the secondary throttle will commence to open causing the main secondary port also to supply fuel mixture to the cylinder. With this improved breathing, a better performance response to medium loads and speeds is achieved

(Fig. 15.8(c)). At very high speed and load the auxiliary secondary port is made to open progressively (Fig. 15.8(c)) by an exhaust-pressure operated rotary valve (Fig. 15.8(b)). Therefore, this excess of mixture charge supplements the normally declining volumetric efficiency at the very top end of the speed range.

15.1.9 Lubrication system
(Fig. 15.9)

Oil is drawn from the sump-pan via the oil strainer and pick-up pipe to the eccentric bi-rotor type oil pump, where it is pressurized before it is circulated through the oil cooler (Fig. 15.9). The cooled oil is then cleaned as it passes through the oil filter media, and it is then delivered to the front and rear main bearings via the main oil gallery. Pressurized oil is supplied via radial drillings intersecting the main bearing journals, so that oil is made to flow along the axial drilling in the eccentric output shaft. Oil is also fed through radial holes to the rotor bearing journals (lobes), and a small quantity of oil is further supplied via drilled passages to the gear side and side faces of each rotor.

The majority of the oil spilling from the rotor bearings enters the internal bore of the rotors and acts as a coolant. Circulation and ejection of oil from the rotor is achieved due to the planetary rotation of the rotor; that is, the rotation about its own centre of gravity, which coincides with the axis of the eccentricities, as well as the axis of the output shaft. Thus, the oil within the rotor is subjected to alternate inward and outward acceleration, which provides a controlled oil flow movement and, at the same time, rotor bearing side spill provides adequate lubrication for the internal rotor gear and the fixed external gear housed inside the rotor casting.

Excessive oil pressure in the oil passages when the engine is cold is controlled by the pressure-relief valve, which will return oil to the sump.

When the engine is cold the thermo bypass valve opens and permits oil to bypass the oil-cooler so that it flows directly from the oil-pump to the oil-filter and, once the operating temperature has been reached, the wax-pellet type bypass valve closes forcing the heated oil to circulate through the oil cooler.

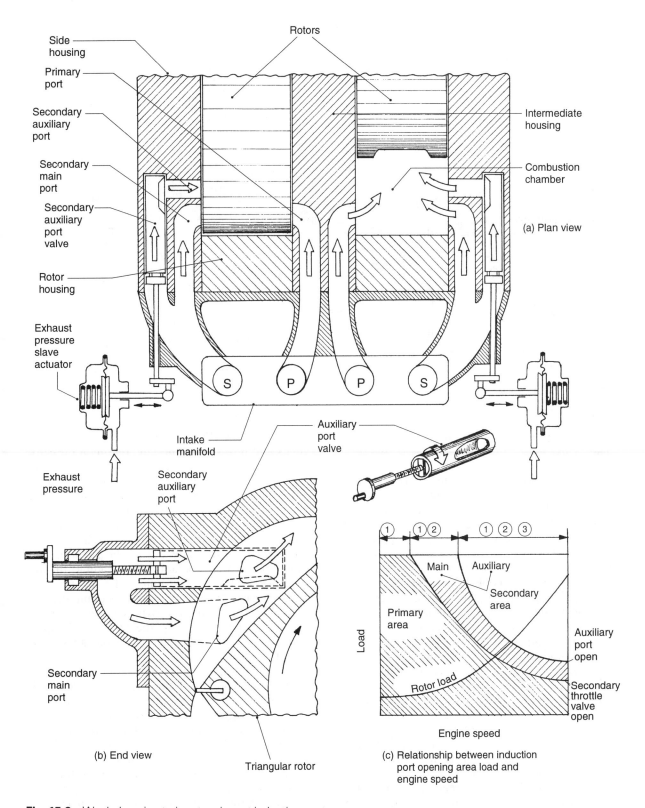

Side housing

Primary port

Secondary auxiliary port

Secondary main port

Secondary auxiliary port valve

Rotor housing

Exhaust pressure slave actuator

Exhaust pressure

Rotors

Intermediate housing

Combustion chamber

(a) Plan view

S P P S

Intake manifold

Auxiliary port valve

Secondary auxiliary port

Secondary main port

(b) End view

Triangular rotor

Load

Engine speed

① | ① ② | ① ② ③

Main Auxiliary

Secondary area

Primary area

Rotor load

Auxiliary port open

Secondary throttle valve open

(c) Relationship between induction port opening area load and engine speed

Fig. 15.8 Wankel engine twin rotor six port induction system

767

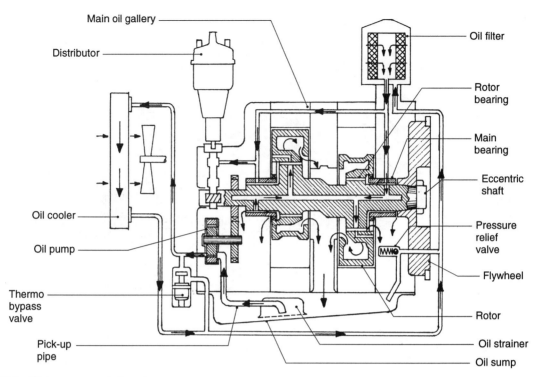

Fig. 15.9 Twin rotor wankel engine lubrication system

15.1.10 Cooling system
(Fig. 15.10)

The engine block is made up of five separate section housings: the front and rear end housings, two rotor bore housings and an intermediate housing, all of which are clamped together by bolts (Fig. 15.10). Coolant circulation is provided around the bore walls by axial passages cast into each housing section, while the side walls of the twin cylinders are cooled by extended coolant passages adjacent to the side walls cast in both the end covers and the intermediate housing.

When the engine is running, a proportion of the heat from combustion will be absorbed by the coolant, and therefore the hot less-dense liquid will tend to rise while the cold more-dense liquid will gravitate to its lowest position.

Hot coolant on the spark-plug side of the cylinder-bore housing will tend to accumulate and, in the process, it will rise and move to the radiator header-tank. Here, it displaces previously heated liquid which is therefore compelled to move down the radiator tubes.

At the same time as the relatively hot coolant flows down the radiator tubes, cold air will be forced through the radiator matrix, consisting of tubes and fins, due to the fan creating a pressure difference across the radiator matrix. The effect of the cold air flowing across and around the tubes will be to extract heat continuously from the hot liquid. Consequently, the liquid cools, becomes heavier and so sinks to the bottom of the radiator. Cooled liquid in the lower portion of the radiator will then tend to come out of the radiator and enter the lower coolant pipe leading to the entrance to the coolant-pump. To increase the rate of heat extraction, the coolant-pump pressurizes the liquid, and this provides a forced convection current to speed up the amount of liquid actually moving through the passages and radiator tubes.

For rapid engine warm-up, a thermostat valve (Fig. 15.10) blocks off the radiator header-tank when the engine is cold, thus preventing the cold air from absorbing the heat from the coolant. As soon as the operating temperature is reached, the thermostat valve automatically opens enabling the normal coolant circulation to recommence. A bypass pipe is incorporated between the thermostat chamber and the inlet to the coolant pump so that about 10% of the liquid coolant is permitted

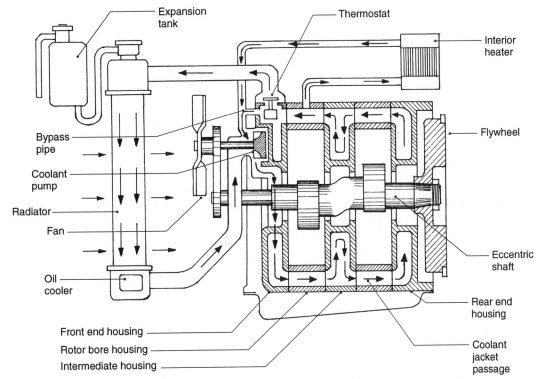

Fig. 15.10 Twin rotor wankel engine liquid cooling system

to circulate at all times, this being a safeguard to overheating and boiling of the coolant in the combustion burning zone.

'O' rings are inserted in grooves formed in the side walls of the housing sections to prevent leakage between the various housings.

15.1.11 Engine displacement, capacity and compression ratio

Rating and comparing the Wankel engine with the reciprocating engine is, in the first instance, based on the maximum cubic capacity of one of the combustion chambers when a rotor flank has moved adjacent to the minor axis (waisted portion) of the epitrochoid bore, but is at its furthest point from it.

Clearance and swept volume

The volume created between the curved chamber bore, flat side walls and one of the rotor flanks when in the nearest-approach position (TDC) adjacent to the waisted portion of the epitrochoid bore, is known as the *clearance volume*. The swept volume of a Wankel engine is obtained by subtracting the minimum volume of the combustion chamber at the point of greatest compression (TDC) from the maximum volume to which the chamber expands (BDC) as the rotor revolves.

The swept volume of the chamber may be given by the following:

$$V_s = 3\sqrt{3RWe}$$

where V_s = swept volume (cm³)
R = radius of rotor from its centre to one of the apex tips (cm)
W = width of rotor (cm)
e = distance between the axes of the output shaft and eccentric lobe, known as the eccentricity (cm)

Compression ratio

The compression ratio will be the conventional formula, as follows:

$$CR = \frac{V_s + V_c}{V_c}$$

therefore

$$CR = \frac{(3\sqrt{3RWe}) + V_c}{V_c}$$

769

where CR = compression ratio
V_s = swept volume (cm^3)
V_c = clearance volume (cm^3)

Engine displacement capacity

There are three power strokes per rotor revolution in a Wankel engine, and the output shaft completes three revolutions for every one revolution of the rotor. Therefore, there is one power stroke per revolution of the output shaft. This, therefore, can be seen to be equivalent to a two-cylinder four-stroke engine which also has a power stroke every crankshaft revolution. Because of this reasoning a two-chamber swept volume equivalence is generally recognized.

The displacement capacity is therefore twice the swept volume of a single chamber multiplied by the number of rotors; that is,

$$V = 2V_s n$$

where V = displacement capacity (cm^3)
V_s = swept volume (cm^3)
n = number of rotors

15.2 The Stirling engine

A Scottish clergyman, Dr Robert Stirling, invented a closed-cycle hot-air engine in 1816. It was an externally heated reciprocating engine in which an enclosed mass of air recirculates each working cycle. In the late 1930s and 1950s, N.V. Philips (a Dutch company) established a research and development programme, and found that, by changing from air to a light gas such as helium and hydrogen, efficiency could be improved and, therefore, instead of being called a hot-air engine it became known as the Stirling engine.

A single-acting Stirling engine has an upper displacer piston and a lower power piston, which are used to divide the cylinder into two regions: an upper hot space and a lower cold space. The two cylinder spaces at different temperature levels are interconnected through heater tubes, a regenerator and cooler, so that their volumes are varied cyclically as the gas is shuttled from one space to the other. The displacement of the gas from one end of the cylinder to the other involves the absorption and rejection of heat. Gas in the hot space will expand and perform more work than that in the cold space; therefore, a net amount of work is gained. There are three main types of drive mechanisms which have proved successful, these are: (1) swashplate; (2) slider crank; and (3) rhombic drive.

15.2.1 Construction and description of the heater unit
(Figs 15.11 and 15.12)

The heat-exchanger which transfers the heat of combustion to the cylinder via the working gas consists of a ring of tubes joined to the cylinder's hot space, which rises from it to a circular gallery, from which a second group of pipes descends to a series of regenerator cups, grouped around the cylinder (Fig. 15.11). The regenerator cups are filled with a matrix of small-diameter crimped wire, which extracts and absorbs a large proportion of the heat energy contained in the gas as it passes from the cylinder's hot space to the cold space (Fig. 15.12). Conversely, when the cooled gas flows in the opposite direction, from the cold space to the hot space, it absorbs heat from the wire mesh. Further pipes from the bottom of the regenerator are joined to the water cooler, and finally the gas circuit between the hot and cold cylinder spaces is completed by a passage formed between the cylinder's cold space and the lower part of the cooler.

Fuel (this can be diesel, paraffin or petrol) is injected through the atomizing nozzle into the combustion chamber mounted above the cylinder where it mixes with the air supply and continuously burns at around 700°C. The hot combustion gases pass between the heat-exchanger heater-tubes to alternate spiral passages of an annular pre-heater unit. After passing through the spaces between the heater-tubes, most of the remaining heat contained in the gas is transmitted by conduction through the metal spiral passage to the inlet air, which is forced into alternate inlet passages of the pre-heater by a blower (not shown). As a result, the spent exhaust gases are reduced to a low temperature before they are expelled from the engine into the atmosphere. It is claimed that the efficiency of the heat transference from the hot exhaust gases to the cool inlet air is something in the order of 90%. A heater-plug is only used to start the combustion process, and the combustion heat exchanger (heater tubes) is then operated at a constant rate by a thermostat which adjusts the heat exchanger output by switching on and off the fuel spray. Thus, the heater output can be varied to meet the output power demands. Since the burning process

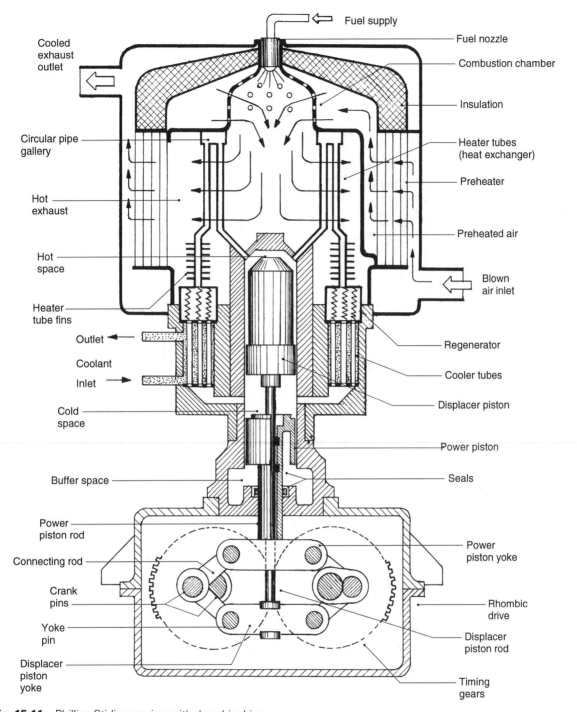

Fig. 15.11 Phillips Stirling engine with rhombic drive

771

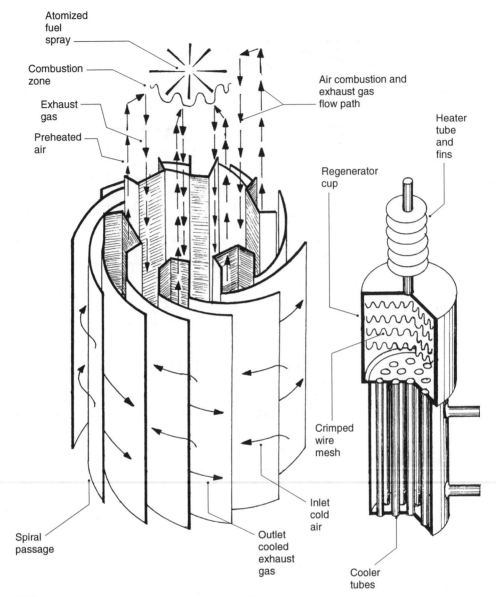

Fig. 15.12 Stirling engine preheater, regenerator and cooler units

is continuous and it is practically completed before the gases leave the engine, levels of exhaust emission are relatively low.

15.2.2 Rhombic drive
(Figs 15.11 and 15.13(a–d))

The function of the rhombic drive is to control both piston movements so that the hot-space volume variation has an approximate 90° phase lead with respect to the cold-space volume change. Only the power piston regulates the total volume of the working gas, which is to be expanded or compressed, whereas the displacer piston controls the transfer of the working gas between the cylinder's hot and cold spaces.

The rhombic drive has two crankshafts (Fig. 15.11) offset from the engine's central axis and geared together by two timing-gears, so that they rotate in opposite directions to each other. The power piston is attached to the top yoke by a hollow-rod while the displacer-piston is joined to the bottom yoke by a solid-rod passing through the centre of the power-piston rod. Both top and bottom yokes are linked to each crankshaft by pairs of equal length connecting rods coupled to a common crank-pin to form a rhombus.

With this form of mechanism, perfect symmetry is provided throughout the working cycle and the inertia forces resulting from this motion can be completely balanced by adding rotating masses equal to the piston and rod mass to each timing-gear diametrically opposite the crank-pins (Fig. 15.13(a–d)). The advantages of this linkage are that it can be fully balanced and that there is no piston side thrust on the rods and, therefore, sealing problems are minimal. To balance the pressure exerted by the working gas, the buffer space is filled with pressurized gas, with seals being provided inside and outside the hollow power piston rod to prevent the working gas and buffer gas (respectively) from escaping.

15.2.3 Single acting Stirling engine cycle of operation
(Fig. 15.13(a–d))

Phase 1
Compression and transference of gas to the cold space commences (Figs 15.13(a) and 15.14)
Both the displacer and power pistons move towards TDC (Fig. 15.13(a)). The gas is therefore displaced from the reducing cylinder hot-space through the heater tubes to the regenerator and the cooler where it gives up some of its heat energy as it passes through to the cylinder's cold-space. The reducing cylinder volume caused by the power-piston moving up its stroke, raises the gas pressure from State 1 to State 2, with the temperature remaining approximately constant (Fig. 15.14).

Phase 2
Compression is completed while the gas is transferred to the hot space (Figs 15.13(b) and 15.14)
The power-piston continues to move up its stroke while the displaced-piston moves in the opposite direction, downwards (Fig. 15.13(b)). The reducing cylinder cold-space therefore forces the cool gas through the cooler, regenerator and heater tubes, so that it absorbs heat stored in the regenerator from the previous cycle and from combustion through the pre-heater tubes as it passes through to the cylinder hot-space. The gas in the

cylinder's hot-space also continues to receive heat energy through the heater tubes due to the burning of the fuel—the gas pressure therefore rises from State 2 to State 3, with a minimal change in the volume of the gas (Fig. 15.14).

Phase 3
Expansion power occurs with gas transference completed (Figs 15.13(c) and 15.14)
Both displacer and power pistons move towards BDC with the effect that the cylinder's cold-space has decreased and the cylinder's hot-space has increased (Fig. 15.13(c)). Heat energy from combustion of the fuel continues to be supplied to the gas in the cylinder's hot-space through the pre-heat tubes. Subsequently, expansion takes place pushing both pistons down their stroke. As a result, the total gas volume increases while its pressure decreases from State 3 to State 4—there being theoretically very little change in temperature during this phase (Fig. 15.14).

Phase 4
Heat rejection while gas is transferred back to the cold space (Figs 15.13(d) and 15.14)
Both displacer and power pistons move away from each other towards TDC and BDC respectively (Fig. 15.13(d)). The upward moving displacer-piston transfers the gas in the cylinder's hot-space through the heater tubes to the regenerator and cooler, giving up some of its heat energy before it passes into the enlarging cylinder's cold-space, created by the power-piston's outward movements. The cooled gas therefore expands to fill the increasing volume of the cylinder's cold-space; consequently, the gas pressure drops from State 4 to State 1, with the total volume of the hot and cold spaces being practically unaltered through this state of change (Fig. 15.14).

Theoretical and actual P–V diagrams
In practice, the power and displacer piston movement does not take place exactly in four distinct phases, so that the pressure rise from State 2 to State 3 will only be partially at constant volume. Likewise, the pressure drop from State 4 to State 1 will not be at constant volume but occurs while the hot and cold spaces are undergoing a volume change. The actual shape of the P–V diagram for a practical Stirling engine is more in the form of a rounded loop, as shown within the theoretical diagram (Fig. 15.14).

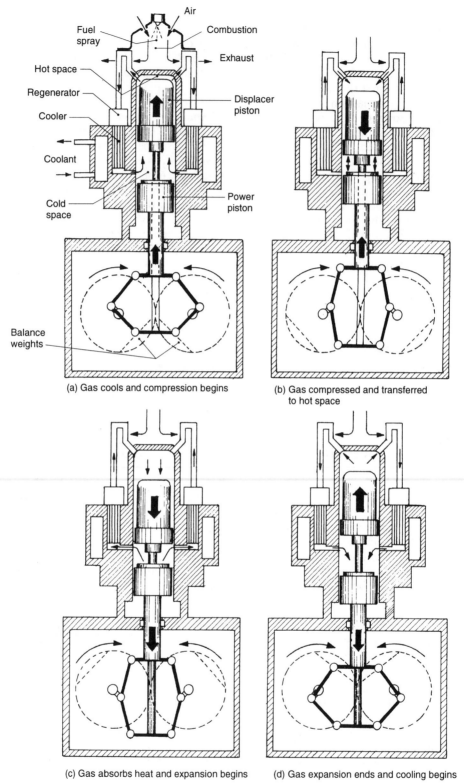

(a) Gas cools and compression begins

(b) Gas compressed and transferred to hot space

(c) Gas absorbs heat and expansion begins

(d) Gas expansion ends and cooling begins

Fig. 15.13 Stirling engine phases of operation

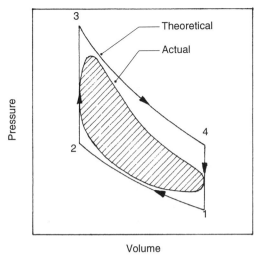

Fig. 15.14 Ideal and actual Stirling engine P–V diagram

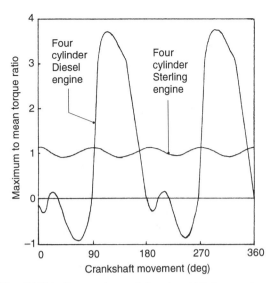

Fig. 15.15 Comparison of diesel and Stirling engine torque variation over cycle of operation

15.2.4 Design and performance comparisons
(Fig. 15.15)

With the externally heated Stirling piston engine, the swept volume is small and the clearance volume is large, which is the opposite to the conventional reciprocating internal combustion engine where the clearance volume is small and the swept volume is much larger. This relationship between swept and clearance volume means that the Stirling engine compression ratio is only about 1.5–2.0 to 1 compared with 8–12 to 1 for petrol engines and 14–24 to 1 for diesel engines.

The output of a Stirling engine depends largely on the mass of working gas transferred between the cold and hot cylinder space in much the same way as the conventional internal combustion engine's power output relies on the mass of air induced into the cylinder per stroke. Raising the working gas pressure will therefore increase its mass and proportionally increase the output. Thus, doubling the pressure doubles the mass of gas which, potentially, very nearly doubles the power output.

If the strength of the engine limits the maximum working gas pressure, this would be in the order of 150 bar whereas 75 bar is considered the minimum operating pressure. Hence, in the extreme, this would only give a compression ratio of 2:1. Because of the high working pressure, the Stirling engine, even with its low compression ratio, develops about the same output per unit cylinder capacity as either the spark or the compression ignition engine, whereas it is heavier than the petrol engine but roughly the same weight as an equivalent output diesel engine.

The pre-heating of the air charge and the continuous combustion process and heat exchange produces a thermal efficiency of the order of 38% at full load, which is very similar to typical values obtained with diesel engines.

The Stirling engine is much quieter than either the petrol or diesel engines, due mainly to the gradual pressure variation over each cycle (Fig. 15.15) compared with a very rapid pressure rise in the conventional petrol or diesel engine, at the point of initial combustion. Another feature with the Stirling engine is that there is no camshaft and valve mechanism, which would otherwise considerably increase engine noise.

15.3 The gas turbine engine

The first patent for a gas turbine was taken out in 1791 by an Englishman named John Barber. Since then, a long list of pioneers, inventors and designers have contributed to the progressive development of the gas turbine engine. Hans Holzwarth, of Mannheim, Germany, is credited for building the first economically practical gas turbine in 1905. Sir Frank Whittle saw the possibility for a gas turbo-jet engine propelling an

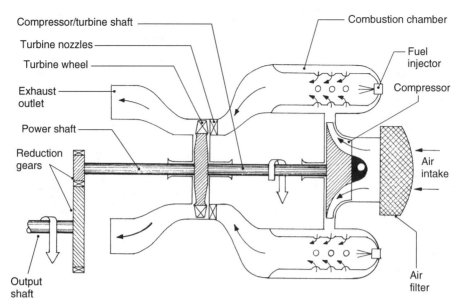

Fig. 15.16 Simple single shaft single turbine engine

aeroplane and applied for a patent in 1930. Ever since then, extensive research and development of gas turbine engines has been undertaken by all major engine and vehicle manufacturers throughout the world. The first turbine-driven car in the world was made by the Rover Company and was submitted to an officially observed RAC test at the proving group of the Motor Industry Research Association (MIRA) on 8 March 1950.

15.3.1 Comparison of the single and two-stage turbine engine
(Figs 15.16, 15.17 and 15.18)

The simplest gas turbine layout with a single centrifugal compressor and axial flow turbine on the same shaft produces no low speed torque and is not self-sustaining much below half maximum shaft speed (Fig. 15.16).

With a split shaft separate power turbine added (Fig.15.17) the compressor assembly can be operated at optimum speed while the power turbine can be run up and down the speed range and will produce torque even at a standstill. This two-stage turbine layout is known as the *free-turbine engine*.

A free-turbine engine basically consists of the air inlet, a compressor, and the combustion chamber which burns and expands the heated gases on their way to the first turbine. This turbine de-

velops only enough power to drive the compressor and hence is known as the *compressor turbine*. The gas is then expanded through a second turbine which develops the power output of the engine and hence is known as the *power turbine*. The gas is then finally expanded to the atmosphere via the exhaust system.

The major difference between the split-shaft two-stage turbine engine and the single-shaft turbine-engine is that the power turbine is not connected by a shaft to the compressor turbine; that is, the turbine is free and, because of this, the two-turbine arrangement can run and will be operated at widely varying speeds.

With the single-shaft gas turbine engine, the available torque falls rapidly with a decrease in shaft speed from the designed maximum, and therefore a large number of gear ratios would be necessary (Fig. 15.18). By using the two-shaft free-turbine layout, the torque rises with decreasing speed (Fig. 15.18). Thus, less than half the number of gear ratios would be needed to propel a vehicle if the engine was of the two-shaft double-stage turbine, as opposed to the single-shaft gas turbine engine. Figure 15.18 also shows and compares the diesel engine torque/speed characteristics, which are seen to be relatively constant with decreasing engine speed.

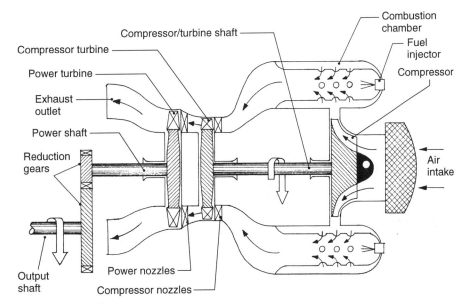

Fig. 15.17 Free turbine split shaft two-stage turbine engine

15.3.2 Gas turbine operating sequence
(Fig. 15.19)

The gas turbine engine performs the same basic operations as in any other internal combustion engine, that is, induction, compression, power (combustion and expansion) and exhaust. With the reciprocating piston engine the phases of operation take place in the cylinder at different time intervals, whereas with the gas turbine, they all occur simultaneously in different parts of the engine. The operating sequence and events can be understood by comparing the diagrammatic view of the gas turbine engine (Fig. 15.19) with the changes in gas speed, temperature and pressure illustrated underneath.

Induction and compression phase
(Fig. 15.19)

When the compressor is rotated, unthrottled air from the atmosphere is drawn through a filter into the intake side of the compressor. As the air enters the vanes of the impellor, centrifugal force will compel it to move radially outwards, thus pressurizing the air leaving the compressor. A pressure ratio in the order of 4:1 is achieved with an operating compressor speed of around 30 000 to 50 000 rev/min. The compressed air then passes through the twin rotary regenerator (heat exchanger) honeycomb matrix, it absorbs heat and continues on its way to the combustion chamber.

Combustion phase
(Fig. 15.19)

While the engine is running, a continuous supply of fuel will be sprayed at a pressure of around 50 bar into the combustion chamber where it will be atomized and burnt with the mass of compressed and pre-heated air charge constantly moving through the chamber.

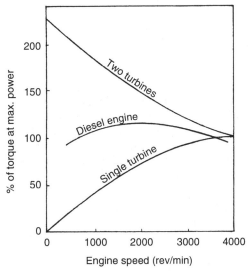

Fig. 15.18 Torque verses output shaft speed for single shaft and two shaft gas turbine engines and diesel engine

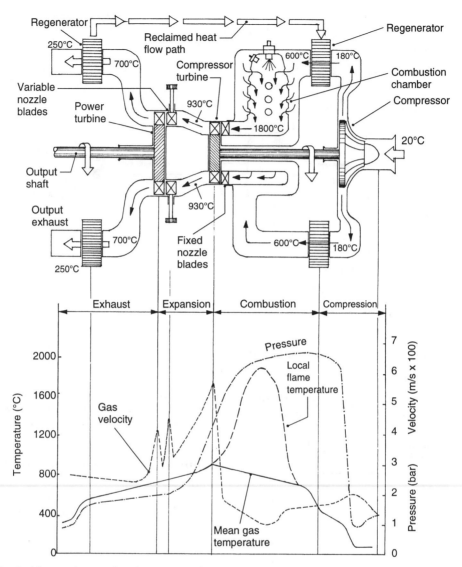

Fig. 15.19 Gas turbine engine combustion process throughout the system

A small proportion of the pre-heated air from the regenerator enters the primary zone of the combustion chamber near the fuel spray, where it is used to burn locally the finely atomized fuel with a slightly rich air–fuel ratio of about 14:1. Some of the remaining air is used for skin cooling the combustion-chamber, while the bulk of it enters the dilution holes at the downstream end of the combustion chamber in order to dilute and lower the air–fuel ratio to roughly 60:1. The effect of the change in air–fuel ratio is to reduce the combustion temperature from about 1800°C near the injector in the primary spray zone, down

to approximately 930°C at the entry to the compressor turbine blades.

The heat released by the combustion expands the charge but, as the chamber has an open end, there is no rise in pressure. In fact, there is a slight fall in pressure from the entry to the outlet at the compressor turbine blades.

Expansion phase
(Fig. 15.19)
The rapid expansion of the gas, caused by the liberated heat of combustion, considerably in-

778

creases its velocity before it forces its way between the guide vanes in the stator-ring of the compressor-turbine. These vanes direct the flow to the appropriate angle of thrust for the blades on the periphery of the compressor-turbine wheel, so that the wheel rotates rapidly. The large quantity of heat energy still remaining (expressed in terms of gas velocity) then passes through the variable power-turbine guide-nozzles (fixed vanes) where it is redirected onto the power-turbine blades at an angle dependent upon the load and speed conditions required. It is this second phase movement of gas, from stationary guide-blades to power-turbine blades, which provides the output shaft with the torque and power. As can be seen at this stage, the gas energy (and hence its velocity) is decreasing, so the small peak velocity increases, indicating the restriction of the gas path between each set of blades.

Exhaust phase
(Fig. 15.19)
The expelled gas coming away from the power turbine then passes through the rotary regenerator (heat exchanger) honeycomb matrix where it releases a major proportion of its remaining heat. The operation of the regenerator is such that the heat stored in the honeycomb matrix is continuously being transferred by the rotation of the regenerator wheel to the input side of the combustion chamber. Finally, the spent gases pass out the exhaust ducts to the atmosphere at a much reduced temperature, pressure and velocity.

15.3.3 Regenerator
(Figs 15.19, 15.20 and 15.21)

To improve the thermal efficiency (Fig. 15.21) of the gas turbine, a regenerator-type heat-exchanger is incorporated to extract the majority of the heat contained in the exhaust gases (after these gases have left the power-turbine) and to feed this heat to the incoming compressed air before the air enters the combustion-chamber (Fig. 15.19). As a result, heat which would normally be wasted is used to preheat the pressurized air charge before it enters the combustion-chamber, thereby reducing the amount of fuel required to raise the temperature of combustion so that the gas temperature at the entrance to the first turbine is well over 900°C.

Twin disc-type regenerators made from a sin-

tered ceramic, bound together with glass, are commonly used (Fig. 15.20). This glass-ceramic material, developed by the Corning Glass Company is sold under the trade name 'Cercor'. The rotary regenerator disc is built like a honeycomb in section by wrapping alternate layers of corrugated and flat unfired Cercor strip around a solid hub. After firing, a solid rim is bonded to the matrix. Sometimes the honeycomb matrix is cut into segments which are then bonded together again such that the flat strips in the matrix are largely in a radial direction. This construction reduces internal leakage between the air and gas sides of the disc.

A drive from the compressor shaft rotates these discs through a reduction gear (Fig. 15.20) with a ratio (depending upon the maximum speed of the compressor shaft) of the order of 2500:1 to 3000:1. It therefore provides disc speeds from idle to maximum power of 9 to 22 rev/min.

The hot exhaust gas stream leaving the power-turbine passes through one side of the regenerator disc matrix to the exhaust outlet and, in the process, gives up as much as 90% of the heat contained in the gas to the glass-ceramic honeycomb matrix. It is claimed that the exhaust temperature is reduced to about 90°C under idling conditions, and to about 260°C when full power is developed. The heat energy is stored in the matrix until the disc has rotated about half a revolution, and the charged portion of the matrix then intersects the passage leading from the compressor to the combustion-chamber. Heat will thus be absorbed by the stream of fresh compressed air moving through the corrugated matrix on its way to the combustion-chamber.

15.3.4 Variable power turbine nozzle control
(Figs 15.20, 15.22(a–c) and 15.23)

The variable turbine nozzles consist of stator blades which are permitted to twist on their pivots so that they can guide the gas passing from the compressor turbine to the power turbine in any direction (Fig. 15.20). Varying the angle of the nozzle blades relative to the turbine's axis of rotation enables the gas to be deflected so that optimum thrust can be produced to meet all normal operating conditions. Nozzle-blade movement also provides a position for a retardation effect on the forward rotation of the power-

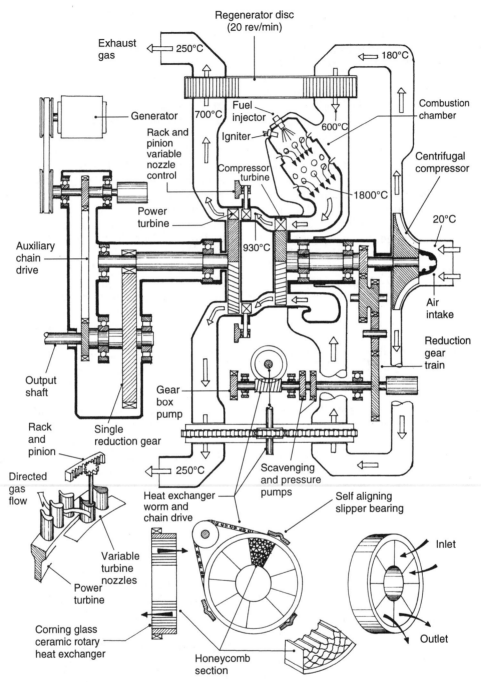

Fig. 15.20 Gas turbine engine layout

turbine, for vehicles which are braking when road speeds are in excess of 25 km/h.

There are normally about 20 blades (vanes) equally spaced around, and adjacent to, the turbine-blades. Each blade is mounted on a spindle pivot which has a quadrant gear attached to it (Fig. 15.22). A large ring-gear meshes with each blade quadrant so that any angular movement of the ring-gear alters the angle of entry of the gas to all the turbine-blades.

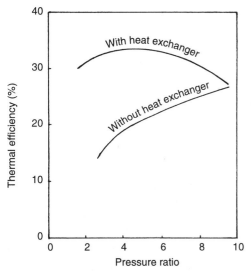

Fig. 15.21 Thermal efficiency verses pressure ratio with and without regenerator

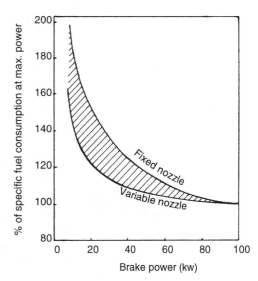

Fig. 15.23 Specific fuel consumption verses brake power for both fixed and variable power turbine nozzles

The ring-gear is linked to the accelerator pedal via a nozzle actuator and transmission governor. When the pedal is in the rest position, the variable nozzles are turned to the ineffective idle and economy position (Fig. 15.22(b)). Closing the nozzles in the forward direction provides the power turbine blades with the greatest gas thrust; that is, the best power or steady-state condition (Fig. 15.22(a)). Closing the nozzles in the reverse

direction at road speeds in excess of 25 km/h (15 mile/h) projects the gas stream in a direction which opposes the rotation of the power turbine (Fig. 15.22(c)), this therefore produces a degree of forced speed retardation.

A comparison of the specific fuel consumption achieved with and without variable nozzles, plotted against the percentage of brake power, is shown in Fig. 15.23.

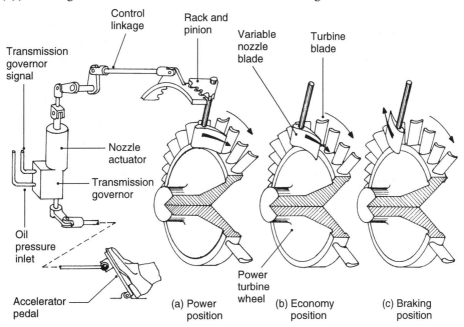

Fig. 15.22 Power turbine variable nozzle blade mechanism

781

15.3.5 Start, idle and output control
(Fig. 15.24)

To start the engine, the starter switch is turned. This energizes the starter-motor, the air assistance pump, the contact breaker and coil for the high tension spark-plug (ignitor) and opens the sprayer solenoid. The starter-motor accelerates the compressor and, therefore, the first turbine-shaft from a standstill, until it reaches a speed of about 5000 rev/min. The fuel supply and the ignitor are then automatically switched on.

When the compressor shaft reaches a self-sustaining speed of around 14000 rev/min (due to the effective commencement of combustion) both the starter-motor and ignitor circuits are de-energized by a signal, whereas the metered starting fuel supply will continue to be delivered permitting the compressor-shaft to accelerate to the idling setting of the governor; that is, around 18000 to 22000 rev/min.

The accelerator pedal position governs the compressor and first turbine-shaft speed, which operates between idle and full speed, while the speed range of the power-turbine and output shaft extends from zero rotation at stall, to the rated power speed. The maximum speed (rated power) setting depends upon the derating of the engine (Fig. 15.24), in some engines it may be as low as 30000 rev/min to prolong engine life whereas other engines may run as fast as 55000 rev/min. In addition to the compressor-shaft governor setting, the accelerator pedal position, combined with speed and temperature signals from the engine all contribute to the controlled operation of the variable power turbine nozzles.

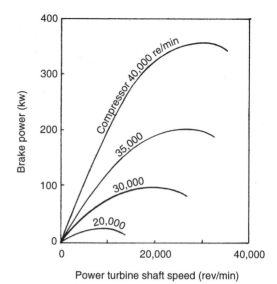

Fig. 15.24 Brake power verses output shaft speed for different compressor shaft speeds

15.3.6 Comparison of gas turbine and diesel engines

The initial cost of a gas turbine engine is relatively high, but it is estimated that maintenance savings of some 50% over the diesel engine are possible. Engine life expectancy before overhaul is in the region of 12000 hours, normally equivalent to 800000 km (500000 miles) which is more than comparable with a quality built diesel engine.

The engine will operate without adjustment on diesel fuel, paraffin or low-lead petrol. Fuel consumption, when tested on heavy-goods vehicles, is comparable with the diesel engine over a wide operating range down to 30% power and the oil consumed is almost nil.

The noise level of the gas turbine power unit is lower than that of a diesel engine of equivalent power output. Due to the process of continuous burning, as contrasted with the intermittent combustion of the diesel engine, the exhaust from the gas turbine engine is smoke-free.

Index